WHAT IS
CHEMISTRY?

WHAT IS CHEMISTRY?
A Chemical View of Nature

Joseph Nordmann
LOS ANGELES VALLEY COLLEGE

Harper & Row, PUBLISHERS

NEW YORK, EVANSTON, SAN FRANCISCO, LONDON

To the students who shared the classroom experience with me. Your efforts and aspirations were all-important. You are the future.

Photograph used on the title page (pp. ii–iii) courtesy Damon Corporation.

Sponsoring Editor: John A. Woods
Project Editor: Lois Wernick
Designer: Rita Naughton
Production Supervisor: Valerie Klima

What Is Chemistry? A Chemical View of Nature

Library of Congress Cataloging in Publication Data

Nordmann, Joseph.
    What is chemistry?

    1. Chemistry.   I. Title.   [DNLM: 1. Chemistry.
QD33 N832w 1974]
QD31.2.N67   540   73-8370
ISBN 0-06-044854-7

CONTENTS

PREFACE

What Is Chemistry? is designed to be used as either a one- or two-term introduction to college chemistry. In brief outline, it consists of nine parts with a few chapters each: one, atoms; two, elements; three, bonding; four, stoichiometry; five, gases; six, solutions; seven, electrochemistry; eight, organic chemistry; and nine, biochemistry. The contents are adaptable to either a qualitative or quantitative presentation by the professor.

Introductory chemistry should, I believe, be an overview of nature from the chemist's standpoint and a description of the efforts of scientists to understand nature's complexities. Because the course is primarily for non-science majors, it should deal with the character as well as the content of science and should concentrate equally on theory and application.

I introduce the material historically. This gives the student a reassuring reference to reality while resisting reduction of the subject to alienating abstraction and dogma. By retracing paths the human mind took in acquiring knowledge, the student sees how problems are chosen, experiments

...structed, data gathered, principles applied, and conclusions drawn. In addition, he may acquire insight into the scientific thought and method, and, as a beginner, this may serve him better than the conclusions themselves.

I begin each subject with background matter which I hope will interest and orient the student and give him reasons for reading further. For instance, instead of discussing the chemical elements solely in terms of the periodic table, gases only with the gas laws, water only with liquid structure theory, and biochemistry just with the structure of biopolymers, I step back for a long-range interdisciplinary view. Synthesis of nuclides in stars and the combination of elements in the earth are described before classifying the elements; atmospheric research and air pollution precede gas laws; the ocean and freshwater problems come before liquid-solution theory; and a hypothesis for the chemical evolution of life leads into biochemistry. Principles that follow are then illustrated with examples from areas of modern research and technology. To cite a few cases, X ray crystallography is used to explain silicates and the structure of the earth's crust, voltaic cells furnish a basis for talking about electric cars and electric eels, and the DNA theory is coupled with theories of biological cell differentiation, cancer, and human aging.

Inclusion of modern research, both pure and applied, is profitable. Immediately obvious is the high motivational value, for results from research laboratories which promise an ever deeper comprehension of nature have long-range implications in all our lives. Opportunity is present to recognize current benefits of science as well as to point out certain dysfunctions of its technology. Further, one can demonstrate the important point that chemistry is an empirical science which progresses with the aid of its advancing implementation. Experimental methodology is very much a part of science, because it is a source of facts — and thus of theories. I do not hesitate, therefore, to bring forth applications of the electron microprobe (rock structure), NMR spectrometer (organic conformers), gas chromatograph (fat analysis), peptide sequencer (enzyme secondary structure), and other instruments.

At the beginning of each chapter I list the important concepts that follow. This outline should guide the student in his reading. After finishing the chapter he will benefit by reviewing the outline to test his ability at defining its scientific terms and explaining each of its statements. Chapters end with sets of questions and problems. I have tried to make these informative, provocative, and nonroutine, with frequent reference to problems that scientists are contemplating today.

A multidisciplinary approach, almost of necessity, generates a relatively large volume of material, and this is true here. It is expected, therefore, that except for a two-semester class in which most of the book might be assigned, the professor will concentrate on those sections he considers essential to his instruction and will assign some of the remainder for collateral reading. Enough is available for the design of three types of one-term courses.

The first type of course is for liberal arts students. It can be broad and descriptive with only occasional reference to mathematics. A basic selection of chapters for this course is 1 and 2 (beginnings of chemistry and atomic structure), 4 (nuclear energy), 5 and 6 (cosmology, geochemistry), 7 (periodic table), 13 (the atmosphere), 15 (the ocean), 20 (organic structure), and 22 and 24 (origin of life, molecular genetics). In addition the descriptive sections of 9 (bonding), 16 (acids, bases), and 18 (electrolytic cells) would be helpful. It is not expected that the laboratory will be used.

A second type of course is for the student who needs to qualify himself for the freshman science majors' general chemistry course, but who has not had high-school chemistry. His fare will be more conceptual, and he should be concurrently enrolled in a mathematics class. Appropriate chapters are 2, 3, and 4 (atomic structure, quantum theory, nuclear chemistry), 7 and 8 (classification and properties of elements), 9 and 10 (bonding), 11 and 12 (chemical equations and stoichiometry), 14 (gas laws), 16 (liquids and solutions), and 19 (redox and voltaic cells). If time allows, the professor might also use 17 (equilibrium) or 20 (organic structure). Two to four hours of laboratory work per week would normally accompany the lectures. (*What Chemists Do*, the lab manual for this text, is available for this purpose.)

A third type of class organization is a modification of either of the other two for students such as nursing majors who need more familiarity with organic chemistry and biochemistry. It would incorporate Chapters 20 through 24, omit some earlier chapters on inorganic chemistry, and have, say, two hours per week of laboratory.

Preparation of this text owes much to the direct assistance of reviewers and to the many authors from whose published material I obtained most of my facts. For reading chapters in their specialties and offering suggestions for improvements, I thank Prof. John Coleman, Prof. Julius Glater, Dr. Ruth Glater, Dr. William Harris, Prof. Myron Mann, and two former students of mine, Dr. Norman Schnautz and Dr. Dennis Smith. I also thank the reviewers whose valued services were arranged for by the publisher. Books that were especially helpful to me were *Introduction to the Meaning and Structure of Physics* by Leon N. Cooper, *Frontiers of Astronomy* by Fred Hoyle, *Biology and the Future of Man* by Philip Handler (editor), *The Development of Modern Chemistry* by Aaron Ihde, *Principles of Geochemistry* by Brian Mason, and *General Chemistry* by Linus Pauling. Essential also were papers, which I credit here only collectively, that appeared in the last few years in *American Scientist, Chemical & Engineering News, Endeavour, Environmental Science and Technology, Journal of Chemical Education, Science,* and *Scientific American.* One suggestion that helped lead to development of the "Important Concepts" sections came from Mrs. Caroline Lanford Eastman of Harper & Row. Finally, special citations for efficiency and patience must go to Miss Diane Reid, who helped with indexing and problems checking, and to Mrs. Johanna Each who, in cheerfully making the manuscript and its revisions legible, typed, at least twice, every word and formula these covers enclose.

Joseph Nordmann
Beverly Hills, California

ONE
WHAT IS THE ATOM?

December 2, 1942, under the University of Chicago stadium, when the controlled release of atomic energy by the fission of atomic nuclei was first demonstrated. Right, the graphite–uranium pile; center, man operating the pile's main control rod; left, the scientists (Fermi with slide rule, Compton standing in middle). (Painting by Gary Sheahan, Atomic Energy Commission.)

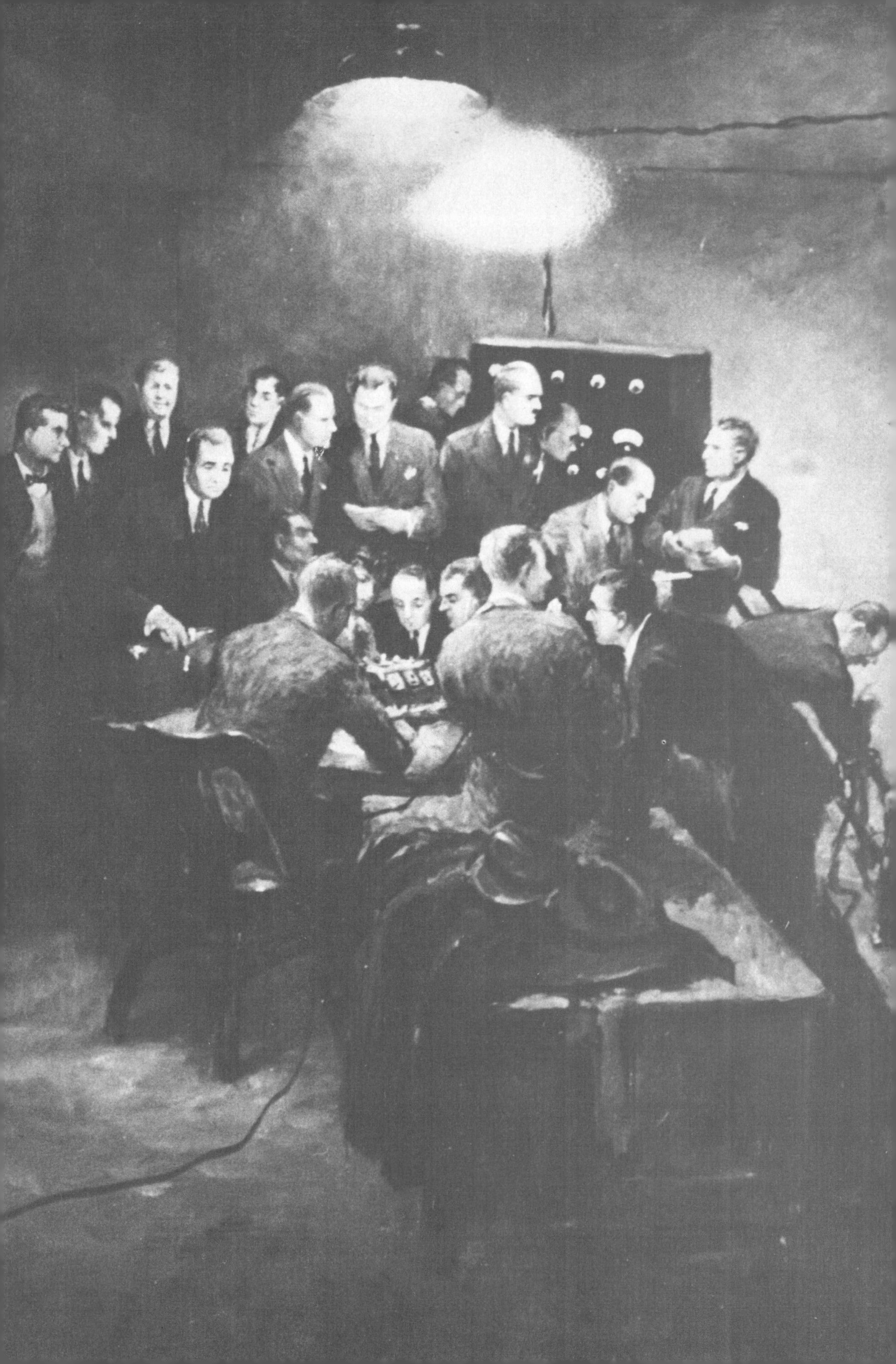

THE IMPORTANT CONCEPTS

1-1. Origins in Greece
1. Science originated in Western culture.
2. The atom concept came from speculation on the composition of matter.

1-2. Alchemy to chemistry in Western Europe
1. Chemistry began in alchemy.
2. Aristotle's influence was broken when inductive logic replaced deductive logic.
3. Chemistry as science began in the seventeenth century with a turn toward principles as established by experiments.
4. Newton revived the idea of atoms.

1-3. The atomic theory established
  1. Three laws of chemical combination suggested the atomic theory to Dalton.
  2. Dalton's important assumption was atoms of different weights for different elements.
  3. Atomic weights could not be found without chemical formulas.
  4. The combining volume law led to the chemical formulas that chemistry needed.
  5. Organic chemistry took impetus from molecular structure theory.
  6. Direct evidence for atoms was found in the nineteenth century.

Science (L. *scientia*, knowledge) is an orderly compilation of verifiable truths in which general laws of behavior can be perceived. A *scientist* is a person who looks for system in the workings of nature.

1-1. ORIGINS IN GREECE

Although the culture of the East predates that of the West by centuries, science is mainly a Western development. Reasons for this have been traced to attitudes in Western religions that pictured a universe with unity and purpose wherein man could expect natural laws to operate.

Science began in ancient Greece (600 B.C. to A.D. 400) where an enlightened civilization evolved a tradition of searching for explanations through logical thought. Science was taught by philosophers who obtained subject matter from casual observations of their environment and created interpretations through discussion.

A central question for speculation concerned the composition of the universe and the character of matter or material in it. Some philosophers thought matter filled all space and could be subdivided without limit. Others said that if a sample of matter could be cut up successively more finely with an infinitely sharp knife, a limit of subdivision must be reached. It is conceivable that the most fundamental proposition in science, *the atomic* (Gk. *átomos*, indivisible) *theory*, originated in that argument.

Stated by Leucippus about 450 B.C. the Greek atomic theory pictured all matter as composed of atoms. It defined atoms as the ultimate state of subdivision—invisible seeds infinite in number that filled all space. It explained creation of the universe as a whirling motion of large atoms from which came the earth and other celestial bodies, whereas smaller atoms of air, fire, and water were set free to circle around them. Democritus, a student of Leucippus, strengthened the theory by proposing that atoms varied in shape, perhaps had hooks for holding on to one another, and that different things were composed of different atoms. Man he imagined to have arisen from a primeval ooze as a mixture of every kind of atom. Change (such as the seasons) he ascribed to atomic combination and separation, adding that since atoms were indestructible, motion in nature was destined to go on forever.

These modern sounding ideas deserved to be guide lines along which science should have grown. They were but philosophical discourse, however, supported only by a proponent's eloquence, and their promise was lost when attacked by more forceful argument.

Aristotle (384–322 B.C.), the greatest scientist of antiquity, rejected atomicity. He believed matter to be continuous and infinitely divisible and that atoms, if they existed at all, were unimportant since the smallest parts of any given substance were probably all alike. He replaced atoms with

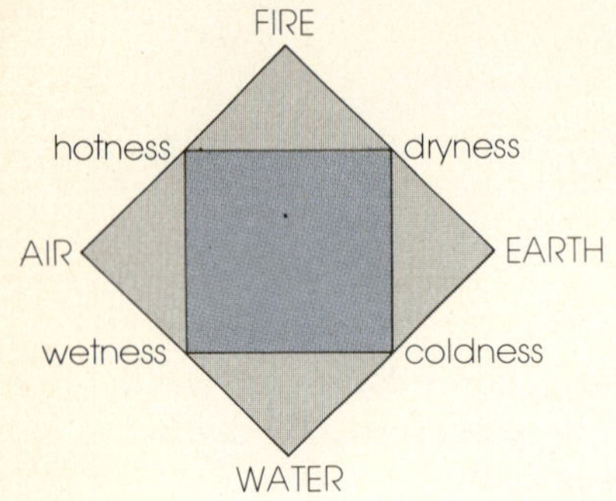

Figure 1-1 The terrestrial (earthlike) elements (in capital letters) and the fundamental properties, as conceived by the ancient Greeks. Man was related to nature because his drink was water, his food came from the earth, he breathed out warmness indicative of internal fire, and breathed in coolness of the air.

prima materia, an ill-defined substance that under various stresses assumed different forms. From Empedocles, an earlier philosopher, he borrowed the idea of the "four elements"—air, earth, fire, and water. The elements supposedly came about by combination of pairs of the "four properties"—hotness, dryness, coldness, and moistness (Figure 1-1)—and were kept in motion by the opposing forces of conflict and love.

Much of Aristotle's reputation as a scientific authority was due to his *empiricist* view that knowledge came from experience and that natural laws were revealed by examining physical things. To foster his philosophy he founded the Lycaeum, a school on which colleges were later modeled. He taught several subjects, organized the library, dissected specimens (he classified 500 kinds of animals), and left a written record that was an encyclopedia of learning. Unfortunately the method he applied so well to biology was lost. Later, when his works were rediscovered, they were taught as dogma instead of being used to encourage a spirit of research.

1-2. ALCHEMY TO CHEMISTRY IN WESTERN EUROPE

The leadership of Athens declined after Aristotle's death and the center of culture moved to Alexandria, Egypt, which had been conquered and partly populated by Greeks. Science became a mixture of Eastern mysticism and practical arts for producing dyes and alloys and objects of porcelain, metal, and glass. *Chemistry* can be said to have started then, although origin of the word "chemistry" is not clear. Its precursor *chemia,* which apparently meant "the Egyptian Art," was in use in A.D. 100. The art was to become known as *alchemy* and its practitioners were *alchemists.*

In A.D. 640 Arab conquerors overran Egypt and the Near East. After preserving via translations much of the science and learning they found, they introduced their culture to southern Europe around A.D. 1150. There the fall of the Roman Empire seven centuries before had left an intellectual wasteland.

Although alchemy soon became a tangled maze that embraced parts of philosophy, religion, astrology, magic, and other subjects, to Europeans it looked like a technological renaissance because the Romans had advanced nothing similar from the heritage of Greek science (Figure 1-2). Metallurgists adopted its implements. Druggists prepared its remedies in the name of medicine. Mystics, interpreting the ancient dictum of a common origin to mean that substances were interconvertible, looked for routes to the "ferment" whose drinking would give everlasting life, and to the "philosopher's stone'' whose touch would transmute common metals to the ''perfect'' metal, gold. Practical men derived laboratory methods and process equipment from alchemy, but apparently nobody advanced chemistry as a science. A useful theory of chemical reactions did not exist, and the concept of atoms was practically forgotten.

Some of Aristotle's lesser and faulty works, especially in physics, discovered in the eleventh century, were revitalized by St. Thomas Aquinas (Italy 1225–1274). The most renowned scholastic theologian of the Middle Ages, Aquinas tried to combine Aristotelian philosophy with Christian thinking and succeeded in establishing the doctrine that Aristotle's works and methodology were the authority from which scientific answers were to be sought. The result was that scholars were satisfied to teach and interpret authority while rejecting experimentation. Emergence of a true science was delayed for another four centuries.

Deductive logic was a Greek method whose influence lasted 20 centuries: A general operating principle was assumed (often on flimsy evidence) and then applied to specific cases according to an advocate's ability at debate. Break with fashion came about 1500, at the end of the Middle Ages, with adoption of *inductive logic* as a more productive method: After individual cases were examined, a general principle was announced.

Figure 1-2 An alchemist's laboratory in Europe in the sixteenth century. The inscription reads, "In the fire is the juice of everything, in short, the distillation of the bodies. Water becomes vigorous, limpid, and volatile." (Edgar Fahs Smith Memorial Collection, University of Pennsylvania.)

Two figures stand out for their criticism of Aristotelian system and advocacy of experimentation as the best means for gathering and testing evidence. One was Italian scientist Galileo Galilei (1564–1642), who combined in the manner of modern science observation, theory, and mathematics; the other was the English philosopher Frances Bacon (1561–1626). Whereas the Greeks believed speculation was more reliable than observation, Bacon believed both were useful. He anticipated that science would come to depend on research for discovery and invention of new knowledge, and that science would be recognized universally for applications that benefited society. He described the true scientist as

> The good man is one who possesses or is capable of exercising the intellectual and ethical virtues demanded by the aims and methods requisite for the discovery of truth in the study of nature, and the good life is the dedication to the improvement through this means of man's lot on earth.

The seventeenth century saw the beginning of systematic experimentation and the emergence of some genuine chemical principles. Chemists characterized three states of matter: *solids*, which were rigid and had

Discovery dates	Elements, in order of discovery												Total known
Prehistoric times	C	S											2
Metals of the ancients	Au	Ag	Cu	Fe	Sn	Pb	Hg						9
Medieval times	Zn	As	Bi										12
1600–1700	Sb	P											14
1701–1750	Co	Pt											16
1751–1775	Ni	N	O	Cl	Mn								21
1776–1800	Mo	Te	W	**U**	Zr	H	Ti	Y	Be	Cr			31
1801–1805	Nb	Ta	Rh	Pd	Ce	Os	Ir						38
1806–1810	Na	K	B	Mg	Ca	Sr	Ba						45
1811–1825	I	Cd	Li	Se	Si								50
1826–1850	Br	Al	**Th**	V	La	Er	Tb	Ru					58
1851–1875	Cs	Rb	Tl	Ga	In								63
1876–1885	Ho	Yb	Sm	Tm	Sc	Gd	Pr	Nd					71
1886–1900	Ge	Dy	F	Ar	He	Ne	Kr	Xe	**Po**	**Ra**	**Ac**	**Rn**	83
1901–1935	Eu	Lu	**Pa**	Hf	Re								88
1936–1950	**Tc**	**Fr**	**At**	**Np**	**Pu**	**Am**	**Cm**	**Pm**	**Bk**	**Cf**			98
1951–	**Es**	**Fm**	**Md**	**No**	**Lw**	**Rf**	**Ha**						105

Figure 1-3 Discovery dates of the chemical elements. Boldface type, radioactive; color, gases; grey, other nonmetals; all others, metals. The first nine elements were those found free (uncombined) in nature or whose oxides were easily reduced to metal by heat. The next few were discovered accidentally in alchemical experiments. Modern chemistry began with the discovery of nitrogen (N) and oxygen (O) in air research. Discovery of elements since then has been made possible by new methods of producing energy for reactions, invention of sensitive analytical instrumentation, elucidation of the structure of the atom, and organization of the natural sequence of the elements called the periodic table (see inside of back cover). The last 17 elements have been synthesized in high-energy experiments.

Figure 1-4 Isaac Newton (England 1642–1727), one of the scientific giants. Newton's guiding ambition was to understand forces. "I wish I could derive all phenomena of nature by some kind of reasoning from mechanical principles," he said, "for I have many reasons to suspect that they all depend upon certain forces by which the particles of bodies are either mutually attracted and cohere in regular figures or are repelled and recede from each other." By offering rational causes for natural phenomena Newton's work gave civilization sight of freedom from superstition. The feeling of hope was epitomized in an epitaph written by Alexander Pope, a contemporary verse satirist,

"Nature and Nature's laws lay hid in night.
God said: 'Let Newton be,' and all was light."

(Edgar Fahs Smith Memorial Collection, University of Pennsylvania.)

definite shape; *liquids*, which flowed to fit containers; and *gases*, which weighed little and filled any container. They distinguished between a *substance* such as chalk, and its *properties* such as color, hardness, and taste. Changes that matter underwent they classified as *physical* (melting ice, boiling water) when only physical form changed, and *chemical* (burning wood, dissolving iron in acid) when entirely new substances appeared for old. They began to recognize nature as a mixture to be understood when sorted out and analyzed. Chemical analysis, though crude, gave new meaning to earlier ideas that substances (later called "elements") existed which were not separable into anything simpler. Experiments with *chemical reactions* showed how elements could be combined to make *compounds* and how compounds could in turn be broken down into elements. By 1700 chemists recognized 14 elements (Figure 1-3) and several hundred compounds.

Attention was called once again to atoms, this time by Newton (Figure 1-4), who set the direction of science for the next 200 years. He discovered the laws of motion, invented the calculus, performed experiments in optics, and created with his ideas on light and gravity the first examples of theories that encompassed phenomena on a scale worthy of the best theories today. From work in both physics and alchemy he formulated a theory of matter:

1. Matter is created of atoms. There may be only one kind of atom. Differences among substances are probably due to different arrangements of atoms.
2. Attractive and repulsive forces (in character perhaps like the force of gravity) operate between atoms. They account for the stability and instability of compounds.
3. An "ether" fills all space; it is the medium that carries light.

Chemists took two directions from Newton's theory of matter. Theorists thought chemistry should look for evidence of atoms and try to measure their affinity so the properties of matter could be expressed in numbers. Experimentalists believed an atomic concept was unnecessary because chemistry could proceed through experimentation alone. Neither group made much progress but the breakthrough to modern science was not far off.

1-3. THE ATOMIC THEORY ESTABLISHED

Laws of combination and Dalton's theory

By the beginning of the nineteenth century the *metric system of measurement* (Appendix A-1) had been invented, laboratory apparatus was standardized to it, improved chemical balances for weighing substances were in use, analytical techniques yielded quantitative data on chemical reactions, and scientists were communicating with one another through publication of technical papers. Three *basic laws of combination* were discovered (of which more will be said in Chapter 12). The first stated that chemical reaction involved no measurable net change in mass: that the total weight of *products* from a reaction was the same as the total weight of *reactants*. The second law said that a chemical compound always had the same properties and composition: that nature combined elements in definite rather than arbitrary proportions. The third law stated that if two elements, A and B, combined individually with a given weight of a third element C, the ratio of the weights of A and B in those two compounds would be the same as, or a multiple of, the ratio by which they combined with each other.

An English chemist, John Dalton (see Figure 12-1) contemplated these chemical combination rules, and it seemed to him they all indicated that elements were composed of particles having individual weights* proportional to the weight ratios found in experiments. If so, a standard was needed against which all elements could be compared. Arbitrarily assigning hydrogen (the lightest element) a weight of one unit, he worked out values for the "atomic weights" of other known elements. The list then inspired him to propose a hypothesis about atoms. His textbook of 1808 contained this summary:

1. Matter is composed of atoms too small to be seen.
2. The atoms of a given chemical element are identical in all properties, including weight.
3. Different elements are made of different atoms which have different individual weights.
4. Atoms are indivisible. During chemical changes, atoms retain their characteristics.
5. Atoms combine chemically, forming *molecules*.
6. Elements combine in simple whole number ratios. In the simplest case, 1 atom of element A + 1 atom of element B gives 1 molecule of AB.

*See Appendix A-1 for usage of the word "weight."

The decisive point Dalton made was that different elements were composed of different atoms. This meant each different atom had a characteristic weight, as did each different molecule formed from atoms in a chemical reaction. From that assumption alone the laws of combination could be deduced. The crude laboratory techniques of the day failed to furnish direct proof, however, and Dalton proposed a test. In effect he stated that if atoms actually did exist, then when two elements, A and B, combined to form several different compounds, AB, A_2B, AB_2, etc., the weights of one element reacting with a fixed weight of the second element should reflect the atom ratio and therefore be in the ratio of small whole numbers. When experiments soon supported this thesis, the atomic hypothesis had its first major success.

To gain acceptance for his theory Dalton saw that he needed to have the *formulas* of compounds; that is, he had to know how many and what kinds of atoms combined in each molecule. No method for finding formulas was known, however, so he assumed a method. Using the "rule of greatest simplicity," he assumed that if two elements reacted to form only one compound, they did so in 1:1 atom ratio $(A + B \rightarrow AB)$. If two elements formed more than one compound, then the atom ratio was 1:1 in the most common compound and 1:2, 2:1, 1:3, 2:3, etc., in less familiar combinations. The method fit some cases but not others, however, and confusion about atoms continued.

Chemical formulas from gas reactions

A solution to the problem of chemical formulas was discovered in 1808 by the French chemist J. Gay-Lussac. He noted that when gases reacted with one another, their volumes were in the ratio of small whole numbers. He correctly interpreted the experiments to mean that a simple relationship existed between the volume of a gas and the number of particles it contained. Not realizing the combining volume rule (later law) was highly significant and would furnish strong support for his atomic theory, Dalton denied Gay-Lussac's conclusions because he could see no connection between gas volumes and weights of atoms.

The complete interpretation of the combining volume law was given in 1811 by Avogadro (see Figure 12-2). He said it meant that at given experimental conditions equal volumes of gases contained equal numbers of particles (molecules). Because small whole number ratios of volumes were found in all gas reactions, it followed that (a) the compounds formed had simple formulas, and (b) weights of equal volumes of individual gases must be in the ratios of the masses of their individual particles. Thus 1 liter of oxygen was 16 times heavier than 1 liter of hydrogen. Therefore the weight ratio of the oxygen atom to the hydrogen atom was 16:1.

Avogadro also showed how to interpret experimental results such as the following:

hydrogen (1 vol) + chlorine (1 vol) = hydrogen chloride (2 vol)

Dalton had used this example to argue against Gay-Lussac, pointing out that two volumes of product were formed although only one volume could be expected:

H (1 vol) + Cl (1 vol) ⟶ HCl (1 vol)

Avogadro, however, assumed that gaseous elements like hydrogen and chlorine occurred as *diatomic molecules* — in modern symbolism,* H_2 and Cl_2. Reaction of 1 molecule of each would then give 2 molecules (or 2 gas vol) of hydrogen chloride:

$$H_2 + Cl_2 = 2HCl$$

Despite the clarity of reasoning and the vital need for a method of determining correct formulas and atomic weights, most chemists looked elsewhere for help. Many remained suspicious of atomicity itself, and neither Avogadro nor Dalton lived to see his ideas vindicated.

The atomic weight problem resolved

Berzelius (see Figure 11-1), whose accurate experimental data substantiated the laws of chemical combination, also supplied support for Dalton's atomic theory by preparing tables of atomic weights and trying to determine the formulas of compounds. He chose oxygen as an atomic weight standard because of its availability in the atmosphere and its ease of chemical combination. In a typical experiment he heated a weighed quantity of a metallic element in air, weighed the metal oxide formed, then calculated the weight of metal that proportionately would have combined with the standard weight of oxygen. If the combining ratio was 1 atom of metal to 1 atom of oxygen, the calculated weight of the metal was its atomic weight. If the ratio was $2:1$, then the atomic weight of the metal was half that computed for the $1:1$ case, etc.

When the atomic weights of Berzelius are recalculated to correspond with today's atomic weight scale, they are in excellent agreement except in those cases in which his assumed oxide formula was incorrect. But precisely because formulas had to be assumed, other chemists remained unconvinced and this careful work also went largely ignored.

A new means of investigating chemical composition had become available about 1800 when chemists learned that substances could be decomposed by electric current. By passing electricity through various compounds and measuring the amount of decomposition, Faraday (see Figure 18-3) arrived at a list of numbers he called *electrochemical equivalents*. He found the same quantity of electricity that would produce 1 gram of hydrogen (arbitrarily called "1 equivalent") from the electrolysis of water, would also liberate 8 grams of oxygen, 36 grams of chlorine, 104 grams of mercury, etc. These values should have soon led to a corresponding list of atomic weights since, as we will see later, a very simple relationship connects an element's equivalent weight and atomic weight. The relationship was not discovered, however, and, because reproducible equivalent weights could be determined easily, most chemists preferred to use them and circumvent argument about atomic theory.

Indecision concerning atoms continued until 1860, when a congress of European chemical specialists was convened to determine if theoretical chemistry, which had been operating haphazardly as a collection of individual preferences, assumptions, and rules of thumb, could be unified

* H and Cl are *symbols* for elements and can be interpreted as individual atoms or many atoms of the elements; H_2, Cl_2, and HCl are formulas for molecular substances. They can be interpreted either as individual molecules containing two atoms each, or many molecules. Formula and equation writing are discussed in Chapter 11.

through agreement on atoms, atomic weights, symbolism, and nomenclature. One notable contribution came from the meeting. Italian chemist S. Cannizzaro showed convincingly how Avogadro's hypothesis, neglected for 50 years, had all the time been the one real clue to atomic weights. He reviewed the reasoning:

1. Gases reacted with one another in small whole-volume ratios, 1:1, 1:2, etc. From that one could assume equal volumes of different gases at the same experimental conditions contained the same number of molecules.
2. Hydrogen, chlorine, and some other gaseous elements had to contain two atoms per molecule to account for the volume relationships in reactions such as

$$H_2 + Cl_2 = 2HCl \qquad \textit{(1 vol + 1 vol = 2 vol)}$$

3. It followed from (1) that if equal volumes of gases were weighed under the same conditions, their weights would be in proportion to the masses of the molecules they contained. Reference to a standard gas would give a table of relative molecular weights. If hydrogen, the lightest element, was given an atomic weight of one, $H = 1$, then the hydrogen molecule was $H_2 = 2$. And since experiment showed that chlorine gas was 35.5 times heavier than an equal volume of hydrogen gas under the same experimental conditions, the molecular weight of chlorine had to be twice its atomic weight, or $Cl_2 = 2 \times 35.5 = 71$, etc.

Acceptance of Cannizzaro's conclusions, which today are taken for granted because all scientists accept the atomic theory, was the rationale chemistry needed to emerge as a science. *Chemists* became scientists who thought in terms of atoms (and later in terms of molecules). *Organic chemistry,* which now encompasses most of the chemical compounds known, started its explosive growth as the atomic theory furnished a way to visualize complex structures built up of carbon atoms. In *inorganic chemistry* the atomic theory spurred the search for new elements. They were found in the earth, the air, even in the gases around the sun, and given sensible arrangement in the periodic table. Correct formulas of compounds, complete chemical equations, and atomic and molecular weights, which followed from atomic theory, established a foundation for analytical chemistry. The ability to quantitatively analyze products from chemical reactions immediately assisted both basic science and industry. For a hundred years *chemical industry* had been limited merely to improving a few simple processes for making acids, alkalis, and bleaches. Given new direction from laboratory research in the last quarter of the nineteenth century, it emerged as a powerful innovative force. With the synthesis of dyes, solvents, photographic chemicals, and pharmaceuticals, it showed the way toward science-based technologies.

Accompanying progress in applied chemistry was a tenuous though persistent advance of atomic theory. Several provocative bits of evidence were found which solidified the belief that atoms and molecules actually existed. The observation under a microscope of random displacements of tiny particles in a drop of water (see Figure 14-10) was interpreted as caused by bombardment of still smaller and invisible water molecules that had ceaseless motion characteristic of molecular-sized matter. In another experiment a film believed to consist of single molecules was prepared and measured; its thickness of one 10-millionth of an inch established the first estimate of molecular diameters. In another experiment the number of molecules contained in a mass equal to the molecular weight of any substance expressed in grams (2 g of H_2, 71 g of Cl_2, etc.) was calculated to be 10^{23}, an enormous number that emphasized the minuteness of individual atoms.

The source of attraction between atoms came under new scrutiny. Scientists recognized that atoms could hardly be homogeneous spheres, although it was handy in developing elementary theories to picture them as that. Dalton had not suggested how atoms joined together to form molecules and nobody seemed to know how to follow up Berzelius' conclusion that since compounds were decomposable by electric current, electric forces must hold them together. Then physicists discovered a few techniques for examining nature at the atomic level, although at first they could not interpret the results. First, observations with a spectrometer (see Figure 2-3) on light emitted from heated substances showed that each element gave a different pattern of so-called spectral lines. Second, experiments in which electricity was sent through gases held at low pressure produced rays that seemed to be composed of charged particles; however, the size of some of the particles was even smaller than that calculated for atoms. In further experiments X rays were discovered and interpreted as another atomic phenomenon. Then it was found that certain atoms in nature were spontaneously unstable, losing huge amounts of energy as rays shot from them. Then, . . .

QUESTIONS

1. Sciences Physics is described as a *laboratory science* because its problems are usually investigated in a controlled manner in the lab. Volcanology (study of volcanoes) is called an *observational science* because many of its questions can only be answered by observing nature as it is. From leafing through this textbook, which type of science do you think chemistry is? Give some examples.

2. Mixtures A *substance* has constant composition and properties throughout. A *mixture* contains two or more substances. Powdered breakfast drinks, gelatin desserts, pesticides, cleaning preparations, and other packaged products found on the grocers' shelves are mixtures. Copy the trade name and list of substances from the label of one such mixture. In alphabetical order list as many of the elements present as you can. Which of the compounds do you think are man-made rather than derived from natural sources? What do you think is the purpose of each substance in the mixture?

3. Properties Properties are the distinguishing qualities of matter. Chemical properties are related to the transformation of matter into other substances via chemical reaction. Physical properties are characteristics observable without chemical change. Classify the following properties accordingly: melting point, electrical conductance, crystal form, flammability in air, digestibility in the stomach.

4. Properties Properties of matter are sometimes described as *intensive* when they are independent of, and *extensive* when they are dependent on the quantity of matter present. Use the two terms to classify the following properties and discuss briefly with reference to some specific substance(s) that you select: hardness, mass, odor.

5. Materials Material is any kind of matter. *Homogeneous* material has uniform properties throughout. *Heterogeneous* material is a mixture of homogeneous materials and has nonuniform properties. Classify the following materials accordingly: the air you breathe, drinking water, aluminum foil, asphalt paving.

6. Scientific method Scientists seek correlations in nature. They do so by gathering facts through experimentation, organizing the facts, generalizing the facts into a hypothesis, then testing the hypothesis. A student dissolves a spoonful of sugar in water. In another experiment he heats a spoonful of sugar and the sugar turns black and finally burns. He announces this hypothesis: Water-soluble white crystalline substances are combustible. What is wrong with his use of scientific method?

7. Aristotle Aristotle established *syllogism* as a method of reaching conclusions about nature. **(a)** Copy a dictionary definition of syllogism. **(b)** Explain how the following example fits the definition. Acids are good solvents for minerals, and liquids that taste sour are acidic; therefore sour-tasting liquids will dissolve minerals. **(c)** Do you think this syllogism is true? Explain. **(d)** What do you think of syllogism as a scientific method?

8. Dalton In Dalton's notation symbols for oxygen and nitrogen were ○ and ⊕, respectively. He represented nitric oxide ⊕ ○ and nitrous oxide ⊕ ○ ⊕. **(a)** How are the compounds written in chemists's shorthand today? What advantage do you see in the modern system? **(b)** If the masses of oxygen and nitrogen atoms are about the same, how do the masses of 1 liter each of nitric and nitrous oxides compare?

9. Equations In 1833 Gaudin, a French chemist, gave the following volume diagrams (each square representing 1 gas vol) to show reactions in which hydrogen chloride gas ⊙ ⊗ and water vapor ⊙ ○ ⊙ were products:

(a) Identify ⊙ ⊙, ⊗ ⊗, and ○ ○. **(b)** Did Gaudin understand Avogadro's hypothesis (as later explained by Cannizzaro)? Discuss.

10. Equations Carbon monoxide gas burns in oxygen gas to given carbon dioxide gas. In modern symbolism Dalton would have described the reaction as $CO + O \rightarrow CO_2$ to represent 1 vol of each gas. Experiments prove, however, that 2 vol of carbon monoxide react with 1 vol of oxygen to give 2 vol of carbon dioxide. How did Avogadro's hypothesis explain the experimental facts? How should the equation be written?

PROBLEMS

11. Elements Plot a graph from the data of Figure 1–3. Turn the paper to make the *abscissa* (horizontal axis) the longer axis. On it put "discovery dates" 1600, 1625, 1650, etc., to 1975. On the *ordinate* (vertical axis) put "total elements known" 0, 10, 20, 30, etc., to 110. Represent each element with a distinct dot. After you have connected all the dots with a continuous line, draw an arrow to each dot and label each arrow with the element's symbol. Use a separate color or other device to designate gases. Do the same for other nometals and also for radioactive elements. Title the graph and add a note explaining the colors.

12. Elements From Figure 1–3 calculate the percentages of elements that are nonmetals, gases, metals, radioactive.

13. Weights **(a)** A sackful of padlocks weighs 4.77 kg. A sack containing an equal number of padlock keys weighs 227 g. How many times heavier is a lock than a key? (See Appendix A for metric measurements.) **(b)** At the same temperature and pressure 1 liter of hydrogen cyanide gas weighs 1.20 g as does 0.61 liter of carbon dioxide gas. What does 1 liter of carbon dioxide weigh? **(c)** How much heavier is a carbon dioxide molecule than a hydrogen cyanide molecule? **(d)** The molecular weight of carbon dioxide is 44. What is the molecular weight of hydrogen cyanide?

14. Atomic weight In an experiment 0.500 g of zinc was heated with sulfur, giving a quantitative yield of zinc sulfide weighing 0.745 g. **(a)** What weight of sulfur combined with the zinc? **(b)** If zinc and sulfur combined in 1:1 ratio ($Zn + S = ZnS$), what is the atomic weight of zinc, given that the atomic weight of sulfur is 32.0?

15. Atomic weight When 0.400 g of sulfur (S) is burned in air, 0.800 g of a sulfur oxide forms. If 1 atom of sulfur has combined with 1 molecule of oxygen ($S + O_2 = SO_2$), what is the atomic weight of S, given that the atomic weight of O is 16.0?

16. Atomic weight In one experiment Berzelius prepared zinc oxide and found it was 20.0 percent oxygen and 80.0 percent zinc by weight. **(a)** Based on his atomic weight standard of $O = 100$ and assuming the oxide formula to be ZnO, what atomic weight of zinc did he obtain? **(b)** Had he used the modern value, $O = 16.0$, what atomic weight of zinc would he have gotten?

17. Molecular weight Avogadro weighed various gases in a flask and obtained data such as these:

flask filled with hydrogen	73.85 g
flask evacuated	73.75 g
flask filled with oxygen	75.35 g
flask filled with chlorine	77.30 g

(a) What weights of oxygen, hydrogen, and chlorine were contained in the flask? **(b)** Given the molecular weight of oxygen, $O_2 = 32.0$, what were the experimental molecular weights of H_2 and Cl_2? **(c)** What were the atomic weights of hydrogen and chlorine from these data?

18. Atomic weight Use Cannizzaro's reasoning to determine the atomic weight of bromine (Br) and the molecular weight of hydrogen bromide from the following data, given that the atomic weight of hydrogen is 1.00: 1 liter of hydrogen gas weighs 0.0893 g; at the same temperature and pressure 1 liter of bromine gas weighs 7.14 g. And experiments show that

bromine (1 vol) + hydrogen (1 vol) = hydrogen bromide (2 vol)

19. Molecular weight **(a)** Name the elements in this molecule and tell how many atoms of each are present: $CaSiF_6$. **(b)** Add up the atomic weights (periodic table) to obtain the molecular weight of the compound.

20. Atomic weight In the early 1800s it was hoped that the density (mass per unit volume) of an element might provide a clue to its atomic weight. Here are the densities in grams per cubic centimeter of these metals: $Be = 1.82$, $Ti = 4.50$, $Fe = 7.86$, $Ag = 10.5$, $Hg = 13.6$, $Au = 19.3$, and as a standard the atomic weight of silver, $Ag = 108$. **(a)** Calculate the atomic weights of the metals assuming density to be directly proportional to atomic weight. Arrange your results in a table. **(b)** Beside each calculated atomic weight write the accepted value (periodic table). Comment on the assumption in (a).

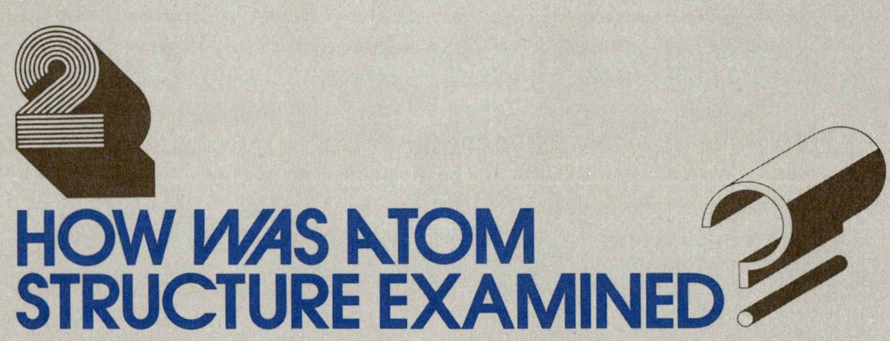

HOW WAS ATOM STRUCTURE EXAMINED?

THE IMPORTANT CONCEPTS

2-1. The spectroscope
1. Electromagnetic radiation in nature exists over a vast range of wavelengths.
2. Light is the visible part of the electromagnetic spectrum.
3. Different wavelengths of radiation can be separated and measured with a spectrometer.
4. A structural feature of the atom is related to the spectrum it gives when heated.

2-2. Electricity and magnetism
1. Rubbing a surface creates charge called "static electricity."
2. Coulomb's law governs force between fixed charges.
3. A force field exists around a fixed electric charge and around a magnetic pole.
4. Charges caused to move by a battery constitute an electric current.

5. Electricity and magnetism are related.
   a. Electric current creates a magnetic field perpendicular to its flow.
   b. A charged particle moving in a magnetic field "feels" a force perpendicular to both its motion and the field.

2-3. Experiments in discharge tubes

1. Passage of electric current through a low-pressure gas produces cathode rays and a flow of positive ions.
2. Cathode rays consist of electrons, small particles from atoms.
3. Electron mass and charge can be measured.

2-4. Discovery of X rays and radioactivity

1. Energetic electrons striking a metal surface produce X rays, a very short wavelength radiation.
2. Some elements are naturally radioactive; they decompose spontaneously, giving off rays.

2-5. The Rutherford atom

1. The scattering technique is a way to examine atom-sized particles.
2. Experiments indicate that the atom consists of negative electrons and mostly empty space around a small, dense positive nucleus.
3. Atom to nucleus radius ratio is at least $10^4 : 1$.

2-6. Z, the atomic number

1. Z, the serial number of an element in the periodic table, is the number of unit positive charges in the atom's nucleus.

A hundred years ago few scientists believed that the existence of atoms could be proved, and fewer yet imagined that a fine structure within the atom would also sometime be revealed. Experiments with electric discharges in gaseous elements were even then, however, providing evidence that would culminate in discovery of the electron, a fundamental subatomic particle. Identification of rays shooting out of spontaneously decomposing (radioactive) elements soon gave additional ideas about atom composition. Then one of the most famous experiments in physics revealed that the atom acts as if it is composed of a tiny dense nucleus and much "empty" space. Another experiment gave a way to determine the electric charge on nuclei and to assign to elements the serial numbering found in the periodic table.

2-1. THE SPECTROSCOPE

Spectral lines

Newton showed that the passage of sunlight through a narrow slit and a glass prism produces the *spectrum* illustrated in Figures 2-1 and 5-1. He correctly interpreted the effect as demonstrating that white light contains visible light of all colors, each component of which is refracted (bent) a different amount by passage through glass.

We will define *radiant energy* as energy carried through space or matter in the form of electromagnetic waves. Electromagnetic waves have both an electric and a magnetic component. The waves are classified according to

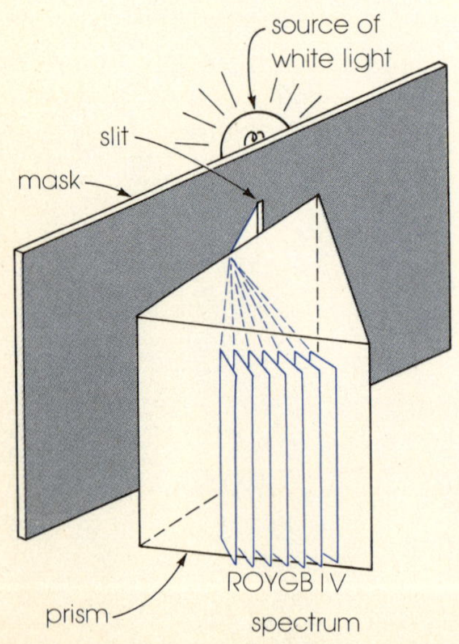

Figure 2-1 Passage of white light through a glass prism yields a continuous spectrum of blending colors. (If the spectrum is focused through a lens into another prism, white light is reconstituted.) VIBGYOR is a nmonic for the colors: violet, indigo, blue, green, yellow, orange, red. Violet light is the most energetic, red the least.

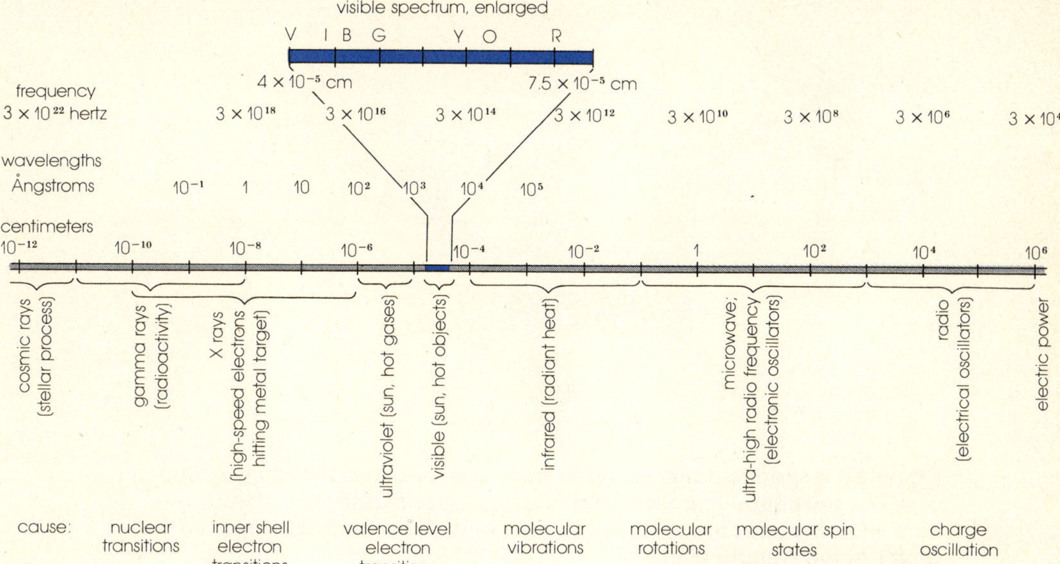

Figure 2-2 The complete electromagnetic spectrum. The longer the wavelength, the less energetic the radiation. In a vacuum all electromagnetic waves travel at the same speed, 3×10^{10} cm/sec. (For a definition of wavelength see Fig. 3-4.)

their *wavelength,* or distance between successive crests, and *frequency,* or number of waves passing a point in unit time. The complete electromagnetic spectrum (Figure 2-2) encompasses energies having wavelengths from less than 10^{-12} cm* to more than 10^6 cm.

Visible light is the small segment of the electromagnetic spectrum perceived by the human eye. Different wavelength regions within that segment are seen as different colors. An instrument that separates radiant energy into its components is called a *spectroscope.* One that also measures the wavelengths is called a *spectrometer* (Figure 2-3). German chemists G. Kirchhoff and R. Bunsen were among the first to make spectrometric measurements to catalogue the spectra of known elements and apply them to the analysis of unknown mixtures. A coincidence of spectral lines identified a known element, an absence of lines proved the corresponding elements were missing, and one or more unidentifiable lines announced the presence of a new element (they discovered rubidium and cesium in this way). In 1859 they discovered the basic *laws of spectroscopy:* (1) Each element when heated displays a characteristic spectrum. (2) The cooler vapor of each element is capable of absorbing that radiation which is characteristic of the element when heated.

For 25 years physicists studied spectra, finding many lines at visible as well as ultraviolet and infrared wavelengths, yet they found no mathematical relationships in the data. Then Swiss schoolmaster J. Balmer deduced a formula that predicted the wavelengths in the hydrogen spectrum (Figure 2-4),

* Abbreviations for measurements are given in Appendix A, Table A-3; cm means *centimeter;* 2.54 cm = 1 in.

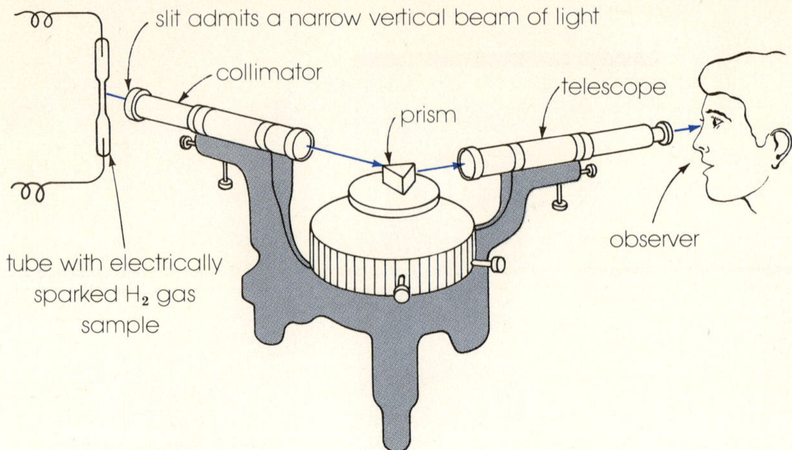

Figure 2-3 A spectrometer, set up to study the spectrum of hydrogen. The observer sees, superimposed on a scale, a series of narrow, colored, vertical lines. The lines are images of the slit and characteristic of excited atoms of the sample element.

$$\frac{1}{\lambda} = R\left(\frac{1}{2^2} - \frac{1}{n^2}\right) \tag{2-1}$$

where λ is the wavelength in centimeters; R is the Rydberg constant $(1.097 \times 10^5 \text{ cm}^{-1})$; and n has integral values 3, 4, 5, ... If one postulated that hydrogen existed as atoms, then the equation indicated some regularity in the hydrogen atom. No theory of atomic structure could account for the

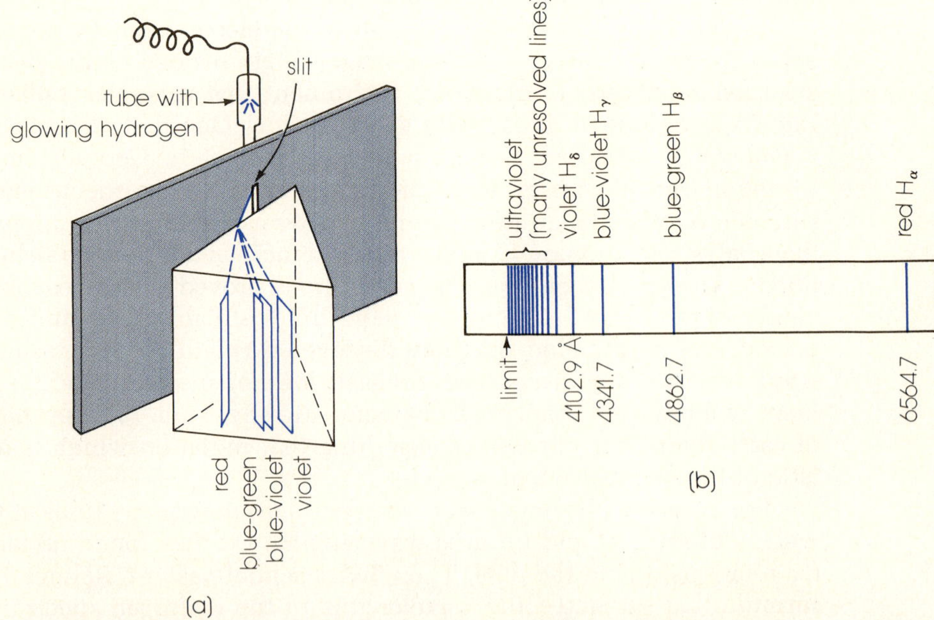

Figure 2-4 The hydrogen spectrum. (a) Producing the spectrum by passing an electric discharge through hydrogen gas. (b) Reproduction of hydrogen spectral lines as photographed through a spectrometer. Numbers are wavelengths in angstrom units (see Figure 2-2) of prominent lines.

equation, however, nor for spectra in general. According to established physical principles then available, spectra meant that some sort of vibration was occurring in matter. But what was vibrating?

2-2. ELECTRICITY AND MAGNETISM

Static electricity

To understand the experiments that preceded formulation of the first good model of the atom, we must briefly review related work in electricity and magnetism.

When the surface of a solid is rubbed, it acquires an electric charge called *static electricity*. Although static electricity was known to the ancient Greeks, who learned that an amber rod rubbed with wool would pick up bits of dry straw, the effect was not investigated until the seventeenth century. A generally accepted theory of electricity was proposed about 1730 by C. Du Fay of France, who described two kinds of electricity, vitreous and resinous (now called *positive* and *negative*). With an *electrometer* (Figure 2-5) he demonstrated how bodies charged with opposite electricities attracted, whereas those with like electricities repelled one another. Du Fay's theory was gradually replaced by the single-fluid theory of Benjamin Franklin, American statesman and scientist. Franklin assumed that electricity was a flow of only one kind of charge. Absence of that charge created charge of the opposite sign.

Quantitative electrical experimentation began during the late 1700s, when French engineer C. Coulomb discovered the fundamental *law of electrostatics*, or the science of electricity at rest. Using a torsion balance (Figure 2-6) he proved that two charges of like signs repelled one another (and two charges of opposite sign attracted one another) with a force, F_{elec},

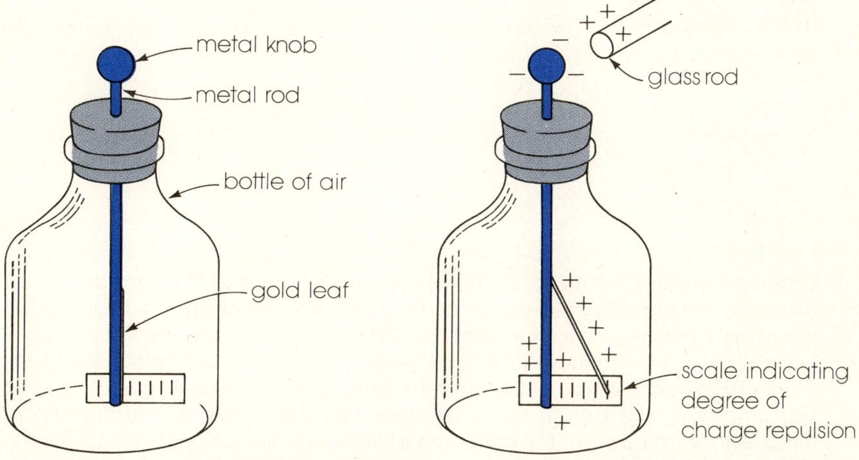

Figure 2-5 An electrometer. (Left) Uncharged. (Right) A glass rod rubbed with a silk cloth becomes positively charged. As it is brought up to the metal knob it attracts negative charges (later called electrons) to the top of the stem, leaving the bottom positively charged. Mutual repulsion of lower stem and thin gold leaf causes the leaf to move a distance apart that is proportional to magnitude of charge on the glass rod. Removing the rod returns the electrometer to its original neutral condition.

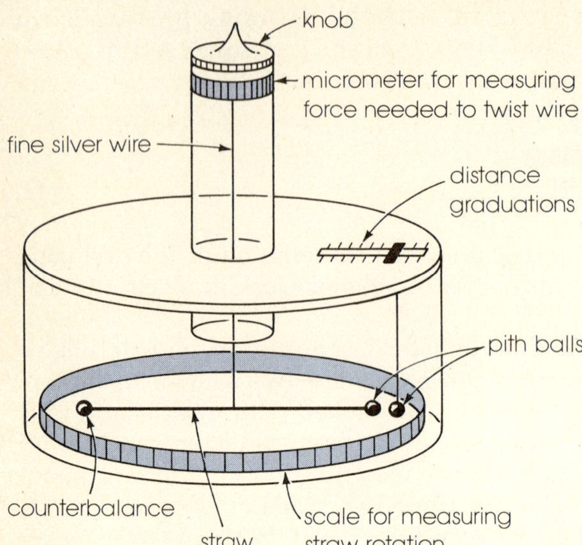

Figure 2-6 Coulomb's torsion balance. As the pith balls are given like charges they repel one another and the straw rotates a number of degrees, as read on the lower scale. The knob is then turned to twist the wire and bring the pith balls closer together against their mutual charge repulsion. The force required is measured on the micrometer. The micrometer is previously calibrated by determining the mechanical force needed to twist the wire.

directly proportional to the product of the magnitudes of the charges q_1 and q_2, and inversely proportional to the square of the distance r between the charges,

$$F_{\text{elec}} = k\,\frac{q_1 q_2}{r^2} \tag{2-2}$$

where k is a proportionality constant.*

Electric fields

To visualize the influence of one electric charge on another, Faraday originated the idea of a *field of force* existing around a concentrated charge fixed in space. He then described interaction of charges as the interaction of fields. We continue to use his concept today. To discuss a field we imagine

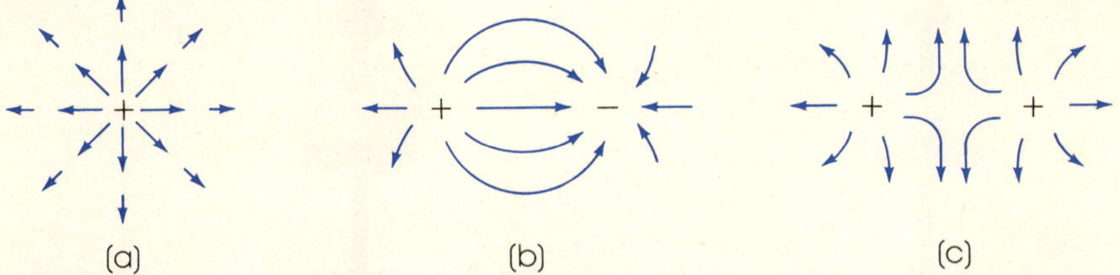

Figure 2-7 Electric fields. (a) Some of the vectors or force lines needed to describe the electric field about an isolated positive charge. Force (vector intensity) varies with the square of the distance in accordance with Coulomb's law. (b) Continuous lines that begin on positive and terminate on negative charges are used to describe the field around two charges of opposite sign. If the two charges lie close together they are called an electric dipole; it has zero total charge. (c) Repulsion of lines of force about two like charges.

---

* In the *centimeter, gram, second (CGS) system of measurement*, force is expressed in dynes, distance in centimeters, and time in seconds. Letting $k = 1$, the equation becomes

$$F_{\text{elec}} = \frac{q_1 q_2}{r^2}$$

This defines q as a unit of charge called the *electrostatic unit* (esu), which we will use later in describing charged atoms. Two points of charge, each of 1 esu and 1 cm apart, exert on one another a force of 1 dyne. No fundamental explanation has been found for the experimental fact stated by Eq. 2-2.

introducing a very small test charge in the neighborhood of a large reference charge, and then ask what force* the test charge experiences.

Electric field intensity is a *vector* quantity. A vector is a physical quantity that requires both magnitude and direction for its description. It is represented by an arrow whose length indicates magnitude. To picture the invisible electric field we draw field vectors about the charges, as illustrated in Figure 2-7.

Magnetic fields

Magnetism was known to the ancients, who found that a natural magnetic ore called *lodestone* or magnetite (a mixture of iron oxides, FeO and Fe_2O_3) attracted iron. They also found that iron rubbed with magnetite became magnetic and that a suspended needle of magnetized iron tended to align itself in a north–south direction. The action as a direction indicator or *compass* was correctly explained by English physician W. Gilbert in 1600 as an attraction of the needle's ends or poles by the earth, which itself must be a huge magnet with north and south poles. Coulomb subsequently proved that (1) magnetic force followed the inverse-square law (Eq. 2-2), and (2) unlike magnetic poles attracted whereas like poles repelled one another.

The battery (see Figure 18-1), a device for producing moving charges or *electric current* in a wire, was invented in 1800. In 1820 during a lecture demonstration Danish physics professor H. Oersted observed that a wire connected through a battery caused the needle of a compass in its vicinity to move in a direction perpendicular to that of the flow of electric current. By contrast, a stationary charge had no effect on the needle.

When Oersted published his observation, physicists everywhere recog-

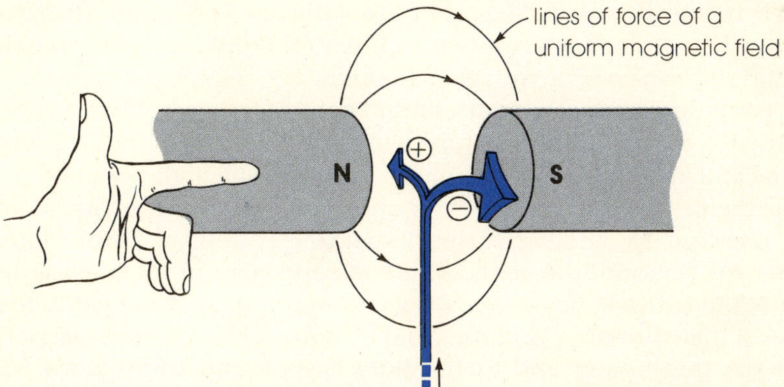

Figure 2-8 Force in a magnetic field. Direction of force on a charged particle moving in a magnetic field is given by the "left-hand rule": when thumb, forefinger, and other fingers of the left hand are held at right angles to one another, if the thumb points in the initial direction of travel of a negative particle and the forefinger points in the direction of the magnetic field (north to south), the other fingers point in the direction of the force. (A positive particle "feels" force in the opposite direction.)

---

* The force F is equal to the product of the magnitude of the test charge q and field strength E:
$$F = qE$$
where F may be expressed in dynes, q in esu, and E in dynes per esu.

nized that he had discovered a new natural force that was directional and associated with motion, and that related electricity and magnetism. Experiments quickly proved that the magnetic field around a current-carrying wire was arranged in concentric circles perpendicular to the direction of the wire. Further experimentation proved that a charged particle moving through a uniform field of a magnet experienced a force perpendicular to both the direction of the particle's motion and the direction of the magnetic field (Figure 2-8).

Analysis of the force–particle motion relationship showed that although the direction of a charged particle changed in a magnetic field and its path became circular, its velocity remained constant. The equation derived to describe these conditions was

$$\frac{m}{q} = \frac{Br}{cv} \qquad\qquad (2\text{-}3)$$

where B is the magnetic field strength; r is the radius of the particle's circular path; c is a constant (3×10^{10} in the CGS system); and v, q, and m, respectively are particle velocity, charge, and mass. When the terms on the right side were known, m/q, the mass-to-charge ratio of a particle acted on by a magnetic field, could be calculated. The equation was soon to prove useful in atomic investigations.

2-3. EXPERIMENTS IN DISCHARGE TUBES

Rays negative and positive

Work done in carrying a unit positive or unit negative charge from one point to another is called *electric potential difference*. It is usually measured in a unit called the *volt* (V) (details are in Section 18-3). When electric current flows between two points, the points are said to exist at a potential difference or a voltage difference.

Gases are normally poor conductors of electricity. However, as a gas is gradually pumped from a tube equipped with a *cathode* (negative electrode) and an *anode* (positive electrode) held at a high potential difference, a sparking or *electric discharge* is observed between the two electrodes. As gas pressure is reduced further, sparking is replaced by a glow. At still lower gas pressure the general glow disappears but the glass at the end opposite the cathode begins glowing (fluorescing) as if the cathode is sending rays in that direction (Figure 2-9a). A solid object placed between cathode and the fluorescent end of the tube casts a shadow (Figure 2-9b), and a magnet causes displacement of the fluorescence pattern (Figure 2-9c).

English physicist W. Crookes, who named the rays *cathode rays*, concluded that this behavior showed that the cathode was giving off negative particles of some sort, perhaps negative ions (negative atoms).

If Crookes' assumption was correct, where were the positive particles that compensated for the negative particles in cathode rays? In 1886 German physicist E. Goldstein found the answer by using a tube with a perforated cathode through which positive rays could travel in a direction opposite that of cathode rays (Figure 2-10). Wilhelm Wien (Germany 1864–1928, Nobel prize in physics 1911), who carefully measured the bending of Goldstein's rays in strong electric and magnetic fields, proved

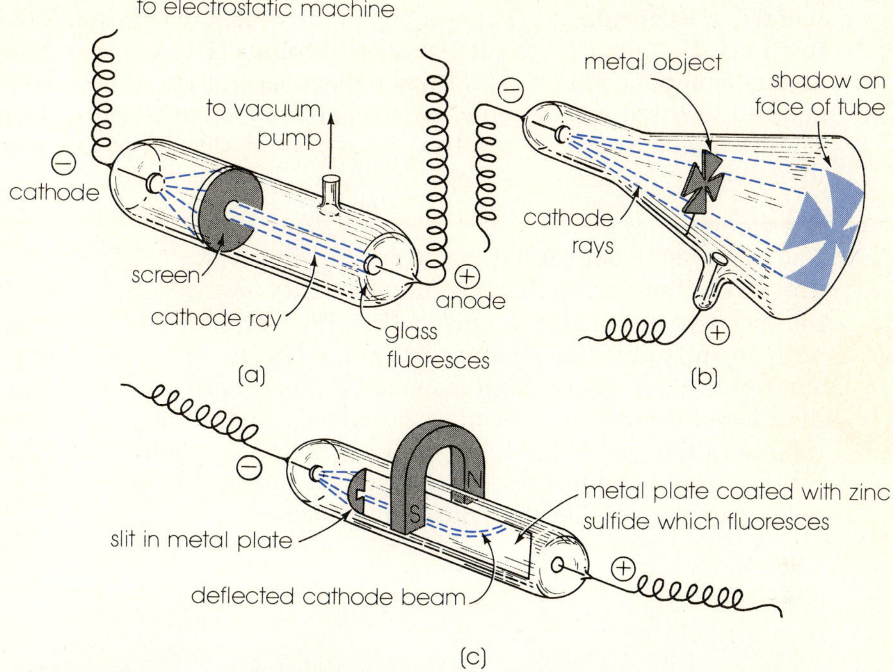

Figure 2-9 Gas-discharge tubes. (a) The basic design. Anode and cathode operate at a high-voltage difference. As air pressure is reduced by a vacuum pump to 1/100 of atmospheric pressure, a silent electric discharge begins to pass between the electrodes, the gas glows, and the glass fluoresces. (b) A Crookes tube, showing cathode rays producing a shadow. (c) Another Crookes tube showing deflection of the cathode beam with a magnet.

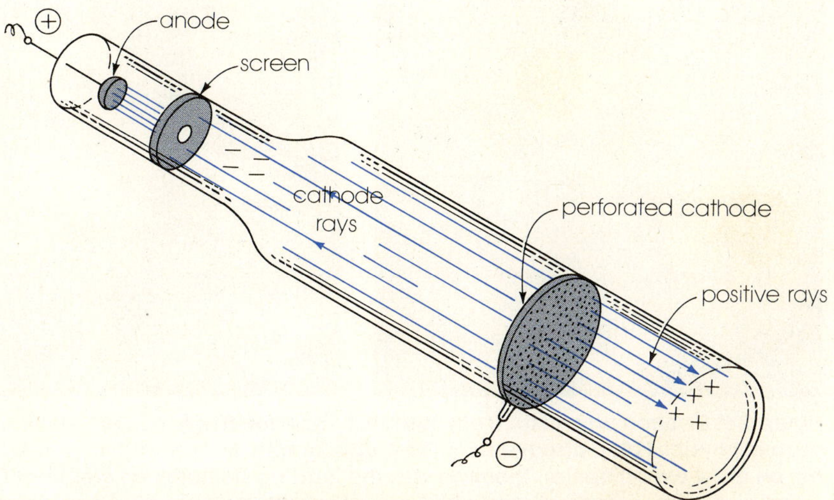

Figure 2-10 Goldstein's tube for examining positive rays. Each gas examined produced positive particles having mass-to-charge ratios characteristic only of that gas. Positive rays also had different colors: pink for hydrogen, yellow for oxygen, etc. Positive rays were later identified as atoms and molecules of the sample gas stripped of one or more electrons.

that when hydrogen gas (H_2) was put in the tube, the rays consisted of positive ions, H_2^+ and H^+, the latter called *protons*. His research renewed the argument about cathode rays. If gas molecules and atoms became positively charged in these experiments, then what were Crookes' negative particles? The dilemma would be resolved by the great discovery that the atom is divisible.

The electron

The mathematician turned experimental physicist, J. J. Thomson (see Figure 2-11), thought discharge tube electricity probably split gas molecules into positive and negative ions as $H_2 = H^+ + H^-$. In 1897 he tested the idea with an apparatus that produced a cathode beam whose deflection could be studied in both electric and magnetic fields (Figure 2-12). He immediately discovered that his assumption was wrong. Instead of giving negative ions unique to the gas in the tube, each gas tested gave identical cathode rays. Further, the cathode rays acted as if they consisted of negative particles having a mass smaller than the atomic mass.

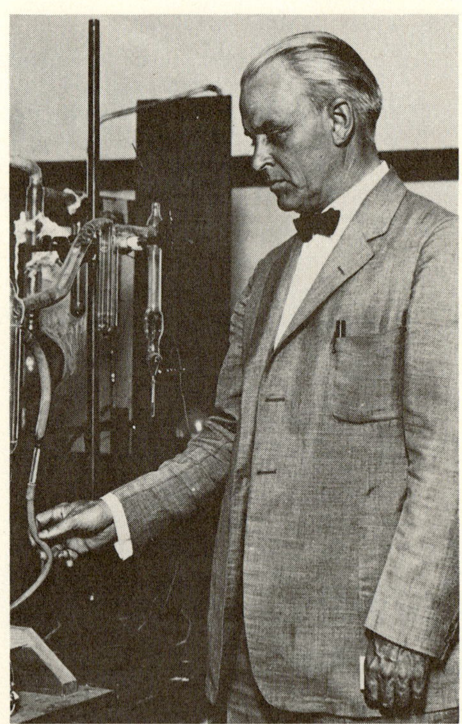

Figure 2-11 Electron physicists. (Left) Joseph J. Thomson (England 1856–1940, Nobel prize in physics 1906), discoverer of the electron, with his cathode ray tube. Though not an artful experimenter, Thomson was the leading authority on electrons for 40 years and director of the Cavendish Laboratory. (Right) Robert Millikan (United States 1865–1953, Nobel prize in physics 1923), discoverer of the electronic charge. Millikan also verified Einstein's photoelectric equation and did research on cosmic rays, which he so named. In later life he was president of the California Institute of Technology. (Photographs: left, Cavendish Laboratory, University of Cambridge; right, courtesy California Institute of Technology.)

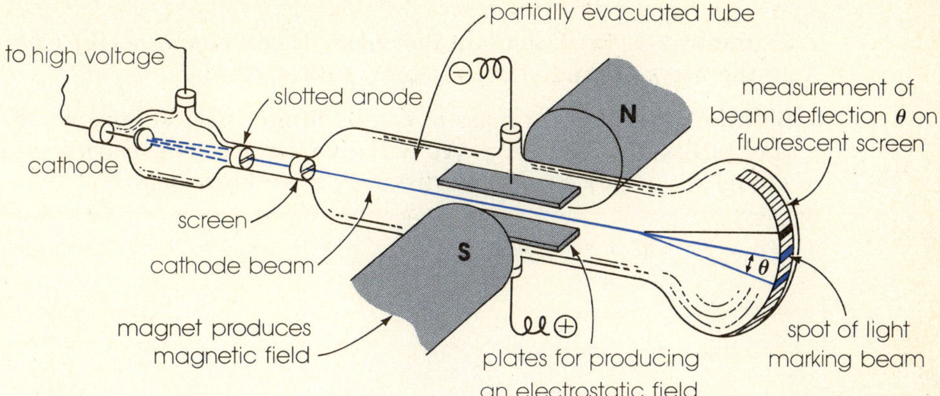

Figure 2-12 The apparatus with which Thomson discovered the electron. His calculations involved plate length, the strengths of the electric and magnetic fields, and the cathode beam deflection angle.

Balancing magnetic and electric fields, Thomson measured beam deflections. Substituting data into an equation similar to Eq. 2-3, he calculated first the velocity of the particles in the beam, then their mass-to-charge (m/q) ratio. He found that

$$\frac{m}{q} = 1.90 \times 10^{-18} \, g/esu = 5.7 \times 10^{-9} \, g/coulomb^*$$

Faraday had established that passage of 96,500 coulombs of electricity through a water solution of acid produced 1.01 g of hydrogen gas. The m/q ratio of hydrogen was therefore known to be 1.01/96,500 or 10.4×10^{-6} g/coulomb. Thomson's experimental measurements meant that either the cathode ray particles (electrons) had a charge roughly 2000 times larger than the hydrogen ion (H^+), or their mass was 1/2000 that of the lightest atom known. Thomson drew the latter conclusion by assuming that the particles were fragments of atoms that carried a charge equal but opposite in sign to the charge of H^+.

The electronic charge

The electron's mass-to-charge ratio had been found. Could the properties of a particle so small be investigated further? In seeking to isolate and measure the electron's charge Thomson and his co-workers set up a chamber in which they allowed tiny water droplets to condense on electrons, giving the droplets negative charge. By measuring the voltage needed to suspend a droplet of known mass between charged plates, they were able to estimate the electronic charge. The experiment was refined by Millikan, who used oil droplets to carry electrons (Figure 2-13). By alternately allowing a charged droplet to fall by gravity and rise by applied electric field, Millikan (Figure 2-11) demonstrated that the minimum charge on the droplet was 4.77×10^{-10} esu. That established the electronic charge, one of nature's fundamental constants.

* The coulomb is a practical unit of electric charge (Section 18-3). One coulomb is equal to 3×10^9 esu.

Example 2-1 **(a)** Calculate the mass of the electron. **(b)** Compare it to the mass of the hydrogen atom, 1.68×10^{-24} g.

Solution **(a)** Electron mass m can be found from Thomson's m/q ratio and Millikan's value of q. We will give the electron its own symbol for charge, e ($q = e$). From text discussion we know that

$$\frac{m}{e} = 1.90 \times 10^{-18} \ g/esu$$

and

$$e = 4.77 \times 10^{-10} \ esu$$

Multiplying m/e by e, gives m, the mass:

$$m = (1.90 \times 10^{-18})(4.77 \times 10^{-10}) = 9.10 \times 10^{-28} \ g$$

(b) The ratio of hydrogen atom mass to electron mass follows:

$$\frac{1.68 \times 10^{-24} \ g}{9.10 \times 10^{-28} \ g} = 1845$$

Thus the electron is 1/1845 as massive as the lightest atom.

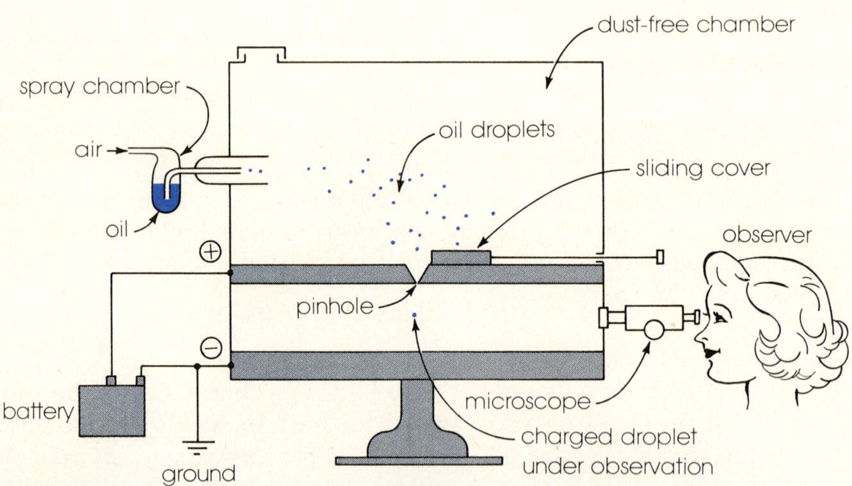

Figure 2-13 Millikan's apparatus. Droplets acquire one or more electrons by friction as they come from the spray chamber. After a droplet falls through the pinhole, the hole is covered and the droplet observed. By knowing the gravitational force and electric field strength, calculating the droplet's mass, buoyancy, and friction, and measuring the droplet's velocity of fall due to gravity and its velocity of rise due to the electric field, Millikan had all the information necessary to calculate the electron charge e the droplet carried. The present value is

$$e = 1.60 \times 10^{-19} \ coulomb = 4.80 \times 10^{-10} \ esu$$

All electric charges in nature are a multiple of this value.

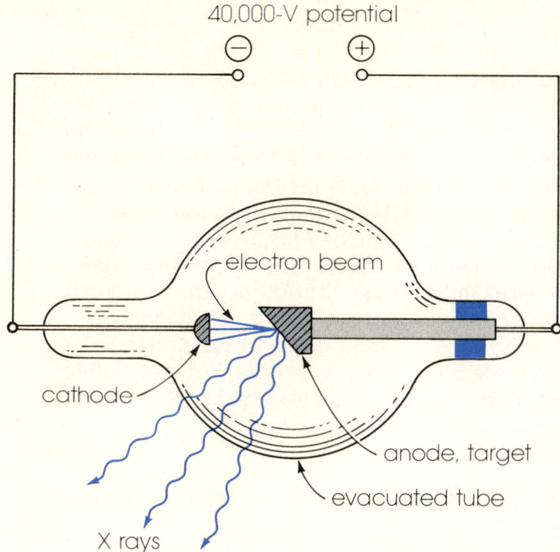

Figure 2-14 An early X ray tube. High-speed electrons striking the metal anode give energy to target atoms which is reemitted as X rays.

2-4. DISCOVERY OF X RAYS AND RADIOACTIVITY

Closely following Thomson's discovery of the electron came two other major discoveries: X rays and radioactivity.

X rays were found accidentally in 1895 by Wilhelm Röntgen (Germany 1845–1923, Nobel prize in physics 1901) while working with a highly evacuated Crookes tube (Figure 2-14). As he increased the potential across the electrodes to several thousand volts, an invisible radiation (which he named *X radiation*) was emitted from the positive electrode, passed through the tube's glass wall, and lit up a fluorescent screen standing a few feet away. Holding his hand in the rays Röntgen was startled to see a shadow of its skeleton on the screen. In further experiments he found that X rays made air electrically conducting by causing ions (N_2^+, N^+, O_2^+, etc.) to form. He also found that the rays were not deflected in a magnetic field. He concluded that X rays could not be composed of charges like cathode rays, but were a new kind of high-energy radiation produced by a then unknown atomic process.

X rays created a worldwide sensation. Within three weeks of their announcement the medical profession was taking X ray photographs of broken bones. And sharp businessmen were advertising "X ray-proof underwear—no lady safe without it."

When Antoine-Henri Becquerel (France 1852–1908, Nobel prize in physics 1903) heard of Röntgen's experiments, he wondered if certain minerals and salts, known to fluoresce in sunlight, might be giving off X rays. He found that a salt of the element uranium not only gave off rays of some sort that left an image of its crystals on a photo plate through several layers of wrapping paper, but did so whether exposed to sunlight or not. Experimentation showed that the rays also ionized surrounding air and that the rate of ray emission could not be changed by cooling, heating, or exerting pressure on the sample. Such substances were described as *radioactive*.

Investigation of radioactivity was undertaken by Marie Curie (Figure

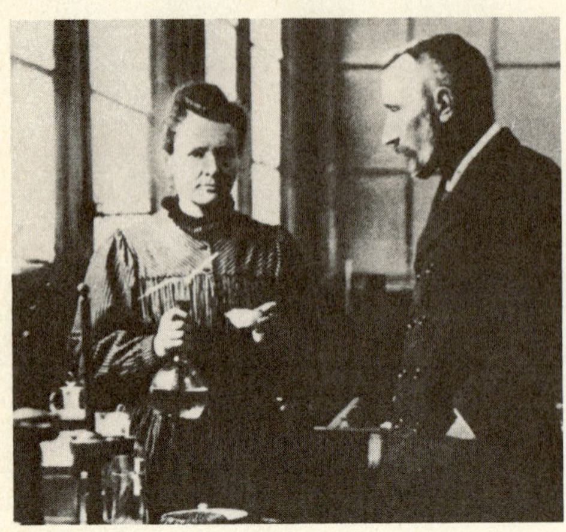

Figure 2-15 Marie Sklodowska Curie (Poland and France 1867–1934) and her husband, Pierre Curie (France 1859–1906). In 1903 the Curie's shared the Nobel prize in physics. In 1911 she won the Nobel prize in chemistry. Commenting that the use of radium in cancer treatment could not have been foreseen when they first isolated the element, Marie Curie said, "This is proof that scientific work must not be considered from the point of view of the direct usefulness of it. It must be done for itself, for the beauty of science, and then there is always the chance that a scientific discovery may become, like radium, a benefit for humanity." (Physics Today.)

2-15). Examining every available element, she found that only uranium and thorium gave the rays Becquerel described. Because radioactivity varied directly with the quantity of radioactive element present, she concluded that it was an atomic process. In pitchblende, a uranium mineral, she made an important discovery: The sample was more radioactive than its uranium content allowed. Suspecting that it contained an unidentified and very active uraniumlike element, she and her husband set out to isolate it. Their approach was to dissolve several tons of pitchblende in acid, then add chemicals known to precipitate heavy metals (see the periodic table). In a bismuthlike residue they found one new highly radioactive element, polonium (element 84). In a bariumlike residue they found a few milligrams of another, radium (element 88). Radium, its intense radiation capable of killing cells, became the basis of cancer therapy.

2-5. THE RUTHERFORD ATOM

The Kelvin-Thomson Model

Scientists at first assumed that radioactive emanations were X rays. Rutherford (Figure 2-16), however, found that uranium emitted two kinds of rays, which he named *alpha* (α) and *beta* (β). (The next year a third emanation called *gamma* (γ) was discovered.) Becquerel proved that the β component was electrons. The α component was much more massive, giving heavier cloud-chamber tracks and requiring larger magnetic and electric fields for deflection. From the experimental evidence, and knowing that helium gas was found in uranium ores, Rutherford concluded that α rays were streams of helium ions, He^{2+}. He proved it by sealing a sample of an α-emitting element in an evacuated tube and later, using a spectrometer, identifying the presence of helium gas.

The discovery of radioactivity furnished direct proof of the existence of atoms and gave scientists pieces from which to build an atom model. Around 1906, in England, Lord Kelvin (William Thompson 1824–1907) and J. J. Thomson proposed a "raisin muffin" model consisting of a spherical mass of positive charge in which negative electrons were emplanted. They assumed that most of the atom's mass and volume resided in the positive portion, and that special electrical effects operated which kept normally repellent like charges together and normally attractant unlike charges

Figure 2-16 Ernest Rutherford (New Zealand and England 1871–1937, Nobel prize in chemistry 1908). As a graduate student Rutherford's first interest was wireless signal sending (he achieved a range of 1000 yards), but when one professor advised him wireless had no future, he turned to the investigation of radioactivity. Rutherford lies buried beside Newton in Westminster Abbey. (Cavendish Laboratory, University of Cambridge.)

apart. Thomson calculated that if such a body was about 2×10^{-8} cm in diameter, its vibration when heated would create a spectrum like that observed from heated elements. This calculation was the model's strongest point because 30 years previous a successful theory of molecular movement in gases (details in Sections 14-3 and 14-4) had predicted a similar value for the atom's size.

Rutherford's scattering experiment

In 1911, when Rutherford decided to test the Kelvin-Thomson atomic model, three indispensable laboratory tools, still used in principle today, were available to him. One was the *cloud chamber* (Figure 2-17) in which paths of high-energy particles became visible. The second was the *scintillation detector,* invented by Rutherford's assistant, H. Geiger, which consisted of a small screen coated with zinc sulfide; every energetic subatomic particle striking the screen gave a tiny flash of light visible under a microscope. The third was an experimental method called the *scattering technique.* A thin foil or section of the material to be studied was prepared. In front was fixed a radioactive substance arranged to give a beam of α particles that would strike the foil. Using a cloud chamber or scintillation detector the investigator then studied the manner in which α particles were scattered (deflected) on passing through the specimen.

The experimental configuration set up by Rutherford's assistants is shown in Figure 2-18. The mass of the α particle had been established as four times that of a hydrogen atom, and its ejection speed from a source like polonium was known to be tremendously fast, 10,000 miles per second. Rutherford felt sure that such "bullets" would go straight through a very thin target, and nobody expected any deflection beyond very small angles. They were all amazed, therefore, when some particles were de-

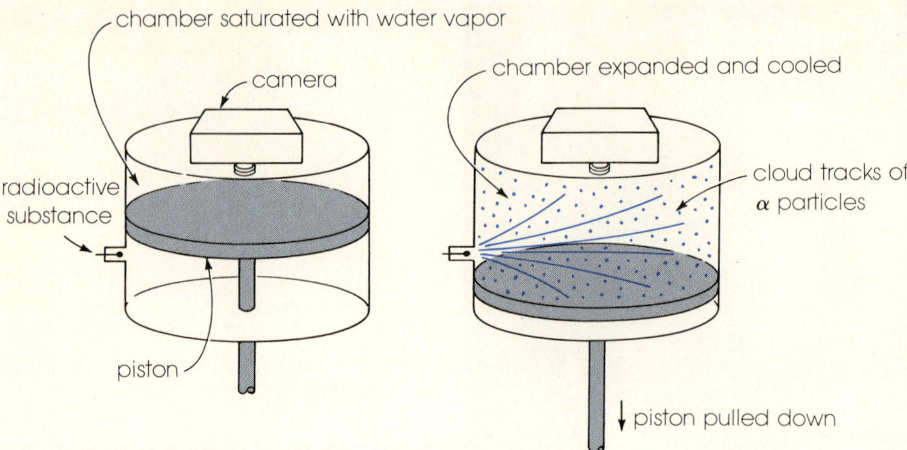

Figure 2-17 Principle of the cloud chamber. The water vapor atmosphere is cooled when expanded as the piston is suddenly lowered. An α particle hurtling through the chamber causes a track of tiny water vapor droplets to condense on the path of ions left in its wake. The chamber was invented by Charles Wilson (Scotland 1869–1959, Nobel prize in physics 1927) whose early interest was natural clouds. Trying to form clouds in the laboratory he discovered that X rays caused fog formation via ionization in his water-vapor apparatus. It occurred to him later that the effect might be a way to study radioactivity.

flected at larger angles and a few returned at greater than 90 degrees. In describing it later Rutherford recalled,

> . . . I remember Geiger coming to me in great excitement and saying, "We have been able to get some of the α particles coming backward. . . ." It was quite the most incredible event that has ever happened to me in my life. It was almost as incredible as if you fired a 15-inch shell at a piece of tissue paper and it came back and hit you.

How could the strange results be reconciled? Electrons were too light to deflect α particles. And the Kelvin-Thomson atom, with its distribution of positive charge over a relatively large volume, had insufficient charge concentration to cause deflection at large angles. Rutherford concluded that most of the atom's mass was its positive material, centralized as a very

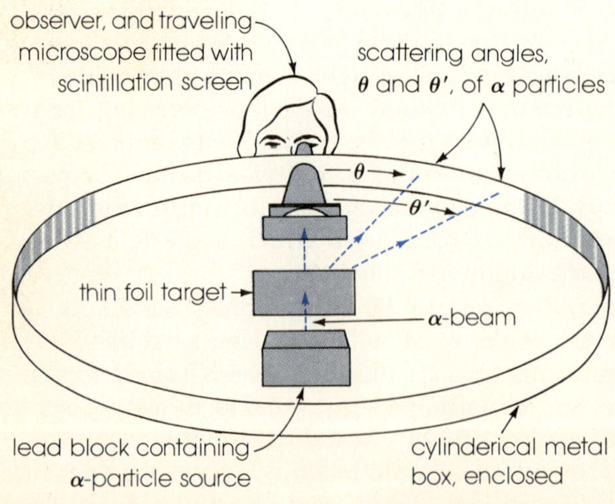

Figure 2-18 The α-scattering experiment that demonstrated atoms have dense, tiny nuclei. Alpha particles from an enclosed radioactive element bombard the very thin target and make individual flashes as they strike the microscope screen. Swinging the microscope in an arc about the target allows the observer to count particles deflected at various angles.

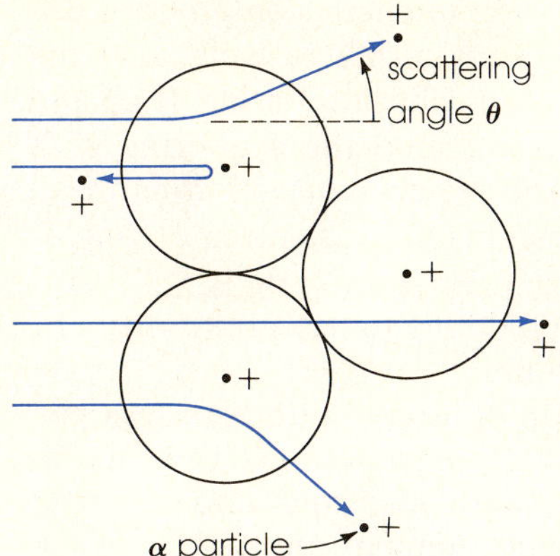

Figure 2-19 Interpretation of the scattering experiment. From the masses of the α particle and gold atom, the number of positive charges they carried, and the number of α particles deflected at a given angle, Rutherford obtained a ratio of atom to nucleus diameters of 10^4:1. (Today's value is about 10^5:1) The size relationship can be appreciated by visualizing an atom cut in half and its cross section enlarged to the size of a football field. An aspirin tablet lying in the center would cover the nucleus.

dense *nucleus*. Outside the nucleus he assumed that negative charges (electrons) were spread over a much larger volume. To get the deflections noted in the experiment (Figure 2-19) he calculated that the nuclear diameter had to be somewhat smaller than 2×10^{-12} cm. That gave a ratio of atom diameter to nucleus diameter of $(2 \times 10^{-8}$ cm$)/(2 \times 10^{-12}$ cm$)$, or at least 10,000:1. Little wonder that only one α particle in 20,000 was observed being turned through an average angle of 90 degrees and that Rutherford would describe the atom as "mostly empty space."

2-6. Z, THE ATOMIC NUMBER

Moseley's experiment

In his scattering experiment Rutherford assumed that nuclei of the bombarded atoms had electric charge and that deflection was due to a nuclear electric field. He further assumed the nuclear charge to be positive rather than negative to help explain the expulsion of positive α particles from radioactive nuclei. The equation he developed to describe scattering contained a term Z that represented the number of positive charges on each target atom. Working with it he obtained values of about 100, 50, and 30 times the unit charge of the hydrogen atom for atoms, respectively, of gold, silver, and zinc. Knowing the atomic masses A of gold, silver, and zinc to be 197, 108, and 65, he noted that the ratio A:Z in each case was roughly 2.

One interpretation of Rutherford's "atomic numbers" was that nuclei were built up of α particles; the α particle mass of 4 and charge of 2+ would give an A:Z ratio of exactly 2. Another guess was that since atomic weights of many elements were close to whole numbers, nuclei might be composed solely of protons (that is, hydrogen nuclei, charge 1+, mass 1), together with enough electrons to neutralize about half of them. Such a scheme would leave a nuclear charge equal and opposite the total negative charge of electrons held outside the nucleus. It would also account for the neutrality of unreacted atoms as well as for the ejection of electrons (β rays) from radioactive atoms.

In 1885 Swedish physicist J. Rydberg, attempting to find a law that would explain the placement of elements in the periodic table, had theorized that the simple serial numbering, H 1, He 2, Li 3, Be 4, etc., possessed some deep yet unrealized significance. After Thomson discovered the electron, the idea was advanced that if each element contained a different number of electrons, perhaps the electron numbers were also the sequence numbers in the periodic table. When Charles Barkla (England 1877–1944, Nobel prize in physics 1917) investigated the scattering of X rays by various elements, he deduced that the number of electrons in an atom was about half the element's atomic weight. Because atomic weights were known to increase by about two units (H 1, He 4, Li 7, Be 9, B 11, etc.) throughout the periodic table, his data lent credence to the possibility of a unit increase in charge from one atom to the next.

The experiment that resolved the question of *atomic numbers* was performed by Henry Moseley (England 1887–1915). In a cathode ray tube he bombarded various metallic element targets with electrons and recorded the X ray spectra produced (Figure 2-20). Each element emitted X rays of many wavelengths featured by several sharp peaks. When Moseley measured the wavelength of the most prominent peak (designated K_α) in each spectrum and arranged the elements in order of decreasing K_α values, he arrived at the periodic table sequence of elements (Figure 2-21). As discussed in Chapter 7, that sequence had been ascertained by chemists only after 80 years of work on atomic weights and chemical reactivities. "We can therefore conclude," wrote Moseley, "from the evidence of the X ray spectra alone, without using any theory of atomic structure, that these integers are really characteristic of the elements."

A few years later, when the atom was interpreted to have electrons in various energy levels about the nucleus, physicists determined that X ray energy was radiated when a low-level electron close to the nucleus was removed and replaced by a higher-energy electron from a level farther out. The greater the positive charge of the nucleus, the more energy needed to remove the low-level electron and therefore the more energetic (shorter wavelength) the X ray emission when an electron dropped back to fill the

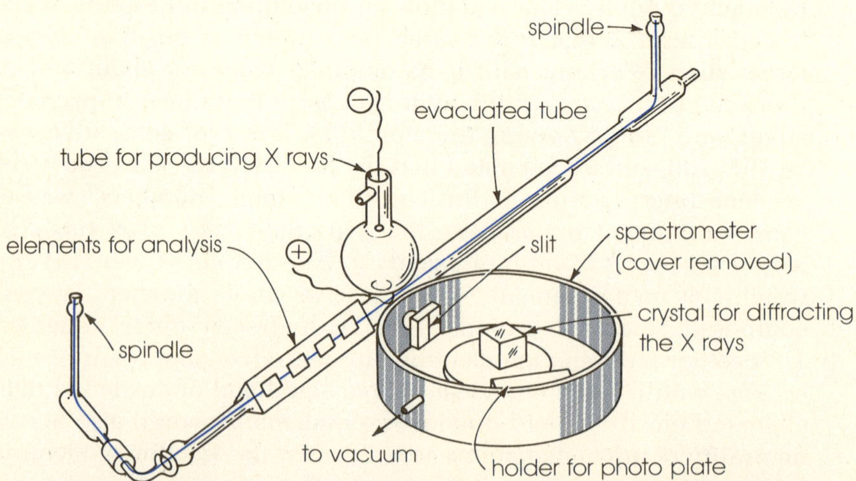

Figure 2-20 Moseley's original apparatus, handmade in the Cavendish Laboratory tradition. Elements were pulled one by one into position for analysis in the X ray tube by manipulating the spindles. X rays coming through the slit fell on the crystal and were diffracted, producing spectra that were recorded photographically.

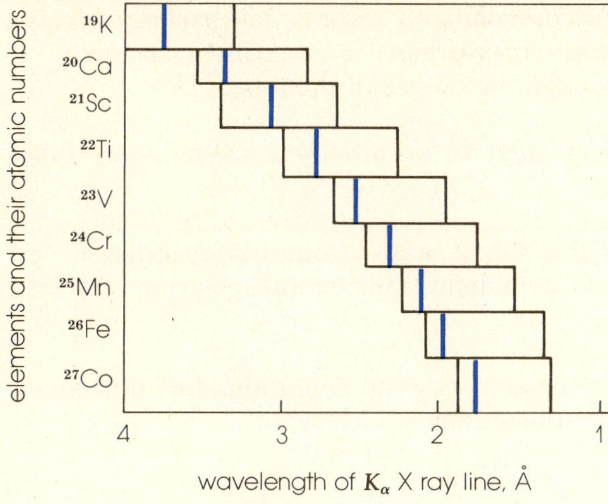

Figure 2-21 X ray spectra and atomic numbers. Moseley discovered the relationship

$$\frac{1}{\lambda} = A(Z - 1)^2$$

where λ is the K_α X ray wavelength, Z the atomic number and A a constant. He found no X ray lines corresponding to elements of Z = 43, 61, or 75. All three were discovered later (see Figure 1-3). X ray spectra are used today to identify elements in mixtures (see Figure 6-2).

"hole." Thus a sequence of elements taken from the periodic table gave a progression of peak X ray wavelengths.

Scientists now feel certain that nature has built up a succession of neutral atoms such that the atom of each next element in the periodic table contains one more unit of positive charge inside the nucleus and one more unit of negative charge outside the nucleus than the atom of the preceding element. The number of unit positive charges in an atom is its atomic number.

QUESTIONS

1. Vectors, scalars Physical quantities may be described as vectors or scalars. A vector has both magnitude and direction. A scalar has magnitude but no direction. Classify the following: **(a)** distance; **(b)** an electric field; **(c)** a magnetic field; **(d)** mass; **(e)** dollars. Explain.

2. Particles **(a)** Atoms are electrically neutral yet contain electrically charged particles. Explain. **(b)** Arrange the following in order of increasing mass: β particle, α particle, hydrogen atom. **(c)** What electric charge does each have?

3. Radioactivity **(a)** Rays from a radioactive sample may ionize the air they pass through. Explain. **(b)** A homemade electrometer may be used to examine minerals for radioactivity. Explain.

4. Spectroscope When Russian test rockets have been fired toward the Pacific Ocean, American planes with cameras mounted on spectroscopes have recorded the firey reentry into the atmosphere of the nose cones. Photographs show a series of short vertical lines of varying intensity. What information do the photos contain? What atomic phenomenon is responsible for the spectra?

5. Cathode rays Why did measurements on cathode ray particles always give the same m/q ratio, whereas measurements on positive ray particles yielded an m/q ratio that depended on the gas in the tube?

6. Cathode ray tube If a neon sign is essentially a cathode ray tube, explain its operation.

7. Definitions (library) With the aid of at least one library reference in chemistry or physics, write your own definitions for **(a)** spectrum; **(b)** field (of force); **(c)** cloud chamber.

8. Definitions (library) As in Question 7, write definitions for **(a)** ionization; **(b)** atomic number; **(c)** radioactivity.

PROBLEMS

9. Force A charge of 10 esu and another of 20 esu are 5 cm apart. What is the magnitude of the force between them if the charges have the same sign? Opposite sign?

10. Spectral lines Find the wavelength of the line in the hydrogen spectrum corresponding to $n = 3$ in the Balmer equation. Express the answer in centimeters and angstroms. Use Figure 2-4b to check your answer.

11. Rutherford's atom If you were asked to construct a scale model of a Rutherford hydrogen atom using a bead $\frac{1}{8}$ in. in diameter to represent the nucleus, at a distance of how many feet would you position the electron?

12. Millikan's experiment **(a)** A certain food market sells eggs by metric weight instead of by number. The eggs marked "Large, Grade AA" contained in several small baskets have these total weights: 1.02 kg, 1.50 kg, 480 g, 1.92 kg. What is the maximum gram weight of one egg? **(b)** If this actually is the weight of one egg, how many eggs are in each basket? **(c)** In the Millikan experiment these electronic charges on oil drops were found: 4.8×10^{-19}, 8.00×10^{-19} 14.4×10^{-19}, 17.6×10^{-19}, and 3.2×10^{-19} coulomb. What is the maximum charge one electron may possess? **(d)** How many electrons were on each of the drops?

13. Moseley's experiment The equation given with Figure 2-21 infers that when $1/\lambda$, obtained by experiment, is plotted against Z^2, the result will be a straight line. **(a)** Obtain Z values from the periodic chart and complete Table 2-1. **(b)** Plot $1/\lambda$ (vertically) versus Z^2 and draw a line through the points. **(c)** Two unknown elements give X rays having K_α wavelengths of 2.30 and 3.37 Å. From your graph deduce Z for each. Name the two elements.

14. Forces Coulomb's law (Eq. 2-2) has the same mathematical form as *Newton's law of gravitation:*

$$F_{\text{grav}} = G \frac{m_1 m_2}{r_2}$$

Table 2-1. K_α X ray wavelengths and atomic numbers

Element	λ	$1/\lambda$	Z	Z^2
Aluminum	8.36 Å			
Potassium	3.76 Å			
Vanadium	2.52 Å			
Iron	1.95 Å			
Copper	1.55 Å			

where m_1 and m_2 are masses of two bodies attracting each other; G is the gravitational constant; and r is the distance between the bodies. **(a)** Use Coulomb's law to calculate the force of attraction between the hydrogen atom's proton and electron. Express the charges in coulombs and their separation in meters (5.3×10^{-11} m); for k use 9×10^9 newton m²/coulomb² (the answer is in newtons). **(b)** Calculate the gravitational attraction between proton and electron using the same r value and masses in kilograms; G is 6.7×10^{-11} newton m²/kg² (again the answer is in newtons). **(c)** How many times larger than the gravitational force is the coulomb force?

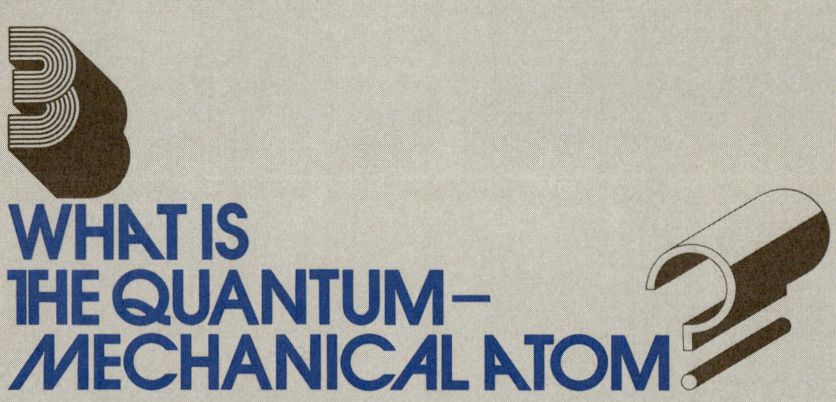

THE IMPORTANT CONCEPTS

3-1. The quantum theory

1. Energy comes in small units called quanta.
2. The photoelectric effect is evidence for quantum energy transfer.
   a. The electron has particle character.

3-2. The Bohr atom

1. A planetary model is a crude atom picture.
2. Quantum theory refines the atom model.
   a. Electrons occupy energy levels.
   b. Spectral lines are related to transitions between electron energy levels; one electron jump is one quantum.
3. Ionization energies are interpretable by the Bohr atom.

3-3. The quantum-mechanical atom
  1. Electron "spin" is evidence of a quantized atom.
  2. The electron has wave character.
     a. Electron waves refine the atom model.
  3. Diffraction and interference are two evidences of wave motion.
  4. Diffraction of electrons is proof of electron waves.
  5. A modern atom model is blurred; electron position is expressed as a probability.
  6. Wave-particle duality, by its very nature, makes precise position-velocity determinations impossible.
  7. Quantum theory has three features.
     a. It defines electron energy and atom size.
     b. It predicts electron wave patterns (which dictate shapes of molecules).
     c. It introduces four quantum numbers that define electron energies.

The physics of Newton dealt with observable objects such as the planets; given their positions and velocities and the forces acting on them, one could calculate the paths they would take in the future. Such was found not to hold in the domain of molecules, atoms, nuclei, and subatomic particles. When classical physics failed to explain their behavior, a more inclusive theory called *quantum mechanics* was developed. Basic to quantum theory are the concepts that energy is absorbed or emitted only in small definite amounts called *quanta*, that events are governed by laws that determine probability not certainty, and that the uncertainty can never be completely removed.

3-1. THE QUANTUM THEORY

Nature's energy ladder

In 1900 one problem confronting science was the origin of "blackbody" radiation, that is, light coming from a little hole in a heated furnace (Figure 3-1). The spectrum observed (1) was continuous (not in the form of lines) and thus contained all frequencies in its range, (2) was independent of furnace material, and (3) depended only on temperature. The best explanation

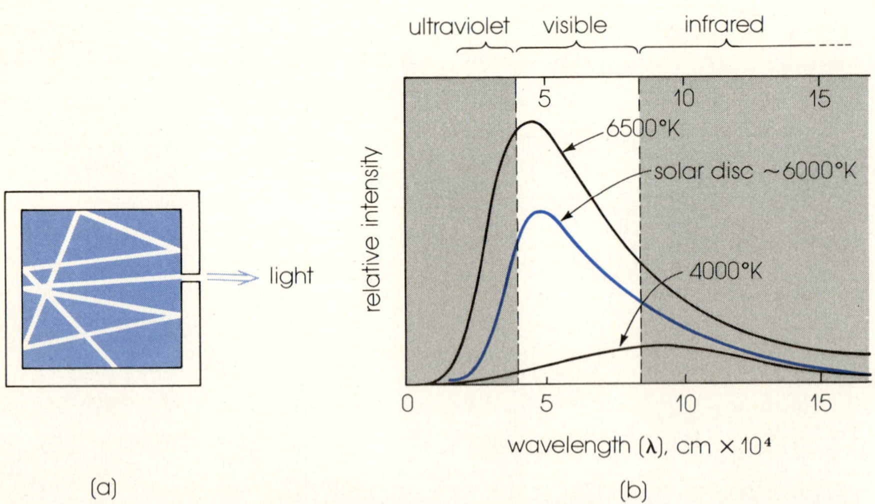

Figure 3-1 Blackbody radiation. (a) A blackbody. Escaping light, having been reflected many times inside, represents radiation having the spectral distribution of a blackbody radiator at that temperature. (b) Blackbody spectra at two different temperatures and, for reference, the sun's distribution of energies. The hotter the body, the more sunlike its emission; that is, the more the wavelengths are shifted from longer values in the infrared region toward values in the visible and ultraviolet (for reference see Figure 2-2).

seemed to be that the walls contained electric oscillators (later identified as atoms) which, when heated, generated a spectrum that came from the spread of intensities of the frequencies at which the oscillators vibrated. The best mathematical approach was to consider a large number of oscillators, allow them to freely exchange energy and vibrate at all frequencies, and to write an equation that would reproduce the spectra.

Calculations were only partially successful until Planck (see Figure 3-3) took the opposite approach by first deriving a mathematical expression to fit the experimental data, then inventing a model that explained it. He made the revolutionary assumptions that (1) atoms (oscillators) could absorb or emit energy only in discrete units or bundles called *quanta,* and (2) the quantity of energy in each quantum was related to the frequency (or the wavelength) of the light by

$$E = h\nu = \frac{hc}{\lambda} \tag{3-1}$$

Equation 3-1 expresses the most fundamental relationship in quantum theory. As we will employ it, E is energy in ergs; ν (nu) is the frequency of radiation in oscillations per second (or hertz); c is the speed of light, 3×10^{10} cm/sec; λ is the wavelength in centimeters (centimeters per oscillation); and h is *Planck's constant:*

$$h = 6.63 \times 10^{-27} \ erg \ sec \tag{3-2}$$

Planck's quantum hypothesis was a new way of viewing matter. Classically, physicists related the energy of a wave (say, a wave at sea) only to its height (amplitude); common sense dictated that a high wave must contain more energy than a ripple. In the atomic domain, however, Planck was saying that (1) the energy of a wave was proportional only to its frequency — that is, to the number of its vibrations per unit of time; and (2) absorption and emission of energy were discontinuous processes. It was as if an energy ladder existed in nature and atoms could only be on rungs of the ladder and never in between. To jump from one rung to the next required absorption or emission of energy equal to that energy gap.*

Planck's ideas were at first regarded as only the answer to the blackbody question. When successfully applied to the photoelectric effect (see below), however, they became recognized as a clear break from classical science. Invention of the quantum theory and the theory of relativity (1905) then marked the beginning of *modern physics.*

The photoelectric effect

Newton had postulated that light consisted of particles, but eighteenth- and nineteenth-century experiments (see Figure 3-13) indicated that light consisted of waves. In 1905 Einstein revived Newton's idea when he said that the *photoelectric effect* — the release of electrons from a metal surface illuminated by light (Figure 3-2) — suggested a particle character for light. Measurements proved that the *kinetic energy* (energy of motion) of the electrons depended not on the intensity of light, as classical theory predicted,

---

* Thus Planck assumed that an atom could only have an energy E that was a whole number multiple of the quantity $h\nu$, that is, $1h$, $2h$, $3h$, etc. He later started the ladder at $\frac{1}{2}h\nu$ instead of $0h\nu$, giving the rungs the values $\frac{3}{2}h\nu$, $\frac{5}{2}h\nu$, ... This agrees with theory today.

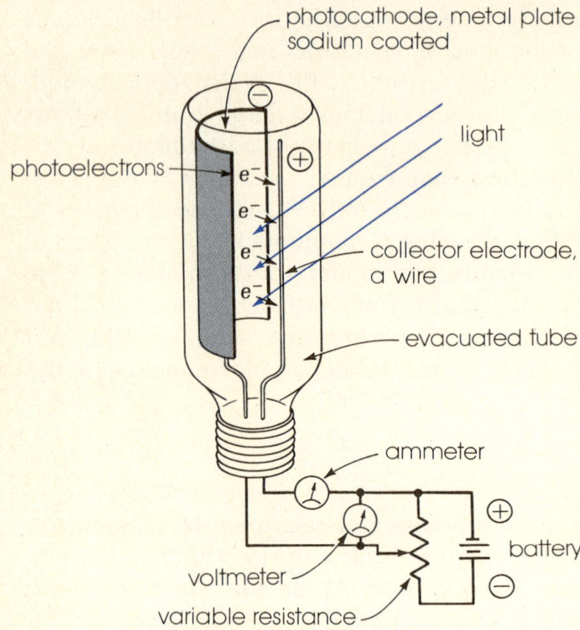

Figure 3-2 The photoelectric effect. Light striking the (negative) cathode of the photoelectric tube energizes surface electrons. As electrons leave and travel to the collector electrode, their flow (electron current) is indicated by the ammeter. To detect the energy of the photoelectrons, connections to the battery are reversed. When the collector is made sufficiently negative by adjustment of the variable resistance, no photoelectrons reach the collector (now called the stopping electrode) and the ammeter reads zero. The corresponding voltage on the voltmeter is a measure of the photoelectron energy. The higher the frequency of light, the greater the photoelectron energy.

but only on the light's frequency. That echoed Planck's conclusion about blackbody radiation.

Einstein (Figure 3-3) explained the photoelectric effect by assuming that light was quantized, meaning that it arrived in quanta or tiny energy units (now called *photons*). Following Planck he assigned an energy of $h\nu$ to the photon and proposed that when one photon of light struck the metal cathode (negative electrode) of a photoelectric tube, its energy was transferred to one electron of charge e. The portion of the energy used by the electron for escape from the metal surface he called the *ionization energy* E_i. The remainder he assigned to the free electron's kinetic energy. This reasoning gave a photoelectric equation in terms of a voltage V measured as described by Figure 3-2:

$$V = \frac{h\nu}{e} - \frac{E_i}{e} \tag{3-3}$$

The equation was later verified experimentally by Millikan.

There is no doubt among scientists today that photons, and all entities of atom size, have properties of both particles and waves. An attempt to represent their peculiar duality is made in Figure 3-4.

3-2. THE BOHR ATOM

Bohr's assumptions

The atom of 1913 consisted of two regions of electricity and much empty space. Its center was a very small, dense nucleus about 10^{-12} cm in radius. The nucleus contained almost all the atom's mass in the form of positive charges equal in number to the atomic mass A. About half the positive charges were assumed combined and neutralized by electrons. Uncom-

Figure 3-3 Founders of the quantum theory. (Left) Max Planck (Germany 1858–1947, Nobel prize in physics 1918), specialist in thermodynamics, the science of heat. Noting the problem older scientists had grasping the uncommon sense of quantum theory, Planck made this observation on human limitations: "I have come to think that when one has reached a certain age, however learned one may be, there are some ideas which are too difficult; one may make use of them, but without understanding them very thoroughly. Fortunately, however, people die; so that, after a time, all who could not understand have disappeared from circulation, and affairs are taken over by those who can." (Right) Albert Einstein (Germany and United States, 1879–1955, Nobel prize in physics 1921), contributor to quantum theory, relativity, and cosmology. Einstein came to the United States in 1933. A staunch pacifist, he believed in world government. (Nobel Foundation.)

bined positive charges, equal in number to the atomic number Z, were each equivalent to a hydrogen ion with a charge of 4.8×10^{-10} esu and a mass of 1.7×10^{-24} g. Out to a distance of about 10^{-8} cm from the center was a collection of free electrons equal in number to Z. Each electron had a mass of 0.91×10^{-27} g and a charge of -4.8×10^{-10} esu, making the atom electrically neutral. Because basic physical laws governing electrical and gravitational forces dictated that charges could not remain fixed around a collection of opposite charges, the electrons were assumed to revolve around the nucleus like planets in a miniature solar system.

Tidy though it was, the model contained a fatal flaw. In 1873, *Treatise on Electricity and Magnetism* had been published. Written by mathematical physicist James Clerk Maxwell (England 1831–1871), it correlated electricity, magnetism, and light with the same authoritativeness that Newton's *Principia* had organized mechanics. One prediction Maxwell's equations made was that an accelerated charge creates an electromagnetic disturbance. That had already led to the discovery of radio waves (charge oscilla-

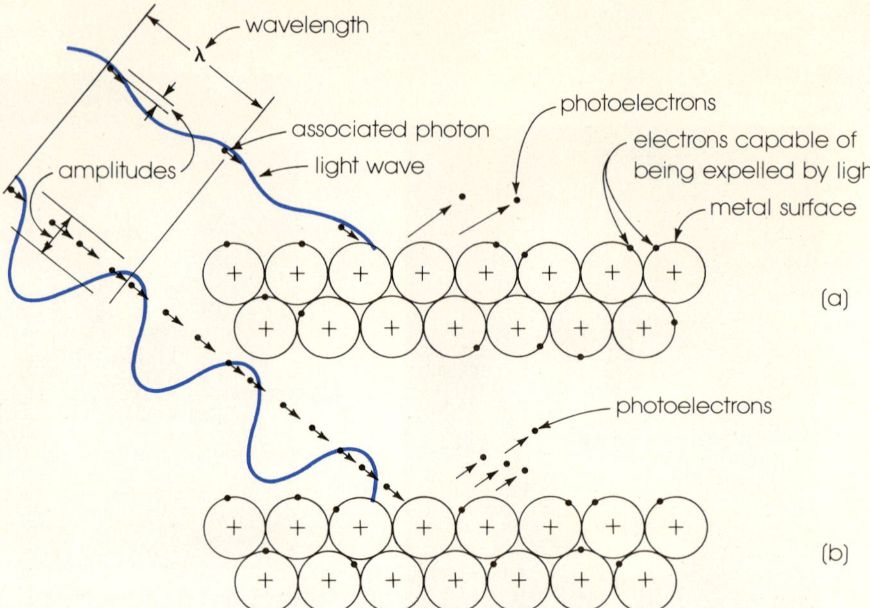

Figure 3-4 The photoelectric effect and the particle nature of light. (a) A light wave of small intensity (amplitude) and its associated photons strike atoms in a metal surface and dislodge electrons. (b) A light wave of large intensity but identical wavelength and its associated photons strike a metal surface and dislodge more electrons. All the freed electrons have the same energy because although more photons are associated with the wave in (b), the photons in both cases have the same energy. Classical physics would lead one to expect the wave of greater amplitude to expel electrons of greater energy. (Adapted from George Gamow, The Principle of Uncertainty, Scientific American, Jan. 1958, 52. Copyright © Scientific American, Inc.)

tion in an antenna makes a radio signal), and physicists asked why the principle should not also apply to the atom. If it did, the orbiting electron would emit electromagnetic radiation, lose energy, and spiral immediately into the nucleus. And if that were true, the electron would have all energies between its initial and final positions, and would yield a continuous spectrum instead of the line spectra (see Figures 2-4, 5-3, etc.) observed with the spectroscope. Furthermore, all atoms measured at least an angstrom unit in radius, which would not be the case if the electron could get arbitrarily close to the center.

The planetary model was rescued by Bohr (Figure 3-5), who took the great intuitive leap of ascribing quantum behavior to the atom. For atom stability he made the following assumptions:

1. Electron orbits about the nucleus are circular and only certain prescribed orbits are permitted.
2. Contrary to Maxwell's theory, the electron moves in an orbit without radiating energy.
3. The electron is quantized. Each electron orbit has different energy. When an electron changes orbits, it acquires or releases a definite amount of energy.

From Newton's second law of motion (a body is accelerated by force

Figure 3-5 Niels Bohr (Denmark 1885–1962, Nobel prize in physics 1922). Bohr was only 28 when he fit quantum theory to the atom by proposing quantized electron orbits. Later he did theoretical research on radioactivity and the atomic nucleus. During World War II he worked in the United States on the atom bomb. (Nobel Foundation.)

applied to it) and Coulomb's law (Eq. 2-2), Bohr derived an equation to describe the energy of the hydrogen atom which consists only of an electron of charge $-e$ circling a nucleus of charge $+e$. To reduce the number of orbits from an infinite number to the few permitted ones (Figure 3-6), he assumed that the electron's angular momentum L (product of particle mass, square of distance from the axis of rotation, and angular velocity about that axis) was quantized. He calculated L to be $h/2\pi$ for the atom's normal or *ground state* in which the electron occupied the orbit closest to the nucleus. For larger orbits of higher energy he calculated L to be $2h/2\pi$, $3h/2\pi$, etc., or in general $nh/2\pi$. The integer n he called the *principal quantum number*.

Bohr next assumed that when the electron moved from an orbit of higher energy E'' to an orbit of lower energy E', the result would be the release of one photon (one quantum) of energy E. Utilizing Eq. 3-1 he wrote,

$$E = E'' - E' = h\nu = \frac{hc}{\lambda} \tag{3-4}$$

Evaluating E in terms of electron mass m and charge e he obtained

$$\frac{1}{\lambda} = \frac{2me^4}{ch^3}\left(\frac{1}{n'^2} - \frac{1}{n''^2}\right) = R\left(\frac{1}{n'^2} - \frac{1}{n''^2}\right) \tag{3-5}$$

where c is the speed of light; h is Planck's constant; n' is an electron orbit of lower energy; n'' is an electron orbit of higher energy; and R is a combination of the previous group of constants. Within limits of measurement, R was identical to the Rydberg constant (Eq. 2-1), an expression derived 30

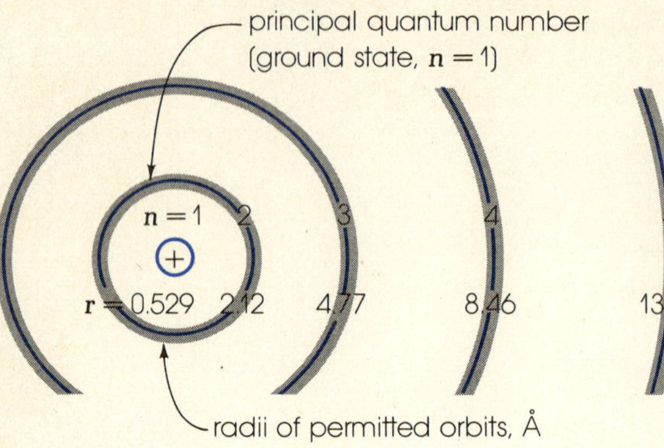

Figure 3-6 The Bohr hydrogen atom. Combining classical and quantum physics, Bohr constructed a remarkably useful model. Calculations based on it gave the following predictions: orbit radii, $0.529 \times 10^{-8} n^2$ cm; atom energies, $-5.21 \times 10^{-19}/n^2$ cal* electron velocities, $2.19 \times 10^8/n$ cm/sec; electron frequencies, $6.58 \times 10^{15}/n^3$ cycles/sec. (n is an integer denoting the orbit or electron energy level: 1, 2, 3, . . .)

years before solely from spectroscopic measurements on heated hydrogen gas.

Of Eq. 3-5 Bohr said,

> We see that this expression accounts for the law connecting the lines in the spectrum of hydrogen. If we put $n' = 2$ and let n'' vary, we get the ordinary Balmer series. If we put $n' = 3$, we get the series in the infrared observed by Paschen. . . . If we put $n' = 1$ we get a series in the extreme ultraviolet which has not been observed, but the existence of which may be expected.

(Two years later the Lyman ultraviolet series was discovered. See Figure 3-7.)

The coincidence of the values of Bohr's R and Rydberg's R has been called the most satisfying and impressive numerical agreement between

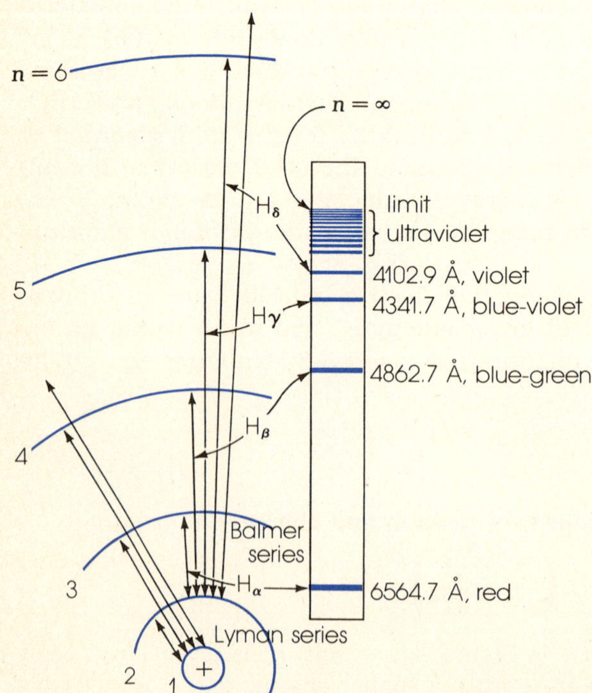

Figure 3-7 An energy-level diagram of the hydrogen atom with particular reference to the (visible) Balmer series of spectral lines. Straight arrows indicate electron transitions between energy levels. If transition is from a higher to a lower level, energy (hν) is emitted and a bright emission line is observed. If transition is from a lower to a higher level, energy (hν) is absorbed and a dark absorption line is observed. Series are named for their discoverers.

* cal is an abbreviation of calorie, a unit of heat. Negative energy means the electron is bound in the atom. (A completely free electron would be assigned zero energy.)

theory and experiment in the history of science. Bohr's model was likewise unusual for the breadth of physics it incorporated, drawing as it did on Newton's laws of motion, Rutherford's nuclear atom, the speed of light determined by Michelson, Thomson's electron mass, Millikan's electronic charge, and the quantum ideas of Planck and Einstein. The model also had the attribute of being readily visualized, so that chemists, for example, were soon using it to discuss the formation of molecules as interactions of electrons which bonded atoms together (see Section 9-1).

Example 3-1 Show that Eq. 3-5 correctly describes the electronic transition responsible for the Balmer blue–violet line.

Solution To use the equation (with R in cm^{-1}) we need first to convert the wavelength 4341.7 Å (Figure 3-7) to centimeters. To do so we rewrite the number exponentially (Appendix B, Section B-2), then use the factor-unit conversion method (Appendix D) with the appropriate conversion factor (Appendix A, Table A-3):

$$? \, cm = 4.342 \times 10^3 \, \text{Å} \left(\frac{10^{-10} \, m}{1 \, \text{Å}} \right) \left(\frac{10^2 \, cm}{1 \, m} \right) = 4.342 \times 10^{-5} \, cm$$

We substitute this value into Eq. 3-5. For the Balmer series, $n' = 2$; n'' is unknown. The answer we seek is merely a small integer (the quantum level). Therefore, we will round off (Appendix B-1) the values of the wavelength and R. Then

$$\frac{1}{4.34 \times 10^{-5} \, cm} = \frac{1.10 \times 10^5}{cm} \left(\frac{1}{2^2} - \frac{1}{n''^2} \right)$$

$$\frac{1}{4.76} = \frac{1}{4} - \frac{1}{n''^2}$$

$$\frac{1}{n''^2} = \frac{1}{4} - \frac{1}{4.76} = \frac{4.76}{19.0} - \frac{4.00}{19.0} = \frac{0.76}{19.0} = 0.03999$$

$$n'' = \sqrt{\frac{1}{0.0399}} = \sqrt{25} = 5.0$$

Thus an electron transition between quantum level 5 and quantum level 2 is responsible for emission (or absorption) of a photon having a wavelength of 4342 Å.

Ionization energies

Experimental support for the Bohr theory appeared quickly. Johannes Stark (Germany 1874–1957, Nobel prize in physics 1919) demonstrated that when a powerful electric field was applied to a glowing gas in a positive ray tube, each spectral line was split into a series of finer lines. He interpreted the result to mean that electron orbits could become asymmetric (elliptical instead of circular), making slightly different quantum jumps possible. A similar effect with an applied magnetic field, discovered in 1896 by Pieter Zeeman (Holland 1865–1943, Nobel prize in physics 1902), was then recalled and interpreted as magnetically induced splitting of quantized electron energy states.

An experiment of more interest to chemists was performed in Germany by Franck and Hertz. Their apparatus (Figure 3-8) was designed to examine the effect of passing an electron beam of variable energy through a gas. As they raised beam energy by increasing tube voltage, they discovered that for each sample there was a series of critical voltages at which the electron beam suddenly lost energy to the gas.

In quantum theory terms the experiment was interpreted as follows. For simplicity one assumed that the tube contained gaseous atomic hydrogen H. Until electrons in the beam attained enough kinetic energy to collide with a hydrogen atom and raise its electron from the ground state ($n = 1$) to the next higher state ($n = 2$), they bounced off with little energy loss as indicated by a steady current reading on the ammeter. However, at the first critical voltage, one quantum of energy was transferred from an electron to an atom upon collision. The electron causing the effect, having given up some of its energy, did not retain sufficient energy to reach the collector, and the ammeter indicated a drop in circuit current. The critical voltage, indicated on the voltmeter, was equivalent to the difference in energy between the ground quantum state and the second quantum state ($E_2 - E_1$). That value was just the energy difference calculated from measurements on the corresponding spectral line.

Subsequent energy absorptions corresponding to successive boosts of the electron to higher allowed quantum states could also be demonstrated,

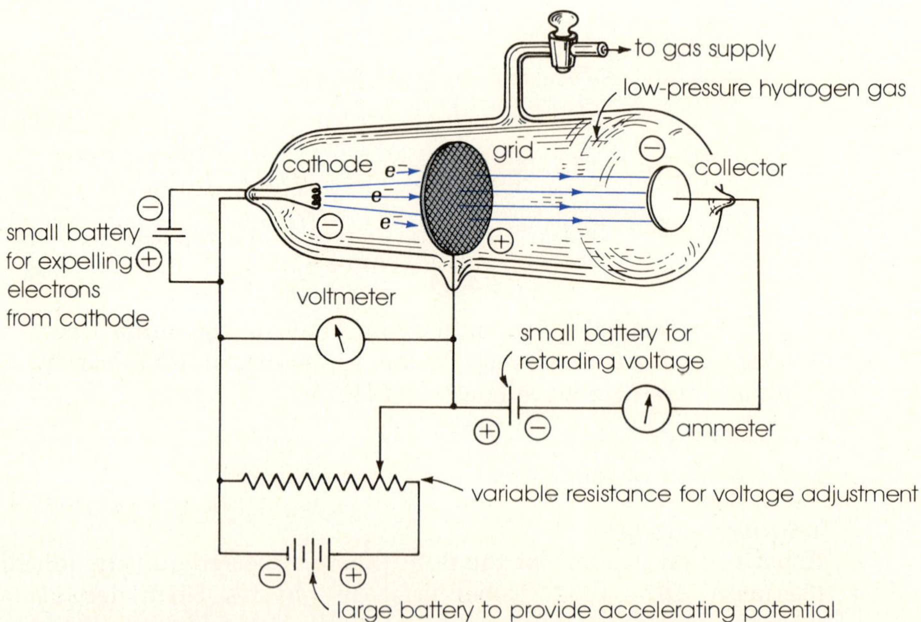

Figure 3-8 The experiment of James Franck (Germany and United States 1882–1964) and Gustav Hertz (Germany and USSR, 1887–   ), who shared the Nobel prize in physics in 1925. They discovered laws governing atom–electron collisions.

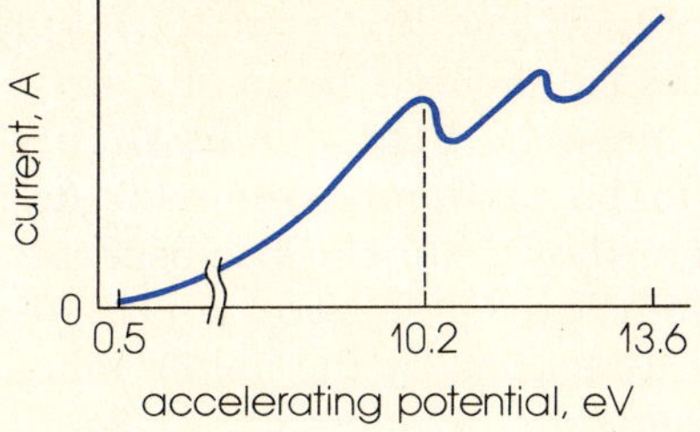

Figure 3-9 Results of the Franck-Hertz experiment with hydrogen atoms. The first critical voltage, 10.2 volts, corresponds to raising the electron from the first Bohr orbit to the second. The limiting value, 13.6 volts, corresponds to ionization of the gas: $H \rightarrow H^+ + e^-$.

as indicated in Figure 3-9. The limit was reached at the atom's *ionization potential*, the voltage at which the electron finally escaped the atom completely.

Chemists have used ionization potential data to make valuable correlations. One example discussed later is a rationalization for the positioning of elements in the periodic table (see Figure 7-9).

3-3. THE QUANTUM-MECHANICAL ATOM

In 1925 two Dutch physicists, G. Uhlenbeck and S. Goudsmit, found indications that the electron not only had motion about the atom nucleus, but also had "spin"; that is, it acted as if it rotated about its own axis. With respect to a given axial orientation (vertical for example), an isolated electron could then be thought of as spinning in a clockwise or counterclockwise direction.

Evidence of electron spin was obtained experimentally (Figure 3-10) using the silver atom, which has a lone electron in its outermost energy

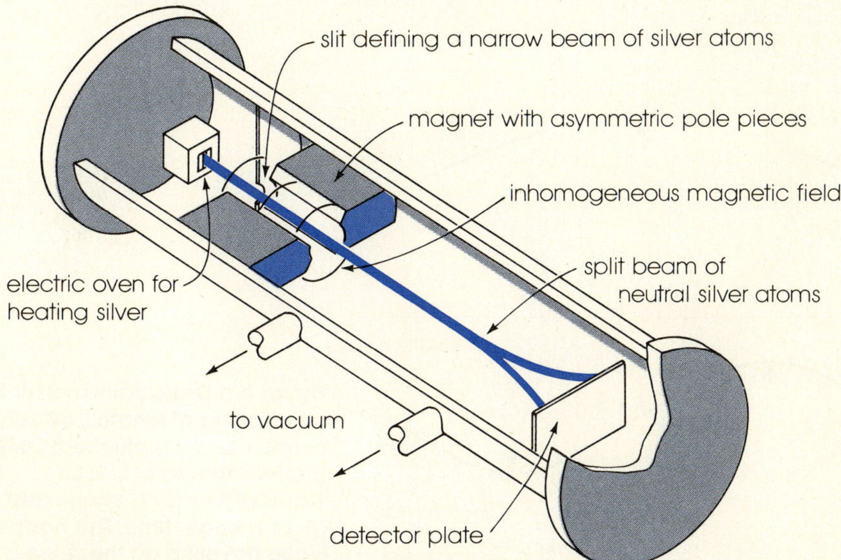

Figure 3-10 Demonstration of electron spin quantization with a molecular beam apparatus built by Otto Stern (Germany and United States 1888–1969, Nobel prize in physics 1943) and W. Gerlach. The beam of neutral silver atoms divides upon passing through the magnetic field because spin of the electron about its own axis gives the atom two spin orientations.

level. (The electron is not paired with another electron, which would cancel evidence of its spin.) According to classical theory a beam of neutral atoms passing through an inhomogeneous magnetic field should simply spread out because atoms were assumed to be randomly oriented. According to quantum theory, however, the beam should split in two because, due to the spin of the lone electron, silver atoms existed in two spin conditions which would be deflected in opposite directions by the field's inhomogeneity. Splitting into two beams was observed.

Electron waves

Despite spectacular initial success, the Bohr theory remained intact only a dozen years. It applied well to hydrogen but only qualitatively to larger atoms. Instinctively scientists did not like its arbitrary combination of classical and quantum physics. A true quantum mechanics was needed, one based upon its own postulates and mathematics.

In 1924 such a science began with Louis de Broglie (France 1892– , Nobel prize in physics 1929), who worked mathematically to understand the particle-wave duality in light and in atoms. He wrote,

> When I began to consider these difficulties I was chiefly struck by two facts. On the one hand the Quantum Theory of Light cannot be considered satisfactory, since it defines the energy of a light corpuscle by the (Einstein) equation $E = h\nu$, containing the frequency ν. Now a purely corpuscular theory contains nothing that enables us to define a frequency; for this reason alone, therefore, we are compelled, in the case of Light, to introduce the idea of a corpuscle and that of periodicity simultaneously. . . . But corpuscles and waves cannot be independent of each other: in Bohr's terms, they are two complementary aspects of Reality: and it must consequently be possible to establish a certain parallelism between the motion of a corpuscle and the propagation of its associated wave.

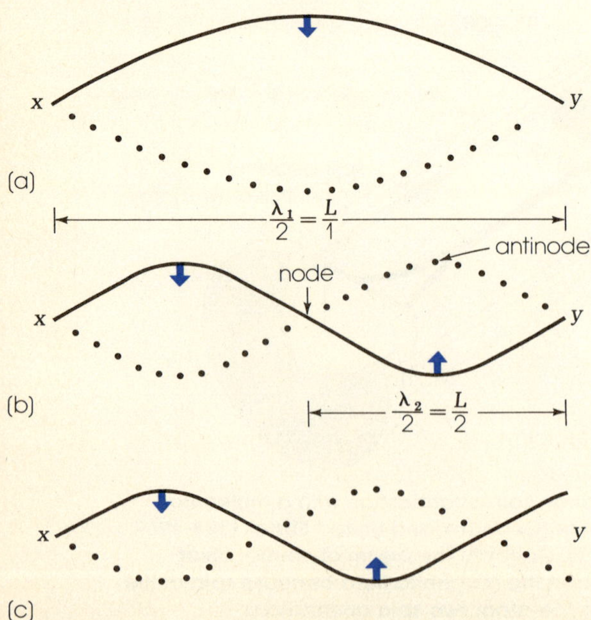

Figure 3-11 Classical wave motion. A guitar string of length L attached at points x and y is plucked. Only those vibrations that total 1, 2, 3, . . . half-wavelengths ($\lambda/2$) occur, and the string gives a single tone. This happens as the wave traveling up the string is reinforced or canceled by the wave traveling down the string. Such a wave is called a standing wave because it seems to stand still; (a) is called a fundamental; (b) is called an overtone; condition (c) does not occur.

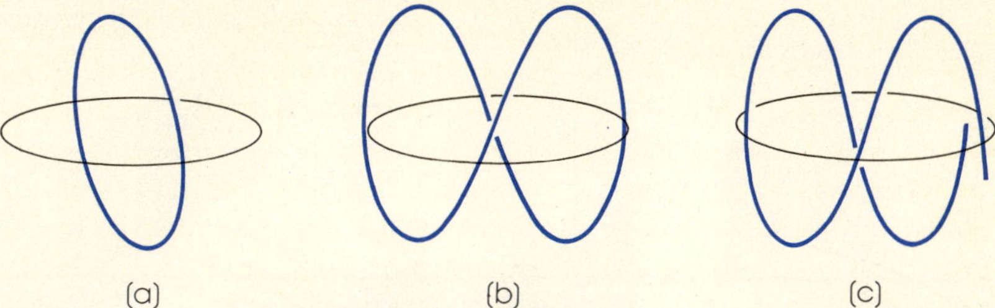

(a)   (b)   (c)

Figure 3-12 A representation of quantum wave motion of the electron about the nucleus. Corresponding to Figure 3-11, (a) is a fundamental, (b) is an overtone, and (c) is not allowed.

From the Einstein equation de Broglie derived a relationship between the wavelength λ of a particle and its momentum (product of its mass m and velocity v):

$$\lambda = \frac{h}{mv} \tag{3-6}$$

He then connected electron waves with atomic properties by letting electrons be guided in orbits via waves related to their motion. Using the experimental observation that only certain wave patterns occur in a vibrating string fixed at its ends (Figure 3-11), he assumed that around an atomic nucleus also only a definite whole number of waves could be accommodated.

In Bohr's concept the quantum number n was associated with the average distance between electron and nucleus. In de Broglie's concept n became the number of whole wavelengths in an orbit. Orbit 1 had one wavelength, orbit 2 two wavelengths, etc. (Example 3-2 illustrates a calculation.) As shown in Figure 3-12, the atom became a vibrating system capable only of those vibrations that gave it stability.

Example 3-2 Bohr calculated the velocity of the electron in the second orbit of the hydrogen atom to be 1.1×10^8 cm/sec. Calculate its de Broglie wavelength, and comment on the relationship to the Bohr model.

Solution We will use Eq. 3-6, substituting into it text values for Planck's constant, electron velocity, and electron mass:

$$\lambda = \frac{6.6 \times 10^{-27} \; erg \; sec}{(1.1 \times 10^8 \, cm/sec)(9.1 \times 10^{-28} g)} = 6.6 \times 10^{-8} \; cm$$

This length of 6.6 Å is about three times the calculated radius of the second orbit (Figure 3-6) and half the circumference c:

$$c = 2\pi r = 2 \times 3.14 \times 2.12 = 13.3 \; \overset{\circ}{A}$$

Thus two de Broglie wavelengths fit into the second Bohr orbit.

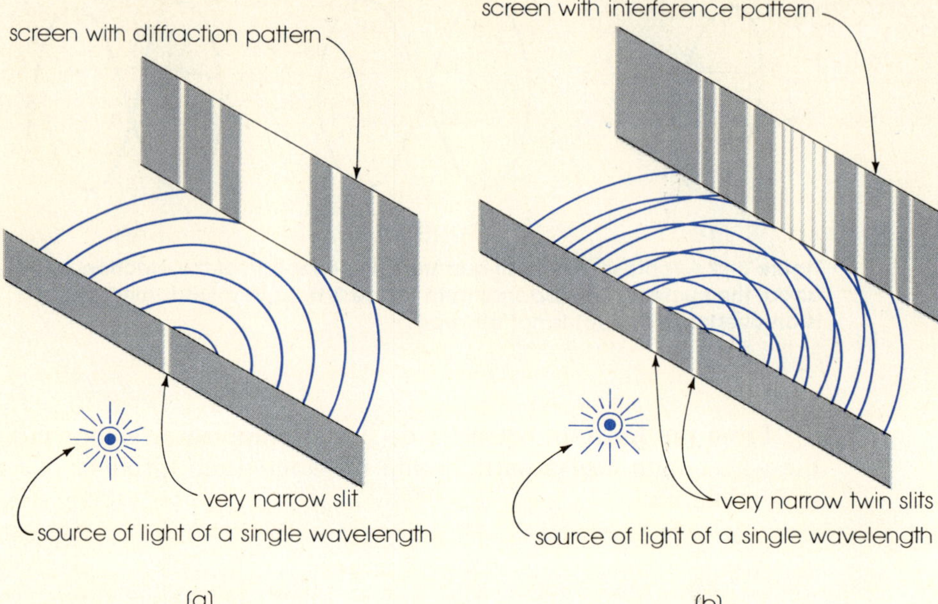

Figure 3-13 Evidence of wave motion (schematic views). (a) Diffraction. As light goes through the slit it "bends" around the edges to spread out in a circular pattern as if its origin was the slit. (Semicircles represent crests of waves, spaces between them are wave troughs.) If light had only particle character and were propagated in a straight line, a single bright band the size and shape of the slit would be seen on the screen. Instead, a light and dark diffraction pattern appears. (b) Interference. Wave motions coming from both slits cross one another at a glancing angle. At certain points waves enhance each other whereas at other points they diminish each other, giving a light and dark interference pattern on the screen. From the distances involved, the wavelength of the light can be calculated.

Electron diffraction

When matter is perturbed, energy is propagated (spread) from the disturbance. Transmittance of energy by propagation is called *wave motion*. Two related tests of wave character, illustrated in Figure 3-13, are *diffraction*, or the bending of waves into an area that would be shadow if light traveled in

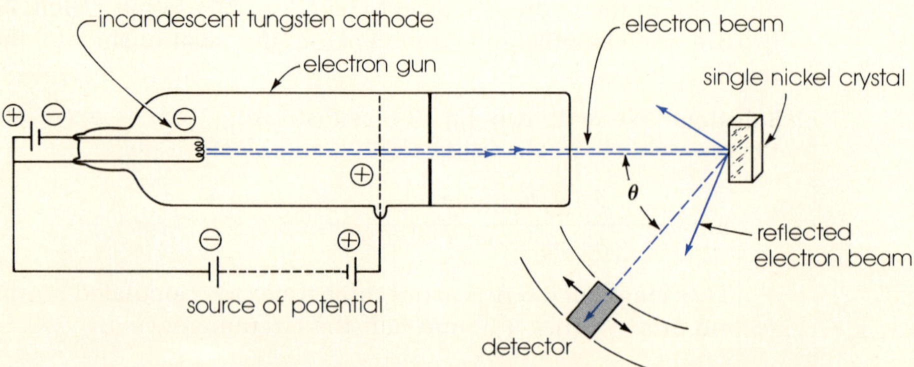

Figure 3-14 The Davisson-Germer experiment (a vacuum chamber surrounding the gun and specimen is not shown). The electron detector records strong reflection at angles corresponding to a diffraction pattern.

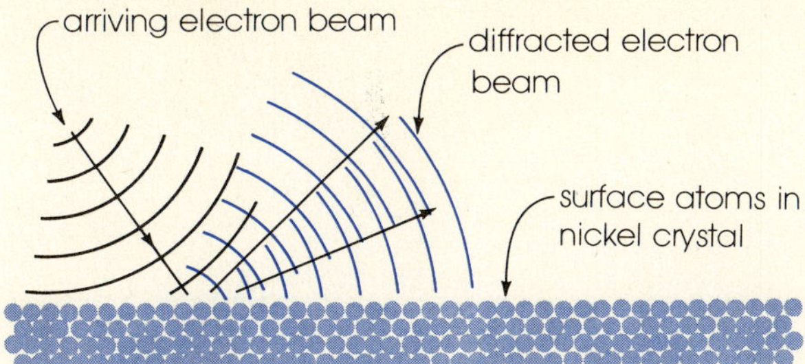

Figure 3-15 A closer look at the reflection of electrons from the top few rows of atoms at a crystal surface. Diffraction occurs there because atom–atom distances are comparable to the calculated wavelength of electron waves. This constructive effect is possible only if electrons have wave properties.

straight lines, and *interference*, or the process by which waves combine either to enhance or to diminish their individual effects.

Proof of de Broglie's electron-wave hypothesis was found in 1926 by Clinton Davisson (United States 1881–1958, Nobel prize in physics 1937) and L. Germer of the Bell Telephone Laboratories. They bombarded a crystal of nickel metal with low-energy electrons and measured the intensity of the reflected beam at various angles (Figure 3-14). The observed distribution of intensities suggested that the crystal had acted as a diffracting medium for the associated electron wave.

Planes of atoms indeed can reflect short wavelength energy, as pictured in Figure 3-15 (and Figure 6-5). Constructive wave interference occurs at reflecting angles that depend on wavelength and atom spacing. When Davisson and Germer calculated the wavelength in their experiment, they found that it was the λ predicted by de Broglie.

All experiments corroborate the wave-particle complementarity of the electron. It transfers its mass and charge as a unit, but its motion is a wave.

Particle waves do not belong to the spectrum of electromagnetic waves (see Figure 2-2). They are never radiated by or lost from the particle, and their speed is variable.

The son of J. J. Thomson, George Thomson (England 1892– , Nobel prize in physics 1937), and co-workers extended electron diffraction to very thin specimens. The principle of their apparatus is illustrated in Figure 3-16.

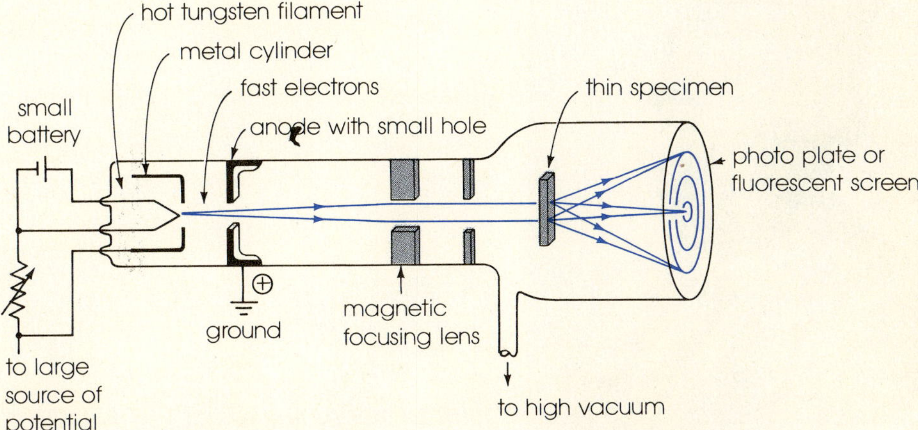

Figure 3-16 Schematic view of an electron diffraction camera. It operates under high vacuum to prevent electron scatter by air molecules. The specimen is only 50 to 1000 Å thick. Distance between specimen and screen, 30 to 50 cm, is such that a ring image a few centimeters in diameter forms.

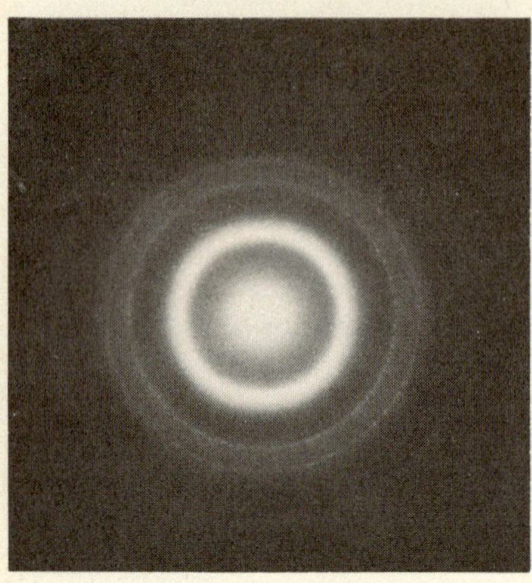

Figure 3-17 An electron-diffraction pattern, created by shooting an electron beam through a very thin sample of electroless nickel (see Eq. 18-12). (Courtesy A. J. Ardell, University of California, Los Angeles, School of Engineering.)

Photographs taken with the Thomson technique consist of a pattern of concentric circles, as shown in Figure 3-17. Circle radii are related to atomic dimensions in the target material. Electron diffraction is a powerful means for determining interatomic distances in matter.

Another valuable research tool, whose invention in 1932 was suggested by the wave character of electrons, is the electron microscope (see Figure 13-13).

Matrix mechanics, wave mechanics

Physicists Max Born (Germany and England 1882–1970, Nobel prize in physics 1952) and Werner Heisenberg (Figure 3-18) incorporated de Broglie's ideas into a new model by replacing the particle electron with a wave electron. Their mathematical solutions, which employed terms in rectangular arrangements called *matrices*, correctly predicted spectral line intensities and other characteristics of the atom.

Figure 3-18 Werner Heisenberg (Germany 1901– Nobel prize in physics 1932). An originator of quantum mechanics, Heisenberg also discovered the uncertainty principle. During World War II he worked on the unsuccessful German atomic reactor. (Nobel Foundation.)

Figure 3-19 Erwin Schrödinger (Austria and Ireland 1887–1961, Nobel prize in physics 1933). Schrödinger interpreted de Broglie's electron wave ideas mathematically. The Schrödinger wave equation produces quantum numbers that describe the atom. (Physics Today.)

Paul Dirac (England 1902–   , Nobel prize in physics 1933) carried matrix mechanics a step further by combining Einstein's special theory of relativity with quantum physics to create expressions that could be applied even to *relativistic particles,* those with very high energies. His mathematics interrelated four equations that conferred on the electron four "internal states." Included therein were right- and left-handed spins (already noted in the Stern-Gerlach experiment) and positive and negative energies. Dirac interpreted negative energy to mean a positive electron might exist.

The positive electron or *positron* was discovered four years later by Carl Anderson (United States 1905–   , Nobel prize in physics 1936). Anderson caught its track on film carried aloft by a balloon to an altitude where high-energy streams of particles called *cosmic rays* caused collisions that produced positrons. The positron was the first "antimatter" particle ever identified. Scientists today believe that for every electrically charged particle there can exist an identical particle of opposite charge.

Schrödinger (Figure 3-19) especially, who believed atomic physics was becoming too purely mathematical, preserved some of its former visual values when with de Broglie he developed *wave mechanics,* an alternate way of dealing with particle waves. Beginning with the atomic nucleus as the force that restrained the electron and hence its associated wave, he derived his now famous *wave equation,* through which can be calculated the patterns that electron waves assume under restraint.

Solutions to the wave equation are called *wave functions* or *orbitals.* Orbitals can be translated into diagrams like those of Figures 3-20 and 3-21, which depict electron distribution in a given region of space. Besides the wave function ψ (psi), terms in the Schrödinger equation include particle mass, potential energy, and permitted (quantized) energies. The equation is

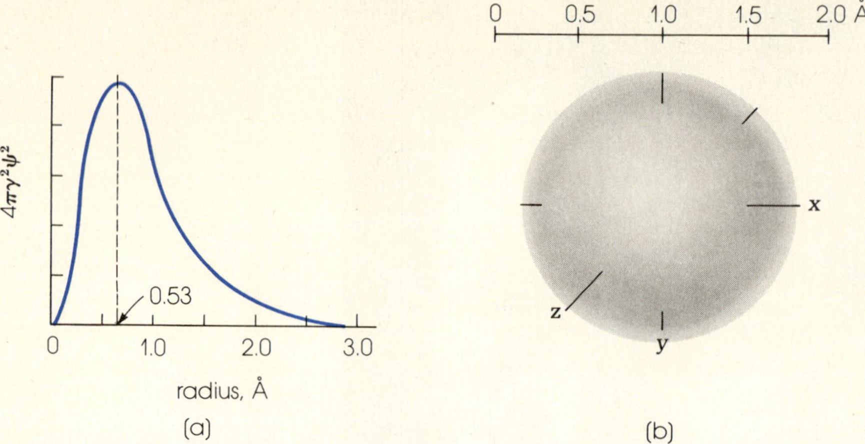

Figure 3-20 Quantum mechanical representation of the ground state (n = 1) hydrogen atom. (a) Radius is plotted against a function called the radial probability distribution. The curve's maximum at 0.53 Å corresponds to the Bohr radius. (b) Distribution of the electron around the atomic nucleus. The figure encloses the volume in which there is a 95 percent chance of finding the electron.

solved by substituting known values for particle mass and potential energy, then finding the permitted energies that satisfy the equation. The energies are related by integers called *quantum numbers*, which describe types of electron motion. Energies calculated for the hydrogen atom are exactly those found by spectral measurements.

The significance of ψ went unrecognized initially. Then Born gave it its present interpretation by suggesting that ψ^2 represented the probability of finding the electron in some volume of space. Unlike the Bohr atom,

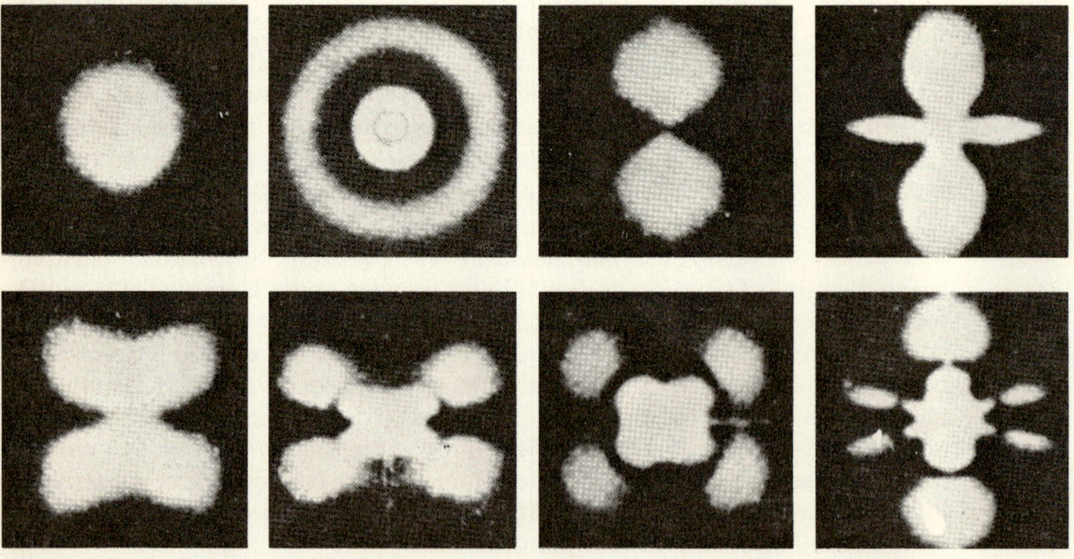

Figure 3-21 Some electron wave patterns (orbitals) for various energy states of the hydrogen atom. These are not models of reality but representations of mathematical solutions of the wave equation. In each pattern the atomic nucleus is to be imagined at the center, surrounded by a three-dimensional electron cloud. (Victor F. Weisskopf, "Physics in the Twentieth Century," Science 168, 923–930, May 1970. Copyright 1970 by the American Association for the Advancement of Science.)

with its precise electron positions and energies, the quantum-mechanical atom became blurred, its electron position expressed as a probability.

The uncertainty principle

Heisenberg reacted to Schrödinger's criticism of too much abstraction by introducing, with visual reference, an idea called the *uncertainty principle*. With it he stated that observational error could never be reduced to zero when one tried to inspect atomic particles closely. In the "thought experiment" which led to this conclusion, Heisenberg imagined having a supermicroscope under which a single electron was to be inspected by illumination with a single photon of light. Location of the very small electron required a photon of very short wavelength λ, meaning one of high momentum or high energy (Eq. 3-6). When the photon collided with the electron, however, the electron would move as momentum was transferred to it, and therefore at the moment of being "illuminated" neither its position nor momentum could be exactly known. Heisenberg showed that the uncertainty in electron position Δx was equal to or greater than a photon wavelength $\pm\lambda$, and the uncertainty in electron momentum $m\,\Delta v$ was equal to or greater than h/λ. The product of the uncertainties is incorporated in the Heisenberg uncertainty equation,

$$m\,\Delta v\,\Delta x \geqslant \lambda\frac{h}{\lambda} \geqslant h \tag{3-7}$$

The uncertainty principle is today one of the basic laws of physics. As illustrated in Example 3-3, it says that matter must remain incapable of precise observation at close range because there is no way to prevent the measuring method from disturbing the subject. The very nature of atomic particles creates this condition.

Example 3-3 Consider the electron of Example 3-2 moving in the second Bohr orbit with its calculated velocity of 1.1×10^8 cm/sec. What uncertainty exists in knowing its position if we wish to measure its velocity to within, say, 10 percent, 0.1×10^8 cm/sec?

Solution Writing Eq. 3-7 to read "approximately equal to" we obtain $m\,\Delta v\,\Delta x \simeq h$. Solving for Δx, the uncertainty in position, gives

$$\Delta x \simeq \frac{h}{m\,\Delta v} \tag{3-8}$$

Substituting the uncertainty in velocity, and text values for h and m gives

$$\Delta x \simeq \frac{6.6 \times 10^{-27}}{(9.1 \times 10^{-28})(0.1 \times 10^8)} \simeq 73 \times 10^{-8}\ cm \simeq 73\ \mathring{A}$$

The second Bohr orbit has a radius of only 2.12 Å. The uncertainty of knowing the electron's position, however, is 73 Å. Thus when we demand to know the electron's velocity to within 10 percent, strange as it may seem the uncertainty in locating the particle itself is about 34 times larger than its average distance from the nucleus.

What does quantum mechanics yield?

Quantum theory is an equation-oriented departure from the traditional physics of experience. It enables us to understand structure and energy at the atomic level in the following way.

First, it interrelates electron–proton attraction and electron charge and mass to define electron energy and electron distance from the nucleus. The relationships account directly for atomic spectra and the size of atoms.

Next the theory develops a wave equation for the calculation of the shapes electron waves assume when subjected to the influence of the nucleus and neighboring electrons. Although the electron's position and momentum cannot be simultaneously determined by the equations, wave patterns or orbitals are aways the same for an atom under specified conditions (Figure 3-21). Orbital symmetries are reflected in the shapes of molecules (see Table 9-2) because it is electrons that establish bonds between atoms.

Finally, quantum mechanics gives us four quantum numbers as solutions of the wave equations. As we will see later (Section 7-2), combinations of the numbers and a rule called the "exclusion principle" specify all the energy states an atom may possess.

Although the quantum state cannot be explained simply or adequately pictured, we will be able, where necessary, to apply generalizations from quantum theory. In the following chapters we deal mostly with large collections of atoms and molecules as we observe them in nature and in the laboratory. The features they have that concern us will be comparatively easy to describe.

QUESTIONS

1. Photoelectric effect The current (ammeter reading) in the photoelectric tube of Figure 3-2 varies with the wavelength of light falling on the cathode. When the cathode is coated with sodium metal in one experiment and cesium metal in another, Figure 3-22 data are obtained. **(a)** How do you interpret the results? **(b)** What happens in each case when light of longer wavelength than λ' falls on the photocell? Explain.

2. Photon **(a)** How does an atom generate photons? **(b)** What determines the energy (and wavelength) of the photons? **(c)** Which has more energy, a photon of visible light or a photon of ultraviolet light? Explain.

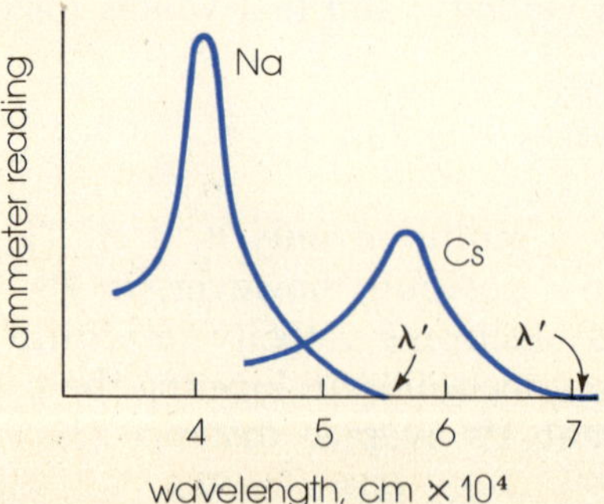

Figure 3-22

3. Quantum number What feature of the hydrogen atom did Bohr relate to the principal quantum number n? What is n in de Broglie's atom?

4. Uncertainty principle **(a)** Does the uncertainty principle mean that our present measurement methods fail because they are relatively crude, and that at a future date more refined methods will make measurements certain? Explain. **(b)** Why is the uncertainty principle meaningful for atoms but not for tennis balls?

5. Definitions Define or explain, and give the significance of **(a)** Planck's constant; **(b)** Stern-Gerlach effect; **(c)** orbital.

6. Electron Cite experimental evidence that indicates the electron is **(a)** a particle; **(b)** a wave.

7. Spectral lines The Paschen series of lines in the hydrogen spectrum is generated by transitions of electrons between the third energy level and higher energy levels. Using Figure 3-7, explain why you would expect these lines to be in the infrared region whereas the Balmer series is in the visible and the Lyman series is in the ultraviolet.

PROBLEMS

8. Bohr atom **(a)** Use Figure 3-6 and its accompanying description to calculate the velocity of the electron in the fourth energy level and the circumference of its path. **(b)** Find the number of orbits it makes per second.

9. Uncertainty Suppose you want to express the position of the electron in the first orbit of the Bohr hydrogen atom to within 1×10^{-8} cm (approximately the orbit's diameter). **(a)** What is the uncertainty in its velocity? **(b)** How does the uncertainty compare to the electron's velocity given by Bohr (Figure 3-6)?

10. Spectra Balmer's original formula describing his (visible) series of hydrogen spectral lines was

$$\lambda = \frac{3645n^2}{n^2 - 4}$$

where n is an integer greater than 2 and wavelength λ is expressed in angstroms. Show that the formula predicts each of the four wavelengths marked in Figure 3–7.

11. Spectra **(a)** Use the above equation to predict λ for energy level 7. **(b)** Why was this line not observed by early experimenters?

12. Radio waves **(a)** Show that Eq. 3-1 can be rearranged to $c = \lambda\nu$. **(b)** In 1887 German physicist Heinrich Hertz discovered radio waves by application of Maxwell's theory that an oscillating charge creates electromagnetic waves. His waves had a wavelength of about 9.7 m at a frequency of

3.1×10^7 cycles/sec. Calculate the speed (in centimeters per second) of radio waves. How does the answer compare with the speed of visible light?

13. Planck's constant In an experiment similar to that of Figure 3-8, the ionization energy of mercury is found to be 10.4 eV (1.60×10^{-12} erg $= 1$ eV). This corresponds to a spectral line having a frequency of 2.53×10^{15} cycles/sec. Show how, from these experimentally determined values, one can calculate a value for Planck's constant.

14. Wavelength A batter who strikes out in a baseball game excuses his performance by saying he forgot to allow for the de Broglie wavelength of the thrown ball. Calculate the ball's wavelength in centimeters, given its mass, 140 g, and velocity, 90 mi/hr (4×10^3 cm/sec). Comment on his excuse.

THE IMPORTANT CONCEPTS

4-1. Natural nuclear transformations
1. Alpha particles, beta particles, and gamma rays have different properties.
2. The displacement law allows prediction of nuclei formed in decay processes.
3. Isotope and isobar are useful words in describing related nuclei.
4. The mass spectrograph separates positive ions precisely, by mass.

4-2. Induced nuclear transformations
1. A nucleus may be transformed by acceptance of a subatomic particle.
2. The neutron is a neutral particle with ability to penetrate nuclei.
3. An atom's proton, neutron, and electron numbers are predictable from its atomic weight and atomic number.
4. Accelerators produce high-energy particle streams.

Study of the atomic nucleus is an active research field today. It began with the discovery of radioactivity, which disclosed the high-energy content of the nucleus, and with Rutherford's discovery of the nucleus itself. It received major impetus in 1932 with (1) discovery of the neutron, which for binding in the nucleus required postulation of a new kind of natural force, (2) announcement of Fermi's β-decay theory, which required postulating another new force, and (3) discovery of the positron, which indicated that antiparticles might exist.

Concurrent work on radioactivity disclosed that (1) radioactive atoms lost particles from their nuclei and became other atoms, (2) radioactivity could be induced by neutron bombardment, and (3) great quantities of nuclear energy were available in the fissioning of heavy nuclei or the fusing of light nuclei. There followed invention of atomic weaponry and attainment of control of atomic energy for peaceful uses.

4-1. NATURAL NUCLEAR TRANSFORMATIONS

Isotopes, isobars

Soon after radioactivity was discovered, simple ways were devised for studying it. Decay products—α particles (helium nuclei), β particles (electrons), and γ rays (short-wavelength radiation)—were separated in a magnetic field and their penetration characteristics determined. Experimental results are summarized in Figures 4-1 and 4-2.

Chemists who first synthesized pure compounds of radioactive elements were puzzled to find that the decay modes changed with time. A freshly prepared uranium salt, for example, emitted only α and γ rays.

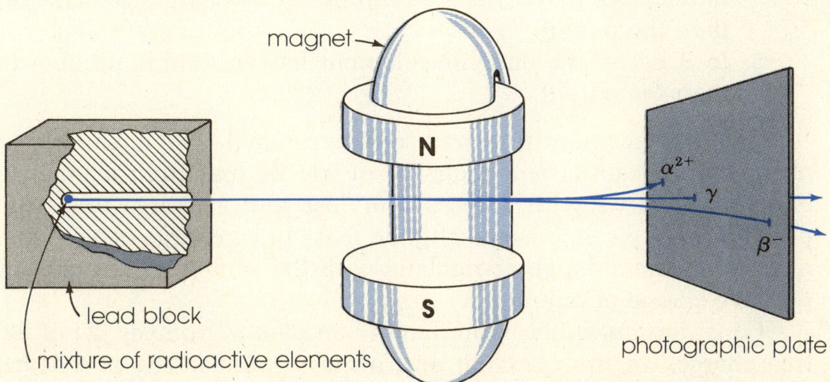

Figure 4-1 Natural radioactivity in a magnetic field. Charged components are deflected in opposite directions. (Refer to Figure 2-8.) The uncharged γ ray, which always accompanies α or β emission, is not deflected.

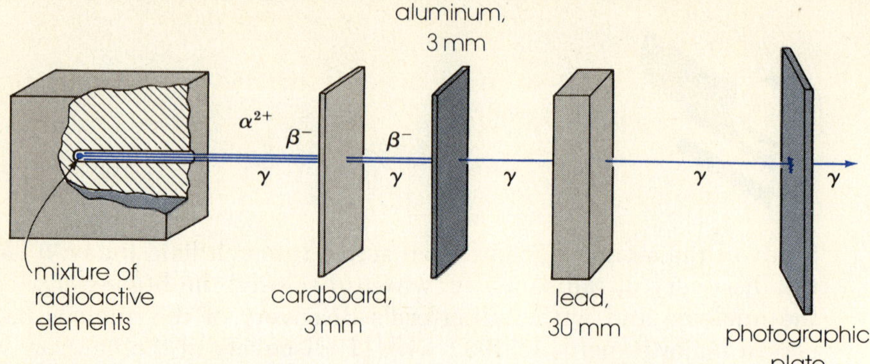

Figure 4-2 Penetration power of natural radioactivity. Alpha particles are stopped by cardboard. Beta particles, about 100 times as penetrating, are stopped by a sheet of aluminum. Gamma rays penetrate even thick lead shielding.

After a few hours, however, it also was found to be emitting β rays. Although α and γ activity remained high, β activity rose to a maximum in a few days, then held at that level.

The changes were explained in 1903 by Rutherford and Frederick Soddy (England 1877–1956, Nobel prize in chemistry 1921), who discovered the following generalizations:

1. A pure radioactive element gives off either α particles or β particles but not both.
2. Radioactivity involves the disintegration of atoms and formation of different kinds of matter, including atoms of other elements that may themselves be radioactive.
3. A law governs radioactive decay: The rate of decay is proportional to the amount of radioactive element present, and the fraction of radioactive element decaying in a given time interval is constant.

Later Soddy and co-workers discovered two more generalizations, which they called the *displacement rule:*

4. In α decay the *daughter element* (major product from decay of a radioactive parent) has an *atomic number* (nuclear charge) two less than the parent.
5. In β decay the daughter element has an atomic number one higher than the parent.

The displacement rule was soon explained. The α particle has a mass of four atomic units and a charge of 2+. Its loss from a nucleus leaves a daughter nucleus with a mass four less and a charge two less than the parent. The β particle has negligible mass but a charge of 1−. Its loss from a nucleus yields a daughter nucleus with the same mass as the parent but a positive charge of one more.

Consider a uranium atom having an *atomic number* (Z) of 92 (92 positive charges in the nucleus) and a *mass number* (A) of 238 (the integer nearest its atomic weight, 238.03). Loss of an α particle ($Z = 2$, $A = 4$) leaves an atom with an atomic number of $92 - 2 = 90$, and a mass number of $238 - 4 = 234$. From the periodic table we identify element 90 as thorium. The uranium atom has changed spontaneously into a thorium atom.

To represent *nuclear reactions* chemists write *nuclear equations*. Conservation of charge (in this case 92+) is carried in subscripts, conservation of mass (238) in superscripts. As described above,

$$^{238}_{92}U = {}^4_2He + {}^{234}_{90}Th \tag{4-1}$$

We will follow the process a few steps further; ^{234}Th is a β emitter ($Z = -1$, $A = 0$ for the electron). According to the displacement rule the resulting (second generation) daughter must be element 91, protoactinium, which has the same mass as its parent but one positive charge more:

$$^{234}_{90}Th = {}^0_{-1}e + {}^{234}_{91}Pa$$

Now ^{234}Pa is radioactive and a β emitter. As before, we can predict its daughter will have an atomic number of 92 and an atomic mass of 234. Any element with an atomic number of 92 is uranium:

$$^{234}_{91}Pa = {}^0_{-1}e + {}^{234}_{92}U$$

^{234}U is an α emitter. It yields thorium again:

$$^{234}_{92}U = {}^4_2He + {}^{230}_{90}Th$$

The problems encountered by early workers in nuclear research can now be appreciated. Uranium-234 and uranium-238 are both uranium, give the same spectral lines, and cannot be separated by chemical methods although they decay at different rates. The same can be said for the two kinds of thorium. The puzzle of reaction sequences was gradually unraveled by chemists who separated the different radioactive elements and catalogued their decay modes and *half-lifes* — that is, the time interval in which half the atoms present at any given instant would decay. They also clarified the displacement law with two terms:

1. *Isotope* (Gk., equal place): an atom having the same atomic number as another atom but different atomic weight.
2. *Isobar* (Gk., equal weight): an atom having the same atomic weight as another atom but different atomic number.

In the examples above, $^{234}_{92}U$ and $^{238}_{92}U$ are isotopes, $^{234}_{90}Th$, $^{234}_{91}Pa$, and $^{234}_{92}U$ are isobars.

Uranium-238, uranium-235, and thorium-232 are the parents of three long decay series that have been active since the earth was formed. The series end, respectively, with the stable isotopes lead-206 (Figure 4-3), lead-207, and lead-208.

Sorting particles by mass

A great aid to the studies just described (and to research ever since) was the mass spectrometer (Figure 4-4). Using electric and magnetic fields it permits precise separations of positive ions according to their masses. The "mass spec" idea, conceived by J. J. Thomson, was made practical by co-worker Francis Aston (England 1877–1945, Nobel prize in chemistry 1922). In early experiments they discovered that neon taken from the air was a mixture of 90 percent neon-20 and 10 percent neon-22. These were the first nonradioactive isotopes known.

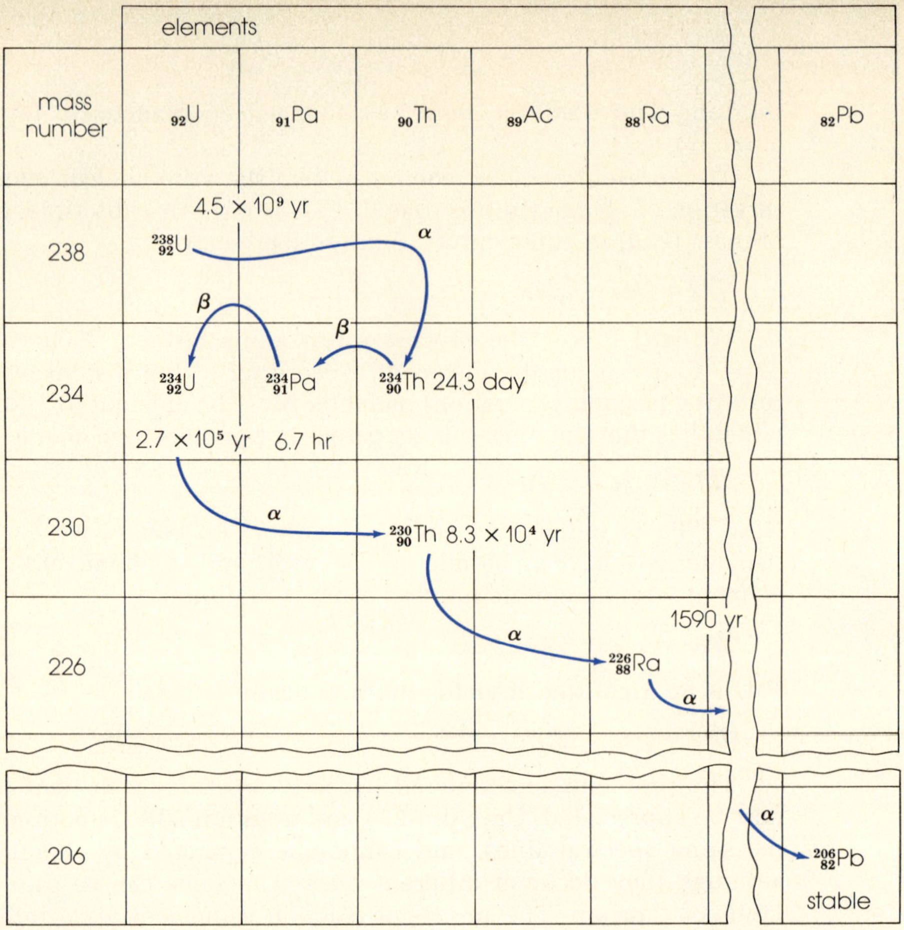

elements

mass number	$_{92}$U	$_{91}$Pa	$_{90}$Th	$_{89}$Ac	$_{88}$Ra	$_{82}$Pb
	4.5×10^9 yr					
238	$^{238}_{92}$U $\quad\alpha$					
	β	β				
234	$^{234}_{92}$U	$^{234}_{91}$Pa	$^{234}_{90}$Th 24.3 day			
	2.7×10^5 yr	6.7 hr				
230		α	$^{230}_{90}$Th 8.3×10^4 yr			
226				α	$^{226}_{88}$Ra $\quad$ 1590 yr $\quad\alpha$	
206					α	$^{206}_{82}$Pb stable

Figure 4-3 A portion of the uranium-238 decay sequence. Times given are half-lifes. Between uranium-238, the original parent, and lead-206, the final daughter element, there is a mass loss of 32 atomic weight units due to emission of eight α particles.

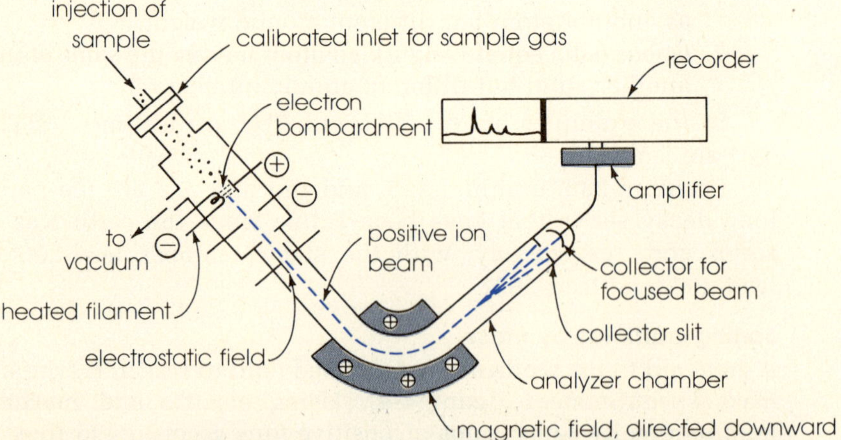

Figure 4-4 Schematic view of a mass spectrometer (Aston's design). Today's best instruments are capable of separating ions differing in mass by only 1 part in 750,000. Applications include analysis of trace (barely perceptible) impurities in samples, determination of isotopic and molecular weights, separation of isotopes in practical quantities, and identification of complex compounds (see Figure 23-9a).

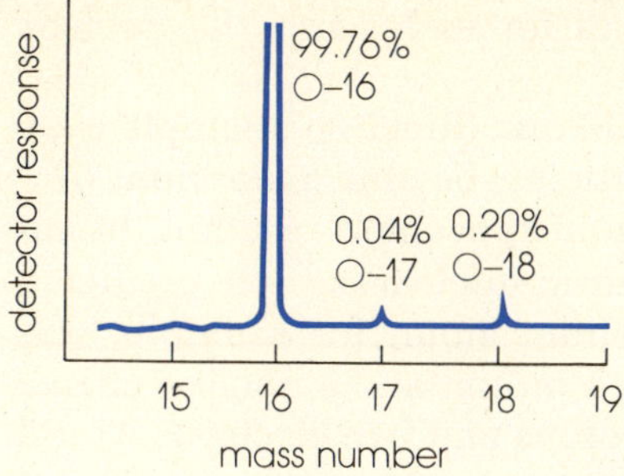

Figure 4-5 Separation by mass spectrometry of the natural isotopic mixture of oxygen taken from the atmosphere.

In mass spectrometry a fraction of a milligram of gaseous sample is admitted to the instrument and subjected to electron beam bombardment. As the beam knocks electrons from the sample, positive ions are produced. An atomic sample gives simple ions like Ne$^+$, whereas a molecular sample gives the parent ion and fragments of it. After acceleration, the ions are passed as a beam between charged plates. Slower ions of a given mass are deflected more than faster ions because they remain longer in the electrostatic field. As the beam then passes through a magnetic field at right angles to the electric field, slower ions are again deflected but this time in the opposite direction. Adjustments therefore permit focusing on the detector all ions of a given mass regardless of velocity. As indicated in Figure 4-5, detector response is proportional to the number of positive ions striking it.

4-2. INDUCED NUCLEAR TRANSFORMATIONS

Discovery of the neutron

In 1919 Rutherford bombarded nitrogen gas with α particles and obtained a strange result: Cloud chamber photos showed that some collisions were followed by proton (H$^+$) tracks. Where had hydrogen come from? "We must conclude," Rutherford wrote, "that the nitrogen atom is disintegrated under the intense forces developed in a close collision with a swift α particle, and that the hydrogen atom which is liberated formed a constituent part of the nitrogen nucleus."

Working with James Chadwick (England 1891–   , Nobel prize in physics 1935), Rutherford concluded that because various light elements could be induced to give up protons, the proton was probably the source of nuclear positive charge. His student Patrick Blackett (England 1897–   , Nobel prize in physics 1948) deduced that the nitrogen nucleus had captured an α particle and then disintegrated, giving a proton. His evidence was those few cases of head-on collision (eight events in the 400,000 cloud chamber tracks he studied) in which the α particle was suddenly halted and the track forked into a long thin proton branch and a short thick branch of a nucleus more massive than the α particle. The pictures could only be explained by assuming that nitrogen had been converted to an isotope of oxygen:

$$^{14}_{7}N + {}^{4}_{2}He = {}^{17}_{8}O + {}^{1}_{1}H \tag{4-2}$$

Not only was it evident then that transmutation of elements was possible (the alchemist's dream), but scientists could begin estimating, from the known α-particle energies, the strength of forces holding the nucleus together.

Several avenues of research were begun. In one direction attempts were made to cause heavy nuclei to accept α particles; beyond potassium (nuclear charge 19+), however, coulomb repulsion prevented reaction. In another direction evidence was sought for a neutral nuclear particle predicted by Rutherford several years before. The helium atom, for example, was known to have a mass four times larger than a proton and a nuclear charge of 2+. Although a nucleus containing *four* protons and two electrons would account for these facts, so would a nucleus containing *two* protons and two equally massive neutral particles.

In 1932 H. Becker and Walther Bothe (Germany 1891–1957, Nobel prize in physics 1954) demonstrated that α-particle bombardment of the light metal beryllium gave protons and a penetrating radiation that they could not identify. Chadwick suspected that Bothe's radiation might be evidence of Rutherford's neutral particle, which had escaped detection because neutral particles did not leave cloud chamber tracks. From experiments in which he allowed the radiation to bombard nitrogen, he was able to calculate the speed of its particles and their mass: about 0.1 percent greater than the proton mass. Chadwick named the particle the *neutron*. Carrying one atomic mass unit but no charge, the neutron's presence meant that beryllium had been converted to carbon by reaction with α particles:

$$\ce{^{9}_{4}Be} + \ce{^{4}_{2}He} = \ce{^{12}_{6}C} + \ce{^{1}_{0}n} \tag{4-3}$$

The atom thus became describable by three particles, a description sufficient for many chemical purposes today. Consider a neutral atom of mass number A and atomic number Z. In the nucleus are Z protons and $A - Z$ neutrons. Outside the nucleus in quantum orbits are Z electrons; $\ce{^{19}_{9}F}$, for instance, stands for a fluorine atom with nine protons, and $19 - 9 = 10$ neutrons in the nucleus, and nine electrons in orbits.

Artificial radioactivity

The discovery that radioactivity could be induced in nonradioactive atoms was made in 1934 by Frederick Joliot (France 1900–1958) and his wife Irene Curie (France 1897–1956), who shared the Nobel chemistry prize in 1935. Bombarding a target of aluminum with α particles and studying the result in a cloud chamber centered in a magnetic field, they found that the target emitted positive electrons (positrons) even after bombardment was halted. In the target they identified a short-lived isotope of phosphorus, apparently formed by neutron loss. Its decay by positron (atomic number 1, mass number 0) emission gave a stable isotope of silicon:

$$\ce{^{27}_{13}Al} + \ce{^{4}_{2}He} = \ce{^{1}_{0}n} + \ce{^{30}_{15}P}$$

$$\ce{^{30}_{15}P} = \ce{^{30}_{14}Si} + \ce{^{0}_{1}e}$$

Within 18 months of the publication of these results, laboratories around the world had reported production of 100 artificial radioactive species.

Today about 1350 isotopes are known, of which some 1080 are radioactive. Of the latter approximately 1025 are man-made. A recent application of induced radioactivity by neutron bombardment is the technique called *neutron activation*. By identifying and counting the radioactivities of isotopes formed, analytical chemists obtain ultrasensitive and nondestructive quantatitive measurements on elements in the sample. Average detection limit for some 65 elements is presently about 5×10^{-8} g, or 5 parts per billion in a 10-g sample.

High-voltage accelerators

For research to continue on induced radioactivity, scientists needed a versatile, mechanical source of high-energy projectiles. The first machine to successfully accelerate particles for "atom smashing" experiments was built during 1929–1932 by John Cockcroft (England 1897–   ), and Ernest Walton (Ireland 1903–   ), who shared the Nobel prize in physics in 1951. In the Thomson tradition of thrift it was made partly of salvage (the operator sat in a curtained packing crate to observe flashes of particles under a microscope) and cost \$2500, the largest sum spent to that date at the Cavendish Laboratory on one experiment. The sudden discharge of a bank of capacitors produced protons of sufficient energy (up to 0.7 MeV*) to produce α particles upon collision with a lithium target:

$$^{7}_{3}Li + ^{1}_{1}H = ^{8}_{4}Be = ^{4}_{2}He + ^{4}_{2}He$$

In 1930 at the University of California, Ernest Lawrence (United States 1901–1958, Nobel prize in physics 1939) and S. Livingstone built a circular accelerator called a *cyclotron*. A cyclotron (Figure 4-6) consists of two

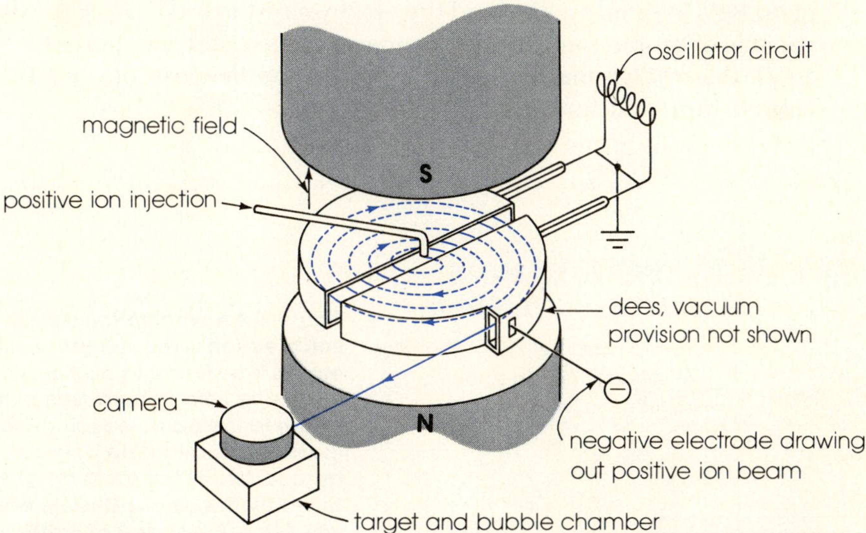

Figure 4-6 The cyclotron. If alternating potential difference across the dees is 100,000 eV, a particle gains 200,000 eV in one revolution by jumping the gap twice. After 100 revolutions (completed in about 0.001 sec), its energy gain is 20 MeV.

Lawrence realized the cyclotron could be useful in medicine through its production of isotopes and neutrons. He treated his own mother, who suffered from cancer, with neutrons from his cyclotron and the cancer became dormant. (Later work showed neutrons less effective than X rays in such treatment.)

* 1 MeV = 1 million eV = 1.6×10^{-6} erg.

hollow and evacuated D-shaped chambers ("dees") above and below which are pole pieces of a large magnet. An alternating voltage difference of about 100,000 volts is impressed across the gap between the dees by a high-voltage oscillator. As a charged particle is introduced in the center of the configuration, it is accelerated by attraction of opposite charge on one dee and repulsion of like charge on the other dee. As the particle enters a dee, the magnetic field, acting at right angles, causes it to take a circular path. Traversing the first dee it jumps the gap into the other dee. As it does so, it gets another boost in energy because at that instant the oscillator has changed the dee voltage. As the particle reaches higher speeds, the radius of its path increases until it exits near an attracting electrode at the dee periphery. Outside, a target and detection apparatus await its impact.

A circular machine of fixed particle path (Figure 4-7) was conceived in 1945 by V. Veksler (USSR) and by Edwin McMillan (United States 1907–   , Nobel prize in chemistry 1951). In their design, bunches of particles shot into a hollow ring are accelerated when they jump a gap between electrodes operated by an oscillator. As predicted by Einstein's relativity theory, the particles gain mass as they gain energy, and they become more difficult to confine. Control is maintained by increasing the deflecting magnetic field in synchronization with the accelerating voltage. By keeping the particles whirling a million times or more, billion electron-volt (BeV) energies are attained.

To attain billion electron-volt energies in a linear design requires a long path length. The world's largest linear accelerator (Figure 4-8) produces bunches of electrons traveling very close to the speed of light. At an energy of 40 BeV their masses exceed the electron rest mass by 50,000. Although incapable of producing particles as energetic as cosmic ray particles (up to 10^{12} BeV), these large machines are giving us the deepest look yet into nuclear composition and forces.

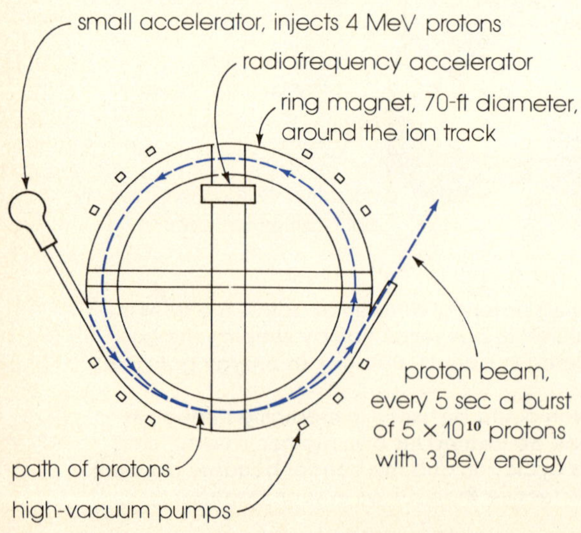

Figure 4-7 A synchrotron design of a billion electron volt accelerator. Pulsed energy is transferred to particles and the magnetic field is increased in synchronization to the particles' relativistic gain in mass. The relationship between rest mass m_0 and relativistic mass m_v at speed v, is given by the Lorenz-Einstein equation in which c is the speed of light):

$$m_v = \frac{m_0}{\sqrt{(1 - v^2)/c^2}}$$

At an energy of 100 MeV an electron has a speed 0.9999 that of light and a mass 200 times m_0. The largest such machine is the 200-BeV proton synchrotron at Weston, Illinois, scheduled for startup in 1975.

Figure 4-8 A linear billion electron volt accelerator. The accelerator is actuated by 240 high-power amplifier (klystron) tubes which transmit electromagnetic waves that carry 20-BeV electrons down a straight 2-mile copper pipe. Electrons are introduced at the far upper end of the tunnel and accelerated to a beam switchyard for magnetic focusing. Beams are directed into the experimental buildings (bottom of photo) where targets are hit and magnetic spectrometers sort out the particle products. (Courtesy Stanford University.)

4-3. NUCLEONS AND NUCLEAR STRUCTURE

Weak and strong forces

Information on nuclear shape and size comes from scattering experiments. The nucleus is spherical to slightly ellipsoidal. *Nucleons* (protons and neutrons) seem to act in it like golf balls in a sack, mobile yet in close contact.

Nuclear volume is proportional to the nucleon number. It therefore is also proportional to the atom's mass number A because each nucleon has a mass of about 1 atomic mass unit (amu*). Nuclear radius R is given by

---

* 1 amu equals one-twelfth of the mass of one ^{12}C atom. The unit is discussed in Section 12-2.

$$R = r_o A^{1/3} \tag{4-4}$$

where r_o is the nuclear constant, 1.3×10^{-13} cm. The unit of nuclear measure is the fermi (F): $1\ \text{F} = 1.0 \times 10^{-13}$ cm.

Nuclear radii measure only a few fermis. With relatively massive particles packed so tightly, nuclear density is stupendous. A cubic centimeter of pure nuclear material would contain 2×10^{38} nucleons and weigh 3×10^{14} g or 330 million tons.

Until methods were available for critically examining the atomic nucleus, the only natural forces known to operate between objects were *electromagnetism* (coulomb force) and the very feeble force of *gravity*. Physicists now recognize that at least two more forces operate in nature: weak interaction and strong interaction between subatomic particles in nuclei. The concept of *weak force* was introduced by Fermi (see Figure 4-12) to explain the ability of a nuclear electron to escape (β decay). *Strong force* is used to explain the binding of nucleons — that is, the binding that keeps the nucleus together despite the very powerful repulsive forces between the nucleons. As shown by studies on "mirror nuclei" such as $^{13}_{6}\text{C}$ ($6p^+$, $7n$) and $^{13}_{7}\text{N}$ ($7p^+$, $6n$), proton–proton, proton–neutron, and neutron–neutron combinations have essentially the same binding characteristics. On a relative scale in which the strong nuclear force is given a value of 1, the coulomb force at proton–proton distance in the nucleus is 10^{-2}, the weak β-decay force is perhaps 10^{-10}, and the force of gravity a very faint 10^{-39}.

Objects are assumed to be attracted or repelled by force fields that are projected by *field carriers*. Physicists recognize magnetic, electric, gravitational, and nuclear fields. Theoretically a particle is associated with a field, and modern theory treats particle and field simultaneously. Electromagnetism is carried by photons. The carrier of gravity is presumably the graviton, a hypothetical particle whose existence was tentatively established in 1970. It is believed the strong nuclear force is carried by *mesons*. The weak force may be carried by the W particle, evidence for which was found in 1971. The two nuclear forces are attractive at a range of 1 fermi but become repulsive at about half that distance. As a consequence the nucleus has small size but does not collapse under strong-force attraction.

Mesons as carriers

In 1935 Hideki Yukawa (Japan 1907–   , Nobel prize in physics 1947) predicted the nuclear existence of a medium-weight (274 times the electron mass) carrier of the field with which nucleons presumably interact. Because its calculated energy was 25 times more than the energies observed in radioactivity, he proposed that the particle might best be sought among fragments produced by the impact of cosmic rays on matter. Twelve years later Yukawa particles, now called *pions* or π (pi) mesons, with all the predicted properties were found in stacks of photographic emulsions exposed to cosmic radiation.

For study purposes today mesons are produced by striking a metal target with very high-energy particles (Figure 4-9).

Three kinds of pions are known; π^+; π^-, mass 273; and π^0, mass 264. As protons and neutrons exchange mesons in a quantum mechanical way,

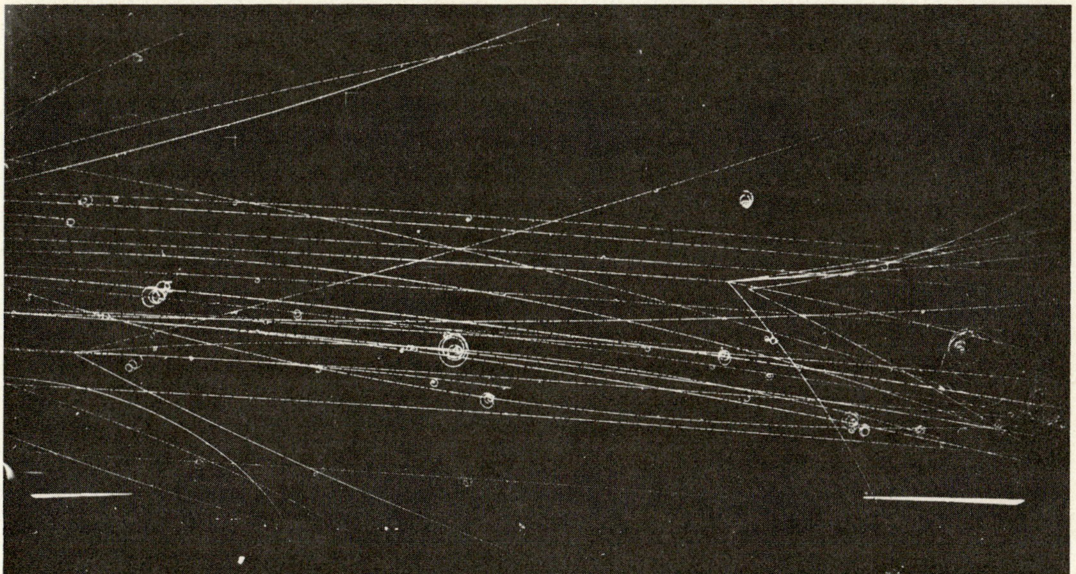

Figure 4-9 Pi-meson events, as recorded in a 20-inch liquid hydrogen bubble chamber. Pions were produced by a high-energy beam of protons striking an aluminum target. About 15 pions enter in a parallel beam from the left. Branching tracks (like those center, right) indicate collision of a pion with a hydrogen nucleus (proton) and production of secondary particles. Small circular tracks are due to low-energy electrons knocked loose by pions. (Courtesy Brookhaven National Laboratory.)

force is transmitted between them and they appear to be merely two conditions for one particle, the nucleon (Figure 4-10). Interchange reactions are

$$n = p^+ + \pi^-$$
$$p^+ = n + \pi^+$$

Nucleon structure

Evidence for a substructure in nucleons is presently being obtained by examining (1) the spectra that nucleons produce on absorption and reemission

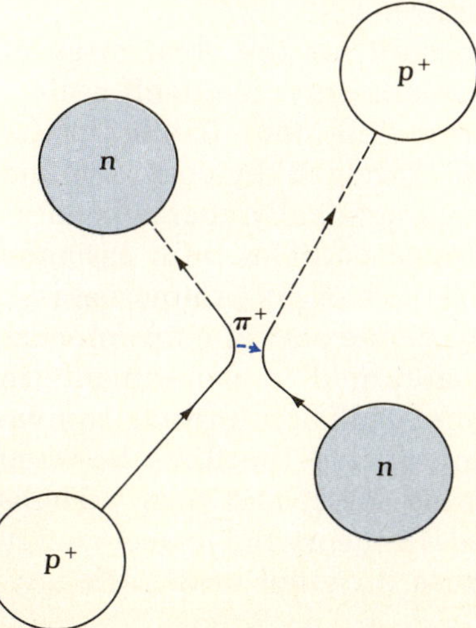

Figure 4-10 Proton-neutron interaction at high energy. The collision is called "inelastic" because the particles change identity. Transfer of a π meson converts a proton into a neutron, and vice versa.

of energy, and (2) the manner in which nucleons scatter an electron beam. Designs of the experiments resemble two earlier investigations on atomic structure—Franck and Hertz's proof of a progression of excited electron states (see Figure 3-8), and Rutherford's scattering technique (see Figure 2-18). The difference today is that targets are about 50,000 times smaller, and require 10 times the excitation energy. Whereas electron shells are perturbed by a few electron volts and nuclei by a few million electron volts, billion electron-volt energy is needed to perturb nucleons. Excited nucleons exist in a whole spectrum of quantum states indicative of an as yet unresolved internal organization that is held together by a force more powerful than the nuclear strong force.

Both neutron and proton appear to contain a core of radius 0.3 fermi and charge $0.5e$ (e the unit electronic charge). Extending about 1 fermi outside the core is a mantle with charge $-0.5e$ for the neutron and $+0.5e$ for the proton. Extending another fermi or so is a diffuse ring, presumably of mesons in the process of being expelled and recaptured. Ring charge is about $+0.1e$ for both proton and neutron. Besides these characteristics nucleons have spin and magnetism. They are clearly not "simple" particles.

Leptons, mesons, baryons

The existence of a variety of subatomic particles and quantum states has been proved, but no theory yet accommodates them all.

The particles are grossly classified by mass. Lightest are the *leptons* (Gk. *leptos*, small). They include two kinds of *neutrinos* (a very small neutral particle), the electron, the muon (a 200-amu electron), and their antiparticles. Of intermediate mass are the mesons (Gk. *mesos*, middle) and antimesons. More than 100 meson energy levels have been found. Although all mesons are unstable, the eight with lowest energy are well characterized. Lowest three of the eight are the pions. Heaviest particles are the *baryons* (Gk. *barys*, heavy) and antibaryons. A baryon is a nucleon in any energy state. As with mesons, over 100 baryon energy levels are known, and again at the bottom levels is a family of eight particles. The ground energy level is occupied by the proton and neutron.

Several ideas have been advanced to account for the great array of *hadrons* (Gk. *adron*, strong), as mesons and baryons are collectively called. One is the *quark hypothesis*, put forth by Murray Gell-Mann (United States 1928–   , Nobel prize in physics 1970) and G. Zweig. They proposed that hadrons are composed of combinations of three fractionally charged quarks and antiquarks. Quarks are hypothetical ultimate particles with assigned spin, charge, and other properties (mass may be five or six proton masses). Whether quarks are real substances or merely mathematical conveniences is not yet known. But since they have not appeared in experiments in which colliding beams of protons have produced collision impacts equivalent to 2000-BeV energies, it has been surmised that the force between quarks is even stronger than the nuclear strong force, and thus perhaps quarks have never been permanently separated. In theory the lowest-energy quark is stable and collectible for study because it cannot decay into anything else.

Liquid-drop or shell?

To help describe induced radioactivity, Bohr in 1936 pictured a nucleus as absorbing a particle, being raised to an excited state, then returning to a lower state by ejecting a particle or γ ray. The discovery of nuclear fission (Section 4-4), in which a heavy nucleus splits into the nuclei of two smaller atoms, sharpened his idea into the *liquid-drop model*. This theory pictures the nucleus as a drop that starts vibrating when its balance of cohesive and repulsive forces is upset and finally breaks into smaller drops (see Figure 4-13).

A rival model began with Fermi when he demonstrated that neutron capture by a nucleus depends on neutron energy, and that only certain sharply defined energies are acceptable to a given nucleus. That suggested a grouping of nucleons in shells, perhaps like the orbital arrangement of electrons in the Bohr atom. The shell theory has since been supported by correlations between nucleon count and nuclear shape and reactivity. A nucleus with a neutron or proton count of 2, 8, 20, 28, 50, 82, or 126 has special stability and spherical shape, and its isotopes are naturally abundant. For lack of an explanation these numbers were originally called "magic numbers." Since then physicists, including Maria Goeppert Mayer (Germany and United States 1906–1972) and Johannes Jensen (Germany 1907–   ), who shared the 1963 Nobel physics prize, have developed a *nuclear shell model* (Figure 4-11) which accounts for most of the known facts and predicts the magic numbers. According to their theory, protons and neutrons are grouped in close-spaced energy shells or nuclear quantum states. A magic number of protons or neutrons establishes filled shells and nuclear sphericity. (Doubly magic nuclei, those containing magic numbers of both protons and neutrons such as $^4_2$He, the α particle, and $^{208}_{82}$Pb, the end product of the thorium decay series, are particularly stable.) If nonmagic numbers of nucleons are present, higher shells are presumably unfilled. Nonmagic nuclei are ellipsoidal and less stable.

As will be discussed in Section 7-2, the periodic table is an ordering of atoms according to similarities that occur periodically in electron-shell

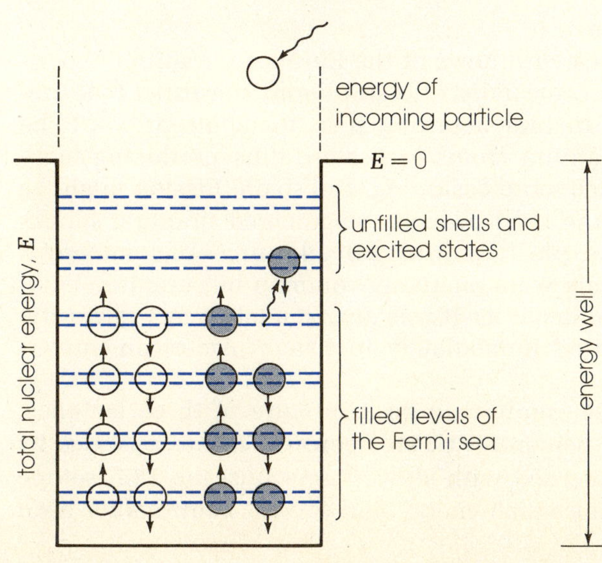

Figure 4-11 The nuclear shell model. Nucleons fill a "Fermi sea" of levels at the bottom of the energy well. Above the sea are unfilled shells and excited energy states. Nucleons may be activated enough to reach those positions when entry of a foreign nucleon creates a disturbance. Pairing by spin (indicated by arrows up and down) of no more than two protons and two neutrons is permitted per level.

structure. We are probably not far also from the publication in textbooks of a similar chart which will group nuclei according to repeating patterns of nucleons.

4-4. RELEASE OF NUCLEAR ENERGY

Atomic fission

Work on neutron-promoted nuclear reactions began immediately after the neutron's discovery. Initially the world leader was Fermi's group in Italy. Their approach was to bombard elements with neutrons slowed by passage through paraffin or water. This increased the (de Broglie) wavelength of the neutron to a value compatible with nuclear vibration, permitting its entry into the nucleus.

When they exposed a uranium foil to neutrons, it acquired β activities indicative of new elements. By dissolving the foil in acid, adding known compounds to "carry" the unknown elements, then isolating the knowns by established chemical precipitation procedures, they got partial separation of the radioactive components. They expected to find elements near uranium in the periodic table but failed to identify any.

German chemist I. Noddack guessed what happened. She wrote,

> It would be equally possible to assume that when a nucleus is demolished in this novel way by neutrons, "nuclear reactions" occur . . . when heavy nuclei are bombarded with neutrons, the nuclei in question might break into a number of larger pieces which would no doubt be isotopes of known elements but not neighbors of the elements subjected to radiation.

In Germany in 1938 Hahn (Figure 4-12) and co-workers finally identified barium, lanthanum, and cerium (elements 56, 57, 58), then krypton, rubidium, and strontium (36, 37, 38) in uranium (92) target material. The uranium nucleus had indeed split into two parts, they announced, one being approximately $Z = 57$, $A = 140$, the other approximately $Z = 37$, $A = 96$. Each splitting event released a huge amount of energy.

Sustaining the fission reaction

In 1939 Bohr came to America with news of the European uranium experiments. At a meeting of atomic specialists he and Fermi suggested that uranium fission products might include neutrons; if so, the neutrons would be capable of causing more uranium atoms to fission, thus producing more neutrons, which would induce more fissioning, and so on. Fission might be made self-sustaining, therefore, and properly engineered uranium blocks could either be used as high explosives or long-lived sources of commercial heat energy. Uranium minerals were relatively common but uranium compounds had few uses and the metal itself was almost a laboratory curiosity. Research on element 92 started immediately in several American universities.

Analysis showed natural uranium to be a mixture of three isotopes: ^{234}U, 0.006 percent; ^{235}U, 0.7 percent; ^{238}U, 99.3 percent. When separated by mass spectrograph and bombarded with slowed neutrons, the 235 isotope proved most reactive, yielding much energy and several neutrons for each nucleus split (Figure 4-13).

Figure 4-12 Pioneers in nuclear energy. (Left) Enrico Fermi (Italy and United States 1901–1954, Nobel prize in physics 1938). Fermi used his Nobel prize trip to Stockholm as a means of leaving Italy and getting to the United States. He was the principal theoretician in the group that built the first self-sustaining nuclear reactor and he later worked on the atom bomb. Following the war he was active in meson research (University of Chicago). (Right) Otto Hahn (Germany 1879–1968, Nobel prize in chemistry 1944). Hahn's position in radiochemistry in Germany was similar to that of Rutherford in England. He discovered protoactinium (element 91), helped elucidate radioactive decay series (Figure 4-3), and proved uranium and thorium could be split by slow neutrons to release nuclear energy. (Photographs: Left, Courtesy The University of Chicago; right, Nobel Foundation.)

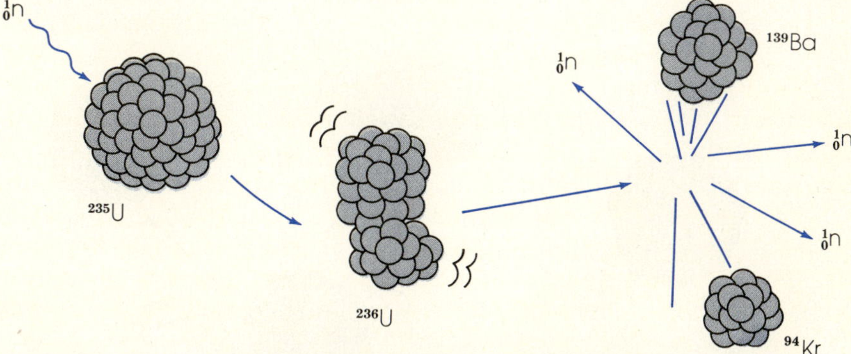

Figure 4-13 Uranium fission. Neutron capture by the uranium-235 nucleus gives a nuclide heavier by 1 amu. (A word association: Nuclide is to nucleus as isotope is to atom.) Uranium-236 is unstable, splitting with the release of nuclear energy into smaller nuclei and several neutrons. Typically,

$$^{235}_{92}U + ^{1}_{0}n = ^{236}_{92}U = ^{139}_{56}Ba + ^{94}_{36}Kr + 3^{1}_{0}n$$

In August 1939 a group of American scientists persuaded Einstein to sign a letter to President Franklin Roosevelt which explained (1) the possibility of self-sustaining fission (a so-called *chain reaction*) for military use, (2) the poor grade of domestic uranium ores, and (3) the halting of the sale of uranium by Germany in international markets. The President appointed a government liaison committee to work with the scientists and appropriated $6000 for their research. The following month Germany invaded Poland from the west and the USSR moved troops to the Polish border from the east. World War II had begun.

At the University of California a team headed by McMillan, then by Seaborg (see Figure 8-4) sought to find what happened when a uranium-238 nucleus captured a neutron. They demonstrated that the first product, uranium-239, was a β emitter which quickly became a new element, number 93, neptunium. In similar fashion 93 became another new element, 94, plutonium (Figure 4-14). Somewhat simplified,

$$^{238}_{92}U + ^{1}_{0}n = ^{239}_{92}U$$

$$^{239}_{92}U = ^{239}_{93}Np + ^{0}_{-1}e \tag{4-5}$$

$$^{239}_{93}Np = ^{239}_{94}Pu + ^{0}_{-1}e$$

Like uranium-235, plutonium-239 was readily fissioned by neutrons. A separate intensive research program focused on its utility in bombs. Plutonium could be made from abundant uranium-238 and separated chemically because it was a different element. Separating uranium-235 from uranium-238 would prove quite another matter.

In 1942 a research team led by Fermi and Arthur Compton (United States 1892–1962, Nobel prize in physics 1927) was assembled to test the idea that if some hundreds to thousands of pounds of natural uranium were brought together and fission of the small amount of uranium-235 therein begun with a few neutrons, it might be possible to produce enough neutrons so that the excess beyond those needed to convert uranium-238 to uranium-239 would sustain the fissioning of uranium-235 for its produc-

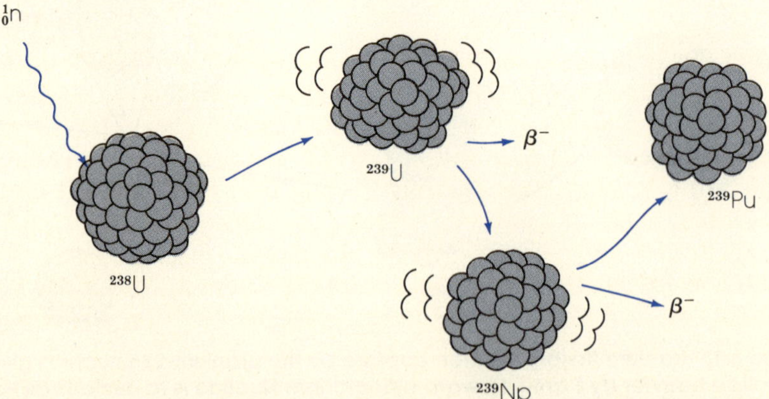

Figure 4-14 Synthesis of neptunium and plutonium. Neutron capture by uranium-238 is followed by loss of two β particles. Half-lifes are ^{239}Np, 2.4 days; ^{239}Pu, 24,000 years.

tion of neutrons. The "laboratory" was a squash court under the University of Chicago stadium. The apparatus was a cubical latticework or "pile" about 20 feet on a side. When completed, it contained 6 tons of uranium in the form of uranium oxide bricks, interarranged with blocks of very pure graphite (carbon) which served to slow down neutrons to enhance their acceptance by nuclei. Into the pile were inserted cadium metal rods to serve as neutron absorbers that would moderate the fission reaction.

As construction proceeded, the pile was tested for neutron production. The day it reached critical proportions, the main control rod was withdrawn slowly as the scientists watched neutron counting instruments. Two men stood ready to pour a cadmium salt solution through the latticework should other control fail. When it became evident that the neutron count would continue rising and not level out as it had previously, Fermi, who was computing with a slide rule, said calmly, "The reaction is self-sustaining." A bottle of Chianti wine was opened and paper cups passed around. Controlled release of nuclear fission energy, one of the most significant discoveries in the history of mankind, had become a reality (see the figure on page 1).

Work at Chicago and Berkeley demonstrated several possible methods for separating uranium-235 from uranium-238 in natural uranium, and plutionium-239 from uranium after its preparation in a pile. Main effort for uranium-235 isolation was based on a physical process in which gaseous compounds of uranium isotopes were to be pumped through porous barriers where separation according to slightly different rates of travel would take place (see Graham's law, Chapter 14). Isolation of plutonium-239 was to be accomplished by chemical methods (solution in acid, extraction, precipitation, etc.). Both efforts were begun at a site near Oak Ridge, Tennessee, where cooling water was available for the piles producing plutonium. When reactor design was worked out, production-scale reactors were erected on the Columbia River near Hanford, Washington. The Hanford plant was constructed successfully to utilize complex chemical procedures, some of which had been studied only under the microscope on microgram samples prepared in a cyclotron. The scaleup was an unprecedented 10 billionfold.

The fission bomb

In 1942 the U.S. Army organized the effort named the *Manhattan Project* to build the atomic bomb. Scientists had to be recruited (renowned names in atomic physics answered the call—Bohr, Fermi, Chadwick, Bethe, Segre); equipment was procured; and a great laboratory was built. Los Alamos, a science city serving 9000 persons, sprang up on an isolated mesa in New Mexico and there the problems related to bomb assembly from Oak Ridge and Hanford isotopes were solved. An American physicist, J. Robert Oppenheimer (1904–1967), who later came to represent the consciences of all scientists who left university laboratories to contribute to nuclear mass destruction, headed the project.

The enormous energy available in the nucleus was predicted from nuclear stabilities and an equation relating mass and energy. The mass of every atom (except ^1H) is smaller than the mass calculated by adding up the masses of its individual particles. As described below, the mass difference, called the *mass defect*, could be interpreted by the law of mass–energy conservation as equivalent to energy lost when in some primeval natural event the atomic nucleus was created from particles.

Example 4-1 By mass spectrograph the 1+ ion of $^{133}_{55}$Cs is found to have a mass of 132.93479 amu. Calculate the mass defect.

Solution The ion consists of 54 electrons, 55 protons, and 78 neutrons. Total particle mass is

$$
\begin{array}{lrl}
electrons: & (54)(0.00055) = & 0.02970 \\
protons: & (55)(1.00728) = & 55.40040 \\
neutrons: & (78)(1.00867) = & 78.67626 \\
\hline
& & 134.10636 \ amu
\end{array}
$$

The difference between the particle sum and the atomic (in this case ionic) mass is the mass defect: $134.10636 - 132.93479 = 1.17157$ amu.

Nuclei are conveniently compared by dividing mass defect by the sum of protons and neutrons, giving the mass defect per nucleon. In our example it is $1.17157/133 = 0.00883$ amu per nucleon. The larger this number, the more stable the nucleus. The smaller the number, the more unstable the nucleus and the more energy theoretically derivable from its nuclear reaction. As shown in Figure 4-15, cesium-133 is a relatively stable element. It is not surprising that cesium is one of the elements in highest yield from fission.

When mass defect is restated as energy, it is called the *binding energy* of the nucleus. Mass and energy are related by a now familiar equation deduced in 1905 by Einstein from relativity theory

$$E = m_0 c^2 \tag{4-6}$$

Given rest mass m_0 in grams, and the velocity of light c (3.00×10^{10} cm/sec), energy E is obtained in ergs. Binding energies, however, are

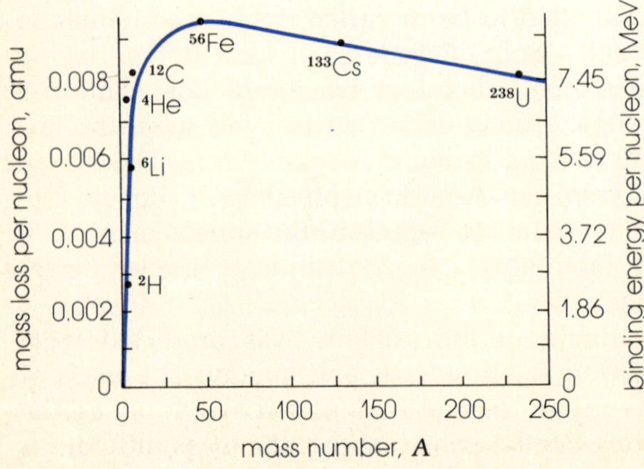

Figure 4-15 The nuclear stability curve. Mass loss and binding energy are plotted as a function of mass number, A. From A = 20 to A = 160, binding energy is about 8 MeV per nucleon; such elements are stable. When A > 160, binding energy decreases because the many protons present are repelling one another. When A < 20, binding energy also decreases because most of the nucleons are partly exposed at the nuclear surface where attractive force is less.

usually expressed in million electron volts. Equivalencies needed in calculations are

$$1\ amu = 1.66 \times 10^{-24}\ g = 1.492 \times 10^{-3}\ erg = 931.2\ MeV \tag{4-7}$$

For cesium-133 the energy equivalent of the mass defect per nucleon is found by converting units as follows:

$$?\ MeV/nucleon = \left(\frac{8.83 \times 10^{-3}\ amu}{nucleon} \right) \left(\frac{9.31 \times 10^2\ MeV}{1\ amu} \right) = 8.22\ MeV/nucleon$$

Example 4-2 If 1.00 g of ^{235}U is converted to energy in a nuclear explosion, how much heat is generated?

Solution The identity of the matter converted to energy need not be specified, for m_0 is simply a mass. Substitution in Eq. 4-6 gives

$$E = (1.00\ g) \left(3.00 \times 10^{10}\ \frac{cm}{sec} \right)^2 = 9.00 \times 10^{20}\ erg$$

Conversion to calories (units of heat) is made using the identity

$$1\ cal = 4.18 \times 10^7\ erg \tag{4-8}$$

$$?\ cal = 9.00 \times 10^{20}\ erg \left(\frac{1\ cal}{4.18 \times 10^7\ erg} \right) = 2.15 \times 10^{15}\ cal$$

10^{15} cal is enough heat to change 9.5 million gal of room-temperature water to steam at 100°C, and is equivalent to the explosive power of 20,000 tons (20 kilotons) of the conventional explosive trinitrotoluene (TNT).

The first atomic bomb was the \$2-billion product of 65 months of effort by 125,000 people. Named *Fat Boy* for its 59-inch girth, it contained at least 11 pounds of plutonium in the form of a hollow sphere surrounded by shaped charges of gunpowder arranged to squeeze (implode) the sphere into supercriticality. Getting compaction within microseconds (1 μsec $= 10^{-6}$ sec) insured a reasonable fission yield from the very fast chain reaction before the mass vaporized. The device was triggered on a tower by observers 10 miles away. The explosion lit the night landscape like day, melting sand to glass for a radius of a half mile. Measuring instruments not destroyed gave its yield as 20 kilotons of TNT.

World War II was almost over. Germany had surrendered to the Allies two months earlier. Decision to use the bomb on the Japanese was made by a committee appointed by President Harry Truman. It included Oppenheimer, Compton, Lawrence, Fermi, General Groves (army coordinator of the Manhattan Project), and Secretary of War Henry Stimson.

Three weeks after the test shot an A-bomb containing a minimum of 33 pounds of uranium-235 was dropped from an airplane on Hiroshima, Japan, a city of 320,000. Exploding at an altitude of 2200 feet, it destroyed two-thirds of all buildings, killed at least 80,000 persons outright, and injured 70,000 more. (The effects are still there; people exposed to the radiation continue to die from various related causes at the rate of about three a day.) When Japan did not heed a demand to quit fighting a plutonium-239 bomb was dropped on the city of Nagasaki. Simultaneously the USSR declared war on Japan, and its armies invaded Japanese-held Korea and Manchuria. The next morning Japan surrendered.

The fusion bomb

The possibility of deriving energy from nuclear fusion, the very high-temperature process in which light nuclei fuse to give heavier nuclei, had been discussed by Oppenheimer's group as early as 1941, but no follow-up was made for eight years. Then dust samples collected by high-flying U.S. Army planes, and rainwater samples collected by the U.S. Navy at various global stations showed the presence of yttrium-91 and cesium-141 in the air; Russia had exploded an unannounced atomic bomb.

Weapons research was resumed at Los Alamos with the "the father of the H-bomb," physicist Edward Teller (Hungary and United States 1908–   ) as a group leader. From theory associated with Figure 4-15 it was known that fusion of hydrogen isotopes would yield pound for pound three times the energy of heavy-atom fission. Furthermore, because fusion does not depend on neutrons in the manner that fission does, a fusion weapon would not be limited by a critical size. The main question to be answered by experiments was one of temperature; could the million-degree temperature required for fusion be produced, and would fusion then take place? More specifically, could an A-bomb, so recently the ultimate weapon that ended the greatest war, be used as a cap or trigger to set off a more powerful explosive?

One isotope needed was "heavy" hydrogen or *deuterium*, abbreviated ^2_1H, or D. One in 5000 hydrogen atoms in nature is deuterium. By carefully distilling ordinary water, heavy water, D_2O, can be concentrated in the residue, and D_2 gas then produced by decomposing the D_2O with electric current. Another necessary element was radioactive "superheavy" hydrogen or *tritium*, ^3_1H (half-life 12.4 yr). It had to be prepared in an atomic reactor.

In 1951 the U.S. Army assembled the first thermonuclear device in the South Pacific. It consisted of a plutonium trigger and a large refrigerator that maintained liquid hydrogen isotopes at −250°C. It was exploded successfully, but the yield was below the theoretical value.

Less than two years later air sampling disclosed that the Russians (physicist I. Kurtchatov was in charge of their bomb project) had found how to make a portable hydrogen bomb of 1 megaton (1 million tons of TNT equivalent) yield. By this time at Los Alamos, Teller and mathematician S. Ulam learned the principle also, and in 1954 the United States exploded a nonportable 15-megaton device. Fusion material in this and subsequent superbombs was lithium deuteride, $^6Li^2H$, a stable, white, salt-like compound made by direct chemical combination of lithium-6 metal and deuterium gas at 600°C. Eventually England, Red China, and France worked out fusion technology and exploded H-bombs in the atmosphere.

Largest of all test shots was set off by the USSR. Fusing a ton of light elements its estimated yield was 60 megatons, equivalent to all the powder used in the entire history of warfare. A 60-megaton bomb detonated at an altitude of 60,000 feet would ignite dry paper over an area comparable to that of the state of New Jersey (about 8000 square miles).

Principal fusion reactions are as follows:

$$_1^2H + {}_1^2H = {}_2^3He + {}_0^1n$$

$$_0^1n + {}_3^6Li = {}_2^4He + {}_1^3H$$

$$_1^3H + {}_1^2H = {}_2^4He + {}_0^1n \qquad (4\text{-}9)$$

$$_1^2H + {}_1^2H = {}_1^3H + {}_1^1H$$

$$_1^1H + {}_3^7Li = {}_2^4He + {}_2^4He$$

Today the world's nuclear arms total perhaps 50,000 megatons. World population is 3.6 billion. Dividing 5×10^{10} by 3.6×10^9 reveals the grim fact that 14 tons of TNT-equivalent is stockpiled for each human being on the planet.

Biological effects of radioactivity

Besides the obvious effects of heat and blast, atomic weapons produce radiation and *fallout*. The latter consists of radioactive isotopes and debris. When thrown into high altitude and carried by wind, fallout can attain global coverage. The most studied fallout product is strontium-90, a β emitter with a half-life of 25 years. It gains access to bones by transfer from soil to green plants to cow's milk to humans. Excluding marrow, bone is 25 percent water, 30 percent organic (protein mostly), and 45 percent inorganic. The latter, mainly calcium phosphate, $Ca_3(PO_4)_2$, accounts for 99 percent of the calcium in the body. Strontium, being chemically similar to calcium, readily takes its place in bone. Strontium-90 bombarding bone marrow, wherein the structures operate that synthesize blood cells, can cause anemia and other serious diseases.

Radiation* affects living matter in several ways. A high level of radiation kills cells (and whole organisms) by ionizing organic molecules in them. Low-level radiation can alter cell functions, including those that reproduce the organism (Figure 4-16). Damage done one generation may thus be evident in the next. Genetic effects of radiation were studied as early as 1920 by Hermann Muller (United States 1890–1967, Nobel prize in medicine or physiology 1946), who found that fruit flies which had been exposed to X rays during the period their reproductive cells were forming produced offspring with more mutations than normal. Since then, many experiments have been done on higher life forms. Because organisms (including man) have become adapted to this earth through a long evolutionary history during which time natural processes randomly produced

---

* *Radiation dosage* is measured in röntgens (R). One röntgen of X rays produces 1.61×10^{12} ion pairs (O$^+$, e$^-$, for example) of 1 g of air. A smaller, more convenient unit is the milliröntgen (1 mR = 0.001 R). *Radiation effect* in living matter is measured in a unit called the röntgen-equivalent-man (Rem), or in a unit called the rad. For X rays, β rays, and γ rays, in air and in soft tissue.

1 Rem = 1 rad = 1 röntgen

All three units are defined on the basis of the effect per gram of medium absorbing the radiation.

Radioactivity is often measured in millicuries (mCi). One millicurie of a radioactive source is equivalent to about 1 mg of radium; it undergoes 3.70×10^7 disintegrations per second.

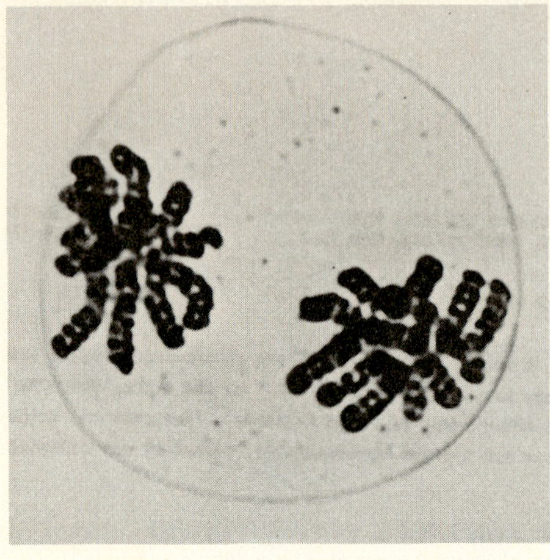
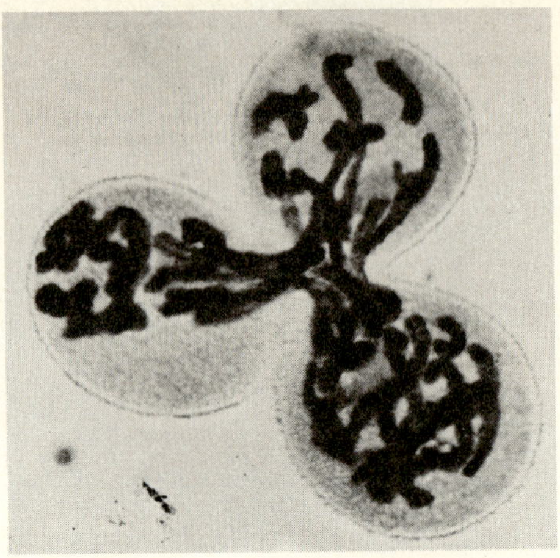

(a)   (b)

Figure 4-16 Effects of radiation (X rays) on genetic material. (a) Chromosomes in a normal plant cell. (As detailed in Chapter 24, cells contain heredity units called genes. In each cell of the body, including the cells used in human reproduction, thousands of genes strung together as threads called chromosomes carry the responsibility of transferring heredity.) (b) Chromosomes rendered abnormal by X radiation. Instead of dividing normally into two cells, the cell is attempting to divide into three. (Courtesy Brookhaven National Laboratory.)

mutations and only the fittest reproduced, it seems certain that anything that accelerates the rate of cellular change increases the probability that future generations will be born defective.

4-5. The Peaceful Atom

Electric power from nuclear fission

Principal peacetime use of nuclear reactions is the production of electricity. In energy potential 1 cubic foot (1080 lb) of uranium is equivalent to 1.7 million tons of coal, 32 billion cubic feet of natural gas or 7.2 million barrels of oil. For home use electricity is measured in kilowatt-hours.* In round numbers the fissioning of 1 gram of uranium releases 1 megawatt-day of heat, which is equivalent to the production of 7000 kilowatt-hours of electricity. Assuming that average family use is 23 kilowatt-hours per day, 1 cubic foot of uranium represents a year's supply of electricity for 420,000 homes.

By century's end world population is expected to double, and demand for electricity may climb sixfold. As traditional fuels are irrevocably depleted, the trend toward use of nuclear fuels will increase (see Table 4-1 and Figure 24-10). Uranium and thorium are about as common in rock as lead and tin; enough is recoverable to satisfy needs for several centuries.

The United States now has 100 nuclear power plants ("nukes") in operation and in various stages of planning and construction. To keep up with energy demands another 70, each of 1000-megawatt capacity, will be

* The kilowatt-hour (kWh) is a unit of energy. (Electrical units are defined in Section 18-3.) A megawatt (MW), a unit of power, is 1000 kilowatts (kW); 1000 kW generated for 24 hours is 1 MW-day.

Table 4-1. U.S. electrical generation (billions of kilowatt-hours)

Energy source	1950		1970		1980, est.	
	Energy	Percent	Energy	Percent	Energy	Percent
Coal	155	47	825	55	1255	45
Hydro	95	29	225	15	320	11
Gas	45	14	290	20	350	13
Oil	35	10	110	7	165	6
Nuclear	0	0	50	3	720	25
Totals	330		1150		2810	

needed by 1980. Present installations are designed to consume only the 0.7 percent of mined uranium that is the readily fissionable uranium-235. Operating principles are shown in Figure 4-17. Fission heat is used to make steam. Steam turns the bladed cylinder of a turbine, which transfers its energy through a rotating shaft to a generator. The generator produces electricity by rotation of a conductor through a magnetic field.

A more efficient reactor is the *breeder reactor*. It produces a fissionable isotope by neutron capture as its initial charge of fuel is being used up. A federally sponsored research goal in the United States is the building of a liquid metal fast breeder reactor (LMFBR) by 1980. (The Soviets started operating theirs in 1973.) Coolant will be liquid sodium or lithium. Generating capacity will be about 400 megawatts. After 10 years of operation enough uranium-238 will have been converted to plutonium-239 (Figure 4-14) to refuel the original reactor and another like it.

Despite obvious advantages (no fossil fuel consumption, high power production), nukes introduce their own kinds of problems. One is the danger of a major release of radioactive material if, in case of overheating, the core cooling system proves to be inadequate. Other problems are the small but continuous losses of radioactive gases and the difficulties in disposing of radioactive solid wastes and large quantities of excess heat (see Section 15-3). Plutonium synthesis in breeders creates still other problems. By 1980 world plutonium production is expected to reach 70 pounds per day. Plutonium is exceptionally toxic. And 70 pounds is enough for several atomic bombs. By international consent inspectors are on the job right now to prevent theft and black marketeering in nuclear explosives.

Electric power from nuclear fusion

In the twenty-first century (see Figure 24-10) the world's energy base may begin to be deuterium. Feasibility of controlling its fusion with tritium,

$$^2H + {}^3H = {}^4He + {}^1n + 17.6\ MeV \tag{4-10}$$

will probably be demonstrated before 1980. Although the "physics" of fusion appears favorable, the engineering problems are formidable. They include (1) the necessity of producing 50- to 100-million degree temperatures, (2) getting adequate plasma* density (at least a millionth that of the

* A *plasma* is a completely ionized yet neutral gas that consists of particles such as nucleons and electrons at high temperature. Plasma properties are unique enough to consider it a fourth state of matter.

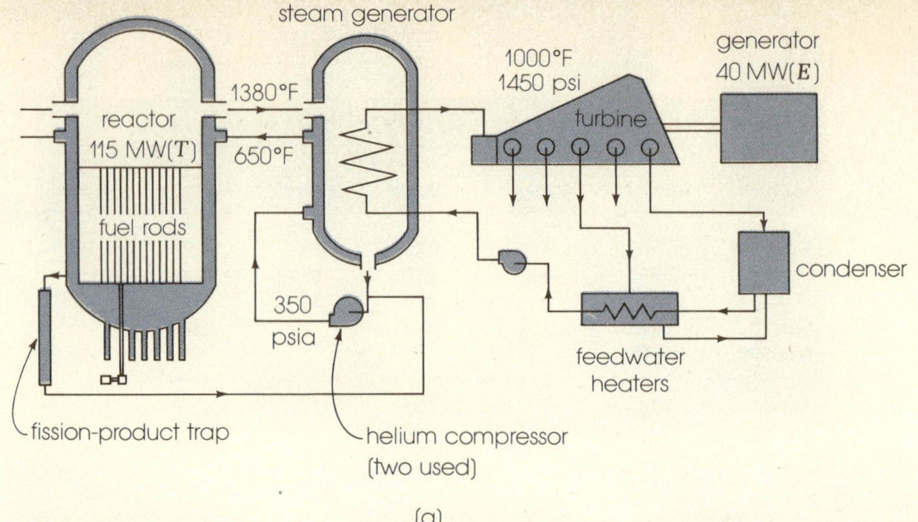

(b)

Figure 4-17 A prototype high-temperature gas-cooled reactor (HTGR) power station. (a) Schematic view. Helium coolant under pressure flows through the reactor core which contains about 1.9 tons of thorium and enriched uranium. After being heated by the core, the helium is passed to the steam generators to heat water to high-temperature steam. After spinning the turbine which spins the electric generator, the steam is condensed and returned to the steam generators. Helium, after going through the steam generators, is returned to the reactor. (Abbreviations used in the figure: psia = pounds per square inch absolute (a unit of pressure), MW(T), and (E) = megawatts thermal and electrical. (b) The reactor operating floor. (Courtesy Gulf General Atomic.)

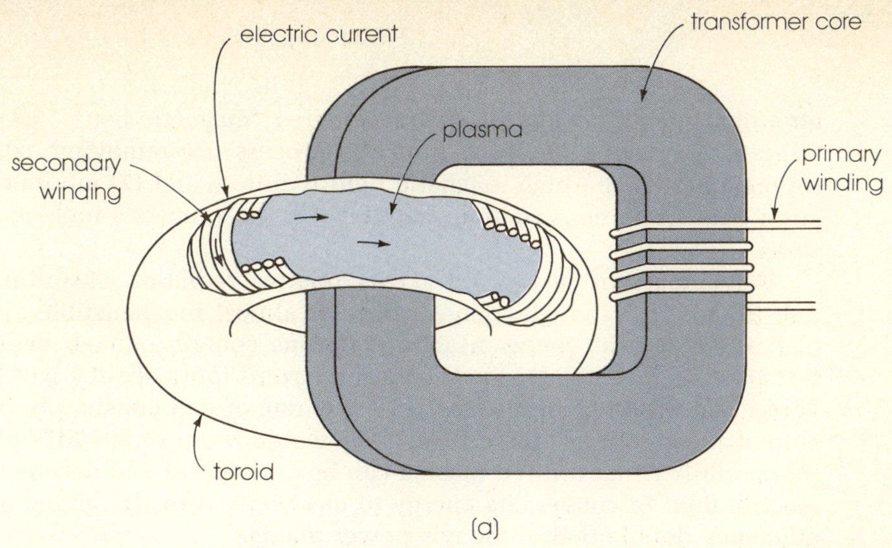

Figure 4-18 (a) Principle of the Tokamak (acronym for Toroidal kamera (chamber) magnetic). A transformer consists of two coils wound on the same ferromagnetic core but insulated from each other. Depending on windings, it transforms a varying voltage, introduced into the primary winding, to a larger or smaller voltage in the secondary winding. Largest Tokamak in operation has these specifications: major toroid diameter, 6 ft; minor diameter, 1 ft; electric current, 10^5 amp; plasma temperature, 10^7°C. For commercial use donut diameter will have to exceed 100 ft. (b) An experimental Tokamak. (Courtesy Princeton University Plasma Physics Laboratory.)

air around us), (3) confining the plasma in a "magnetic bottle," (4) regenerating tritium using neutrons from the process, (5) removing excess heat, (6) maintaining the high magnetic field required, and (7) minimizing radiation damage to construction materials by the process's high-energy neutrons.

Advantages of fusion are (1) its relative radiation cleanliness (short half-life of ^3H), (2) high-energy yield, (3) almost inexhaustible availability of fuel (at present energy usage the oceans contain enough deuterium to last 40 trillion years), (4) low explosion hazard (only about 1 g of hydrogen is reacting at any given instant), (5) slow rate of fuel consumption (conversion of a few pounds of hydrogen a day would run a 10,000-MW plant), and (6) possibility that unused plasma can be drained off and decelerated in an electric field to convert its energy to electricity directly, giving an overall efficiency double that of today's power plants.

One fusion design being implemented in the 1970s is the Tokamak, invented in the USSR and now also under development in the United States. Plasma is contained within a toroid ("donut") which in effect is a secondary winding of a heavy transformer (Figure 4-18). A strong current pulse introduced into the transformer primary winding ionizes gas in the toroid, causing current there to flow. The current both heats the plasma and produces a second magnetic field around the toroid which confines the plasma within its circular path until fusion takes place.

Another fusion design uses high-powered *lasers* (optical devices giving intense beams of coherent light) to compress and heat fusion fuel. Lasers bypass the need for a magnetic field.

Earth moving with nuclear explosives

The U.S. Atomic Energy Commission (AEC) is attempting to commercialize atomic explosives through a program called *Project Plowshare*. Applications are in large-scale jobs for which other methods seem too puny—digging canals and harbors, for example, and fracturing tight geological formations to get at natural gas, oil, copper, and steam (Figure 4-19).

Major interest since 1967 has been in natural gas recovery with the simultaneous creation of underground caverns for gas storage. Test shots in Colorado have produced gas that contains some radioactivity due to krypton-85 and hydrogen-3. Diluted with gas from normal gas wells it is advertised as safe because it exposes the average consumer to only 0.5 millirem yearly of extra radiation. (The Federal Radiation Council has set 170 millirems per year as an overall recommended maximum exposure.) Critics say that any extra radiation is harmful. Other hazards in the program are the possibilities of causing damaging seismic shocks and contaminating groundwater.

Unproven gas reserves beneath the United States and its coastal waters are estimated to be 1200 trillion cubic feet. Occurring mostly at depths of 2 to 3 miles, perhaps no more than two-thirds of the gas can be recovered by conventional drilling. Proven gas fields contain less, only about 280 trillion cubic feet. At the present annual consumption rate of 25 trillion cubic feet, these wells will be empty before 1990.

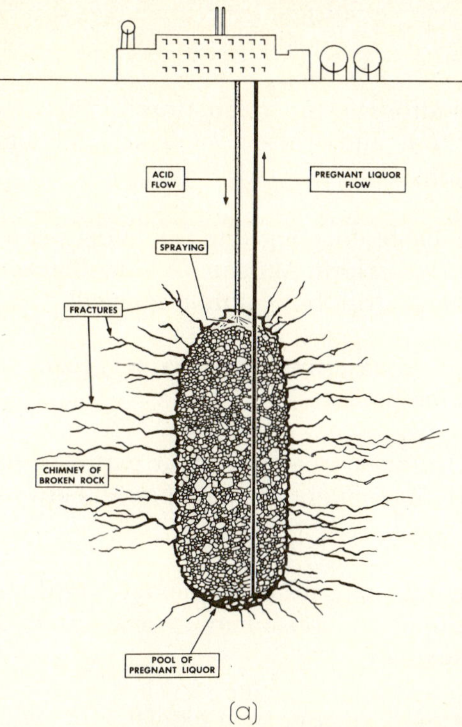

Figure 4-19 Proposed peaceful uses of nuclear explosives. (a) Leaching an underground copper ore deposit. An underground nuclear explosion characteristically creates a "chimney" of broken rock. Acid sprayed into the chimney leaches out copper. Leachate from the bottom is pumped up to a plant where copper ions are reduced to copper metal. (b) Proposal for creating steam-generating cavities. After cavities are created (left), water is introduced into the fracture area and the chimney bottom. Steam that comes from the chimney top is piped to a plant for the generation of electricity. Condensate water is recycled. Hot, dry rock formations thought suitable for the process exist under parts of the western United States. [(a) Courtesy CER Geonuclear Corporation; (b) courtesy Lawrence Radiation Laboratory, University of California.]

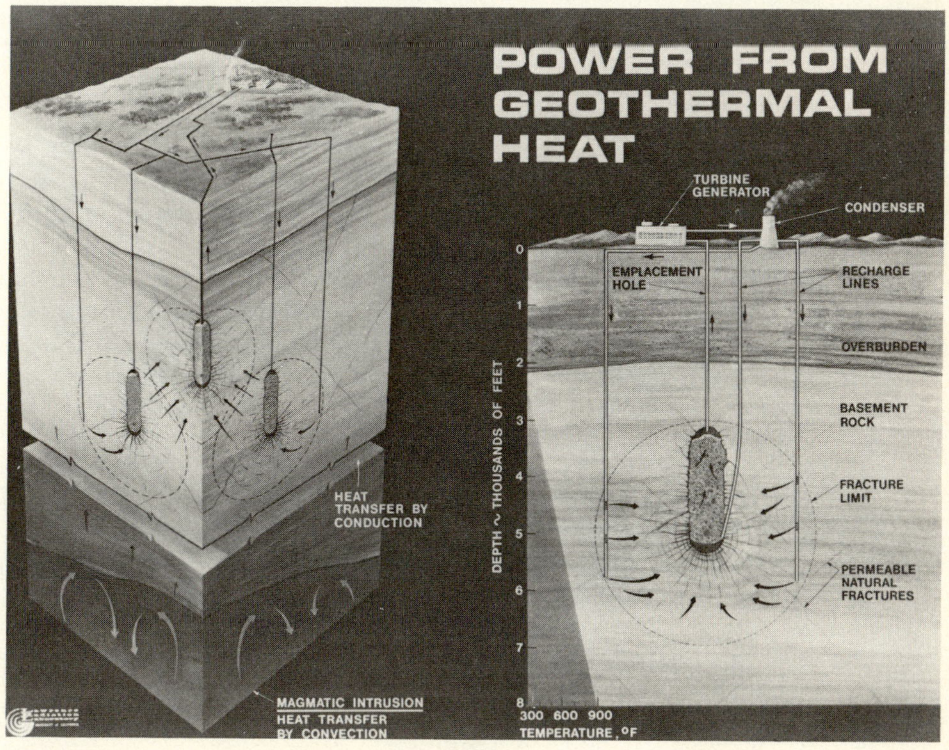

(b)

---

1. Neutron Discovery of the neutron afforded an explanation for the existence of isotopes such as $^{16}_{8}O$ and $^{17}_{8}O$, and isobars such as $^{210}_{82}Pb$ and $^{210}_{83}Bi$. Define all new terms, and explain.

2. Physicists In 1932 the Cavendish Laboratory had the following people working: Chadwick, J. J. Thomson, Rutherford, Wilson, Aston, Blackett, Cockcroft, and Walton. What did each contribute to atomic science?

3. Nuclear equations **(a)** Equation 4-2 is described as an (α, p^+) reaction. Explain. **(b)** In like symbolism describe Eq. 4-3.

4. Radioactivity You are given two isotopes, one emitting α particles and γ rays, the other β particles and γ rays. How might you distinguish between the two samples?

5. Radiation Flying coast to coast in a jet exposes one to about 4 millirem of radiation. Living a year in a brick or stone house may expose one to 50 millirem. Define the new terms, and explain.

6. Decay A nuclide whose p^+/n fraction is smaller than required for stability may become stable if one of its neutrons decays to a proton. A nuclide deficient in neutrons may become stabilized if a p^+-to-n transformation occurs. Which case is illustrated in each of the following? Explain.

$$^{13}_{7}N = {}^{13}_{6}C + {}^{0}_{1}e$$

$$^{28}_{13}Al = {}^{28}_{14}Si + {}^{0}_{-1}e$$

7. Isotopes **(a)** In 1965 after Red China detonated a nuclear bomb, the United States gathered air samples. Analysis of them showed the presence of lithium-6. What did that mean? **(b)** In 1958 the last of a series of American hydrogen bomb tests was concluded at Bikini in the Marshall Islands. In 1967 a survey by the AEC showed that residual radiation was no longer a hazard in the ground and that the Bikinians could return if certain conditions were met. One condition was that all coconut crabs, a local food source, had first to be eliminated because they contained strontium-90. Explain.

8. Definitions What is the difference between **(a)** atomic weight and atomic number? **(b)** nucleon and meson? **(c)** natural and induced radioactivity? **(d)** matter and antimatter? **(e)** cyclotron and linear accelerator?

9. Radioactivity Carbon-10 decays as follows:

$$^{10}_{6}C = {}^{10}_{5}B + {}^{0}_{1}e$$

(a) What is $^{0}_{1}e$? **(b)** Describe the behavior you would expect of the particle if subjected to the experiments described in Figures 4-1 and 4-2.

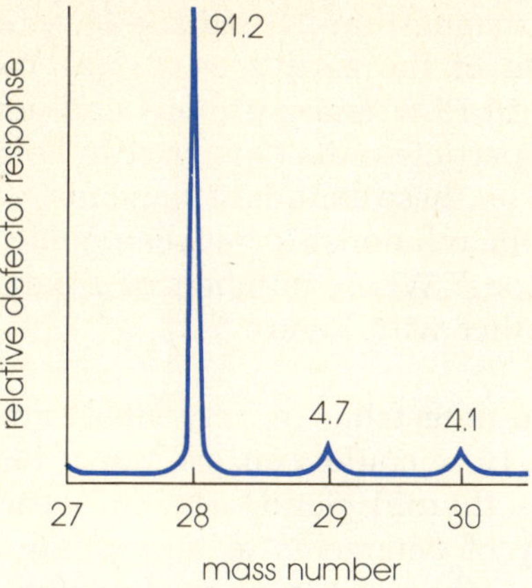

Figure 4-20

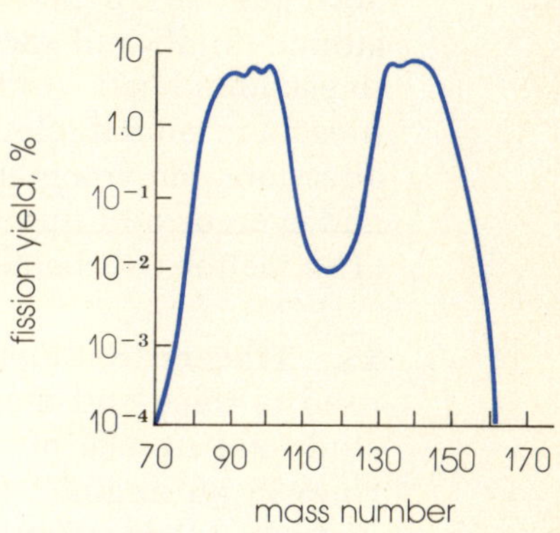

Figure 4-21

10. Mass spec Mass spectrograph analysis of the element silicon taken from sand (SiO_2) is shown in Figure 4-20. Interpret the result.

11. Fission Fission of uranium-235 by slow neutrons gives the data shown in Figure 4-21. Explain, and tentatively identify some of the more abundant products.

12. Nuclear power Figure 4-22 is a sketch of a proposed fission-powered agricultural-industrial complex now under study for possible construction in Puerto Rico and the Middle East. As visualized here it could produce up to 10^9 gal of water daily and 2000 MW of electricity, feed 6 million persons from several hundred thousand acres of irrigated land and run satellite industries making fertilizers, metals, and chemicals. What advantages and disadvantages do you think such plans have?

13. Reaction energy Why does reaction (b) release about 30 million times more energy than reaction (a)?

(a) $C + O_2 = CO_2 + 0.7\ eV$
(b) $C + He = O + 20\ MeV$

Figure 4-22 (Courtesy United States Atomic Energy Commission.)

14. Nuclear stability Oxygen-16, magnesium-24, silicon-28, and calcium-40 comprise 70 percent by weight of the earth's crust. **(a)** What is the atomic number of each element? **(b)** How many protons and neutrons are in each nucleus? **(c)** How many α-particle units $(2p^+,2n)$ can be considered present in each nucleus? **(d)** Based on these abundant elements, what generalizations can you make regarding the relationship between nuclear stability and even or odd numbers of nucleons? Whole numbers of α-particle units? (The matter will be considered further with Figure 5-2).

15. Tracers Radioisotopes, being detectable in very small amounts, are used to trace and measure things. How could you use tracer technique to study **(a)** uptake of iodine by the thyroid gland? **(b)** the wear of piston rings in an engine? **(c)** the ability of detergents to take grease off plastic dishes? **(d)** the flow rate of gases exhausted from an automobile engine?

PROBLEMS

16. Mass spec Mercury in nature consists of seven isotopes. Mass numbers and approximate abundances are: 202, 30 percent; 200, 23 percent; 199, 17 percent; 201, 13 percent; 198, 10 percent, 204, 7 percent; 196, 0.1 percent. In the manner of Figure 4-20, sketch the readout one would obtain from a mass spectrograph.

17. Nuclear reactions In Figure 4-3 these decay reactions follow $^{226}_{88}\text{Ra}: \xrightarrow{\alpha} Q \xrightarrow{\alpha} X \xrightarrow{\alpha} T \xrightarrow{\beta} Z$. In the manner of Eq. 4-1 write the four reaction equations. Give the correct symbols for elements Q, X, T, Z.

18. Nuclear equations Write equations like Eq. 4-1, giving correct symbols for all elements. **(a)** A uranium-235 nucleus captures a neutron and splits into a xenon-133 nucleus, a strontium-90 nucleus, and several neutrons. **(b)** A uranium-238 nucleus bombarded with a carbon-12 nucleus gives four neutrons and a heavy nucleus.

19. Nuclear equations As above. **(a)** Cobalt-60 has replaced radium in cancer therapy. It is prepared when another cobalt nuclide captures a neutron. **(b)** Gold has been synthesized by causing a platinum-198 nuclide to capture a neutron. The nuclide formed emits a β particle, giving a gold nuclide.

20. Nuclear size Calculate the nuclear radii and volumes of these atoms: ^{64}Cu, ^{125}Sb, ^{256}Md. Use the fermi as the unit.

21. Decay series Thorium-232 changes to lead in stepwise fashion, during which sequence these particles are ejected: α, β, β, α, α, α, α, β, β, α. By inspection, what lead isotope is the end product? Explain briefly.

22. Mass defect Based on the mass-defect principle (Example 4-1), which nuclide is the more stable: ^6Li, 6.01496 amu; or ^7Li, 7.01602 amu?

23. Mass defect The mass of one $^{238}_{92}U$ nucleus is 238.123 amu. Calculate the mass defect and the mass defect per nucleon.

24. Mass and energy Electron–positron collision results in their disappearance and replacement by two photons of light:

$$_{-1}^{0}e + _{1}^{0}e = 2\,\gamma$$

The photons have short wavelength (0.0122 Å) and together possess an energy of 5.46×10^{-7} erg. Show that this experimental measurement amounts to a test of the mass-energy conservation equation. (*Hint:* Eqs. 3-1 and 4-8 both define energy; consider equating them.)

25. Nucleons From the text description of nucleon structure, draw pictures of a neutron and a proton. Label the parts, give charges and dimensions.

26. Energy Prepare a chart from the data of Table 4-1. Plot energy vertically from 0 to, say, 1500 billion kWh, versus years. Put each of the five curves in a different color or style of line. **(a)** Which energy source is growing fastest? **(b)** In what years will nuclear energy probably overtake oil, hydroelectricity, and gas?

27. Nuclear explosives A problem of antiballistic missile (ABM) defense in which radar is used to seek out enemy missiles is complicated because nuclear explosions disrupt communications by ionizing the air. Calculation of disruption volume is based on fireball diameter D (in kilometers) which is related to megaton yield Y of the warhead exploding: $D = (Yd_o/d)^{1/3}$. In this equation d_o is air density at sea level and d is air density at the explosion altitude. For a 1-megaton explosion, find the fireball diameter at **(a)** sea level; **(b)** an altitude of 30 miles, where air density is only one-thousandth that at sea level.

TWO

WHAT ARE THE CHEMICAL ELEMENTS?

Formation in 1963 of the island of Surtsey, near Iceland. Lightning crackles through clouds of ash and volcanic gases, and magma flows into the ocean. The earth's early history may have been dominated by such events, which created landmasses and the atmosphere and filled depressions with water in which life eventually began. (Almenna Bokafelagid, photo by S. Jonasson.)

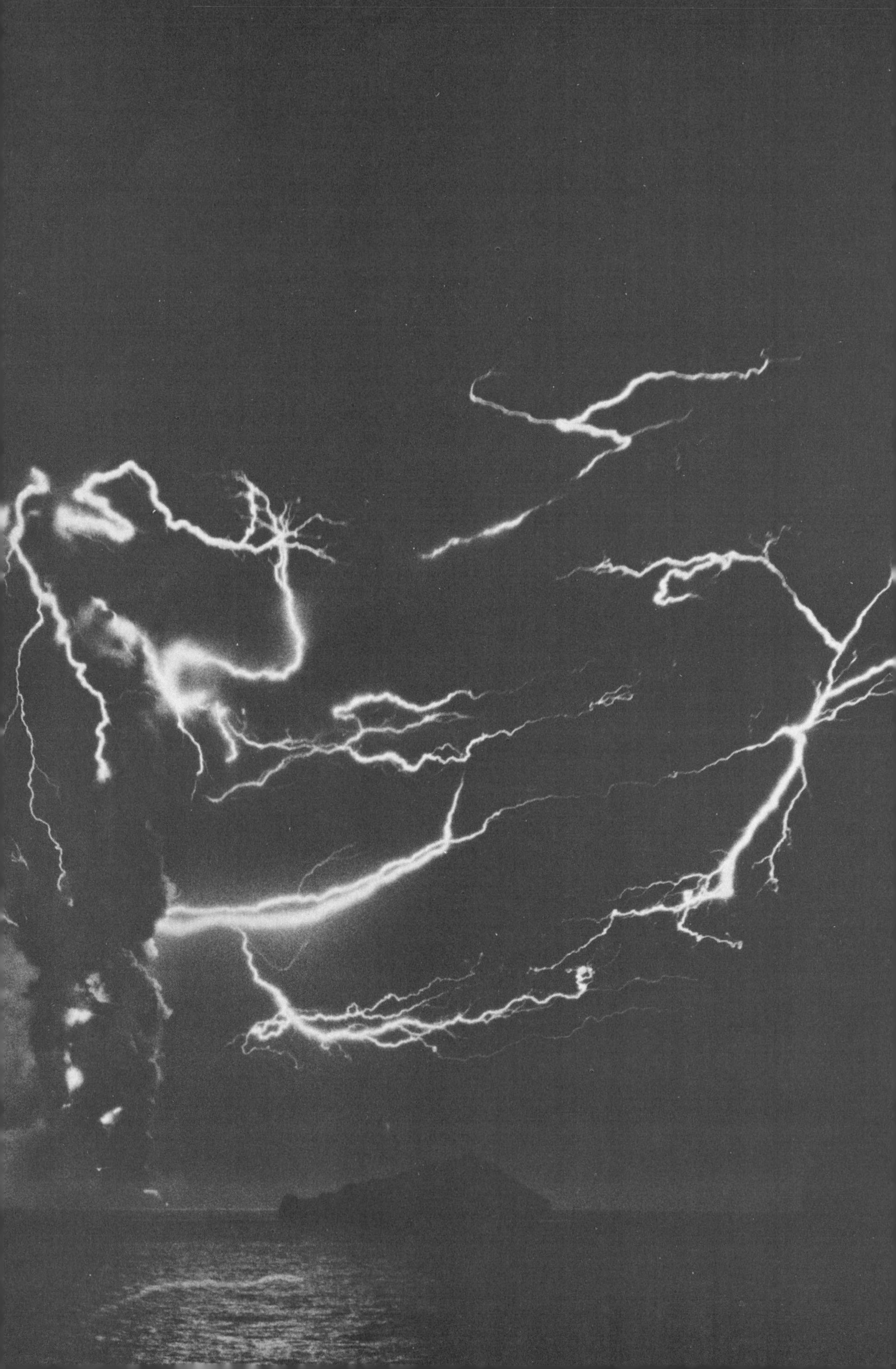

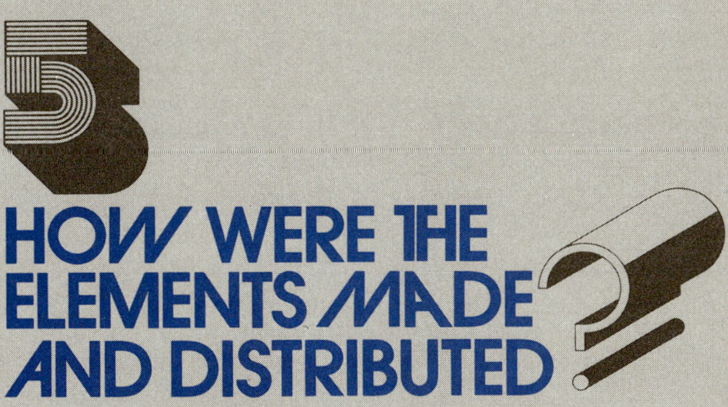

HOW WERE THE ELEMENTS MADE AND DISTRIBUTED?

THE IMPORTANT CONCEPTS

5-1. Analyzing starlight

1. The identity and abundance of elements in bright stars can be determined from their spectra.
   a. The observable universe is almost entirely composed of two elements.
2. Theories of origin of the universe must explain its apparent expansion.

5-2. Synthesis of the chemical elements

1. The universe may have begun as a fireball in which nuclear reactions produced light elements.
2. Stars evolve through thermonuclear reactions that first yield helium, then heavier elements.

5-3. The solid earth

1. The planets probably formed from solar disc material.
2. Meteorites and the earth are chemically related.
3. Earth age can be determined by atomic clocks.
4. Laws of radioactive decay are reducible to a few mathematical expressions.
5. The earth is differentiated into several zones.
6. Composition of the earth's zones can be approximated.
7. Earth conditions favored life.

Ninety different naturally occurring elements have been discovered (see Figure 1-4). Where did they come from and how did they build the earth? Astronomers believe that everywhere in the universe elements are being continuously created in stars, then scattered through space. Many of the elements on earth, including the atoms in our bodies, originated long ago in these thermonuclear furnaces.

To investigate stellar (star) processes we will apply principles considered in nuclear chemistry. To obtain an overview of our planet's structure and composition we will make acquaintance with methods of geochemistry (Gk. ge, earth). And to discuss the history of the universe and the solar system we will reach out from facts at hand to engage in some scientific speculation.

5-1. ANALYZING STARLIGHT

Stellar composition

Modern astronomy began in 1865 when English astronomer W. Huggins installed in his private observatory a telescope fitted to a *spectrograph*. (The principle of the latter is given with Figure 6-3.) Photographing spectra of the sun and other stars, he demonstrated that quantitative information could be obtained on stellar composition, physical state, and movement. Spectra he observed were absorption spectra, that is, dark line spectra created by electronic transitions of elements absorbing energy in stellar gas atmospheres. Element identification he made by reference to laboratory spectra. Elemental abundance values followed from temperature measurement (with a heat-sensing instrument attached to the telescope) and a proportionality constant mathematically relating the intensity of a spectral line to the number of excited atoms causing it.

Typical spectra, including the solar spectrum, are reproduced in Figure 5-1 (facing p. 102). Most of the elements known on earth have been identified in stars—63 of them in the sun.

What is the universe made of? Based on star spectra, 92 percent of all atoms in existence are hydrogen, 7 percent are helium, and all the other elements make up a mere 1 percent. Figure 5-2 is a chart of stellar elemental abundances. Using an abundance scale that is logarithmic, it compares all elements to silicon, an element common to rocks in the earth's crust. We can see that a smooth curve drawn through the data would slope downward from elements numbered 1 to about 44 (atomic weights 1 to 100), then level out through the heavier elements. It is evident also that elements with even atomic numbers are generally some 10 times more abundant than their neighbors with odd atomic numbers. Elements with a neutron-to-proton ratio of unity are especially abundant (C, N, O, Ne, Mg, Si, S, Ca),

Figure 5-1 Spectra. (Top) A beam of white light produces a spectrum called continuous because it contains no breaks. (Middle) 1 is a continuous spectrum; 2 is the sun's (discontinuous) absorption spectrum. The dark (Fraunhofer) lines are produced by elements absorbing energy at the same wavelength they would emit energy if at a higher temperature (see caption, Figure 3-7.) The next three spectra are emission spectra produced for reference in the laboratory by heating sodium, hydrogen, and calcium. (Bottom) Light from stars in the group called the Hyades, as viewed through a telescope equipped with a prism. Redder, darker stars are cooler than bluer, brighter stars at the same distance. Most of the elements known on earth have been identified in starlight. (Bottom: Courtesy University of Michigan, Department of Astronomy and Eastman Kodak Company.)

and elements having even numbers of nucleons and even mass numbers are more abundant than those with corresponding odd numbers.

Star types

Color is an indication of a star's surface temperature: white means hot, red means cool. Brightness is related to a star's total energy output. When corrected for distance to make comparisons among stars meaningful, bright means high-energy output, dull means low-energy output per unit of time. In 1913 E. Hertzsprung in Holland and H. Russell at Princeton University discovered a relationship between color and brightness. It is shown in Figure 5-3.

The ordinate of the diagram is a logarithmic scale of brightness relative to the sun's brightness. The abscissa is a sequence of spectral classes, designated by letters, which defines surface temperatures. When data are plotted for a related group of stars equidistant from the earth, three star types become evident. When information from their spectra is added on, the distinctions sharpen. The great majority of stars, rich in hydrogen, lie along a middle curve called the *main sequence*. Above the main sequence are some larger, cooler, redder stars that contain considerable helium. Below are a few tiny, very dense, white stars composed of heavier elements.

The *Hertzsprung-Russell correlation* is one of the most valuable discoveries in astronomy for its suggestion that stars change character with time; that in main sequence, giant, and dwarf stars we are not looking at unrelated phenomena but at bodies in three stages of life. Theory of stellar processes and astronomical observations today agree so well that stellar evolution is considered to be a fact. Scientists believe that stars change color, size, and brightness through a series of thermonuclear reactions in which the chemical elements are synthesized.

The spectral red shift and Hubble's law

When a light-emitting object is moving away from an observer at a reasonable fraction of the speed of light, light from the object appears redder because fewer waves per second reach the observer's eye. If the object is moving rapidly toward the observer, its light appears bluer because more waves per second reach his eye (Figure 5-4). All bodies in the universe are moving. By measuring the wavelength shift of reference lines in starlight, astronomers have determined star velocities relative to the earth.

Until the 1920s scientists assumed that our galaxy, the *Milky Way*, was the only collection of objects in the universe. Then spectral measurements,

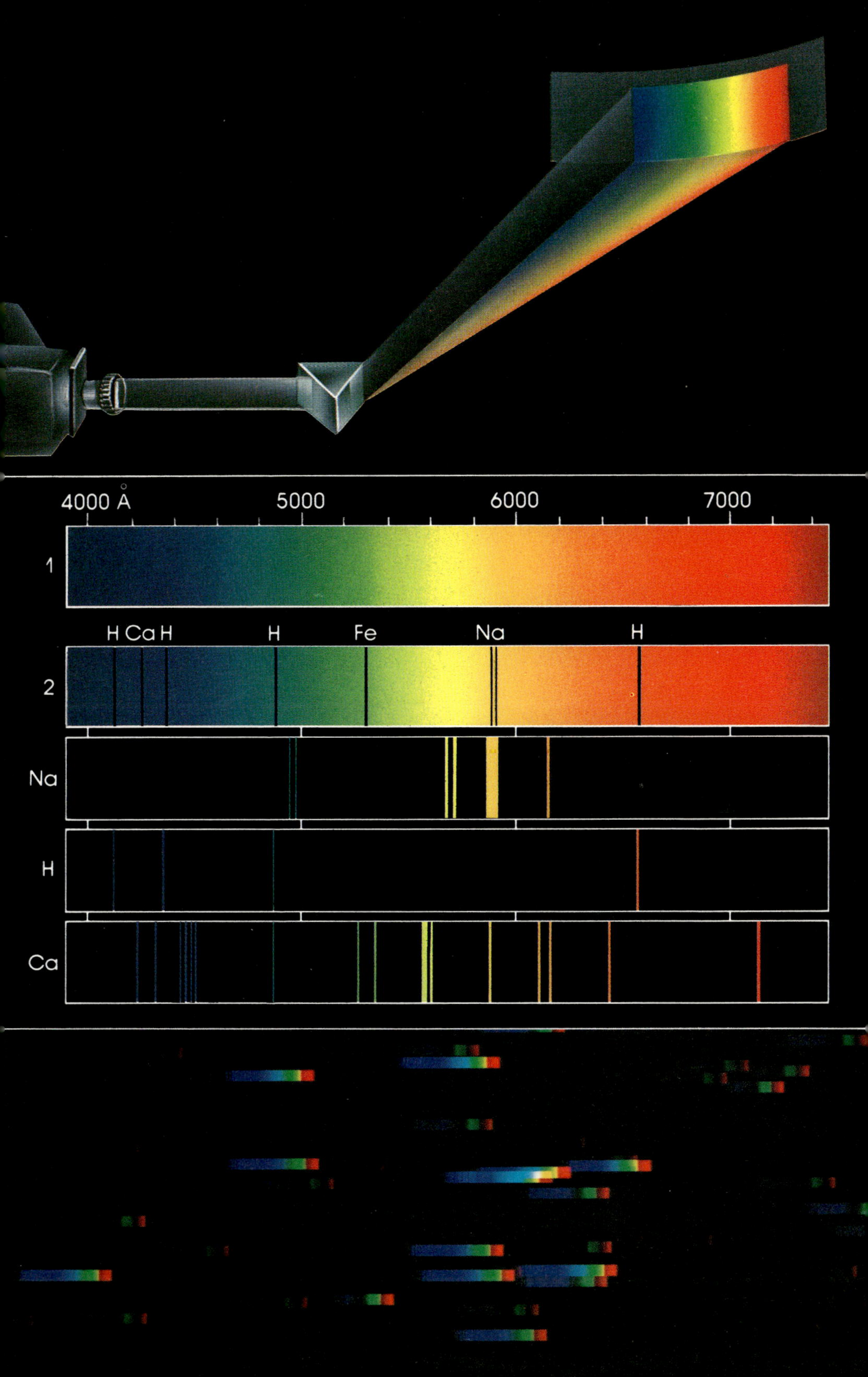

4000 Å
5000
6000
7000
1
H Ca H
H
Fe
Na
H
2
Na
H
Ca

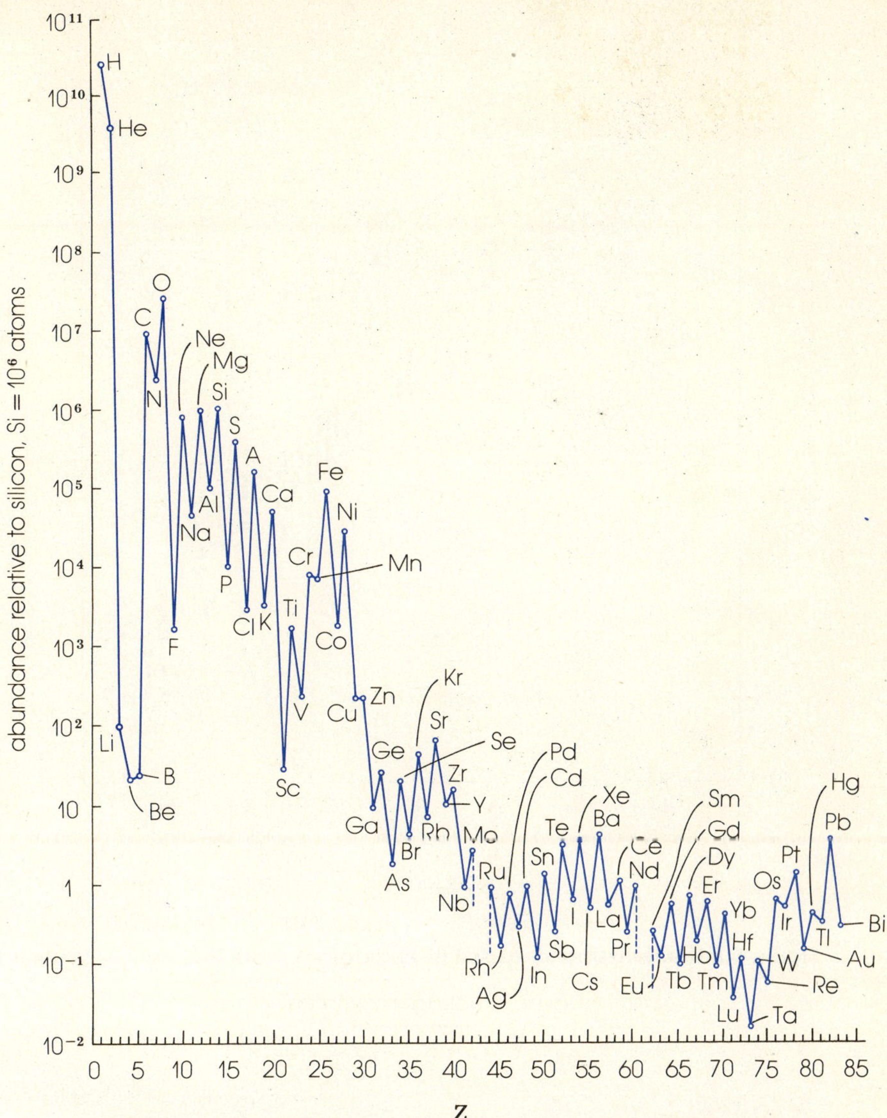

Figure 5-2 Abundance of atoms in the observable universe, relative to silicon. To make a representative mixture of elements starting with 10⁶ atoms of silicon, one would add 3 × 10¹⁰ atoms of hydrogen, 6 × 10⁹ atoms of helium, etc. (From L. H. Ahrens, Distribution of Elements in our Planet. Copyright 1965, McGraw-Hill. Used with permission of McGraw-Hill Book Company.)

like those illustrated in Figure 5-5, of some 40 nebulae (luminous distant clouds) revealed 95 percent of them to be shifted toward the red, indicating a general retreat of matter from our part of the sky. Edwin Hubble, at the Mt. Wilson Observatory, showed that the "nebulae" were actually far-off galaxies. Determining their distances by brightness relationships (light intensity diminishes as the square of distance; thus a star is one-ninth as bright when three times more distant), he found some of them millions of light-years* away, the first objects proved to be outside the Milky Way.

* A light-year (lt-yr), the distance light travels through space in a year, is equal to 5.9×10^{12} mi. (The speed of light is 186,000 mi/sec or 300,000 km/sec.)

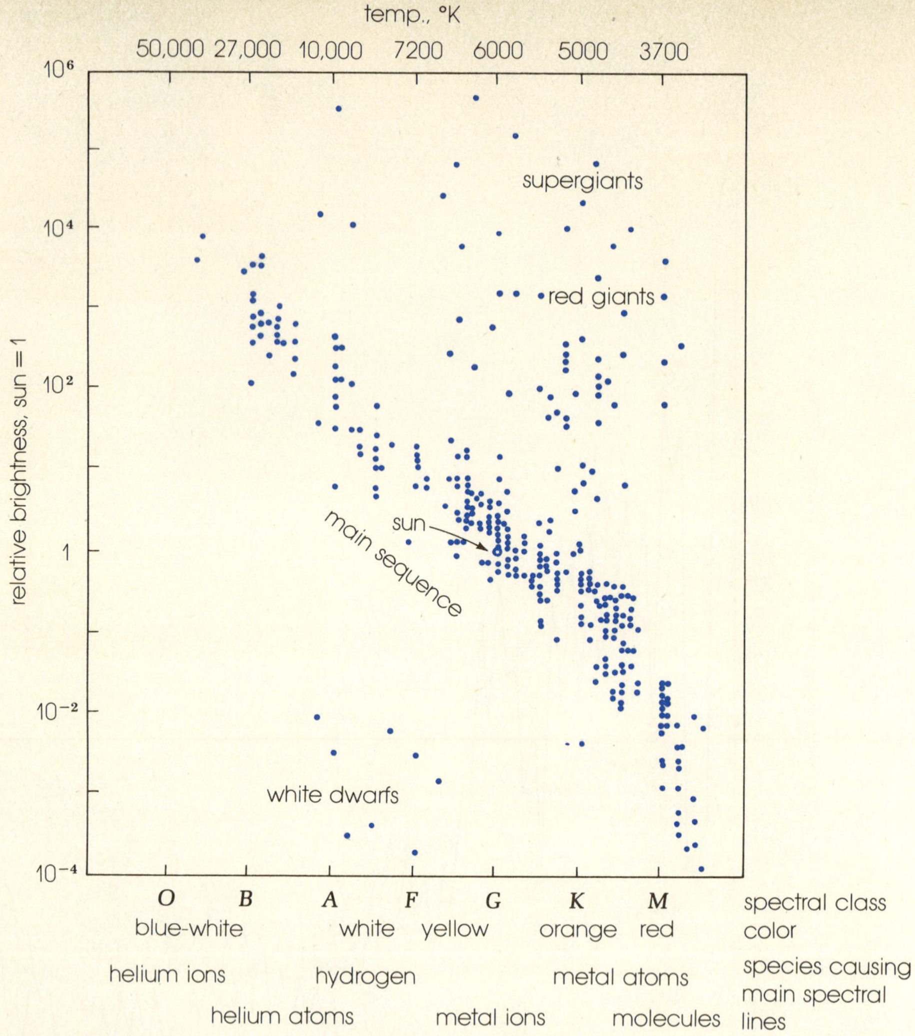

Figure 5-3 The Hertzsprung-Russell (H-R) diagram. Note that neither scale is linear.

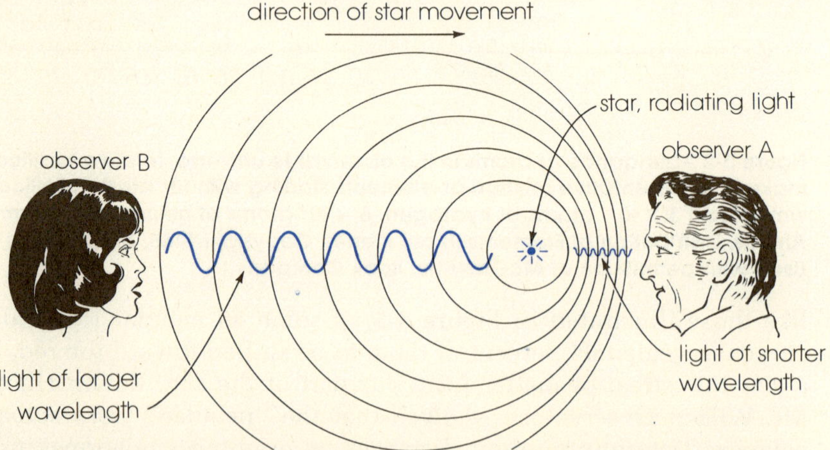

Figure 5-4 Change in wavelength with motion, a phenomena called the Doppler effect. As the star moves toward observer A and away from B, A sees a shift of spectral lines toward the blue end of the spectrum (shorter wavelength due to higher frequency), whereas B sees a shift of lines toward the red end (longer wavelength due to lower frequency). (Excerpt from Fred Hoyle, Astronomy. Copyright © 1962 by Rathbone Books, Limited. (Reprinted by permission of Doubleday and Company, Inc.)

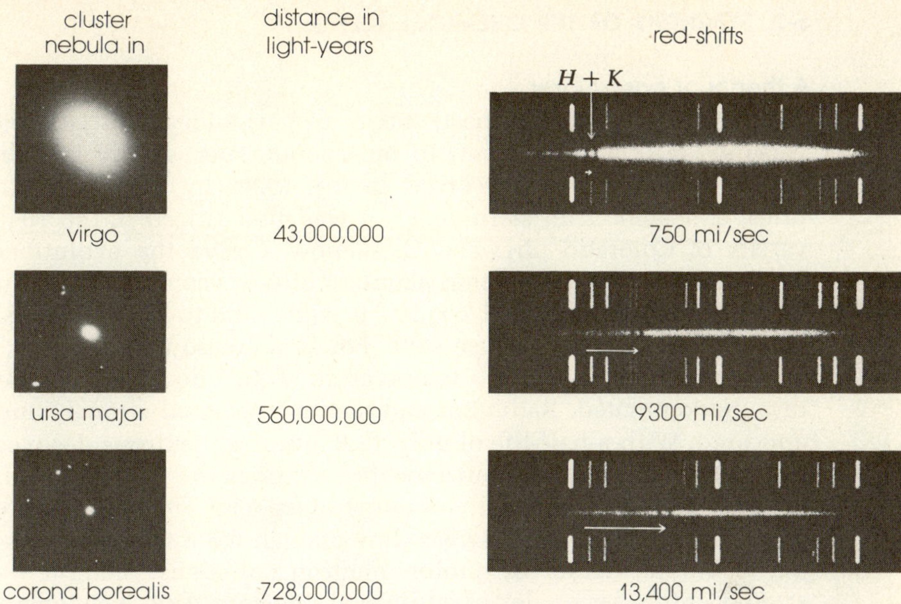

Figure 5-5 Red shifts in three remote galaxies. The galaxies are centered in the left series of the photos. Centered at the right are film strips of their spectra. Displacements toward longer (redder) wavelengths, as measured by shifts of the common calcium H and K reference lines, are indicated by the arrows. Above and below each spectrum the well-known spectrum for iron is reproduced for wavelength comparisons. (Courtesy Mt. Wilson and Palomar Observatories.)

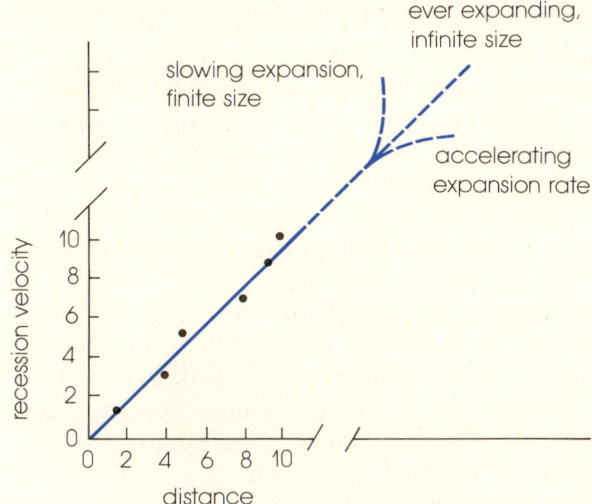

Figure 5-6 The expanding universe. Hubble found recession velocity v of galaxies to be directly proportional to their distance r from the earth,

$$v = Hr$$

where H is Hubble's constant. The slope of the line, v/r or H, expresses the rate of velocity increase with distance. The present value for H is 10.3 mi/sec per million light-years. This means velocity increases by 10.3 mi/sec for each distance increase of 10^6 light-years (5.9×10^{18} mi). The units of 1/H are time,

$$\frac{1}{H} = \frac{5.9 \times 10^{18} \ mi}{10.3 \ mi/sec} = 5.7 \times 10^{17} \ sec = 18 \times 10^9 \ yr$$

Eighteen billion years is called the Hubble age of the universe. Presumably it is when expansion began with the "big bang."
Very long distance measurements on the faintest galaxies obviously lack precision, but astronomers continue to make them, hoping to ascertain the shape of the graph's upper region which predicts the future of the universe.

In 1929 Hubble announced that the (red shift) velocity of galaxies increased in direct proportion to distance (Figure 5-6). This conclusion became one of astronomy's cardinal principles. After Hubble, all theories proposed for the creation and structure of the universe had to account for the red shift, meaning a universe that is expanding.

5-2. SYNTHESIS OF THE CHEMICAL ELEMENTS

A theory of cosmology

The first (and today the best) theory of cosmology (Gk. *kosmos,* the universe as an ordered system) to incorporate successfully the observations described above was conceived in the 1920s by G. Lemaître, a Belgian priest educated as an astronomer. It was modernized in the 1940s by University of Colorado physicist G. Gamow. Called the evolutionary or *big bang theory*, it envisions that about 18 billion years ago the entire mass of the universe (an estimated 2×10^{56} g, equivalent to 10^{80} nucleons) was contained in an infinitely dense state. For some reason it exploded, creating a billowing "fireball" with a temperature of 10^{12} degrees or higher. Gamma rays predominated. Radiation and matter interacted strongly and neutrons appeared. With a half-life of only 12.8 minutes, neutrons decayed into protons, electrons, and antineutrinos. In 5 minutes the fireball had thinned out to approximately the density of our stratosphere. The temperature then was a few hundred million degrees, low enough for nucleogenesis — the building of atomic nuclei by proton–neutron collisions. Helium was synthesized. Capture by nuclei of additional neutrons gave heavier nuclei. By β decay nuclei overloaded with neutrons gave elements of higher atomic number. After a half hour, nucleogenesis was probably finished, halted by neutron disappearance and the cooling and diluting effects of the expansion. Perhaps 100 million years later clouds of hydrogen, helium, and other atoms, still moving away from the center of creation and rotating, condensed into star clusters called *galaxies*. The Milky Way is a typical galaxy. It is spiral shaped and rotating, contains 200 billion stars, has a diameter of 10^5 light-years, and is 10^6 light-years from the next galaxy. Ten billion galaxies are within range of our telescopes.

In the mathematics of the big bang theory an eventual reversal is predicted which will bring all matter together again if enough mass exists to exert sufficient gravitational force to halt the expansion. Not all the required matter has been found but, as we will learn, much may be invisible. Evidence for contraction would be a general violet shift as matter rushes together to the state from which another explosive expansion would restart the universe.

Use of the radio telescope (Figure 5-7), which in effect is a very long-range spectrometer for the microwave region (see Figure 2-2), has in the last two decades helped cosmology become a science. One detail it reveals is that space is uniformly filled with microwave energy whose spectrum and intensity correspond to the radiation of a blackbody (see Figure 3-1) at a temperature of about 3°K (-270°C). Big bang theory predicts that space will possess this property as a vestige of the original fireball, and its discovery has given the theory some of its most convincing support.

Synthesis of helium on the main sequence

A star is born as gravitational contraction brings together a vast cloud of hydrogen and other interstellar matter. At a density 1000 times that of ordinary water (10^3 g/cm³), core temperature of the mass rises to 10 million degrees or more, turning it into a plasma of high-velocity protons and electrons. As hydrogen begins fusing to helium, the star becomes bright and joins the main sequence (see Figure 5-3).

Figure 5-7 The largest radio–radar telescope, at Arecibo, Puerto Rico. It is used to study (a) very distant radio sources (the range is at least 6 billion light-years, twice that of the largest optical telescope), (b) nearby planets by bouncing radar signals off them, and (c) the earth's ionosphere (see Figure 13-7). The parabolic dish (1000-foot diameter), shaped in a natural bowl and covered with wire mesh, focuses radio waves on the antenna (center). The antenna then sends an electrical signal to instruments for interpretation. Twenty different atoms, molecules, and radicals (H, NH_3, $NO\cdot$, etc.) have been identified in space from the distinctive way in which each absorbs radio wavelengths. Discovery in 1971 of molecules in galaxies besides our own carries implications that molecular formation processes go on everywhere, and that the chemical evolution necessary to the emergence of living things may be universal (see Figure 22-4). (Courtesy Cornell University.)

Core reactions begin with proton–proton interactions. Ninety percent of the material probably follows the sequence $^1H \rightarrow {}^2H \rightarrow {}^3He \rightarrow {}^4He$, the net result being

$$4{}^1_1H \longrightarrow {}^4_2He + energy$$

As expansion due to heating is balanced by radiation loss and the force of gravity, the star settles down to a relatively quiet life some tens of millions to a few billion years long, during which it "burns" hydrogen. Each 4 g of hydrogen burned to helium is equivalent to 6×10^{18} kcal of heat energy.

Synthesis of intermediate elements in giants

When the core of a star has become at least 30 percent helium, proton-chain reactions come to an end for lack of fuel. Instead of dying out, how-

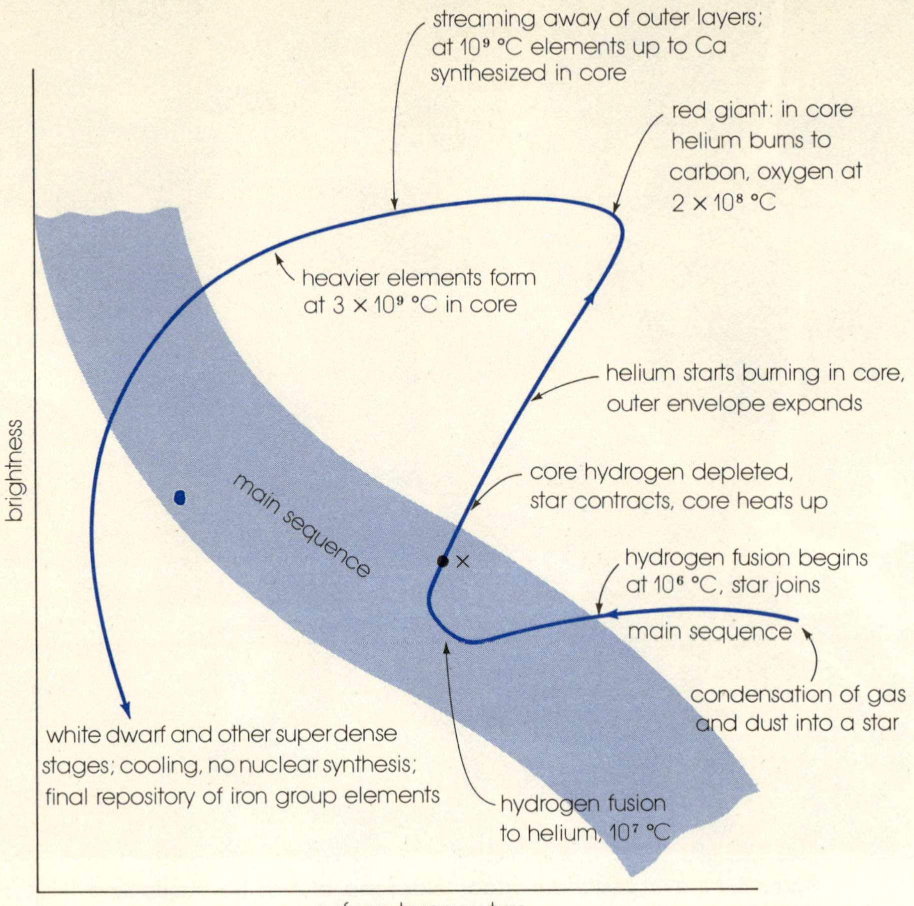

Figure 5-8 **Theoretical evolutionary route of a star like the sun. The sun may presently be at point x. Its age has been sought by experiments designed to count neutrinos coming from its nuclear reactions.**

ever, the star contracts under its gravitational force, and core temperature rises. The sudden rise heats the rest of the star, causing it to expand several hundredfold. The surface is now cooler as radiation comes from a much larger area, so the star appears red. Gas streams out into space because gravity at the surface is much weaker than before. The star has left the main sequence and joined the giants (Figure 5-8).

At a core temperature of 200 million degrees and a density of 10^5 g/cm³, mutual repulsion of α particles is overcome. Fusion of helium to carbon begins, releasing energy by the so-called triple α process in which beryllium-8 may be a short-lived intermediate:

$$^4_2He + {}^4_2He \longrightarrow {}^8_4Be$$

$$^8_4Be + {}^4_2He \longrightarrow {}^{12}_6C \tag{5-1}$$

When the helium is gone, the core contracts again, giving still higher temperatures. Nuclides formed by helium fusion now decompose and their α particles react with other nuclei to form more massive nuclides:

$$^{12}C \xrightarrow{^{4}He} {}^{16}O \xrightarrow{^{4}He} {}^{20}Ne \xrightarrow{^{4}He} {}^{24}Mg \xrightarrow{^{4}He} {}^{28}Si, \text{ etc.}$$

Element building probably carries at least to ^{56}Fe, the nuclide with maximum binding energy (see Figure 4-15).

The synthesis of other elements takes place between 100 million and 1 billion degrees by neutron capture followed by β decay, as

$$^{35}_{17}Cl + {}^{1}_{0}n \longrightarrow {}^{36}_{17}Cl \longrightarrow {}^{36}_{18}Ar + {}^{0}_{-1}e \qquad (5\text{-}2)$$

Once the proper temperature is reached, these processes may go in only a few hours. They are terminated with production of the heaviest elements, most of which are probably destroyed by neutron fission and α decay. Reaction velocities soar as temperature continues to climb to an estimated 3.8 billion degrees, and density increases to a fantastic 3 million g/cm³.

Stellar graveyards

With the synthesis of iron in the giant, fusion energy ceases to exist. Thermal pressure of the plasma and radiation pressure of photons can no longer balance the force of gravity, and the core begins to collapse. As the foundation beneath outer gas layers falls away, gravitational energy heats them up. The surface brightens once more and the star begins moving toward the left in the H-R diagram. Material expelled at high velocity, including some from the core, scatters a variety of elements into space. The decomposing star is called a *nova*. It may increase 100,000 times in brightness and burn for a week or two before suddenly dying down.

If the giant contains 10 to 50 solar masses of material, it has evolved quickly and may now be unstable enough to expire by supernova explosion. Over a period of a few days the brilliance of a *supernova* can be 100,000 times that of a nova and the equal of an entire galaxy.

The hot remnant of a star is compressed in proportion to the mass remaining. Carrying with it most of the intermediate-weight elements synthesized, it will slowly and quietly cool to the temperature of space. A sun-mass star becomes an earth-size white dwarf. Its hot atoms are completely ionized so electron orbits do not exist. Electrons and protons are crowded together as a very dense (10^6 g/cm³) gas, a state known as *degenerate matter*. A star containing a few sun masses is compressed into a ball less than 50 miles in diameter. Called a *neutron star*, it is in effect one big atomic nucleus of neutrons. Unless gravity becomes repelling at extremely short range, a still larger star is squeezed to the limit—its "gravitational radius"—which theoretically is only a point. Because the gravitational force of several solar masses focused at a point is so extreme that not even light can escape, it is called a *black hole*. Black holes are invisible.

A significant fraction of the entire mass of the universe may reside in the superdense cinders of dead stars.

5-3. THE SOLID EARTH

Origin of the sun and its planets

Some astronomers theorize that the solar system began as a cool cloud of hydrogen, helium, dust, small molecules such as H_2O and NH_3, and chunks ·

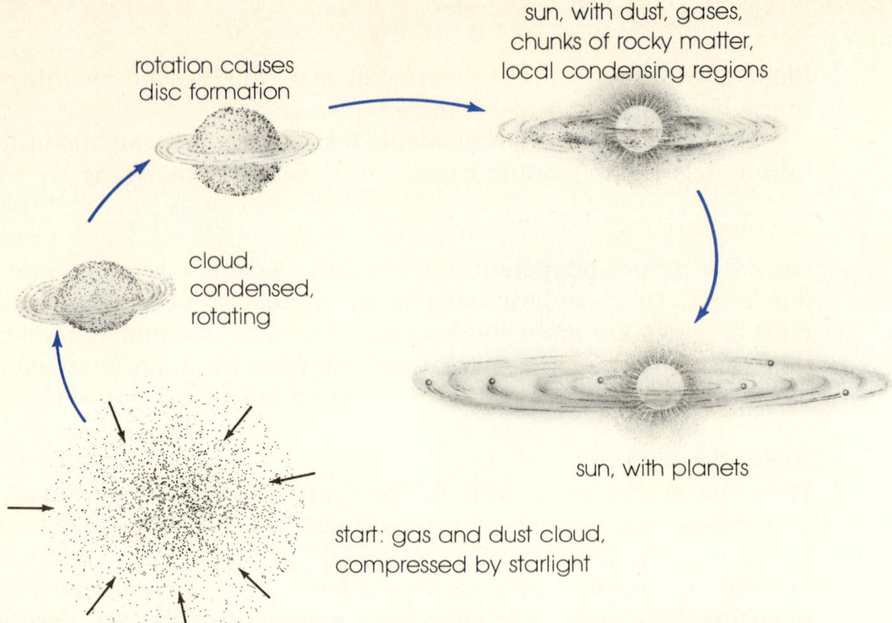

Figure 5-9 Formation of the solar system. This evolutionary route is suggested because (a) the axis of the system's rotation is perpendicular to the elliptical planetary orbits that lie in one plane, and (b) all planets move in the same direction about the sun. The sun contains more than 99.8 percent of the system's mass.

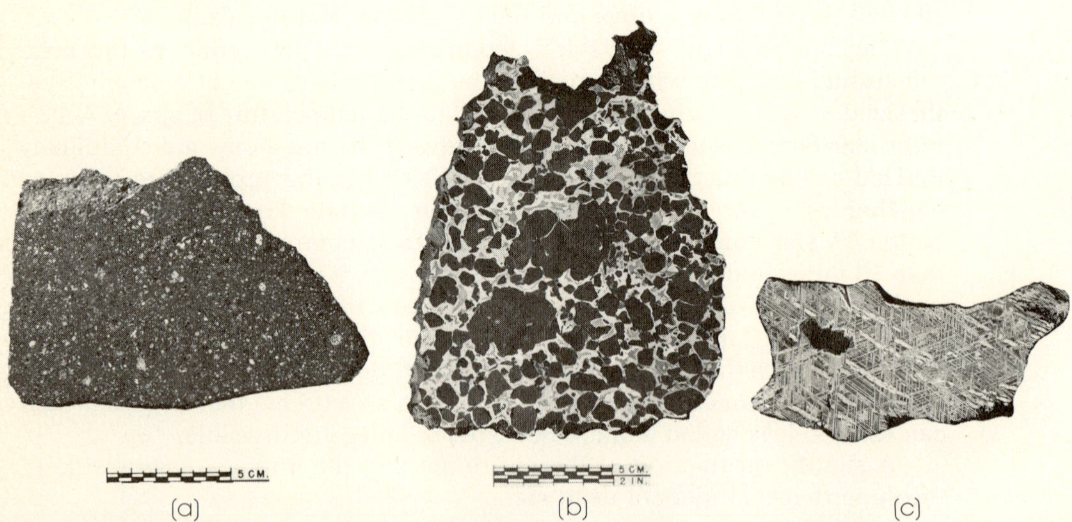

Figure 5-10 Meteorites. Meteorites are classed as stones, stony-irons, and irons (a) A stone. It consists mainly of silicates of iron and magnesium. In addition this specimen contains some carbonaceous material. (b) A stony-iron. It is made of iron and magnesium silicates in a matrix that includes sulfides and phosphides of iron, cobalt, and nickel and an iron–nickel alloy. (c) An iron. About 94 percent iron and 6 percent nickel, the specimen was cut and polished to show the unique grain structure that identifies it as a meteorite. Irons can be located on the ground with a magnet. (Courtesy The Smithsonian Institution.)

of heavier elements from exploded stars (Figure 5-9). Compressed by starlight, then contracting by gravity, the cloud began to rotate. Becoming elliptical, it threw off a flat equatorial disc of gas and solid particles. As the disc swirled, local areas began to condense. In the center, where most of the mass remained to create the sun, nuclear fire lit up and the disc spread out.

Dust, metals, and rocky particles, being least volatile, aggregated relatively close to the sun. The sun's radius and luminosity were greater then, and some melting and vaporization of the material possibly took place. After a brief hot phase, cooler fragments may have formed larger bodies, from which by collisions the inner planets grew. The remaining debris may have been blown out to space by gas bursts from the sun.

Evidence that the inner planets (see Figure 5-11) are composed mostly of combinations of oxygen, silicon, iron, and magnesium comes from the chemical examination of *meteorites* (Figure 5-10), which are probably samples of original disc material. Infall of meteoric material (mainly dust) on the earth is estimated to be a few thousand tons a day.

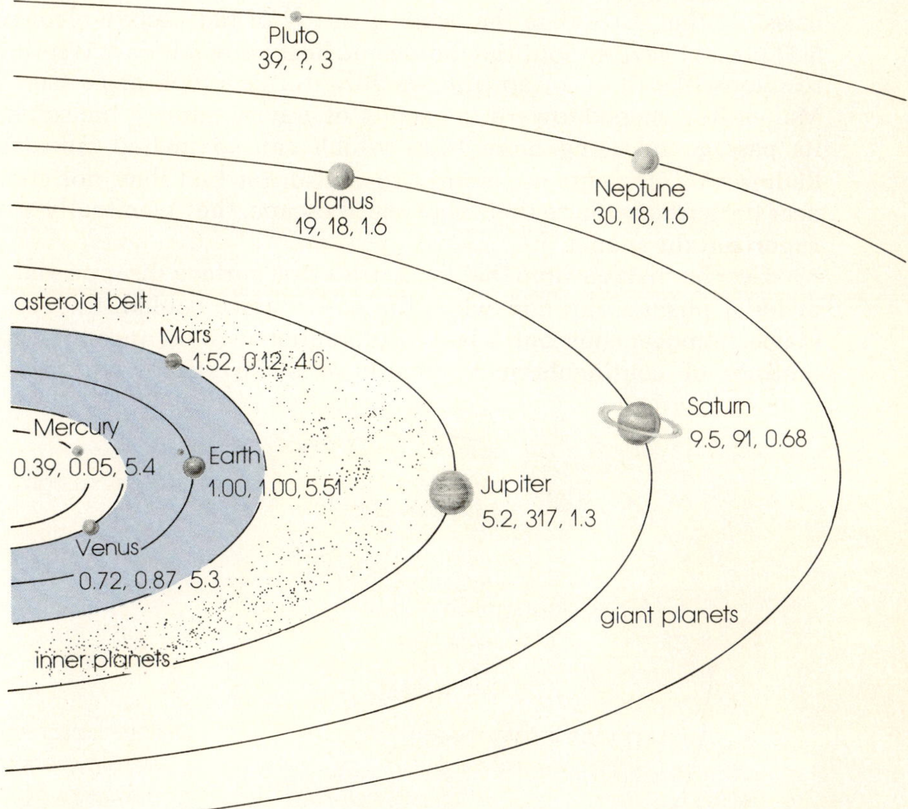

Figure 5-11 The solar system. Numbers given with planet names are, respectively, relative average distance from the sun (earth = 1), relative mass (earth = 1), and density (grams per cubic centimeter). The sun (not shown, to the left) is 864,000 miles in diameter (109 earth diameters), and its mass is 2.2×10^{27} tons (334,000 earth masses). The earth's average distance from the sun is 93 million miles. Earth mass is 6.59×10^{21} tons. The shaded area is that region where the presence of water and a moderate temperature make life feasible. The asteroid belt may be unconsolidated matter left over from the original solar disc.

The giant outer planets (Figure 5-11) probably consolidated from dust, hydrogen, and other gases. The extreme pressures inside Jupiter and Saturn will compress hydrogen to a solid resembling a metal where in all atoms are equidistant and electrons are uniformly distributed (see Table 10-1). Forty percent of the planetary system's mass may be metallic hydrogen.

If the planets had not formed and inherited most of the solar system's momentum, the sun might be spinning 100 times faster than it is. Because most stars also rotate slowly, astronomers assume that they too evolved with planetary systems, although their planets are too dimly lit to be observed. In the Milky Way alone there may be a billion or more stars with habitable planets. It seems possible, therefore, that other "earths" exist.

Formation of the earth

The classical working hypothesis for discussing evolution of our planet is illustrated in Figure 5-12. Swirling dust in the solar disc is assumed to have condensed into chunks and then into the earth. Solar radiation, conversion of gravitational energy, and radioactivity heated the roughly homogeneous mass. At that early date the larger quantity of radioactive elements (Table 5-1) present gave sixfold the radiogenic heat known today. Within the earth reactions like those in an iron smelting furnace took place (see Eq. 8-13). Molten iron seeped toward the center of gravity through the softened mass, its passage releasing more heat, which caused melted silicates to rise. Radioactive elements apparently rose also, for had they not concentrated near the surface where their heat could escape, they eventually would have vaporized the planet.

Geochemists assume that the earth's first surface disappeared as molten material pushed out, and when the new surface cooled, the earth was no longer homogeneous but a body differentiated into concentric zones. The outlines of continents may then have taken shape due to radioactive

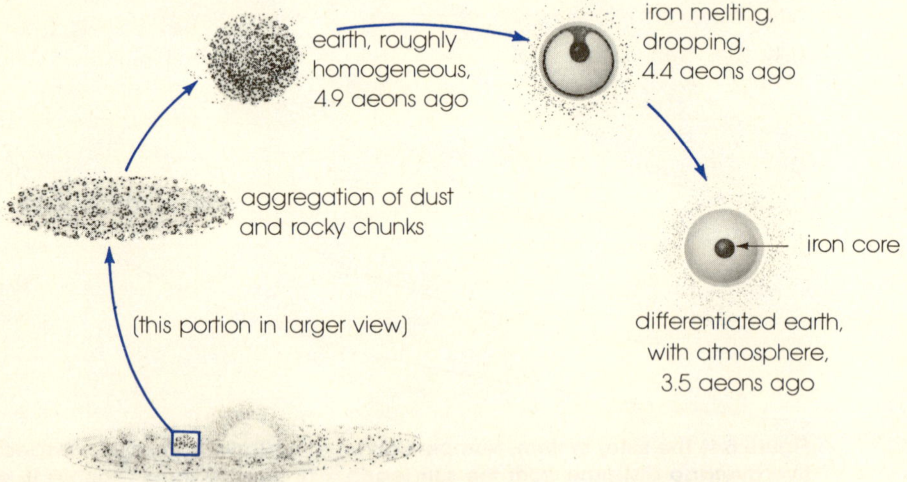

Figure 5-12 Model for the earth's formation. Dust, gases, and perhaps snow segregated in the sun's disc. Fusion of aggregates then gave the planet. Heating caused a shell of molten iron to accumulate and drop to the center, forming a semiliquid core. Simultaneously lighter silicates and radioactive elements rose to form the crust. The result, a differentiated body.

Table 5-1. Radionuclides useful in dating the earth's distant past

Nuclear reaction	Half-life aeons
$^{40}K + \beta \longrightarrow {}^{40}Ar^{a}$	1.3
$^{232}Th - \cdots \rightarrow {}^{208}Pb + 6\alpha + 4\beta$	14.1
$^{235}U - \cdots \rightarrow {}^{207}Pb + 7\alpha + 4\beta$	0.731
$^{238}U - \cdots \rightarrow {}^{206}Pb + 8\alpha + 6\beta$	4.51

[a] This process is called *electron capture*. The nucleus assimilates an electron from the nearest electron level; ^{40}K also undergoes β decay, forming ^{40}Ca.

elements melting their way through the crust, followed by a series of remeltings and recrystallizations. With volcanic action came gases that would make the atmosphere (see photo on page 97 and Chapter 13) and water that would fill the basins (Chapter 15). Chemicals produced by atmospheric reactions collected in lakes where interactions gave increasingly complex molecules and finally the first stirrings of life (Chapter 22). Perhaps 1.5 *aeons* (1 aeon = 1 billion years) elapsed between dust cloud and life.

Age of the earth

Prior to 1900 the age of the earth was a highly speculative matter. Kelvin, who discovered some of the basic laws of heat, calculated an age of 20 to 40 million years based on the measured heat flow to the surface (50 cal/cm²/yr) and the assumption that the planet was cooling normally after a molten beginning. Discovery of heat from radioactivity invalidated his estimate. From measurements on the rates at which rocks are weathered and sediments built up, geologists arrived at a planetary age of more than a billion years. Then in 1908 B. Boltwood, an American working with Rutherford, suggested that the energies of α and β emissions were so large that radioactive decay must be unalterable in any environment; hence radioactive elements with very long half-lifes should be usable as clocks to measure geologic time. His suggestion became practicable when invention of the mass spectrograph made routine the determination of nuclide distribution in radioactive minerals.

Consider a mineral as old as the earth and originally containing 235 mg of uranium-235. After one half-life, 0.731 aeon (see Table 5-1), half (118 mg) of the uranium-235 remains. The other half has been converted to lead-207 (104 mg) and helium (13 mg). The uranium-to-lead weight ratio of 118:104 (atom ratio 1:1) found by mass spec analysis indicates that the mineral's age is 0.731 aeon. Suppose that the mineral had been twice that old. Then half the uranium remaining after one half-life would have decayed, leaving 59 mg of uranium-235. And an additional 52 mg of lead-207 and 6.5 mg of helium-4 would have been created. The uranium-to-lead weight ratio of 59:156 (atom ratio 1:3) would then indicate that the age is two half-lifes or 1.46 aeons. The way radioactivity diminishes with time is shown in Figure 5-13.

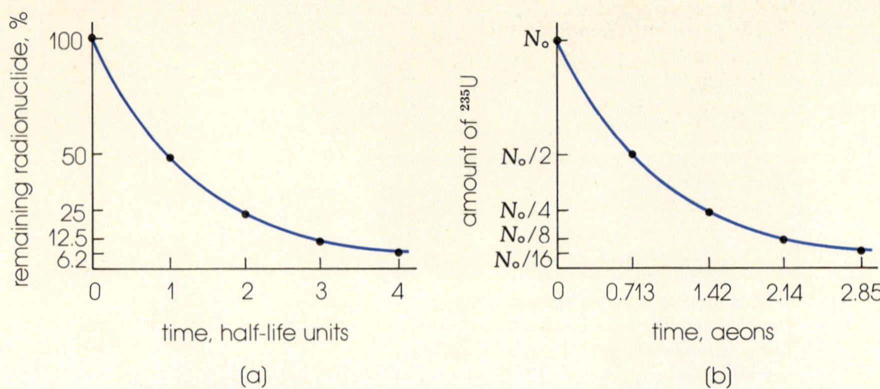

Figure 5-13 Radioactive decay. (a) The general form of the decay curve. During each half-life interval, half the radionuclide present at the beginning of that interval decays. (b) The decay curve for ^{235}U. N_o is the amount (number of atoms, their weight or radioactivity) at the start of counting. After one half-life, $\frac{1}{2}$ remains; after two half-lifes, $\frac{1}{2} \times \frac{1}{2}$ or $\frac{1}{4}$ remains, etc. After 10 half-lifes, $(1/2)^{10}$ or about 1/1000 remains, and after 20 half-lifes, $(1/2)^{20}$ or about 1/100,000 remains. Radioactivity is then essentially gone. In general,

$$N = \frac{N_o}{2^n}$$

where N is the amount of radionuclide left after n half-lifes.
For further analysis two more equations are useful,

$$2.3 \log \frac{N_o}{N} = kt$$

where t is the time elapsed during which N_o is reduced in amount to N, and k is the decay rate constant, a proportionality constant having the units 1/time. The value of k is different for each isotope. A k value for use in the equation above is calculated from

$$k = \frac{0.693}{t_{1/2}}$$

where $t_{1/2}$ is the nuclide's half-life. In the symbolism of the equation, half-life is the time needed for N to become $N_o/2$.

Nuclear clock measurements indicate the oldest crustal rocks are 3.5 aeons old. However, meteorites as well as the oldest moon rocks, which presumably formed approximately when the earth did, have an age of 4.5 aeons. The 1-aeon difference is considered to be the time between the earth's consolidation and the establishment of a stable crust. The best estimate is that the earth was born 5 billion years ago. If it were much older, radionuclides like potassium-40 and uranium-235 would no longer be found.

Composition of the earth's zones

Our knowledge of the earth's skin comes from analyzing surface rocks and material brought up from a few miles depth by drilling and by volcanoes. The other 99.5 percent of the earth remains unexplored by direct observation.

The cross-sectional picture of the earth presented in Figure 5-14 has been established by following the paths of seismic (Gk. *seismos*, shock) waves created by earthquakes and nuclear explosions (Figure 5-15). In general the less dense and more rigid and incompressible the material a seismic wave encounters, the faster the wave moves. Rigidity and incompressibility increase markedly with depth; so if the earth was homoge-

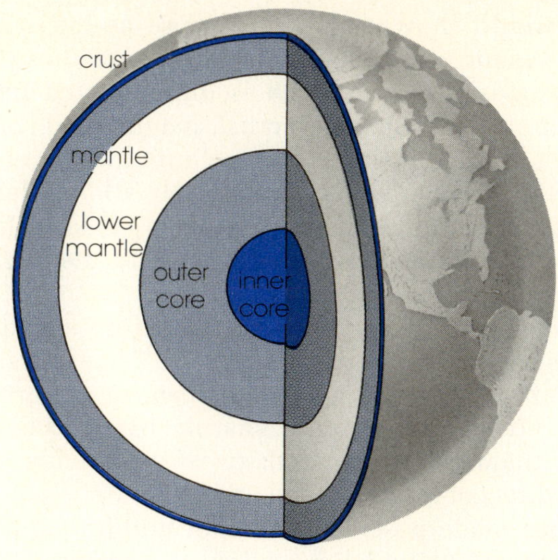

0 17 1000	2900	5000 6371,	distance from surface, km	
1200 1500		5000,	estimated temperature, °C	
5 400 1400		3200 3600,	estimated pressure, kilobars (1 kb = 1013 atm)	
2.8 4.8 5.8		12.0 13.0,	estimated density, g/cm³ (overall = 5.52 g/cm³)	

Figure 5-14 Zones of the earth's interior.

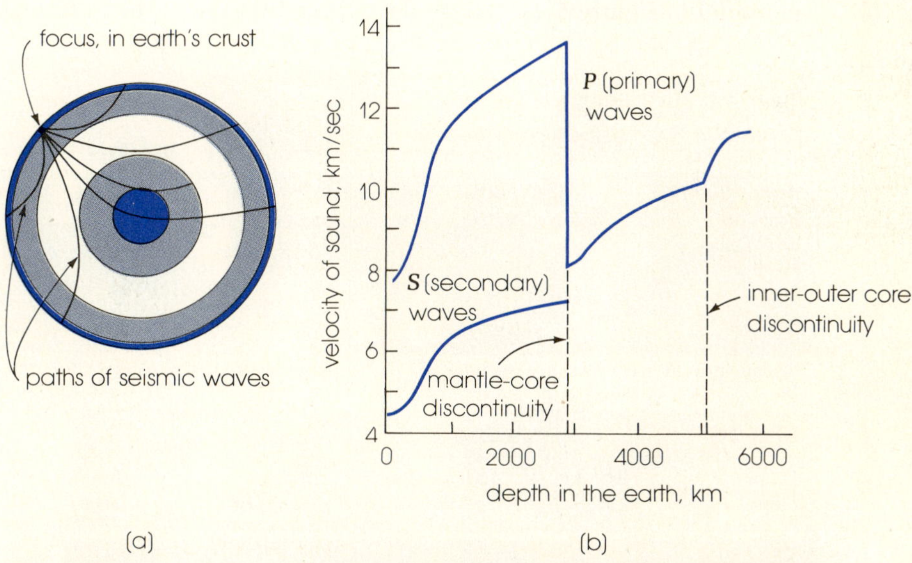

Figure 5-15 Seismic studies. (a) An earthquake creates seismic waves. Electronic stations coordinated to listen for the waves record the time required for them to surface. (b) Plots of wave velocity verses earth depth show two deep zones inside the planet. Wave pattern discontinuities mark mantle-core and inner-outer core boundaries; P (primary) waves travel like sound waves in air, by alternately compressing and stretching the medium; S (secondary) waves travel at right angles to P waves and do not penetrate liquids, which the outer core is thus assumed to be. (Adapted from S. P. Clark, Jr. and A. E. Ringwood, Reviews of Geophysics, 2 (1), 67.)

neous, wave velocity would show a regular increase as the center is approached. Results of seismic experiments, however, are those of Figure 5-15b. The interpretation is that the planet is differentiated into several subregions of three main concentric zones: *crust, mantle,* and *core.*

If to the solid zones we add three more—the *atmosphere,* or gas envelope about the planet, the *hydrosphere,* or collectively all the planet's readily available water, and the *biosphere,* or all organisms living and dead—we have compartmentalized everything physical there is to study. The six zones are characterized in Tables 5-2 and 5-3.

The composition of the crust has been calculated by averaging many chemical rock analyses and estimates of rock masses. The eight elements in greatest abundance (Table 5-4) make up 98.5 percent of the crustal mass.

Crust and mantle are separated quite sharply by a mile-thick zone called the *Moho.* The mantle's greater density is attributable to both its greater iron content and its compaction. The upper mantle is hot enough to be plastic. Plasticity has made possible mountain building and spread of the ocean floor (see Figure 15-2), which has allowed the continents to move apart (see Problem 16).

The core of the earth is believed to be similar in composition to iron meteorites (see Figure 5-10c). The exceptionally sharp core demarcations indicated in Figure 5-15b probably mean that the outer core is liquid metal, whereas the inner core, under greater pressure, may be solid metal. The magnetic field about the earth is possibly caused by electric currents associated with movements of molten core iron caused by temperature differentials and the planet's rotation.

Estimates of whole-earth composition given in Table 5-5 are based on the model of Figure 5-14. These data show that the earth is 92 percent iron,

Table 5-2. Constitution of the earth's zones[a]

Zone	Chemistry	State
Atmosphere	Nitrogen, oxygen, other gases	Gas
Biosphere	Organic matter, water, calcium phosphate	Solid, liquid
Hydrosphere	Salt water, fresh water, ice, snow	Liquid mostly
Crust	Rocks; silicates of abundant metals	Solid
Mantle	Dense silicates	Solid
Core	Iron mostly; nickel, cobalt, possibly sulfur, silicon	Liquid, solid

[a] Adapted from Brian Mason, *Principles of Geochemistry* (3rd ed.). New York: Wiley, 1966, Table 3.3.

Table 5-3. Physical measurements of the earth's zones[a]

Zone	Thickness km	Mean density g/cm^3	Mass percent	Volume percent
Atmosphere	—	—	9×10^{-5}	—
Hydrosphere	3.80	1.03	0.024	0.1
Crust	17	2.8	0.4	1
Mantle	2883	4.5	67.2	85
Core	3471	11.0	32.4	14
Whole earth	6371	5.52	100	100

[a] Adapted from Brian Mason, *Principles of Geochemistry* (3rd ed.). New York: Wiley, 1966, Table 3.3.

Table 5-4. Average abundances of commonest elements in crustal rocks

Element	Atomic number	Percent by weight	Element	Atomic number	Percent by weight
Oxygen	8	46.6	Calcium	20	3.6
Silicon	14	27.7	Sodium	11	2.8
Aluminum	13	8.1	Potassium	19	2.6
Iron	26	5.0	Magnesium	12	2.1

Table 5-5. Estimated composition of the whole earth

Element	Percent by weight	Element	Percent by weight
Iron	34.6	Sulfur	1.9
Oxygen	29.5	Calcium	1.1
Silicon	15.2	Aluminum	1.1
Magnesium	12.7	Sodium	0.4
Nickel	2.4	All others	1.1

oxygen, silicon, and magnesium. By comparison, when we translate the data of Figure 5-2 into weight percentages, we find that the (spectral) universe at large is 98.9 percent hydrogen and helium (oxygen is 0.7 percent, iron a mere 0.009 percent).

The tiny spaceship we cling to may be only one of billions of planets in the Milky Way alone, but in the vastness of space it is a most unusual body. Its history has given it the chemical composition, mass, and temperature that make the life we know possible. How singular an event life is we may never find out; the nearest planet like ours is probably thousands of light-years away, so we will never explore it. In the universe as a whole, compact collections of molecules necessary for life are a rarity because almost everywhere temperatures and pressures are either too high or too low for their formation. Nevertheless, it is natural to wonder how many other habitable planets exist and what societies there might tell us about our future. Our planetary system is unusually young. Most others are probably twice as old.

QUESTIONS

1. Stars Stars burn out and die. Physically and chemically what is a dead star? Why isn't the sky observed to be full of dead stars now?

2. Meteorites Why do meteorites provide better samples for studying cosmic elemental abundances of nonvolatile elements than do earth rocks?

3. Sun The sun's spin (1 revolution every 25 days) was first measured from the movement of sun spots (cool gas zones). Rotation time can also be established by using a spectrometer to measure violet shift at the approaching edge and red shift at the receding edge. Explain.

4. Star, planets Compare the sun and earth on the basis of size, chemical composition, structure, age, surface temperature.

5. Red shift Suppose that atomic processes, including the electronic vibrations that produce light, were once slower than they are today, and that they have progressively speeded up. That would mean light emitted long ago by stars, and which we now view arriving from many light-year's distance, originated with lower than expected frequency. How would this theory, proposed in 1932 by British astronomer E. Milne, furnish an explanation alternate to Hubble's for the red shift?

6. Planets The earth has a magnetic field, a gas atmosphere, and much surface water. Mars has little or no magnetic field and only small amounts of gas and surface water (ice). Suggest differences in composition or planetary histories, or both, that might account for the facts.

7. Elemental syntheses **(a)** Figure 5-16a is an abbreviation of a reaction going on in the sun. Describe it in words. Tell what the net reaction is, and its significance. **(b)** Figure 5-16b is the Crab Nebula, the remainder of a supernova explosion reported in its present position in A.D. 1054 by oriental astronomers. What is a supernova? What causes it? Could elements in the earth have come from ancient supernovae explosions? Explain.

8. Elemental abundances **(a)** How were the data of Figure 5-2 obtained? **(b)** Name the four most abundant elements in the universe; the whole earth; the earth's crust.

9. Dating Tritium (T or ^{3}H, half-life 12.3 yr) is formed like ^{14}C in the upper atmosphere (see Problem 21) and reaches the earth's water supplies in very small quantities by falling as T_2O snow and rain. Because decay and natural replenishment of tritium are in global equilibrium, the tritium content in surface waters is constant. **(a)** How can water from one well contain tritium activity whereas water from another does not? **(b)** A 200-year-old bottle of brandy is discovered. Can it be tritium dated? Explain.

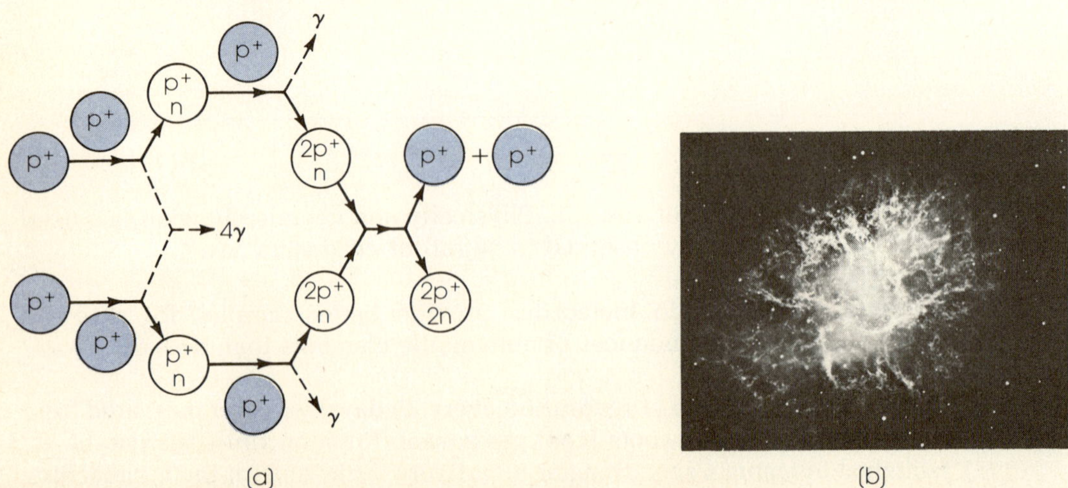

Figure 5-16 [(b) Courtesy Mt. Wilson and Palomar Observatory.]

10. Depletion processes Imagine lighting a 12-in. candle and watching it burn. Let t be the time it takes to burn down to the 6-in. mark. **(a)** When another interval t elapses, will the candle be at the 3-in. mark, some other mark, or all burned up? **(b)** Why is the burning of candle wax the same as or different from the decay of a radioactive element (Figure 5-13)?

11. Core **(a)** What is the Curie point of iron (Chapter 8)? **(b)** Two hundred years ago scientists believed the earth to be completely solid and its magnetism due to a core of permanently magnetic iron. From data in Figure 5-14 and your definition in (a), what can you say about this theory? **(c)** How is the earth's magnetic field explained today?

12. Moon During the Apollo moon missions, part of the used equipment was allowed to fall and crash on the moon's surface. Impact was recorded by a seismometer. **(a)** What could scientists hope to learn about the moon's structure from this experiment? **(b)** In order for the moon to have a structure like the earth (Figure 5-14), what processes would have had to take place?

PROBLEMS

13. Universe's age According to the big bang theory, the universe has been cooling since the original fireball began expanding. The theory offers this equation relating time t in seconds and temperature K in absolute degrees: $K = 1.5 \times 10^{10}/\sqrt{t}$. If $K = 3$ degrees (the temperature of outer space today), calculate the **(a)** seconds and **(b)** years that have presumably elapsed since expansion began.

14. Hubble's law The "Hubble radius" of the universe is that distance at which receding galaxies attain the speed of light, 1.86×10^5 mi/sec (and beyond which we could therefore never see them). Calculate this distance in light-years and in miles, from Hubble's law.

15. Solar system If our planetary system formed from a disc of gas and dust around the sun, temperature distribution in the disc may have followed the law $K = k/r$, where K is absolute temperature, k is a constant, and r is the relative distance from the sun (Figure 5-11). **(a)** Assuming that the temperature at the earth's distance was $220°K$ ($-50°C$) at that time, calculate k. Then calculate the temperatures at which Uranus and Neptune presumably formed. **(b)** Could solid ammonia and solid methane (melting points, respectively, $200°K$ and $90°K$) have been available for incorporation into Uranus? Neptune?

16. Oceans' age (library) Using an atlas, estimate average distance between the coast lines of South America and Africa. **(a)** Convert the distance from miles to centimeters (1 mi = 1609 m). **(b)** Assume that the two continents were once in contact but since then have been moving apart at a constant rate of 2 cm/yr (see Figure 15-2). How long ago were they presumably in contact? **(c)** The oldest fossils found in Atlantic Ocean sediments, taken by oceanographers who cored the ocean floor, appear to be no more than 200 million years old. How does this compare with your calculation in (b)?

17. Uranium–lead dating Consider a meteorite formed 4.51 billion years ago and containing at that time 100 mg of pure ^{238}U. What weight of ^{206}Pb would it contain today? (See Table 5-1.)

18. Mineral age A mineral containing a single radioactive element has an activity measured with a Geiger counter as 5000 cpm (counts per minute). Use the first equation of Figure 5-13 to calculate the count after two, four, and six half-lifes.

19. Radioactive decay Consider 1000 atoms of a radioactive element. How many atoms of parent and of daughter remain after **(a)** one half-life; **(b)** two half-lifes; **(c)** three half-lifes; **(d)** four half-lifes? **(e)** Calculate the daughter-to-parent ratio in each case. Then plot the ratios versus time (in half-life units). Plot time on the abscissa.

20. Decay curves (a) Using the numbers calculated in the first four parts of Problem 19, plot a graph of the parent atoms remaining versus time (in half-life units). Plot time on the abscissa. **(b)** In another color, plot the number of daughter atoms. Label the curves.

21. Carbon-14 dating In the 1950s Willard Libby (United States 1908–   , Nobel prize in chemistry 1960) found that cosmic ray collisions with atoms in the upper atmosphere produce neutrons. When a neutron is

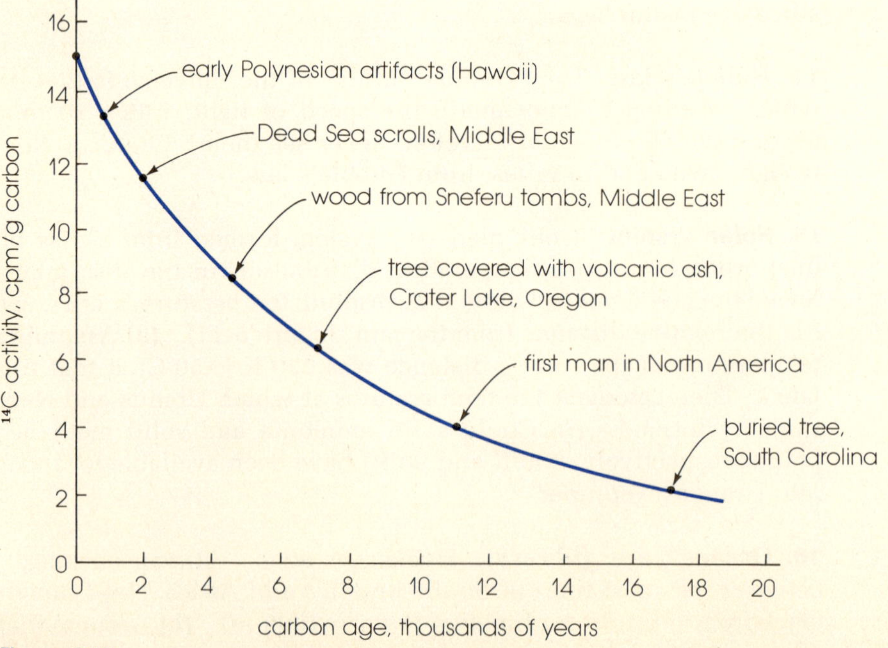

Figure 5-17

captured by a nitrogen-14 atom, a proton is emitted, leaving a radioactive carbon-14 atom (half-life, 5730 yr). Oxidation of carbon gives a molecule of $^{14}CO_2$. Living things take in the $^{14}CO_2$. As long as they are alive, they contain an equilibrium concentration of radioactive carbon (intake is balanced by its decay) equivalent to 12.5 cpm/g of total carbon. When they die, intake stops. The ^{14}C decay then constitutes an atomic clock by which ages up to about 50,000 years can be determined. Figure 5-17 shows some determinations that have been of value to archeologists. **(a)** Use the last equation of Figure 5-13 to calculate the rate constant of ^{14}C. **(b)** Wood taken from a boat in an Egyptian tomb showed an activity of 7.00 cpm/g of total carbon. Use the second equation of Figure 5-13 to calculate its age.

22. Coral dating Certain contemporary corals add a readily observed ridge each year (comparable to a tree's annular ring). They also add a microscopic line each day. Similar markings are found on fossil corals that died millions of years ago. When they were alive, the earth is thought to have rotated about its axis faster and days were shorter because drag on the earth's rotation caused by ocean tidal friction had operated for less time. If true, then geologic age is proportional to the number of days in a year. And "coral age" becomes an independent method by which radioisotope dating may be checked. Assuming that the tidal effect has been constant in causing an increase in the day of 2 seconds every 100 years, calculate the age of a fossil coral that shows 400 microscopic lines between ridges.

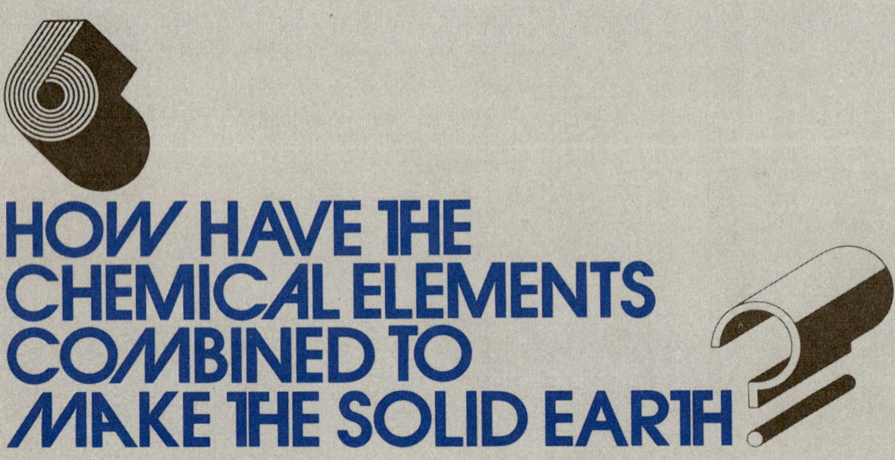

HOW HAVE THE CHEMICAL ELEMENTS COMBINED TO MAKE THE SOLID EARTH?

THE IMPORTANT CONCEPTS

6-1. Minerals and rocks

1. Minerals are classified by their negative groups.
2. Rocks are classified by origin.
3. The geochemical cycle correlates migration of crustal elements.
4. Modern instruments make geochemical analyses possible.

6-2. Crystals

1. Solids may be crystalline or amorphous.
2. X ray diffraction establishes distances between atom planes in crystals.

3. Crystals are described by the disposition and length of their crystallographic axes.
4. A lattice is built of unit cells.
5. Simple geometric figures describe the six lattice types.

6-3. Allotropy
1. Some elements exhibit allotropy.
2. Physical properties of substances can be correlated with atomic arrangement.

6-4. Structure and composition of common minerals
1. Silicate mineral structures are explained by the several ways in which the orthosilicate group condenses.
2. Ion exchange properties of framework silicates are explained by the open structure.
3. A few metal–oxygen–silicon combinations account for most of the earth's crust.

We have seen that much is speculative about the earth's history and much remains to be learned. Even a single rock can represent a complex study because of the great variability possible in its composition, structure, and mode of formation. Earth scientists believe that our planet continues to evolve through processes explainable by natural laws. Insight into earth science, including wise management of earth resources, is gained by chemically examining, materials called *minerals* and *rocks* because their properties are related to the transformations that have taken place at the earth's surface. There would be no life as we know it if water and atmospheric gases had not come out of the planet, rock erosion had not produced soil, and minerals had not concentrated in usable deposits.

6-1. MINERALS AND ROCKS

Eight kinds of minerals

A *mineral* is a naturally occurring, homogeneous crystalline inorganic solid that may have definite chemical composition. Minerals generally originate in solutions and crystallize from water, molten rock, or volcanic vapors. Some 2000 minerals have been found in nature but only 1 percent of them are common. About 25 new minerals are discovered each year.

Mineral classification began 200 years ago when Scottish physician James Hutton (1726–1797), the "father of geology," showed the value of examining the earth as an evolving body. In 1818 Berzelius (see Figure 11-1) categorized minerals according to the chemical character of their negative groups. His method was refined in 1837 by J. Dana of Yale University. Dana's system is still in use today.

The eight Dana mineral classes are given in Table 6-1. A mineralogist who has learned the ordinary physical properties of minerals can classify and identify many minerals on sight. For others he uses chemical tests or instrumental means such as the spectroscope.

Three kinds of rocks

A *rock* is a naturally occurring mixture of minerals. Geologists classify three kinds of rocks according to the processes that deposited them. In the earth's crust 80 percent of the rocks are termed *igneous,* 15 percent *metamorphic,* and 5 percent *sedimentary.* On the earth's surface 75 percent of all rocks are sedimentary.

Igneous (L. *ignis,* fire) rocks are products of melting and crystallization. Continental crusts (and apparently the moon's surface, too) are mainly basalt and granite, which are silicates of calcium, magnesium, iron, aluminum, sodium, and potassium. Their origin is *magma* or molten silicate rock material brought up by volcanism. As magma cools, compounds crystallize

Table 6-1. A classification of minerals

No.	Class	Mineral example	Formula	Chemical name
1.	Native elements	Gold	Au	Gold
2.	Sulfides	Galena	PbS	Lead sulfide
3.	Oxides, hydroxides	Hematite	Fe_2O_3	Ferric oxide
4.	Halides	Sylvite	KCl	Potassium chloride
5.	Carbonates, borates, nitrates	Calcite	$CaCO_3$	Calcium carbonate
6.	Sulfates, chromates, molybdates, tungstates	Barite	$BaSO_4$	Barium sulfate
7.	Phosphates, vanadates, arsenates	Pucherite	$BiVO_4$	Bismuth vanadate
8.	Silicates	Willemite	Zn_2SiO_4	Zinc orthosilicate

from it in a sequence that depends on solubility in the melt and the melting points of the compounds. In general, minerals rich in calcium, magnesium, and iron crystallize first, whereas those rich in aluminum, potassium, sodium, and silica (SiO_2) crystallize last. Figure 6-13d illustrates the intimate mixing found in rocks and points up the problem that geologists have in characterizing their materials.

Sedimentary (L. *sedimentum*, settling) rocks form at or near the earth's surface. They come from the cementation (consolidation) of sediments created mechanically, chemically, or biochemically from older rocks. Their composition may be even more varied than that of igneous rocks. They are called *clastic* (see Figure 6-13b) if most of the fragments in them were transported piece by piece to the compaction zone by water, wind, mud flow, or ice, and *chemical* if a majority of the material was derived from water solution. Because the sequence of weathering, transport, deposition, condensation, and cementation varies in rate, sedimentary rocks occur in layer formations such as the bands of limestones and sandstones seen in the Grand Canyon. Iron ore deposits are products of precipitation processes that may go back to very early earth history when iron was less oxidized and more soluble.

Metamorphic (Gk. *meta + morphe*, altered form) rocks are made from other rocks by mechanical effects of heat and pressure that break and deform them, and by chemical effects. The latter at higher temperatures may include reactions with neighboring material that cause new crystalline masses to appear. Metamorphic processes take place at depths too shallow for complete melting but too deep for weathering. The chemical reactions are often attended by loss of volatile substances such as water. A typical metamorphosis is the recrystallization of limestone to denser marble (both $CaCO_3$).

The geochemical cycle

Over long periods of time, events that crustal materials experience are repetitious. The sequence, called the *geochemical* or *rock cycle,* is described by Figure 6-1. The cycle begins at the lower left of the figure with molten silicates intruding from the mantle into the crust. They may solidify in the crust and be uplifted later, or pour out of volcanoes and solidify as surface igneous rock. Igneous rock may be changed into metamorphic rock or un-

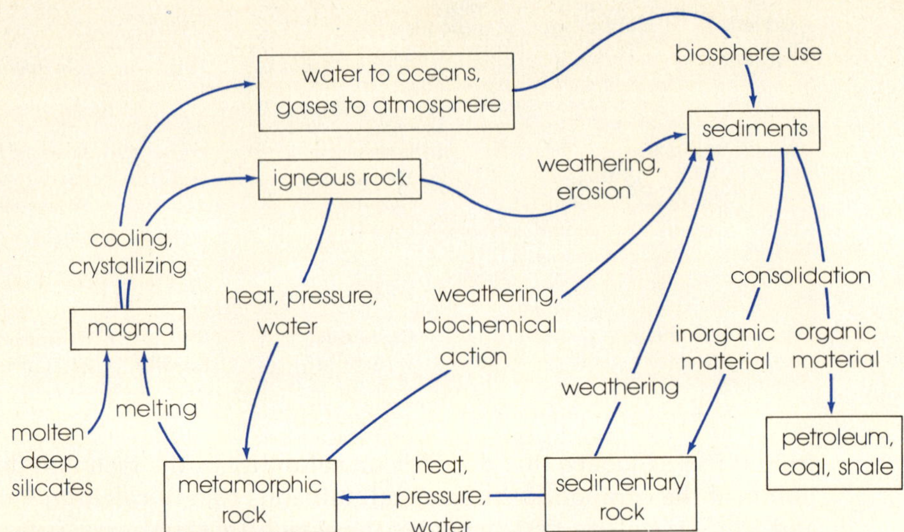

Figure 6-1 The geochemical cycle. The cycle provides a basis for discussing many aspects of geology. One aim of geochemical research is to follow individual elements through the cycle. (Adapted from Brian Mason, Principles of Geochemistry, 3rd ed. Wiley, New York, 1966, p. 286.)

dergo breakup due to weathering. Weathering and biochemical action on metamorphic rock give sediments. Sediments rich in organic material, if covered sufficiently deep and long, are converted to oil shale, petroleum, and coal. By contrast, inorganic sediments, subjected to pressure and to water that contains cementing chemicals, are compacted into sedimentary rock. Mechanical rearranging and recrystallization of sedimentary rock produce metamorphic rock, and weathering remakes sediments. When metamorphic rock is remelted in the lower crust, the cycle has come full turn.

The cycle also explains *soil* formation. Soil is sediment that has been decomposed further by water, air, and the action of vegetation and microorganisms. Inspection of the sides of trenches cut through the ground to bedrock reveals that soils are generally differentiated into three layers or *horizons*. The *A* horizon on top is depleted by rain of the more soluble elements, sodium, potassium, calcium, and magnesium. Middle horizon *B* contains fine clay and some elements leached from *A*. Bottom horizon *C* is rocky particles in the process of becoming *A* and *B*. Soil fertility depends on chemical composition, moisture, permeability, microorganisms, other organic constituents, and the particle-size distribution.

The geochemical cycle also helps explain how minerals become concentrated. An example on a globe-girdling scale is found where excessive rainfall has so depleted soil that only a clay binder with oxides of silicon, aluminum, iron, and titanium remains. The richest aluminum and iron mines operating today are in countries around the equator. A local example of mineral concentration is found in the Green River area of Colorado, Utah, and Wyoming where black shale deposits exist. They were formed by sedimentary processes that compressed mud and organic material into beds containing petroleum hydrocarbons. Locked in an area of about 16,000 square miles are deposits estimated at 2 trillion barrels of oil. The deposits contain 6 times the known oil reserves of the Middle East and enough to supply the United States' demand for 150 years. Exploitation awaits practical methods for getting the oil out (using nuclear heat for distillation?) and managing attendant environmental problems.

Mineral analysis by instrumental means

Mineral analysis is complicated because minerals commonly are *solid solutions*. A solid solution is a homogeneous crystalline solid of variable composition. Thus olivine, $(Mg,Fe)_2SiO_4$, can be considered as having formed when molten Mg_2SiO_4 and Fe_2SiO_4 mutually dissolved in one another, giving a series of solid solutions with compositions between those of the two pure components. Some solid solution tendency is evidenced in practically all minerals. A chemical analyst therefore is always faced with trying to find minor constituents in the presence of much larger amounts of the more common elements. Three instruments that have aided in solving these, as well as other chemical problems, are discussed below.

With the *electron microprobe* (Figure 6-2), spot-by-spot analysis of tiny volumes (typically 10^{-15} cm³) can be made in sample surfaces, thus eliminating the laborious task of physically separating tiny grains for examination. As an electron beam excites specimen atoms, each element heavier than carbon yields an X ray spectrum which is analyzed with a spectrometer. The detection limit of a given element is about 100 parts per million parts of sample. In the small volume analyzed this may correspond to measurement of only 10^{-14} g of element. Few instruments have such great sensitivity.

The *spectrograph* (Figure 6-3), in principle a spectrometer equipped for picture taking, has already been mentioned (with Figure 5-1). As used in mineral analysis, a few milligrams of sample is burned between carbon electrodes of a high-temperature electric arc. Light from the incandescent sample is passed through a slit, its components separated by means of a prism or other dispersing element, then focused on a strip of photographic film. Line positions identify an element, and density of a line gives the element's concentration. All metals, carbon, silicon, boron, and phosphorus are detectable, most at concentrations down to 10^{-3} percent.

The *flame spectrophotometer* (Figure 6-4) is an instrument used for analyzing elements whose spectra are excited when dilute sample solutions are sprayed into a small flame. In the *emission-flame* mode of operation, atoms that have absorbed energy reemit the energy as characteristic wavelengths when they return to their normal unexcited state. For each element, a wavelength isolated by a prism-slit combination is allowed to fall on a photoelectric tube whose response is proportional to the element's concentration. In the *absorption-flame* mode, better known as *atomic absorption* (or AA), sample solution is drawn into a bushy flame. A light beam produced in a lamp containing the *analate* (element to be analyzed) is sent through the flame toward the detection system. The beam is characteristic only of the analate. Therefore, electrons in analate atoms absorb its energy. The difference in intensity between the beam entering and leaving the flame is read out as analate concentration.

Metallic elements respond best in flame analysis. In the atomic-absorption mode, sensitivities are not uncommonly in the parts per billion range (1 ppb = 1 μg of metal ion in 1 liter of water).

(a)

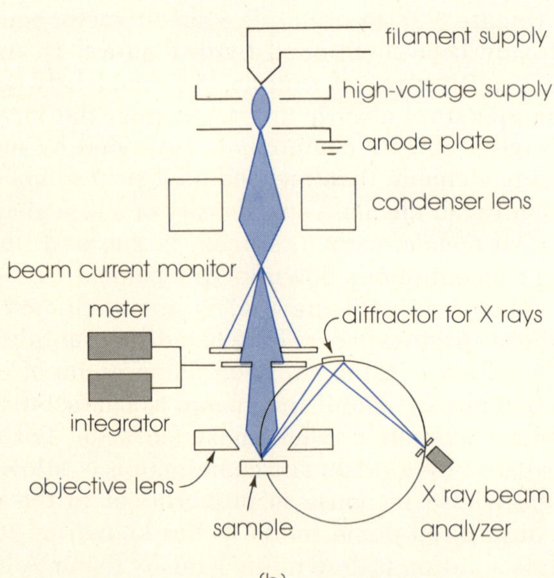

(b)

Figure 6-2 The electron microprobe X ray analyzer. (a) A modern instrument. In the middle are the electron gun, electron detector, light microscope, specimen chamber, scanning mechanisms, and their controls. At left is the readout and display console which includes recorder, programing controls, and power supplies. At right is the control console which houses high-voltage power supplies and the vacuum control. (b) Schematic of the electron optical system. The electron beam from the filament is focused by electromagnetic lenses on a spot of the sample's surface chosen by the operator using the ordinary light microscope. The beam excites X ray spectra which are analyzed. (Courtesy Applied Research Laboratories.)

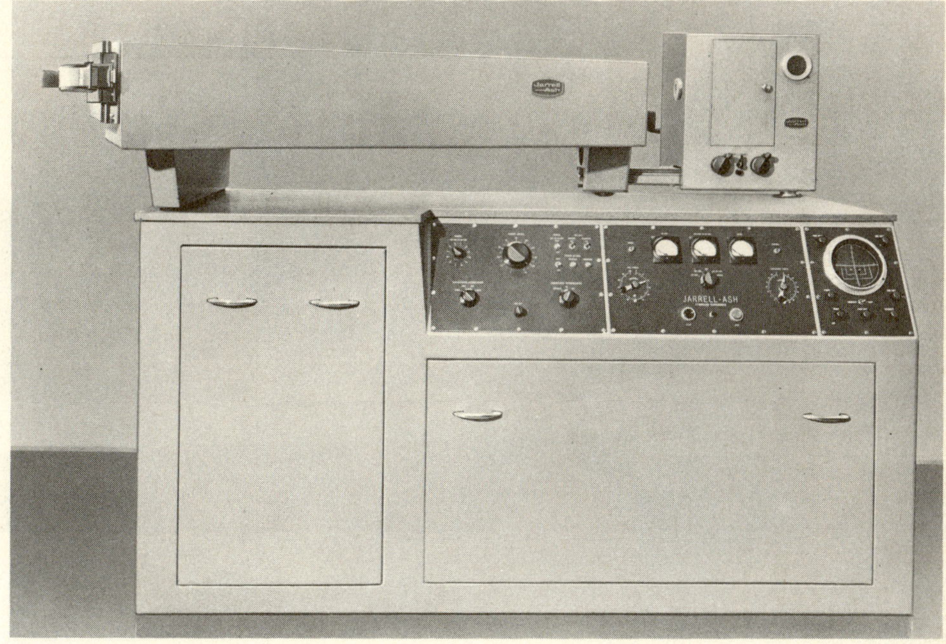

(a)

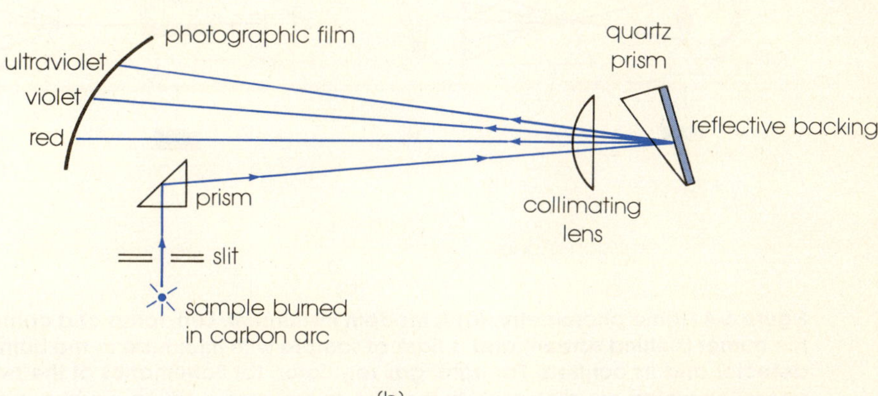

(b)

Figure 6-3 The spectrograph. (a) A modern instrument. Top left, film holder; top right, housing for electrodes where sample is burned; bottom, control cabinet. (b) Schematic view of the optics in a prism spectrograph. (Courtesy Jarrell-Ash Division, Fischer Scientific Company.)

6-2. CRYSTALS

Solids are distinguished by rigidity and frequently by crystallinity. By their nature, *crystals* (Gk. *krystallos*, ice) have sharp melting points and symmetrically arranged plane surfaces that intersect at definite angles. Crystal sizes vary with crystallization conditions. Small crystals may be obtained in the laboratory by mixing concentrated solutions that produce many crystal nuclei but little crystal growth. Larger crystals are obtained by mixing dilute solutions and allowing time for the fewer nuclei formed to increase

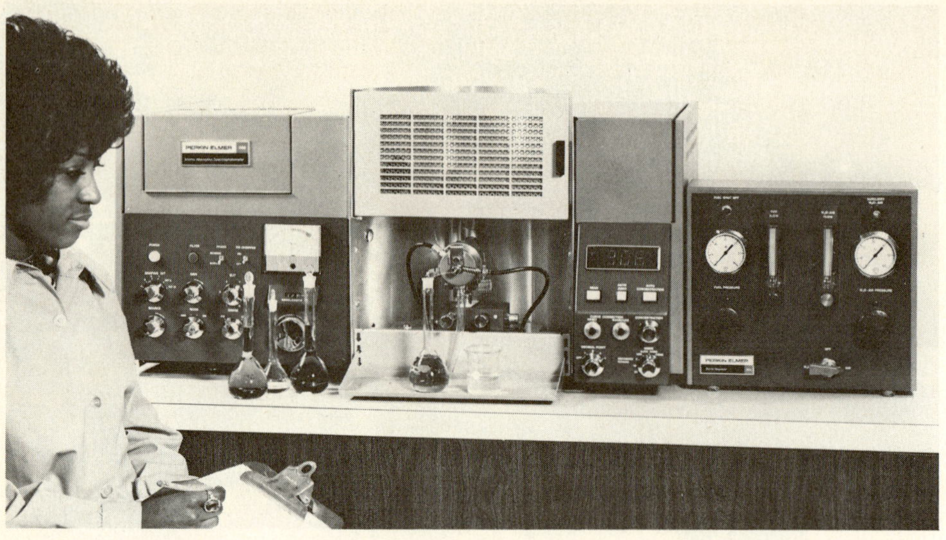

(a)

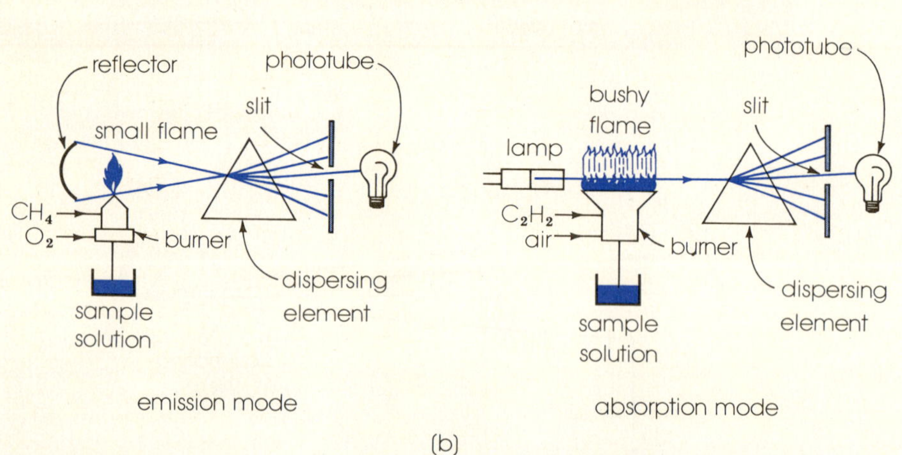

(b)

Figure 6-4 Flame photometry. (a) A modern instrument. Left, lamp and controls; center, the burner (behind screen) and a flask of sample with inlet tube to the burner; right, detector and its controls. Far right, gas regulator. (b) Schematics of the two modes of operation which are discussed in the text. In emission work an oxygen–acetylene flame is often used; in absorption work the flame is usually air–acetylene. Wavelength range in the latter mode is 2000 to 8500 Å, which is somewhat broader than the former. (Courtesy Perkin-Elmer Corporation.)

in size. Crystals found in the earth vary from microscopic grains to specimens with dimensions measured in feet, the latter having come from slowly evaporating solutions or slowly cooling melts.

Solids such as glass and plastics that lack characteristic form and soften over a temperature range are called *amorphous*. They are considered liquids cooled below the point where they cease to flow.

Various physical properties of crystals, such as ability to cleave (split) along certain planes and mechanical resistance to compression, depend on the crystal direction on which the stress is placed. (In amorphous materials these properties are the same in all directions.) The inference is that the readily observed geometric shapes of crystals are indicative of a repeating structural order down to the tiniest units from which they are built.

Examining crystals with X rays

Before much chemistry was known, mineralogists had worked out schemes of identifying minerals based on form, color, hardness (see Question 1), cleavage, luster, and density. Around 1800 the first instruments were invented for measuring crystal surface angles. Crystallographers thought geometric form was related to both chemical composition and atomic arrangement, but they had no means of proving it. The discovery of X rays gave them the tool they needed.

In 1914 Max von Laué (Germany 1879–1960, Nobel prize in physics 1914) and his students showed that a single crystal could act as a diffraction grating for X rays because parallel planes of atoms in the orderly internal arrangement lay at distances comparable to X ray wavelengths. (The diffraction effect has been illustrated in Figures 3-13 to 3-17.) X ray diffrac-

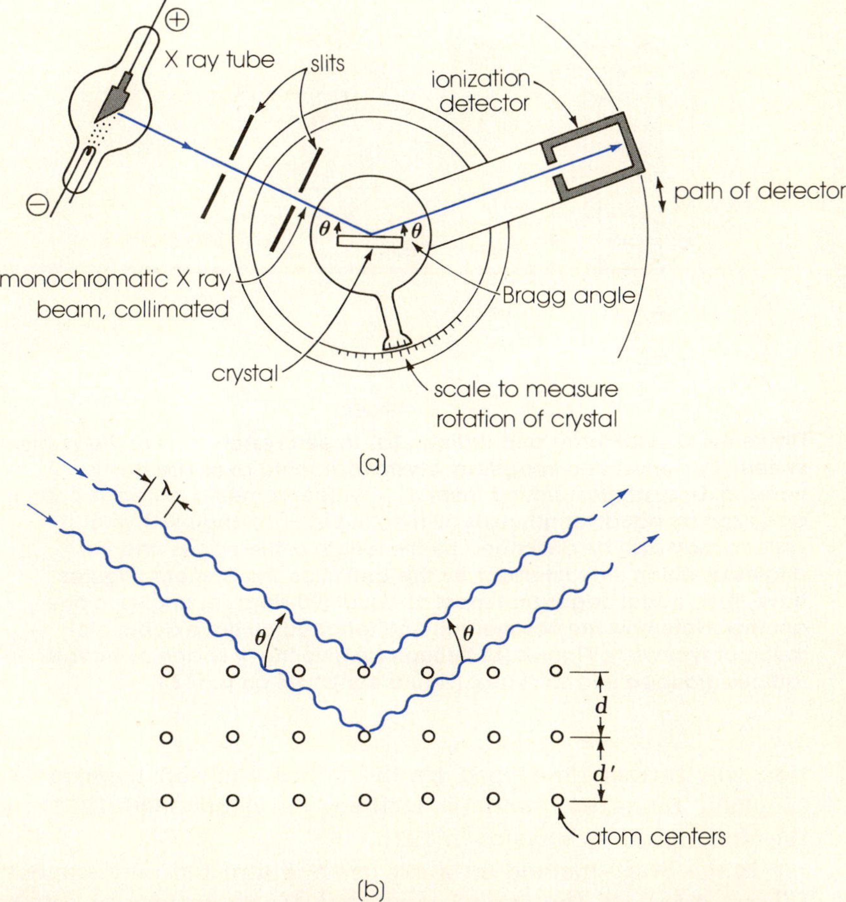

Figure 6-5 The Bragg method. (a) Schematic view of the equipment. (b) X ray reflectance (scatter) from planar arrays of atoms in a crystal lattice. As the crystal is turned the strength of the reflected beam suddenly grows and fades depending on the effectiveness of reinforcement from successive planes. (See Problems 12 and 18.)

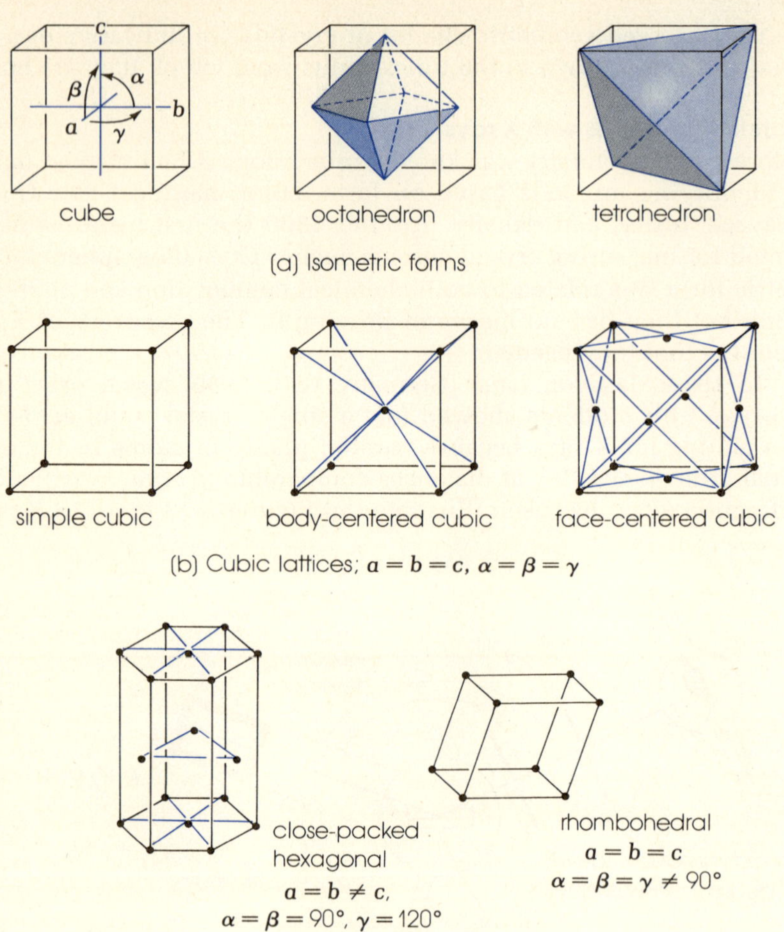

Figure 6-6 Crystal forms and lattices. (a) Three crystal forms of the cubic system. By convention imaginary crystallographic axes are designated a, b, c, angles around them α, β, γ. The isometric system is characterized by equal length axes all meeting at right angles. Crystal systems may also be described by the length of the edges and the angles at which they intersect. By this definition the isometric figures have three equal edges that meet at equal (90-degree) angles to one another. Note how the octahedron and tetrahedron have cubic elements of symmetry. Figures (b) through (g) give the 14 space or Bravais lattices grouped into six types. (Figure continues on p. 133.)

tion was further developed by the father and son team of W. H. Bragg (England 1862–1942) and W. L. Bragg (England 1890–1971), who shared the Nobel prize in physics in 1915.

In the Bragg method an X ray beam is directed on a single crystal face (Figure 6-5a). As the crystal is rotated, its characteristic pattern of X ray reflectance is followed with an ionization detector such as a Geiger counter. At various angles reinforcement of reflectance from a particular series of parallel atomic planes in the crystal takes place, giving a peak signal. At other glancing angles X ray waves cancel and the signal is weak.

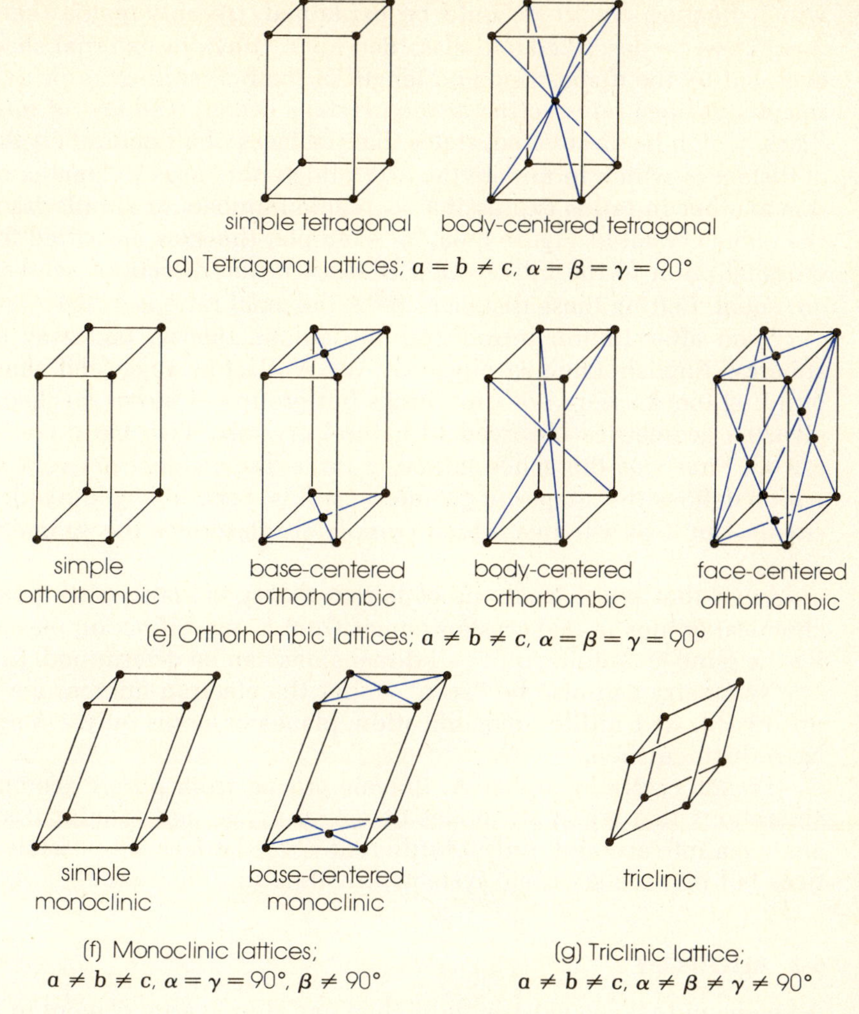

(d) Tetragonal lattices; $a = b \neq c$, $\alpha = \beta = \gamma = 90°$

(e) Orthorhombic lattices; $a \neq b \neq c$, $\alpha = \beta = \gamma = 90°$

(f) Monoclinic lattices; $a \neq b \neq c$, $\alpha = \gamma = 90°$, $\beta \neq 90°$

(g) Triclinic lattice; $a \neq b \neq c$, $\alpha \neq \beta \neq \gamma \neq 90°$

For the condition of reinforcement it can be shown by trigonometric construction related to Figure 6-5b that

$$n\lambda = 2d \sin \theta \tag{6-1}$$

where n is called the X ray order (1, 2, 3, . . .), λ is X ray wavelength, θ is the reflecting angle, and d is the distance between planes of atoms doing the reflecting. After d values are found, a computer is often used today to correlate them to obtain the desired three-dimensional *conformation* (spatial arrangement) of atoms in the crystal. X ray diffraction is one of the most powerful means we have for investigating structure. (For a biochemical example see Figure 23-17.)

Crystal systems

Twenty years before Dalton, French Abbé R. Haüy founded crystallography, the science of crystal forms. Haüy reasoned that if nature constructed crystals from tiny multifaced units (polyhedra) stacked in space-filling parallel fashion, then interfacial angles, faces, symmetry, and the planes along which fracture occurred could be explained. He anticipated that crystals would best be described and classified not by obvious external shape, however, but by the disposition and length of their *crystallographic axes*, three imaginary lines intersecting at the crystal's center. The *law of rational indices*, which he discovered, states that the faces of all natural crystals occur at distances which terminate the crystallographic axes in lengths related to one another in ratios expressible as whole numbers or simple fractions. In the cubic crystal of Figure 6-6a, for example, the axes measured from their crossing point to the intersection of faces in the directions marked a, b, c are equal. Letting these distances be 1, the axial ratio is $1:1:1$.

Soon after Dalton introduced the atomic theory, picturing atoms as spheres, English chemist-physicist W. Wollaston suggested that Haüy's building blocks were not tiny bricks but groups of atoms stacked so as to give the geometries observed in natural crystals. Two basic concepts followed. First was the *space lattice*, a three-dimensional network of points which defines the shapes of geometric solids. Second was the *unit cell*, the smallest unit of a lattice which completely describes the geometry of the array.

Proof that point locations of lattice theory actually mark positions of chemical groups in real crystals comes from X ray diffraction measurement. By the same technique, unit cell dimensions can be determined. Crystal lattice symmetry can also be "seen" under the electron microscope (see Figure 13-13). At 1 million magnification, planes of atoms only 2 Å apart have been distinguished.

French crystallographer A. Bravais proved from purely geometric considerations that when all modes of mirror-image and rotational symmetry are taken into account, only 14 different space lattices are possible. The lattices fall into the six basic systems illustrated in Figure 6-6.

6-3. ALLOTROPY

An element that can exist in more than one atomic arrangement in the same physical state is called *polymorphous* ("several forms") or *allotropic* (Gk. *allos* + *tropos,* different direction). Physical properties of allotropes may differ greatly. Two allotropic elements found in the earth are carbon and iron. Their properties reflect their crystal structure.

Structures of carbon

Carbon occurs in two allotropic forms, *diamond* and *graphite*. (*Charcoal,* which is seemingly different, has the graphite structure.) In the form of diamond, carbon is the hardest natural mineral; as graphite, it is one of the softest.

Diamond is made up of tetrahedrally oriented carbon atoms. As shown in Figure 6-7a internal structure consists of puckered rings containing six carbon atoms, each carbon of which is attached to four other carbons. In effect every diamond crystal is one giant molecule of carbon. Hardness is due to a compact structure in which strong bonds of equal length exist between atoms.

Graphite (Figure 6-7b) is representative of substances having *layer lat-*

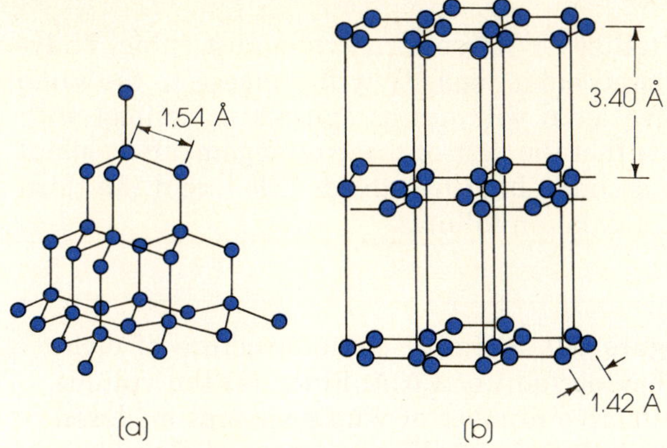

Figure 6-7 Carbon structures. (a) Diamond, a tetrahedral arrangement. (b) Graphite, showing alignment of alternate sheets of atoms. In both (a) and (b) each carbon is bonded to four other carbons.

tice structure. Instead of forming puckered hexagons, the carbon atoms form flat hexagons. Carbon–carbon distance in the plane of each layer is comparatively small and attachment is strong. Distance between layers, however, is more than twice as great and attachment between layers is weak. The slippery feel of graphite is due to slippage of its crystals one over another. Graphite is found in older rock formations and in some meteorites. It is prepared industrially in an electric furnace.

Diamond is more dense than graphite. Natural diamonds are found in tubular formations called *kimberlite pipes,* whose roots probably reach into the mantle. Diamonds can be synthesized by putting graphite under high pressure.

Structures of iron

Up to a temperature of 912°C pure iron has a body-centered cubic lattice called *alpha-iron.* Its unit cell contains a central atom surrounded by eight other atoms (Figures 6-6b, 6-8a). At 912°C alpha-iron changes to gamma-iron, an allotrope having the face-centered arrangement (Figures 6-6b, 6-8b). Atoms are more closely packed in gamma-iron and the allotropic change is marked by a change in density (7.9 to 8.5 g/cm^3).

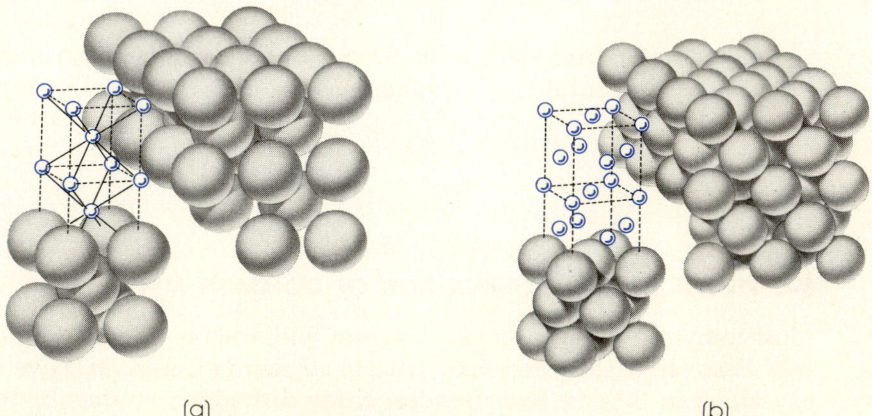

Figure 6-8 Iron structures. (a) Ordinary or α iron, with its body-centered cubic lattice. (From COLLEGE CHEMISTRY, 3rd. edition, by Linus Pauling, W. H. Freeman and Company. Copyright © 1964.) (b) With its face-centered lattice, γ iron.

Structure and density

If an allotropic element crystallizes in three forms, simplest cubic, body-centered cubic, and face-centered cubic, density will increase in the same order. If we represent effective atom volumes as spheres in contact with one another, it is easy to show that the first of these arrangements is about half filled with "atoms," the second about two-thirds filled, and the third about three-fourths filled. A calculation follows.

Example 6-1 Consider Figure 6-9, a simple cubic structure of eight identical spherical atoms having radii of 1.0 Å. Find **(a)** the volume of the unit cell, **(b)** the effective number of whole spheres enclosed by the cube that defines the cell, and **(c)** the percentage of the cube's volume occupied by spheres.

Solution **(a)** Because sphere radius is 1.0 Å and the spheres touch each other, the distance between sphere centers is 2.0 Å. This is the length of the side of the unit cube or cell. Cell volume is

$$V = (2.0 \text{ Å})^3 = 8.0 \text{ Å}^3$$

(b) Imagine the reference cell of Figure 6-9 as a small interior part of a stacked structure of similar cells that completely enclose it. To see that any corner of the reference cell is also the corner of seven other identical cells that intersect at that lattice point, consider atom x in Figure 6-10. It is shared by the four cells pictured, as well as by four other cells that have been removed from the front of this structure. Each sphere of the reference cell, therefore, is shared equally among a total of eight cells. In effect then only one-eighth of each sphere in any cell is inside the cube that describes that cell. The effective number of whole spheres enclosed by the unit cube is

$$\tfrac{1}{8} \times 8 \text{ spheres} = 1 \text{ sphere}$$

(c) The volume of one sphere is

$$V = \frac{4\pi r^3}{3} = (4)(3.14)\frac{(1.0 \text{ Å})^3}{3} = 4.2 \text{ Å}^3$$

Cube volume, from part (a), is 8.0 Å³. Sphere volume within the cube is 4.2 Å³. Percent filling by spheres is

$$\frac{4.2}{8.0} \times 100 = 52 \text{ percent}$$

6-4. STRUCTURE AND COMPOSITION OF COMMON MINERALS

Most minerals contain metals, oxygen, and silicon. Silicon (L. *silex*, flint) was discovered 150 years ago. Silicate structures, however, have only been explained in the last few decades. X ray diffraction studies by the Braggs, Pauling, and others have shown that the unit in silicate chemistry is the *orthosilicate ion*, SiO_4^{4-}. It consists of a central silicon atom surrounded tetrahedrally by four larger oxygen atoms. The wide variety of mineral sili-

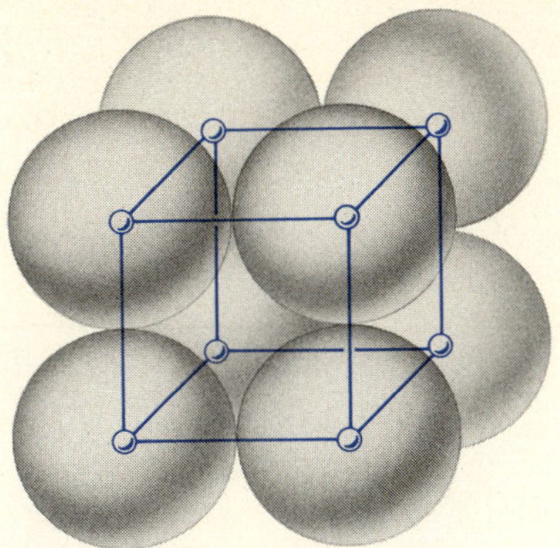

Figure 6-9

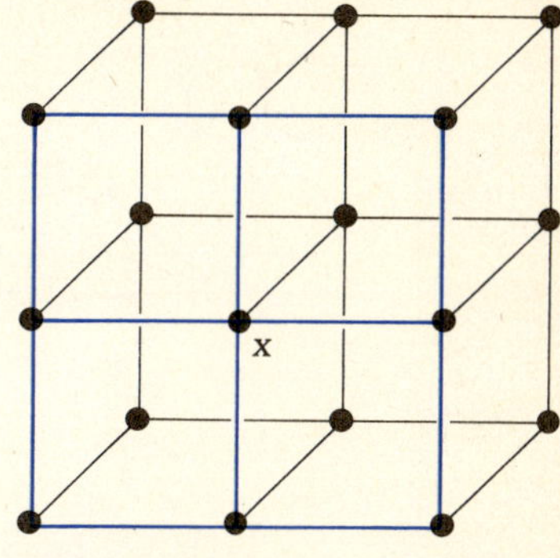

Figure 6-10

cates is explained first by the ability of SiO_4 units to condense (link up) to build large structures by sharing oxygen atoms, and second by the ability of Al^{3+} to substitute for Si^{4+}, giving $(AlO_4)^{5-}$ units that also condense.

Simpler silicates

Simplest silicates, called *orthosilicates*, contain individual SiO_4 tetrahedra. The SiO_4 unit is strongly bonded (Figure 6-11a).

The pairing of two tetrahedra via the sharing of one oxygen atom gives disilicates containing the $Si_2O_7^{6-}$ grouping (Figure 6-11b).

Condensation of six tetrahedra gives a $Si_6O_{18}^{12-}$ group (Figure 6-11c).

Many tetrahedra, each sharing two oxygen atoms, can form very long single chains (Figure 6-11d). The repeat unit is SiO_3^{2-}.

When tetrahedra share three oxygens, long double chains form (Figure 6-11e). The repeat unit is $Si_4O_{11}^{6-}$.

When chains are joined in parallel fashion, they give a sheet structure (Figure 6-11f). The repeat unit is $Si_4O_{10}^{4-}$.

More complex silicates; zeolites

When every oxygen atom in the SiO_4 unit is shared, the three-dimensional network formed is called a *framework silicate*. Most rocks have this structure.

As pictured in Figure 6-12, framework silicates are porous. Depending on relative hole size and chemical characteristics they can permit or block the passage of molecules and accept or not accept ions from solution. Framework aluminosilicates, called *zeolites* (both natural and synthetic) and having the capability of exchanging ions with solutions, are used to "soften" water. Exposed to concentrated brine (NaCl solution), the many holes in the zeolite take up sodium ions. Exposed to "hard" water containing Ca^{2+} and Mg^{2+}, the structure releases sodium ions while acquiring calcium and magnesium ions, and the water is softened. When the holes become saturated with calcium and magnesium, the useful ion exchange capacity is used up. By soaking the zeolite in brine once more, exchange is reversed and the zeolite is again loaded with Na^+ and ready to soften more water:

$$zeolite \cdot 2Na + Ca^{2+} \rightleftharpoons zeolite \cdot Ca + 2Na^+$$
$$\qquad (hard) \qquad\qquad\qquad (soft)$$

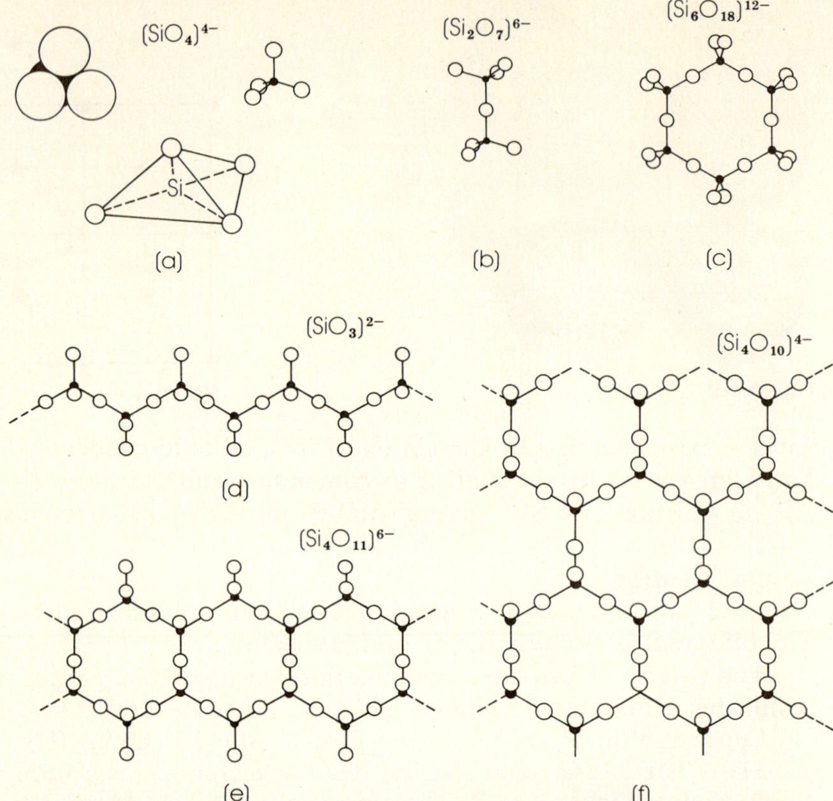

Figure 6-11 Simpler silicate structures. (a) Three equivalent ways to represent the tetrahedral SiO_4 unit. (b) The double-tetrahedra structure. (c) A ring structure in which each silicon atom shares two oxygen atoms. (d) The single-chain structure. (e) The double-chain structure. (f) The sheet structure, a flat hexagonal network.

In each structure the Si—O—Si distance is 3 Å.

Important minerals

Nine minerals form the bulk of the earth's crust. They are listed in Table 6-2.

Feldspars comprise almost half the content of crustal rocks, making them the most common minerals. Feldspar structure is a framework con-

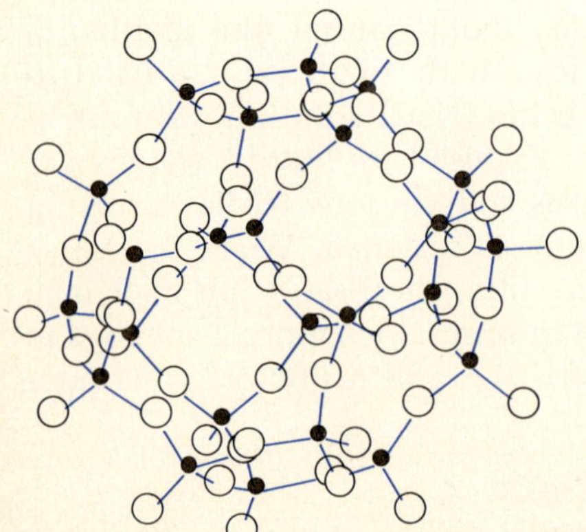

Figure 6-12 A framework silicate. The oxygen atom of every SiO_4 unit is shared between two silicon atoms. With their bonds extending in three dimensions, framework structures are harder and much more difficult to break than chain or sheet structures. As in Figure 6-11, Si atoms are the small dark circles, O atoms, the open circles.

Table 6-2. Commonest rock-forming minerals of the crust

Name	Formula[a]	Name	Formula
Feldspars		**Micas**	
Orthoclase	$KAlSi_3O_8$	Biotite	$(Mg,Fe)_3AlSi_3O_{10}(OH)_2$
Plagioclase	$NaAlSi_3O_8$	Muscovite	$Al_2AlSi_3O_{10}(OH)_2$
Microcline	$KAlSi_3O_8$	**Pyroxenes**	$CaMgSi_2O_6$
Quartz	SiO_2	**Amphiboles**	$Ca_2Mg_5Si_8O_{22}(OH)_2$
Olivines	$(Mg,Fe)SiO_4$		

[a] Except for quartz the formulas are variable. Microcline and orthoclase are different crystal forms of the composition given.

sisting of SiO_4 and AlO_4 tetrahedra with alkali metal ions occupying holes.

Quartz is next to feldspar in abundance (Figure 6-13a). Quartz also has framework structure.

Olivines are magnesium–iron silicates. Containing individual SiO_4 tetrahedra, they belong to the orthosilicate class. Their appearance is glassy, like quartz.

Micas are sheet silicates, as is evident from their cleavage characteristics (Figure 6-13h).

Pyroxenes are chain silicates. Silicon–oxygen chains are parallel to the vertical axis of their crystals. Relatively weak structural links occur at regular intervals where metal atoms hold chains together. Pyroxenes cleave at the weak points to give prism-shaped fragments (Figure 6-13d).

Amphiboles are also chain silicates but their composition is often not simple. Besides calcium and magnesium they may contain iron and aluminum.

Some rock and mineral examples

Representative minerals and rocks are pictured in Figure 6-13.

Granite (a) may be igneous or metamorphic, having crystallized from a magma or from remelted igneous rock. It contains randomly oriented grains of several common minerals, typical of melting–crystallizing processes.

Chert breccia (b) is a sedimentary rock. Light-colored areas are fragments of hard, older rock that is a form of silica called *chert*; the matrix is calcium carbonate.

Gneiss (c) is a metamorphic rock. Light bands are quartz and feldspars; dark bands are minerals like biotite.

Moon rocks (d) indicate that extensive melting took place on that body. The percentages of titanium, zirconium, and yttrium are higher than expected, assuming that the moon formed with the earth (see Figure 5-12).

Halite (e), or sodium chloride, crystallizes in cubes. When a large cube is struck, it cleaves into smaller cubes.

Quartz (f) is the most common simple mineral known and a major constituent in many igneous and metamorphic rocks. It crystallizes in the hexagonal system with ends terminated by rhombohedra; SiO_4 units are arranged in right- or left-handed spirals which give quartz special optical properties (see Figure 20-2). Because quartz passes ultraviolet light more readily than does glass, components are cut from it for use in scientific instruments (Figure 6-3b).

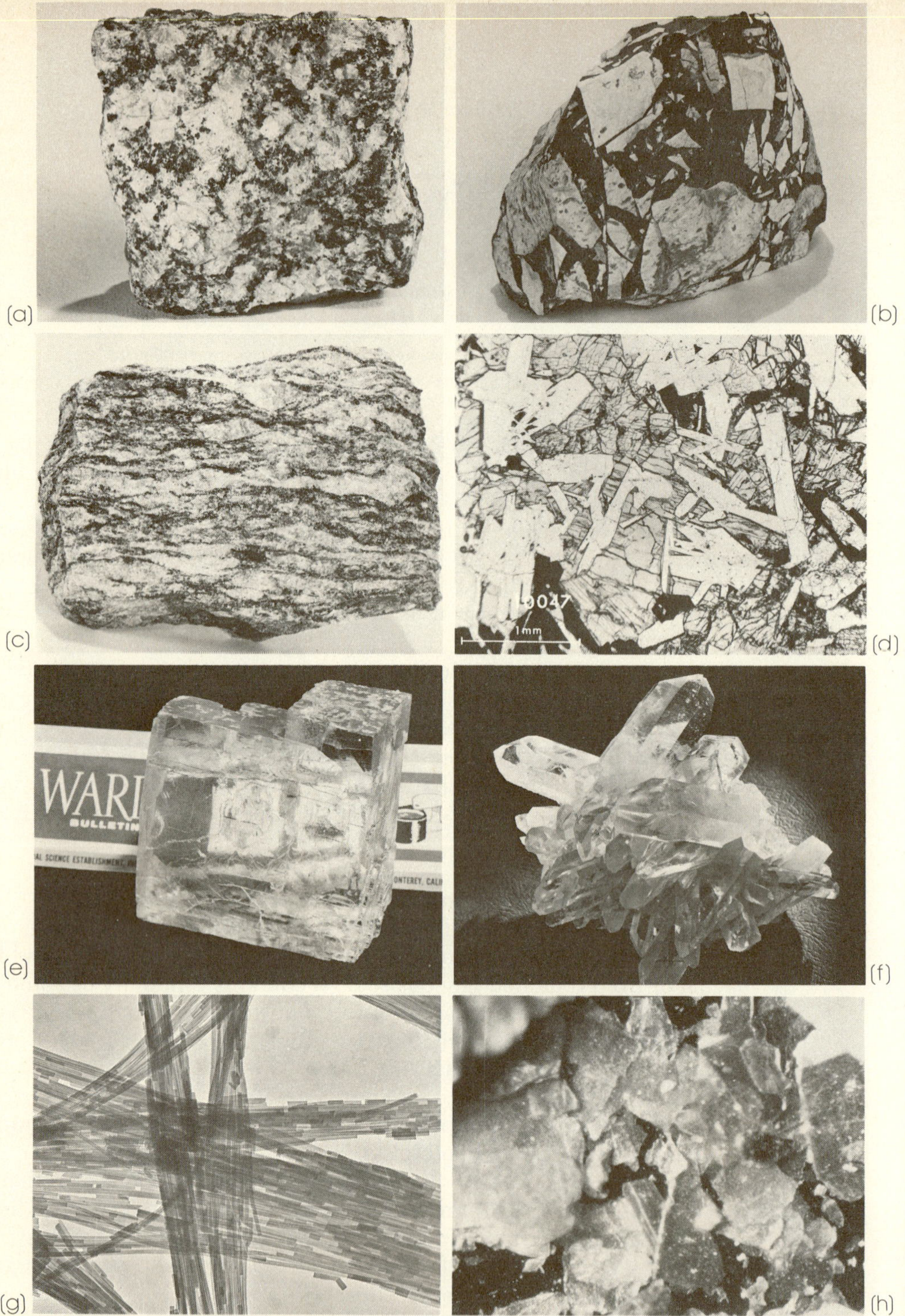

Figure 6-13 Representative minerals. (a) Granite. The individual grains of quartz, biotite, and feldspar (Table 6-2) can be seen with a hand lens. (b) Chert breccia. (c) Gneiss. (d) Lunar rock ($\times 65$), showing light-colored feldspar, grey and fractured pyroxene, and black ilmenite $FeTiO_3$. (e) Halite. (f) Quartz. (g) Chrysotile asbestos. (h) Mica. (Photographs: a, b, c, e, f, Ward's Natural Science Establishment, d, Courtesy NASA; g, Courtesy Calidria Asbestos, Union Carbide Corporation, Mining and Metals Division.)

Asbestos (g) is an amphibole. Silicate chain structure explains its fibrousness. Asbestos is used to wrap furnace pipe and make roofing shingles. Asbestos dust may cause lung cancer.

Mica (h) is a sheet silicate. The clear, flexible sheets produced by cleavage are used to insulate electrical elements.

QUESTIONS

1. Hardness Table 6-3 shows Mohs' scale of mineral hardness. Experiment shows that a fingernail will barely scratch gypsum, a copper penny will barely scratch fluorite, a piece of broken glass will easily scratch apatite but not orthoclase. Assign Mohs hardness numbers to fingernail, copper, and glass.

Table 6-3. Mohs' scale

Hardness	Mineral
1	Talc, $Mg_3Si_4O_{10}(OH)_2$
2	Gypsum, $CaSO_4 \cdot 2H_2O$
3	Calcite, $CaCO_3$
4	Fluorite, CaF_2
5	Apatite, $Ca_5(PO_4)_3(OH)$
6	Orthoclase, $KAlSi_3O_8$
7	Quartz, SiO_2
8	Topaz, $Al_2SiO_4(OH)_2$
9	Corundum, Al_2O_3
10	Diamond, C

2. Graphite Graphite is not a good lubricant in the vacuum of outer space. A theory advanced to account for this assumes that air molecules normally occupy space between layer lattice sheets and the layers slide over them (like sheets of plywood rolling on ball bearings). Describe experiments that might be conducted to test the theory.

3. Definitions What is the difference between **(a)** minerals and rocks? **(b)** crystalline and amorphous solids? **(c)** igneous and sedimentary rocks? **(d)** a body-centered and a face-centered lattice? **(e)** silica and a silicate?

4. Crystals **(a)** You are given a solid to examine. What tests and observations might you make to find if it is crystalline or amorphous? **(b)** One granite sample consists of fine crystals, another of large crystals. Which probably cooled more slowly when it was formed? Explain.

5. Crystal system **(a)** The mineral albite ($NaAlSi_3O_8$) occurs in crystals that yield these data: $a:b:c = 0.63:1:0.56$, $\alpha = 94$ degrees, $\beta = 116$ degrees, $\gamma = 89$ degrees. To what crystal system does it belong? Explain. **(b)** A zircon ($ZrSiO_4$) crystal gives these data: $a:b:c = 1:1:0.64$, $\alpha = \beta = \gamma = 90$ degrees. Explain.

6. Mineral classes To what class (Table 6-1) does each of the following minerals belong? **(a)** scheelite $CaWO_4$; **(b)** graphite C; **(c)** wollastonite $Ca_3Si_3O_9$; **(d)** uranite UO_2; **(e)** cryolite Na_3AlF_3; **(f)** olivine $(Mg, Fe)_2SiO_4$; **(g)** smithsonite $ZnCO_3$; **(h)** pyrrhotite FeS. **(i)** colemanite $Ca_2B_6O_{11} \cdot 5H_2O$; **(j)** Chile saltpeter $NaNO_3$.

7. Lava **(a)** If given a sample of lava, how could you determine at what temperature it quit flowing and solidified? **(b)** How could you determine by inspection that the lava is neither a sedimentary nor a metamorphic rock? **(c)** How can weathering make soil from lava?

8. Sand Both talc and sand are composed primarily of silicon and oxygen. Explain what structural differences cause the former to be soft and slippery, the latter hard and gritty.

9. Structure Chemists have synthesized solids, called *molecular sieves,* from the oxides of silicon, aluminum, and sodium. These materials selectively pass small gas molecules but not larger gas molecules (thus helium can be separated from methane with which it occurs naturally). What is probably the structure of the sieve? Explain.

10. Silicates These minerals are all described in general structural terms in Figure 6-11. What structure do you expect in each? zircon $ZrSiO_4$, tremolite $Ca_2Mg_5Si_8O_{22}(OH)_2$, diopside $CaMgSi_2O_6$, beryl $Be_3Al_2(Si_6O_{18})$, hemimorphite $Zn_4Si_2O_7(OH)_2 \cdot H_2O$.

PROBLEMS

11. Soil Draw a cross-sectional picture of a typical soil from surface to bedrock. Label the layers as described in the chapter and indicate a few chemical components.

12. Atom planes **(a)** Using graph paper or a metric ruler, prepare an accurate enlargement of Figure 6-14, making distance d_2 twice the length of d_1. In other colors draw three other differently spaced sets of parallel straight lines (two lines per set are sufficient) through the lattice to indicate various rows of atoms which in a crystal lattice would be planes for X ray reflection. Label the sets d_3, d_4, d_5. **(b)** List the five sets in decreasing order

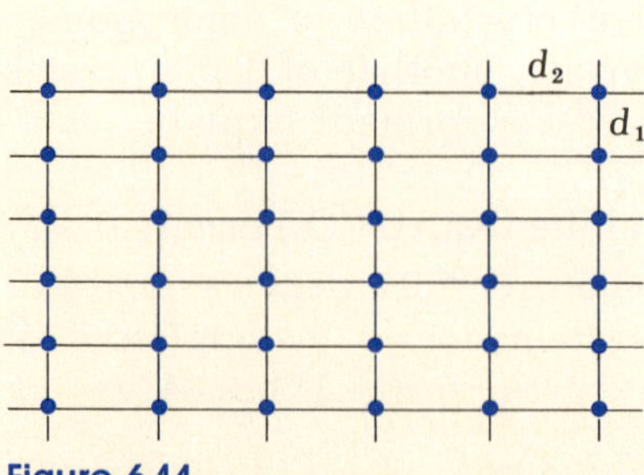

Figure 6-14

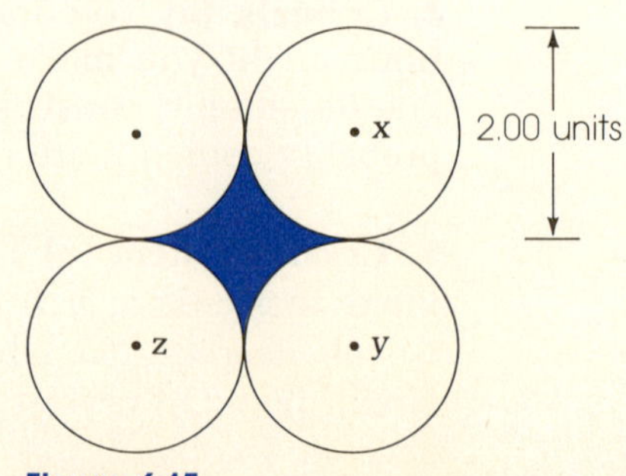

Figure 6-15

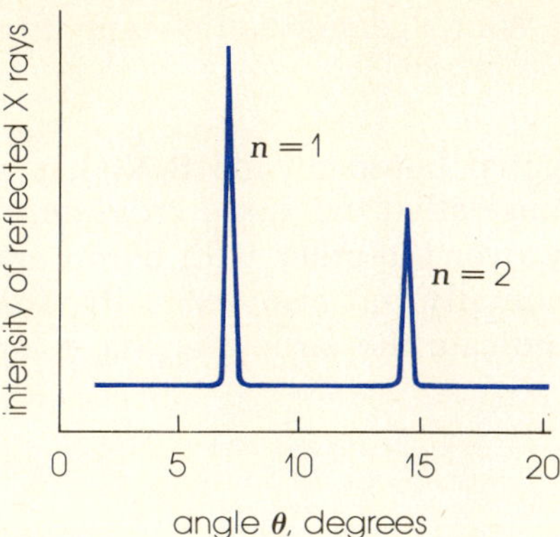

Figure 6-16

of atom population (which in X ray diffraction analysis would mean decreasing intensity of reflection). With the ruler measure the centimeter distance between lines in each set. Give the measurements.

13. Two-dimensional crystal Figure 6-15 shows four circles of equal size in contact. Aided as needed by geometric construction on an enlarged reproduction of the figure, calculate **(a)** the area of one circle; **(b)** distance xy; **(c)** the area of the shaded space.

14. Two-dimensional crystal As in Problem 13, calculate **(a)** the area of the largest circle inscribable in the shaded area; **(b)** distance xz.

15. Sphere packing Use the method of Example 6-1 to determine the number of sphere equivalents contained within the unit cells of the following lattices of Figure 6-6: **(a)** simple orthorhombic; **(b)** base-centered orthorhombic.

16. Shale A certain shale deposit contains hydrocarbons equivalent to 25 gal of crude oil per ton of shale. One cubic foot of shale weighs 125 lb. If 1 barrel of crude oil (1 barrel = 45 gal) in the ground is valued at $1.25, what is the oil value of land covering 10^4 acres (1 acre = 43,560 ft²) in which a uniform shale bed 100 ft thick exists?

17. Silicates Ring silicates are known that contain these units: $Si_3O_9{}^{6-}$ and $Si_4O_{12}{}^{8-}$. Draw each structure, following Figure 6-11 as a model.

18. Bragg method Results related to the Bragg method are shown in Figure 6-16. The X rays used had a wavelength of 0.708 Å. The sample was a crystal of sodium chloride (Figure 6-11e) with a cube face exposed for analysis. The angle of reflection for first order (n = 1) X rays was 7°14′. Calculate the distance between planes of atoms parallel to the cube face.

19. Lattice type Sulfur crystals found around volcanoes have crystallographic axes that all intersect at right angles, and length ratios $a:b:c$ of 0.813:1:1.90. **(a)** Draw the axes to scale, making b 5 cm long, and indicate the angles about them. **(b)** Which general lattice type (Figure 6-6) is represented?

20. Carbonlike structures Hexagonal boron nitride B_3N_3 has a structure similar to graphite. (It is sometimes called *inorganic graphite*.) The rings, however, are packed one directly over the other with boron and nitrogen atoms alternating, not only horizontally but also vertically. Draw several sheets of several rings each to indicate the structure. Make boron atoms dark, nitrogen atoms light.

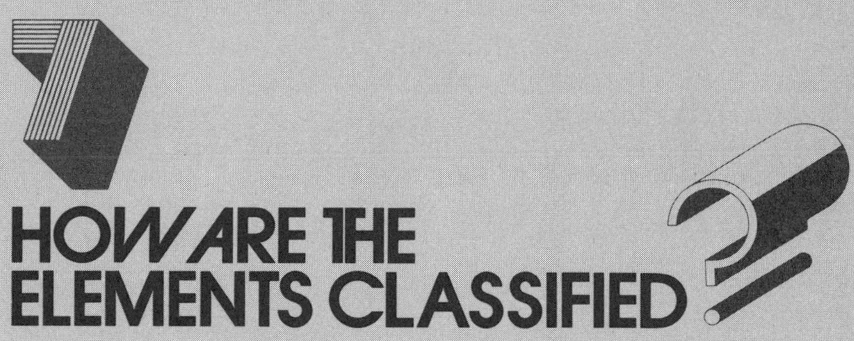

HOW ARE THE ELEMENTS CLASSIFIED

THE IMPORTANT CONCEPTS

7-1. The periodic law
1. A repetition of properties is evident when elements are arranged in order of their atomic numbers.
2. Periodic table trends are noticeable in both families and periods.
   a. The most obvious and useful generalization: Family members are chemically similar.
3. Metals, nonmetals, and metalloids are identifiable in the table.

7-2. Electronic basis of the periodic table
1. There are four quantum numbers.
   a. Three of the numbers are interrelated.
   b. Any electron in any atom is uniquely designated by a combination of the four numbers.
2. An orbital contains a maximum of two electrons.
3. The orbital filling sequence is known.

4. Orbitals have three-dimensional shapes.
5. Orbital filling order is logically represented by blocks of elements in the periodic table.
6. Properties of elements can be correlated with electron structure and periodic table position.

Discovery of the elements (see Figure 1-4) was accompanied by the challenge of cataloguing their properties and finding a rational way to correlate their chemistry. The scheme that does this has been under development for the past century. Called the *periodic table,* it is the most useful compilation chemists have.

7-1. THE PERIODIC LAW

The forerunners

The idea that a natural order might exist for listing the chemical elements was first published in 1829 by the German chemist J. Döbereiner. He stated that in certain three-element groups ("triads") the combining weight (weight combining with a fixed weight of some reference element) of the middle element was approximately midway between the combining weights of the other two. The examples of triads he gave included chlorine–bromine–iodine, and calcium–strontium–barium.

In 1854 Harvard professor J. Cooke published a paper in which he attempted to show that all elements could be classed in six groups if certain algebraic formulas were used to interrelate atomic weights. Döbereiner's triads appeared as members of groups.

In France in 1862 geology professor A. de Chancourtois made a three-dimensional arrangement of the elements, placing them on a line sloping at 45 degrees which wound down a cylinder of paper. Horizontally he arranged 16 columns, vertically he wrote 1, 2, 3, 4, ... Using atomic weights newly established by Cannizzaro he marked the positions of elements on the line in the order of atomic weights. Once again the triads appeared, one below the other. (See Question 1.)

In 1865 English analytical chemist J. Newlands published a seven-column table of the 62 elements then known. He numbered them 1 through 62 and arranged them in order of ascending atomic weight. A repetition of chemical properties often appeared, either in each eighth element, or in each sixteenth. When Newlands named his system *the law of octaves,* he was ridiculed for trying to relate chemistry to the octave interval in music. Discouraged, he quit the work. The same approach by more forceful scientists then produced the organization chemists sought.

The organizers

During 1868–1871, Mendeleev (Figure 7-1) in Russia and J. L. Meyer in Germany, independently of one another and also apparently unaware of Newlands' work, each created a useful correlation table of the elements. Both are credited with discovering the principle of the scheme used ever since, *the periodic law.*

Figure 7-1 Dimitri Mendeleev (Russia 1834–1907). Mendeleev was born in Siberia, the youngest of 17 children. His mother died bringing him to western Russia for an education. He taught at St. Petersberg University where he wrote several textbooks and did research on the properties of elements and the origin of petroleum. Following a dispute with the administration (he supported student demands for university reforms) he resigned his professorship. In later studies on world population trends he concluded that science, technology, and education would rescue mankind from overpopulation problems and that all nations would combine into one confederation. He called the United States "a nation of the future" for its understanding of industrialization. (Picture of the postage stamp first printed in Chemical and Engineering News, Feb. 23, 1970.)

The periodic law states that the properties of the chemical elements are related to their atomic structure and are a periodic function of atomic number. Mendeleev got hints of the law from a study of the chemical and physical properties of elements which he compiled on file cards. Putting the cards in order of increasing elemental atomic weight, he found that properties systematically repeated. Writing the elements in tabular form (omitting hydrogen because it did not fit the scheme), he obtained the eight-column arrangement reproduced in Figure 7-2 which he called the "natural system of elements." He listed his conclusions as follows:

1. The elements if arranged according to their atomic weights exhibit an evident *periodicity* of properties.
2. Elements that are similar as regards their chemical properties have atomic weights which are either of nearly the same value (for example, platinum, iridium, osmium) or which increase regularly (for example, potassium, rubidium, cesium).
3. The arrangement of the elements, or of groups of elements in the order of their atomic weights, corresponds to their so-called *valencies** as well as, to some extent, to their distinctive chemical properties.
4. The elements which are the most widely diffused (in nature) have *small* atomic weights.
5. The *magnitude* of the atomic weight determines the character of the element just as the magnitude of the molecule determines the character of a compound.
6. We must expect the discovery of many yet unknown elements, for example, analogous to aluminum and silicon, whose atomic weights should be between 65 and 75.
7. The atomic weight of an element may sometimes be amended by a knowledge of the contiguous elements.
8. Certain characteristic properties of the elements can be foretold from their atomic weights.

* Valency or chemical combining capacity is explained in Section 9-1.

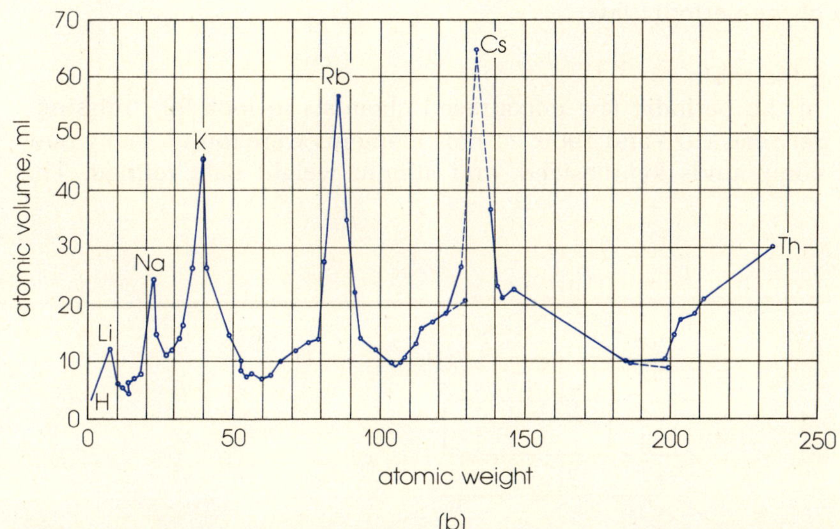

Reihen	Gruppe I — R²O	Gruppe II — RO	Gruppe III — R²O³	Gruppe IV RH⁴ RO²	Gruppe V RH³ R²O⁵	Gruppe VI RH² RO³	Gruppe VII RH R²O⁷	Gruppe VIII — RO⁴
1	H = 1							
2	Li = 7	Be = 9,4	B = 11	C = 12	N = 14	O = 16	F = 19	
3	Na = 23	Mg = 24	Al = 27,3	Si = 28	P = 31	S = 32	Cl = 35,5	
4	K = 39	Ca = 40	− = 44	Ti = 48	V = 51	Cr = 52	Mn = 55	Fe = 56, Co = 59, Ni = 59, Cu = 63.
5	(Cu = 63)	Zn = 65	− = 68	− = 72	As = 75	Se = 78	Br = 80	
6	Rb = 85	Sr = 87	?Yt = 88	Zr = 90	Nb = 94	Mo = 96	− = 100	Ru = 104, Rh = 104, Pd = 106, Ag = 108.
7	(Ag = 108)	Cd = 112	In = 113	Sn = 118	Sb = 122	Te = 125	J = 127	
8	Cs = 133	Ba = 137	?Di = 138	?Ce = 140	−	−	−	− − − −
9	(−)	−	−	−	−	−	−	
10	−	−	?Er = 178	?La = 180	Ta = 182	W = 184	−	Os = 195, Ir = 197, Pt = 198, Au = 199.
11	(Au = 199)	Hg = 200	Tl = 204	Pb = 207	Bi = 208	−	−	
12	−	−	−	Th = 231	−	U = 240	−	− − − −

(a)

(b)

Figure 7-2 Relationships among the elements. (a) Reproduction of Mendeleev's table, published in German in 1872. The R is the general symbol for an element as given in formulas of characteristic compounds at the tops of groups. Missing elements hindered development of the bottom of the table. (b) Reproduction of Meyer's plot of atomic volume versus atomic weight, published in 1870. This was one of several graphs Meyer developed to demonstrate periodicity of various elemental properties. Peak positions are held by the active metals of Group I in Mendeleev's table. (See Problems 11, 12, 14, 15.)

So sure was Mendeleev of his system that he changed the order of a fourth of the elements from that dictated solely by the atomic weights then established to one consistent with chemical properties. Later work vindicated his bold judgments that the atomic weight data had been wrong. Where no known element fit a table position, he left a blank space, assuming that element had not yet been found.

By successfully predicting the properties of unknown elements he removed all skepticism from the system's workability. An example was "ekasilicon" (Sanskrit *eka*, first), the element below silicon in his table

Table 7-1. Mendeleev's predictions for ekasilicon (germanium)

Property	Ekasilicon (theory)	Germanium (found)
Atomic weight	72	72.59
Specific gravity	5.5	5.47
Melting point	High	958°C
Combination with oxygen	EO_2	GeO_2
Oxide melting point	High	1100°C
Oxide specific gravity	4.7	4.70
Combination with chlorine	ECl_4	$GeCl_4$
Chloride boiling point	100°C	83°C

(germanium was discovered 15 years later). A comparison of properties (Table 7-1), predicted and found, shows how well Mendeleev understood the utility of the periodic law.

Completing the table

Discovery of the periodic law encouraged chemists to look for "missing" elements. Between 1870 and 1900, 21 new elements were found, many new inorganic compounds synthesized, and atomic weight data refined. The

Figure 7-3 Pyramid form of the periodic table. This was a 1900 version, when 83 elements were known. It was less popular than the horizontal form.

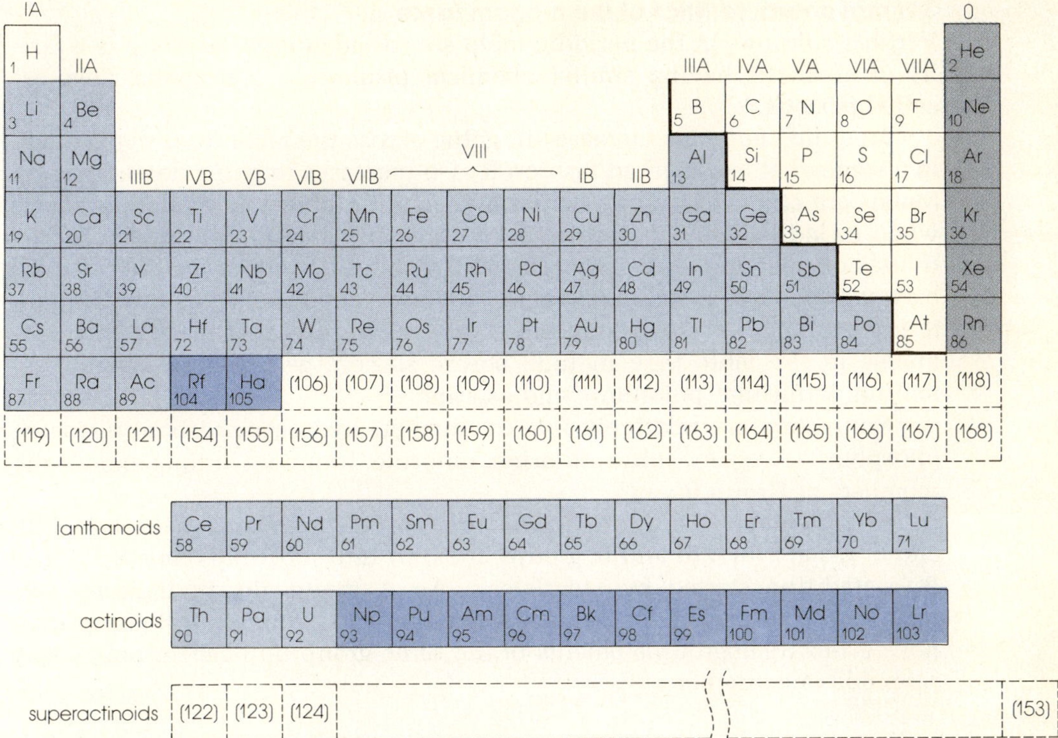

Figure 7-4 A modern long-form table. Spaces are provided (numbers in parentheses) for superheavy elements beyond 105 that may someday be synthesized (see Section 8-4). Color code: light blue, metals; dark blue, transuranium metals; white, active nonmetals; light grey, noble gases.

noble gases, discovered in 1894, were accommodated readily in the Mendeleev table by adding the "zero" column.

A table of pyramidal shape was developed to organize the elements into six completely defined horizontal rows having places for 2, 8, 8, 18, 18, and 32 elements. As shown in Figure 7-3, a seventh period in this configuration was incomplete.

After Moseley's work on X ray spectra (Figure 2-21), *atomic number* was used to order elements in the table. This removed the discrepancies evident when *atomic weight* was used. For example, according to chemical reactivity, tellurium obviously belonged with the sulfur group and iodine with the chlorine group, although atomic weights indicated the reverse order: iodine, 126.9; tellurium, 127.6. The correct atomic numbers are, respectively, 53 and 52.

In 1920 Bohr made a long-form version of the periodic table based on completed shells of electrons. It provided for 2, 8, 8, 18, 18, 32, and 32 elements in successive horizontal rows. To maintain compact organization, elements 58 through 71, and later elements 90 through 103, were put into separate groupings below the table's main body. Following 1939, when the radioactive elements beyond $Z = 92$, and elements 43, 61, and 85 were synthesized by nuclear reactions, they were incorporated to give the table we use today (see Figure 7-4 and inside the back cover).

General characteristics of the modern table

Vertical columns in the periodic table are called *groups*. A group is a family of elements having similar chemical properties. Horizontal rows are called *periods*.

Metallic character increases in going across the table from right to left. In a somewhat less marked fashion it also increases in going down a group. Metals are characterized by luster, malleability (ability to form thin sheets when rolled), ductility (ability to be drawn into wire), and high conductivities of heat and electricity. Elements exhibiting properties intermediate between those of metals and nonmetals are called *metalloids*, or *amphoteric* elements. Located along the stairstep line dividing metals and nonmetals in the table they include boron, silicon, germanium, arsenic, selenium, tellurium, polonium, and astatine.

Nonmetals are relatively few in number: hydrogen, carbon, nitrogen, phosphorus, oxygen, sulfur, fluorine, chlorine, bromine, iodine, and all the members of Group 0.

The elements are divided into nine groups numbered I through VIII, and 0. Except for VIII and 0, groups are split into *subgroups marked A and B*, a tradition started by Mendeleev. An A group always includes two members from periods containing eight elements, whereas a B group does not. A and B subgroup elements of the same group number do not closely resemble one another.

7-2. ELECTRONIC BASIS OF THE PERIODIC TABLE

Quantum numbers

We discussed in Chapter 3 how numbers called *quantum numbers*, which describe electron energy and motion, were assumed by Bohr and generated mathematically by Schrödinger using the wave equation. We will now see how they form a theoretical foundation for the periodic law.

There are four quantum numbers, designated by letters.

Principal quantum number n denotes main energy levels or quantum shells in which the electron is permitted. Numbering from the level closest to the nucleus, n has integral values:

$$n = 1, 2, 3, 4, 5, \ldots \tag{7-1}$$

Quantum number l is related to electron *angular momentum*; l determines the shape of the volume called an *orbital* in which the electron has reasonable probability of being found as calculated by the Schrödinger equation. Angular momentum varies with energy level so that l depends on n as follows:

$$l = 0, 1, 2, \ldots, (n-1) \tag{7-2}$$

The electron moving about the atomic nucleus generates a magnetic field. The effect is accounted for by m, the *magnetic quantum number*; m depends on l. For a prescribed value of l, m has $2l + 1$ different values; m may have any whole number value between l and $-l$, including zero:

$$m = l, l - 1, l - 2, \ldots, 0, -1, -2, \ldots, -l \tag{7-3}$$

As noted in Chapter 3, the electron acts as if it spins on its own axis while moving about the nucleus. Spin is defined by *spin quantum number* s. Spin is unrelated to the other three quantum numbers and has only two values, depending on the direction of spin in relation to the direction of orbital motion:

$$s = +\tfrac{1}{2} \quad or \quad -\tfrac{1}{2} \tag{7-4}$$

Each electron in an atom is describable by a different combination of the four quantum numbers. This statement is called the Pauli *exclusion principle,* after Wolfgang Pauli (Austria and Switzerland 1900–1958, Nobel prize in physics 1945). The principle is one of the basic rules of quantum mechanics although only theoretical reasons have been found for it. It simply makes atomic structure theory fit experimental facts.

The filling of electron energy levels and sublevels may now be summarized.

1. Quantum number n refers to main levels. In each main level there are n sublevels. Each sublevel has a different quantum number l.
2. In each sublevel there are $2l + 1$ orbitals. Each orbital has a different quantum number m.
3. Each orbital can hold only two paired electrons. Paired electrons have opposite spin.

s, p, d, and f electrons

As stated in Eq. 7-2, l may have values from 0 to $n - 1$. When $l = 0$, the electrons are called s electrons; when $l = 1$, p electrons; when $l = 2$, d electrons; and when $l = 3$, f electrons. (The abbreviations s, p, d, f come from the words *sharp, principal, diffuse,* and *fundamental* which early spectroscopists used to describe line spectra.)

When $n = 1$, l can only have a value of 0. This limits electron energy to a single main level. When $n = 2$, l may have values of 0 and 1, meaning two subshells of electrons are possible within the second main shell. When $n = 3$, three subshells are possible; and when $n = 4$, four subshells. If a main shell is large enough to accommodate s, p, d, and f electrons, their order of increasing energy is $s < p < d < f$. These generalizations are summarized in Figure 7-5.

When $l = 0$, quantum number m is also 0 and only one orbital is possible. Such an orbital contains a maximum of two s electrons in a spherical volume of space. An s orbital is formed from the first two electrons to occupy a main energy level. An s orbital's radius is proportional to n as indicated in Figure 7-6a.

When $l = 1$, quantum number m can have three values, 1, 0, -1, making possible three orbitals. Each contains a maximum of two electrons, giving a total of six p electrons. The shape is that of double lobes having direction along three mutually perpendicular axes in space as shown in Figure 7-6b.

When $l = 2$, m has values 2, 1, 0, -1, -2, giving five orbitals of two electrons each, a total of 10 d electrons. Five space orientations are possible.

When $l = 3$, m has values 3, 2, 1, 0, -1, -2, -3, giving seven orbitals of two electrons each, a total of 14 f electrons.

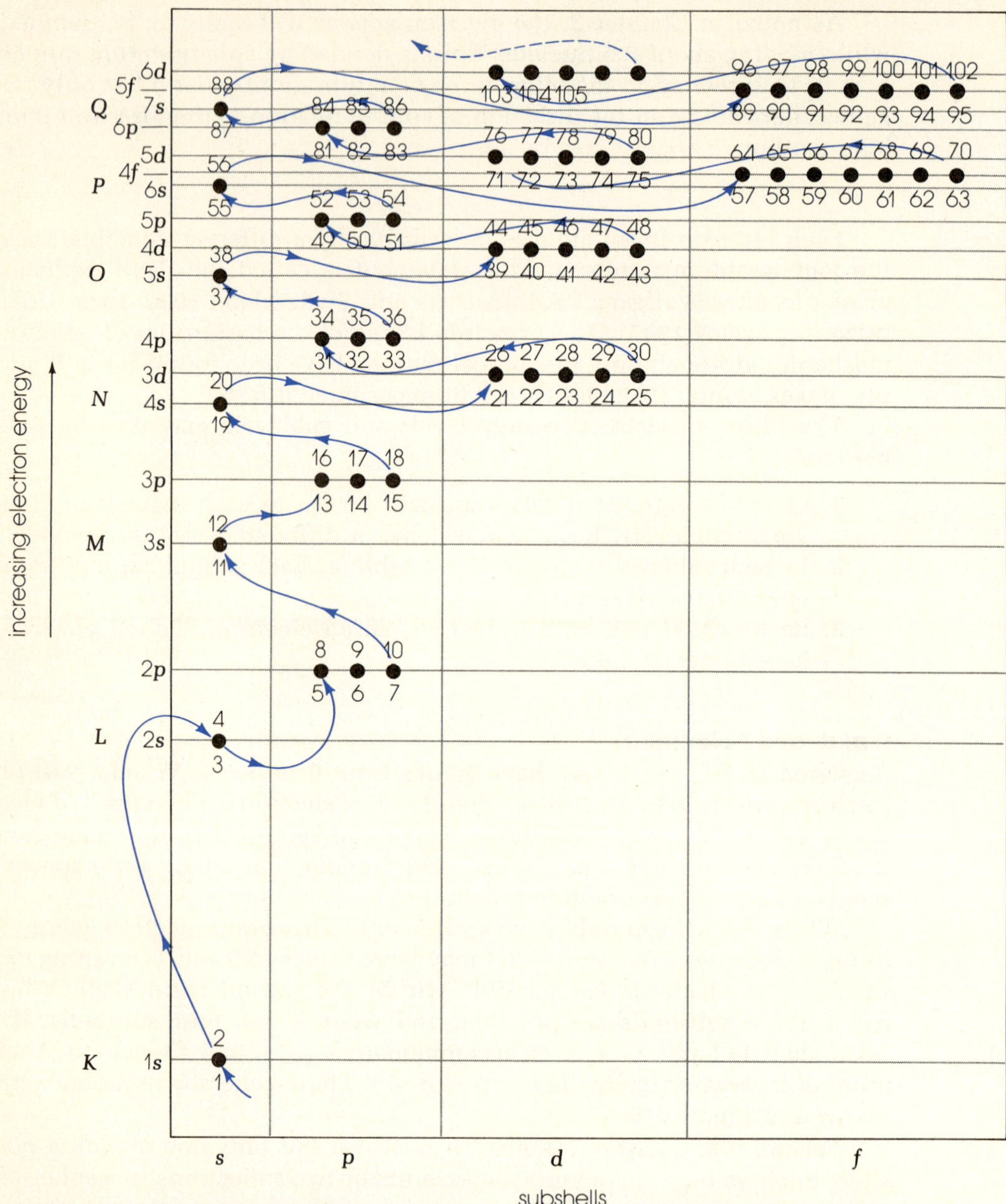

Figure 7-5 Schematic view of electron energies and order of filling of atomic orbitals. Each circle represents an orbital in which a maximum of two electrons may be found. Two electrons in an orbital have identical n, l, and m quantum numbers but opposite values of s. The curving line traces the idealized order in which orbitals are filled with electrons. Beginning with hydrogen the order is 1s, 2s, 2p, 3s, etc. Each new element is assumed built from the preceding one by the addition of one positive charge to the nucleus and one electron to the lowest-energy orbital that is not yet filled. Originated by Bohr, this theoretical way of arriving at electron configurations is called the Aufbau (Gr. development) process. Irregularities in filling order are known in some real atoms.

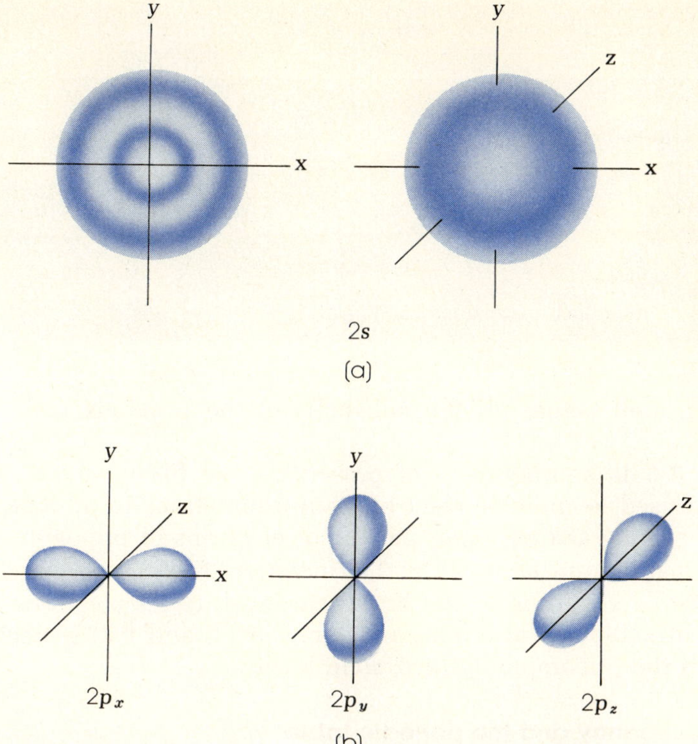

Figure 7-6 Atomic orbital shapes. (a) The s orbital of the second electron shell (n = 2, l = 0). (Left) A cross section showing two regions of electron density, corresponding to two s electrons in the first level and two more s electrons in the second level. (Right) The spherical 2s orbital. (b) p orbitals in the second electron shell (n = 2, l = 1). The three mutually perpendicular orientations along the x, y, and z axes are a consequence of quantum number m; p orbitals in larger shells are similar in shape and correspondingly larger.

These figures are calculated to represent boundaries within which there is 95 percent probability of finding the electron.

Shapes of orbitals for several n, l, and m values have been indicated without specification in Figure 3-20.

In general, for any main energy level n, the number of orbitals it contains is equal to n^2. The total maximum number of electrons therein is equal to $2n^2$. The outermost level of any neutral atom contains a maximum of eight electrons.

Table 7-2 summarizes conditions in the first four main electron shells.

Atomic orbital occupancy

With few exceptions Figure 7-5 can be used to write the orbital filling order and electron configuration about the nucleus of any neutral atom. The diagram also indicates that as electron energy increases, energy difference between electron levels narrows. Beginning with the $4p$ electrons, sublevel overlap occurs, so that the filling order thereafter is not a simple progression. As illustrated by the $4s$ orbital filling before $3d$, filling can begin in a

Table 7-2. Shells, subshells, and electrons

	1	2	3	4
Main electron shell	1	2	3	4
Number of subshells (n)	1	2	3	4
Number of orbitals (n^2)	1	4	9	16
Kind of orbital	s	s p	s p d	s p d f
Number of orbitals	1	1 3	1 3 5	1 3 5 7
Maximum number of electrons	2	2 6	2 6 10	2 6 10 14
Total maximum number of electrons ($2n^2$)	2	8	18	32

new main shell before all the subshells of the previous main shell are complete.

Table 7-3 is another way of presenting the filling order. Arrows in circles are used to indicate electron spin orientations in orbitals. Two oppositely pointing arrows mean pairing of electrons of opposite spin. Consider the magnesium atom $_{12}$Mg. The notation $1s^2 2s^2 2p^6 3s^2$ means that it contains two s electrons in the first main level, two s electrons and six p electrons in sublevels of the second main level, and two s electrons in a sublevel of the (incomplete) third main level.

Orbital occupancy and the periodic table

The filling order of orbitals that one can trace in Figure 7-5 may be divided into segments as follows:

$$1s \quad 2s, 2p \quad 3s, 3p \quad 4s, 3d, 4p \quad 5s, 4d, 5p$$

$$6s, 4f, 5d, 6p \quad 7s, 5f, 6d, 7p \ldots \qquad (7\text{-}5)$$

Since s orbitals are filled with 2 electrons, p with 6, d with 10, and f with 14, completed segments represent 2, 8, 8, 18, 18, 32, 32 electrons. Logically

Table 7-3. Electron levels and sublevels of light elements

Period	Element	Electron configuration	Schematic orbital representation
			1s 2s 2p 3s 3p
1	$_1$H	$1s^1$	(↑)
	$_2$He	$1s^2$	(↑↓)
2	$_3$Li	$1s^2 2s^1$	(↑↓) (↑) (○)(○)(○)
	$_4$Be	$1s^2 2s^2$	(↑↓) (↑↓) (○)(○)(○)
	$_5$B	$1s^2 2s^2 2p^1$	(↑↓) (↑↓) (↑)(○)(○)
	$_6$C	$1s^2 2s^2 2p^2$	(↑↓) (↑↓) (↑)(↑)(○)
	$_7$N	$1s^2 2s^2 2p^3$	(↑↓) (↑↓) (↑)(↑)(↑)
	$_8$O	$1s^2 2s^2 2p^4$	(↑↓) (↑↓) (↑↓)(↑)(↑)
	$_9$F	$1s^2 2s^2 2p^5$	(↑↓) (↑↓) (↑↓)(↑↓)(↑)
	$_{10}$Ne	$1s^2 2s^2 2p^6$	(↑↓) (↑↓) (↑↓)(↑↓)(↑↓)
3	$_{11}$Na	$1s^2 2s^2 2p^6 3s^1$	(↑↓) (↑↓) (↑↓)(↑↓)(↑↓) (↑) (○)(○)(○)
	$_{12}$Mg	$1s^2 2s^2 2p^6 3s^2$	(↑↓) (↑↓) (↑↓)(↑↓)(↑↓) (↑↓) (○)(○)(○)
	$_{13}$Al	$1s^2 2s^2 2p^6 3s^2 3p^1$	(↑↓) (↑↓) (↑↓)(↑↓)(↑↓) (↑↓) (↑)(○)(○)

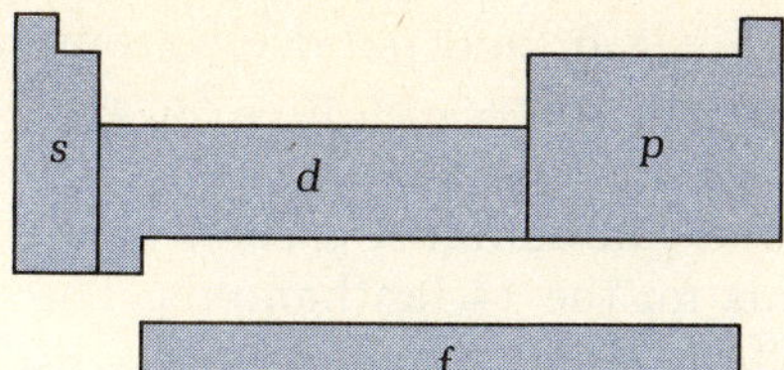

Figure 7-7 The periodic table, divided into blocks of elements on the basis of the filling of orbitals.

these are, in order, the numbers of elements in periods 1 through 7 of the periodic table. Also, elements in which s, p, d, and f orbitals are being filled fall into natural table groupings (Figure 7-7). The s block contains two families, the p block, six. Together they are called the *representative elements*; d-block families are called *transition elements*; f-block families are *inner transition elements*.

An expansion of Figure 7-7 is Figure 7-8. We will read through this scheme to point out some of the details.

Period 1 consists of two elements, $_1$H and $_2$He, whose atoms, respectively, contain one and two s electrons.

Periods 2 and 3 consist of eight elements each, $_3$Li to $_{10}$Ne and $_{11}$Na to $_{18}$Ar. In them one s and three p orbitals are filled.

Periods 4 and 5 both contain 18 elements: $_{19}$K to $_{36}$Kr, and $_{37}$Rb to $_{54}$Xe.

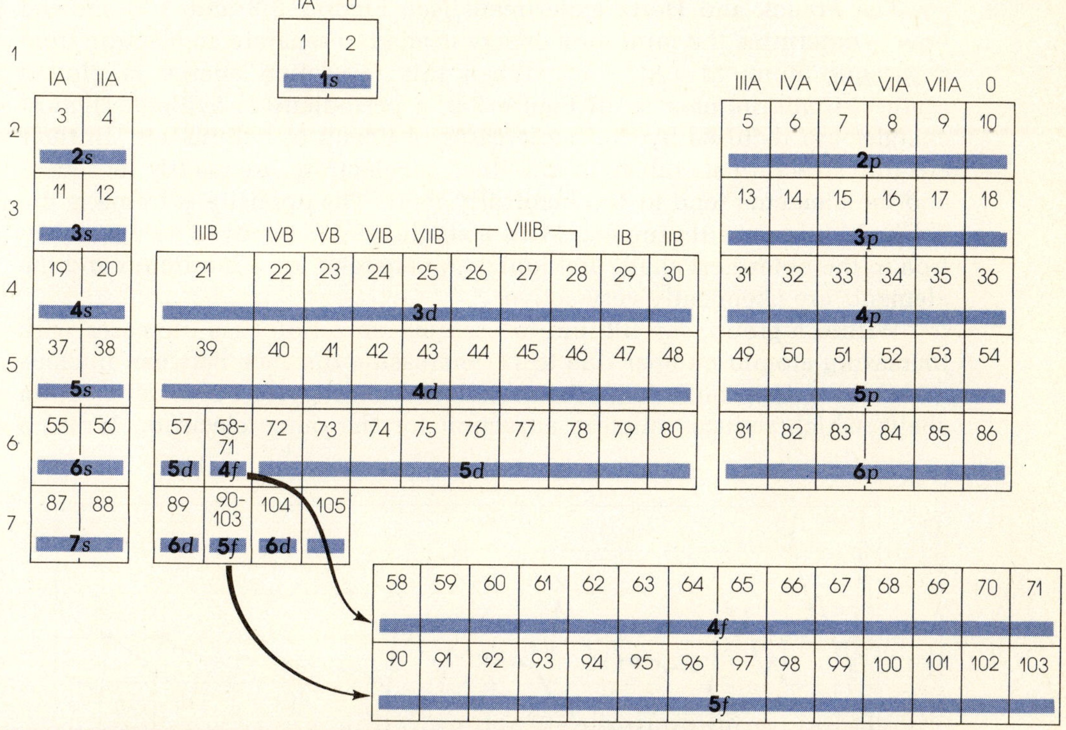

Figure 7-8 Division of the periodic table into blocks of elements, which emphasizes both the order of filling of atomic orbitals and the theoretical basis for the periodic law. Periods begin with an element having one valence s electron and end with an element having filled orbitals. The arrows point to period 6 and 7 continuations in the two inner transition series. (Adapted from Charles W. Keenan and Jesse H. Wood, General College Chemistry (3rd ed.). New York: Harper & Row, 1966, pp. 62–63.)

In each sequence orbital filling requires two *s*, six *p*, and ten *d* electrons. Buildup of the 3*d* and 4*d* subshells, respectively, spans $_{21}$Sc to $_{30}$Zn, and $_{39}$Y to $_{48}$Cd. These are the first two rows of the transition metals.

Period 6, $_{55}$Cs to $_{86}$Rn, which includes the lanthanoids, $_{58}$Ce to $_{71}$Lu, consists of 32 elements. The 4*f* filling accounts for the 14 lanthanoids. The 5*d* filling occurs in the third row of transition elements, $_{57}$La, and $_{72}$Hf through $_{80}$Hg.

Period 7, which includes the actinoid series, is incomplete. (Nineteen of 32 possible elements are known.) Starting with $_{89}$Ac, 6*d* filling of a fourth row of transition elements is begun. Within it are the 14 actinoids, $_{90}$Th through $_{103}$Lr. In that sequence 5*f* filling is completed. The 6*d* filling would be completed with element 118.

Chemists are mainly concerned with the outer electron structure of atoms. Elements can be placed in the periodic table according to the orbital characteristics of the last electron added in the building process described by Figure 7-5. And, as will be explained in the next chapter, the outer electrons include the valence electrons that are used in chemical reactions.

Periodicity and ionization energies

Chemical and physical properties of elements are related to their electron structure. Properties of elements can consequently be understood from periodic relationships. We will look at one example.

The Franck and Hertz experiment (see Figures 3-8 and 3-9) showed how to determine the minimum energy needed to separate an electron from a gaseous atom: $M = M^+ + e^-$. When this ionization energy is plotted against atomic number as in Figure 7-9, a periodicity is evident. Highest energies are required by the noble gases of Group 0: helium, etc. In their electron structure all subshells are filled, no electrons are readily loosened, and the elements tend to be chemically inert. The opposite extreme is the Group IA metals: lithium, etc. With a single, easily removed valence electron in the outermost shell, first ionization energies are a minimum and the elements are chemically very active.

Within a group, say lithium to cesium, ionization ease increases with increasing atomic number due to (1) increasing distance between nucleus and valence electron, and (2) the so-called shielding effect of inner electron shells which reduces nuclear attraction for the outer electron. Across a

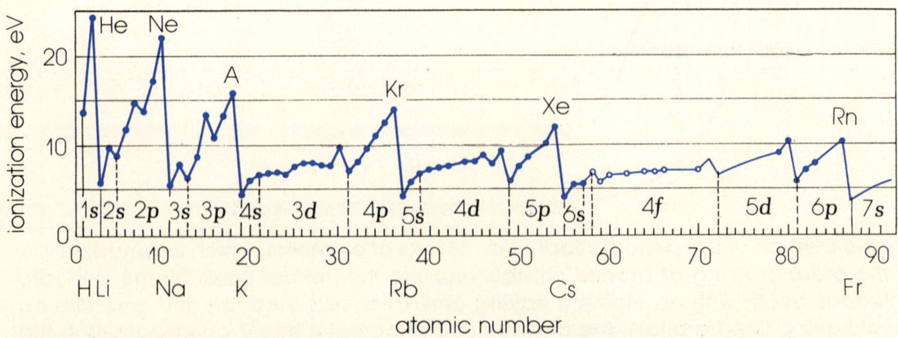

Figure 7-9 First ionization energies of the elements plotted against atomic number.

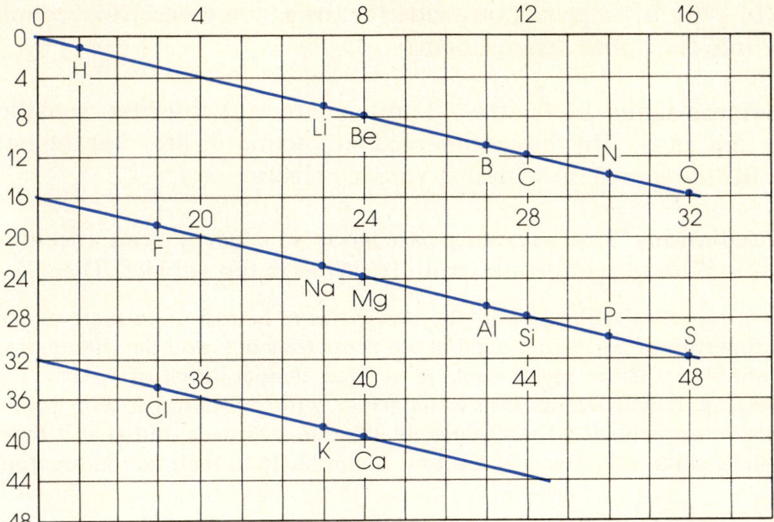

Figure 7-10

period (for instance, from Li through Be, B, C, N, O, F, to Ne) a general increase in ionization energy is noted. It is due to the increase in nuclear charge while electrons are being added at about the same distance from the nucleus (no new main shells are begun).

Extremes in ionization energies are found in francium (lowest) and helium (highest). Defined by ease of electron release, francium has the most metallic character, helium the most nonmetallic character.

QUESTIONS

1. Periodic table Figure 7-10 shows the *telluric screw* of de Chancourtois (see page 147). **(a)** Find a dictionary definition of "telluric." **(b)** Explain the meaning of the scales and how the chart operates. **(c)** To construct a complete chart through element 20 (Ca) today, what additions would be needed?

2. Periodic table Figure 7-11 shows part of the spiral system of H. Baumhauer (1870). **(a)** What elements occupy the positions labeled Q, X, and

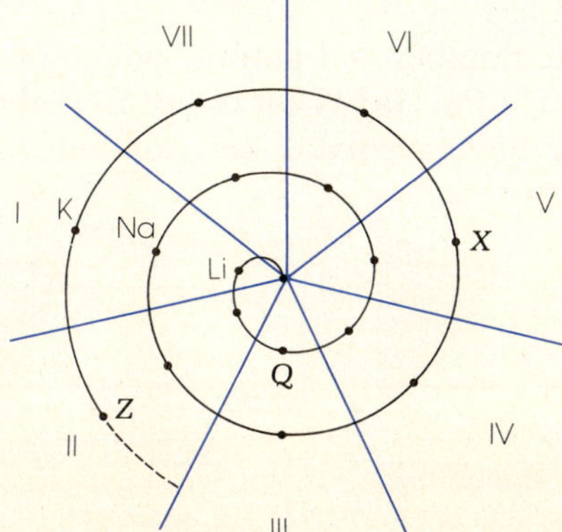

Figure 7-11

Z? **(b)** Why is no provision made for the noble gases? How would you fit them into the spiral arrangement?

3. Periodic table In Figure 7-5 tell where **(a)** the first transition series starts and ends; **(b)** the lanthanoids and actinoids are; **(c)** the noble gases are; **(d)** the elements with one valence electron are.

4. Predictions The following passage is an excerpt of an article published in 1871. Who do you think wrote it? What is the subject? Discuss.

> Its properties will resemble the properties of Si and As to such an extent as the properties of As itself resemble the properties of P and Se; that is, it will at any rate be a fusible metal capable at high temperatures of volatilizing and oxidizing. It will decompose water vapor with difficulty. It will have almost no action on acids; that is, it will not liberate hydrogen and it will form very unstable salts. Alkalies will react on it similarily to their action on zinc and arsenic.

5. Predictions Use the periodic table to make logical deductions about the following: **(a)** Pt and Ni are used as catalysts (see Figure 17-4). Name one other element that might have similar activity. **(b)** Name the following and classify each as a metal, nonmetal, or metalloid: Tl, Br, W, Ge, V. **(c)** The densest element is iridium. Name three other elements of probable high density. **(d)** Most potassium compounds are water soluble. What other metals probably form similarily soluble compounds? **(e)** What is the likelihood that a new element will be discovered between numbers 1 and 105?

6. Groups, periods What does each colored area in Figure 7-12 denote? Explain.

7. Metals, nonmetals Electronically, how would you classify or describe metals? Nonmetals?

8. Electron arrangement **(a)** What is the greatest number of electrons found in the outer shell of any neutral atom? **(b)** Is the number characteristic of any particular family of elements? Explain.

9. Electron arrangement **(a)** What atom is represented by $1s^2 2s^2 2p^4$? **(b)** What does the symbolism mean?

10. Atomic number List the atomic number and atomic weight of each element in these pairs: Ar, K; Co, Ni; Th, Pa. **(a)** What is peculiar about these data? **(b)** How would Mendeleev have resolved the problem of placing

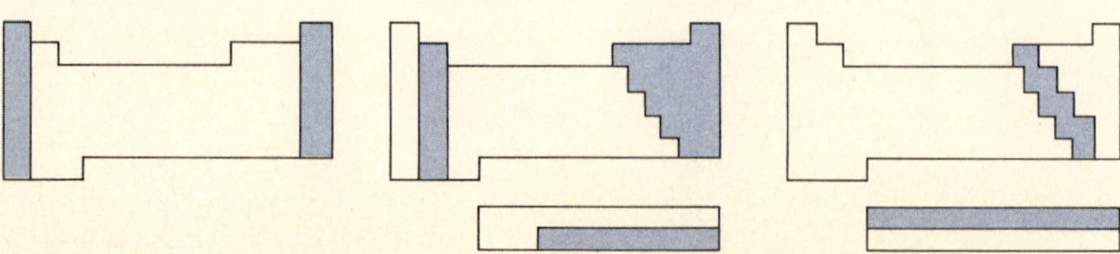

Figure 7-12

Table 7-4.

Element	At. no.	Melting point °C
Na	11	?
K	19	63.7
Rb	37	38.9
Cs	55	28.7

(Accepted values are 97.8, 0.218 for Tables 7-4 and 7-5, respectively.)

Table 7-5.

Element	At. no.	Electrical conductance $1/\mu ohm$ 20°C
Mg	12	0.224
Ca	20	?
Sr	38	0.043
Ba	56	0.016

Table 7-6.

Compound	At. no.	Specific gravity
$SiCl_4$	14	1.50
$GeCl_4$	32	1.85
$SnCl_4$	50	?
$PbCl_4$	82	5.80

(Accepted values are 2.23, 2.67 for Tables 7-6 and 7-7, respectively.)

Table 7-7.

Element	At. no.	Atom radius Å
Li	3	1.55
K	19	2.35
Rb	37	2.48
Cs	55	?

these elements in the periodic table with respect to one another? **(c)** How would Moseley have done it?

PROBLEMS

11. Predictions Prepare graphs from the numerical data in Tables 7-4 and 7-5. From the graphs predict the missing values. Record each one. Check your predictions against the given experimental values.

12. Predictions Follow instructions as in Problem 11 (use Tables 7-6 and 7-7).

13. Periodic chart Make an enlargement of Figure 7-11. **(a)** Add the symbols of all elements up to and including Ca. **(b)** Suggest how the chart might be modified at this point to accommodate the next 10 (transition) elements before elements 31 through 35 are fitted in normally.

14. Predictions **(a)** Use the data in Table 7-8 to calculate the atomic volumes in milliliters of the first five metals. **(b)** Comment on the agreement of your results with those of Meyer, Figure 7-2b. **(c)** Predict the density of francium. (Graph the table data, and extrapolate.) From that calculate its atomic volume.

15. Predictions Plot atomic number (horizontally) versus melting point (take data from periodic table, inside back cover) for the first 36 elements. List evidences of periodicity in the graph.

16. Orbitals **(a)** Use Figure 7-5 to complete the orbital structure of element 19: $1s^2 2s^2$. . . . **(b)** As in Table 7-3, write the electron configurations and orbital representations of $_{16}S$ and $_{18}Ar$.

Table 7-8.

Element	At. wt.	Density g/ml, 20 °C
Li	6.94	0.53
Na	23.0	0.97
K	39.1	0.86
Rb	85.5	1.53
Cs	133	1.90
Fr	233	?

17. Quantum numbers   Consider the last two electrons in the magnesium atom, Table 7-3. Assign them quantum numbers, n, 1, m, s. Give a brief explanation of how you arrived at each value.

18. Electrons   **(a)** What is the maximum number of s, p, d, and f electrons that can be present in a given main level?   **(b)** What is the total number of electrons in the $n = 2$ level?   **(c)** In the $n = 4$ level?

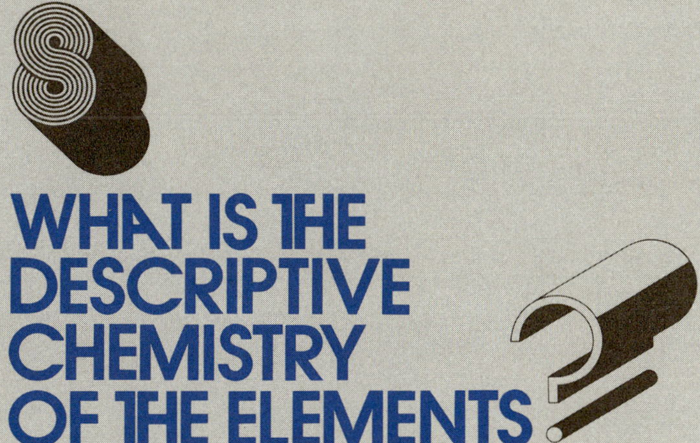

WHAT IS THE DESCRIPTIVE CHEMISTRY OF THE ELEMENTS?

THE IMPORTANT CONCEPTS

1. Elements are studied via the periodic table.
2. Elements and their compounds have both chemical and economic significance.

8-1 A subgroup elements

1. Electron filling in main group elements is in s and p levels.
2. Group number, outermost level electron number, and oxidation state of an element are related.
3. Metallic character changes in going across a period and down a group.

8-2. B subgroup elements and group VIII

1. Electron filling is in the outermost d level.
2. Properties, including oxidation states, do not follow simple rules.
3. Magnetic properties and color of ions are related to unpaired d electrons.
4. Common oxide and sulfide minerals are often readily reduced to the metals.
5. Iron is the most familiar and useful transition element.

8-3. The inner transition elements

1. Electron filling is in the outermost f level.
2. Ion exchange is used to separate the lanthanoids.
3. Lanthanoid metals are chemically similar and quite active.
4. Most of the actinoids are man-made; all are radioactive.

8-4. Extending the periodic table

1. Elements up to $Z = 168$ may be synthesized and accommodated in an expanded table (Figure 7-4).
2. Doubly magic nuclei are expected to be islands of stability amidst short-lived neighbors.

We have seen that the periodic table is the embodiment of the observation that the properties of the elements are recurring functions of their atomic numbers. Logically it is our best means of organizing the elements for the study of their descriptive chemistry. Thus far we have perhaps emphasized chemical theory more than fact. One purpose of theory is to express useful generalizations based on facts that are too numerous to memorize and manage effectively. Theories, however, are not necessarily more useful than facts, nor do they last as long. We therefore will find it profitable to develop some factual background by taking a brief tour through the families of elements and their common compounds.

In discussing the winning of elements from natural sources and the preparation of compounds from them, it is also appropriate to point out attendant economic values and the contributions to society of the mining, refining, and chemical industries. Without their mass production of low unit-cost commodities, modern civilization would not be possible. Consider the data in Table 8-1. The most obvious conclusion is that among the common *metals,* the United States is self-sufficient only in iron. Of 40 minerals essential to our present high level of industrial activity, we import all but seven, and the grade of domestic deposits is growing poorer.

Table 8-2 gives production figures for the dozen most common *inorganic chemicals.* Though their manufacture and use are generally inconspicuous to the public, the quantities handled are huge. To carry 30 million tons of sulfuric acid at one time, for instance, would require a solid

Table 8-1. Metals: estimated present and future supply and demand in the United States

Typical metals[a]	Approx. present primary[b] production million tons	Approx. present primary demand million tons	Est. cumulative 1975–2000 primary demand million tons	Est. reserves million tons
Chromium	none	0.5	20	2
Iron	58	86	3300	110,000
Manganese	0.05	1.2	47	0.3
Nickel	0.02	0.2	8	0.2
Aluminum	0.4	4	350	9
Copper	1.2	1.5	100	80
Lead	0.3	0.8	40	32
Magnesium	0.1	0.1	40	15
Zinc	0.5	1.5	65	33

[a] The first four elements listed are called *ferrous metals* because they are alloyed in steel; the others are called *nonferrous metals.*

[b] *Primary* means metal production from virgin ores, as opposed to *secondary* metals from sources such as scrap. Primary plus secondary production equals total production. Present total U.S. iron production, for example, is annually about 140 million tons.

Table 8-2. Biggest volume inorganic chemicals—approximate 1972 U.S. production[a]

Chemical, 100 percent pure	Production million tons	Chemical, 100 percent pure	Production million tons
Sulfuric acid, H_2SO_4	30	Nitric acid, HNO_3	7
Oxygen, O_2	15	Ammonium nitrate, NH_4NO_3	7
Ammonia, NH_3	14	Nitrogen, N_2	7
Sodium hydroxide, NaOH	10	Phosphoric acid, H_3PO_4	6
Chlorine, Cl_2	10	Hydrochloric acid, HCl	2
Sodium carbonate, Na_2CO_3	7	Ammonium sulfate, $(NH_4)_2SO_4$	2

[a] Inorganic industrial products that come directly from the mining industry such as salt and limestone are omitted. Their annual tonnages are immense (for example: NaCl, 40 million tons).

line of railroad tank cars stretching from Los Angeles to Chicago (2200 miles). We should know how such compounds are made and used.

In the pages that follow we will make some use of formulas and molecular equations to describe compounds and chemical reactions. In previous chapters we acquired enough feel for this symbolism to appreciate its serviceability. In Chapter 11 we consider the chemist's shorthand in detail.

8-1. A SUBGROUP ELEMENTS

Elements called *main group*, *A subgroup*, or *representative* have already been singled out in Figure 7-7 as *s*- and *p*-block elements. Their symbols are given in Figure 8-1.

IA: The alkali metal group

The alkali (Arabic, *al-qily*, ashes) metals (Table 8-3) are soft enough to be cut with a knife. Their melting points and densities are low, their electrical and heat conductivities are excellent. The most abundant, sodium and potassium, were first isolated in 1807 through the decomposition of their fused (melted) compounds by electric current. The method is still used today (see Figures 18-2 and 18-4). The metals are so reactive they are kept in airtight containers or under oil.

IA							0
1 (H)	IIA	IIIA	IVA	VA	VIA	VIIA	2 He
3 Li	4 Be	5 B	6 C	7 N	8 O	9 F	10 Ne
11 Na	12 Mg	13 Al	14 Si	15 P	16 S	17 Cl	18 Ar
19 K	20 Ca	31 Ga	32 Ge	33 As	34 Se	35 Br	36 Kr
37 Rb	38 Sr	49 In	50 Sn	51 Sb	52 Te	53 I	54 Xe
55 Cs	56 Ba	81 Tl	82 Pb	83 Bi	84 Po	85 At	86 Rn
87 Fr	88 Ra						

Figure 8-1 The A subgroup section of the periodic table. (Transition elements have been removed.)

Table 8-3. IA elements

Element	Electron shells	Oxidation states	Density g/ml	Melting point °C
Lithium, Li	2,1	1[a]	0.53	180
Sodium, Na	2,8,1	1	0.97	98
Potassium, K	2,8,8,1	1	0.86	64
Rubidium, Rb	2,8,18,8,1	1	1.53	39
Cesium, Cs	2,8,18,18,8,1	1	1.90	29
Francium, Fr	2,8,18,32,18,8,1	1		27[b]

[a] Most important oxidation states are italicized.
[b] Most common isotope.

Sodium is the most familiar alkali metal. It is employed in atomic reactors to carry heat, and in cables to carry electricity. It is also used (1) to make the sodium–lead alloy from which the gasoline additive tetraethyl lead is prepared, (2) to coat photoemissive tubes (see Figure 3-2), and (3) in certain chemical reactions in which reducing power (electron release) is needed (see Figure 19-8),

$$Na = Na^+ + e^-$$

When combining chemically, the alkali metals lose the outermost electron and form 1+ ions. (That is the meaning of "oxidation states" in Table 8-3.) Colorless in solution, the ions give *flame tests*; that is, they impart distinctive color to a flame that serves to identify them. The Li^+ flame is red, Na^+ is yellow, K^+ is reddish violet, Rb^+ is red, Cs^+ is bluish. Most alkali metal compounds are quite soluble in water.

The alkali metals are the most electropositive elements known (see Figure 7-9). With water, they react violently to give the hydroxide and hydrogen,

$$2Na + 2H_2O = 2NaOH + H_2 \tag{8-1}$$

With hydrogen, they form hydrides (hydrogen is the negative ion),

$$2K + H_2 = 2KH$$

And with Group VIIA elements (halogens), they give halides,

$$2Rb + Cl_2 = 2RbCl$$

With oxygen, lithium gives an oxide, sodium a peroxide, and the higher members give superoxides: Li_2O (white), Na_2O_2 (yellowish), CsO_2 (yellow).

The most common alkali metal compound is sodium chloride, ordinary table salt or rock salt (Table 8-2). Its recovery from salt mines, brine (salt) wells, and the ocean is among the oldest of mining operations. The use pattern is 40 percent for the manufacture of sodium hydroxide or lye (Table 8-2) and chlorine via electrolysis (Table 8-2 and Figure 18-7); 25 percent for making sodium bicarbonate (baking soda, $NaHCO_3$) and sodium carbonate or soda ash (Table 8-2); 12 percent for highways (deicing, etc.); and 23 percent for miscellaneous uses—in the home, to preserve meats, to manufacture other products.

Some other compounds whose large-scale preparation is related to sodium chloride supplies are (1) sodium chlorate ($NaClO_3$), a bleaching agent; (2) sodium chromate (Na_2CrO_4), an intermediate in the manufacture of pigments and chromium chemicals; (3) various sodium phosphates such as sodium tripolyphosphate ($Na_5P_3O_{10}$), a "builder" that increases the efficiency of detergents (but which is being replaced by sodium carbonate); (4) various sodium silicates such as sodium tetrasilicate ($Na_2Si_4O_9$), which is an intermediate for the preparation of oil industry catalysts, cardboard box adhesives, and nonphosphate detergent builders; and (5) sodium thiosulfate 5-hydrate (hypo, $Na_2S_2O_3 \cdot 5H_2O$), the "fixer" used to dissolve untreated silver salts in photography.

Of the many other useful alkali metal compounds two are worthy of mention. The first is potassium chloride, which is used in large tonnages in fertilizer preparation (green plants need potassium, phosphorus, and nitrogen). The second is lithium carbonate, which finds application in ceramics and the making of such compounds as lithium hydroxide, a reactant in the manufacture of automotive greases and in alkaline batteries.

IIA: The alkaline earth group

The IIA metals (Table 8-4) are in general harder, denser, and less active chemically than the IA metals.

The discovery of radium was mentioned earlier (see Figure 2-15). The other IIA metals were known for centuries in the form of oxides that gave an alkaline reaction in water (hence "alkaline earths"). Davy was the first to obtain the pure metals, preparing magnesium, calcium, strontium, and barium in 1808 by electrolyzing their melted hydroxides at a mercury electrode, then distilling off the mercury.

The great abundance of magnesium and calcium in the earth have been noted (see Table 5-4). Other members of the group are less common, especially radium.

Group IIA metals lose their two outermost s electrons during chemical combination and form dipositive ions. In reactions with oxygen, water, sulfur, and halogens they form, respectively, oxides (general formula MO, where M = metal), hydroxides ($M(OH)_2$), sulfides (MS), and halides (MX_2, where X = halogen). When heated with hydrogen, carbon, and nitrogen, they yield hydrides (MH_2), carbides (MC_2), and nitrides (M_3N_2). The IIA ions are colorless in aqueous solution. Like the alkali metals they give flame tests: Ca^{2+} is reddish orange, Sr^{2+} and Ra^{2+} are red, Ba^{2+} is yellowish green.

Beryllium is obtained from the mineral beryl ($Be_3Al_2Si_6O_{18}$) by first preparing beryllium chloride ($BeCl_2$) then electrolyzing its melt. Beryllium is the strongest of the light metals; pure beryllium is used to make missile and aircraft parts. Copper–beryllium alloys are used in electronic products,

Table 8-4. IIA elements

Element	Electron shells	Oxidation states	Density g/ml	Melting point °C
Beryllium, Be	2,2	2	1.85	1277
Magnesium, Mg	2,8,2	2	1.74	650
Calcium, Ca	2,8,8,2	2	1.55	838
Strontium, Sr	2,8,18,8,2	2	2.60	768
Barium, Ba	2,8,18,18,8,2	2	3.61	714
Radium, Ra	2,8,18,32,18,8,2	2		700

nonsparking tools, and other hardware. Beryllium compounds have few uses. Like the powdered metal itself, they are poisonous.

Magnesium is obtained both from dolomite ($Ca,Mg(CO_3)_2$) by reduction with a carbon–silicon alloy and from seawater. Magnesium is used in castings for automobile wheels, in anticorrosion applications (see Figure 19-17b), and in alloys with aluminum for making transportation and housing products and packaging items such as pop-tops for cans. The metal burns brightly in air, forming a mixture of the white oxide and yellowish nitride. Familiar compounds include milk of magnesia, a stomach antacid made from MgO and water, and epsom salt, a laxative, magnesium sulfate 7-hydrate ($MgSO_4 \cdot 7H_2O$). (*Hydrates* are discussed later, with Table 11-4.)

Calcium metal is used as a hardener in some lead alloys and as an oxygen "scavenger" in the making of steel and copper alloys. (The alloys become embrittled if air forms oxides in them.) Two important calcium compounds are the mineral gypsum, calcium sulfate 2-hydrate, used to make plasterboard, and the half hydrate called "plaster of Paris." They are interconvertible,

$$2CaSO_4 \cdot 2H_2O \xrightleftharpoons{\Delta} 2CaSO_4 \cdot \tfrac{1}{2}H_2O + 3H_2O$$

Portland cement is primarily a mixture of calcium silicates and aluminates ($CaSiO_4$, $Ca_3Al_2O_6$, etc.) made by heating a mixture of limestone and clay to 1500°C then grinding the resulting "clinker" to a powder. When water is added, the compounds react to form long needle-shaped calcium aluminate crystals that interlock to give a rigid structure. Concrete, used in road making and other construction, is a mixture of cement, sand, and crushed rock.

Lime (calcium oxide), used in steel making and other processes, is prepared by heating limestone:

$$CaCO_3 \stackrel{\Delta}{=} CaO + CO_2 \tag{8-2}$$

Slaked lime (calcium hydroxide) is made by adding water to lime:

$$CaO + H_2O = Ca(OH)_2 \tag{8-3}$$

Calcium carbide is made by heating lime and coke in an electric furnace:

$$CaO + 3C \stackrel{\Delta}{=} CaC_2 + CO$$

Calcium carbide is used to make acetylene gas for welding as well as calcium cyanamide, a source of nitrogen for fertilizer manufacture:

$$CaC_2 + 2H_2O = C_2H_2 + Ca(OH)_2$$

$$CaC_2 + N_2 = CaCN_2 + C$$

Another industrial calcium compound is calcium chloride ($CaCl_2$). It is used to deice roads, to control roadside dust (it gathers moisture from the air, forming hydrates, $CaCl_2 \cdot H_2O$, etc.), and to modify setting time of concrete. Still another common compound is calcium phosphate, a mineral from which phosphorus compounds are manufactured.

Strontium and barium metals have few uses. Strontium nitrate ($Sr(NO_3)_2$) gives red color to flares and fireworks. Strontium carbonate ($SrCO_3$) is incorporated during manufacture in television tube glass for its

Table 8-5. IIIA elements

Element	Electron shells	Oxidation states	Density g/ml	Melting point °C
Boron, B	2,3	3	2.34	2030
Aluminum, Al	2,8,3	3	2.70	660
Gallium, Ga	2,8,18,3	3	5.91	30
Indium, In	2,8,18,18,3	3	7.31	156
Thallium, Tl	2,8,1,32,18,3	3, 1	11.85	303

ability to absorb X rays that might be produced in normal television set operation. White, insoluble barium sulfate ($BaSO_4$) is added to plastics to increase their density. It is also administered internally to make the alimentary canal opaque in diagnostic X ray pictures; unlike soluble barium compounds, it is nonpoisonous.

IIIA: The boron or aluminum group

Aluminum is the most abundant metallic element in the earth's crust (see Table 5-4). The other IIIA elements are rare by comparison (Table 8-5).

Group IIIA elements lose three valence electrons and therefore have an oxidation state of 3+. The usual oxides, hydroxides, and halides are M_2O_3, $M(OH)_3$, and MX_3. The oxides tend to be hard (Al_2O_3 is used as an abrasive). As precipitated in solution the hydroxides are gelatinous. The halides *hydrolyze*; that is, they react to split water into H and OH,

$$AlCl_3 + 3H_2O = Al(OH)_3 + 3HCl \tag{8-4}$$

Boron is a metalloid. Impure amorphous boron, a black solid, can be made by heating the oxide with magnesium. Pure boron is made by passing a gaseous mixture of boron bromide and hydrogen through an electric arc:

$$B_2O_3 + 3Mg \overset{\Delta}{=} 3MgO + 2B$$

$$2BBr_3 + 3H_2 \overset{\Delta}{=} 6HBr + 2B$$

Pure boron crystals resemble diamonds: transparent, brittle, very hard, high melting, and unreactive. Filaments made of boron are unusually rigid. They are employed to stiffen plastic aircraft components. (Fiber technology originated 30 centuries ago with the Israelites, who reinforced bricks with straw.)

The source of boron chemicals is borax, sodium tetraborate 10-hydrate, found in Searles Lake, California. Borax is used directly in detergents, in the manufacture of special glass, and as a *flux* for dissolving oxide films before metals are joined by welding; for example,

$$Na_2B_4O_7 \cdot 10H_2O + NiO \overset{\Delta}{=} 2NaBO_2 + Ni(BO_2)_2 + 10H_2O$$

Boric oxide is made by treating borax with a strong acid. With water, B_2O_3 gives the very weak acid, boric acid (H_3BO_3). Boron reacts directly with fluorine to give boron trifluoride, a petroleum industry catalyst. With lithium hydride, boron trifluoride gives lithium tetrafluoborate and diborane, a gas:

$$8BF_3 + 6LiH = 6LiBF_4 + B_2H_6$$

Heated above 100°C, diborane gives a series of higher boranes (B_4H_{10}, etc.), most of which are spontaneously flammable in air. For a given weight they yield about 50 percent more heat when burned than kerosene. They have been tested as rocket fuels.

Aluminum metal (Table 8-1) is produced by electrolysis (see Figure 18-5). The method was invented in 1886 at Oberlin College by chemist Charles Hall, and independently in France by P. Heroult. Alumina (aluminum oxide) consumed in the process is prepared from bauxite, a mineral containing about 55 percent Al_2O_3. Treatment with sodium hydroxide solubilizes only the aluminum oxide, as sodium aluminate, which is filtered free of impurities. Treatment of the clear solution with carbon dioxide gives aluminum hydroxide, which is filtered off and dried:

$$Al(OH)_3 + NaOH = NaAl(OH)_4$$

$$NaAl(OH)_4 + CO_2 = Al(OH)_3 + NaHCO_3$$

$$2Al(OH)_3 \overset{\Delta}{=} Al_2O_3 + 3H_2O$$

Aluminum metal is used in many structural applications because some of its alloys (nominally 95 percent Al, with Cu, Mg, Mn, Si) rival low-carbon steel in strength at about one-third the density. Today's medium-size automobile contains about 80 pounds of aluminum. Pure aluminum is chemically active yet corrosion resistant due to acquisition of an oxide coating that protects the metal's interior.

Aluminum dissolves in bases and most strong acids, releasing hydrogen and forming salts. In hydrochloric acid and sodium hydroxide, aluminum chloride and sodium aluminate are products:

$$2Al + 6HCl = 2AlCl_3 + 3H_2$$

$$2Al + 2NaOH + 6H_2O = 2NaAl(OH)_4 + 3H_2 \tag{8-5}$$

Aluminum burns in oxygen with an evolution of a great deal of heat, indicating Al_2O_3 is an unusually stable compound. This fact is taken advantage of in the thermite process wherein aluminum is oxidized while the oxide of a less active metal is reduced. In the reaction with iron oxide enough heat is released to give molten iron:

$$2Al + Fe_2O_3 = Al_2O_3 + 2Fe$$

An important industrial chemical is aluminum sulfate 18-hydrate ($Al_2(SO_4)_3 \cdot 18H_2O$), made by heating bauxite with sulfuric acid. It is used as a source of aluminum hydroxide in sizing paper (filling around the fibers) and in settling particulate matter in sewage water treatment.

Gallium, thallium, and indium are soft, low-melting metals. They find minor applications in electrical devices and telecommunications. Gallium occurs in bauxite and its chemistry is like that of aluminum. The pure metal melts slightly above room temperature. Thallium has oxidation states of 3+ and 1+. Its compounds are poisonous.

IVA: The carbon or silicon group

Carbon is a nonmetal, silicon and germanium are metalloids, and tin and lead are metals (Table 8-6). Carbon, lead, and tin were known to the an-

Table 8-6. IVA elements

Element	Electron shells	Oxidation states	Density g/ml	Melting point °C
Carbon, C	2,4	4, 2	2.26	3727[a]
Silicon, Si	2,8,4	4	2.33	1410
Germanium, Ge	2,8,18,4	4	5.32	937
Tin, Sn	2,8,18,18,4	4, 2	7.30	232
Lead, Pb	2,8,18,32,18,4	4, 2	11.4	327

[a] Graphite

cients. Silicon was discovered by Berzelius, who heated potassium hexa-fluosilicate with potassium metal:

$$K_2SiF_6 + 4K \overset{\Delta}{=} Si + 6KF \tag{8-6}$$

The chemistry of carbon contains many details (Chapters 20, 21) because carbon atoms can bond to one another to form indefinitely extended structures. The same ability is not carried lower in the IVA group. The limit of —Si—Si— bonding seems to be six atoms, that of —Ge—Ge— only three. The occurrence of elemental carbon as graphite and diamond has been mentioned (see Figure 6-7). Carbon is found chemically combined in atmospheric carbon dioxide, in large sedimentary deposits of limestone, and in all living things.

Silicon is a gray solid. It and carborundum, silicon carbide (SiC), are made by heating sand and carbon together in an electric furnace. Elemental silicon is used in electronics. Carborundum, a blue-black solid with the structure of diamond, is an industrial abrasive.

Ordinary window glass, also called *soft glass* or lime glass, is made by melting together sand, lime, sodium carbonate, and aluminum oxide. The product is about 75 percent SiO_2, 23 percent $CaO + Na_2O$, and 2 percent Al_2O_3. Pyrex *glass*, also called hard or "borosilicate glass," is made by substituting boron oxide for calcium oxide. Pyrex has a higher softening point, greater resistance to attack by alkalis, and smaller change in dimensions with temperature.

Silicones are a class of synthetic liquids and solids consisting of strong —Si—O—Si—O— chains to whose silicon atoms are bonded organic units such as methyl groups, —CH_3. Silicones are used in making sealants, adhesives, hydraulic fluids, lubricating oils, plastics for artificial anatomical parts, etc.

Germanium is obtained by heating GeO_2 with carbon; it is purified by *zone refining*. Refining is done by passing the crude germanium bar slowly lengthwise through a ring-shaped hot zone that melts the bar only in that locality. Impurities move with the melt to the end of the bar. After several passes, the main part of the bar has become one of the purest substances known, its impurities typically reduced to 0.1 part per billion.

The atomic structure of germanium and silicon allows them to be modified with traces of impurities so they become semiconducting. Semiconducting crystals (see Figure 19-10) have properties that make them equivalent to vacuum tubes.

Tin has three allotropic forms. When common white tin (tetragonal) is heated, it changes to brittle tin (rhombic). When white tin is cooled below 13°C, it slowly changes to gray tin (cubic) that can be crumbled in the hand. Tin metal is a base for low-melting alloys such as pewter (86 percent Sn, 6 percent Cu, 6 percent Bi, 2 percent Sb) and babbitt or bearing metal (90 percent Sn, 7 percent Sb, 3 percent Cu); it is also used alone as a rust-preventive coating on iron and steel. Tin-coated steel cans are used for foods because tin resists corrosion and is nontoxic.

Tin is obtained from the mineral cassiterite by reduction with carbon at 1300°C:

$$SnO_2 + C \overset{\Delta}{=} CO_2 + Sn \tag{8-7}$$

Tin compounds find largest usage in tin electroplating and in making glass. Stannous fluoride (SnF_2) is incorporated in some toothpastes to inhibit decay in children's teeth. Organic tin compounds are used to stabilize plastics and kill microorganisms (example of a fungicide: tributyltin hydroxide, $(C_4H_9)_3SnOH$).

Lead (Table 8-1) is a soft, gray metal. During the Roman Empire it was used to make tableware and water pipes. (Life expectancy was less than 30 years then, due in part possibly to lead poisoning.) The metal is used in lead storage batteries, radiation shielding, electric cable sheathing, sound-proofing, and such low-melting alloys as plumber's solder (67 percent Pb, 33 percent Sn). The biggest growth in use appears to be in battery manufacture; a small electric car (see Figure 19-7) can use several hundred pounds of batteries.

Lead is obtained by roasting the mineral galena (PbS) in air to lead oxide, then reducing the oxide with carbon. Lead compounds and their uses include tetraethyl lead for gasoline, lead dioxide (PbO_2) for the positive plates of batteries, lead monoxide (PbO) for rubber processing and glass and ceramic glaze manufacture, and red lead (Pb_3O_4) for undercoat paints on structural steel.

The chemistry of lead in relation to air pollution is described in Section 13-4.

VA: The nitrogen or phosphorus group

The same change from nonmetallic to metallic properties with increasing atomic number noted among the IVA elements is evident in VA elements (Table 8-7): Nitrogen and phosphorus are nonmetals, arsenic is a metalloid,

Table 8-7. VA elements

Element	Electron shells	Oxidation states	Density g/ml	Melting point °C
Nitrogen, N	2,5	5,4,3±,2±,1±	0.81[a]	−210
Phosphorus, P	2,8,5	5,3±,2−,1	1.82[b]	44[b]
Arsenic, As	2,8,18,5	5,3±	5.72	817
Antimony, Sb	2,8,18,18,5	5,3±	6.62	630
Bismuth, Bi	2,8,18,32,18,5	5,3±	9.8	271

[a] Liquid, at its boiling point.
[b] White allotrope.

antimony is more metal than metalloid, bismuth is definitely a metal. Each element is capable of forming two or more series of compounds as a consequence of variable usage of valence electrons in chemical bonding.

Nitrogen (N_2) is a colorless, odorless gas. It comprises 78 percent by volume of the atmosphere and also occurs with methane in the ground; N_2 is the least active of the VA elements and the most stable diatomic molecule known. Nitrogen is obtained commercially (Table 8-2) by liquefying air then separating nitrogen from oxygen by distillation. Liquid nitrogen (b.p. $-196°C$) is used to displace air from oxygen-sensitive processes, freeze-dry foods, cool high-voltage electric cables to increase power-carrying capacity, cool patients in low-temperature surgery, suffocate insects in grain silos, and refrigerate long-distance shipments of perishable goods.

Main nitrogen minerals are nitrates: saltpeter (KNO_3) and Chile saltpeter ($NaNO_3$). The four most important nitrogen compounds (Table 8-2) are ammonia, ammonium nitrate, and ammonium sulfate, all used as fertilizers, and nitric acid used to manufacture nitrates. Ammonia is made commercially by direct combination of N_2 and H_2 (see Section 17-1). Nitric acid is synthesized by oxidizing ammonia (see Question 6, Chapter 17.)

Complex organic nitrogen compounds are found in every form of life. Their source is nature's nitrogen cycle which operates both through the ocean (Figure 15-3) and the earth. In the latter atmospheric nitrogen is "fixed" (caused to combine chemically) by soil bacteria associated with the roots of plants such as clover, and by electrical storms which produce nitrogen oxides that are carried down from the atmosphere by rain. Nitrogen as nitrate ion (NO_3^-) is taken up by plants and formed into proteins that are eaten by animals. Animals excrete nitrogen compounds, and both plants and animals eventually die and by decay return their nitrogen to the cycle.

Phosphorus can be prepared in several allotropic forms, the main ones being the stable red form (P_x, where x is a large number), and the poisonous and spontaneously flammable white form (P_4). White phosphorus changes slowly to the red form when heated in the absence of air. The main phosphorus mineral is apatite, a calcium phosphate, whose origin was biological organisms. Elemental phosphorus is obtained by heating the mineral with sand and carbon in an electric furnace. Ninety percent of phosphorus production is used to make phosphoric acid (Table 8-2) via tetraphosphorus decoxide:

$$2Ca_3(PO_4)_2 + 6SiO_2 + 10C \overset{\Delta}{=} P_4 + 10CO + 6CaSiO_3$$

$$P_4 + 10CO + 10O_2 \overset{\Delta}{=} P_4O_{10} + 10CO_2$$

$$P_4O_{10} + 6H_2O = 4H_3PO_4 \tag{8-8}$$

The H_3PO_4 is used to make fertilizers such as "triple superphosphate" (calcium dihydrogen phosphate, $Ca(H_2PO_4)_2$) and alkaline phosphates such as sodium metaphosphate ($NaPO_3)_6$, for use in water softening and detergents (see Section 15-3). Other inorganic phosphorus compounds include tetraphosphorus trisulfide (P_4S_3), used in the heads of matches, and sodium hypophosphite (NaH_2PO_2), a reducing agent for the electroless plating of nickel and cobalt (Eq. 18-12).

Two synthetic organic phosphorus derivatives are the powerful insecticide parathion and the nerve gas tabun, respectively,

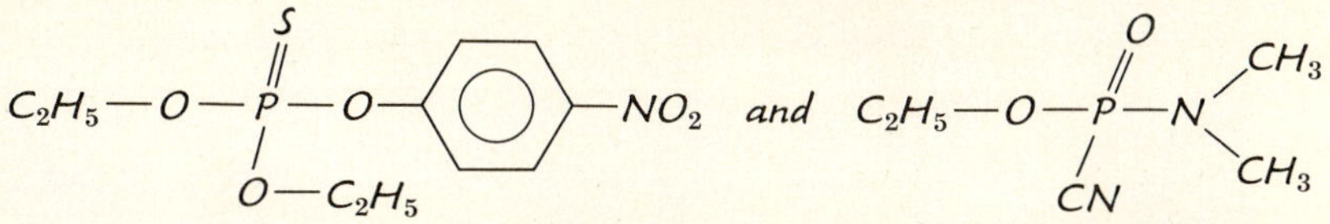

$$C_2H_5-O-\overset{\overset{S}{\parallel}}{\underset{\underset{O-C_2H_5}{|}}{P}}-O-\!\!\langle\bigcirc\rangle\!\!-NO_2 \quad and \quad C_2H_5-O-\overset{\overset{O}{\parallel}}{\underset{\underset{CN}{|}}{P}}-N\overset{CH_3}{\underset{CH_3}{<}}$$

These compounds can cause death by inhibiting the biological catalyst (enzyme) system necessary for normal nerve and muscle function.

Like nitrogen, phosphorus passes through living things as part of a massive natural cycle (see Figure 15-4).

Arsenic, antimony, and bismuth are more similar to one another than to nitrogen and phosphorus. All show allotropy. Both arsenic and antimony can be prepared in gray form—brittle metalloids that conduct electricity—and in yellow form—soft waxy solids. Arsenic, antimony, and bismuth are incorporated in low-melting alloys. The elements are prepared by roasting the sulfide ores to oxides, then reducing the oxides with carbon. Typical minerals are orpiment (As_2S_3), stibnite (Sb_2S_3), and bismite ($Bi_2O_3 \cdot H_2O$).

Compounds of arsenic, antimony, and bismuth are poisonous. Cacodylic acid (dimethylarsinic acid, $(CH_3)_2AsO_2H$) is an effective defoliant. Lead hydrogen arsenate ($PbHAsO_4$) controls fruit tree insects. Sodium arsenite ($NaAsO_2$) has been used to kill trees, aquatic weeds, and termites. Use of heavy metal poisons is being curtailed. Principal antimony chemical is the trioxide (Sb_2O_3); it is the preferred flame retardant for all major plastics.

Two organic arsenicals of historical interest are salvarsan, the first drug for the treatment of syphilis, synthesized by Paul Ehrlich (Germany 1854–1915, Nobel prize in medicine or physiology 1908), and lewisite, a toxic gas invented in the United States in 1918 for military use:

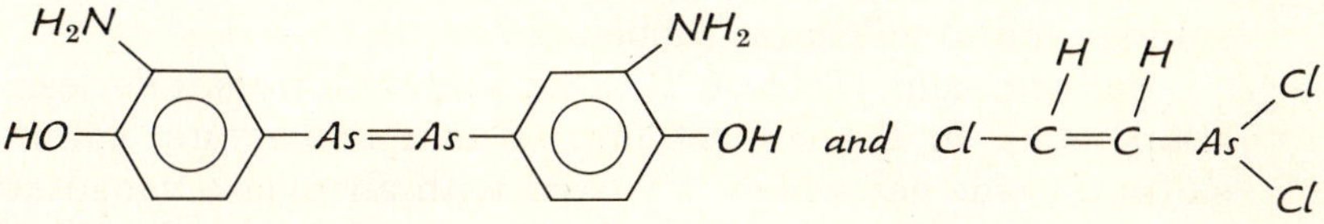

$$HO-\!\!\langle\bigcirc\rangle\!\!\overset{H_2N}{-}As=As-\!\!\langle\bigcirc\rangle\!\!\overset{NH_2}{-}OH \quad and \quad Cl-\overset{\overset{H}{|}}{C}=\overset{\overset{H}{|}}{C}-As\overset{Cl}{\underset{Cl}{<}}$$

VIA: The chalcogen group

Oxygen and sulfur are abundant nonmetals; selenium, tellurium, and polonium are scarce metalloids (Table 8-8). Of the group, oxygen is the most electronegative; that is, it tends to gain electrons in chemical reactions.

Oxygen is a colorless, odorless, tasteless gas at ordinary temperatures. It is the most widely used industrial gas (Table 8-2) and the most abundant element in our environment. By weight, oxygen comprises 50 percent of silicate rocks, 89 percent of water, and 23 percent of air. It is obtained as a light-blue liquid (b.p., $-183°C$) from liquid air.

Two-thirds of all industrial oxygen goes into iron and steel making; air enriched a few percent with oxygen gives higher furnace temperatures and thus conserves fuel. Largest future markets for oxygen will be in rocket propulsion, chemical processing, firing glass-making furnaces and cement kilns, welding, bleaching paper, and treating sewage and polluted water.

Table 8-8. VIA elements

Element	Electron shells	Oxidation states	Density g/ml	Melting point °C
Oxygen, O	2,6	2−	1.14[a]	−219
Sulfur, S	2,8,6	6,4,1−,2−	2.07	119
Selenium, Se	2,8,18,6	6,4,2−	4.79	217
Tellurium, Te	2,8,18,18,6	6,4,2−	6.24	450
Polonium, Po	2,8,18,32,18,6	4,2	9.2	254[b]

[a] Liquid O_2 at the boiling point.
[b] Best known isotope.

Atmospheric oxygen is diatomic O_2. Its poisonous allotrope ozone is triatomic O_3. Ozone is produced in the laboratory by passing a silent electric discharge through oxygen; O_3 is also produced naturally in the atmosphere (Eq. 13-5). The use of oxygen by all animals in the process called "respiration" is described in Section 22-5.

Oxygen is chemically active, especially at elevated temperature. It combines with all elements except helium, neon, and argon. Water, carbon dioxide, sand, and rust are familar oxides.

Sulfur was known to early civilizations (sulfur matches were used in ancient Rome), who found it in elemental form around volcanoes. Today most sulfur is produced by the Frasch process in which compressed air and superheated water (150°C) are forced into underground sulfur deposits to bring molten sulfur to the surface through a system of concentric pipes. Secondary sulfur sources, those requiring chemical reactions to obtain the element, include furnace gases, mineral sulfides and sulfates, coal, and "sour" gas — natural gas containing hydrogen sulfide (H_2S). Ordinary sulfur is a bright-yellow crystalline solid with orthorhombic symmetry (see Figure 6-6). Monoclinic and amorphous allotropes are also known. Principal use of sulfur is in sulfuric acid synthesis. Elemental sulfur is also used to make sulfides and to vulcanize rubber.

Sulfuric acid (Table 8-2) is an oily, colorless, odorless liquid, and a strong acid. In the contact process of manufacture, sulfur is burned to sulfur dioxide gas which is passed with air over a vanadium or platinum catalyst to make sulfur trioxide. The latter is absorbed in dilute sulfuric acid to make concentrated acid:

$$S \xrightarrow{O_2} SO_2 \xrightarrow[cat.]{O_2} SO_3 \xrightarrow{H_2O} H_2SO_4 \tag{8-9}$$

One-third of all sulfuric acid is used to make phosphate fertilizers. The remainder goes to synthesize chemicals, pigments, and fibers; process crude oil; leach nonferrous metals from ores (Figure 4-19a); make paper, explosives, and alcohols; and pickle (clean) metals before plating.

Sulfur dioxide is employed as a bleach and disinfectant in the dried fruit industry and to synthesize sulfites. Calcium hydrogen sulfite ($Ca(HSO_3)_2$) is used to dissolve lignin (a natural binder that is 10 to 20 percent of the weight of wood) so the remaining cellulose fibers can become paper pulp.

Carbon disulfide (CS_2) is made by direct combination of carbon and sulfur in an electric furnace at about 900°C. It is used in the viscose process

(1892) in which cellulose is dissolved in a mixture of sodium hydroxide and carbon disulfide, coagulated with sodium chloride, then forced as a fiber from a small hole or as a sheet from a narrow slot into an acid bath for hardening. The product is regenerated cellulose; the fiber is called "rayon," the sheet "cellophane." Synthetic organic sulfur compounds include drugs, insecticides, and dyes.

Selenium and tellurium have semiconducting properties. They occur with sulfur and with copper minerals from which they are obtained as by-products.

Plants such as locoweed, in which selenium becomes concentrated from the soil, often have a garlic odor and can be toxic to grazing animals. Some selenium is needed in the diet, however, to prevent serious diseases. The electrical conductance of gray selenium increases with the intensity of visible light that falls on it, making it useful in photoelectric cells. It is also used in steel alloying and to impart red color to glass and ceramics. Tellurium is used to toughen rubber for cable insulation and in steel and copper alloys to increase their machinability.

Polonium is a radioactive element in the uranium decay series (see Figure 4-3).

VIIA: The halogen group

The halogens (Gk. *halo + gene*, salt producing) (Table 8-9) are chemically active nonmetals. They are volatile (easily vaporized), form volatile compounds, have irritating odors, and are poisonous. In general they have odd-number oxidation states. In nature they occur as halide ions (X^-), but in the free state their form is diatomic molecules (X_2). Seawater is 1.9 percent Cl^- by weight. In the earth's crust the percentages are as follows: F^-, 0.1; Cl^-, 0.2; Br^- and I^-, 10^{-3}.

Fluorine is a pale yellow gas (b.p., $-187°C$). It was prepared first by Henri Moissan (France 1852–1907, Nobel prize in chemistry 1906) via electrolysis of an anhydrous mixture of potassium hydrogen fluoride (KHF_2) and hydrogen fluoride (HF). Essentially the same method is used today. The main mineral source is fluorspar (CaF_2).

Fluorine is the most electronegative element known. Combustibles ignite spontaneously in it. Products from the reaction of fluorine and wood include HF, O_2, and C. Fluorine is used in rocket propulsion ($H_2 + F_2 = 2HF + heat$) and in the preparation of various fluorides. Fluorine can be

Table 8-9. VIIA elements

Element	Electron shells	Oxidation states	Density g/ml	Melting point °C
Fluorine, F	2,7	1—	1.50[a]	−220
Chlorine, Cl	2,8,7	7,6,5,4,3,1,1—	1.56[a]	−101
Bromine, Br	2,8,18,7	5,4,1,1—	3.12	−7.2
Iodine, I	2,8,18,18,7	7,5,4,1,1—	4.94	114
Astatine, At	2,8,18,32,18,7	7,5,3,1,1—		302[b]

[a] Liquids, at the boiling point.
[b] Best known isotope.

stored and shipped in steel tanks because inert iron fluoride forms on the surface.

Hydrogen fluoride gas is used to make refrigerant gases like freon 12 (dichlorodifluoromethane CCl_2F_2), high-melting plastics such as teflon $(—CF_2—CF_2—CF_2—)_x$ and synthetic cryolite (Na_3AlF_6) for aluminum metal production. A water solution of hydrogen fluoride is hydrofluoric acid. A corrosive liquid, the acid dissolves glass and other silicates. It is stored in certain kinds of plastic bottles.

Chlorine (Gk. *chloros*, yellow–green) is a yellowish gas. It was discovered by K. Scheele in 1774, who prepared it by oxidizing muriatic acid (hydrochloric acid). The main mineral is halite (NaCl), found in salt-dome formations and salt water.

Chlorine is not quite as active as fluorine. Most commercial chlorine comes from the electrolysis of brine (see Table 8-2 and Figure 18-7). Chlorine consumption is a measure of activity in the chemical industry. The four most important uses are (1) preparation of chlorinated organic compounds (Chapter 21) such as ethyl chloride (ethyl gasoline), vinyl chloride (plastics), and chloroform (insecticides); (2) bleaching paper pulp; (3) disinfecting public water supplies; and (4) making inorganic chemicals such as household bleach (sodium hypochlorite),

$$2NaOH + Cl_2 = NaOCl + NaCl + H_2O$$

Hydrogen chloride gas dissolved in water is hydrochloric acid (HCl). It is a fuming, corrosive liquid used in chemical manufacture, metal cleaning, and oil well treatment (to stimulate oil flow through tight formations).

Bromine (Gk. *bromos*, stench) is a red–brown fuming liquid at room temperature. It is recovered by displacement with chlorine from natural salt waters that contain bromide ion:

$$Cl_2 + 2Br^- = Br_2 + 2Cl^- \tag{8-10}$$

Three-fourths of all bromine produced has in the recent past been used to make dibromoethane ($C_2H_4Br_2$), a constituent of leaded gasoline. The rest is utilized directly in disinfecting water, in synthesizing methyl bromide (CH_3Br) and other fumigants, and in making flame-retardant chemicals, hydrobromic acid (HBr, a strong acid like HCl), and lithium bromide (LiBr), the latter a water absorber employed in air conditioning systems.

Iodine (Gk. *iodes*, purple) is a black crystalline solid that gives purplish vapor when heated. It is obtained by chlorine displacement from seawater, oil well brines, and kelp (a seaweed). In human beings iodine concentrates in the thyroid, a gland, located in the neck, that regulates body growth.

In organic solvents iodine is purple, indicative of I_2 molecules; I_2 is nearly insoluble in water but quite soluble in iodide solutions, where it forms brown triiodide ion I_3^-. Largest markets for iodine are in compounds serving as catalysts, medicinals, and feed supplements. Iodoform (CHI_3) is a disinfectant. Some photographic film contains AgI. Alcoholic solutions (tinctures) of iodine and KI are used as antiseptics. Hydriodic acid (HI) is a strong acid.

Astatine is a synthetic radioactive element. Its most stable isotope has a half-life of 8.3 hours. It is more metallic than iodine but has some similar properties; taken internally it migrates to the thyroid gland.

Table 8-10. 0 elements

Element	Electron shells	Oxidation states	Density g/ml	Melting point °C
Helium, He	2	0	0.126	−270
Neon, Ne	2,8	0	1.20	−249
Argon, Ar	2,8,8	0,2	1.40	−189
Krypton, Kr	2,8,18,8	0,2,4	2.6	−157
Xenon, Xe	2,8,18,18,8	0,2,4,6	3.06	−112
Radon, Rn	2,8,18,32,18,8	0,2,4,6		−71[a]

[a] Best known isotope.

Group 0: The noble gas group

The relationships of helium and radon to nuclear processes have been noted in previous chapters. Discovery of the Group 0 elements (Table 8-10) is described with Figure 13-6.

Helium forms no chemical compounds. The other noble gases, so named for their reluctance to combine, form a few compounds. Inertness is due to their closed electron shells (see Figure 7-5) which contain totals of 2, 10, 18, 36, 54, and 86 electrons. The first noble gas compound, xenon hexafluoroplatinate ($XePtF_6$), was not synthesized until 1962 (by Canadian chemist N. Bartlett), although 30 years before that Pauling (see Figure 9-3) predicted such compounds could exist. Since then XeF_2, XeO_3, RnF_4, and others have been prepared. Some are explosively unstable.

Helium is found in natural gas wells in concentrations up to 5 percent. It can be separated as a gas (b.p., −269°C) after liquefying the other gases by cooling and compressing. The National Aeronautics and Space Administration (NASA) takes about half the helium produced for rocket fuel pressurization. The gas is also utilized in balloons and as an additive to oxygen in breathing apparatus. Liquid helium is a valuable coolant in laboratory work.

Neon, argon, and krypton are used in certain electric light bulbs and as inert atmospheres in the processing of active metals.

8-2. B SUBGROUP ELEMENTS AND GROUP VIII

The transition metals

The 33 elements occupying a central block between Groups IIA and IIIA in the periodic table are called *transition elements* (Figure 8-2). "Transition" describes elements that bridge the gap between the highly metallic elements of Groups IA and IIA and the elements on the table's right side

IIIB	IVB	VB	VIB	VIIB	VIII			IB	IIB
21 Sc	22 Ti	23 V	24 Cr	25 Mn	26 Fe	27 Co	28 Ni	29 Cu	30 Zn
39 Y	40 Zr	41 Nb	42 Mo	43 Tc	44 Ru	45 Rh	46 Pd	47 Ag	48 Cd
57− La	72 Hf	73 Ta	74 W	75 Re	76 Os	77 Ir	78 Pt	79 Au	80 Hg
89− Ac	104 Rf	105 Ha							

Figure 8-2 The B subgroup and Group VIII section of the periodic table.

whose trend from IIIA to 0 is toward nonmetallic character. Some transition elements are common (Fe, Ti) in nature, whereas others are rare (Rh, Ir). Properties vary irregularly from group to group and also within a group, although hardness and high melting point are typical. All are metals.

Chemical behavior of the transition elements is related to the filling of d orbitals (see Figure 7-8). Because five *d* orbitals require 10 electrons for filling, each complete series — Sc through Zn, Y through Cd, and La through Hg — contains 10 members. Indicative of electron-level buildup, the first series begins

$$_{21}Sc: \; 2, \, 8, \, 9, \, 2; \qquad _{22}Ti: \; 2, \, 8, \, 10, \, 2; \qquad _{23}V: \; 2, \, 8, \, 11, \, 2$$

Or in orbital notation,

$$_{21}Sc: \; 1s^2 2s^2 2p^6 3s^2 3p^6 3d^1 4s^2; \qquad _{22}Ti: \; 1s^2 2s^2 2p^6 3s^2 3p^6 3d^2 4s^2; \qquad etc.$$

Most of the elements use various combinations of valence electrons from their two unfilled shells (shells 3 and 4 in the first long period) in forming compounds, and thus they exhibit more than one oxidation state. The states cannot be predicted by a simple rule.

Transition metal ions generally have one or more electrons unpaired (one or more orbitals only half-filled), which makes them paramagnetic. Consider manganous chloride. In forming $MnCl_2$ the manganese *atom*, 2, 8, 13, 2, donates two 4s electrons to the chlorine atoms, giving the Mn^{2+} *ion* with the electronic arrangement $1s^2 2s^2 2p^6 3s^2 3p^6 3d^5$. Paramagnetism increases with increasing numbers of unpaired electrons, and measurement shows Mn^{2+} has a paramagnetism corresponding to five unpaired *d* electrons. Such observations have led to the conclusion that orbitals in a sublevel usually are singly occupied before electron pairing in them commences, a generalization known as *Hund's rule of maximum multiplicity*.

Color and ease of complexation of transition metal ions are also related to incomplete d levels. Color results mainly from the absorption of specific quanta of energy in the visible region by the easily excited unpaired electrons. As will be explained in Section 10-1 incomplete sublevels favor formation of complexes like the dark-blue tetrammine copper(II) ion:

$$Cu^{2+} + 4NH_3 = Cu(NH_3)_4^{2+} \tag{8-11}$$

Complexes that are more deeply colored than the simple metal ions help analysts identify the metals.

Group IIIB: Sc, Y, La, Ac

The IIIB metals are scarce and presently of little use. They are strongly electropositive (oxidation number 3+). Actinium is radioactive.

Group IVB: Ti, Zr, Hf

Titanium is the tenth most abundant element in igneous rocks. Zirconium is less common and hafnium is rare. The group's most important oxidation number is 4+. The free metals are obtained from their oxides by the Kroll process in which conversion to tetrachlorides is followed by reduction with a magnesium–sodium alloy in an inert atmosphere of argon,

$$ZrO_2 \xrightarrow{\; Cl_2 \;} ZrCl_4 \xrightarrow[Ar]{\; Mg\text{-}Na \;} Zr + MgCl_2 + NaCl$$

Titanium is a tough, relatively light (sp. gr., 4.49) metal used in jet aircraft and other construction. The important mineral is rutile (titanium dioxide, TiO_2). Refined TiO_2 has exceptional whiteness and "hiding" ability — properties that make it useful in paper and paint manufacture.

Zirconium metal is used in some steel alloys and by itself in constructing atomic reactors (it absorbs few neutrons). One mineral is zircon ($ZrSiO_4$). Zirconium dioxide is employed as a furnace lining.

Hafnium accompanies zirconium in its ores. Its alloys are high melting.

Group VB: V, Nb, Ta

The VB metals are fairly scarce. Their most important oxidation state is 5+.

Vanadium is incorporated in steels to strengthen them. Vanadium pentoxide (V_2O_5) is a catalyst in the manufacture of sulfuric acid and other compounds. One mineral is vandinite, a lead chloride vanadate, $Pb_5Cl(VO_4)_3$. With oxidation numbers ranging from 2+ to 5+ vanadium forms an array of colored compounds.

Niobium (formerly called "columbium") is alloyed in steels.

Tantalum has outstanding resistance to such corrosive substances as hydrochloric acid. The pure metal is used in the manufacture of chemical process equipment and high-temperature electronic devices.

Group VIB: Cr, Mo, W

The VIB metals are used to harden steel. The principal oxidation state is 6+. Chromium also exists in the 2+ and 3+ states. Molybdenum and tungsten oxidation states are 2+, 3+, 4+, 5+, and 6+. The oxides are not soluble in acid but they can be put into solution as Na_2MO_4 (M the 6+ metal) by heating with an alkali:

$$M_2O_3 \xrightarrow[air, \Delta]{Na_2CO_{3)}} Na_2MO_4$$

Chromium is incorporated in *stainless steels*, the most widely used corrosion-resistant alloys. Chromium is obtained from the mineral chromite ($Fe(CrO_2)_2$). Pure chromium is electroplated from chromic acid (CrO_3) solutions onto automobile bumpers, household appliances, and brass plumbing fixtures to give them a shiny, durable surface. Most chromium chemicals are manufactured from sodium chromate (Na_2CrO_4) prepared as above. Acid solutions of sodium dichromate 2-hydrate ($Na_2Cr_2O_7 \cdot 2H_2O$) are used to tan leather, a process in which chromium compounds formed with proteins in the animal skin improve its strength and resistance to bacterial attack. Green chromic oxide (Cr_2O_3) and yellow lead chromate ($PbCrO_4$) are paint pigments.

Sources of molybdenum and tungsten metals are the minerals molybdenite (MoS_2) and scheelite ($CaWO_4$). Tungsten has the highest melting point of any metallic element (3410°C); molybdenum, the fifth highest (2610°C). The metals are prepared first as powders by reduction of their trioxides with hydrogen at high temperature; they are then compressed and hot rolled to produce compact metal. Their very hard carbides (MoC, WC) are used in the cutting edges of metal shaping tools and as studs embedded in snow tires. Lamp filaments, spark plug contacts, and X ray tube anodes are made from tungsten. Lubricants, catalysts for petroleum refining, and organic phosphors for television screens are prepared from molybdenum chemicals. Molybdenum orange, a mixture of lead molybdate $Pb(MoO_4)$ and other compounds, is the most important inorganic pigment developed in the last 50 years.

The simple chromates, molybdates, and tungstates (MO_4^{2-}) have similar properties. In acid solution the ions condense by water elimination as illustrated by conversion of yellow chromate ion to orange dichromate ion:

$$2CrO_4^{2-} + 2H^+ \rightleftharpoons Cr_2O_7^{2-} + H_2O \tag{8-12}$$

This tendency makes molybdenum and tungsten chemistry complicated, for the condensation continues:

$$MoO_4^{2-} \xrightleftharpoons{H^+} Mo_3O_{11}^{4-} \xrightleftharpoons{H^+} HMo_6O_{21}^{5-}, \text{ etc.}$$

Reducing 6+ chromium to a lower oxidation state gives chromic ion (Cr^{3+}). Reducing 6+ molybdenum and 6+ tungsten, however, yields a mixture of hydrated oxides $(Mo_8O_{23} \cdot xH_2O$, etc.) in which the metal appears in two oxidation states. Deep colors of such compounds identify the metals.

Group VIIB: Mn, Tc, Re

Manganese is the only important VIIB element. Technetium is produced in uranium fission. It is not found naturally, having decayed since the earth's formation ($t_{1/2}$ for ^{99}Tc is 10^6 yr), but its spectral lines are detectable in some stars where heavier elements are being made. Rhenium was discovered in ores of metals near it in the periodic table — an application of the periodic law. Only carbon and tungsten among the elements have a higher melting point (3180°C). Rhenium is rare.

Manganese is found on the continents and on the sea floor as the mineral pyrolucite (manganese dioxide). Carbon reduction of MnO_2 gives the metal. Manganese toughens steel. Hadfield's manganese steel (10 to 14 percent Mn) is used to make railway crossings, rock-crusher jaws, tractor shoes, etc., because it work hardens and resists wear. Dry cell battery electrodes are made with MnO_2 (Figure 19-5). The important oxidation state of manganese is 2+; other states are 3+, 4+, 6+, and 7+. A representative synthetic manganese compound is potassium permanganate $(KMnO_4)$. It is a dark purple substance used in laboratory analysis as an oxidizer. It is manufactured via potassium manganate:

$$MnO_2 \xrightarrow[air, \Delta]{KOH,} K_2MnO_4 \xrightarrow[CO_2]{H_2O} KMnO_4 + K_2CO_3 + MnO_2$$

Group VIII: Fe, Co, Ni; Ru, Rh, Pd; Os, Ir, Pt

The convention begun by Mendeleev (Figure 7-2a) of classifying Group VIII elements in a block continues to be observed. For our discussion we will consider the trio iron, cobalt, and nickel as one unit and the other six elements as another.

Iron is the second most abundant metal in the earth's crust. World iron reserves total 750 billion tons, with plenty on all continents. Iron is obtained from hematite (Fe_2O_3) and magnetite (Fe_3O_4) ores by smelting (fusing) with a mixture of coke and limestone in a blast furnace. A hot air–oxygen oxidizing mixture keeps the furnace operating at 1300°C. While carbon monoxide reduces iron, silica from the ore and lime from the limestone react to give a molten silicate slag:

$$\left\{ \begin{array}{ll} 2C + O_2 \overset{\Delta}{=} 2CO & 3CO + Fe_2O_3 \overset{\Delta}{=} 3CO_2 + 2Fe \\ CaCO_3 \overset{\Delta}{=} CO_2 + CaO & CaO + SiO_2 \overset{\Delta}{=} CaSiO_3 \end{array} \right\} \tag{8-13}$$

Slag and molten iron in a 1:2 weight ratio are tapped separately from the furnace bottom. The iron, containing embrittling impurities of phosphorus, carbon, silicon, oxygen, nitrogen, and sulfur, is cast in large molds to produce ingots. Traditionally, refining has then been done simply by remelting the ingots to oxidize impurities and bring them to the surface. In the newer electroslag process the impure ingot is suspended vertically with its tip resting on a slag bed in a furnace. By passing a high-density electric current through the assembly, the slag is kept molten and the tip is melted. As iron drips through the slag, sulfur, nitrogen, and oxygen are removed therein by chemical reactions. Purified steel solidifies below the bed while impurities concentrate in the uppermost, liquid part (see Question 7).

When iron is heated, its properties change. At 768°C, called the *Curie temperature*, the ordered electron spins responsible for ferromagnetism become random and the metal loses its strong magnetism. Ordinary or alpha-iron has the body-centered cubic lattice (see Figure 6-6). At 912°C alpha-iron changes to another allotropic form, gamma-iron, which has a face-centered cubic lattice (Figure 6-8). At 1410°C gamma-iron changes to delta-iron, which again has the body-centered lattice. That arrangement is retained to the melting point 1535°C.

Steels are iron alloys containing 0.1 to 1.5 percent carbon and usually other elements as well. When molten carbon steel is quenched (cooled quickly), allotropic change is impeded and remains incomplete. The result is martensite, an alloy of carbon and gamma-iron that is hard and brittle. When martensite is heat treated—that is, heated and held at a temperature somewhat below the melting point—cementite or iron carbide (Fe_3C) appears as separate granules in crystals of alpha-iron. In this disordered structure atom layers will not readily slide by one another when stressed. The steel therefore has the desirable characteristics of being harder than pure iron and tougher than martensite.

Principal iron oxidation states are 2+ and 3+. Bivalent ferrous compounds are usually green. An example is ferrous sulfate 7-hydrate ($FeSO_4 \cdot 7H_2O$), a by-product from the pickling of iron. Tervalent ferric compounds are usually brownish due to surface formation of red–brown ferric hydroxide ($Fe(OH)_3$). Two common compounds in which ferrous and ferric iron appear in the negative ion are potassium ferrocyanide 3-hydrate ($K_4Fe(CN)_6 \cdot 3H_2O$) and potassium ferricyanide ($K_3Fe(CN)_6$). Ferric compounds give Prussian blue with the former whereas ferrous compounds give Turnbull's blue with the latter. The formation of these pigments, both approximately $KFe(Fe(CN)_6) \cdot 6H_2O$, identifies ferric and ferrous ions in solution.

Cobalt is less active than iron; it is also harder and stronger and ferromagnetic below 1150°C. One mineral is the arsenide, smaltite ($CoAs_2$). Chief use of cobalt is as an alloying element. Alloy examples are alnico (a Co–Fe–Al–Ni mixture) from which powerful electromagnets are made, and the Haynes stellites (Co–Cr–W), useful in making cutting tools. Cobalt oxidation states are 2+ and 3+. The latter is stable in water solution only in the form of such complex compounds as sodium cobaltinitrite ($Na_3Co(ONO)_6$). Cobaltous oxide (CoO) imparts blue color to glass and ceramics.

Main uses of nickel metal are in coinage and corrosion-resistant alloys like stainless steel and monel (75 percent Ni, 25 percent Cu). Largest deposits known are the nickel–iron sulfides such as pentlandite ($2FeS \cdot NiS$) found in New Caledonia and Ontario, Canada. Impure nickel is obtained by oxidizing the sulfides in air and reducing the oxide with carbon. Nickel free of iron and cobalt is then obtained by converting it to nickel carbonyl, separating the compound (a toxic gas), then heating to decompose it,

$$NiS \xrightarrow[\Delta]{air} SO_2 + NiO \xrightarrow[\Delta]{C} CO_2 + Ni \xrightarrow[\Delta]{CO} Ni(CO)_4 \xrightarrow{\Delta} CO + Ni$$

In nickel carbonyl the metal's oxidation number is 0. More familiar nickel oxidation states are 2+ and 4+. Representative compounds are nickel sulfate 6-hydrate ($NiSO_4 \cdot 6H_2O$), a source of the metal for electroplating, and hydrated nickel dioxide ($NiO_2 \cdot xH_2O$), a coating for the anode in some alkaline storage batteries.

Ruthenium, rhodium, palladium, osmium, iridium, and platinum are called the "noble" or platinum metals. They are rare (Pt sells for more than \$100/oz), high melting, dense (Os and Ir are the densest of all elements), ductile, malleable, and resistant to chemical attack. They are found chemically uncombined in nature but physically mixed with one another as alloys that require multistep chemical processing to separate.

Platinum was the first of the group known, its decorative use by Central and South American indians having been described by sixteenth century European explorers. Platinum is especially valuable as a catalyst in industrial gas reactions (see Figure 17-4). If 0.1 ounce must be used in an automobile exhaust system, a 10-million car production year will require 1 million ounces, or roughly 25 percent of the world's supply. Platinum is used to catalyze oxidations (nitric acid manufacture), reductions, and molecular rearrangements (producing high-octane gasoline from low-octane hydrocarbons). Other applications that take advantage of its inertness and high melting point (1769°C) are found in the manufacture of temperature-measuring instruments, laboratory crucibles, dental bridges, and jewelry. Platinum is readily sealed in glass to make equipment such as electrodes because platinum and soft glass have about the same coefficient of cubical expansion.

Oxidation states of the platinum metals are commonly 2+ and 4+.

Group IB: Cu, Ag, Au

Copper, silver, and gold, known as *coinage metals* in the era of hard currency, are exceptionally ductile and malleable (gold outranks all elements in these respects). The pure metals are soft, dense, melt at about 1000°C, and have high electrical and thermal conductivities. In nature, copper and silver are found both free and chemically combined, whereas gold is found free only—evidence that IB elements are reluctant to react.

Copper, "the red metal," is obtained from such minerals as chalcopyrite ($CuFeS_2$) and malachite ($CuCO_3 \cdot Cu(OH)_2$) via roasting and reducing with carbon; it is then refined electrolytically (Figure 18-6). Copper is fabricated into electric wiring and switch gear, heat exchange tubing, roofing sheets, and water pipe. It is also used in alloys such as brass (Cu–Zn) and bronze (Cu–Sn) which are corrosion resistant and easily formed. Bronze implements 5000 years old have been found in the artifacts of Mesopotamia. (The Bronze Age followed the Stone Age and preceded the Iron Age.)

The oxidation states of copper are 1+ (cuprous) and 2+ (cupric). Compounds of the former are generally white and easily oxidized to cupric compounds. The latter, which are more important, are generally blue or green. Copper metal dissolves in nitric acid giving cupric nitrate ($Cu(NO_3)_2$). Sulfuric acid converts copper oxide to cupric sulfate. Evaporation yields blue crystals of cupric sulfate 5-hydrate ($CuSO_4 \cdot 5H_2O$), a familiar compound used to make copper-containing antifouling paints, insecticides, and fungicides. Copper, its compounds, and its alloys give a green flame test.

Silver and gold alloys are fabricated into jewelry, dental work, and tableware. Sterling silver is 92.5 percent silver, 7.5 percent copper. Twenty-four-carat gold is 100 percent gold, 12-carat gold is an alloy containing 50 percent gold. Silver and gold are electroplated from their cyanide solutions onto objects such as electronic parts to protect surfaces from corrosion. Metallic silver is chemically deposited from alkaline silver solutions onto glass to make mirrors. Silver is said to tarnish when its surface becomes coated with black silver sulfide (Ag_2S).

Native silver and gold are recovered from crushed ores by water washing (denser metal settles as gravel is carried away). Further concentration is made by dissolving the metals either in metallic mercury (which is later distilled off) or in aqueous sodium cyanide. Cyanide products are decomposed with zinc:

$$Au \xrightarrow{NaCN} NaAu(CN)_2 \xrightarrow{Zn} Au$$

Oxidation states are Ag^+, Au^+ (aurous), and Au^{3+} (auric).

The most important silver compound is the nitrate,

$$3Ag + 4HNO_3 = 3AgNO_3 + NO + 2H_2O \tag{8-14}$$

Silver nitrate is unusually soluble in water and used to prepare other silver chemicals.

Large quantities of silver compounds are used in photography, including some photocopy processes. Photographic film is made by thinly coating a sheet of plastic with an aqueous mixture of silver nitrate, potassium bromide, and gelatin (a protein). The layer produced, called a *photographic emulsion*, contains a dispersion of tiny silver bromide crystals. As spread on 35 mm film, several billion crystals are available for each exposure (about 1.3 in.^{2}). Silver bromide is sensitive to photons. When struck by light, it yields a latent image of the subject photographed. Imaging apparently begins with electrons (activated by photons) migrating in crystals until trapped at active crystal sites. Sites become activated when they acquire traces of silver sulfide by reaction with sulfur atoms from the gelatin. It is there that electrons and silver ions interact to give nuclei of metallic silver,

$$Ag^+ + e^- = Ag$$

When the film is treated with an alkaline reducing solution called a "developer," further reduction to silver starts at the nuclei and the latent image darkens as a metallic silver negative. It is densest where exposed most to light, least dense where exposed least.

Unreacted silver halide is then removed as complex ions by washing in a solution containing sodium thiosulfate:

$$AgBr \xrightarrow{Na_2S_2O_3} Ag(S_2O_3)_2{}^{3-}$$

Finally, the film is water washed and dried. A positive print is made by shining a light through the negative and focusing the image for a few seconds onto a piece of paper coated with photographic emulsion. The paper is then subjected to the process of developing, fixing, washing, and drying.

Silver halide particles about 10^{-5} cm in diameter incorporated in window glass give a product called "photochromic" glass which is being installed in office buildings. Exposed to sunlight, it darkens because metallic silver forms

$$AgCl \xrightarrow{h\nu} Ag + Cl$$

When sunlight no longer strikes the glass, it lightens as the reaction reverses itself.

Gold compounds are used mainly in the electroplating of electronic hardware.

Group IIB: Zn, Cd, Hg

Zinc, cadmium, and mercury are soft metals with relatively low melting and boiling points. Their common oxidation state is 2+. Sources are sulfide minerals: white zinc blende (ZnS), yellowish greenockite (CdS), and dark red cinnabar (HgS). Roasting in air gives the oxides; HgO decomposes as it is formed and mercury metal distills (b.p., 357°C). Mercury and cadmium vapors are toxic; ZnO and CdO must be reduced by carbon before the metals are distilled.

Iron coated with zinc is called *galvanized iron*. In moist air the zinc acquires an adherent hydroxide and carbonate coating ($Zn(OH)_2$ and $ZnCO_3$) which protects the iron from rusting. Galvanized iron is employed in roofing, siding, highway construction, and fencing. Zinc is also used in the die casting of automotive and appliance parts, in dry cell batteries, and to make alloys such as brass.

Zinc is chemically active. It dissolves in both dilute acids and bases to form salts:

$$ZnSO_4 \xleftarrow{H_2SO_4} Zn \xrightarrow{NaOH} Na_2Zn(OH)_4$$

Zinc oxide (ZnO), made by burning zinc vapor in air, is used in ceramic ware, as a pigment in white paint, a filler in rubber, and an antiseptic in ointments. Zinc chloride ($ZnCl_2$) is used as a flux in soldering and in pulping wood for wallboard manufacture.

Cadmium metal is used as (1) an electroplated coating (from cyanide solution) to protect iron, (2) the negative electrode in some alkaline batteries, (3) control rods in nuclear reactors, and (4) a component of low-melting alloys. Cadmium sulfide is a yellow pigment.

Mercury is the only metal that is liquid at room temperature (m.p.,

$-38.9°C$). It is used in electrolysis cells (see Figure 18-7), certain lighting fixtures, electrical switches, thermometers, and other laboratory equipment. Mercury dissolves many metals by forming mixtures called *amalgams*. An amalgam of silver and tin is used in dental fillings.

Mercury forms two series of compounds, mercuric (Hg^{2+}) and mercurous (Hg_2^{2+}); the latter ion contains two metal atoms bound together. Structures for mercuric and mercurous chlorides are Cl—Hg—Cl and Cl—Hg—Hg—Cl, the lines representing chemical bonds. Mercury oxidizes slowly when heated in air to give yellow and red forms of mercuric oxide that differ apparently only in particle size. Mercuric chloride can be made by reaction of hydrochloric acid on mercuric oxide. Mercurous chloride is prepared by adding mercury to mercuric chloride:

$$Hg \xrightarrow{O_2} HgO \xrightarrow{HCl} HgCl_2 \xrightarrow{Hg} Hg_2Cl_2$$

The common name for mercurous chloride, calomel (Gk. *kalos* + *melas*, beautiful black), refers to the black precipitate of finely divided mercury it gives when identified by reaction with ammonia,

$$Hg_2Cl_2 + 2NH_3 = HgNH_2Cl + Hg + NH_4Cl \tag{8-15}$$

Environmental problems connected with mercury and cadmium are discussed in Section 15-3.

8-3. THE INNER TRANSITION ELEMENTS

Within the transition elements just discussed are two series of metals called *inner transition elements* (Figure 8-3). Both involve filling of electrons in the quantum shell second from the outermost shell. In the lanthanoids this is the $4f$ level, in the actinoids the $5f$ level. Most of the actinoids are man-made.

The lanthanoids: elements 58 through 71

The lanthanoids were originally called *rare earths* because of their apparent scarcity. Awareness of their existence began about 1790 with the discovery in Sweden of two new minerals from which oxide mixtures named *ceria* and *yttria* were obtained. It was more than a hundred years, however, before all the metals in these minerals were cleanly separated.

The following were key developments in rare earth research:

1. The periodic table enabled chemists to determine the number of lanthanoids possibly in existence.
2. Moseley's X ray spectra method established the order of the lanthanoids in the table.

lanthanoids

58 Ce	59 Pr	60 Nd	61 Pm	62 Sm	63 Eu	64 Gd	65 Tb	66 Dy	67 Ho	68 Er	69 Tm	70 Yb	71 Lu
90 Th	91 Pa	92 U	93 Np	94 Pu	95 Am	96 Cm	97 Bk	98 Cf	99 Es	100 Fm	101 Md	102 No	103 Lw

actinoids

Figure 8-3 The inner transition metal section of the periodic table.

3. Knowledge of electronic structure prepared chemists for the chemical similarities among the lanthanoids that would make separation of them difficult.
4. Development of the ion-exchange method (described below) gave a way to separate lanthanoid ions.
5. Invention of the spectrograph gave means for determining purity of samples.

Ceria was found to contain lanthanum and six lanthanoids: cerium (Ce), praseodymium (Pr), neodymium (Nd), samarium (Sm), europium (Eu), and gadolinium (Gd). Analysis of yttria proved the presence of the IIIB metals scandium and yttrium and eight lanthanoids: gadolinium, terbium (Tb), dysprosium (Dy), holmium (Ho), erbium (Er), thulium (Tm), ytterbium (Yb), and lutetium (Lu). Only promethium (Pm) was not discovered in these unusual deposits. Evidence for promethium in the form of spectral lines derived from various rare earth fractions was sought for many years without success. Finally, in 1947 its separation was made from uranium fission products at the Oak Ridge atomic plant and the half-life of its stablest isotope established at 2.6 years.

The main source of lanthanoids today is monazite sand, a mixed phosphate in which thorium is usually the most abundant metal. The components are separated by heating with sulfuric acid, diluting, treating with various chemicals to precipitate thorium, iron, and titanium, then pouring the solution through columns packed with ion-exchange resins. The resins are synthetic high-molecular weight organic compounds containing groups that hold or release species in a manner similar to that described previously for zeolites. After selective adsorption on the columns, ions are eluted (washed out) into separate containers.

As shown in the periodic table, the lanthanoids constitute a series between lanthanum ($Z = 57$) and hafnium ($Z = 72$). They possess in common three incomplete electron shells, $4f$, $5d$, and $6s$. Filling of the seven $4f$ orbitals with two electrons each accounts for the series' 14 members.

Pure lanthanoids are quite electropositive and about as active chemically as Group IIA metals. With acids they give hydrogen and M^{3+} salts. Less familiar oxidation states are 2+ and 4+. With moist air they give oxides (M_2O_3) and hydroxides ($M(OH)_3$). Above 200°C, they burn with halogens, giving halides (MX_3). Above 300°C, they give hydrides with hydrogen (MH_3). At still higher temperatures, they yield nitrides with nitrogen (MN) and carbides with carbon (MC_2).

Free lanthanoids are prepared by electrolysis of their molten chlorides. Small amounts of the metals are used to modify properties of nonferrous alloys. For example, 1 percent mischmetal (a mixture of rare earths that is about 50 percent cerium) improves the strength of aluminum used for high-voltage transmission lines.

Lanthanoid oxides are incorporated in ceramics and optical glass; in glass blowers' goggles, for instance, they eliminate sodium flame glare. Various oxide mixtures have also been tried with cobalt as catalysts in petroleum refining and in automotive exhaust systems as a substitute for platinum. Several lanthanoid compounds are used as phosphors to improve color television screens. They work by furnishing to the operating electron beam a relatively large number of easily excited electrons that emit visible light on returning to less energetic states.

The actinoids; elements 90 through 103

Prior to 1940 thorium, protoactinium, and uranium were placed, respectively, below hafnium, tantalum, and tungsten in the main body of the periodic table because chemists thought they represented a filling of $6d$ orbitals. When, however, neptunium and plutonium were synthesized, the possibility became evident of a series of 14 elements corresponding to the lanthanoid series in which filling would be in the $5f$ orbitals. These elements are now called the "actinoid" series. It seems certain the $5f$ subshell is filled at lawrencium ($Z = 103$) and that $6d$ filling proceeds with element 104 which chemically resembles hafnium.

Thorium, uranium, protactinium, neptunium, and plutonium (the latter two in trace amounts) occur naturally. Since 1940 Lawrence Radiation Laboratory scientists (Figure 8-4) have synthesized all the elements beyond uranium. Generally, half-life decreases while fissionability by neutrons increases with increasing atomic number. For example, the half-life of

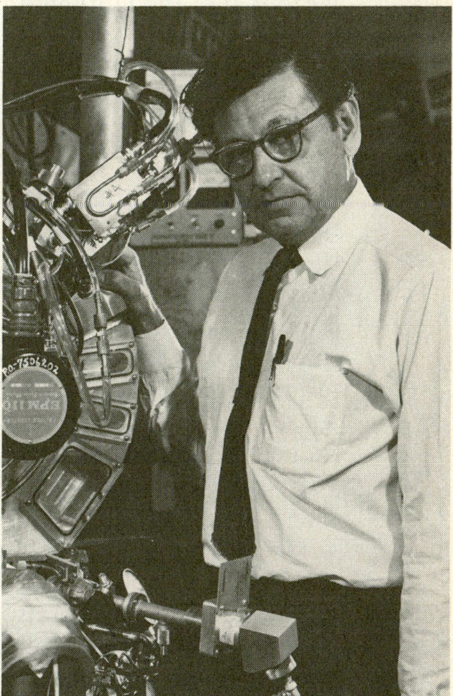

Figure 8-4 Synthesizers of transuranium elements. (Left) Glenn Seaborg (United States 1912–   , Nobel prize in chemistry 1951). Seaborg helped synthesize plutonium and seven other transuranium elements. Later he was chairman of the Atomic Energy Commission. In the background of this 1957 photo is the once popular short form of the periodic table. (Compare to Figures 7-2a and 7-4.) (Right) Albert Ghiorso. Codiscoverer of nine synthetic elements, Ghiorso stands in his laboratory with equipment used to identify elements 104 and 105. The device is a wheel containing radiation detectors. Below is the target bombarded by the beam of the heavy ion linear accelerator (HILAC). The accelerator is located behind shielding at the rear. (Photographs courtesy Lawrence Radiation Laboratory, University of California.)

$^{246}_{96}$Cm is 5000 years but the half-life of $^{246}_{100}$Fm is only 3 seconds. The heaviest actinoids are therefore best made not by successive neutron additions but by bombardment with heavy particles. For instance,

$$^{253}_{99}Es + ^4_2He = ^{256}_{101}Md + ^1_0n$$

The intense radioactivity of short-lived isotopes requires shielding for personnel and creates unusual experimental difficulties. It causes compounds to heat up and decompose, crystals to deform, and water solutions to break down to hydrogen and oxygen with simultaneous changes in metal-ion oxidation states. Studies are aided, however, by the readily measured paramagnetism of the ions and the well-defined spectra related to energy absorption by f electrons. Techniques for separating mixtures of the ions include ion exchange and extraction with organic solvents that are immiscible (incapable of solution) with water.

Free actinoid metals are obtained by heating the fluorides with an alkaline earth or alkali metal under an argon atmosphere,

$$NpF_3 + 3Na \overset{\Delta}{=} Np + 3NaF \tag{8-16}$$

The actinoids have a silvery appearance. Their chemical reactivity is similar to that of the lanthanoids. All but two form 3+ ions. More than half also form 4+ ions.

Military and electric power station uses of uranium, neptunium, and plutonium were outlined in Chapter 4. Possible uses of the other actinoids are various, but as yet incompletely realized. Easily shielded α emitters such as curium-244 (half-life, 17.6 yr), whose heat output is equivalent to 3 watts per gram, are being tried as power sources for satellites and navigation buoys. An americium isotope was used in the Surveyor 5 spacecraft that soft landed on the moon in 1967 and carried out the first chemical analysis of the lunar surface. As its α particles bombarded the soil and bounced back with energies proportional to the masses of nuclei struck, the impacts were translated into electronic signals and sent to earth for interpretation. Plutonium-238 was used to run a moon station left by the Apollo 12 mission. The same isotope is used to power a pacemaker for human heart implantation; expected working life of the device is 5 to 10 years. (At least 10,000 new patients a year in the United States are presently being fitted with pacemakers powered by conventional batteries.)

Californium-252 (half-life 2.6 yr), a spontaneously fissioning element that gives off great quantities of neutrons, is being considered for applications in (1) cancer therapy (insertion of a californium needle could direct a highly localized dose of neutrons to cancer tissue), (2) measuring the density of engineering structures, and (3) *neutron activation analysis*. In the latter technique samples are bombarded with neutrons and their induced radioactivities measured with a γ-ray spectrometer that identifies the elements. Neutron activation is the most sensitive method in general analytical use today. Under ideal conditions, detection sensitivities are in the nanogram (10^{-9} g) range.

Radioisotopes are being put to many uses in applied medicine (cobalt-60 treatment of cancer, etc.) and in research. The latter is typified by work that elucidated the mechanism of photosynthesis (Figures 22-14, 22-15).

8-4. EXTENDING THE PERIODIC TABLE

Elements synthesized since 1969 are numbers 104 and 105. Claims of originality made in the United States seem more convincing than those made in the USSR. Names given here, therefore, are those suggested by Americans: rutherfordium (Rf) after Rutherford, and hahnium (Ha) after Hahn. Elements beyond 105 are expected to be prepared in particle accelerators and to have properties appropriate to spaces provided for them in an extension of the periodic table. Using computer programs to predict electronic structures and nuclear characteristics, Seaborg, Ghiorso, and colleagues have estimated the stabilities of superheavy nuclides and the possibility of carrying nuclear synthesis as far as $Z = 168$. That would complete period 7, add period 8, and put a row of 32 superactinoids below the actinoids (see Figure 7-4.)

For several years scientists have unsuccessfully sought naturally occurring superheavy elements in such hiding places as platinum minerals, ocean sediments and lead from windows of the oldest churches. Indication that superheavy elements are formed in stellar processes comes from mass-spectrum analyses of stony meteorites. Xenon isotopes have been found in some samples in ratios that apparently cannot be accounted for as daughter species unless parents, which fissioned long ago, had atomic numbers between 112 and 119. The very heavy atoms that will contain magic number nuclei and therefore possess the stability ascribed to closed nucleon shells have been predicted from nuclear structure theory (Section 4-3). The nucleus for which $Z = 114$, $N = 184$ will be doubly magic like lead 208 ($Z = 82$, $N = 126$). Although the same principles that determine electronic arrangement (hence placement in the periodic table) do not apparently also determine nuclear stability, element 114 would be a Group IVA element like lead. Also in that group would come element 164, the next element to have a doubly magic nucleus. It is estimated that element 114 will have a half-life of at least 1000 years, whereas half-lifes of neighboring elements will probably be measured in fractions of seconds. Synthesis of 114, "eka-lead" in Mendeleev's nomenclature, is expected to require isolation of quantities of special neutron-rich isotopes for reactions. Also needed will be modification of present equipment to give more massive projectiles sufficient energy to overcome nuclear repulsive forces. One possible route, using an actinoid element for a target, is

$$^{244}_{94}Pu + ^{48}_{20}Ca = ^{288}_{114}X + 4^{1}_{0}n$$

where X is the arbitrary symbol for the new element.

Such reactions give a variety of products. To separate and prove the existence of a given element requires techniques of both nuclear science and analytical chemistry. The latter especially depend on predictions of chemical behavior related to position in the periodic table.

QUESTIONS

1. Industrial chemicals Without the use of equations, tell how each chemical in Table 8-2 is obtained. Give one industrial use of each.

2. Supply and demand Following are estimated total primary reserves of some metals in the world (including the United States); values are in millions of tons: Al, 1200; Cu, 300; Pb, 90; Mg, 2600; Zn, 120. **(a)** Using Table 8-1 data, calculate the percentage of each that the United States will presumably show demand for during the last quarter of this century. **(b)** Steel production is a measure of a country's muscle because it relates directly to growth in building and transportation. In a poor country steel production might be a few pounds per person per year. What is it in the United States? (See footnote, Table 8-1.)

3. Metal properties **(a)** Iridium and platinum, two of the densest elements, crystallize with cubic symmetry. Which of the three cubic lattices of Figure 6-6b do you expect for them? Explain. **(b)** Archimedes reported that during one cold winter in ancient Greece, statues made of tin weakened and fell. Explain. **(c)** A silver spoon tarnishes when used with soft boiled eggs. Explain. **(d)** Electron levels in the Cu^+ ion are $1s^2 2s^2 2p^6 3s^2 3p^6 3d^{10}$. Do you expect the ion to be diamagnetic? To be colored? Explain. **(e)** What element will element 106 probably most resemble? For 106 predict an important oxidation state, and the metal's approximate melting point.

4. Definitions Define or explain **(a)** galvanizing; **(b)** Frasch process; **(c)** zone refining; **(d)** neutron activation analysis; **(e)** amalgamate; **(f)** Pyrex glass; **(g)** soft glass; **(h)** stainless steel.

5. Diagonal relationships When heated with nitrogen, both lithium and magnesium give nitrides: Li_3N, Mg_3N_2. Sodium does not form a nitride. This behavior is called a *diagonal relationship* in the periodic table. Beryllium and boron also show diagonal simililarities to neighboring elements. Which neighbor is involved in each case? From the descriptive chemistry given in the text, cite one similarity shown by each of these two pairs.

6. Transition elements An alloy that is 82 percent tungsten, 18 percent rhenium is used in jet engine applications. A chemical analysis is to guarantee its composition. Acid treatment of the alloy followed by heating gives a mixture of WO_3 and Re_2O_7, which must be separated. Assume data are lacking on the two oxides. However, the following melting point data are known: CrO_3, 197°C; MoO_3, 795°C; Mn_2O_7, −20°C. Based on periodic table relationships, describe how the analysis might proceed.

7. Iron (library) From at least one reference book gather information that enables you to discuss the process illustrated in Figure 8-5.

8. Lanthanoids (library) From reference reading, describe one commercially useful compound of each of five different lanthanoids that you select. (Indexes may list the elements as lanthanides or rare earths.)

9. Electrical conductance **(a)** How do you think electrical conductance of elements might vary within a periodic table group, say group IVA? **(b)** Within a period, say period 3? Explain.

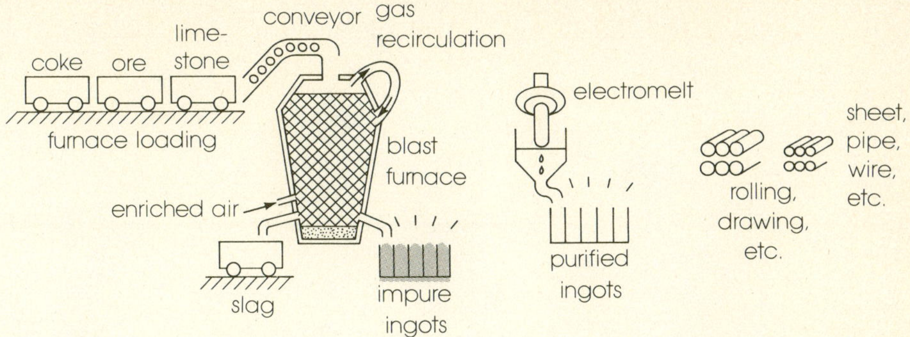

Figure 8-5

10. Superheavy elements Aided by Figure 7-4, predict some properties of elements 117 and 118.

PROBLEMS

11. Ion properties Vanadous ion, V^{2+} is formed when a vanadium atom loses two electrons. The ion's electron arrangement is 2, 8, 11 or $1s^2 2s^2 2p^6 3s^2 3p^6 3d^3$. **(a)** Give the latter type of notation for Fe^{3+} (2, 8, 13) and Zn^{2+} (2, 8, 18). **(b)** Use Hund's rule to predict how many unpaired d electrons are in each of the three ions. **(c)** Which of the ions do you expect to show color? Explain. **(d)** Which of the ions do you expect to be paramagnetic? Explain.

12. Orbital filling **(a)** The chromium atom has this electron configuration: $1s^2 2s^2 2p^6 3p^6 3d^5 4s^1$. In what respect is it an exception to the idealized filling order of Figure 7-5? **(b)** In the same notation give the expected electronic structure of chromic ion Cr^{3+}.

13. Orbital notation **(a)** There is a certain element that is described as $K_2 L_8 M_{18} N_{32} O_{18} P_8 Q_2$. What element is it? Explain. **(b)** The outermost electron structure of a certain atom is . . . $5s^2 5p^6 5d^{10} 6s^2 6p^2$. What are two probable oxidation states? Explain.

14. Nitrogen From description of the natural environmental circulation of the element nitrogen, prepare a diagram of a cycle similar to that of Figure 15-3.

15. 104, 105 **(a)** Hahnium-260 was prepared by the Ghiorso team by bombarding californium-249 with nitrogen-15 nuclei; four neutrons were the by-product. **(b)** Rutherfordium-257 was made by bombarding californium-249 with carbon-12. Write the equations, using Eq. 4-3 as a model.

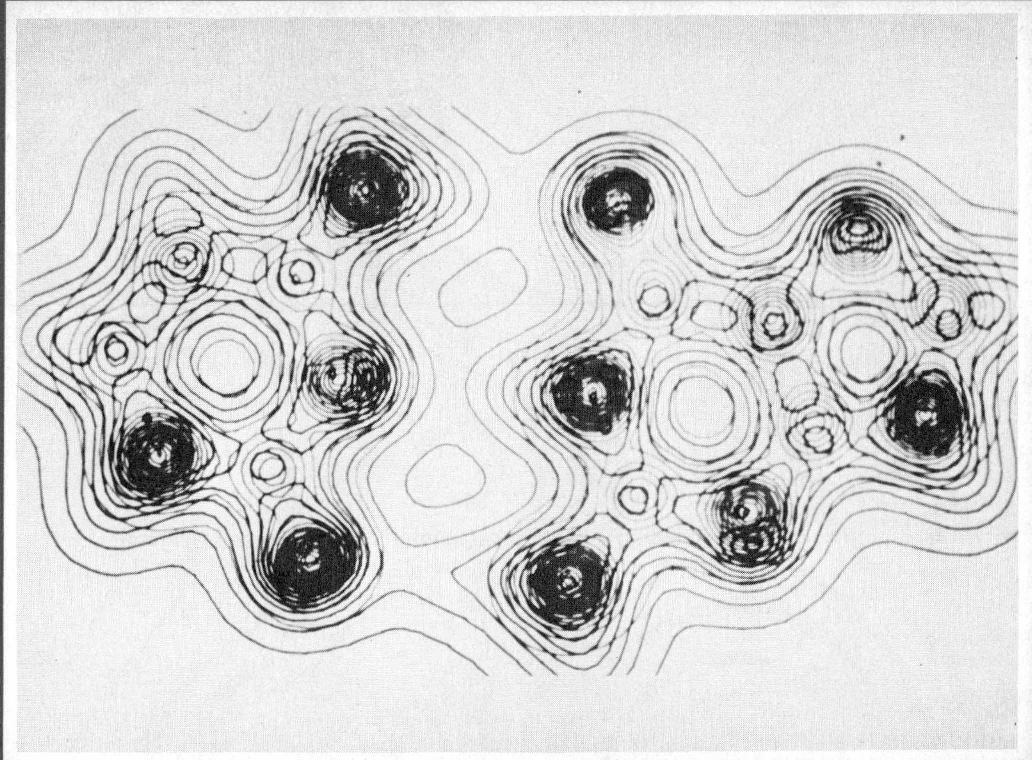

THREE

WHAT IS CHEMICAL BONDING?

A computer-generated electron-probability contour map of the hydrogen-bonded cytosine-guanine base pair found in the material of heredity in biological cells. The plotted electron densities lie 1 bohr radius (0.53 Å) above the plane of the molecules. The wider spaced contour lines refer to the valence σ-electron density. The denser contours refer to π-electron density of electronegative oxygen, and nitrogen atoms. To produce the plot 400 hours of computer time, representing 5 trillion mathematical operations, was needed. By connecting a televisionlike screen to a computer that is programed to solve quantum-mechanical expressions, scientists can study images of simulated atoms in step-by-step interactions. (Correlate the map with Figure 24-2b.) (Courtesy John Bailey, IBM Research Laboratory.)

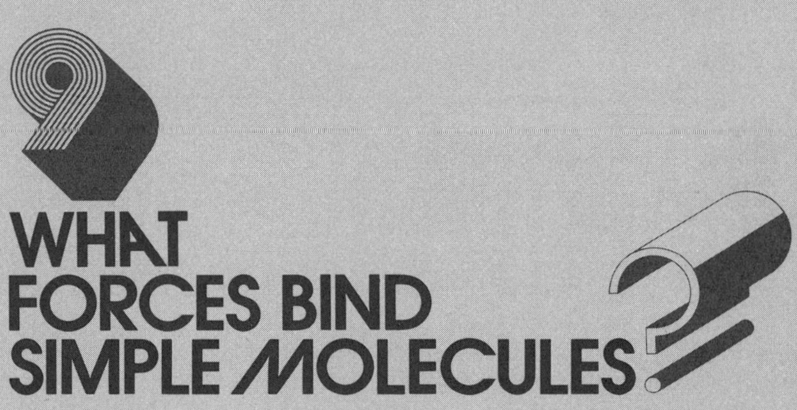

WHAT FORCES BIND SIMPLE MOLECULES?

THE IMPORTANT CONCEPTS

9-1. The ionic bond
1. Electrical (electronic) forces operate between atoms.
2. The octet rule helps to interpret reactions between main group metals and nonmetals.
3. The electrovalent bond is found in ionic structures.

9-2. The covalent bond
1. Electron sharing gives covalent bonds.
   a. Single, double, and triple bonds are known.
   b. Bond length, strength, and multiplicity are related.
2. Electron dot and valence bond formulas are practical representations of molecules.
3. Molecules that resonate are pictured by several different bond structures.

4. A coordinate covalent bond is formed between electron donor and acceptor.
5. Exceptions to the octet rule are known.
    a. Some valence shells hold more than eight electrons.
    b. Free radicals are odd-electron species.
    c. An atom with a closed valence shell may form bonds.
6. Covalent bonds may be polar or nonpolar.
7. Bond character may be predicted from electronegativity values.
8. Molecules may be polar or nonpolar.
9. Dipole moment is related to molecular shape.

9-3. The orbital interpretation of covalence

1. The molecular-orbital method gives a picture of bonding and antibonding orbitals.
2. Hybridized orbitals have shapes that give maximum separation of electron pairs.
    a. A few simple geometries describe the common hybridizations.

9-4. Bonding from computers

1. Computer-generated electron-density maps of simple species aid in visualizing molecules and bond formation.
    a. Change in system energy with atom–atom distance can be calculated.

A molecule is an electrically neutral aggregate of atoms that possesses distinctive chemical properties. It is held together by one or more chemical bonds of electronic origin. Although for our purposes the chemical bond is describable by simple diagrams, its details must be dealt with quantum mechanically. A knowledge of bonding gives insight into some of chemistry's most fundamental questions: What interactions occur and how much energy is involved when atoms are brought close to one another? Why do atoms stick together, forming molecules? Why do molecules have definite composition? Why do molecules have characteristic shape?

9-1. THE IONIC BOND

Recognition of electrical forces between atoms

The notion of chemical bonding is an old concept, for if one postulates atoms too small to see, then one must also offer an explanation for the buildup of strong and massive structures. To Greek atomists, chemical bonds consisted of atomic hooks and interlocking shapes—and the idea of a mechanical linkage between atoms persisted into the nineteenth century.

Natural force acting between atoms was foreseen by Newton, even the need for postulating special effects at interatomic distances. He wrote,

> We must learn from the Phaenomena of Nature what Bodies attract one another and what are the Laws and Properties of Attraction, before we enquire the Cause by which the Attraction is performed. The attractions of Gravity, Magnetism, and Electricity reach to very sensible distances . . . and there may be others which reach to so small distances as hitherto escape Observation; and perhaps electrical Attraction may reach to such small distances, even without being excited by Friction. . . . I infer that Particles attract one another by some Force, which in immediate Contact is exceedingly strong, at small distances performs chemical operations, and reaches not far from the Particles with any sensible Effect.

When Davy used a battery in 1807 to decompose solutions and molten salts in electrolysis experiments, he concluded that (1) particles in compound substances that conducted electric current had charge (and migrated to oppositely charged electrodes), (2) electric forces held compounds together, and (3) bonding energy was probably an intrinsic property of matter.

Berzelius used the results of electrolysis experiments to discuss compound formation. Proposing, "that in all chemical combinations there is neutralization of opposing electricities," he sought to classify elements in two groups based on electric charge. In 1831 he wrote,

> If these electrochemical views are correct, it follows that all chemical combination depends solely on two opposing forces, positive and negative electricity,

and that thus each combination should be composed of two parts united by the effect of their electrochemical reaction, provided that there exists no third force. Whence it follows that each compound substance, regardless of the number of its constituent principles, may be divided into two parts, of which one is electrically positive and the other negative.

This so-called *dualistic theory* successfully described the condition of strong alkalis such as NaOH and typical salts like KCl, which today we know consist of oppositely charged ions. It failed, however, to explain how electronegative atoms like chlorine could be substituted for electropositive atoms like hydrogen in organic (carbon) compounds.

Around 1845 English chemist E. Frankland showed *substitution reactions* were common in organic chemistry. Chlorinating the hydrocarbon ethane in the presence of light, he produced chloroethane, dichloroethane, etc., eventually substituting halogen atoms for all the hydrogens;

$$C_2H_6 \xrightarrow[light]{Cl_2} C_2H_5Cl \xrightarrow[light]{Cl_2} C_2H_4Cl_2 \xrightarrow[light]{Cl_2} \cdots \longrightarrow C_2Cl_6$$

Combining his results with evidence of fixed atomic combinational ratios that were evident in such compounds as NH_3, NI_3, PH_3, SbH_3, and $SbCl_3$, Frankland concluded that " . . . no matter what the character of the uniting atoms may be, the combining power of the attracting element, . . . is always satisfied by the same number of these atoms."

Combining power came to be called *valence* (L. *valentia*, capacity). Today the word "valence" is used as an adjective, as in "valence electron." In place of valence as a noun we employ the terms *valency, valence number, oxidation number,* or *oxidation state.*

Chemists began writing chemical formulas with lines or dots linking symbols of elements, each link inferring one unit of valency for an atom. The univalence of hydrogen and chlorine, bivalence of oxygen and sulfur, tervalence of nitrogen, and quadrivalence of carbon (to use Latin prefixes appropriate to the origin of the word "valence") were depicted as follows in *structural formulas* for water, ammonia, hydrogen sulfide, and chloroethane:

$$H\cdots O\cdots H \qquad H\cdots \overset{\displaystyle \vdots}{\underset{\displaystyle H}{N}} \cdots H \qquad H\cdots S\cdots H \qquad Cl\cdots \overset{\displaystyle \overset{H}{\vdots}}{\underset{\displaystyle \underset{H}{\vdots}}{C}}\cdots \overset{\displaystyle \overset{H}{\vdots}}{\underset{\displaystyle \underset{H}{\vdots}}{C}}\cdots H$$

Structural formulas clarified chemical function so well that they launched organic chemistry as a whole new chemical field. Inorganic chemists, however, had difficulty agreeing on even the formulas of ordinary salts. Whereas organic chemists dealt mainly with a few elements — carbon, hydrogen, oxygen, nitrogen — whose valencies appeared invariant, inorganic chemists dealt with all the other elements, many of which had variable valency. They also could not explain the easy passage of electric current through solutions of most salts, acids, and bases; how could molecules composed of neutral atoms conduct electricity?

An answer to the electrical conductance question was given by Arrhenius (see Figure 16-7) in 1887 when he assumed that conducting substances gave *ions* (Gk. *ienai*, to move) in solution. Ions he defined as positively or negatively charged atoms or groups of atoms (called *radicals*) having properties different from those of neutral atoms. In sodium nitrate solution, for example, he visualized that Na^+ and NO_3^- carried current between charged electrodes. Molten salts were known to conduct electricity, and Arrhenius concluded that they too were composed of ions held together by mutual attraction of opposite charges (coulomb forces). A salt crystal was therefore always neutral because according to the valence rules there could only be equal numbers of positive and negative charges.

These conclusions are still valid. However, a comparable insight into the bonding in organic compounds, most of which are nonconductors in both solution and the pure state, was not developed for another 30 years.

Lewis, Kossel, and Langmuir

In 1916 Lewis (see Figure 9-2) in the United States and W. Kossel in Germany each proposed that chemical bonding could be explained in terms of electron transfer and electron sharing between atoms. Their ideas came from the Bohr atom model and from certain experiments. First, calculations indicated chemical reaction energy was enough to perturb only an atom's outermost (valence) electrons without affecting the remaining (core) electrons. Second, spectroscopic observations established the number of electrons surrounding each nucleus and also demonstrated that the number of valence electrons of main group elements in the periodic table corresponded with their group numbers. (Group IA elements, 1 valence electron; IIA elements, 2 valence electrons; etc.)

Kossel's work dealt mainly with ionic compounds. Noting the noble gas group lay between the highly electronegative halogens and the highly electropositive alkali metals, he suggested that reactions between Group VIIA and IA atoms gave ions of noble gas electron configuration through a gain and loss of electrons. Like noble gas atoms, the resulting ions would be expected to be unreactive because nuclei surrounded by 2, 10, 18, 36, 54, or 84 electrons were inert. Because reactions between metals and nonmetals of periodic table groups near the noble gases left nuclei with an outside shell of eight electrons, Kossel's generalization became known as the *octet rule*.

The typical metal–nonmetal reaction between sodium and chlorine could then be explained as follows. The neutral sodium atom has the electron configuration 2, 8, 1, and the neutral chlorine atom 2, 8, 7. As the two atoms approach each other, a vigorous reaction occurs. Sodium gives up its single valence electron and becomes sodium ion with the neon arrangement 2, 8. The chlorine atom adds the electron to its valence shell, giving the chloride ion with the argon configuration 2, 8, 8. The sodium ion, with 11 protons in the nucleus and only 10 electrons in shells, has a net charge of 1+ (ionic valency 1+), whereas the chloride ion with 17 protons and 18 electrons has a net charge of 1− (ionic valency 1−). The product is a salt,

$$Na° + \cdot \ddot{\underset{\cdot\cdot}{Cl}} : \; = \; Na^+ : \ddot{\underset{\cdot\cdot}{Cl}} : ^- \tag{9-1}$$

Salts, a product of metal–nonmetal reactions, were verified by Bragg's earliest X ray diffraction work to be latticeworks of ions. Electrostatic force unmistakably held ionic structures together and a name was given to it—*the ionic or electrovalent bond*.

In discussing bonding Kossel used for the atom a stationary structure with rings of electrons around a positive nucleus (Figure 9-1a). Lewis ar-

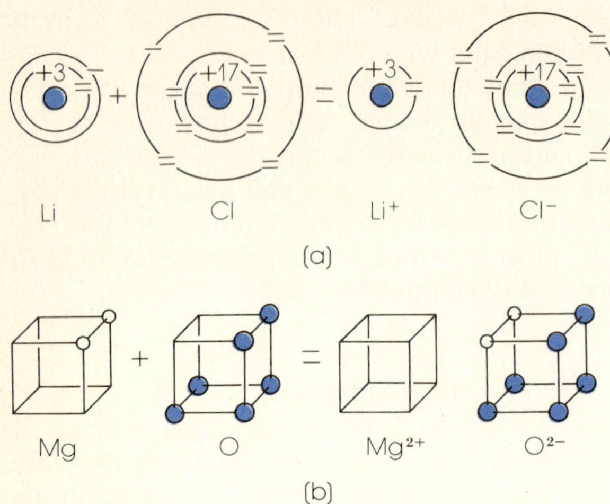

Figure 9-1 First versions of the ionic bond. (a) Kossel's picture of the formation of lithium chloride, published three years after Bohr proposed an atom with planetary electrons. (b) Lewis' picture for the behavior of valence electrons in the formation of magnesium oxide.

Figure 9-2 Pioneers in electron-bond theory. (Left) Gilbert N. Lewis (United States 1875–1946). Lewis was dean of the chemistry college at the University of California. His idea that a pair of shared electrons constituted a chemical bond gave chemists a simple way to think about molecular structure. He also developed acid–base theory (Section 16-6). (Right) Irving Langmuir (United States 1881–1957), Nobel prize in chemistry 1932). A physical chemist, Langmuir worked on catalyst mechanisms (See Figure 17-5), molecular films, weather modification by cloud seeding, and electric discharge in gases. He invented numerous devices, including the mercury diffusion pump required in high-vacuum research. A sample of his philosophy was that overplanning inhibits creativity, and that too much theorizing hinders experimental work by creating preconceptions that block chances of benefiting from serendipity. (Photographs: left, Edgar Fahs Smith Memorial Collection, University of Pennsylvania; right, courtesy Research and Development Center, General Electric Company.)

rived at the same conclusions about bonding with a cubic atom model in which corners represented the octet of valence electron positions (Figure 9-1b). In Lewis' visualization noble gas configurations were obtained when all eight corners became either empty or filled.

Lewis' thoughts on bonding and the electron octet were extended by Langmuir (Figure 9-2). Langmuir abandoned models with fixed electrons (which physicists abhorred) for a model with oscillating electrons that anticipated the quantum-mechanical atom.

9-2. THE COVALENT BOND

Electron pairs

Lewis initially thought separate theories might be needed to describe the linkages in inorganic (primarily ionic) compounds and organic (primarily nonionic) compounds. Subsequently he decided that the two cases were simply extremes of the general phenomena of bonding that might be explained by one theory. Realizing that most compounds were diamagnetic and that practically all contained an even number of electrons, he concluded that it was electron pairing that in some way gave chemical bonds their cohesion. In ionic compounds he assumed that pairing took place through electron transfer. In nonionic compounds he thought that pairing occurred through a mutual sharing of the two electrons. Langmuir named the shared-pair linkage the *covalent bond*.

Lewis wrote,

> Two electrons thus coupled together, when lying between two atomic centers, and held jointly in the shells of the two atoms, I have considered to be the chemical bond. . . . The pair of electrons which constitutes the bond may lie between two atomic centers in such a position that there is no electric polarization, or it may be shifted toward one or the other atom in order to give to that atom a negative, and consequently to the other atom a positive charge. But we can no longer speak of any atom as having an integral number of units of charge, except in the case where one atom takes exclusive possession of the bonding pair, and forms an ion.

Simple covalent molecules

For our purposes it will be sufficient to regard the covalent bond as an electron pair that occupies a normal orbital in each of two sharing atoms. In representing molecules we will continue to use the Lewis-Langmuir convention that the symbol of an element stands for its atomic nucleus and core electrons and that dots represent its valence electrons. In an *electron dot formula* a pair of dots between two symbols will mean a covalent bond, as in the hydrogen molecule H:H. Alternatively we will use a short line for a covalent bond, as in H—H, and call such a formula a *valence bond formula*.

Formation of a hydrogen molecule can be written as

$$H \cdot + H \cdot = H : H \qquad (or\ H—H)$$

The two 1s electrons are shared in H_2, allowing each hydrogen atom to attain the closed electron configuration of helium. Mutual repulsion of the nuclear protons is overcome by a screening effect of the electron pair whose experimentally measured density is greatest between the two nuclei.

A more complex molecule for consideration is formaldehyde (H_2CO). Like H_2 it consists of nonmetallic elements only, and is covalently bonded. To write any correct bonding formula we need to know first the arrangement of atoms in the molecule, and second the number of valence electrons about each atom. In formaldehyde, carbon is the central element and the other three atoms are bonded to it. For valence electron information we refer to main group numbers in the periodic table (carbon = Group IV = 4 valence electrons, etc.) To follow the electrons we will designate the atoms $H\cdot$, $H\cdot$, $\cdot\overset{\circ}{\underset{\circ}{C}}\cdot$, and $:\overset{\cdot\cdot}{O}:$. The 12 valence electrons we now distribute in such a way that pair sharing gives hydrogen the helium electron structure and both carbon and oxygen the neon structure. It turns out that two pairs of valence electrons are not used in bonding:

$$\begin{array}{c} H \\ \\ H \end{array}\overset{\circ}{\underset{\circ}{C}}:\overset{\cdot\cdot}{\underset{\cdot\cdot}{O}}: \quad \text{or} \quad \begin{array}{c} H \\ \diagdown \\ \diagup \\ H \end{array}C=\overset{\cdot\cdot}{\underset{\cdot\cdot}{O}} \quad \text{or} \quad \begin{array}{c} H \\ \diagdown \\ \diagup \\ H \end{array}C=O$$

The four-electron bond represented by four dots (or two parallel lines) is called a *double covalent bond*.

Hydrogen cyanide (HCN) contains one single and one *triple*, or six-electron, covalent bond. Writing the neutral atoms $H\cdot$, $\cdot\overset{\circ}{\underset{\circ}{C}}\cdot$, and $\cdot\overset{\cdot\cdot}{N}:$, the structure is

$$H:C:::N: \quad \text{or} \quad H-C\equiv N: \quad \text{or} \quad H-C\equiv N$$

Again, through electron pair sharing, hydrogen attains the helium configuration and carbon and nitrogen the neon configuration. One pair of nitrogen electrons is not used in bonding.

Covalent bond length, multiplicity, and strength are related. As multiplicity increases, length decreases and strength increases. Carbon–carbon single, double, and triple bonds, respectively, have lengths 1.54, 1.33, and 1.20 Å. The energy required to break them is in the ratio $1:1.8:2.3$.

Some substances are held together by combinations of ionic and covalent bonds. An example is sodium nitrite NaONO. Bonding between the sodium and nitrite ions is essentially ionic. Within the nitrite ion itself, however, bonding is primarily covalent. Representing the neutral atoms as Na , $:\overset{\cdot\cdot}{O}:$, $\overset{\circ\circ}{N}:$ and $:\overset{\cdot\cdot}{O}:$, the formula for the compound is

$$Na^+:\overset{\cdot\cdot}{\underset{\cdot\cdot}{O}}:\overset{\circ\circ}{N}:\overset{\cdot\cdot}{O}:^- \quad \text{or} \quad Na^+:\overset{\cdot\cdot}{\underset{\cdot\cdot}{O}}-\overset{\circ\circ}{N}=\overset{\cdot\cdot}{\underset{\cdot\cdot}{O}}^- \quad \text{or} \quad Na^+O-N=O^- \qquad (9\text{-}2)$$

Resonance

When no single electron dot or valence bond formula is adequate to define a molecule, and several approximately equally plausible bonding arrangements can be written, the molecule may be described as a *resonance hybrid*. It is then represented as being a composite of two or more formulas that have (1) the same relative positions of atoms, (2) similar or identical energies, and (3) the same number of electron pairs. The concept of resonance originated in Heisenberg's quantum mechanics. Since then it has been widely applied in simplified form especially by Pauling (Figure 9-3) and organic chemists.

Figure 9-3 Linus Pauling (United States 1901–   , Nobel prize in chemistry 1954, Nobel prize in peace 1962). Pauling is one of the most wide-ranging intellects in modern chemistry. The guiding principle in his research has been to apply quantitative measurements (especially X ray and electron diffraction) on molecular structure to important theoretical problems. In the field of inorganic chemistry he deduced a set of easily understood rules that explains the architecture of silicates (Figures 6-11, 6-12). From studies on molecules of biological origin he advanced theories about immunology, anesthesia, memory, genetic diseases (sickle-cell anemia, etc), mental processes, and behavior. From data on magnetic and electric properties of molecules, their shapes and dimensions, he developed a quantum-mechanical view of chemical bonding that incorporated his own concepts of resonance, orbital hybridization, and directional valence. His proposed linear helix structure for protein chains (see Figure 23-18) anticipated the discovery of evidence for the complementary chain structure in genetic material (see Figure 24-3). He won the Noble peace prize for calling attention to the dangers of radiation released during the atmospheric testing of atom bombs. (Courtesy California Institute of Technology.)

An example of a resonating molecule is carbon dioxide. The formula expected is one showing oxygen atoms linked with double covalent bonds to carbon, $\ddot{O}{=}C{=}\ddot{O}$, giving an electron octet about each atom. The carbon–oxygen distance, however, is 5 percent shorter than the $C{=}\ddot{O}$ in formaldehyde, $H_2C{=}\ddot{O}$. That experimental fact is taken to indicate contributions by triply bonded resonance forms which we enclose in braces and picture as follows:

$$\left\{ :\ddot{O}-C{\equiv}O\colon \qquad \ddot{O}{=}C{=}\ddot{O} \qquad :O{\equiv}C-\ddot{O}\colon \right\} \tag{9-3}$$

The CO_2 molecule is said to resonate among the three hypothetical structures, the real molecule being a composite or hybrid of them all.

Though qualitative as most chemists use it, resonance theory is a powerful structure bonding concept. Its value will become apparent later when we look into some of its applications.

Octet rule exceptions

The octet rule was helpful in developing valence theory related to main group atoms that use s, or s and p valence electrons (see Figure 8-1). Standing outside the rule, however, are the transition elements that can also use d electrons in bonding, and other exceptions are known as well.

One exception is found in substances like boron trifluoride, which are stable although one atom is deficient an electron pair,

$$B\,{\circ} + 3\cdot\ddot{F}\colon = \quad \begin{matrix} :\ddot{F}: \\ B\,{\circ}\,\ddot{F}: \\ :\ddot{F}: \end{matrix}$$

The reactivity of BF_3 is understood by recognizing its tendency to fill out the octet. Thus it forms a bond with fluoride ion to give the inert tetrafluoroborate ion,

$$\overset{\displaystyle :\ddot{F}:}{\underset{\displaystyle :\ddot{F}:}{B\!:\!\ddot{F}:}} + :\ddot{F}:^{-} = \overset{\displaystyle :\ddot{F}:}{\underset{\displaystyle :\ddot{F}:}{:\ddot{F}:B\!:\!\ddot{F}:^{-}}} \quad or \quad \overset{\displaystyle F}{\underset{\displaystyle F}{F\!\rightarrow\!B\!-\!F^{-}}} \qquad (9\text{-}4)$$

The new bond is called a *coordinate covalent bond*. The only difference between it and a covalent bond is the mode of formation — in coordinate covalent bonding one atom donates both the electrons that will be shared. The direction of the arrow used to indicate the bond is toward the electron acceptor atom and away from the donor.

A second general exception is found in substances like sulfur hexafluoride (SF_6), in which the central element's valence shell holds more than an octet; $:\ddot{S}:$ is large enough and $\cdot\ddot{F}:$ small enough so that six fluorine atoms can be clustered in a structure containing six covalent bonds. Electronically, $_{16}S$ is $1s^2 2s^2 2p^6 3s^2 3p^4 3d^0$. By utilizing the six 3s and 3p electrons in a way that will be described later as *hybridization,* the sulfur atom binds six fluorines to give a molecule with octahedral symmetry (see Table 9-2).

Another octet rule violation is found in odd-electron substances like nitric oxide; $:\dot{N}\cdot$ and $\overset{\circ}{\underset{\circ}{:}}\ddot{O}\overset{\circ}{\underset{\circ}{:}}$ combine to give the following resonance hybrid in which a deficiency of one electron exists:

$$\left\{ :\ddot{N}\!=\!\overset{\circ}{\underset{\circ}{O}}\overset{\circ}{\underset{\circ}{}} \qquad :\dot{N}\!=\!\overset{\circ\circ}{\underset{\circ}{O}}\overset{}{\underset{\circ}{}} \right\}$$

Species with one or more electrons unpaired are called *free radicals*. Free radicals are paramagnetic and colored (NO is one exception; it is colorless). They are common in the upper atmosphere (see Figure 13-7) and in polluted air (see Eqs. 13-7 and 13-8), where sunlight energy creates them by splitting electron pairs.

Still another exception was discovered in 1962 when, as mentioned in Section 8-1, it was shown that the heavier noble gases can form compounds although their outer-level electrons comprise a "closed" octet. When a mixture of xenon and fluorine is heated at 400°C for an hour then cooled to −80°C, a stable compound (m.p., 90°C), xenon tetrafluoride (XeF_4), forms. The valence shell of xenon now contains 12 electrons. The geometry of the bonding is octahedral like that in SF_6. The quartet of fluorine atoms with eight bonding electrons lies in a plane. Above and below the plane are two pairs of nonbonding electrons.

Electronegativity and bond character

The relative power of an atom to attract electrons in a covalent bond is termed its *electronegativity*. In a *homonuclear molecule* like H—H or F—F, there is always equal electron sharing and the covalent bond is described as *nonpolar*. In a *heteronuclear molecule* like H—F, the electrons are displaced toward the more electronegative element (F) and the covalent bond is called *polar*. Because bond formation relates to the ability of atoms to donate and to accept electrons, we can anticipate that electronegativity trends follow trends we have already recognized in the periodic table. Thus elements with low ionization energy (Groups IA, IIA) donate readily and have low electronegativity, whereas elements with high ionization energy and high electron affinity (Groups VIIA, VIA) have high electronegativity.

Table 9-1. Electronegativities of main group elements

IA	IIA		IIIA	IVA	VA	VIA	VIIA
H 2.1							
Li 1.0	Be 1.5		B 2.0	C 2.5	N 3.0	O 3.5	F 4.0
Na 0.9	Mg 1.2		Al 1.5	Si 1.8	P 2.1	S 2.5	Cl 3.0
K 0.8	Ca 1.0			Ge 1.8	As 2.0	Se 2.4	Br 2.8
Rb 0.8	Sr 1.0			Sn 1.8	Sb 1.9	Te 2.1	I 2.5
Cs 0.7	Ba 0.9			Pb 1.7	Bi 1.8		

Electronegativities are given in Table 9-1 for main group elements arranged in periodic table order. The numbers were calculated by Pauling from *bond energy* data. Bond energy is the energy required to create neutral atoms by breaking a bond in a molecule. Bond energies can be obtained experimentally by measuring the heat involved when compounds are synthesized.

On the Pauling scale, fluorine is arbitrarily given the value 4.0 units and other elements assigned values relative to it. As expected, electronegativity increases with atomic number across a period and decreases in going down a group. The lowest value is thus found in the lower left corner of the periodic chart and the highest in the upper right. Cesium is the least electronegative (or most electropositive) element, fluorine the most electronegative. The transition elements (omitted here) all have values of about 1.7 units.

The electronegativity *difference* between two elements influences the character of their bond. When the two electronegativities are quite different, the bond is primarily ionic. Ionic bond character is enhanced when the more positive element is univalent and has large size, and the more negative element is univalent and has small size. If two elements have about the same electronegativity, the bond between them is probably covalent.

Pauling has estimated that when the electronegativity difference is 0.2 unit the bond has 1 percent ionic character; when 1.0 unit, 22 percent; 1.7, 50 percent; 2.2, 70 percent; 3.2, 92 percent. A plot of these figures is a curve from which the character of bonds can be assessed (see Problem 18). For example, the electronegativity difference between the atoms in sodium bromide is $2.8 - 0.9 = 1.9$. On the curve 1.9 units corresponds to about 60 percent ionic (and 40 percent covalent) character. The formula Na^+Br^- therefore best describes the compound. By contrast, the electronegativity difference in HBr is $2.8 - 2.1 = 0.7$ unit, corresponding to a bond with about 14 percent ionic character. Hydrogen bromide is essentially covalent and best represented by a molecular formula, HBr, or H—Br, or H:Br.

Polar molecules

Although hydrogen bromide is a neutral molecule, its two dissimilar atoms lead to a noncoincidence of centers of positive and negative charge. We can emphasize this condition by writing the formula for HBr with *partial charges* denoted by the symbol δ (delta). Thus $H^{\delta+}Br^{\delta-}$ means that the atoms carry charges smaller than would be indicated by unqualified + and − signs which symbolize full ion charge. Charge displacement may also be indicated by placing the bonding pair nearer the more electronegative atom, as H :B̈r:.

An essentially covalent molecule that, like HBr, has permanent charge separation is called a *polar molecule* and its bond is a *polar covalent bond*. Such a molecule behaves like an electric dipole—that is, a system containing two equal but opposite separated charges. By contrast nonpolar molecules, those that have charge symmetry, do not act like electric dipoles. As shown in Figure 9-4, polar molecules can be distinguished from

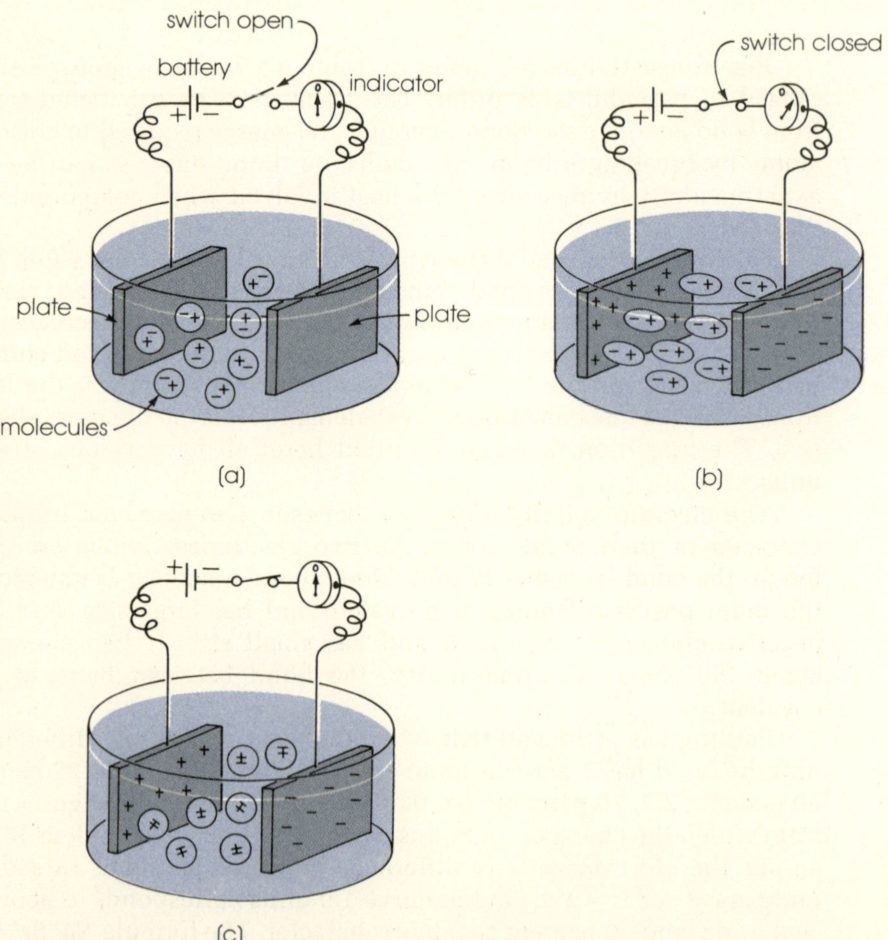

Figure 9-4 Effect of an electrostatic field on molecules. (a) Without the field, molecules have random orientation. (b) With the field, polar molecules orient themselves as electric dipoles, giving a measurable neutralization of the applied field. (c) Nonpolar molecules remain unoriented in an electrostatic field.

Figure 9-5 Peter Debye (Netherlands and United States 1884–1967, Nobel prize in chemistry 1936.) Educated in Germany as a physical chemist, Debye was director of the Kaiser Wilhelm Institute for theoretical physics before coming to the United States and Cornell University in 1940. Debye worked on X ray diffraction techniques, dipole moments, electrolytic conduction, and ionic attraction in solutions (see Figure 16-3). He also applied quantum mechanics to the relationship between heat effects and atomic vibration in solids. (Courtesy American Chemical Society.)

nonpolar molecules because they align themselves in the electric field created by charging the parallel plates of a device called a *capacitor*. For a given operating voltage the plates hold more charge with a polar substance between them than they hold with a nonpolar substance.

A molecule that assumes a preferred orientation in an electric field is described as possessing a *dipole moment*. Dipole moment μ is equal to the product of charge q and distance d separating the centers of plus and minus charge,

$$\mu = qd \tag{9-5}$$

The greater the difference between the electronegativities of two bonded atoms the larger the magnitude of μ.*

Dipole moment and molecular shape

Chemists use dipole moments as clues to the shapes of small molecules. For example, CO_2, CS_2, and $HgCl_2$ have zero dipole moment. This experimental fact indicates linear structure, for only with that configuration can the molecules have symmetrical charge distribution,

$$\ddot{O}{=}C{=}\ddot{O} \qquad \ddot{S}{=}C{=}\ddot{S} \qquad :\ddot{Cl}{-}Hg{-}\ddot{Cl}:$$

On the other hand SO_2, H_2O, and H_2O_2 have dipole moment values, respectively, of 1.6, 1.84, and 2.1 debyes. They are charge-asymmetric, and therefore must have "bent" structures that act like dipoles:

The dipolar character of water is discussed further with Figure 16-3.

---

* Dipole moments are expressed in electrostatic units or debye (Figure 9-5) units (10^{-18} esu $= 1$ D). Values of this magnitude are obtained because the electronic charge is 4.8×10^{-10} esu, and atom–atom distances are about 10^{-8} cm.

9-3. THE ORBITAL INTERPRETATION OF COVALENCE

The molecular-orbital method

The most comprehensive view available today of covalent bonding is a quantum-mechanical approximation that considers how electron density is affected by all atoms present and spread over a molecule as a whole.

In the *molecular-orbital* (MO) method of description developed by Robert Mulliken (United States 1896–   , Nobel prize in chemistry 1966) and others, atoms are considered fixed in positions required by molecular geometry whereas their electrons occupy energy states called *molecular orbitals*. Molecular-orbital expressions are arrived at by mathematically combining atomic orbitals such as those of Figure 7-6. Like atomic orbitals, molecular orbitals are describable by the Schrödinger wave equation, and quantum numbers are associated with them. Each orbital can contain two electrons of opposite spin. When molecular orbitals are built up in Aufbau fashion (see Figure 7-5), beginning with the one of least energy, the result is the molecular equivalent of the 1s, 2s, 2p, 3s, etc., structure of atomic orbitals.

After taking into account atomic-orbital interactions, the total energy of the molecule is found by summing the orbital energies. A molecular-orbital energy that is less than the sum of the energies of the isolated atoms indicates bond stability. Electrical repulsion is overcome as the bonding electron pair occupies a volume extending around its two bonded nuclei, and each electron attains greater mobility than it had when bound to a single atom. Wave-mechanical calculations show that as the volume available to electrons increases, less energy is associated with their wave motion. This is interpreted to mean that bond energy is derived principally from the wave nature of electrons rather than from their electrical properties.

We will use a description of the formation of the hydrogen molecule to indicate how the molecular-orbital method works. Quantum chemists first assume that the mathematical expressions that describe the two individual 1s atomic orbitals should be combined linearly to arrive at two molecular orbitals called *sigma* (σ) *orbitals*. Linear combination means either addition or subtraction of wave functions. Subtraction is visualized as subtraction of the volumes of the two atomic orbitals that overlap. The result is an *antibonding* orbital (designated σ^*1s) of slightly higher energy. Addition, visualized as addition of the overlap volumes, gives a *bonding* orbital ($\sigma1s$) of slightly lower energy. The operations are summarized as follows:

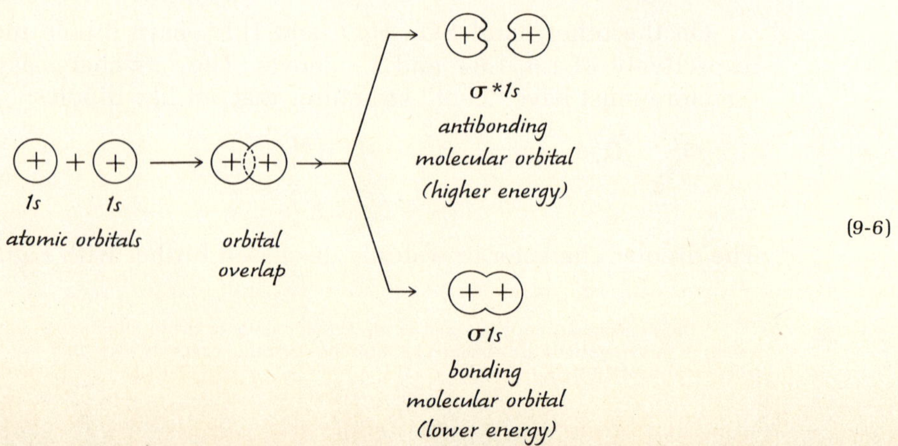

$$(9\text{-}6)$$

Detail is added to the process by employing the symbolism of Table 7-3 to designate electron occupancy of orbitals. Both electrons, their spins paired, are found in the bonding orbital because it is the region of lower energy. The higher-energy antibonding orbital is vacant;

$$(9\text{-}7)$$

Larger atoms require more complex diagrams.

The atomic-orbital method and molecular geometry

A simpler way to interpret the covalent bond is by the atomic-orbital (AO) method. With the exception that bonding electrons are considered to occupy a stable orbital in each of two bonded atoms, the atomic-orbital method treats atoms as if they were isolated. Despite the limitations imposed by using energy levels that are associated with separate atoms rather than with atoms under each other's influence, the method correctly forecasts molecular shapes and is satisfactory for many purposes.

For illustration we will describe the formation of a water molecule from its atoms. We know each of the two hydrogen atoms has one electron in a spherical 1s orbital. We also know that the oxygen atom has in its valence shell two electrons occupying a 2s orbital and four electrons occupying 2p orbitals. The nature of orbitals in atoms more complex than the hydrogen atom is not certain, but we will assume that the same principles govern all atoms. We therefore describe the oxygen p orbitals directionally with x, y, and z rectangular coordinates as in Figure 9-6a. In one p orbital we will position two of the four available p electrons and in the other two orbitals we will place one electron each. Bonding might take place by the combination (overlap) of the hydrogen 1s orbitals and the two half-filled oxygen 2p orbitals. Due to the right-angle orientation of the p orbitals, a water molecule bent at 90 degrees would then be expected. The measured average bond angle, determined by X ray diffraction, however, is slightly more than 104 degrees.

The discrepancy has been rationalized by wave-mechanical calculations which predict that the electron distribution needed for maximum orbital overlap (greatest bond stability) is not as described above but a tetrahedral bond arrangement—that is, one with four available bonding locations 109 degrees apart. These bonds are generated by a process called *orbital hybridization,* a concept originated by Pauling in 1931. Hybridization is a quantum-mechanical means for combining normal atomic orbitals to give a new set of atomic orbitals with orientation idealized for bonding.

In the electron clouds of the oxygen atom, combination of the single 2s and three 2p orbitals gives four equivalent hybrid orbitals. To specify their origin the hybrids are designated sp^3. Two of the new orbitals are filled with two electrons, the other two are singly occupied. Formation of water from its atoms is then described as overlap of the 1s hydrogen orbitals with the two half-filled hybrid orbitals to give two covalent bonds, leaving the filled, nonbonding hybrid orbitals unaffected. Repulsion of electrons in the bonding orbitals by the nonbonding electrons is believed responsible for decreasing the expected bonding angle from 109 to 104 degrees (Figure 9-6c,d).

The formation and shape of the ammonia molecule, NH_3, is explained

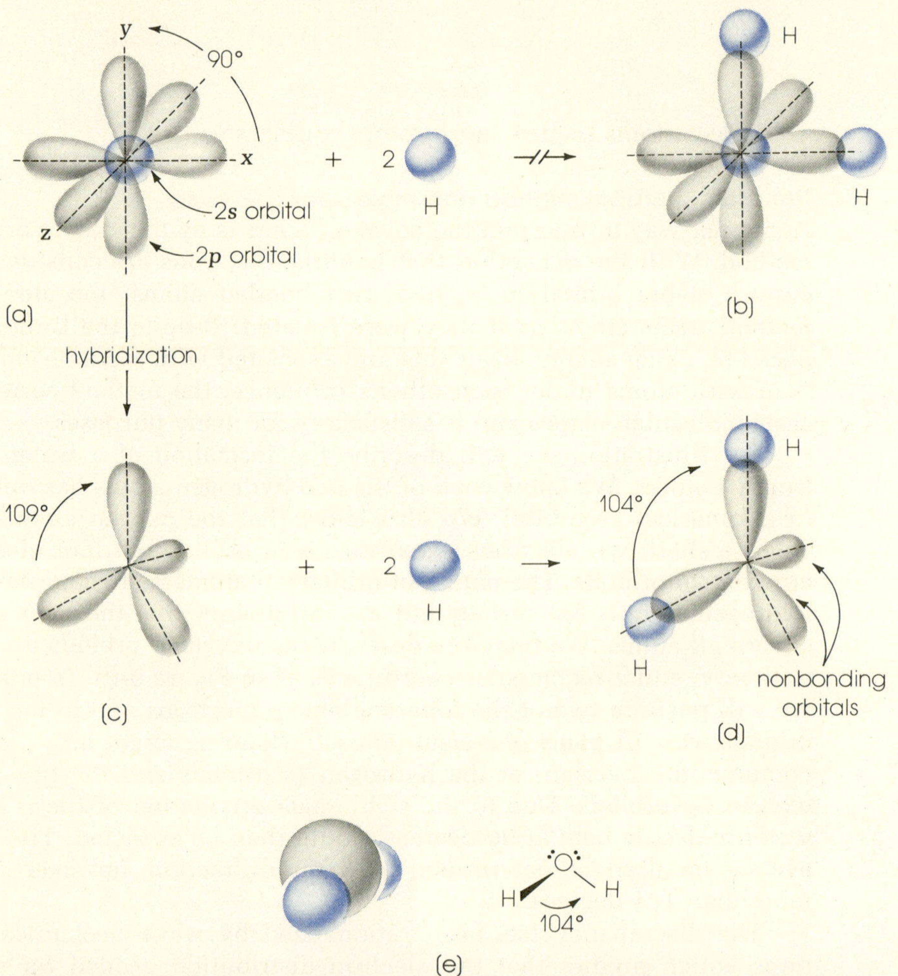

Figure 9-6 Hybrid atomic orbitals and the formation of H_2O. (a) An atomic-orbital model of the oxygen atom, showing separately the spherical 2s orbital and three 2p orbitals. The latter have direction along mutually perpendicular axes. (b) An incorrect assumption that the water molecule is formed directly from (a) by overlap of the two hydrogen 1s orbitals. This view correctly says H_2O will be dipolar but wrongly predicts a bonding angle of 90 degrees, 13 percent smaller than measured. (c) sp³ hybridization in the oxygen atom produces four new orbitals from the 2s and 2p orbitals. They are tetrahedrally directed in space. (d) The accepted view of water molecule formation based on interaction of two hydrogen 1s orbitals with two hybridized oxygen orbitals. The nonbonding orbitals each contain two "extra" electrons. (e) Two other models of the water molecule. The space-filling model on the left is constructed to reflect measured bonding angles and relative atomic sizes.

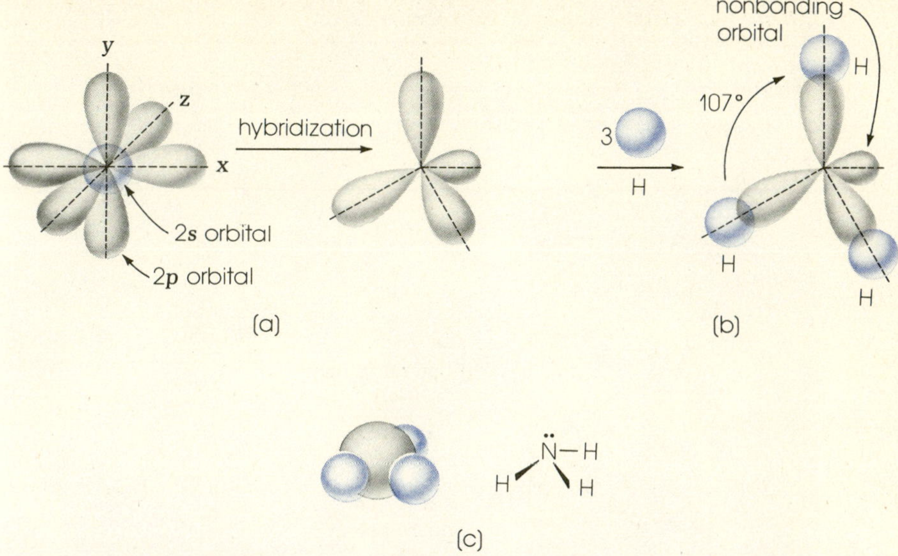

Figure 9-7 Formation of NH₃. (a) Hybridization of the 2s and 2p orbitals of the nitrogen atom. (b) Interaction of the hybrid orbitals with hydrogen 1s orbitals gives NH₃. (c) Two other models of the ammonia molecule.

in the same way (Figure 9-7). The nitrogen atom has the configuration $1s^2 2s^2 2p^3$. Quantum-mechanical combination of the three unpaired p electrons with the pair of 2s electrons gives, as in the oxygen example, four sp^3 hybrid orbitals. Three of the new orbitals contain one electron each, the other is filled with two electrons and is nonbonding. The molecule is formed as overlap occurs between the half-filled hybrid orbitals and 1s orbitals of the three hydrogen atoms.

Hybrid bonding in organic compounds is considered in Section 20-3.

Hybridization and space orientation

The examples of hybridization just cited are not unusual. Actually, most covalent bonding involves hybrid orbitals and, as demonstrated above, the shapes of molecules can be inferred from hybrid-orbital orientation. More than 40 different hybridization modes can in theory be produced from combinations of s, p, d, and f electrons, but real molecules are describable by only a few.

The common hybrid-orbital arrangements are listed in Table 9-2. The first three are flat structures, the others are three dimensional. Study of them has given the useful conclusions that (1) hybridization provides in each geometry the maximum spatial separation of bonding electron pairs, and (2) where no atom is bonded, we can expect unshared electrons to take up the bonding position. These two generalizations, which allow intelligent surmise about the structure of simple species when little else is known, are called the *Sidgwick-Powell rules* (see Question 9). They are a corollary of the *Pauli exclusion principle*, which says that electrons of like spin avoid each other. The exclusion principle is of prime importance in determining molecular shapes.

Without the Sidgwick-Powell rules, quantum-mechanical manipulation

Table 9-2. Hybrid atomic orbital geometries

Hybrid type	Number of orbitals	Geometry	Examples
sp	2	Linear (flat)	$HgBr_2$, $Ag(NH_3)_2{}^+$
sp^2	3	Equilateral triangle (flat)	BF_3, BCl_3
dsp^2	4	Square planar (flat)	$Ni(CN)_4{}^{2-}$, $HgI_4{}^{2-}$
sp^3	4	Tetrahedral (3 dimensional)	H_2O, NH_3, CH_4, $SO_4{}^{2-}$, $SiO_4{}^{2-}$, $Cu(NH_3)_4{}^{2+}$
d^2sp^3 or sp^3d^2	6	Octahedral (3 dimensional)	SF_6, $Fe(CN)_6{}^{4-}$, $Co(NH_3)_6{}^{3+}$

of mathematical functions describing atomic orbitals would be needed to predict hybridized orbital shapes. (In Figure 9-6, for example, there is nothing to indicate that the combination of one s and three p orbitals should yield four tetrahedrally oriented orbitals.) For our purposes, questions about hybrid orbital shapes will be answered simply by reference to Table 9-2.

9-4. BONDING FROM COMPUTERS

Plotting molecular-orbital densities

Visualization of molecules has always been hampered by our inability to solve the complicated quantum-mechanical equations that would describe them. Recently, however, with the aid of computers quantum chemists have been able to refine the approximations and, by using computer output to drive an automatic plotter, have produced quantum-mechanically correct contour maps of electron density in atoms and molecules. The computer program that produced the results shown in Figures 9-8 and 9-9 included the internuclear distances, the contours desired, and equations that described orbital number and symmetry.

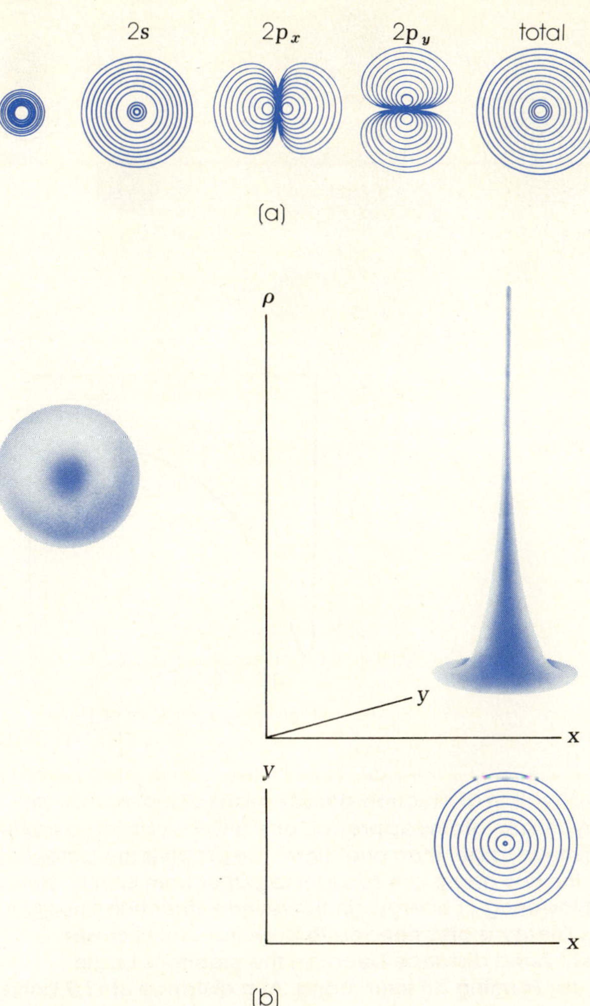

Figure 9-8 Computer–plotter construction of electron-density maps of atoms. In the $2p_x$ and $2p_y$ diagrams of (a), the innermost contours correspond to a probable electron density of 0.125 electronic charge (e⁻) per cubic bohr (1 bohr = 0.529 × 10⁻⁸ cm = 0.529 Å). In all other diagrams (including Figure 9-9) the innermost contour corresponds to a probable electron density of 1.0 e⁻/bohr³. Each successive outer contour decreases by a factor of 2, down to 4.9 × 10⁻⁴ e⁻/bohr³. (a) A sequence showing the building up of the carbon atom. The structures are contour maps of probable electron density ρ (rho) for the four occupied orbitals. (b) The bottom structure in the x-y plane is the result of combining all the structures in (a). It is called a map for the total probable density. The top structure is an x-y-ρ plot of electron density in the x-y plane passing through the nucleus. (Adapted from Film Series of Loops by A. Wahl and U. Blukis, McGraw-Hill, New York, 1969; courtesy also of the Argonne National Laboratory, Argonne, Illinois.)

In the sequence labeled (a) in Figure 9-8, the architecture of a carbon atom is described in terms of probable electron density in its occupied orbitals. In (b) all the orbitals are merged to give an orbital picture of the whole atom. The lower figure in (b) is the two-dimensional electron-density map obtained from the plotter. The upper figures in (b) are drawings of its three-dimensional equivalent. Electron density is seen to decrease rapidly in all directions away from the center.

Electron-density plots representing bond formation are given in Figure 9-9. In (a) the making of a covalent bond is illustrated by the reaction, H + H = H—H. In the accompanying graph of the system's potential energy the two hydrogen atoms can be thought of as being brought together from far apart, starting at the graph's right side. At a distance of 10 bohrs (about 5 Å) they begin to feel mutual attraction. The system loses energy, attaining greatest stability at a covalent bonding distance of about 1.2 bohrs. At shorter distances the atoms repel one another and energy is required to push them closer together (the energy curve rises). The molecule (contour

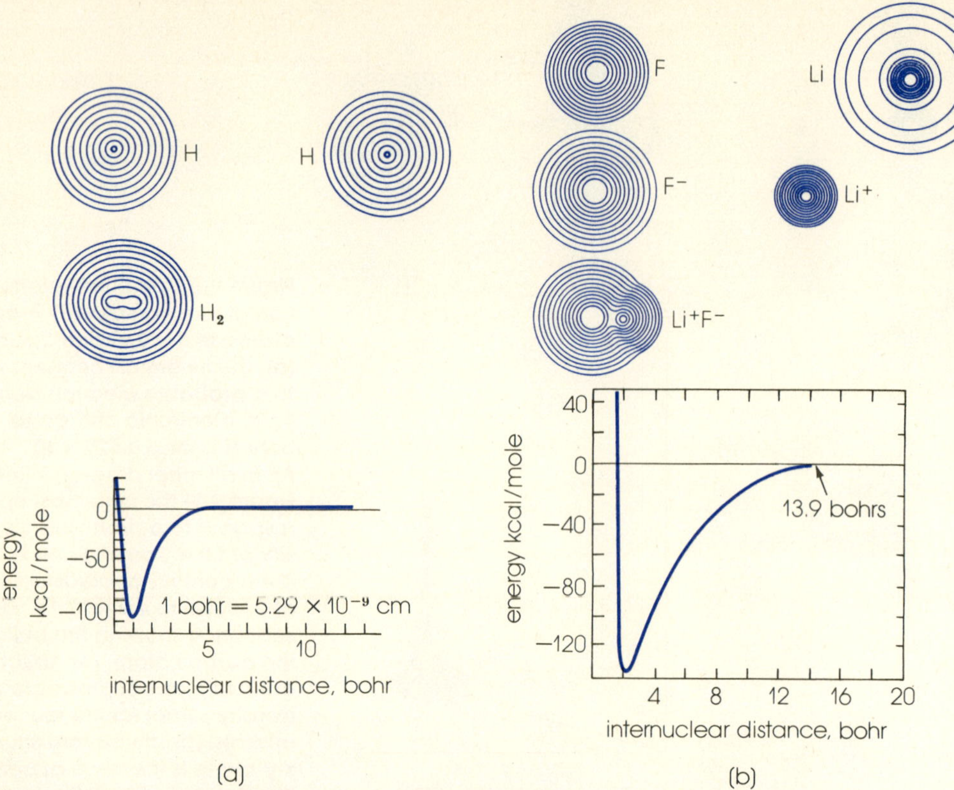

Figure 9-9 Computer–plotter construction of electron-density maps of molecules. (a) Forming a covalent bond. Two hydrogen atoms approach one another, giving a molecule of H_2 that turns out to be not much larger than one atom. The graph is the potential energy curve for the process. As the two atoms are brought together from infinity, bond formation is accompanied by a lowering of energy. (In the reverse direction energy is required to separate the atoms.) Energy is also needed to push the atoms closer together than the normal covalent bond distance because the electrons begin crowding into the same volume. (b) Forming an ionic bond. At a distance of 13.9 bohrs the two atoms become ions. The Li⁺F⁻ unit is about the same size as F⁻. The graph indicates that the optimum ionic bonding distance is about 2.7 bohrs. (Adapted from Film Series of Loops by A. Wahl and U. Blukis, McGraw-Hill, New York, 1969; courtesy also of the Argonne National Laboratory, Argonne, Illinois.)

map just above the graph) is not much bigger than an atom. This means that when a molecule forms, the volume occupied by the electrons shrinks.

In (b) an ionic bond is established between lithium and fluorine atoms. At a distance of 13.9 bohrs (7.35 Å) the ionic state is more stable than the free atom state. At closer distances energy falls rapidly as the two particles, now ions, experience coulomb attraction. The fluoride ion is somewhat bigger than the fluorine atom due to addition of an electron to an orbital. The lithium ion, however, is considerably smaller than the lithium atom due to loss of its valence electron in the $2s$ orbital. Core electron orbitals are relatively unaffected by the ionization process.

Both graphs in Figure 9-9 show that bond formation is accompanied by an energy decrease which allows the system to overcome three obstacles:

electron — electron repulsion, nucleus — nucleus replusion, and the motion (kinetic) energy of the electrons.

As computers yield representations of more complex molecules (page 195), chemists will learn more about atomic interactions, and perhaps be able to derive from them physical constants related to such features as dipole moments and binding energies. We will also have a means for studying reactions under simulated conditions that might be difficult or impossible to observe in actual experiments. In the computer laboratory we can stop a "reaction" at any time. A look at the display screen will show how charge density, and therefore bonding, is shaping up as atoms are brought together.

QUESTIONS

1. Bonds In general what kind of bond do you expect to form between **(a)** a metal and nonmetal? **(b)** two nonmetals? **(c)** a species with a complete octet and a species with an incomplete octet? **(d)** In answering (a) and (b), what aid can you get from Table 9-1?

2. Bonding models The structures of Figure 9-10 were proposed in 1916 by Kossel and Lewis, respectively, to illustrate their theories. Identify the substances and explain.

3. Isoelectronic species **(a)** How many electrons are in the first and second shells of each of the following: O^{2-}, F^-, Ne, Na^+, Mg^{2+}? **(b)** The five substances are *isoelectronic*. What does the term mean?

4. MO The following are MO representations of the formation of (a) Li_2 and (b) Be_2. Interpret them. Tell which molecule you think is more stable and why.

(a)

(b)

5. Electronegativity The energy needed to separate a gaseous molecule into gaseous atoms is called the *bond dissociation energy*. Consider these

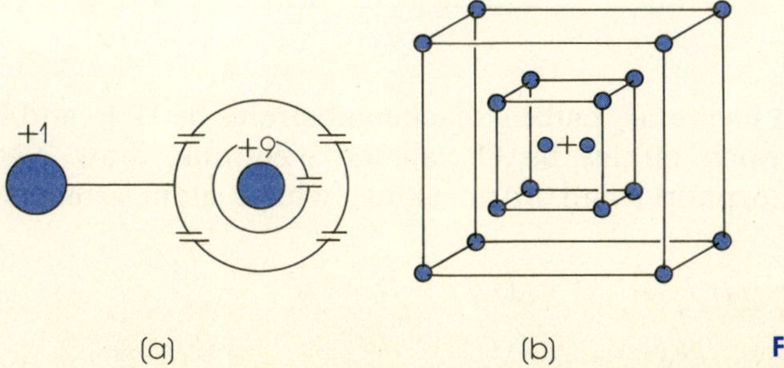

(a)

(b)

Figure 9-10

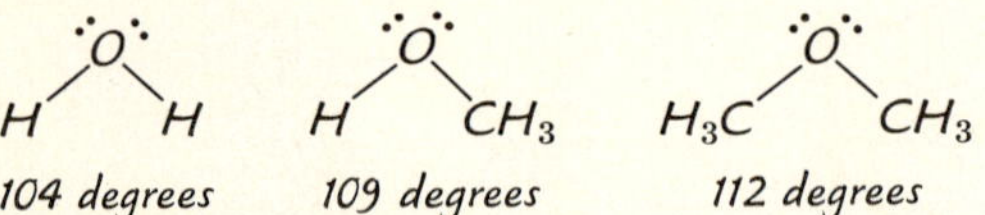

Figure 9-11

relative bond dissociation energy values: HCl, 103; HBr, 87; HI, 71. Is the trend that which might be expected from electronegativity data? Explain.

6. Polar, nonpolar How does the action of dipoles in an electric field permit chemists to distinguish experimentally between a polar substance such as PF_3 and a nonpolar substance such as BF_3? Predict the shapes of these molecules. Explain.

7. Dipole moment **(a)** Arrange the following in order of increasing dipole moment and explain: HBr, HCl, HI. **(b)** SnI_4 and CCl_4 are molecules with zero dipole moment. Explain.

8. Salt **(a)** When lithium metal and fluorine gas combine, the result is the white crystalline salt lithium fluoride (LiF). Why can we assume that it contains equal numbers of positive and negative ions? **(b)** Solid LiF does not conduct an electric current. Fused LiF (m.p., 870°C) does conduct. Explain.

9. Sidgwick-Powell rules The generalized molecules of Figure 9-11 are, respectively, MX_2, MX_3, MX_3e (e = an unshared electron pair), and MX_6. Interpret them with the aid of the Sidgwick-Powell rules. (In the first two structures, X atoms lie on circles; in the last two they lie on spheres.)

10. Bonding angle Following are structures of water, methyl alcohol, and dimethyl ether:

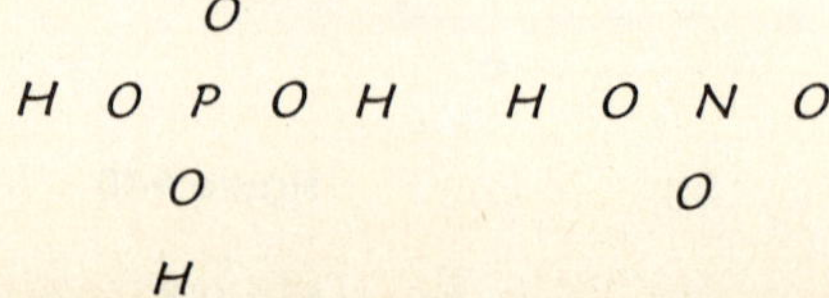

104 degrees   109 degrees   112 degrees

(a) Why are the bonding angles somewhat similar? **(b)** Why does the angle increase as shown? (Propose a theory.)

PROBLEMS

11. Electron dot formulas Letting blackened circles be H, P, and N valence electrons, and open circles be O valence electrons, draw electron dot formulas for phosphoric and nitric acids, whose atom arrangements are

O

H O P O H   H O N O

O   O

H

12. Electron dot equations Write electron dot equations (similar to Eq. 9-1) to show formation of the following ionic compounds from their elements: K_2O, BaS, $AlBr_3$.

13. Electron dot formulas Write electron dot covalent structures for **(a)** HOCl (two single bonds); **(b)** $POCl_3$ (one double, three single bonds); **(c)** N_2 (one triple bond).

14. Electron dot formulas Write electron dot formulas for the following formulas. Show electrons from carbon and phosphorus as open circles, all others as darkened circles. **(a)** S=C=S; **(b)** H—C≡C—CH_3; **(c)** PCl_3.

15. Dipoles **(a)** Draw tetrahedral structures (Table 9-2) for the following molecules. **(b)** Tell briefly how to interpret their measured dipole moment values (debyes): CH_4, 0.0; $ClCH_3$, 1.9; Cl_2CH_2, 1.6; Cl_3CH, 1.1; CCl_4, 0.0.

16. Odd molecule Draw a valence bond formula for chlorine dioxide OClO. Predict some of its properties.

17. Polar, nonpolar To represent polarity in a molecule (as deduced from the electronegativity table) an arrow may be drawn from the less electronegative to the more electronegative element: $\overrightarrow{H-Cl}$. When like forces oppose one another, the molecule is nonpolar: $\overleftarrow{O}=C=\overrightarrow{O}$. Add arrows to the following and tell in each case whether the molecule is polar or nonpolar:

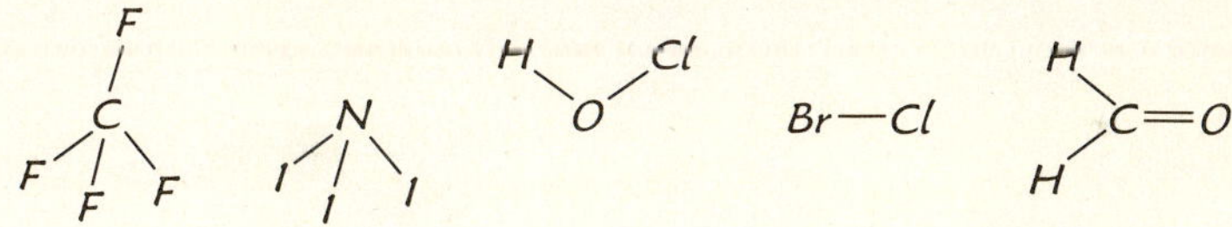

18. Electronegativity Text data accompanying Table 9-1 relate electronegativity differences to bond character. **(a)** Plot the data, with electronegativity on the ordinate. **(b)** Use the table and graph to estimate the percentage of ionic character in bonds of the following compounds: KBr, MgS, H_2O, CO_2, CH_4.

19. Electronegativity **(a)** Copy Figure 9-12 on graph paper. Using Table 9-1 continue the graph to $Z = 20$. Connect all points with a line. **(b)** Write down the generalizations your graph makes evident.

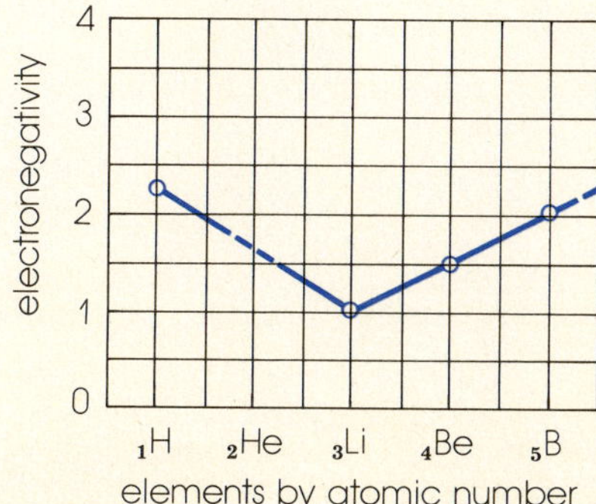

Figure 9-12

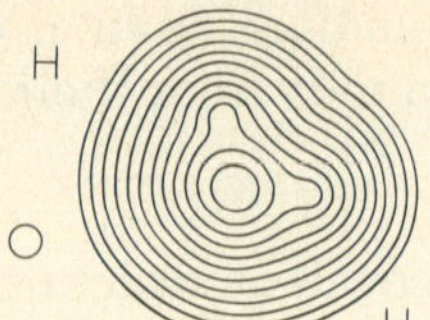

Figure 9-13

20. Computer plot Figure 9-13 is a Wahl electron-density contour diagram of the H_2O molecule. Translate it approximately into an x-y-ρ plot like the upper drawing in Figure 9-8b. Show where the three atoms are located.

WHAT IS THE BONDING IN LARGER STRUCTURES?

THE IMPORTANT CONCEPTS

10-1. Bonding in coordination compounds

1. Neutral and negative groups form complexes with transition metal ions.
2. Complex ions have basically elementary geometries.
3. A few simple laboratory techniques give information on complexes.
4. Ligand-metal ion bonds are coordinate covalent.
5. Through complexation the central metal tends to attain the electron number of the next noble gas.
6. Atomic-orbital theory correlates ligancy, color, magnetic properties, and geometry of complexes.
7. Coordination compounds are important in nature.

10-2. Bonding in solids
1. Ionic crystals, held together by coulomb forces, are hard, high melting, and composed entirely of ions.
   a. Bonding force varies with ion size and charge.
   b. Crystal ligancy varies with ion size.
   c. Ion radius varies with nuclear charge and number of electron shells.
2. Molecular crystals, containing dipole–dipole attractions, are soft, low melting, and composed entirely of molecules.
   a. Van der Waals and London forces are recognized.
   b. Hydrogen bonding, both inter- and intramolecular, may operate in solids, liquids, and gases.
   c. Covalent and van der Waals radii are used in discussing molecular structure.
3. Covalent network solids resembling giant molecules are often hard and high melting; their melts do not conduct electricity.
4. Metals usually crystallize in one of three lattice arrangements.
   a. The Lorentz model qualitatively explains many metallic properties.
   b. The band model distinguishes among conductors, nonconductors, and semiconductors.

In Chapter 9 we looked at atoms, ions, and molecules as structural units. In this chapter we examine how larger structures are built up from the units. We consider two cases: first, coordination compounds, which are molecules containing a grouping such as $Cu(NH_3)_4^{2+}$; and second, crystals. In Figure 6-6 we described crystalline solids according to their symmetry. Here we classify them by the species that occupy the lattice points and the bonding that locks the lattice together.

10-1. BONDING IN COORDINATION COMPOUNDS

The coordination theory

Nineteenth-century valence and structure theory treated organic molecules reasonably well but fit only the simplest inorganic compounds. Particularily difficult to explain were "molecular compounds," later called *coordination compounds,* formed from the combination of certain ionic or nonionic molecules with compounds whose normal valency was already satisfied. Thus mercuric iodide, a red solid insoluble in water, was known to dissolve in potassium iodide to give a pale yellow product, K_2HgI_4, in which seemingly the valency of mercury expanded to accommodate more atoms:

$$I-Hg-I + 2K-I = \quad \begin{array}{c} K \\ | \\ I \diagdown | \diagup I \\ Hg \\ I \diagup | \diagdown I \\ | \\ K \end{array}$$

Reactions of this kind were clarified by Werner (Figure 10-1) during 1890–1918. The *coordination theory* he developed so satisfactorily correlated structure–reaction phenomena it started a renaissance in inorganic chemistry that still guides research today. Its main points were as follows:

1. Coordination compounds are built around a central atom that is typically a transition metal. Two kinds of valency apparently operate. In the first "sphere of attraction," ions or molecules are "coordinated" to form a "complex" that may be quite stable. In the second sphere ions or molecules are less tightly held and quickly lost in solution.

2. Each central atom coordinates a characteristic number of ions or molecules (both called *groups,* later called *ligands*). *Coordination numbers* (*ligancy*) from 1 to 9 are known, 4 and 6 being most common. Coordination structures (*complexes*) have geometric symmetries.

3. Typical molecules that coordinate are H_2O, NH_3, NO, and CO. Typi-

Figure 10-1 Alfred Werner (France and Switzerland 1866–1919, Nobel prize in chemistry 1913). Werner studied in Zurich and spent his career at the university there. His approach to chemistry was largely qualitative (he had earned failing marks in mathmatics). His ideas, however, were highly original, imaginative, and farsighted. His laboratory work consisted of simple physical measurements of solution conductance, freezing point, and optical rotation (see Figure 20-3), and synthesis and analysis of compounds. (Reproduced from the frontispiece of the book of George B. Kauffman, Alfred Werner—Founder in Coordination Chemistry, Springer, New York, 1966.)

cal coordinating ions are CN^-, NO_2^-, $C_2O_4^{2-}$, OH^-, F^-, Cl^-, Br^-, SCN^-, and I^-.

Coordination with a neutral molecule gives a complex having the same valency as the central atom,

$$Cd^{2+} + 4NH_3 = Cd(NH_3)_4^{2+}$$

Coordination with a negative ion gives a complex whose valency is the sum of the valence numbers,

$$Sb^{3+} + 6Cl^- = SbCl_6^{3-}$$

Coordination numbers are not infrequently twice the central atom's valency,

$$W^{4+} + 8CN^- = W(CN)_8^{4-}$$

The complex species may be positive or negative, as shown above, and it is sometimes neutral,

$$Co^{3+} + 3NH_3 + 3Cl^- = Co(NH_3)_3Cl_3$$

4. Complex formation, geometry, and stability depend on such characteristics of central atom and coordinated groups as size and charge (and electron structure).

Werner's methodology

Werner was but 26 years old and with no experience in inorganic coordination chemistry when he chose to examine some complex compounds synthesized by Danish chemist S. Jørgensen. An example of Jørgensen's preparations was a pair of compounds with different properties but the same formula, $Co(NH_3)_4Cl_3$. In solution each was known to give a complex ion and one chloride ion. Jørgensen wrote them,

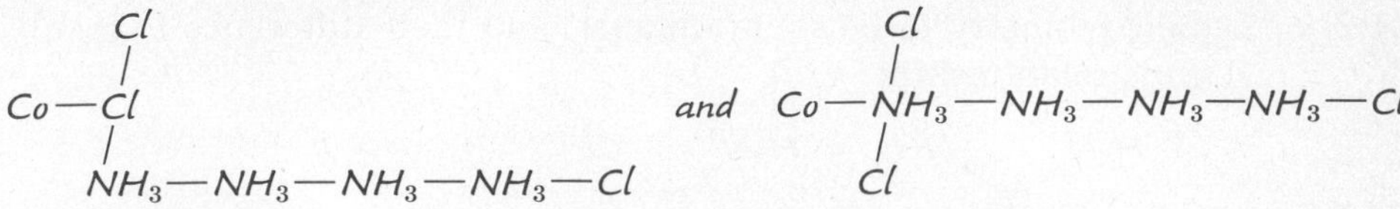

$$Co \underset{\underset{NH_3—NH_3—NH_3—NH_3—Cl}{|}}{\overset{\overset{Cl}{|}}{—Cl}} \qquad and \qquad Co \underset{\underset{Cl}{|}}{\overset{\overset{Cl}{|}}{—NH_3—NH_3—NH_3—NH_3—Cl}}$$

He considered the "closer" chlorine atoms to be nonionizable, the "farther" chlorine (on the ammonia chain) to be the one that ionized.

Unsatisfied with this seemingly weak argument, Werner sought a theory that would include all coordination compounds. In a midnight flash of inspiration (a welcome though little understood ally of the researcher) he conceived of a geometric explanation, and in a few hours worked out the details. Knowing nothing about bonding, but having unusual capacity for visualization in three dimensions, he assumed the geometries of complexes to be those given in Table 9-2. He guessed Jørgensen's compounds represented octahedral symmetries of groups about a central cobalt ion. He further assumed that the difference between them lay in a *cis* (reference groups on the same side) and a *trans* (reference groups on opposite sides) orientation of two of the chloride ions:

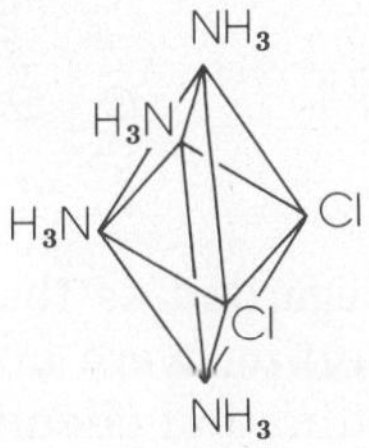

cis

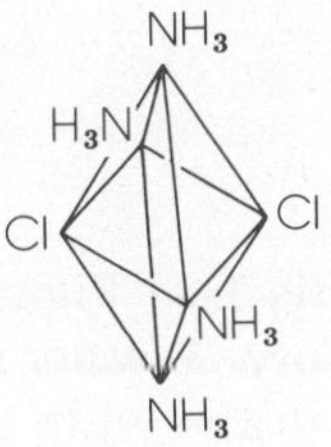

trans

Werner confirmed his ideas with readily understood methodology that featured structure proofs of new compounds he synthesized. Of great assistance was electrical conductance measurement. It allowed him to determine the number of electrically conducting ions a compound gave in solution and thus, indirectly, the number of ligands held tightly in the complex. For example, in one investigation he had prepared two platinum compounds, $Pt(NH_3)_6Cl_4$ and $Pt(NH_3)_3Cl_4$, and found their electrical conductances in water solution to be in the ratio of $5:2$. He concluded that the first compound gave five ions, the second two:

$$Pt(NH_3)_6Cl_4 = Pt(NH_3)_6{}^{4+} + 4Cl^-$$

$$Pt(NH_3)_3Cl_4 = Pt(NH_3)_3Cl_3{}^+ + Cl^-$$

He proved his conclusions by adding silver nitrate to equal molecular quantities of both compounds and obtained AgCl, an insoluble product, in a weight ratio of $4:1$, the ratio predicted for free chloride ions.

In another investigation he had prepared two compounds with the formula $Pt(NH_3)_2Cl_2$. Because they were both nonconducting, he reasoned they were nonionic with all four ligands tightly coordinated. He guessed their geometry to be square planar, and their difference to reside in a *cis–trans* relationship:

cis   trans

To differentiate between the structures he first added silver oxide to the solutions, converting Cls to OHs. Then he added oxalic acid. Only one of the two compounds gave an oxalate complex. He designated that compound *cis*, correctly reasoning that the oxalate ion, $^-$—O—C—C—O—$^-$, could be coordinated only at *cis* positions because it was too short to span opposite corners of the *trans* form:

$$(10\text{-}1)$$

cis   cis   cis

Werner and Kekulé (see Figure 20-4) are regarded as the founders of modern structure theory. Werner's ideas of bonding were crude but they helped call attention to the wide scope of coordination chemistry. Werner lived to see X ray diffraction pictures confirm geometries of complexes that he had deduced intuitively and chemically.

Bonding in complexes

Ideas about bonding in complexes during the 1920s came from Lewis and from English chemist N. Sidgwick. Lewis identified a ligand as a group having an unshared electron pair available for coordinate covalent bond formation with an electron–deficient metal ion. Sidgwick noted that metal ions tended to accept enough electrons during complex formation to acquire the electron configuration of the next noble gas in the periodic table. Thus Co^{3+} with 2, 8, and 14 electrons in its K, L, and M shells, could accommodate four more electrons in the M shell and eight in an N shell. The 12 electrons could come from, say, six neutral $:NH_3$ molecules, giving $Co(NH_3)_6^{3+}$. The metal would then be surrounded by 36 electrons or have an "effective atomic number" of 36, the same as krypton, the next noble gas.

In the 1930s Pauling and others upgraded bonding ideas when they developed atomic-orbital theory. Although extended since then into a more powerful theory, the atomic-orbital theory is still useful for interpreting complexes. To apply the atomic-orbital method to complexes, one first assumes hybridization of the central metal's orbitals, followed by interaction of its empty orbitals with ligand orbitals containing two electrons each.

Consider formation of the ferricyanide ion $Fe(CN)_6^{3-}$. A typical transition metal, iron has available unfilled s, p, and d orbitals. The atom's electron structure is

$$1s^2 2s^2 2p^6 3s^2 3p^6 3d^6 4s^2 4p^0$$

In the symbolism introduced in Table 7-3 we will represent the last three sets of orbitals as

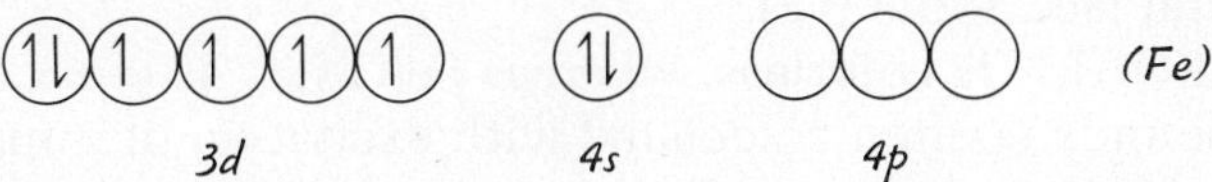

Loss of the two $4s$ and one $3d$ electrons gives Fe^{3+}. Remaining electrons then occupy the d orbitals singly (Hund's rule):

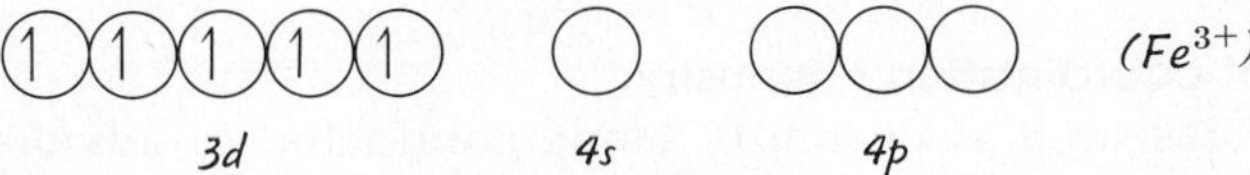

Bonding with a strongly electronegative group such as cyanide is believed to create enough electron repulsion to cause electrons in the $3d$ orbitals to pair up, leaving open two d, one s, and three p orbitals. Their mixing, called d^2sp^3 hybridization, gives six new orbitals having octahedral orientation (Table 9-2). As the six empty hybrid orbitals interact with bonding orbitals of six $:C\equiv N:^-$ groups, the complex forms

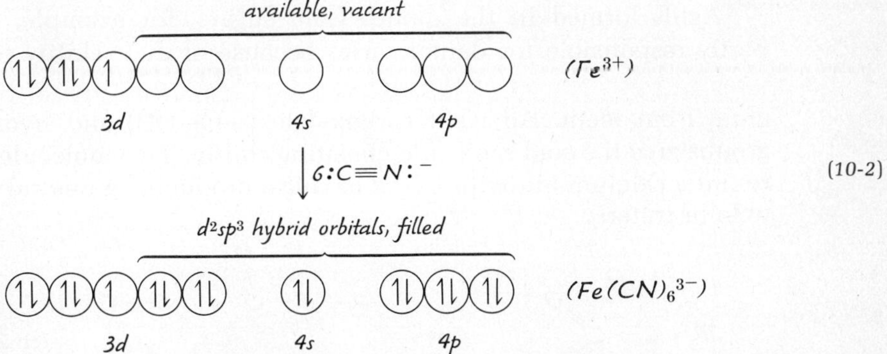

$$(10\text{-}2)$$

Evidence for this scheme is contained in X ray diffraction data that show octahedral symmetry around iron, and in magnetic moment measurement that confirms the complex has one unpaired electron.

When d orbitals are not available, metals may use a combination of s and p orbitals for bonding. The outer electron arrangement in cadmium ion Cd^{2+}, for example, is

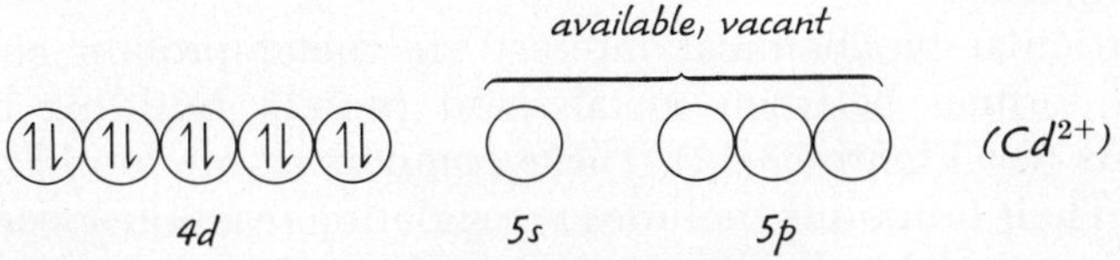

The vacant s and p orbitals can accommodate a total of eight electrons, from, say, four $:OH^-$ ions. The electron arrangement in the complex would then be

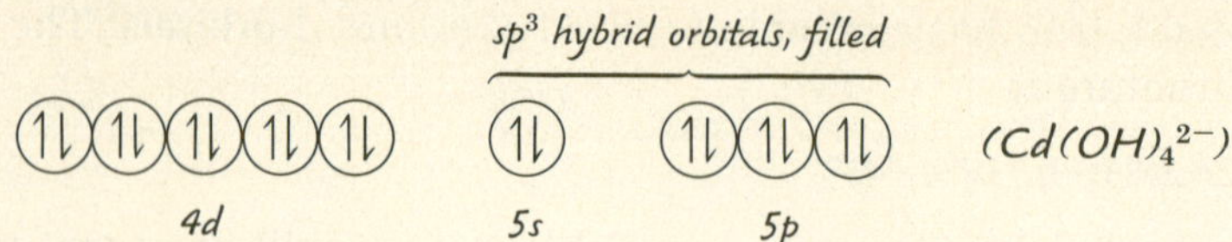

With all electrons paired, no magnetic moment is expected or found. The ion is tetrahedral (see Table 9-2).

The ion $Cd(OH)_4^{2-}$ is colorless, whereas $Fe(CN)_6^{3-}$ is red. *Color* in coordination compounds is often associated with excitation of unpaired d electrons in the transition metal ion. Radiation having energies between about 1.7 and 3.1 electron volts per photon is sufficient to excite d electrons. That happens to be the energy range of visible light (see Figure 2-2).

Importance of coordination chemistry

Considerable research is currently being conducted to advance coordination theory, understand the role of complexes in nature, and apply complexation to practical problems.

Most ligands found in natural complexes are organic *chelating* (Gk. *chela*, claw) *agents*. A chelating agent is a substance with more than one binding site per ligand. Electron donor atoms are most commonly $-\ddot{O}-$, $-\ddot{S}-$, and $-\ddot{N}-$. Like the oxalate group (Eq. 10-1), they enclose the central metal ion in stable ring structures usually made up of five or six atoms.

Acids formed in the mouth from sugars, for example, are probably partly responsible for dental caries because of their ability to extract calcium from teeth. Adjacent carboxylate ($-\overset{O}{\overset{\|}{C}}-OH$) and hydroxyl ($-OH$) groups give the acid molecule chelating ability. Two molecules of acid surround a calcium ion with a pair of rings, producing a neutral complex soluble in water:

$$2R-\underset{H}{\overset{OH}{\underset{|}{\overset{|}{C}}}}-C\underset{O-H}{\overset{O}{\diagup}} + Ca^{2+} = \text{calcium complex} + 2H^+ \qquad (10\text{-}3)$$

sugar acid   *calcium complex*

(R is the remainder of the sugar molecule.) Similar complexing facility is exhibited by mosses that cling to rocks, and microorganisms that grow on iron pipes. In both cases excretion of chelating agents solubilizes metal the organisms need.

Of particular biochemical interest are metal–protein complexes and complexes formed between metals and protein building blocks called *amino acids* (see Figure 23-12). These compounds are found in a variety of processes. Their functions include (1) regulating reaction speeds in cells, (2) transferring material to build up giant molecules such as proteins, and (3) facilitating the passage of material through membranes. A few vital natural chelates are shown in Figure 22-7.

Commercially chelation is used to (1) complex undesirable ions (Ca^{2+},

Mg^{2+}, Fe^{3+}, etc.) in solution during the manufacture of textiles, paper, food products, and pharmaceuticals; (2) introduce micronutrient metals (Fe^{2+}, Zn^{2+}, Mn^{2+}, Cu^{2+}, etc.) into agricultural land; (3) stabilize cosmetics and other sensitive products; (4) fortify detergents; (5) wash up spills of radioactive ions; and (6) clean metal surfaces prior to electroplating.

10-2. BONDING IN SOLIDS

Ionic crystals

In classifying solids by the character of their bonds, chemists recognize ionic, molecular, covalent network, and metallic types of substances.

Crystals of substances such as rock salt are called *ionic crystals* because they contain only anions and cations. *Molecules* of sodium chloride therefore do not exist. Ionic composition is inferred by the good electrical conductance of salt solutions and more emphatically by the electrical conductance of molten salts themselves. The presence of ions is proved by X ray determination of electron densities which reveal that in NaCl, for example, only spherical assemblages of 10 electrons (Na^+) and 18 electrons (Cl^-) can be found.

Forces acting between ions are mainly electrostatic (coulombic). Whether attractive or repulsive, the force F is given by Coulomb's law (Eq. 2-2) which we will write this time as

$$F = \frac{e_1 e_2}{\epsilon r^2} \tag{10-4}$$

where e_1 and e_2 are the electronic charges on the ions, r is the distance between them, and ϵ is a term called the *dielectric constant* of the medium ($\epsilon = 1$ in a vacuum). When e_1 and e_2 have opposite sign, F is negative. Negative F means work is needed to separate the ions, so the force is attractive. When e_1 and e_2 have the same sign, F is positive and the ions repel one another.

As ionic crystals are built up, each ion repels its own kind while attracting ions of opposite charge. The result is an orderly lattice of positive ions surrounded by negative ions, and negative ions surrounded by positive ions at distances dictated by their effective sizes. Assuming ions are charged spheres, each one acts like a point charge from which bonding forces are exerted equally in all directions. Comparing the energy of atoms with the energy of the crystal formed from their ions shows that the crystal has lower energy (greater stability) than the atoms. The source of stability is the ionic bonds or coulomb forces acting between ions.

Bonding force is affected by ion size and charge. As is evident from Eq. 10-4, when distance r between centers increases, attractive force F decreases. And as ion charge increases, attraction increases. We would expect, therefore, that a collection of relatively large singly charged ions such as Cs^+ and Br^- would be less tightly bonded than a lattice of small doubly charged ions such as Mg^{2+} and O^{2-}; CsBr is indeed softer, lower melting, and much more water soluble than MgO — all indications of weaker bonds that permit easier separation of ions.

Ion size also determines the way ions will pack together. Because anions are generally larger than cations, an ionic solid can be thought of as

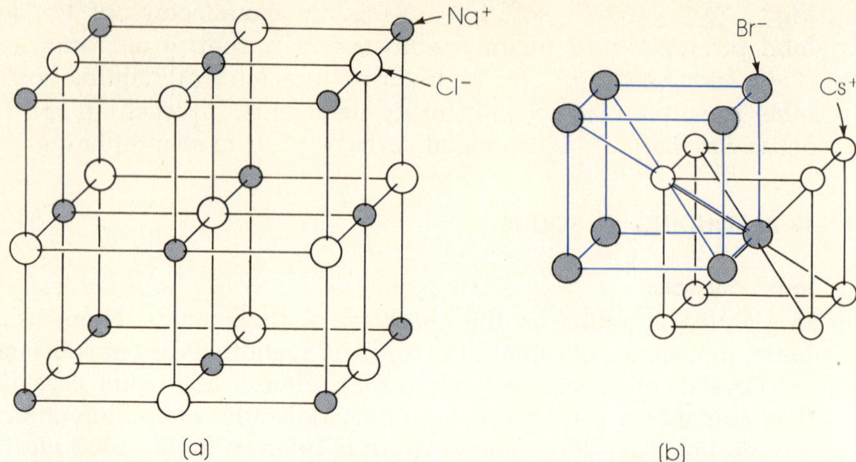

Figure 10-2 Crystal lattices. (a) The face-centered sodium chloride or rock salt lattice. The six Na+ nearest the central Cl− are positioned to form the corners of an octahedron cage. In a massive NaCl crystal each interior ion is surrounded in this manner by six ions of opposite charge (crystal coordination number is 6). (b) The body-centered cesium bromide lattice. The interpenetration of a positive-ion and a negative-ion lattice gives each ion eight oppositely charged near neighbors (crystal coordination number is 8).

a regular array of somewhat bigger spheres (anions), among which smaller spheres (cations) fit. The number of equidistant *nearest neighbors* of opposite charge an ion has in such arrangement is called its *crystal ligancy* or *crystal coordination number*. Ligancy can be discussed in terms of the *radius ratio* of packed spheres, which is defined as the ratio of cation radius r_+ to anion radius r_-. In binary salts coordination numbers of 6 or 8 are common.

When r_+/r_- lies between 0.414 and 0.732, packing of the two different spheres is such that the crystal coordination number is 6. An example is NaCl, Figure 10-2a, in which the ratio is $0.95/1.81 = 0.525$. When r_+/r_- is larger than 0.732, the spheres pack so as to give a coordination number of 8. This is illustrated in CsBr, Figure 10-2b, where the ratio is $1.65/1.95 = 0.846$. The maximum coordination number in ionic crystals seems to be 8, although 12 is the theoretical limit with identical spheres in contact. Coordination number is limited by the radius ratio effect and the repulsion of like charges.

Ionic radii

We have already noted how X ray diffraction is used to find distances (designated d) between planes of atoms in a crystal (see Figure 6-5). X ray measurements made on crystals of simple ionic substances can be used to establish distances between the centers of ions in contact. Assuming ions are spherical, d values represent the sum of two ionic radii. How the sum is divided between cation and anion is determined by reference to crystals such as LiBr in which anion–anion contact seems certain to exist because the cation is too small to prevent it ($r_+/r_- = 0.60/1.95 = 0.308$). This condition is shown in Figure 10-3, where d' is both the distance between Br− centers and the sum of two bromide radii. By assuming that Br− has the

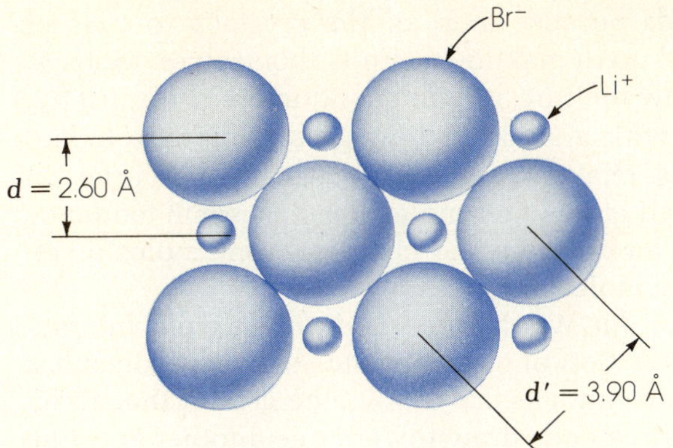

Figure 10-3 A face of a LiBr crystal. Distance d' is the diameter of a bromide ion or the sum of two bromide radii because the anions are in contact. Distance d is only approximately the sum of cation and anion radii because Li^+ and Br^- in a static crystal model do not quite contact one another. Subtraction of d'/2 from d gives an estimate of the Li^+ radius: $2.60 - 3.90/2 = 0.65$ Å. (Compare with Li^+ in Figure 10-4.)

same radius in other crystals, its contribution there is subtracted from the d value corresponding to bromide-cation contact to obtain the probable cation radius.

Some ionic radii established by X ray diffraction are given in Figure 10-4 where relative sizes of isoelectronic species are pictured in rows. Two generalizations are obvious from that figure. First, ionic radius decreases across a row as nuclear charge increases. This is because more attractive force is spread over the same number of electrons. Second, ionic radius increases down a column. Expansion is expected as another electron shell is added in each new row.

Molecular crystals

A *molecular crystal* is a solid whose lattice units are neutral molecules. (In frozen noble gases the "molecular" units are atoms.) Attraction between molecular units is small. It is, in fact, nearly offset by mutual repulsion of

anions		cations		electrons in shells	total electrons
		Li^+ 3+ 0.60	Be^{2+} 4+ 0.31	2	2
O^{2-} 8+ 1.40	F^- 9+ 1.36	Na^+ 11+ 0.95	Mg^{2+} 12+ 0.65	2, 8	10
S^{2-} 16+ 1.84	Cl^- 17+ 1.81	K^+ 19+ 1.33	Ca^{2+} 20+ 0.99	2, 8, 8	18
Se^{2-} 34+ 1.98	Br^- 35+ 1.95	Rb^+ 37+ 1.48	Sr^{2+} 38+ 1.13	2, 8, 18, 8	36

Figure 10-4 Crystal radii of some representative ions. Values given are in angstroms.

interpenetrating electron clouds because most of the crystal's sources of bonding have already been used up in making the units themselves. Molecular crystals are therefore soft, low melting, and nonconductors of electricity.

From research that began with a study of how gases condense to liquids, Johannes van der Waals (Netherlands 1837–1923, Nobel prize in physics 1910) concluded that all atoms and molecules had some tendency to stick together if undisturbed because nuclear charges in one molecule attracted electrons of neighboring molecules.

One component of this *van der Waals force* is *dipole–dipole interaction* — that is, intermolecular attraction of opposite ends of polar molecules. The greater the molecule's permanent electric dipole, the greater the attractive force. As positive and negative ends draw toward one another and line up, "molecular bonds" characteristic of molecular crystals form:

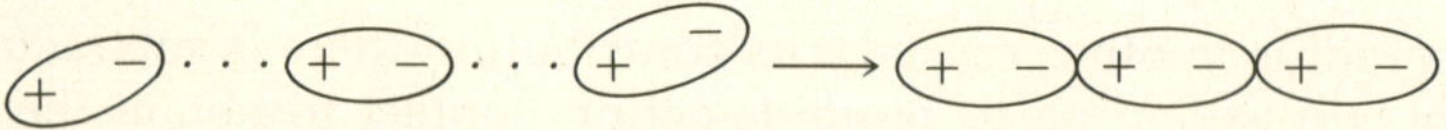

These forces are responsible for many properties of molecular substances, including crystal structure, compressibility, density, and vapor pressure.

In 1929 quantum mechanicist F. London reinterpreted the van der Waals force. He calculated that contributions to cohesion described above were minor compared to another binding effect operating between molecules where permanent dipoles do not exist. London showed that electrons may at any moment be symmetrically or asymmetrically disposed about a nucleus, giving the effect of nuclear displacement relative to the electron cloud:

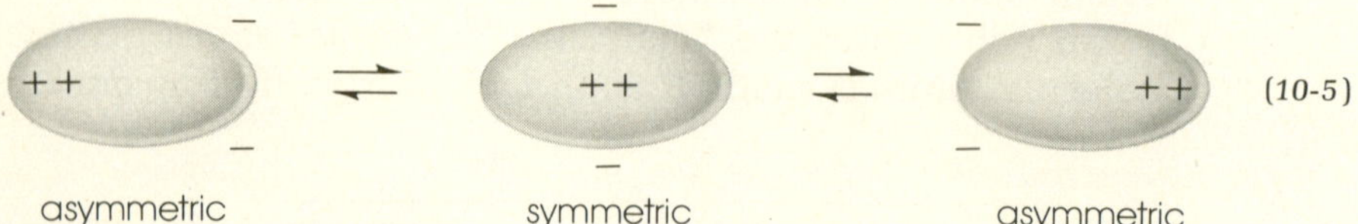

$$(10\text{-}5)$$

Seemingly nonpolar units can thus attract one another in molecular crystals because the asymmetric condition predominates. In crystals of noble gases the *London force* is the only bonding effect operating.

Van der Waals force usually increases with the number of electrons present. In general the larger the molecule, the higher the melting (and boiling) point. An illustration is found in the distillation of crude oil, a mixture of hydrocarbons of various molecular weights (Eq. 21-2). To boil out larger and larger molecules the temperature must be raised higher and higher.

The hydrogen bond

A special directional dipole–dipole effect operates in molecular crystals when the molecules contain a hydrogen atom attached to fluorine, oxygen, or nitrogen. These atoms are so electronegative that their excessive hold on the electron pair they share with hydrogen "uncovers" the hydrogen atom, giving it a partial positive charge attractive to any negative charge. As a result the hydrogen atom is able to act as a bridge between the octets of two electronegative atoms, holding them together with an attraction intermediate in strength between the stout force of a covalent bond and the feeble van der Waals force.

The *hydrogen bond* was first clearly recognized about 1920 by two of Lewis' students, W. Latimer and W. Rodabush. They described how water could attain a certain degree of order through attraction of an oxygen electron pair for a hydrogen on the next molecule. Loose chains of H_2O molecules (hydrogen bonds indicated below by dots) would then build up:

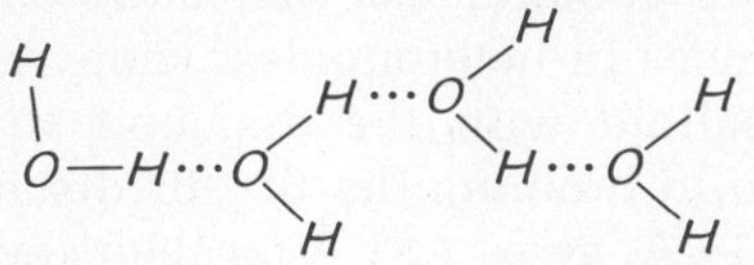

$$
\begin{array}{c}
H \\
| \\
O\!-\!H\cdots O \quad\cdots\quad H\cdots O \\
\end{array}
$$

The relatively high boiling points of water, ammonia, and hydrogen fluoride are evidence of their hydrogen bonding. The boiling points of pure HF, HCl, HBr, and HI are, respectively, 19.4°, −85°, −66°, and −36°C. Intermolecular bonding in the last three is obviously weak, the rising trend in their boiling points being an example of increasing van der Waals force with increasing number of electrons per molecule. On the other hand the condition in hydrogen fluoride is a consequence of the character of the H—F bond. Based on electronegativity difference of the two atoms, the bond is predictably more ionic than covalent, and a single molecule might well be written $H^+\!-\!F^-$. In the gaseous state, clusters of molecules exist of the type $H^+\!-\!F^- \cdots H^+\!-\!F^-$. In solution, unlike the other hydrogen halides which dissociate quite completely, hydrogen fluoride gives relatively few hydrogen ions. Its fluoride ion is hydrogen bonded to HF, giving $F^- \cdots H^+\!-\!F^-$, the hydrogen difluoride ion. In the solid state HF exists as long zigzag chains of hydrogen-bonded molecules:

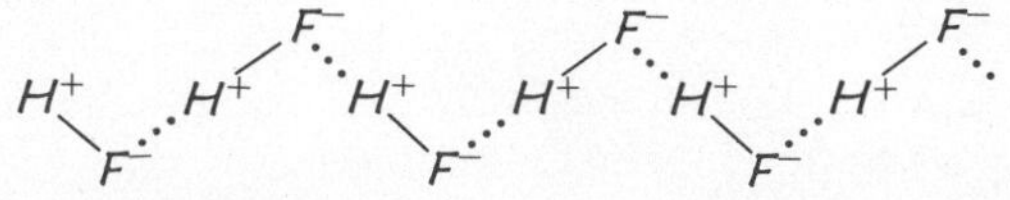

$$
\cdots F^- \cdots H^+ \cdots F^- \cdots H^+ \cdots
$$

*Intra*molecular hydrogen bonds, those within a given molecule, may be found in organic molecules containing two or more oxygen or nitrogen atoms. Their presence leads to ring formation that affects molecular properties. In α-hydroxy acids (like those of Eq. 10-3), for instance, a five-membered ring can form:

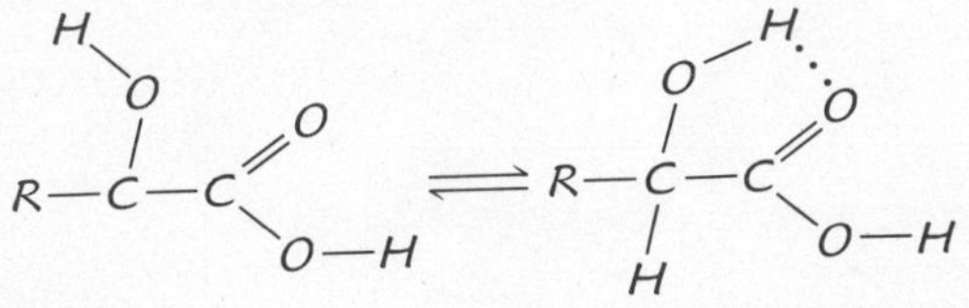

$$
\begin{array}{ccc}
H & & H \\
\backslash & & \cdots \\
O & & O \\
| & \rightleftharpoons & | \\
R\!-\!C\!-\!C\!\!\begin{array}{c}\diagup O\\ \diagdown O\!-\!H\end{array} & & R\!-\!C\!-\!C\!\!\begin{array}{c}\diagup O\\ \diagdown O\!-\!H\end{array} \\
& & |\\
& & H
\end{array}
\qquad (10\text{-}6)
$$

Hydrogen bonding sites occur periodically on the long chains of proteins and other giant molecules found in living cells. As indicated in Figure 23-18, intramolecular hydrogen bonding between N—H and O=C groups is a major factor in determining the stiffness and orientation of protein chains. Molecular orientation is intimately related to biochemical function.

van der Waals radii and covalent radii

The dimensions of covalent molecules are determined by X ray and electron diffraction. The distance measured between the centers of two atomic nuclei joined in a molecule by a single covalent bond is called the *single covalent bond length*. In effect it is twice the average distance between a nucleus and the bonding pair of electrons. In a homonuclear molecule such as Br—Br, half the internuclear distance is called the *single-bond covalent radius* of that atom. (Some values are given in Problem 16.)

Covalent radii determined on homonuclear molecules are added to get approximate internuclear distances in heteronuclear molecules. Thus silicon, a covalently bonded metalloid with the diamond structure, has a Si—Si distance of 2.34 Å. In solid bromine the Br—Br distance is 2.28 Å. Adding the radii, 1.17 Å and 1.14 Å, gives 2.31 Å for the expected length of a bond forming between silicon and bromine. Measurements on solid silicon tetrabromide ($SiBr_4$) give a bond length of 2.32 Å.

In the formation of solid bromine, the molecules arrange themselves into crystals of orthorhombic symmetry (see Figure 6-6), contacting one another at a distance balanced by van der Waals forces and the repulsion of electron shells. The equilibrium distance between the nuclei of two atoms in adjacent molecules is characteristic of the substance. Half this distance is called the *van der Waals radius*. Extending from the nucleus to the effective outside of the atom in directions away from the covalent bond, it is the dimension that describes the volume within which the electrons of an atom are found. Figure 10-5 illustrates the distinction between the covalent radius and van der Waals radius. In general the former is about 0.8 Å shorter.

Chemists use the two kinds of radii to build scale models of molecules as visual aids when examining questions of structure and reaction. Current research on very complex molecules is being aided considerably by the

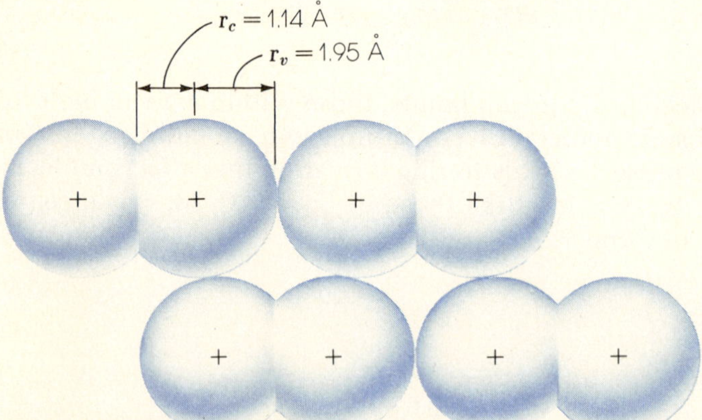

Figure 10-5 A packing of four Br_2 molecules with the covalent and van der Waals radii marked. Length r_c is the covalent radius, and $2r_c$ is the length of the molecule's covalent bond. Length r_v is the van der Waals radius; r_v is essentially the radius of bromide ion :Br:⁻ (as in Na^+Br^-), assuming Br^- to be spherical. Molecules of Br_2 aggregate by van der Waals attraction. The low melting point of solid bromine, −73 °C, shows the intermolecular attraction is weak.

construction of models to match positional assignments inferred from diffraction pictures (see Figures 23-13 and 23-17).

Covalent network solids

Structures in which atoms are linked by covalent bonds to make each crystal one giant rigid molecule are called *covalent network solids*. Examples mentioned earlier include diamond and quartz, which are three-dimensional nets, and graphite, a two-dimensional net.

The strength of network bonding is comparable to that in ionic crystals. Network crystals are often very hard, incompressible, and high melting. Unlike ionic compounds, however, they do not become better conductors of electricity when melted, because no ions are present to carry current.

Metallic crystals

Pure metals generally crystallize in one of three arrangements introduced in Figure 6-6: body-centered cubic, in which each interior atom contacts eight other atoms, and two close-packed structures: face-centered cubic; and hexagonal closest packing, in which 12 atoms contact any interior atom. It is immediately evident that in no case does the number of valence electrons account for the crystal coordination number. This must mean that each bonding electron is shared among more atoms than two and that octets of valence electrons do not become filled.

Consider magnesium, which crystallizes in the hexagonal close-packed arrangement. The pair of valence electrons $(3s^2)$ of any magnesium atom is spread among its 12 near neighbors. To any one of the 12 the reference atom can donate only $\frac{2}{12}$, or $\frac{1}{6}$, of a valence electron. In turn the 12 neighbors share $\frac{1}{6}$ of an electron each with the reference atom, a total of $12 \times \frac{1}{6}$, or two electrons. Any given atom consequently has at its disposal only two valence electrons of its own plus two from other atoms, or a total of four electrons to bond 12 atoms. Four orbitals are usable, the 3s and three 3p, but with only four electrons available they will always remain four electrons short of being filled.

The electron-gas model of metals

We know that metals exhibit the photoelectric effect (see Figure 3-2), have low ionization energies (see Figure 7-9), and form positive ions easily in chemical reactions. Metals also form few diatomic metal molecules (like Li_2) and those that are known readily break apart. All these facts indicate that metal atoms hold their valence electrons loosely and do not seek additional electrons. To make a bonding theory for metals we are forced to deal with mobile electrons in an electron-deficient environment.

In 1916 Lorentz proposed that a metal be considered a lattice of positive atomic kernels (nuclei plus core electrons) surrounded by a gas or sea of free electrons belonging to the entire structure.

The Lorentz model is still useful in accounting qualitatively for metallic character. Metals are ductile and malleable because regardless of lattice deformation the electron gas is still capable of binding the structure as strongly as ever. Good electrical conductivity is due to easy electron movement when an electric potential is applied across the crystal. Good heat conductivity is ascribed to motion imparted to the free electrons by application of heat and the transmission by electrons of energy through collisions. Decreasing electrical conductivity with increasing temperature is explained as a vibration of atomic nuclei which distorts the electron gas and hinders its mobility. Opacity and reflectivity are a result of electrons at all energy levels intercepting light. And, finally, *the metallic bond* can be described as electrostatic attraction between positive nuclei and the electron gas.

The band model of metals

Basis for a quantum-mechanical theory of metals was established in 1927 by Pauli. For the electron gas he substituted electrons quantized into approximately continuous sets of energy levels or "bands." By quantum rules movement of electrons within a band was permitted, but forbidden between bands separated by an energy gap.

Modern metal theory is a molecular-orbital extension of Pauli's ideas. When two metal atoms come together, each valence electron energy level is considered split into two new levels from which molecular orbitals form. As a crystal is built up through addition of many atoms, many levels come into being, arranged in bands of slightly different energies. The bands, a feature of the whole crystal, are either separated by a forbidden region or they overlap. Various degrees of the two conditions make possible the differences observed in properties of different metals. In the series of fourth period elements, from $_{19}K$ to $_{24}Cr$, for example, there is an increase from one to six valence electrons. In the same sequence there is observed an increase in melting point, hardness, and brittleness. The interpretation is that as

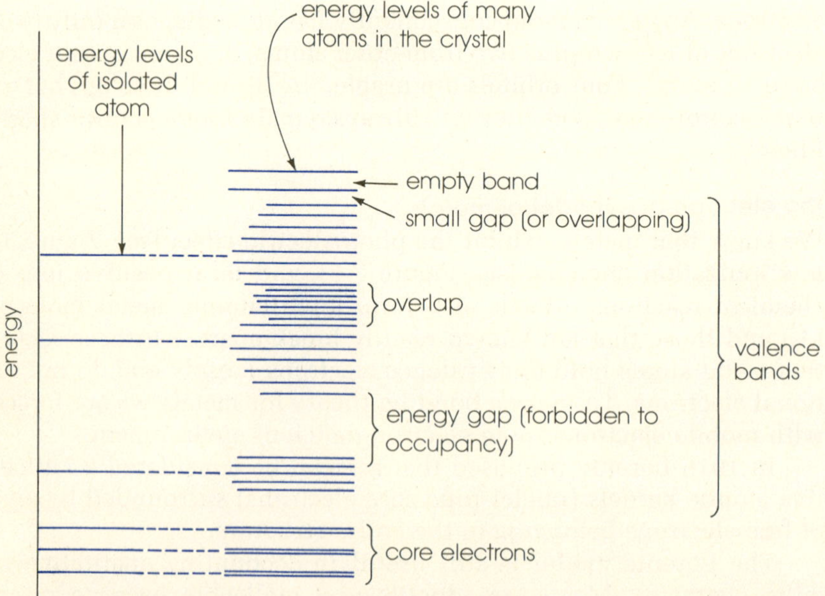

Figure 10-6 Schematic view of energy levels in a typical metal and conductor. In isolated atoms electronic energy levels are distinctly separated, indicative of one center of attraction. In the crystal, however, valence electrons are spread into bands of many closely spaced energy levels. Conducting electrons in uncrowded levels are able to drift through the lattice.

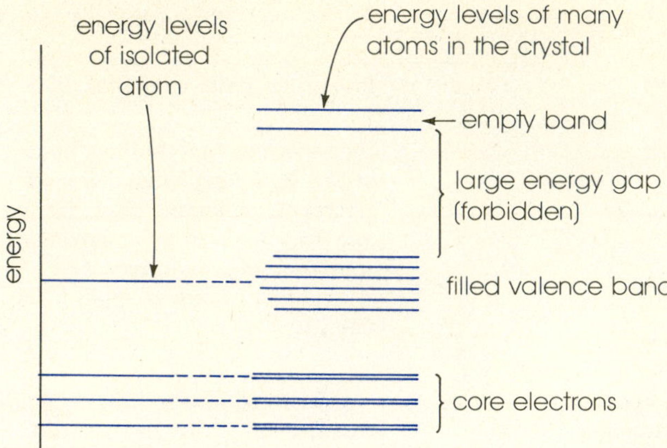

Figure 10-7 A typical insulator (non-conductor). The valence band is filled and a large energy gap exists between it and the next band which is empty. Electrons cannot jump the gap.

bands become more crowded with valence electrons, binding becomes stronger.

Outermost bands in a typical metal are empty of electrons or only partially occupied (Figure 10-6). When outer bands overlap lower bands or lie very close to them in energy, they are accessible to valence electrons promoted upward by application of, say, an electric potential across the metal crystal. In these freer locations electrons flow as charges belonging to the entire lattice and the crystal acts as a conductor.

When, as in diamond and most molecular lattices, the last occupied orbital is filled with paired electrons and a large energy gap lies between it and an empty band beyond, electrons cannot be promoted (Figure 10-7). The crystal is then an insulator.

In substances like silicon, selenium, germanium, and graphite only a small gap exists between filled and empty bands, so their crystals are natural *semiconductors*. A small increase in temperature or exposure to light makes them conducting by elevating electrons into conduction bands. In semiconductors manufactured for use in various electronic devices, impurities are added to aid conduction by introducing empty bands at accessible energy levels (see Figure 19-10).

Bonding in solids, a summary

The electronic nature of a substance dictates the kind of forces that will hold its crystal together. A crystal's lattice geometry is established by the directionality of the forces and the size of the units packed together. These factors in turn are related to the crystal's physical and chemical properties. A summary of types of solids and their bonding is presented in Table 10-1.

QUESTIONS

1. Ions Tell whether the following are true or false, then discuss. **(a)** Melting and boiling points of simple compounds may be used as approximate measures of their ionic character. **(b)** LiF is expected to form harder crystals than CsI. **(c)** The solubility of CaO is expected to be less than that of KCl.

Table 10-1. Bonding in solids

Type/Example	Interunit bonding	Energy[a]	Properties, other examples
Ionic, LiF	Coulombic (electrostatic) attraction	247	High-melting, high-boiling, hard, brittle, poor heat conductors, fair electrical conductors when melted, may be soluble in polar solvents; composed of ions in repeating array; other examples: NaCl, KBr, CaF_2, Cs_2O.
Polar molecular, H_2O	van der Waals forces and dipole-dipole attraction	13.0	Low-melting, low-boiling, generally volatile, soft, easily compressed; insulators, no net electric charge on molecules; molecules themselves bonded by polar covalent bonds; dissolve in polar solvents; other examples: NH_3, HBr, HCN, H_2SO_4, SO_2.
Nonpolar molecular, Cl_2	van der Waals forces and London forces	4.9	Low-melting, low-boiling, volatile, soft, easily deformed; electrical insulators; usually dissolve in nonpolar solvents; molecular symmetrical and covalently bonded; other examples: H_2, He, Br_2, CH_4, CO_2.
Covalent network, SiO_2	Covalent	433	May be very hard, high-melting, high-boiling; electrical insulators; strong, directional bonds; examples found in Group IVA elements and binary compounds where valence electrons total 8: C, Si, BN, SiC, BeO.
Metallic, Fe	Metallic	99	Often high-melting, high-boiling, hard, dense, lustrous, reflective, malleable, ductile; excellent electrical and thermal conductors; valence electrons not localized; other examples: Al, Na, Ni, Cu, Ag, U.

[a] Energy in kilocalories needed to separate 1 mole (6.02×10^{23}) of particles (see Section 12-2).

2. Lattice What is meant by a lattice in which the coordination number is 8?

3. Nitrogen bonding Before 1900 the nitrogen atom in ammonium chloride, NH_4Cl, was thought to hold five atoms. Today we believe only four of the atoms are bonded directly to nitrogen. Explain.

4. Werner's theory Jørgensen prepared two compounds having this formula: $[Co(NH_3)_4(NO_2)_2]NO_3$. Werner said they represented two different octahedral arrangements of 6 coordination around cobalt. Explain.

5. Sidgwick's rule **(a)** Briefly describe Sidgwick's theory of electron structure in complex ions. **(b)** Apply it to these complex ions: $ZnCl_4^{2-}$ and $Cd(NH_3)_4^{2+}$.

6. Complex How many different octahedral structures can theoretically be made from $Co(NH_3)_3Cl_3$?

7. Carbon dioxide Solid carbon dioxide $O{=}C{=}O$ is called Dry Ice. Dry Ice readily sublimes; that is, its molecules evaporate from the crystal surface directly as a gas instead of first forming a liquid. Describe the obviously weak forces that hold a carbon dioxide crystal together.

8. Hydrogen fluoride Hydrogen iodide gas at 30°C consists of HI molecules. Hydrogen fluoride gas at 30°C consists mostly of H_2F_2, H_3F_3, H_4F_4, H_5F_5, and H_6F_6 molecules. Why do HI and HF behave differently?

9. Acetic acid Measurements on acetic acid molecules in the vapor state at 40°C signify that they are roughly twice as heavy as expected from the acid's formula CH_3CO_2H. The best explanation is that the following structure is involved. Interpret it.

$$H-\overset{\overset{\displaystyle H}{|}}{\underset{\underset{\displaystyle H}{|}}{C}}-C\overset{\nearrow O\cdots H-O}{\underset{\searrow O-H\cdots O}{}}C-\overset{\overset{\displaystyle H}{|}}{\underset{\underset{\displaystyle H}{|}}{C}}-H$$

10. Atom radius Why is the covalent radius of an atom smaller than the van der Waals radius?

11. Hydrogen bond Why is the hydrogen bond strong in NH_3 but weak in PH_3?

12. Crystal types Ionic crystals and covalent network crystals share some properties (Table 10-1). They can be distinguished by melting and measuring the electrical conductance of the melts. Explain.

13. Chelation **(a)** What is a chelating group? **(b)** A chelating group having two donor atoms (Eq. 10-1) is said to be bidentate (L. *dens*, tooth), one having three donor atoms terdentate, etc. Why is the term "dentate" appropriately descriptive?

14. Ionic radii The following are isoelectronic ions with their ionic radii in angstroms: Te^{2-}, 2.21; I^-, 2.16; Cs^+, 1.69; Ba^{2+}, 1.35. Why is the size decrease expected?

15. Solid types Classify the following solids by types given in Table 10-1: **(a)** graphite, Figure 6-7c; **(b)** alpha-iron, Figure 6-8a; **(c)** diamond, Figure 6-7a.

PROBLEMS

16. Bond lengths The following are some covalent radii in angstroms: H, 0.30; Li, 1.22; Be, 1.06; B, 0.88; C, 0.77; N, 0.70; O, 0.66; F, 0.64; Al, 1.25; Si, 1.17; P, 1.10; S, 1.04; Cl, 0.99; Ge, 1.22; As, 1.21; Se, 1.17; Br, 1.14; Sn, 1.39; Sb, 1.41; Te, 1.37; I, 1.33. **(a)** Arrange them in the form used for Table 9-1. **(b)** Calculate the distances expected in the following covalent bonds: B—F, C—Cl, Si—Br, Ge—C, N—H, S—Cl. Compare the results to bond distances (angstroms) found by X ray measurements on these compounds: BF_3, 1.30; CCl_4, 1.77; $SiBr_4$, 2.15; $Ge(CH_3)_4$, 1.98; NH_3, 1.01; SCl_2, 1.99.

17. Coordination compounds **(a)** Use Sidgwick's rule and the periodic table to predict the number of ligands (each carrying one pair of bonding electrons) that the following transition metal ions will probably coordinate: Pd^{2+}, Ru^{2+}, Rh^{3+}. **(b)** Aided by Table 9-2 draw geometric structures of HgI_4^{2-}, $Cu(NH_3)_4^{2+}$, $Co(NH_3)_6^{3+}$.

18. Chelation **(a)** Add all unshared electrons as dots to this formula for glycine, the simplest amino acid:

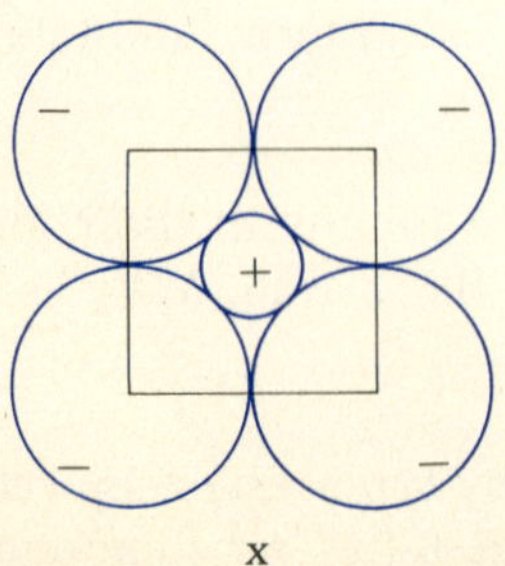

(b) A solution of glycine forms a dark blue square planar chelate with Cu^{2+}. In the manner of Eq. 10-3 show the reaction. Show the product's geometry and all unshared electrons. (Assume bonding of copper is to —O— from —OH, and to the unshared electron pair of —$\ddot{N}H_2$.)

19. Boiling points Plot the following boiling points (vertically) vs. the atomic numbers of the Group VI elements, and connect the points with straight lines: H_2O, 100°C; H_2S, −60°C; H_2Se, −41°C; H_2Te, −2°C. **(b)** On the same sheet, in the same manner, plot these boiling-point data for a series of Group IV hydrides: CH_4, −183°C; SiH_4, −112°C; GeH_4, −90°C; SnH_4, −55°C. **(c)** Explain, aided by an appropriate diagram, why the lightest of the Group VI hydrides has the highest boiling point. **(d)** Explain the rising trend from H_2S to H_2Te. **(e)** Why are the shapes of the two graphs not the same?

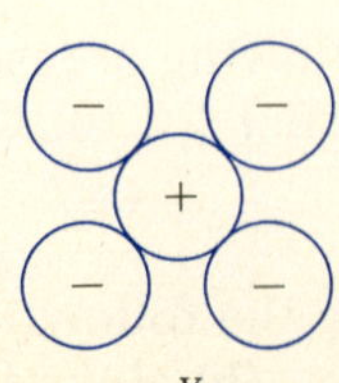

x y **Figure 10-8**

20. Radius ratio The ratio between the radius of a cation r_+ and radius of an anion r_-, can be used to determine when anion–anion contact is attained (as in x) or is not attained (as in y) in crystal planes (Figure 10-8). In x the side of the square is readily seen to be $2r_-$. By the Pythagorean theorem, it is also $(r_- + r_+)\sqrt{2}$. **(a)** Draw an enlarged view of x and show the geometric construction and arithmetic that proves the last statement. **(b)** Solve the equation, $2r_- = (r_- + r_+)\sqrt{2}$ to obtain a numerical value for the radius ratio r_+/r_-. **(c)** Explain, in terms of the ratio, when condition y is expected to arise. **(d)** From Figure 10-4 data, calculate the radius ratio for RbCl which crystallizes in the arrangement designated y. Predict whether or not the chloride ions touch each other.

FOUR

WHAT ARE CHEMICAL EQUATIONS AND STOICHIOMETRY?

An eighteenth-century laboratory and table of affinities published in 1718 by French professor E. Geoffroy. Symbols are a collection from ancient Greek science, alchemy, and Geoffroy. In modern symbolism the third vertical column, for example, in descending order, reads HNO_3, Fe, Cu, Pb, Hg, Ag. "By this table," Geoffroy wrote, "those who are beginning to learn chemistry may form in a short time an adequate idea of the affinities which exist between different substances. Chemists will thereby find an easy method to determine what takes place in many operations which are otherwise difficult to disentangle, and to predict what happens when different bodies are mixed." (Edgar Fahs Smith Memorial Collection, University of Pennsylvania.)

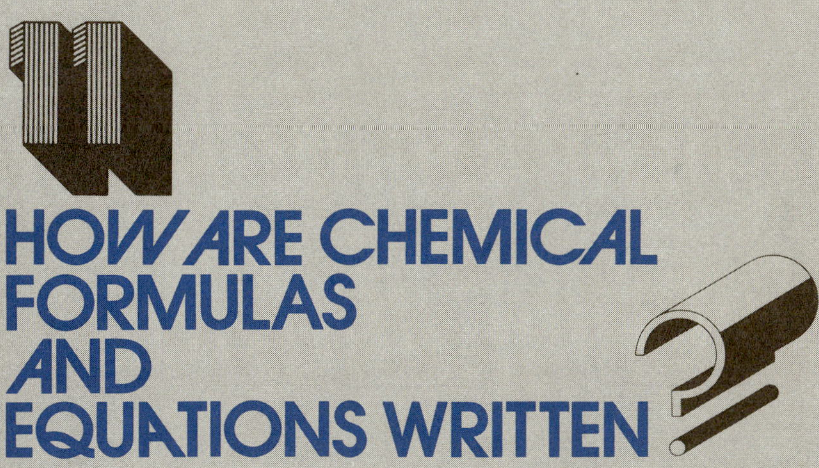

HOW ARE CHEMICAL FORMULAS AND EQUATIONS WRITTEN?

THE IMPORTANT CONCEPTS

11-1. Writing chemical formulas
1. A formula is written from a knowledge of symbols of elements, formulas of radicals, and valence numbers.

11-2. Writing chemical equations
1. Equation writing involves reactants, products, coefficients, and reaction conditions.

11-3. Classification of simple reactions
1. Products may be predicted by analogy from an outline of generalized equations.

2. The vocabulary for discussing reactions includes such words as acidic and basic anhydride, hydrate, complete combustion, halogen activity, replacement, metathesis, and neutralization.
3. A variety of reactions can be correlated around the metal activity series.
4. Molecular equations can be converted into ionic and net ionic equations.

We have seen how the chemical elements are classified (Chapter 7), and learned something about their occurrence, appearance, uses, and physical and chemical properties (Chapter 8). In this chapter we look further into the descriptive chemistry of elements and compounds by outlining their simpler reactions.

11-1. WRITING CHEMICAL FORMULAS

Symbols, formulas, and oxidation numbers

In a paper published in 1814 Berzelius (Figure 11-1) spelled out the basis of chemical symbolism that has been used ever since. Slightly modernized this is what he wrote:

> My symbols are destined solely to facilitate the expression of chemical proportions, and to enable us to indicate the relative number of atoms of the different elements contained in each compound. By determining the weight of the elementary atoms, these figures will enable us to express the numeric result of an analysis as simply as the algebraic formulas in engineering.
>
> The chemical symbols ought to be letters, for the greater facility of writing. I shall take, therefore, for the chemical symbol, the *initial letter of the Latin name of each elementary substance.* 1. In the class which I call *metalloids,* I shall employ the initial letter only. 2. In the class of *metals,* I shall distinguish those that have the same initials with another metal, or a metalloid, by writing the first two letters of the word. 3. If the first two letters are common to two metals, I shall, in that case, add to the initial letter the first consonant which they have not in common: for example, S = sulfur, Si = silicon, Sn = stannum (tin).
>
> The chemical symbol expresses always one atom of the substance. When it is necessary to indicate several atoms, it is done by adding the number of atoms: for example, the oxide of copper is composed of an atom of oxygen and an atom of metal; therefore its formula is CuO. When we express a more complicated compound, we indicate the number of atoms with their letter. For instance, in the synthesis of copper sulfate, which contains four oxygen atoms, we will write $CuO + SO_3 = CuSO_4$.

Chemists today write correct *chemical formulas* from a knowledge of *valence numbers,* also called *oxidation numbers* or *oxidation states.* The theory of valence or combining power of atoms was announced in 1852 by Frankland and Kekulé. Mendeleev adopted valence as a guide to periodicity and suggested that some elements had fixed and others variable valency. As analytical work verified formulas, elements were assigned valence numbers. Positive numbers were given to metals and hydrogen, negative numbers were given to nonmetals and units called *radicals* (SO_4^{2-} = the sulfate radical, for example) that formed acids. A formula for a compound was written by combining symbols of elements and radicals to give

Figure 11-1 Jons Berzelius (Sweden 1779–1848). Berzelius introduced modern chemical symbolism and nomenclature, gave mineralogy a chemical foundation, established reliable atomic weight values, promoted the theory that compounds are composed of plus and minus parts, and discovered the elements cerium, thorium, selenium, and silicon. His annually published review of world chemical progress gave him international recognition as the premier judge of quality in chemical research. (Edgar Fahs Smith Memorial Collection, University of Pennsylvania.)

an algebraic valence number sum of zero. All molecules were treated as neutral.

Oxidation numbers today may be deduced from electron shell structure, but it is handier, with the aid of the periodic table, to memorize symbols for elements and formulas for radicals, and the oxidation numbers that go with them. As we already know (see Figure 7-2a), the A subgroup elements have as one of their valence numbers (it may be the only one) their group number in the periodic table. The valence number of Group IIIB elements is 3+ and that of Group IVB elements is 4+. Valence numbers of the other transition elements are variable; 2+ is most common, followed by 3+. In ionic compounds a valence number is the charge on an ion. In covalent compounds a valence number is simply a convenience to use for formula writing.

Table 11-1. Oxidation numbers of some positive ions (cations)

Univalent		Bivalent			
NH_4^+	Ammonium	Ba^{2+}	Barium	Hg^{2+}	Mercuric
Cu^+	Cuprous	Cd^{2+}	Cadmium		Mercury(II)
	Copper(I)	Ca^{2+}	Calcium	Hg_2^{2+}	Mercurous
H^+	Hydrogen	Co^{2+}	Cobaltous		Mercury(I)
Li^+	Lithium		Cobalt(II)	Ni^{2+}	Nickelous
K^+	Potassium	Cu^{2+}	Cupric		Nickel(II)
Ag^+	Silver		Copper(II)	Pb^{2+}	Plumbous
Na^+	Sodium	Fe^{2+}	Ferrous		Lead(II)
			Iron(II)	Sn^{2+}	Stannous
		Mg^{2+}	Magnesium		Tin(II)
		Mn^{2+}	Manganous	Sr^{2+}	Strontium
			Manganese(II)	Zn^{2+}	Zinc

In writing the formula of an ionic compound and giving it a name, we consider the positive part first, then the negative. We use *subscripts* as multipliers to attain a balance between total positive and total negative valence number, giving an electrically neutral substance.

Combination of Al^{3+} and Br , for example, requires three of the univalent negative bromide ions to balance the plus three charge of the aluminum ion. The correct formula for aluminum bromide is $AlBr_3$.

The sulfate radical (a combination of five atoms acting as a unit) is bivalent, SO_4^{2-}. Forming a neutral "molecule" of aluminum sulfate requires two Al^{3+} ($2 \times 3+ = 6+$) and three SO_4^{2-} ($3 \times 2- = 6-$). The correct formula is $Al_2(SO_4)_3$. The radical is enclosed in parentheses, so the subscript 3 multiplies each atom in it. One molecule of aluminum sulfate contains 17 atoms.

Positive ions

Table 11-1 is an alphabetical listing of common positive ions or *cations*. Hydrogen, Group IA and IIA metals, and ammonium (which acts like a metal), silver, cadmium, zinc, and aluminum normally show only one oxidation state. The names of their ions are the same as the names of the free metals. Thus Ca is calcium, and Ca^{2+} is calcium ion. The remaining metals have at least two oxidation numbers. For a given pair such as Fe^{2+} and Fe^{3+}, names have been formed since the time of Berzelius by adding to the atom's root name the suffix -*ous* for the lower state, and -*ic* for the higher state; thus *ferrous* and *ferric* ions.

A more modern designation of the two iron ions is iron(II) and iron(III), a combination of the element's name and oxidation numbers written as Roman numerals. Worked out from some of Werner's ideas 50 years ago by German chemist A. Stock and expanded in 1940 by the International Union of Pure and Applied Chemistry, it is called the Stock or *IUPAC system of nomenclature*. Chemists use both systems.

Negative ions

Table 11-2 is a listing of common negative ions or *anions*. Those that are single atoms are understandably nonmetals from periodic table Groups

Table 11-1. *(continued)*

	Tervalent				Quadrivalent		Quinquivalent
Al^{3+}	Aluminum	Cr^{3+}	Chromic	Mn^{4+}	Manganese(IV)	Sb^{5+}	Antimonic
Sb^{3+}	Antimonous		Chromium(III)	Pb^{4+}	Plumbic		Antimony(V)
	Antimony(III)	Co^{3+}	Cobaltic		Lead(IV)	As^{5+}	Arsenic
As^{3+}	Arsenous		Cobalt(III)	Sn^{4+}	Stannic		Arsenic(V)
	Arsenic(III)	Fe^{3+}	Ferric		Tin(IV)	Bi^{5+}	Bismuthic
Bi^{3+}	Bismuthous		Iron(III)				Bismuth(V)
	Bismuth(III)						

Table 11-2. Oxidation numbers of some negative ions (anions)

Univalent				Bivalent	
$C_2H_3O_2^-$	Acetate	OH^-	Hydroxide	CO_3^{2-}	Carbonate
Br^-	Bromide	I^-	Iodide	CrO_4^{2-}	Chromate
Cl^-	Chloride	NO_3^-	Nitrate	$C_2O_4^{2-}$	Oxalate
ClO^-	Hypochlorite	NO_2^-	Nitrite	O^{2-}	Oxide
ClO_2^-	Chlorite	MnO_4^-	Permanganate	O_2^{2-}	Peroxide
ClO_3^-	Chlorate	SCN^-	Thiocyanate	SO_4^{2-}	Sulfate
ClO_4^-	Perchlorate			S^{2-}	Sulfide
CN^-	Cyanide			SO_3^{2-}	Sulfite
F^-	Fluoride				

IVA, VA, VIA, and VIIA, whose oxidation numbers can be predicted from the group number. With hydrogen ion or metal ions they form *binary* (two-element) compounds. Names of binary compounds are formed from the names of the two elements and the suffix *-ide*. Calcium sulfide is CaS, for example. When the binary compound is composed of two nonmetals, prefixes are used to designate numbers of atoms in the molecule: N_2O_5 is dinitrogen pentoxide; P_4O_6 is tetraphosphorus hexoxide. (For prefixes, see footnote with reaction type 5.)

Of the radicals listed in Table 11-2 only four do not contain oxygen: CN^-, SCN^-, $Fe(CN)_6^{3-}$, and $Fe(CN)_6^{4-}$. The last two are complex ions. There is a Stock-IUPAC system for naming complex ions, but we will not go into its details.

Some anions not listed are derived by adding hydrogen ions to polyvalent anions. Each H^+ changes the ion's charge by one as

$$\underset{\text{phosphate}}{PO_4^{3-}} \; \underset{-H^+}{\overset{H^+}{\rightleftarrows}} \; \underset{\substack{\text{hydrogen} \\ \text{phosphate}}}{HPO_4^{2-}} \; \underset{-H^+}{\overset{H^+}{\rightleftarrows}} \; \underset{\substack{\text{dihydrogen} \\ \text{phosphate}}}{H_2PO_4^-} \qquad (11\text{-}1)$$

Ions derived from bivalent ions by addition of a hydrogen ion are also given the prefix "hydrogen" or bi-:

$$\underset{\text{carbonate}}{CO_3^{2-}} \; \underset{-H^+}{\overset{H^+}{\rightleftarrows}} \; \underset{\substack{\text{hydrogen carbonate} \\ \text{(bicarbonate)}}}{HCO_3^-}$$

Table 11-1 lists 35 cations; Table 11-2 lists 30 anions. The number of possible plus-minus combinations is $35 \times 30 = 1050$. Most represent stable compounds.

11-2. WRITING CHEMICAL EQUATIONS

Reactants, products, coefficients

Formulas for *reactants* are put on the left side of a *chemical equation*, formulas for *products* on the right. On each side a + sign is placed between any two formulas. To *balance* the equation—that is, to obtain an overall atomic equalization in accord with the conservation of mass law —integers called *coefficients* are inserted before formulas as needed. A coefficient

Table 11-2. (continued)

Tervalent		Quadrivalent	
AsO_4^{3-}	Arsenate	C^{4-}	Carbide
$Fe(CN)_6^{3-}$	Ferricyanide	$Fe(CN)_6^{4-}$	Ferrocyanide
PO_4^{3-}	Phosphate	SiO_4^{4-}	Silicate
N^{-3}	Nitride		

multiplies every atom in the formula following it. When the coefficient is 1, it is simply understood and not written. Once correct symbols and formulas have been written, only coefficients may be adjusted for balance.

Equation balance is signified by employing an "equals" sign between reactants and products. When an equation is not balanced, or when a reaction goes essentially to completion, the symbol $\rightarrow$ pointing toward the products may be used. Catalyst and *reaction conditions* (temperature, pressure, energy input, etc.) may be indicated with the equality symbol.

The method of balancing an equation by trial and error or "inspection" is illustrated in the following example.

Example 11-1 Arsenic(III) fluoride is a liquid boiling at 63°C. It can be prepared by heating a mixture of calcium fluoride, arsenic(III) oxide, and sulfuric acid. By-products are water and calcium sulfate. Write the balanced molecular equation.

Solution Using Tables 11-1 and 11-2 we first write correct formulas for the substances named and arrange them in equation form:

$$CaF_2 + As_2O_3 + H_2SO_4 \longrightarrow H_2O + CaSO_4 + AsF_3$$

We now balance the equation atom by atom. Arsenic is a suitable starting atom because it occurs in only one substance on each side. Two arsenic atoms are on the left, one on the right; therefore we put the coefficient 2 in front of AsF_3. The coefficient 2 also multiplies fluoride in AsF_3, however, giving six fluorine atoms on the right side. To obtain six fluorine atoms on the left, we place the coefficient 3 before CaF_2. This gives three calcium atoms on the left, meaning we must multiply $CaSO_4$ on the right by three to balance calcium; $3CaSO_4$ demands $3H_2SO_4$ on the left to balance sulfate. That gives six hydrogen atoms on the left, requiring $3H_2O$ on the right. The coefficient 3 gives three oxygen atoms on the right (besides sulfate oxygens). Three is just the number available from the one As_2O_3 molecule with which we began. Putting in the symbol for heat over the "equals" sign, we write the balanced equation,

$$3CaF_2 + As_2O_3 + 3H_2SO_4 \overset{\Delta}{=} 3H_2O + 3CaSO_4 + 2AsF_3$$

11-3. CLASSIFICATION OF SIMPLE REACTIONS

Predicting reaction products

Given one or more familiar reactants and asked to predict quickly the products of their reaction under specified conditions, a chemist is not apt to use much theory in giving a reply. He probably first recognizes chemical "types" in the substances. He recalls specific facts he has learned concerning their chemistry and the chemistry of substances like them. From there he may answer by applying a few simple reaction rules or by reasoning from analogy.

The paragraphs below are an abbreviated classification of simple, mostly inorganic reactions, and a summary of chemical activity principles that apply to them. Exercises that follow can be answered by referring to this body of empirical knowledge. Because the facts of chemistry are many, there are exceptions to most rules, analogies are not always justified, and sometimes a wrong conclusion is reached. As beginners with limited time, however, we can attain a modest success in predicting reaction products and writing balanced chemical equations by imitating the examples given, realizing that behind the outline are theories to be considered later.

For simplicity we will start by writing *molecular equations*, in which every substance is represented by a molecular formula. Molecular equations are necessary to certain calculations (see *stoichiometry*, Chapter 12). However, we can already anticipate they will be inadequate to describe reactions of ionic substances in solution. For the latter cases we write equations to contain ions. *Ionic equations* are illustrated at the end of this discussion.

Direct union

We will define a *direct union* as a reaction between two elements, an element and a compound, or two compounds to give a single product,

$$A + B = AB$$

Reactions with the noble gases are not considered. Direct unions are common at elevated temperature in the absence of water.

1. Nonmetallic element + oxygen = nonmetal oxide

An example is the burning of phosphorus in air,

$$P_4 + 5O_2 = P_4O_{10}$$

Oxygen gas normally consists of molecules and is written in equations as O_2. Oxide examples are given in Table 11-3. Other *diatomic elements* are H_2, N_2, F_2, Cl_2, Br_2, I_2. Elemental sulfur, which occurs at ordinary temperatures as S_8, may be written as S in equations. Phosphorus, which can occur as P_4, may be written as P.

2. Metal + nonmetal = binary compound

Letting M be a bivalent metal, some possibilities are

(a) $2M + C = M_2C$ (carbide)
(b) $3M + N_2 = M_3N_2$ (nitride)
(c) $2M + O_2 = 2MO$ (oxide)

Table 11-3. Acids from nonmetal oxides

Oxide	Acid	Acid name	Oxide	Acid	Acid name
As_2O_5	H_3AsO_4	Arsenic	N_2O_5	HNO_3	Nitric
Cl_2O	$HClO$	Hypochlorous	N_2O_3	HNO_2	Nitrous
Cl_2O_3	$HClO_2$	Chlorous	P_4O_{10}	H_3PO_4	Phosphoric
Cl_2O_5	$HClO_3$	Chloric	P_4O_6	H_3PO_3	Phosphorous
Cl_2O_7	$HClO_4$	Perchloric	SO_3	H_2SO_4	Sulfuric
CO_2	H_2CO_3	Carbonic	SO_2	H_2SO_3	Sulfurous

(d) $M + S = MS$ (sulfide)
(e) $M + X_2 = MX_2$ (halide)

Metals with two oxidation states form two series of compounds, for example, CuS and Cu_2S. Generally speaking higher temperatures, as well as reactions with the stronger nonmetals (O_2, F_2, and Cl_2) favor higher oxidation states of metals. Very active metals form peroxides with oxygen, less active metals form oxides (see columns A and L, Table 11-5.)

3. Nonmetal oxide + water = acid

We will define an acid here as a substance that gives hydrogen ions in solution. Nonmetal oxides are called *acidic oxides* or *acidic anhydrides* (acids without water). Tetraphosphorus decoxide, for example, is the anhydride of phosphoric acid, for when water is added to it,

$$P_4O_{10} + 6H_2O = 4H_3PO_4$$

Acid names (Table 11-3) are somewhat different than the names of the corresponding radicals (Table 11-2). When an element forms only one oxygen-containing acid, the acid name ends in -*ic*, for example, carbonic acid, H_2CO_3. When the element forms two acids, the name of the acid with the greater number of oxygen atoms ends in -*ic*, the other in -*ous*. Thus HNO_3 is nitric acid, HNO_2 is nitrous acid. When more than two acids of an element are known, the suffixes -*ic* and -*ous* and the prefixes *per-* and *hypo-* are used in the names. Note the chlorine acids, Table 11-3.

When an element (nonmetal or metal) forms several oxides, those containing more oxygen tend to be most acidic; those with less oxygen, weakly acidic or basic; those between, moderately acidic. Thus in the four oxygen acids of chlorine, $HClO$ is the weakest, $HClO_4$ is the strongest.

4. Soluble metal oxide + water = hydroxide

Most soluble metal oxides give hydroxides whose reactions are described as *alkaline* or *basic*. Metal oxides are therefore called *basic anhydrides* (bases without water). Best examples come from oxides of Group IA and IIA metals. (Other oxides may be too insoluble to react.) A typical reaction is

$$CaO + H_2O = Ca(OH)_2$$

5. Salt (anhydrate) + water = hydrate

Water dipoles are attracted to both plus and minus ions in solution. If the attraction is strong, water can be retained in definite amounts as part of the

Table 11-4.   Familiar hydrates

Formula	Common name	Formula	Common name
$KAl(SO_4)_2 \cdot 12H_2O$	Alum	$K_2Mg(SO_4)_2 \cdot 6H_2O$	Schönite
$CaSO_4 \cdot \frac{1}{2}H_2O$	Plaster of Paris	$Na_2B_4O_7 \cdot 10H_2O$	Borax
$CaSO_4 \cdot 2H_2O$	Gypsum	$Na_2CO_3 \cdot 10H_2O$	Sal soda
$CuSO_4 \cdot 5H_2O$	Blue vitriol	$NaHCO_3 \cdot Na_2CO_3 \cdot 2H_2O$	Trona
$FeSO_4 \cdot 7H_2O$	Copperas	$Na_2SO_4 \cdot 10H_2O$	Glauber's salt
$MgSO_4 \cdot 7H_2O$	Epsom salt	$Na_2S_2O_3 \cdot 5H_2O$	Hypo

structure when the salt crystallizes. This water is called *water of hydration*, and the compound is a *hydrate*. Hydrate water is generally associated with the cation which, due to comparatively small size and high charge, coordinates H_2O molecules. The formula for a hydrate can sometimes be guessed from the cation's known ligancy, but most formulas are simply memorized (Table 11-4). An odd number of water molecules in a hydrate formula is usually indicative of some water being associated with the anion, or filling otherwise unoccupied lattice positions. Thus in cupric sulfate 5-hydrate* four water molecules are coordinated by copper ion as expected (Table 9-2), whereas the other is associated with sulfate:

$$CuSO_4 + 5H_2O = Cu(H_2O)_4SO_4 \cdot H_2O$$

6. Metal oxide + nonmetal oxide = salt

Union of a basic oxide and an acidic oxide gives a salt. The salt's cation will be the metal. The anion will typically be a radical from Table 11-3. For example,

$$CaO + CO_2 = CaCO_3$$

Decompositions

Decompositions of compounds are often more complex than one might expect, giving intermediate products that may be difficult to predict. We will consider only the straightforward cases, those that are essentially the reverse of the direct unions described above. A decomposition may be defined, therefore, by the general equation,

$$AB = A + B$$

For instance, the reverse of type 5 reaction is

$$hydrate \overset{\Delta}{=} anhydrate + water$$

A specific example is

$$CaSO_4 \cdot 2H_2O \overset{\Delta}{=} CaSO_4 + 2H_2O$$

Decomposition of a reactant can be caused by the input of heat, light, or electrical energy. Examples are found in columns J, P, Q, and R, Table 11-5.

---

* The number of water molecules in a hydrate may also be designated with the Greek prefixes *hemi-*, *mono-*, *di-*, *tri-*, *tetra-*, *penta-*, *hexa-*, *hepta-*, *octa-*, *nona-*, *deca-*, *undeca-*, *dodeca-*, for $\frac{1}{2}$ through 12. A hemihydrate may also be written as a monohydrate by doubling the salt.

Organic combustions

When organic compounds containing carbon and hydrogen (and oxygen) are burned to *completion* in air or oxygen, the products are CO_2 and H_2O (and heat). An example is the combustion of decane:

$$2C_{10}H_{22} + 31O_2 \overset{\Delta}{=} 20CO_2 + 22H_2O$$

When combustion is *incomplete* (insufficient oxygen), the products are H_2O and CO, or H_2O and C.

Single replacements

In a single replacement (displacement) reaction one element replaces another:

$$A + BC = B + AC$$

Table 11-5. The metal activity series and related reactions

A	B	C	D	E	F	G	H	I	J	K
React with oxygen, giving peroxides			React with cold water, giving a hydroxide and hydrogen	Chemical activity increases →	Li	Ease of reduction of compounds increases →	Oxides not reduced by hydrogen	Oxides not reduced by carbon	Electrolysis of molten salts yields metals	Only compounds occur in nature
					Cs					
					Rb					
					K					
					Ba					
					Sr					
					Ca					
					Na					
L					Mg					
		React with steam, giving a hydroxide and hydrogen			Be					
					Al					
					Mn			O	P	
					Zn			Oxides reduced by carbon	Electrolysis of aqueous solutions yields metals	
					Cr					
		React with nonoxidizing acids, giving salts and hydrogen			Fe		N			
					Cd		Oxides reduced by hydrogen			
					Co					
					Ni					
					Sn					
					Pb					S
					H					Found native and combined
React with oxygen, giving oxides					Cu					
					Sb				R	
					As		Q			
					Ag		Oxides reduced by heat		Compounds reduced to metals by electrolysis or heat	
M					Hg					T
Oxides must be formed indirectly					Pt					Found native only
					Au					

Reactions of active metals furnish most of the examples, as illustrated below.

7. More active metal + water = hydroxide + hydrogen

Metals more active than cadmium (columns C, D, Table 11-5) decompose water or steam. For instance,

$$2Li + 2H_2O = 2Li(OH) + H_2 \uparrow$$

(The arrow is optional. It indicates hydrogen is a gas.)

8. Active metal + nonoxidizing acid = salt + hydrogen

Metals above hydrogen in the activity series (column F, Table 11-5) replace hydrogen in nonoxidizing acids: HCl, HBr, HI, H_2SO_4, H_3PO_4, and $HC_2H_3O_2$. Metals uppermost in the list react violently, those near hydrogen slowly. If two valence states are possible, the lower is favored:

$$Mn + 2HCl = MnCl_2 + H_2$$

We will not consider more complex reactions given by oxidizing acids such as HNO_3, $HClO_3$, and $HClO_4$.

9. More active metal A + salt of metal B = less active metal B + salt of metal A

The order of replacement of metals presented in Table 11-5 was originally established by running just such reactions. Chemical activity of an element is described by its tendency to form chemical compounds.

The less active metal is displaced as the more active takes its place in a reaction. Zinc, for instance, is more active than mercury but less active than calcium; in solution, therefore, we can predict that

$$Zn + HgCl_2 = ZnCl_2 + Hg$$

$$Zn + CaCl_2 \longrightarrow N.R. \text{ (no reaction)}$$

10. More active halogen A + halide B = less active halogen B + halide A

The *halogen activity order* is predictable from periodic table position: $F_2 >$ $Cl_2 > Br_2 > I_2$. Thus iodine can be recovered from its compounds by chlorine displacement whereas fluorine cannot:

$$Cl_2 + 2KI = 2KCl + I_2$$

$$Cl_2 + KF \longrightarrow N.R.$$

11. Metal oxide + carbon (or hydrogen) = metal + carbon dioxide (or water)

The possibilities of reducing oxides with carbon and H_2 are indicated in columns H, I, N, and O of Table 11-5. Depending on reaction conditions CO might be a product instead of CO_2, and more or less heat will be required depending on the metal's activity. For example, tin can be recovered from its mineral oxide by carbon reduction, but aluminum cannot,

$$SnO_2 + C \overset{\Delta}{=} Sn + CO_2$$

$$Al_2O_3 + C \overset{\Delta}{\longrightarrow} N.R.$$

Metathesis

A *metathesis* is a reaction in which two substances yield two other substances of similar type. In general, if AB is one reactant and CD the other, then cation–anion exchange gives

$$AB + CD = AD + CB$$

Such reactions "go" in water solution when ions are effectively removed from further action because at least one of the products is (1) a weak acid or weak base; (2) a gas such as CO_2, SO_2, or NH_3, which escapes; or (3) insoluble and precipitates. A few rules of solubility are needed to explain (3).

1. When the cation is NH_4^+ or a IA metal ion, or the anion is $C_2H_3O_2^-$, ClO_3^-, NO_3^-, NO_2^-, Cl^-, Br^- or I^-, a salt is usually soluble. (Exceptions are the Cl^-, Br^-, and I^- salts of Ag^+, Hg_2^{2+}, and Pb^{2+}.)
2. Sulfates are soluble except those of Ca^{2+}, Sr^{2+}, Ba^{2+}, and Pb^{2+}.
3. All other simple salts can be expected to have low solubility.

12. Salt A + salt B = salt C + salt D

The following example of a reaction between two salts goes to the right because zinc chromate is insoluble (indicated by a downward-pointing arrow):

$$ZnCl_2 + Na_2CrO_4 = 2NaCl + ZnCrO_4\downarrow$$

13. Salt of weak acid + strong acid

= salt of strong acid | weak acid

Acid strength and weakness are defined by the extent to which the acid dissociates in water, giving ions. At some reasonable dilution a *weak acid* dissociates slightly (say 1 percent) due to strong bonding between H^+ and the anion. A diluted *strong acid* is almost completely dissociated.

Strong *inorganic acids* include

HCl	hydrochloric	$HClO_4$	perchloric
HBr	hydrobromic	HNO_3	nitric
HI	hydriodic	H_2SO_4	sulfuric

Organic acids are generally weak. Examples of weak acids are

$HC_2H_3O_2$	acetic	$H_2C_2O_4$	oxalic
H_3BO_3	boric	H_3PO_4	phosphoric
H_2CO_3*	carbonic	H_2SO_3*	sulfurous
HCN	hydrocyanic	H_2S	hydrosulfuric
HF	hydrofluoric	$H_2C_4H_4O_6$	tartaric
HNO_2*	nitrous	$H_2S_2O_3$*	thiosulfuric

Because of its high boiling point (340°C), pure sulfuric acid is used to displace both weak and strong acids from their salts; heating then drives off

---

* Weak acids that are unstable and decompose spontaneously:

$$H_2CO_3 = H_2O + CO_2\uparrow$$

$$H_2SO_3 = H_2O + SO_2\uparrow$$

$$3HNO_2 = HNO_3 + 2NO\uparrow + H_2O$$

$$H_2S_2O_3 = H_2O + SO_2\uparrow + S\downarrow$$

volatile substances. The by-product may be a sulfate or hydrogen sulfate salt:

$$H_2SO_4 + CaF_2 \overset{\Delta}{=} CaSO_4\!\downarrow + 2HF\!\uparrow$$

14. Ammonium salt + strong base

= ammonia + salt of strong base

For our present purposes, we define a *strong base* as a hydroxide of Group IA metals or of Ca^{2+} or Ba^{2+}. Strong bases displace ammonia gas from ammonium compounds. Ammonia, a weak base, may be written as ammonium hydroxide in solution:

$$NH_4Cl + NaOH \overset{\Delta}{=} NaCl + NH_3\!\uparrow + H_2O$$
$$NH_3 + H_2O = NH_4OH$$

Neutralization

We will define a *neutralization* as a reaction between an acid and a base to give a salt and water. In general,

$$HA + BOH = BA + H_2O$$

Defining acids as giving H^+ and bases as giving OH^- in solution, the essence of neutralization is

$$H^+ + OH^- = H_2O \tag{11-2}$$

Neutralizations go because water is a weak electrolyte.

Group IA metal hydroxides are water soluble, hydroxides of Ba^{2+}, Sr^{2+}, and Ca^{2+} are slightly soluble, and other hydroxides are mostly insoluble but may still react with acids.

15. Acid + base = salt + water

An example is

$$2HCl + Ba(OH)_2 = BaCl_2 + 2H_2O$$

(Acid–base theory is developed further in Section 16-6.)

Ionic equations

Molecular equations are written when it is necessary to specify the chemicals used or produced in reactions. However, to describe the species (some or all of which may be ions) that actually undergo reaction, chemists write ionic equations like Eq. 11-2. Some generalizations follow.

1. In ionic equations weak electrolytes, insoluble compounds, gases, and a few soluble but only slightly ionized heavy metal salts such as $HgCl_2$ and $Pb(C_2H_3O_2)_2$ are written as molecules; everything else is written as ions.
2. Identical ions on both sides of the equation are canceled.
3. The hydrogen ion is written as H^+ or H_3O^+. The latter is a combination of H^+ and H_2O, which indicates how the polar water molecules foster the acid's ionization.
4. When an ionic equation is balanced, there is both an atom and a charge balance. The total charge on the two sides may be zero or some other number, but it will be the same on both sides.

Example 11-2 In the nitrosyl chloride process for synthesizing chlorine gas, Cl_2 is produced by heating a mixture of concentrated nitric acid and sodium chloride to about 110°C. The products besides nitrosyl chloride gas are sodium nitrate and water. Write the balanced ionic equation.

Solution First we write the molecular equation from the description of the process,

$$HNO_3 + NaCl \longrightarrow Cl_2 + NOCl + NaNO_3 + H_2O$$

Next we balance it by inspection, as illustrated in Example 11-1,

$$4HNO_3 + 3NaCl = Cl_2 + NOCl + 3NaNO_3 + 2H_2O$$

Then we reason out which substances should be written ionically, which molecularily.

Nitric acid, sodium chloride, and sodium nitrate are soluble ionic substances in water solution. Chlorine and nitrosyl chloride are gases. Water is a weak electrolyte. The balanced ionic equation therefore is

$$4H^+ + 4NO_3^- + 3Na^+ + 3Cl^- = Cl_2\uparrow + NOCl\uparrow + 3Na^+ + 3NO_3^- + 2H_2O$$

We see that $3Na^+$ and $3NO_3^-$, which are on both sides, do not undergo chemical change. We will cancel them. The result is the *balanced net ionic equation*,

$$4H^+ + NO_3^- + 3Cl^- = Cl_2\uparrow + NOCl\uparrow + 2H_2O \tag{11-3}$$

Total charge on both sides is zero.

Redox Reactions

Redox (a contraction of the words *reduction* and *oxidation*) reactions are those involving changes in oxidation states. They are considered in Section 19-6.

QUESTIONS

1. Definitions Define and give one or more examples: acid, acid anhydride, acid salt, inorganic acid, organic acid, weak acid, base, base anhydride, very active metal, binary compound.

2. Definitions Explain the difference between **(a)** symbol and formula; **(b)** coefficient and subscript; **(c)** Berzelius and Stock nomenclature as applied to the ions V^{2+} and V^{3+}; **(d)** 4P and P_4; **(e)** solvent water and hydrate water.

3. Equations **(a)** Briefly, what is the difference between a molecular and an ionic equation? **(b)** How does a knowledge of covalence and electrovalence help one write an ionic equation when the molecular equation is given? Illustrate your remarks, given this reaction in water solution,

$$2HCl + Na_2SO_3 = 2NaCl + SO_2 + H_2O$$

4. Reaction types Fifteen types of reactions are described in the text. Classify each of the following commercial syntheses as one of them, and explain. **(a)** Hydrogen chloride gas is made by burning hydrogen in chlorine (in a silica combustion chamber). **(b)** Scrap iron is added to a copper solution (from copper mining operations) to recover the copper. **(c)** Plaster of Paris is made by heating gypsum. **(d)** Sodium nitrate is made by mixing nitric acid and sodium hydroxide solutions and evaporating water so the product crystallizes. **(e)** Dichlorine monoxide is prepared by warming hypochlorous acid.

5. Erroneous equations Point out the errors:

(a) $Fe^{3+} + Sn^{2+} = Fe^{2+} + Sn^{4+}$
(b) $Sr(OH)_3 + AlBr_3 = SrBr_3 + Al(OH)_3$
(c) $BaSO_4 + 2HCL = BaCL_2 + 2H^+ + SO_4^{2-}$
(d) $Al_2 + S_3 \rightarrow Al_2S_3$
(e) $HFl_2 + Sb = H\uparrow + SbFl_2$

PROBLEMS

6. Oxidation numbers Figure 11-2 is the beginning of a useful chart originally published in 1918. Copy it lengthwise on a piece of graph paper. Using the periodic table (inside back cover) continue the chart to at least $Z = 36$. Connect points that give lines parallel to the lines shown. Point out some generalizations you detect.

7. Writing formulas Copy Table 11-6. Add molecular formulas.

8. Writing formulas Give formulas for

cuprous sulfide   stannic sulfide
ferrous bromide   nickelous nitrite
mercuric chloride   cupric fluoride
cobaltic carbide   stannous hydroxide
plumbic oxide   manganous cyanide

9. Nomenclature Give the IUPAC name for each of the metal ions in the previous problem.

Table 11-6.

	Acetate	Sulfate	Sulfide	Phosphate	Carbide
Silver(I)					
Copper(II)					
Chromium(III)					
Tin(IV)					
Antimony(V)					

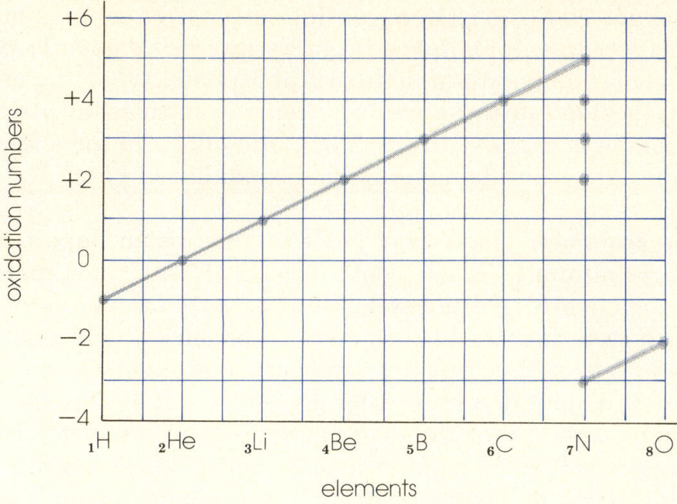

Figure 11-2

10. Writing formulas Give formulas for

potassium acetate	zinc phosphate
silver thiocyanate	bismuth(V) oxide
sodium permanganate	permanganate ion
calcium chlorate	bisulfite ion
lithium peroxide	dihydrogen arsenate ion

11. Writing formulas Give formulas for

magnesium carbonate	sodium hypochlorite
cadmium hydrogen sulfate	acetic acid
manganese(IV) oxide	potassium ferrocyanide
ammonium hydroxide	calcium orthosilicate
nitrous acid	mercurous chloride
mercuric nitrate	

12. Molecular equations Write balanced molecular equations for the following:

(a) hydrogen sulfide + sulfur dioxide = water + sulfur
(b) sodium hydroxide + sulfur trioxide = water + sodium sulfate
(c) sodium oxide + dichlorine pentoxide = sodium chlorate
(d) trona $\overset{\Delta}{=}$ sodium carbonate + water + carbon dioxide

13. Acids Reasoning by analogy from Table 11-3, **(a)** give formulas for the anhydrides of these acids: H_3AsO_3, HBrO, HIO_3; **(b)** name the acids; **(c)** write molecular equations showing how the acids are prepared from their anhydrides; **(d)** write molecular equations showing how each of the anhydrides would react with K_2O; **(e)** name each salt formed in (d).

14. Reaction sequence Write a balanced molecular equation for each step of the following reaction sequence.

Nitrogen can be fixed (chemically combined) by heating it with magnesium metal above 300°C to give magnesium nitride. Water plus magnesium nitride yields ammonia plus magnesium hydroxide. Ammonia plus phosphoric acid gives diammonium hydrogen phosphate. When the magnesium hydroxide is heated, steam is evolved, leaving magnesium oxide. When magnesium oxide is dissolved in hydrochloric acid and the solution evaporated, magnesium chloride monohydrate crystallizes.

15. Reaction sequence The Solvay process (1869) is an important method for converting limestone ($CaCO_3$), coke (C), salt (NaCl), water and ammonia into baking soda ($NaHCO_3$), and soda ash (Na_2CO_3). Overall yield is 75 percent. For each step described below, write a balanced molecular equation.

Limestone is heated, giving lime and carbon dioxide. Coke is burned in air to carbon dioxide. Lime plus water gives calcium hydroxide. Ammonia plus water plus carbon dioxide gives ammonium bicarbonate. Ammonium bicarbonate plus salt gives sodium bicarbonate plus ammonium chloride. Upon heating, sodium bicarbonate gives sodium carbonate plus carbon dioxide plus water. Ammonium chloride plus calcium hydroxide gives ammonia plus calcium chloride plus water.

16. Hydrates *Bieberite* and *bischofite* are common names, respectively, for cobaltous sulfate heptahydrate and magnesium chloride hexahydrate. Write balanced molecular equations showing the result of heating these hydrates.

17. Combustions Write three balanced molecular equations showing the complete combustion in air of propane (C_3H_8), methyl alcohol (CH_3OH), and acetic acid.

18. Ionic equations Predict the products and write net ionic equations; indicate precipitates (↓) and gases (↑): **(a)** dichlorine pentoxide + water →; **(b)** lithium oxide + water →; **(c)** plumbous hydroxide + sulfuric acid →.

19. Ionic equations As in Problem 18: **(a)** strontium metal + hydrochloric acid →; **(b)** magnesium metal + cupric nitrate →; **(c)** silver acetate + sodium bromide →.

20. Ionic equations As in Problem 18: **(a)** sodium thiosulfate + hydrochloric acid →; **(b)** ammonium sulfate + barium hydroxide →; **(c)** mercuric acetate + potassium sulfate →.

12

WHAT ARE STOICHIOMETRIC AND THERMOCHEMICAL CALCULATIONS

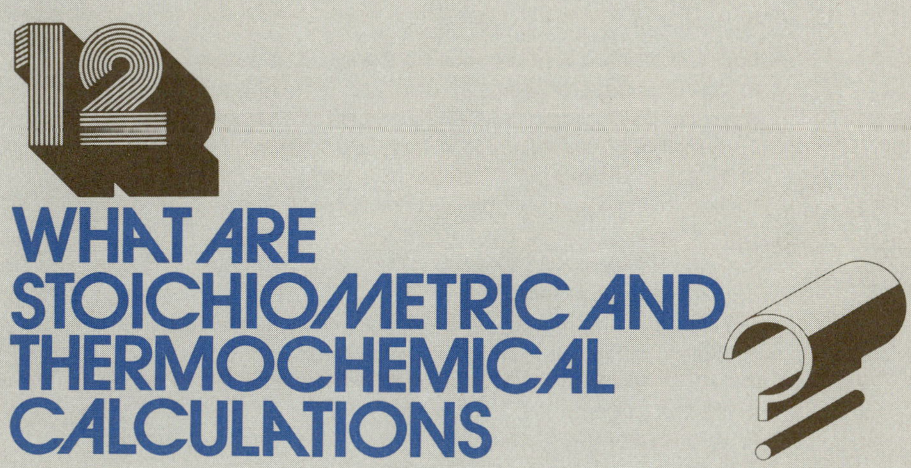

THE IMPORTANT CONCEPTS

12-1. The laws of stoichiometry

1. Matter can be neither created nor destroyed.
2. A pure compound always gives the same chemical analysis.
3. The weights of element A that combine with a fixed weight of element B in forming several compounds are in a ratio of small whole numbers.
4. If elements A and B combine with a fixed weight of element C, the weight ratio A : B is a simple multiple of that found when A and B combine with each other.
5. Equivalent weight was originally defined as the gram weight of an element combining with 8.00 g of oxygen.

6. Elements combine with one another in proportion to their equivalent weights.
7. The atomic theory rationalizes the laws of stoichiometry.

12-2. The problem of atomic and molecular weights

1. At the same conditions of temperature and pressure, gases react with one another in small whole number volume ratios.
   a. At the same conditions, equal volumes of gases contain equal numbers of molecules.
   b. At the same conditions, the weights of the same volume of different gases are proportional to their molecular weights.
   c. Reactive elemental gases such as oxygen are diatomic.
2. The atomic weights of some solid elements can be approximated from heat capacity data.
3. Atomic weights today are referred to carbon-12 and determined with the mass spectrograph.
4. The volume of 1 mole of gas at standard conditions is the molar volume, 22.4 liters.
   a. Determination of gas molecular weights follows directly from definition of the molar volume.
5. The mole is used in dealing with numbers of particles.
6. Avogadro's number N is the number of particles in a mole.

12-3. Stoichiometric calculations

1. Stoichiometric calculations deal with numbers and masses of atoms and molecules.
2. If a precise equivalent weight and an approximate atomic weight of an element are known, a precise atomic weight can be calculated.
3. Percentage composition of a compound can be found, given its formula and the atomic weights.
4. Given the percentage composition of a compound, its simplest formula can be found.
5. If a compound's molecular weight and simplest formula are known, the molecular formula can be found.
6. Avogadro's number can be found experimentally in several ways.
7. Given a balanced equation and atomic weights, one can calculate yield and percentage yield.
8. In gravimetry the analate is put into a pure form for weighing.

12-4. Thermochemistry

1. Heat is conserved in chemical processes.
2. Reaction heats can be measured in a calorimeter.
3. Enthalpy change is related to stability of reactants and products.
4. Enthalpy change depends only on the initial and final states of a reaction; individual enthalpies can be added to calculate the overall enthalpy change.

Calculations dealing with the numbers and masses of atoms and molecules are the subject of *stoichiometry*, a word whose literal meaning is "to measure the parts" (Gk. *stoicheion*, component). The basis of stoichiometry is a few laws discovered around 1800 by work with simple compounds.

Related calculations are found in *thermochemistry*, which concerns itself with the energy change that accompanies almost any chemical change. The quantity of energy absorbed or released during a chemical reaction is exactly proportional to the number of molecules reacting.

12-1. THE LAWS OF STOICHIOMETRY

The law of conservation of mass

In the late 1700s Scottish chemist J. Black was one of the first to investigate chemical reactions through use of the analytical balance. In a typical study he showed that after a measured weight of calcium carbonate was dissolved in hydrochloric acid and the carbon dioxide driven off, the same weight of calcium carbonate could be recovered on addition of sodium carbonate. He described the reactions as

$$limestone + muriatic\ acid \longrightarrow calcium\ salt + water + fixed\ air$$
$$mild\ alkali$$

Lavoisier (see Figure 13-5), who adopted Black's methods, concluded that such results pointed to the permanence of matter. In 1789 he wrote,

. . . nothing is created in operations either of art or of nature, and it can be taken as an axiom that in every operation an equal quantity of matter exists both before and after the operation, that the quality and quantity of the principles remain the same and that only changes and modifications occur. The whole art of making experiments in chemistry is founded on this principle: We must always suppose an exact equality or equation between the principles of the body examined and those of the products of its analysis.

The law of conservation of mass is a condensation of Lavoisier's statement: *The mass of products equals the mass of reactants in any ordinary chemical reaction.*

The law of definite proportions

From analytical comparisons of synthetic and natural minerals, French chemist J. Proust concluded that chemical compounds having fixed composition were the rule. Writing in 1799 he stated,

. . . we must recognize that invisible hand which holds the balance for us in the formation of compounds and fashions properties according to its will. We must conclude that nature operates not otherwise in the depths of the world than at

Figure 12-1 John Dalton (England 1766–1844). At age 12 Dalton began teaching in a one-room country school, and later by self-study became a college professor. He discovered principles of gas behavior (see Dalton's law, Chapter 14), invented symbols for writing chemical equations, developed techniques for weather observation, and made medical discoveries that included description of his own color blindness (daltonism). Results of quantitative mineral analyses and study of air as a mixture of gases started him thinking about particles and eventually led to his great scientific contribution, the atomic theory. (Edgar Fahs Smith Memorial Collection, University of Pennsylvania.)

its surface or in the hands of man. These ever-invariable proportions, these constant attributes, which characterize true compounds of art or of nature . . . (are) no more at the power of the chemist than the law of election which presides at all combinations.

Another French chemist, C. Berthollet, disputed Proust's conclusions. He believed compound composition could vary, and he gave as examples the irregular results when copper and sulfur were combined, and the arbitrary composition of alloys. Proust explained the first example as the formation of mixtures of two different copper sulfides, and the latter as solution mixtures of one metal in another and therefore not compounds at all. Experiments eventually settled the argument in favor of Proust.

The generalization Proust put forth became the *law of definite proportions: The composition of a pure compound is invariant; its elements are combined in fixed proportions by weight.*

The law of multiple proportions

In working on his atomic theory Dalton (Figure 12-1) ran experiments to compare the different weights of one element combining with a fixed weight of a second element in a series of their binary compounds. In nitrous oxide, nitric oxide, and nitrogen dioxide (now known to have the formulas N_2O, NO, and NO_2) he found the weights of oxygen per 1.0 weight unit of nitrogen to be 0.65, 1.3, and 2.6, respectively. Dividing each by the smallest, 0.65, he obtained a whole number ratio of 1:2:4.

When other chemists also discovered uncomplicated weight relationships to exist among related compounds they analyzed, *the law of multiple proportions* took form. Its modern statement follows: *The various weights of one element combining with a fixed weight of another element in forming a series of simple compounds are related by small whole number ratios.*

The law of reciprocal proportions

During the 1760s, Englishman H. Cavendish had measured the quantity of different alkalis needed to react with a fixed amount of an acid and he described the weights of alkalis "as equivalent in neutralizing capacity." Furthermore he found the ratio of weights of various acids reacting with a fixed amount of an alkali to be constant. The significance of his observations was not appreciated until the 1790s when J. Richter, a German assayer, deduced from similar experimentation that "The elements must have among themselves a certain fixed proportion by mass."

Data that Richter examined resembled the following. Starting with 5.6 g of iron, an iron oxide weighing 7.2 g or an iron sulfide weighing 8.8 g could be prepared. The weight of combined oxygen was $7.2 - 5.6 = 1.6$ g, and the weight of combined sulfur $8.8 - 5.6 = 3.2$ g, giving an oxygen-to-sulfur ratio of $1.6:3.2 = 0.50$. In another experiment 3.2 g of sulfur was found to combine with 3.2 g of oxygen to form a sulfur oxide. In this compound the ratio of oxygen to sulfur was $3.2:3.2 = 1.00$, a simple multiple of the oxygen-to-sulfur ratio found in the iron compounds.

The conclusion drawn from many such experiments became *the law of reciprocal proportions: If two elements combine with a fixed weight of a third element, the ratio of their weights in those compounds is a simple multiple of the weight ratio found when they combine with each other.*

Equivalent or combining weights

Chemists systematized reaction-weight ratio experiments by adopting oxygen as a standard because it formed stable compounds with other elements. *The gram weight of an element combining with 8.00 g of oxygen they defined as the element's combining or equivalent weight.*

It was found for instance that 8.00 g of oxygen combined with 1.00 g of hydrogen to form water, and also combined with 35.5 g of chlorine to form chlorine oxide. In turn, 1.00 g of hydrogen and 35.5 g of chlorine reacted to form hydrogen chloride. Being connected through chemical reactions to 8.00 g of oxygen related the equivalent weights of hydrogen and chlorine to one another. It also related their equivalent weights to those of all other elements similiarly defined. This work was summarized in *the law of equivalent proportions: Elements combine with one another in proportion to their equivalent weights, or small whole number multiples thereof.*

For a half century after Dalton most chemists quietly disregarded the atomic theory and atomic weights, using instead equivalent weights that were readily determinable in the laboratory and required no speculation about atoms. The significance of equivalent weight, however, became apparent when atomic weight and valency were understood. Consider the data in Table 12-1. The first three columns of the table list a few representative elements, their atomic weights, and compounds they form with chlorine. In the next column are weights of the elements combined in the compounds with a fixed weight of chlorine. Because 35.5 g of chlorine is 1 equivalent weight in grams, (*a quantity called 1 gram-equivalent or simply 1 equivalent*), the values calculated in column 4 are by definition the gram-equivalent weights of the combining elements. A comparison of the atomic and equivalent weights shows the former are simple multiples of the latter.

Table 12-1.   Equivalent and atomic weights

Element	Atomic weight	Chloride	Grams combining with 35.5 g of chlorine	Valency of element
Hydrogen	1.00	HCl	1.00	1
Oxygen	16.0	Cl_2O	8.00	2
Aluminum	27.0	$AlCl_3$	9.0	3
Carbon	12.0	CCl_4	3.0	4
Antimony	122	$SbCl_3$	40.7	3
Antimony	122	$SbCl_5$	24.4	5

The ratio of the two is the number of equivalents in 1 g-atomic weight or 1 g-*atom* (later defined as 1 mole) of the element in question,

$$\frac{gram/gram\text{-}atomic\ weight}{gram/gram\text{-}equivalent\ weight} = \frac{gram\text{-}equivalent\ weights}{gram\text{-}atomic\ weight} = \frac{equivalents}{gram\text{-}atom} \tag{12-1}$$

The small whole numbers obtained by the division indicated in Eq. 12-1 are the valencies or valence numbers of the elements. In another form,

$$valency = \frac{atomic\ weight}{equivalent\ weight} \tag{12-2}$$

As shown by the two antimony chlorides in the table, an element has a different equivalent weight for each of its valence numbers, and vice versa.

Atomic rationale for the laws of stoichiometry

Experimental results like those just described helped suggest the atomic theory to Dalton. Later the atomic theory rationalized the laws of stoichiometry as outlined below.

Except for nuclear reactions, atoms are as indestructible as Dalton imagined them to be. Chemical reactions take place between atoms or groups of atoms; therefore the mass of a reacting system must remain constant. This is a restatement of the conservation of mass law.

Dalton's assumptions that (1) molecules consisted of fixed ratios of atoms and (2) all atoms of a given element had identical mass made understandable the constancy of results when compounds were analyzed for their percentage composition. By the time refined chemical analyses showed anomalies in atomic weights due to variations in natural isotopic mixtures, the atomic theory was mature enough to be strengthened by the discovery.

The laws of multiple proportions and reciprocal proportions were interpreted from Dalton's assumption that every molecule was a definite combination of small whole numbers of atoms. An illustration with experimental data is useful.

Consider the information in Table 12-2. In the second column of the table are weights of three carbon oxides taken for analysis. In the next two columns are the weights of carbon and oxygen combined in each case. In the last column is the weight of oxygen in each oxide combined with a fixed weight (12 g) of carbon. Dividing each of the numbers in the last column by the smallest, 10.7, gives, in order, 1.5, 3.0, and 1.0. Multiplying by

Table 12-2. Stoichiometry of three carbon oxides

Oxide	Grams of oxide taken	Grams of carbon	Grams of oxygen	Grams of oxygen per 12 g of carbon
Carbon monoxide	28.0	12.0	16.0	16.0
Carbon dioxide	44.0	12.0	32.0	32.0
Carbon suboxide	68.0	36.0	32.0	10.7

2 gives numbers standing in the ratio of small whole numbers: $3:6:2$. This is an illustration of the law of multiple proportions.

We must now note that correct formulas cannot be assigned directly from multiple-proportion ratios. The three oxides just described are not CO_3, CO_6, CO_2. Rather they are CO, CO_2, C_3O_2. Both sets of formulas fit the calculated atom ratios but only the latter will agree with experimental molecular weight data. Until methods for determining atomic and molecular weights were discovered, chemical progress was severly hampered.

12-2. THE PROBLEM OF ATOMIC AND MOLECULAR WEIGHTS

Early work with atomic weights

When Dalton proposed that each kind of atom had its characteristic weight, he also sought to establish a table of relative atomic weights. As a standard he chose $H = 1.0$, and by chemical analysis determined weights of elements combined with 1.0 weight unit of hydrogen. In the compound water he found 8 g of oxygen combined with 1 g of hydrogen. If the oxygen atom was eight times as heavy as the hydrogen atom, that is $O = 8$ on his atomic weight scale, then the formula for water was HO. If oxygen was 16 on the atomic weight scale and 8 g of it combined with 1 g of hydrogen, then the formula for water was H_2O. If $O = 24$, then water was H_3O, etc.

Faced with the dilemma that without atomic weights, formulas could not be found and without formulas, atomic weights could not be found, Dalton made an assumption. He assumed that two elements combined in $1:1$ ratio most commonly, and combined in other ratios with decreasing frequency as molecular complexity increased. His assumed formula for water, HO, was therefore incorrect.

Help came from measurements on gas reactions. French chemist J. Gay-Lussac had found that *when measured at the same temperature and pressure, volumes of gases reacting with one another could be expressed in ratios of small whole numbers.* His observation, later called the *law of combining volumes,* was based on results like the following:

$$\begin{array}{c} 2 \text{ vol of} \\ \text{nitric oxide} \end{array} + \begin{array}{c} 1 \text{ vol of} \\ \text{oxygen} \end{array} = \begin{array}{c} 2 \text{ vol of} \\ \text{nitrogen dioxide} \end{array}$$

and

$$\begin{array}{c} 2 \text{ vol of} \\ \text{hydrogen} \end{array} + \begin{array}{c} 1 \text{ vol of} \\ \text{oxygen} \end{array} = \begin{array}{c} 2 \text{ vol of} \\ \text{steam} \end{array}$$

The small whole number ratios hinted strongly of a direct relationship between gas volumes and the number of contained particles. Dalton, how-

Figure 12-2 Amadeo Avogadro (Italy 1776–1856). Avogadro's interpretation of the combining volume law was the key to solving the all-important problem of determining relative weights of atoms and molecules. (Edgar Fahs Smith Memorial Collection, University of Pennsylvania.)

ever, believed the experiments inaccurate. He claimed that each of these reactions would yield one volume of product not two, because they were limited by one reacting volume of oxygen,

$$NO + O = NO_2 \quad and \quad H + O = HO$$

The correct explanation was offered in 1811 by Avogadro (Figure 12-2), whose hypothesis about reacting gas volumes later acquired the status of a law: (1) *Under identical temperature and pressure conditions equal volumes of all gases contain equal numbers of molecules,* and (2) *reactive elemental gases consist of diatomic molecules rather than single atoms.* The previous examples could then be reinterpreted:

$$2 \; molecules + 1 \; molecule \longrightarrow 2 \; molecules$$
$$2NO \quad + \quad O_2 \quad = \quad 2NO_2$$
$$2H_2 \quad + \quad O_2 \quad = \quad 2H_2O$$

The importance of Avogadro's law went unappreciated until 1860 when Italian chemist S. Cannizzaro showed how it could be used to get atomic and molecular weights. Taking $H_2 = 2.0$ as a molecular weight standard, he compared the weights of equal volumes (and therefore equal numbers of molecules) of various gases under identical conditions. Oxygen and chlorine he found to be, respectively, 16 and 35.5 times heavier than hydrogen. Assuming all were diatomic gases, his experiments gave O_2 a molecular weight of $16 \times 2 = 32$, and Cl_2 a molecular weight of $35.5 \times 2 = 71$. Atomic weights were half the molecular weights, 16 and 35.5.

Atomic weights from heat capacities

In 1819 French physicists P. Dulong and A. Petit measured the *specific heats* (calories needed to raise the temperature of 1 g of substance 1°C) of solid elements. Assuming an atom of one element to have the same capacity to absorb heat as an atom of another element, they multiplied specific heats by the respective atomic weights to obtain "atomic heats," or what are now called *molar heat capacities,*

$$molar \; heat \; capacity \cong specific \; heat \times atomic \; weight \qquad (12\text{-}3)$$

The units are calories per mole per degree Celsius. Again, a *mole* of an atomic substance is one atomic weight of it measured in grams.

Experimenting with a number of metallic elements at room temperature Dulong and Petit obtained roughly a constant, 6.2 cal/mole °C, for the molar heat capacities. Using that number and laboratory determinations of specific heats, they proceeded to check atomic weights with the following relationship, derived by rearranging Eq. 12-3:

$$\frac{6.2}{specific\ heat} \cong atomic\ weight \tag{12-4}$$

By assuming the rule held strictly, they "adjusted" atomic weights that did not fit it and succeeded in showing that some values Berzelius was using were multiples of correct values. Further work, especially on lighter elements, showed that the rule did not fit all cases. Some modern data are given in Table 12-3. (See also Example 12-3 and Problems 36 and 37.)

Heat capacity is related to the frequencies with which atoms vibrate. Around 1910 Debye and also Einstein successfully treated the phenomena mathematically. Their work was a major triumph for quantum theory because it used ideas about excitation of vibrational quanta that classical theory did not recognize.

Toward the ^{12}C standard

During the late 1800s the best atomic weight values of certain elements were those of Belgian chemist J. Stas. Stas' methods of determining combining proportions were refined by Harvard professor Theodore Richards (United States 1868–1928, Nobel prize in chemistry 1914), whose atomic weight values still stand as some of the best ever obtained by purely chemical means (see Problem 20). In 1903 an international committee established the first official table of atomic weights based on the assignment $O = 16.0000$ atomic mass units (amu).

In 1929 research at the University of California by William Giauque (Canada and United States 1895–   , Nobel prize in chemistry 1949) proved that naturally occurring oxygen was not a pure substance, but an isotopic mixture containing mostly ^{16}O with small amounts of ^{17}O and ^{18}O (see Figure 4-5). Following a period of confusion, physicists decided to use ^{16}O = 16.0000 amu as their standard. Chemists decided to continue using the original assignment because the natural oxygen isotopic mixture is essentially invariant in water and air.

Table 12-3. Heat capacities of some solid elements, 20°C

Element	Atomic weight	Specific heat cal/g °C	Molar heat capacity cal/mole °C
Aluminum	27.0	0.217	5.9
Iron	55.8	0.113	6.3
Copper	63.5	0.092	5.8
Cadmium	112	0.055	6.2
Iodine	127	0.052	6.6
Lead	207	0.030	6.3
Uranium	238	0.028	6.7

For precise work the two scales were different enough (the natural oxygen mixture is 16.0044 amu on the physicist's scale) to be inconvenient. In 1961 international agreement was reached on using the carbon-12 isotope $= 12.00000$ amu as the standard. This required only minor adjustment of chemist's values downward (see Problem 19).

As methods for determining atomic weights have increased in sophistication (see Figure 4-4), uncertainties in the results have paradoxically increased. This was evident in 1969 when a revised international table was published. An element's atomic weight is the average relative weight of its natural isotopes, and modern techniques have demonstrated that greater variance in isotope distribution exists than was previously established. In the case of lead even classical chemical analysis can prove that the atomic weight varies slightly from sample to sample. Atomic weight differences in uranium, boron, and lithium samples can also readily be demonstrated because processing related to atomic energy applications has significantly altered the natural abundance ratios of their isotopes.

Molecular weights

Work on determining the weights of molecules relative to the atomic weight standard paralleled work on atomic weights. As already noted, the first molecular weights found were those of gases.

Chemists today define a quantity called the *molar volume* for gases. One molar volume is the volume occupied by 1 mole (1 molecular weight in grams) of a gas at standard temperature and pressure (STP $= 0°C$, 1 atm). Experiments show that

$$\textit{1 mole of gas at STP = 22.4 liters} \tag{12-5}$$

Using this very simple relationship, the molecular weight of a gas or vapor can be calculated without knowing anything about its chemical properties (see Example 12-5.)

Methods for obtaining the molecular weights of nonvolatile compounds were not discovered until the 1880s when chemists found relationships between the molecular weight of a dissolved substance and its solution boiling point, freezing point, and osmotic pressure (details are described in Chapter 16). Techniques for getting molecular weights of very large molecules were worked out in the present century. They involve observation of the viscosity, turbidity, and light-scattering properties of solutions, and the sedimentation rate of suspended particles during centrifugation (details in Chapter 23).

The mole concept

The equation $C + O_2 = CO_2$ can be interpreted to mean 1 atom of carbon and 1 molecule of oxygen react to give 1 molecule of carbon dioxide. It can also be read in terms of gram-atomic and gram-molecular weights: 12 g of C plus 32 g of O_2 yield 44 g of CO_2. Chemists have found it so convenient to talk about reactions in the latter way that a unit, the mole, has been introduced.

The IUPAC definition is

The mole is the amount of substance of a system, which contains as many elementary units as there are carbon atoms in 0.012 kilogram of carbon-12. The elementary unit must be specified and may be an atom, a molecule, an ion, an electron, etc., or a specified group of such entities.

The definition says that the mole is characterized by the number of particles in it. Relationships that hold for relative numbers of particles therefore also hold for relative numbers of moles. The equation $C + O_2 = CO_2$ can therefore also be interpreted as 1 mole of carbon atoms plus 1 mole of oxygen molecules give 1 mole of carbon dioxide molecules.

In engineering applications pound-moles or ton-moles of substances may be used. One ton-mole of carbon is 12 tons of carbon. Oxidized with 1 ton-mole of O_2 (32 tons) the result is 1 ton-mole of CO_2 (44 tons).

Avogadro's number, N

From the preceding discussion we can see that there are exactly as many carbon atoms in 1 mole of carbon as O_2 molecules in 1 mole of oxygen, or CO_2 molecules in 1 mole of carbon dioxide. This number is called Avogadro's number (symbol N). By definition, *Avogadro's number is the number of carbon atoms in exactly 12 g of ^{12}C. One mole of ^{12}C contains Avogadro's number of atoms.* Chemical operations may thus be thought of as "counting" particles by weighing substances. Referring once more to the synthesis of CO_2, we can say that when N atoms of carbon react with N molecules of O_2, the product is N molecules of CO_2. The accepted value of Avogadro's number is

$$N = 6.022 \times 10^{23} \tag{12-6}$$

The enormity of N is difficult both to appreciate and to illustrate meaningfully. It is impossible to visualize, for example, how very small a molecule of water must be such that 6.02×10^{23} of them are contained in only 18 ml (1 mole) of liquid H_2O. If we use units that can be visualized, then the accumulation of N of them becomes unimaginably large. Suppose 6.02×10^{23} drops of rain fell. How much rain would that be? The answer is it would be enough water to submerge the entire continental United States (total land area 3.63 million square miles) to a depth of 10,000 feet.

Avogadro's number has been determined in a variety of independent ways. The first determination was made in 1865 from a theoretical model of molecules moving in a gas. Jean Perrin (France 1870–1942, Nobel prize in physics 1926) calculated a value in 1908 based on Brownian movement (see Figure 14-10). Rutherford and Boltwood obtained a value by literally counting atoms related to α decay (see Problem 34). Millikan obtained a value by relating the charge on the electron to the quantity of electricity needed to produce 1 g of hydrogen during electrolysis (see Problem 33). Another determination of N was made by the Braggs using X ray measurements to calculate the number of particles in 1 mole of rock salt (see Example 12-6). Still another determination came from Langmuir's study of surface films having one-molecule thickness. All methods gave essentially the same answer.

The laws of chemical combination plus the definition of the mole are tools that chemists use in making stoichiometric calculations. Illustrative problems follow.

12-3. STOICHIOMETRIC CALCULATIONS

Multiple proportions

The law of multiple proportions was established by Dalton partly from his own work and partly from experiments by Proust. Proust had demonstrated the principle of definite proportions. He did not, however, follow it up to find the relation between the amounts of one element combined with a fixed amount of another in a series of their binary compounds. The following example shows how some of his data were later interpreted.

Example 12-1 Tin metal oxidized under different conditions gave Proust two tin oxides:

	black oxide	white oxide
percent Sn	88.1	78.4
percent O	11.9	21.6

Show that the data suggest the law of multiple proportions.

Solution For convenience we will calculate the weight of oxygen combined in each oxide with 100 g of tin. To do so we convert percentages to grams:

$$? \; g \; O = 100 \; g \; Sn \left(\frac{11.9 \; g \; O}{88.1 \; g \; Sn} \right) = 13.5 \; g$$

$$? \; g \; O = 100 \; g \; Sn \left(\frac{21.6 \; g \; O}{78.4 \; g \; Sn} \right) = 27.5 \; g$$

The ratio 27.5:13.5 is, within the experimental error of Proust's day, 2:1. The small whole number ratio is the essence of the multiple proportion law.

Avogadro's hypothesis and a molecular formula

Referring to experiments which showed that gases reacted in the ratio of small whole numbers of volumes, Avogadro wrote,

> It must be admitted that very simple relations exist between the volumes of gas and the numbers of molecules which form them. The first hypothesis to present itself in this connection, and apparently even the only admissible one, is the supposition that the number of integral molecules in all gases is always the same for equal volumes.

From this reasoning came the correct molecular representation for reactive elemental gases (O_2, not O, etc.) and the clarification of experiments like the following.

Example 12-2 A sample of calcium metal is heated in nitrous oxide gas. Reaction gives calcium oxide (CaO) and nitrogen gas. When the apparatus has cooled and returned to its original temperature and pressure, it is found that the nitrogen occupies the same volume as did the nitrous oxide. In a separate experiment 1 liter of nitrous oxide is found to be 22 times as heavy as 1 liter of hydrogen. Aided by Avogadro's law find the formula of nitrous oxide.

Solution The observations can be summarized by an equation,

$$\text{1 vol of nitrous oxide} + Ca = CaO + \text{1 vol of nitrogen}$$

From Avogadro's law we can say that because under identical conditions the two gas volumes are equal, the numbers of molecules in them must also be equal. Or

$$\text{1 molecule of nitrous oxide} \longrightarrow \text{1 molecule of nitrogen}$$

Using Avogadro's hypothesis that reactive gaseous elements are diatomic, we write nitrogen N_2. Nitrous oxide therefore is N_2O_n, where n is assumed to be a small whole number. We also know that for each oxygen atom one calcium atom is required—to form CaO. The equation then becomes

$$N_2O_n + nCa = nCaO + N_2$$

The atomic weights are $N = 14$, $O = 16$. If $n = 1$, the molecular weight of nitrous oxide is $(2 \times 14) + 16 = 44$. If n is 2, the molecular weight is $(2 \times 14) + (2 \times 16) = 60$, etc. The actual molecular weight is found from the hydrogen–nitrous oxide comparison. Given that hydrogen $= H_2 = 2.0$, the molecular weight of nitrous oxide is $2 \times 22 = 44$. Its formula must be N_2O.

The Dulong and Petit rule; equivalent and atomic weights

Although approximate, the Dulong and Petit rule was historically important for helping establish precise atomic weights from combining weight experiments. It also was useful in determining correct formulas for compounds and valence numbers of elements.

Example 12-3 A 0.2994-g sample of nickel metal powder reacts with chlorine gas to give 0.6610 g of nickel chloride. In another experiment it is found that 4.4 cal is needed to raise 10 g of nickel 4°C. Calculate **(a)** the precise equivalent weight, **(b)** the approximate atomic weight, and **(c)** the precise atomic weight and valence number of nickel.

Solution **(a)** The weight of chlorine in the chloride is 0.6610 g − 0.2994 g = 0.3616 g. The precise combining weight of an element is that weight which will combine with 1 g-atomic weight (35.45 g) of chlorine:

$$? \ g \ Ni = 35.45 \ g \ Cl \left(\frac{0.2994 \ g \ Ni}{0.3616 \ g \ Cl} \right) = 29.36 \ g$$

(b) The metal's specific heat is computed from the heat data,

$$\frac{4.4 \; cal}{10 \; g \times 4°C} = 0.11 \; cal/g \; °C$$

The approximate atomic weight is found by dividing 6.2 by the specific heat (Eq. 12-4, Dulong and Petit's rule),

$$\frac{6.2}{0.11} = 56$$

(c) By inspection we see the approximate atomic weight of nickel is twice the precise combining weight. The precise atomic weight therefore is

$$2 \times 29.36 = 58.72$$

The integer 2 is the valence number of nickel in the compound. (The compound's formula is $NiCl_2$.)

Percentage composition

The percent by weight of each element in a compound is obtained experimentally by analytical laboratory procedures. When the compound's formula and the atomic weights are known, percentage composition can be calculated from them.

Example 12-4 The molecular formula of the insecticide DDT is $C_{14}H_9Cl_5$. Calculate the percentage composition.

Solution The gram-molecular weight or gram-formula weight is the sum of the gram-atomic weights:

$$(14 \times 12.0 \; g) + (9 \times 1.01 \; g) + (5 \times 35.5 \; g)$$

$$= 168 \; g \; of \; C + 9.09 \; g \; of \; H + 178 \; g \; of \; Cl$$

$$= 355 \; g$$

The elemental weight percentages are

$$percent \; C = (168 \; g/355 \; g) \times 100 = 47.3 \; percent$$
$$percent \; H = (9.09 \; g/355 \; g) \times 100 = 2.6 \; percent$$
$$percent \; Cl = (178 \; g/355 \; g) \times 100 = 50.1 \; percent$$

The sum of the percentages is, of course, 100 percent.

Empirical and molecular formulas

An *empirical* or *simplest formula* expresses the simplest whole number ratio of atoms in a substance. The *molecular formula* of a compound is the same as or a multiple of the empirical formula of that compound. The molecular formula is determined by comparing the empirical weight to the molecular weight and calculating the number of empirical units in the molecular unit. Empirical formulas can be found experimentally.

Example 12-5 One powerful oxidizer used to burn hydrogen for rocket propulsion is a fluorine oxide containing, according to chemical analysis, 70.5 percent fluorine and 29.5 percent oxygen by weight. The compound is volatile and at STP 1.00 liter weighs 2.41 g. Calculate **(a)** the simplest formula, and **(b)** the molecular formula.

Solution **(a)** For simplicity we will consider having 100 g of the compound. Using known atomic weights, $F = 19.0$, $O = 16.0$, we first calculate the number of gram-atomic weights (gram-atoms) or moles present in this amount, and their ratio:

$$? \text{ g-atom of } F = 70.5 \text{ g } F \left(\frac{1 \text{ g-atom } F}{19.0 \text{ g } F} \right) = 3.71 \text{ g-atoms}$$

$$? \text{ g-atom of } O = 29.5 \text{ g } O \left(\frac{1 \text{ g-atom } O}{16.0 \text{ g } O} \right) = 1.84 \text{ g-atoms}$$

The gram-atom ratio is $3.71:1.84 = 2.02$. The empirical formula is therefore OF_2.

(b) One molar volume of any gas occupies 22.4 liters at STP (Eq. 12-5). The molecular weight of a gas is determined by finding the weight of 22.4 liters of the gas at standard conditions:

$$? \text{ g} = 22.4 \text{ liters} \left(\frac{2.41 \text{ g}}{1 \text{ liter}} \right) = 54.0 \text{ g}$$

The molecular weight is 54.0. The weight corresponding to the empirical formula is $16.0 + (2 \times 19.0) = 54.0$. The empirical and molecular weights are the same. The simplest formula is consequently also the molecular formula. The compound is oxygen difluoride OF_2.

A value for N

One of the best ways to determine Avogadro's number comes from X ray examination of crystals. From interatomic distances the number of atoms per mole is calculated.

Example 12-6 In rock salt (see Figure 10-2a) the average distance between Na^+ and Cl^- centers is 2.819×10^{-8} cm, as determined by X ray diffraction. The density of the crystal is 2.165 g/cm³ and the formula weight of NaCl is 58.44 g. Use these data to find a value for N.

Solution We first calculate the volume of 1 mole of solid NaCl from the formula weight and density:

$$? \text{ cm}^3 = 58.44 \text{ g} \left(\frac{1.000 \text{ cm}^3}{2.165 \text{ g}} \right) = 27.00 \text{ cm}^3$$

Rock salt has cubic symmetry, We will imagine a perfect cube of it having a 27-cm³ volume. Volume V of a cube is the length, l, of its edge cubed, $V = l^3$. Therefore

$$edge\ length = \sqrt[3]{27.00\ cm^3} = 3.000\ cm$$

Along this edge is a row of alternating Na^+ and Cl^- ions which we will imagine as spheres in contact with one another. X ray measurement has given their effective diameter or length:

$$ions\ at\ edge = \frac{3.000\ cm}{2.819 \times 10^{-8}\ cm} = 1.064 \times 10^8\ ions$$

The 27.00-cm³ cube contains that number cubed:

$$total\ ions = (1.064 \times 10^8\ ions)^3 = 1.205 \times 10^{24}$$

Half that number is the number of Na^+Cl^- "molecules" in one mole of sodium chloride, or Avogadro's number:

$$\frac{12.05 \times 10^{23}}{2} = 6.025 \times 10^{23}$$

Yield and percentage yield from chemical equations

The yield of a chemical process can be calculated from the chemical equation(s) that describe it. Given a weight of substance A and asked to find the weight of substance B related to it by chemical reaction, a convenient procedure is to

1. Write the balanced chemical equation.
2. Use the periodic table to find the atomic (or molecular) weights of A and B.
3. Convert the weight of A to moles.
4. Use the balanced equation to relate moles of B to moles of A.
5. Convert moles of B to weight units.

Example 12-7 **(a)** What weight of silver chromate can theoretically be prepared from 5.00 g of silver nitrate reacting with sufficient sodium chromate? **(b)** If 4.85 g of pure product is obtained in the experiment, what is the percentage yield?

Solution The balanced equation is

$$2AgNO_3 + Na_2CrO_4 = 2NaNO_3 + Ag_2CrO_4$$

Molecular weights we need are sums of known atomic weights:

$$AgNO_3 = 108 + 14 + (3 \times 16) = 170$$

$$Ag_2CrO_4 = (2 \times 108) + 52 + (4 \times 16) = 332$$

(a) 5.00 g of silver nitrate is

$$\frac{5.00\ g}{170\ g/mole} = 0.0294\ mole$$

The balanced equation shows that for each 2 moles of silver nitrate used, 1 mole of silver chromate is formed. Therefore 0.0294 mole of silver nitrate should give 0.0147 mole of silver chromate. This is

$$0.0147 \ mole \times \frac{332 \ g}{mole} = 4.88 \ g$$

(The same result can be obtained by a factor-unit solution. An example is shown in Appendix 3.)

(b) The result is called the *theoretical yield*; the actual *or experimental yield* is often expressed as a percentage of it:

$$percent \ yield = (actual \ yield/theoretical \ yield) \times 100 \tag{12-7}$$

$$percent \ yield = (4.85 \ g/4.88 \ g) \times 100 = 99.4 \ percent$$

Gravimetric analysis

The method just illustrated is applicable to handling calculations that come from gravimetric analyses. *Gravimetry* is the science of precisely separating a substance of analytical interest (the *analate*) in a form for weighing. As shown in the next example, if the analate is completely accounted for, a stoichiometric calculation can be made even when the reaction equation is incompletely known.

Example 12-8 Rubber is *vulcanized* by heating with sulfur to improve its physical properties. A sample of vulcanized rubber weighing 1200 mg is oxidized completely, converting all sulfur to sulfate ion. Barium chloride solution is added, giving a quantitative precipitate of barium sulfate that is filtered off and dried. The barium sulfate weighs 87.5 mg. Calculate the percentage sulfur in the rubber.

Solution Rubber is a complex mixture, so it is impossible to know all the reactions that go on. We do know, however, that sulfur, the analate, is completely converted to barium sulfate, and that there is one sulfur atom in one $BaSO_4$ molecule. The following statement, with its atomic and molecular weights, is therefore sufficient to solve the problem:

$$S \longrightarrow BaSO_4$$
$$32.0 \qquad\quad 233$$

It is handy when dealing with milligram quantities to define a *millimole*:

$$1 \ mmole = 0.001 \ mole \tag{12-8}$$

One millimole of sulfur is 32.0 mg, 1 mmole of $BaSO_4$ is 233 mg. In the barium sulfate precipitate there is, therefore,

$$\frac{87.5 \ mg}{233 \ mg/mmole} = 0.375 \ mmole \ of \ BaSO_4$$

Since 1 mmole of barium sulfate comes from 1 mmole of sulfur, 0.375 mmole of sulfur must have been present in the rubber sample. Its weight is

$$0.375 \ mmole \times \frac{32.0 \ mg}{mmole} = 12.0 \ mg \ of \ S$$

The percentage sulfur in the rubber sample is

$$\frac{12.0 \ mg}{1200 \ mg} \times 100 = 1.00 \ percent \ of \ S$$

The problem can also be solved by factor-unit calculation (see Appendix 3).

12-4. THERMOCHEMISTRY

Heats of reaction

Reactions giving stable products generally proceed with the evolution of heat (*exothermic reactions*), although some are known that absorb heat (*endothermic reactions*). The quantitative study of chemical reactions from the standpoint of heat transfer is called *thermochemistry.*

Thermochemistry's first law was a forerunner of the law of energy conservation (energy can neither be created nor destroyed). It came from work by Lavoisier and French mathematician P. Laplace in 1780: *The heat needed to decompose a compound is equal to the heat produced in its formation.*

Chemical equations that include a value for the heat liberated or absorbed per mole of reactant are called *thermochemical equations.* The thermochemical equation

$$S\,(s) + O_2\,(g) = SO_2\,(g) + 71.0 \ kcal \tag{12-9}$$

signifies that when 1 mole of solid sulfur and 1 mole of gaseous oxygen react, the products are 1 mole of gaseous sulfur dioxide and 71 kcal of heat energy. According to the first law of thermochemistry an input of 71.0 kcal would be needed to reconvert 1 mole of sulfur dioxide to sulfur and oxygen under the same initial and final experimental conditions.

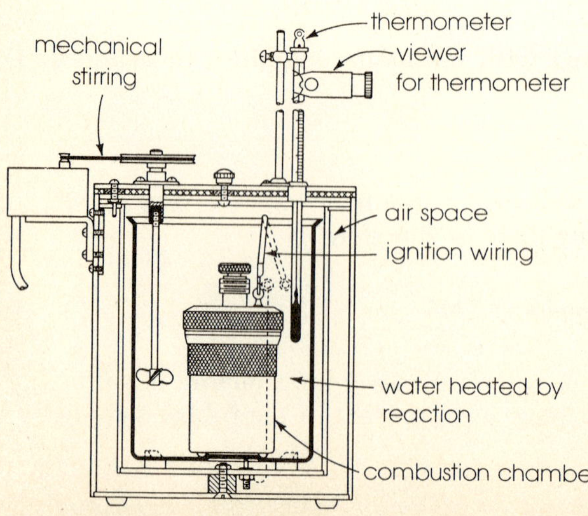

Figure 12-3 Cross-sectional view of an oxygen-bomb calorimeter, an instrument originated by M. Berthelot in 1881. A weighed sample of combustible material is placed in the bomb, oxygen gas introduced under pressure, and the mixture ignited electrically. The heat evolved is absorbed by the known quantity of water in which the bomb is immersed and by all the metal it contacts. From the temperature rise and the weights of water and metal and their specific heats, the total heat of the chemical reaction can be found. The heat of combustion in kilocalories per mole can then be computed. (Courtesy Parr Instrument Company.)

Values for heats of reaction are obtained experimentally by use of a *calorimeter*. A calorimeter (Figure 12-3) is a device in which reaction heat is transferred to a known quantity of water whose change in temperature is noted. Because 1 g of water is changed in temperature 1°C by 1 cal of heat (the specific heat of water being 1 cal/g °C), the total number of calories involved in a reaction is easily determined (see Problem 38).

Enthalpy

Heat change during reaction means that the chemical system contains a different amount of energy than before reaction took place. A system's total energy is termed its *heat content* or *enthalpy* (Gk. *enthalpein*, to warm in). The difference between the enthalpy of reaction products (H_p) and enthalpy of reaction reactants (H_r) is the enthalpy change (ΔH) of the whole system:

$$H_p - H_r = \Delta H \tag{12-10}$$

Chemists have agreed to call 25°C and 1 atmosphere pressure *standard conditions* in thermochemical calculations. At standard conditions all substances are said to be in their *standard states* and the enthalpy change is designated ΔH^0. When the arrangement of atoms in the products is more stable, and thus possesses less energy than the arrangement of atoms in the reactants, the reaction is exothermic. By convention ΔH^0 is then called negative. When the converse conditions hold, the reaction is endothermic and ΔH^0 is positive.

The heat of reaction is a measure of ΔH. In Eq. 12-9 heat of reaction was given a positive sign to indicate it may initially be thought of as a reaction product along with SO_2. By the more useful conventions attached to enthalpy, the equation is written

$$S(s) + O_2(g) = SO_2(g) \qquad \Delta H^0 = -71.0 \text{ kcal} \tag{12-11}$$

When a compound is synthesized from its elements as in Eq. 12-11, ΔH^0 for the reaction is called the *enthalpy of formation* of that compound. Table 12-4 lists a few of the several thousand such values known. Every compound has a standard enthalpy of formation that characterizes it. A compound with a positive ΔH^0 value has a larger heat content per mole than its elements. A compound with a negative ΔH^0 value has a smaller heat content per mole than its elements.

Table 12-4.   Standard enthalpies of formation

Compound	ΔH^0 kcal/mole	Compound	ΔH^0 kcal/mole
$CH_4(g)$	−17.9	$H_2O(g)$	−57.8
$C_2H_2(g)$	54.2	$H_2O(l)$	−68.4
$C_2H_6(g)$	−20.4	$HCl(g)$	−22.0
$CCl_4(l)$	−33.4	$SO_3(g)$	−94.5
$NH_3(g)$	−11.0	$HCl(aq)^a$	−39.6
$NO(g)$	21.5	$NaCl(aq)$	−97.3
$NO_2(g)$	8.1	$NaOH(aq)$	−112
$CO(g)$	−26.6	$H_2SO_4(aq)$	−217
$CO_2(g)$	−94.1	$Fe_2O_3(s)$	−197

^{a} (aq) denotes a very dilute aqueous solution; (g), (l), and (s) stand for gas, liquid, and solid.

Enthalpies are also known for gaseous atoms of the elements. With these values enthalpy changes attending the breaking of bonds in gaseous molecules to form gaseous atoms can be calculated; ΔH values so obtained are called *bond energies* (see Table 13-2). All pure elements at standard conditions are assigned zero enthalpy.

Calculating enthalpies of reaction

A second rule of thermochemistry, called the *law of constant heat summation*, was discovered in 1840 by Russian chemist G. Hess: *The heat liberated (or absorbed) at constant pressure in a chemical process is the same whether reaction takes place in one or several steps.* In other words, when initial and final states of a system are fixed, the overall enthalpy change is fixed.

The thermochemical scheme given in Figure 12-4 is an illustration of Hess' law. At standard conditions, when 2 moles of liquid water is formed from the elements (route 1), 136.8 kcal of heat is liberated. When 2 moles of steam is formed instead (route 2), 115.6 kcal is liberated. When the steam is condensed to liquid water (route 3), an additional 21.2 kcal is evolved. A combination of routes 2 and 3 gives a total of $115.6 + 21.2 = 136.8$ kcal. Thus by either route 1 or routes $2 + 3$ the same quantity of heat is produced.

Hess' law permits arithmetic combination of enthalpies of formation to arrive at enthalpies of reaction. The procedure is as follows:

1. Write the balanced chemical equation.
2. Insert the table value enthalpy of each substance and multiply it by its equation coefficient.
3. Add up the enthalpies on both sides of the equation. Subtract the sum of reactant enthalpies from the sum of product enthalpies to obtain the enthalpy of the reaction.

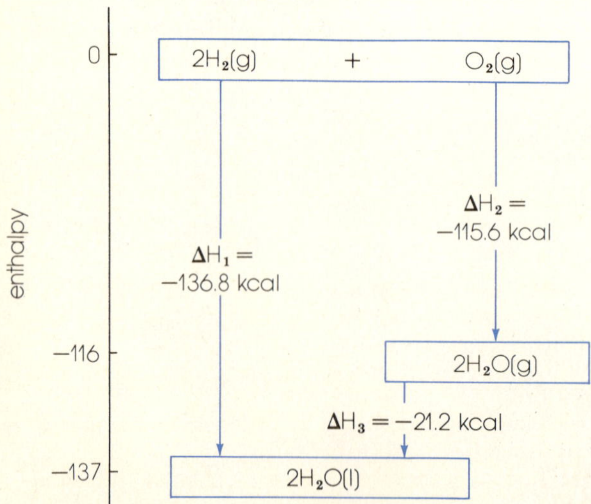

Figure 12-4 An enthalpy diagram. Two equivalent routes are described for the formation of liquid water from its elements.

Example 12-9 In the first step of nitric acid manufacture ammonia gas is oxidized to nitric oxide gas and steam. Assuming constant pressure, calculate the heat released per mole of ammonia used up.

Solution The balanced equation and standard enthalpies (kilocalories per mole) from Table 12-4 are

$$4NH_3(g) + 5O_2(g) = 4NO(g) + 6H_2O(g)$$
$$-11.0 \times 4 \qquad 0 \times 5 \qquad 21.5 \times 4 \qquad -57.8 \times 6$$

$$(12\text{-}12)$$

The total enthalpy of the 10 moles of products is

$$\Delta H^0\, products = 86.0 + (-347) = -261\ kcal$$

The total enthalpy of the 9 moles of reactants is

$$\Delta H^0\, reactants = -44.0 + 0 = -44.0\ kcal$$

The enthalpy of the reaction is

$$\Delta H^0\, reaction = \Delta H^0\, products - \Delta H^0\, reactants$$
$$= -261 - (-44.0) = -217\ kcal$$

$$(12\text{-}13)$$

The minus sign indicates the reaction is exothermic; 217 kcal is evolved when 4 moles (4×17 g) of NH_3 react. Reaction of 1 mole of NH_3 would evolve $217/4 = 54.2$ kcal.

The reaction is a combustion and the answer is called an *enthalpy of combustion* (of ammonia).

Summary

The principles of thermochemistry are important to both theoretical and applied work. In industrial processes in which tons of chemicals may be in a reaction vessel there is obvious need for economical utilization of energy and for equipment appropriate to the quantities of heat involved. In the research laboratory much fundamental understanding of chemical reactivity, bonding, and molecular structure has come from studying the relation of heat energy to chemical change.

In the nineteenth century emphasis was on understanding chemistry through the properties of relatively massive quantities of matter. This century interest has shifted toward seeking explanations based on the energetics of molecules. Among recent advances has been the use of energetics principles to interpret biochemical processes such as oxidation reactions in living cells.

QUESTIONS

1. Mass conservation Why does Einstein's $E = m_0c^2$ (Eq. 4-8) not invalidate the conservation of mass law as applied to chemical reactions?

2. Numbers of particles Avogadro's hypothesis gave a means for measuring out gas samples containing the same number of molecules. Dulong and Petit's rule gave a means for measuring out solid samples containing the same number of atoms. Explain.

3. Heat capacity In 1819 Dulong summarized the result of his work with Petit with this statement, "Atoms of all simple substances have the same capacity for heat." Explain.

4. Dulong and Petit Consider Table 12-3 data. **(a)** What is the shape of a plot of specific heat (ordinant) versus atomic weight? Explain. **(b)** What is the shape of a plot of molar heat capacity versus atomic weight? Explain.

5. Atomic weight Suppose you are given a sample of an unknown metal and asked to determine by simple means its approximate atomic weight. Describe the steps you would take.

6. Atomic weights Twenty-one elements occur on earth in only one isotopic form. Their atomic weights may gradually change as material brought back from space exploration adulterates the environment. Explain.

7. Composition **(a)** Various samples of lead oxide can have different molecular weights. Explain. **(b)** Does that fact upset the law of definite proportions? Explain.

8. Moles Tetranitrogen tetrasulfide is prepared by the reaction,

$$10S + 4NH_3 = N_4S_4 + 6H_2S$$

If 20 moles of N_4S_4 is prepared, tell the numbers of moles of the other substances involved. Explain how the answers are obtained by merely inspecting the equation.

9. Definitions What is the difference between **(a)** an equivalent weight of an element and its atomic weight? **(b)** an empirical formula and a molecular formula? **(c)** heat of reaction and enthalpy of reaction as indicated in Eqs. 12-9 and 12-11?

10. Isomorphism *The rule of isomorphism* (Gk. equally shaped) was discovered in 1812 by English scientist W. Wollaston. Its statement that *compounds with similar formulas can be expected to have similar crystal form* aided early chemists in deducing molecular formulas. Chemical analysis proves a hydrate contains sodium, ferric, and sulfate ions. Examination of its crystals shows they are isomorphous with alum crystals. What is its probable formula?

11. Enthalpy For $AgNO_3$ (aq) and $AgNO_3(s)$, ΔH^0 values are -24.1 and -29.4 kcal/mole, respectively. Several hundred pounds of silver nitrate is to be dissolved daily in a tank of room-temperature water for use at 25°C. Should the tank be provided with means for heating or cooling? Explain.

12. Molecular weight In 1843 French chemist C. Gerhardt proposed that molecular weight be defined as, " . . . that gram weight of a substance

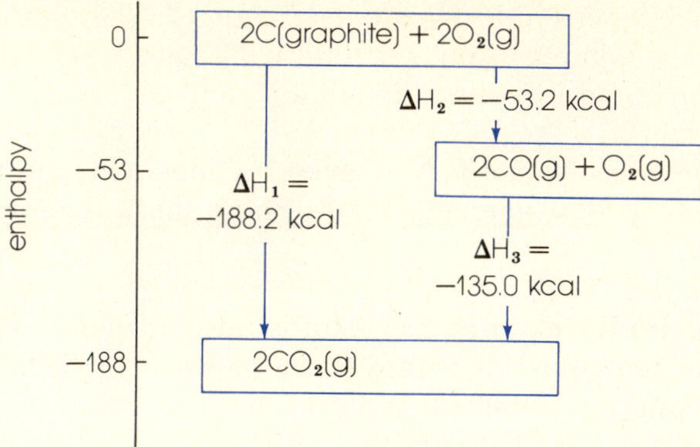

Figure 12-5

which, in the vapor state, occupies the same volume as 2.0 g of hydrogen."
Is his definition still useful? Explain.

13. Molecular weight A liter of fluorine gas weighs 19 times as much as a liter of hydrogen gas under identical conditions. Explain by the reasoning of Cannizzaro and by Avogadro's law how this simple laboratory determination leads to the molecular and atomic weights of fluorine.

14. Hess' law **(a)** Rephrase the text statement of Hess' law in your own words. **(b)** Use it as an aid in fully explaining the scheme of Figure 12-5 (at standard conditions).

15. Molecular formula Analysis proves that phenanthrene, a hydrocarbon found in coal, is 94.4 percent carbon and 5.6 percent hydrogen by weight. **(a)** Is any other information needed before its simplest formula can be determined? Explain. **(b)** The simplest formula is C_7H_5. What data are needed in order to obtain the correct molecular formula (which is $C_{14}H_{10}$)?

PROBLEMS

16. Constant composition In discussing the constancy of composition of minerals taken from various localities, Proust wrote, "We find in the bosom of the earth copper oxide containing 20 percent of oxygen, arsenic with 35, lead with 7, antimony with 16. . . ." Calculate the empirical (simplest) formula for each of these oxides.

17. Atom ratios Three replicate experiments are run in which copper combines with chlorine:

Experiment	Copper	Copper chloride
1	0.375 g	0.795 g
2	0.455	0.965
3	0.415	0.880

For each sample of copper chloride calculate **(a)** grams of chlorine combined; **(b)** percentage composition; **(c)** the Cu:Cl ratio. **(d)** What law of stoichiometry is illustrated in your calculations?

18. Multiple proportions Nickel forms several oxides. One contains 71.0 percent nickel, another 78.6 percent nickel. Show these data suggest the law of multiple proportions.

19. Atomic weights **(a)** If a chemist has an older atomic weight table based on $O = 16.00000$, he can adjust its values to the newer $^{12}C = 12.00000$ scale by dividing each value by 1.000043. What is the new value of oxygen to seven significant figures? **(b)** Bromine in nature consists of 50.54 percent ^{79}Br, isotopic weight 78.92; and 49.46 percent ^{81}Br, isotopic weight 80.92. What number should appear in an atomic weight table for bromine?

20. Atomic weight Richards quantitatively converted a sample of pure silver weighing 0.5394 g to pure AgCl weighing 0.7167 g. Given 107.88 as the atomic weight of silver, calculate to the proper number of significant figures **(a)** the Cl:Ag weight ratio in AgCl, and **(b)** the atomic weight of chlorine. **(c)** In another experiment he converted 0.7192 g of pure zinc to 1.4992 g of pure $ZnCl_2$. Use the atomic weight of chlorine found in (b) to calculate the atomic weight of zinc.

21. Equivalent weight When 300 mg of lanthanum metal (at. wt. 139) is heated with oxygen, the result is 351 mg of oxide. Calculate **(a)** the weight of oxygen in the oxide; **(b)** the weight of lanthanum that would combine with 8.00 g of oxygen (the equivalent weight of lanthanum); **(c)** the oxide's simplest formula.

22. Molecular weights Three isotopes are known for both hydrogen and oxygen. **(a)** Write formulas for all the isotopically different kinds of "water." **(b)** Calculate the molecular weight of each to four significant figures given, $^{1}H = 1.008$, $^{2}H = 2.014$, $^{3}H = 3.016$, $^{16}O = 16.00$, $^{17}O = 17.00$, $^{18}O = 18.00$. Arrange the results in order of increasing molecular weight. **(c)** Circle the most common one.

23. Molecular formula A sample of nicotine is isolated from tobacco. A molecular weight determination gives a rough value of 160. Elemental analysis gives the following percentages by weight: C, 74.1 percent; H, 8.65 percent; N, 17.3 percent. Calculate the simplest formula and molecular formula of nicotine.

24. Molecular formula When 300 mg of a certain metal M is heated with chlorine, 504 mg of a white chloride is formed. **(a)** Assuming the chloride formula may be MCl, MCl_2, MCl_3, or MCl_4, and given the atomic weight of Cl as 35.5, calculate four different possible atomic weights for M. **(b)** Later analysis identifies M as gadolinium. Which is the correct formula for the compound? **(c)** What is the oxidation number of gadolinium?

25. Percentage composition S,S,S-Tributyl phosphorotrithioite is a chemical used to defoliate cotton to facilitate machine picking. Calculate its percentage composition, $P[S(CH_2)_3CH_3]_3$.

26. Percentage composition The following are compounds once used as weed-control agents. Calculate the percentage of metal in each:

$$Mg(ClO_3)_2 \cdot 6H_2O \qquad HO-As \overset{O}{\underset{CH_3}{\overset{\|}{<}}} \overset{CH_3}{} \qquad C_6H_5-Hg-O-\overset{O}{\overset{\|}{C}}-CH_3$$

| magnesium chlorate hexahydrate | dimethyl arsinic acid | phenylmercuric acetate |

27. Molecular formula Methyl parathion is the common name for one of the powerful insecticides related to nerve gases developed during World War II. **(a)** From its analysis derive the empirical formula: 36.6 percent C, 3.80 percent H, 30.4 percent O, 11.8 percent P, 12.2 percent S, 5.32 percent N. **(b)** The molecular weight is about 260. What is the molecular formula?

28. Hydrate formula A barium chloride hydrate, $BaCl_2 \cdot xH_2O$, weighing 674 mg is heated to drive off the water. The residue of anhydrous $BaCl_2$ weighs 573 mg. Calculate **(a)** the milligram weight of hydrate water; **(b)** the millimoles of $BaCl_2$ and of H_2O; **(c)** the simplest whole number ratio of millimoles of $BaCl_2$ and of H_2O; **(d)** the simplest formula for the hydrate.

29. Reaction yield Photographic hypo (mol. wt. 248) can be prepared by the reaction,

$$2Na_2S + 4SO_2 + Na_2CO_3 + 15H_2O = 3Na_2S_2O_3 \cdot 5H_2O + CO_2$$

How many moles of hypo can be made from **(a)** 100 moles of sodium sulfide? **(b)** 640 g of sulfur dioxide?

30. Reaction yield Tetraethyl lead (mol. wt. 323) is synthesized by the reaction,

$$NaPb + C_2H_5Cl \longrightarrow Pb(C_2H_5)_4 + NaCl + Pb$$

(a) Balance the equation. **(b)** Theoretically, to prepare 0.0400 mole of tetraethyl lead, how many moles of ethyl chloride are needed? **(c)** How many grams of ethyl chloride (mol. wt. 64.5) is this? **(d)** If one used that weight of ethyl chloride and obtained 11.0 g of tetraethyl lead, express the actual yield in moles. **(e)** What is the percentage yield?

31. Gravimetric analysis A sample of fertilizer weighing 1.250 g is dissolved and analyzed gravimetrically for its potassium content by converting potassium to insoluble $KClO_4$ (mol. wt. 138) that weighs 0.305 g. **(a)** How many moles of $KClO_4$ is this? **(b)** How many moles of potassium were in the sample? **(c)** How many grams of potassium were in the sample? **(d)** Calculate the percent potassium in the fertilizer.

32. Gravimetric analysis A sample of detergent weighing 0.500 g is treated to convert all phosphorus to $Mg_2P_2O_7$ (mol. wt. 223) weighing 0.150 g. Calculate **(a)** the moles of $Mg_2P_2O_7$; **(b)** the moles and grams of phosphorus in the sample; **(c)** the percent phosphorus in the sample.

33. Avogadro's number Millikan obtained the value 1.602×10^{-19} coulomb for the charge on the electron. One mole of electrons will displace 1 equivalent weight of an element during electrolysis. In the electrolysis of water 1.008 g of hydrogen is displaced by the passage if 96,494 coulombs of electricity. Show how to get a value for Avogadro's number from these facts.

34. Avogadro's number Rutherford and Boltwood found that during a period of a year 2.57×10^{20} α particles were emitted by a sample of radium and formed, at standard conditions, 9.56 ml of helium gas ($\alpha + 2e^- \rightarrow$ He). Use the data to calculate Avogadro's number.

35. Avogadro's number The composition of a certain planet is the equivalent of Avogadro's number of 20-lb rocks. How does it compare in mass to the earth, whose mass is 6.59×10^{21} tons?

36. Specific heat **(a)** A sample of pure tin weighing 0.713 g is converted to an oxide weighing 0.905 g. Calculate the precise combining (equivalent) weight of tin. **(b)** The specific heat of tin is found in another experiment to be 0.0541 cal/g °C. Calculate the approximate and precise atomic weights of tin.

37. Specific heat Elements X and Y are chemically somewhat similar to platinum. Their specific heats are 0.0323 and 0.0586 cal/g °C. Calculate the approximate atomic weights. Name the probable elements.

38. Enthalpy A sample of benzene, C_6H_6, weighing 0.780 g is burned to $CO_2(g)$ and H_2O (l) in an oxygen-bomb calorimeter (Figure 12-3). Bomb and water absorb 7820 cal. Calculate the enthalpy of combustion of benzene in kilocalories per mole.

39. Enthalpy **(a)** Calculate the enthalpy of combustion for **(a)** CH_4 (methane) and **(b)** C_2H_2 (acetylene), assuming in each case that the products are gaseous CO_2 and liquid H_2O. **(c)** Calculate the enthalpy of neutralization for

$$HCl(aq) + NaOH(aq) = NaCl(aq) + H_2O(l)$$

40. Enthalpy Calculate the enthalpy of **(a)** the reduction of iron ore,

$$Fe_2O_3(s) + 3CO(g) = 3CO_2(g) + 2Fe(s)$$

(b) the hydrogenation of acetylene to ethane,

$$C_2H_2(g) + 2H_2(g) = C_2H_6(g)$$

(c) the chlorination of methane to carbon tetrachloride,

$$CH_4(g) + 4Cl_2(g) = CCl_4(l) + 4HCl(g)$$

FIVE

WHAT ARE GASES?

A portion of the earth's atmosphere, as seen from the moon. The earth is half in daylight, half in darkness. The photo was taken in natural far-ultraviolet illumination. The camera was equipped with a filter that blocked the glow caused by atomic hydrogen but transmitted glow caused by atomic oxygen and molecular nitrogen. Such pictures prove that photodecomposition of water vapor in the upper atmosphere is a major source of the earth's supply of free oxygen. (Apollo 16 mission, National Aeronautics and Space Administration.)

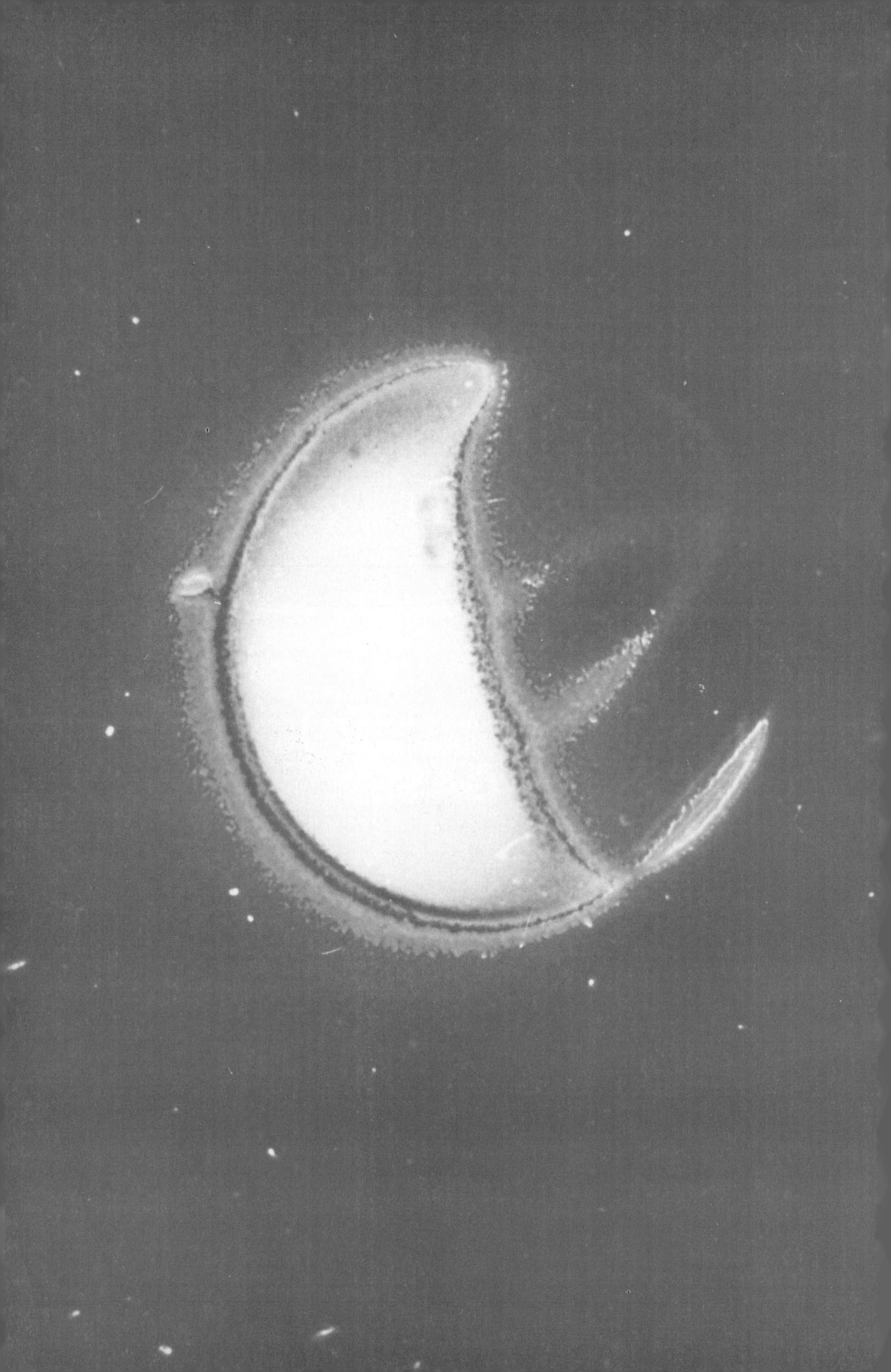

WHAT IS THE ATMOSPHERE?

THE IMPORTANT CONCEPTS

13-1. Origin and evolution of the atmosphere
1. The earth developed an atmosphere through volcanism, gas reactions, and photosynthesis.

13-2. Atmospheric research
1. The atmosphere's mass can be calculated from a single barometric reading.
2. Combustion, originally explained by the phlogiston theory, was correctly explained after oxygen was discovered.
3. One can make elementary gases by simple laboratory procedures.
4. Most of the noble gases were discovered in the atmosphere.
5. The atmosphere has been divided for study into spheres, separated by pauses.
    a. Ozone and ions are constituents of the upper atmosphere.

13-3. Photochemical air pollution
  1. Ultraviolet and visible light can break covalent bonds.
  2. Hydrocarbons, nitrogen oxides, and sunlight give free radical reactions whose products are photochemical smog.
  3. Infrared spectrometry, which measures molecular vibrations, can be used to study air pollution.

13-4. The chemistry of individual atmospheric contaminants
  1. Aerosols come from many sources.
     a. They may be studied under the electron microscope.
     b. They can be trapped with an electrostatic precipitator.
  2. Carbon dioxide, CO_2, is part of nature's carbon cycle.
  3. Combustion processes will never use up our oxygen supply.
  4. Atmospheric CO_2, H_2O, and O_3 cause the greenhouse effect.
  5. Carbon monoxide, CO, comes mainly from the air oxidation of methane produced by organic decay; a natural balance maintains its concentration.
  6. Carbon monoxide adversely affects humans in several ways.
  7. Sulfur dioxide, SO_2, comes partly from coal combustion.
  8. In cities NO_x comes from combustion processes.
  9. Biological systems can be affected by NO_x at the molecular level.
  10. Lead levels in the environment can be estimated back to prehistoric times.

13-5. Air contamination and solid wastes
  1. Solid waste disposal schemes include incineration and conversion to aggregate and fuels.
  2. Any change made on the environment creates a change somewhere else.

An envelope of gases surrounds our planet. Held by gravity, it is concentrated below an altitude of 20 miles. Though churned by heating and cooling, it maintains a layered structure that makes radio communications and other phenomena possible. Its oxygen burns up meteors. As its atoms absorb high-energy radiation from space, they undergo reactions among themselves and with pollutants swept up from the planet's surface. By distributing heat and water through great cycles of evaporation and condensation, the gas envelope regulates temperature and controls weather. This is the air we breathe. Without it there might be no life on earth.

13-1. ORIGIN AND EVOLUTION OF THE ATMOSPHERE

The model we will use to discuss the development of the atmosphere is presented in Figure 13-1.

As in Chapter 5 we will assume that the earth formed from dust and debris, and that it began with an atmosphere (Atmosphere I) which was lost due to lack of gravitational attraction of a large solid body.

Once the planet became consolidated, partial melting started the volcanic action that continues today. Water and other volatiles held chemically and physically in rock then came out of the interior. A typical (Hawaiian) volcano today puts out a vapor that is, by volume, about 85 percent H_2O, 7 percent CO_2, 3 percent SO_2, 2 percent N_2, and has small amounts of HF, HCl, H_2, S, CH_4, CO, H_3BO_3, and NH_4Cl. In the early earth, free iron was more common than it is now. Taking oxygen from its surroundings, it left a gas rich in hydrogen compounds. For instance,

$$H_2O + Fe \overset{\Delta}{=} FeO + H_2$$

$$H_2 + S \overset{\Delta}{=} H_2S$$

$$H_2S + FeCl_2 \overset{\Delta}{=} FeS + 2HCl \tag{13-1}$$

When enough water vapor accumulated to fall as rain, the acidic gases (H_2S, HCl) were washed from the sky, leaving an atmosphere that could have contained H_2 and CH_4 (Atmosphere II).

Atmospheric oxygen is believed to have come from two chemical reactions. Photography from the moon (see the figure on page 289) proves that one source of oxygen is the decomposition of water vapor by solar ultraviolet rays. As explained in Chapter 22 the other source is *photosynthesis* by green plants, the process in which carbon dioxide and water are converted into sugars and starches:

$$xCO_2 + xH_2O \xrightarrow{sun} (CH_2O)_x + xO_2 \tag{13-2}$$

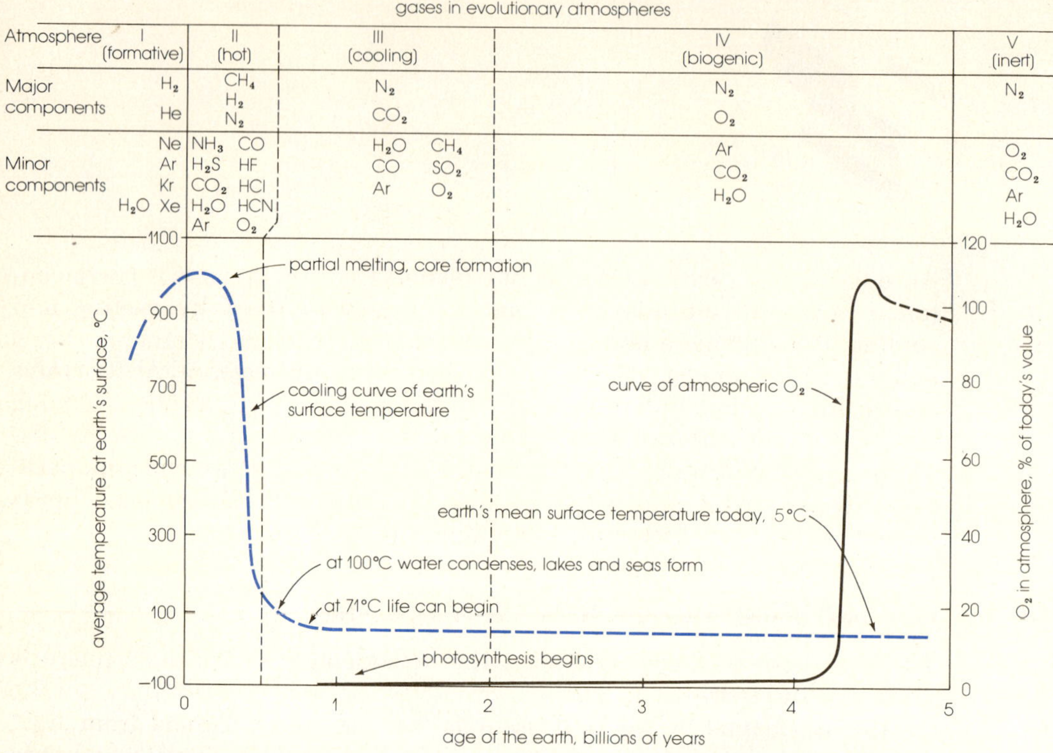

Figure 13-1 **Possible evolution of the atmosphere. Appearance of oxygen and an assumed cooling curve of the earth's surface are plotted in relation to atmospheric compositional changes. Oxygen content increased as oxidizable substances were used up and green plant life grew.**

As oxygen oxidized hydrogen species, Atmosphere II became Atmosphere III. In lakes large quantities of ferrous iron were oxidized to the ferric iron mined today.

When the atmosphere's oxygen content reached 1 percent of today's value, its energy-absorbing capacity was great enough to block out lethal ultraviolet rays, and the planet's surface became hospitable for life. With life forms emerging from the water to proliferate on land, oxygen concentration increased toward the value known today (Atmosphere IV).

How will the atmosphere change further with time? If only natural evolution operates, we could be moving toward Atmosphere V. Crustal rock still contains 15 to 20 times as much nitrogen as is in the air, and N_2 will gradually be freed by erosion and denitrifying bacteria. In the still more distant future, several billion years hence when the sun becomes a red giant and heating evaporates the surface water, the atmosphere will be mostly steam, at a few hundred times the pressure of the present atmosphere.

13-2. ATMOSPHERIC RESEARCH

Investigations by pneumatic chemists

Serious inquiry about the atmosphere and gases in general began around 1600 as the influence of alchemy waned. First of the "pneumatic chemists"

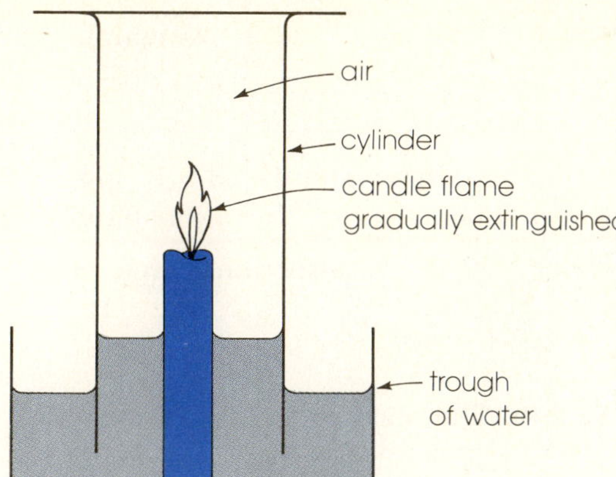

Figure 13-2 One of Van Helmont's experiments. He demonstrated that some of the air is used up as the candle burns, and water rises in the cylinder to take its place. The gas remaining will extinguish a flame and suffocate a mouse.

was Belgian physician J. Van Helmont who coined the word *gas* (Gk. *chaos*, chaos). He claimed to be the inventor of gas because he studied the air (Figure 13-2) and found how to prepare gases (CO_2, NO_2, SO_2, Cl_2) by the action of acids on various substances.

In 1643 Italian physicist E. Torricelli, one of Galileo's students, invented the *barometer* (Figure 13-3), the first instrument for measuring air pressure. Noting approximately 34 feet was the maximum lift of water by suction pumps, he theorized that, "We live immersed at the bottom of a sea of elemental air . . . " (and the pressure of the sea can support a 34-foot column of water). Using mercury, which is 13.5 times more dense than water, he correctly predicted that the air would support a mercury column $34/13.5 = 2.5$ feet high, or about 30 inches.

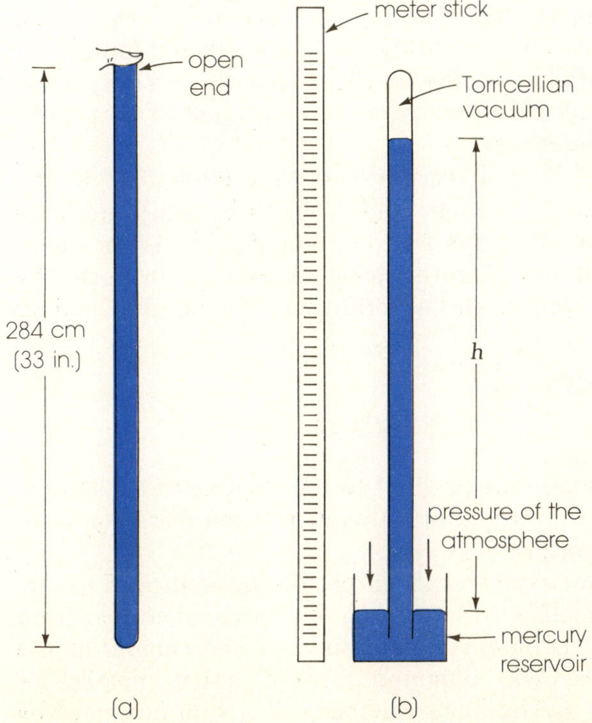

Figure 13-3 The barometer. (a) A thick-walled glass tube is filled with mercury. (b) The open end is held shut until the tube is inverted in a cup of mercury. Standing free, the mercury column then drops a few centimeters, creating a vacuum at its top. Length h is the height of mercury column supported by the pressure of the atmosphere. Pressure is read as a length on the meter stick. Average barometric pressure near sea level is about 760 mm of mercury. This is simply called 760 mm or 760 torr. The torr is a unit named after the barometer's inventor.

As shown in Example 13-1 the barometer permitted a "weighing" of the atmosphere.

Example 13-1 At sea level a water barometer stands 33.6 ft high. Show this leads to determining the total mass of the atmosphere.

Solution Consider a column of water 1 ft² in cross-sectional area and 33.6 ft high. Its volume is

$$(1\ ft^2)(33.6\ ft) = 33.6\ ft^3$$

One cubic foot of water weighs 62.4 lb. The mass of water in the column is

$$(33.6\ ft^3)(62.4\ lb/ft^3) = 2090\ lb$$

On each square foot of the earth's surface, therefore, rests 2090 lb or 1.045 ton of air. The planet's diameter is about 7900 miles. Its surface area is 1.97×10^8 mi², or 5.48×10^{15} ft². Total mass of the atmosphere is thus

$$(1.045\ ton/ft^2)(5.48 \times 10^{15}\ ft^2) = 5.73 \times 10^{15}\ tons$$

Torricelli's theory of the atmosphere was soon validated by B. Pascal, French mathematician who had invented the hydraulic press and discovered the principle of the calculating machine. He had barometers carried up a mountain for the reading of pressures at various heights. As expected for a sea of air, he found that pressure decreased with altitude.

Atmospheric experiments carried aloft by kites began in 1749 when Scottish scientist A. Wilson sent up recording thermometers and found that air temperature decreased with height. Because his balloons rose only a few thousand feet, Wilson concluded that temperature continued to drop until it reached the deep cold of outer space.

D. Cavendish, a wealthy English eccentric with a laboratory in his home, was the first to analyze air brought back to earth by sampling traps attached to balloons. Manipulating gas samples via liquid displacement over troughs of water or mercury (Figure 13-4), he obtained nitrogen by passing air through heated charcoal and absorbing the "fixed air" (carbon dioxide) in aqueous alkali:

$$N_2 + O_2 + C \overset{\Delta}{=} N_2 + CO_2$$

$$N_2 + CO_2 + KOH = N_2 + KHCO_3$$

By comparing the volume of gaseous product to the volume of air sample taken, he arrived at excellent oxygen and nitrogen volume percentages in the air, 20.83 and 79.17 percent, respectively.

Cavendish also carried out experiments in producing hydrogen gas by the action of acids on active metals. Not knowing the chemical composition of acids, he assumed that the "inflammable air" he collected came from the metals because the same gas was obtained from different metals. He thought this gas might be the mysterious substance *phlogiston* proposed by

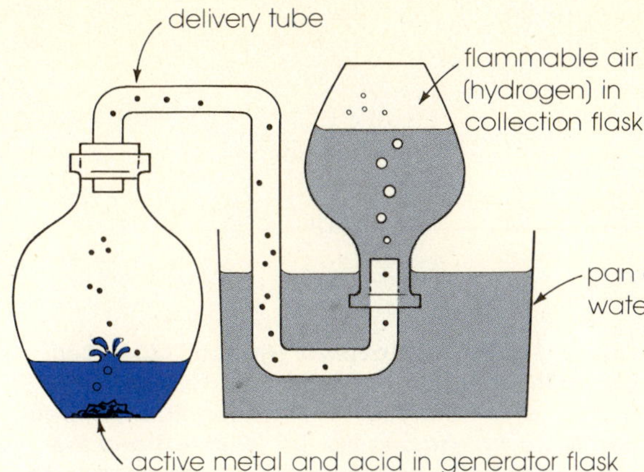

Figure 13-4 Method used by Cavendish, Priestly, and others to collect gases. The collection bottle is filled with water and inverted over the end of the delivery tube under water in the trough. Metal is added to the acid, the delivery tube quickly inserted in the generator, and the hydrogen gas produced displaces its volume of water in the collector. For water-soluble gases like HCl the investigators used mercury in place of water.

German alchemists to account for combustibility. According to their theory combustible substances and metals contained phlogiston (ϕ) which was lost on burning and which was restored when an oxide was heated with a phlogiston-rich material like charcoal:

$$iron\,(\phi) \xrightarrow[air]{\Delta} iron\ oxide\ +\ \phi$$

$$iron\ oxide\ +\ charcoal\,(\phi) \xrightarrow{\Delta} iron\,(\phi)$$

Oxygen (Gk. *oxy-*, sharp or acid) in the air was discovered by C. Scheele (Sweden) in 1770, and independently by J. Priestley (England) in 1774. When Sheele allowed a sample of air to stand several days with wet iron filings in a sealed bottle, he found the iron rusted, approximately a fifth of the air was used up, and a candle would not burn in the remaining air. When he used a chemical balance to compare the weight of remaining air with an equal volume of ordinary air, he found the former slightly lighter. Repeating the experiment with phosphorus in place of iron, he obtained similar results. He concluded that ordinary air was composed of two gases: "foul air" (nitrogen) and "fire air" (oxygen). We now know that both iron and phosphorus combine with oxygen leaving inert nitrogen, and that nitrogen is less dense than oxygen.

Priestley prepared oxygen from both sodium nitrate and mercuric oxide:

$$2NaNO_3 \overset{\Delta}{=} 2NaNO_2 + O_2$$

$$2HgO \overset{\Delta}{=} 2Hg + O_2 \tag{13-3}$$

Feeling exhilarated when he breathed the gas, and finding a candle burned brilliantly in it, he wrote that he had found an air five times as good as the atmosphere.

The correct interpretation of combustion was made by M. Lomonosov (Russia) in 1756, and independently by Lavoisier (Figure 13-5) in 1776. Their work discredited the phlogiston theory and marked the end of alchemy. Lavoisier proved that air consisted of oxygen and nitrogen, and that metals heated with air gained weight equal to the weight of the oxygen taken up. In other important gas experiments he explained the utilization of oxygen in animals as biological oxidation.

Figure 13-5 Antoine Lavoisier (France 1743–1794), one of the founders of chemistry. Lavoisier was a theorist who looked for laws in quantitative experiments. He was among the first to understand the difference between elements and compounds, to use uniform chemical nomenclature, and to publish results free of alchemy's misconceptions. A member of the aristocracy and of a tax collection bureau, he was arrested after the French Revolution for conspiracy and guillotined. (Edgar Fahs Smith Memorial Collection, University of Pennsylvania.)

Discovery of the noble gases

During most of the nineteenth century, little progress was reported in air analysis. Then in 1892 Lord Rayleigh (born John Strutt, England 1842–1919, Nobel prize in physics 1904) found another gas in the atmosphere, and that led to the finding of four more.

Rayleigh first set out to determine accurately the density of oxygen. A careful man, he prepared the gas by several methods:

$$2NaClO_3 \overset{\Delta}{=} 2NaCl + 3O_2$$

$$2H_2O = 2H_2 + O_2 \qquad \textit{(electrolysis at Pt electrodes)}$$

Within the limits of experimental error he found all O_2 samples had the same density. When he repeated the measurements on samples of nitrogen prepared by the following methods, however,

$$\textit{air (free of } H_2O, CO_2) + Cu \overset{\Delta}{\longrightarrow} CuO + N_2$$

$$4NH_3 + 3O_2 = 6H_2O + 2N_2 \tag{13-4}$$

he found something peculiar; N_2 derived from air was always slightly more dense than N_2 from a pure chemical source. His first thought was that air contained a nitrogen allotrope, N_3, corresponding to oxygen's allotrope ozone, O_3. He asked chemist William Ramsay (England 1852–1916, Nobel prize in chemistry 1904) to help him find it.

Rayleigh suggested adapting a method described a century before by Cavendish for removing the principal gases so the remainder could be examined. Enriching air with oxygen he electrically sparked the mixture and absorbed the resulting nitrogen oxides in sodium hydroxide solution (Figure 13-6):

$$N_2 + 2O_2 = 2NO_2$$

$$2NO_2 + 2NaOH = NaNO_2 + NaNO_3 + H_2O$$

About 1 percent of the original air volume remained after treatment. Examining it with a spectrometer (Figure 2-3), the two men found several unclassified spectral lines indicative of an undiscovered element. They confirmed their find by preparing a larger sample of the gas and determining its properties. From heat capacity measurements, they found the element to be monatomic, unlike any normal gas then known except mercury vapor. From density measurement, they established an atomic weight of 38.17. From unsuccessful attempts at combining it with other elements, they declared it chemically inert and named it *argon* ("the lazy one").

Ramsay thought argon might be but one of several unknown inert gases and suggested adding a column (Group 0) for their family in the periodic table between the halogens and alkali metals. While wondering where he might look for them, he learned that a similar unreactive gas had been isolated during treatment of a uranium mineral with acid. Repeating the experiment, he obtained a mixture of nitrogen and a lighter gas having argonlike properties. The prominent yellow line in the latter's spectrum matched a line observed in 1868 by Frenchman P. Janssen, first astronomer to use a spectroscope to study the sun's chromosphere (gas envelope) during an eclipse. The element was helium.

By the time Ramsay returned to air analysis in search of other inert gases, technology for the large-scale liquefaction of air had been worked out by J. Dewar (England) and K. von Linde (Germany). Now he had the experimentalist's luxury of starting with several liters of liquid air, material 600 times more concentrated than the gas he used before. From it he separated by distillation, and identified by spectroscopic means, krypton ("hidden"), neon ("new"), and xenon ("stranger"). When radon, the heaviest of the noble gases was discovered three years later as a product of radium decay, all members of the noble gas family were known.

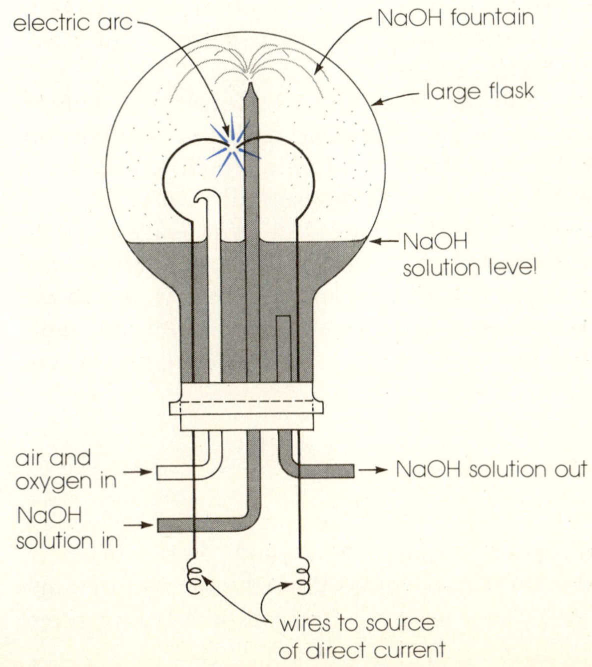

Figure 13-6 Rayleigh's equipment for air analysis. Nitrogen oxides formed at the electric arc are dissolved in the alkaline spray. As oxygen, nitrogen, and carbon dioxide are removed, the gas mixture becomes enriched in inert gases.

Table 13-1.   Composition of dry, clean air near sea level

Gas	Percent by volume	Gas	Percent by volume
Nitrogen, N_2	78.084	Nitrous oxide, N_2O	5×10^{-5}
Oxygen, O_2	20.9476	Carbon monoxide, CO	10^{-5}
Argon, Ar	0.934	Xenon, Xe	8.7×10^{-6}
Carbon dioxide, CO_2	3.14×10^{-2}	Sulfur dioxide, SO_2	$0 - 1 \times 10^{-4}$
Neon, Ne	1.82×10^{-3}	Ozone, O_3	$0 - 7 \times 10^{-6}$
Helium, He	5.24×10^{-4}	Nitrogen dioxide, NO_2	$0 - 2 \times 10^{-6}$
Methane, CH_4	2.0×10^{-4}	Iodine, I_2	$0 - 1 \times 10^{-6}$
Krypton, Kr	1.14×10^{-4}	Ammonia, NH_3	$0 - $ trace
Hydrogen, H_2	5×10^{-5}	Hydrogen peroxide, H_2O_2	$0 - $ trace

Composition of the lower atmosphere

A modern analysis of clean air near sea level is given in Table 13-1. In one
way or another all the gases originated in the earth. Argon-40 is the product
of radioactive potassium-40 decay. Helium is the result of α emission from
uranium and thorium minerals. Gases containing oxygen in chemical combination come from oxidation processes. Methane comes from gas wells
and anaerobic processes going on in marsh lands, rice fields, rubbish fills,
and in the intestinal tracts of animals, particularly cattle.

The upper atmosphere

High-altitude research has proved that in cross section the atmosphere is a
series of overlapping layers (called *spheres*) separated by boundaries
(called *pauses*). In ascending order from the earth's surface, out to an altitude of about 500 miles, the four main layers are the troposphere, the
stratosphere, the mesosphere, and the ionosphere. Each is defined as having
some differentiating property such as a particular range of temperatures.
There is also some stratification by chemical composition with height.
Above the normal atmosphere are zones relatively rich in atomic or ionic
oxygen, helium, and hydrogen (Figure 13-7).

Upper-air investigation started in the 1890s when L. Teisserenc de Bort
(France) discovered the stratosphere with unmanned balloons that carried
recording thermometers and other instruments 10 miles high. In this cold
bottom of the outer air he found an average temperature of $-50°C$.

The stratosphere is noted as the region where O_3 is synthesized. By its
powerful absorption of solar ultraviolet rays, ozone helps both to shield the
earth and to maintain the atmospheric heat balance. Ozone is produced
when ultraviolet light splits molecular oxygen, and oxygen atoms and molecules collide with some third body M (such as a nitrogen molecule),
which removes excess energy:

$$O_2 + h\nu = O + O$$

$$O_2 + O \overset{M}{=} O_3 \tag{13-5}$$

When in 1901 Guglielmo Marconi (Italy 1874–1937, Nobel prize in
physics 1909) sent radio signals 2000 miles across the Atlantic despite popular belief that signals could not travel over the horizon, it was suggested

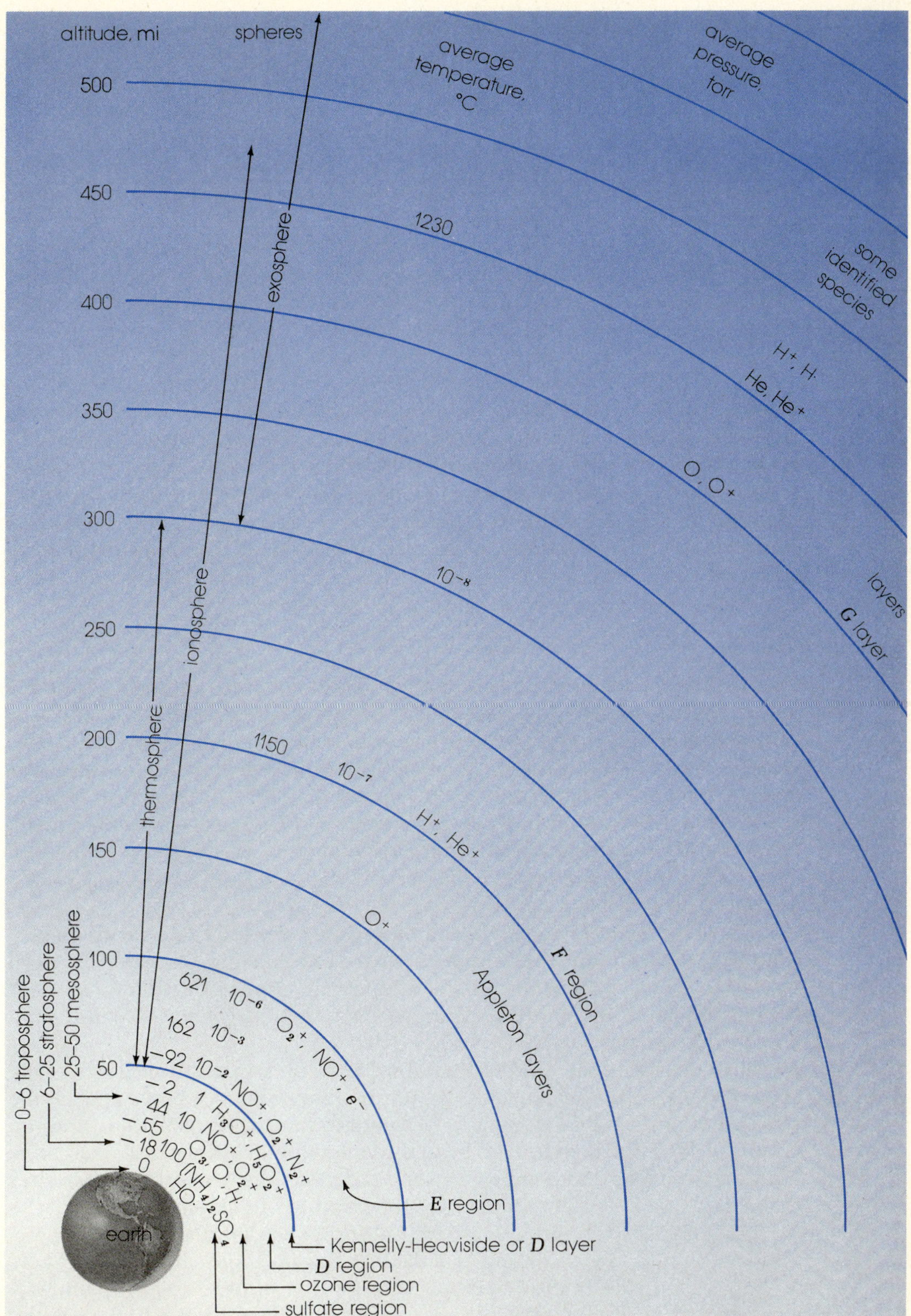

Figure 13-7 Structure and composition of the earth's atmosphere. Species present have been studied by the mass spectometer (Figure 4-4), carried aloft by rocket.

by English physicist O. Heaviside and American engineer A. Kennelly that signal reflection had occurred from a high layer of ionic gas. In the 1920s the Kennelly-Heaviside layer was confirmed by English physicist Edward Appleton (Nobel prize in physics 1947) and M. Barnett, who received radio-wave reflections from an ionic layer beginning at a height of about 60 miles. Using experimental techniques that foreshadowed the invention of radar 20 years later, they identified still higher bands both rich and poor in charged particles, including electrons.

High-altitude research began earnestly after World War II in both the United States and the Soviet Union. Since October 1957, when the Soviets put up the first artificial earth satellite, Sputnik I, more has been learned about the atmosphere than in all previous time.

As indicated by Figure 13-7, the chemical characteristics of the upper air are those of minor constituents and of the solar energy available for reactions at a given altitude. Ions are common. Also found are subatomic particles, which come from particle interactions and *cosmic rays.* Primary cosmic rays are approximately 91 percent protons, 8 percent α particles, and 1 percent a mixture in which electrons and heavy nuclei in the 50 to 60 atomic weight range predominate. Low-energy cosmic rays come from the sun. High-energy cosmic rays may originate in spinning neutron stars (pulsars).

13-3. PHOTOCHEMICAL AIR POLLUTION

The air has never been "clean." From three modern volcanic eruptions alone (Krakatoa 1883, Mt. Katami 1912, Hekla 1947) more gases and particles were added to the air than man has contributed throughout history. On an annual basis total emissions entering the atmosphere are estimated to be 10^{12} tons. Perhaps only 0.06 percent (6×10^8 tons) of this is man-made. That component, however, is a major contributor to reduced visibility, eye and lung irritation, and property damage.

Smog

For seven centuries air pollution in cities has been associated with smoke, soot, fly ash, and sulfurous fumes from the combustion of coal. The word "smog" was coined to describe a disaster in London, England, in 1952 when smoke and fog, undiluted for six days by the absence of wind, combined to cause 4000 deaths.

In the early 1940s a more insidious form of air pollution had already been recognized in Los Angeles, California. For want of a better term it too was eventually called "smog." It smelled faintly of gasoline, and on windless days was a yellowish haze through which landmarks looked blue. Eyes and mucous membranes became irritated, the walls of rubber tires developed cracks, and cellular damage was noted in green plants.

Los Angeles lies on the Southern California coastal plain where a meteorologist would not recommend concentrating millions of people (Figure 13-8). On one side is the Pacific Ocean. On the other side are mountains rising to 10,000 feet. Persistent winds blow landward from a high-pressure area that is fed cool air by the prevailing pattern of air circulation. Cool air is more dense than warm air and its sinking causes compressional heating

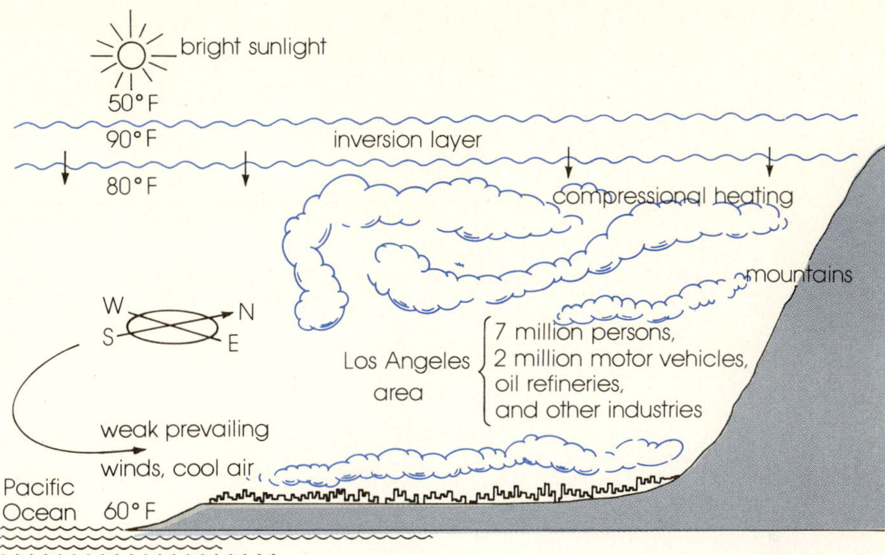

Figure 13-8 Profile of the Los Angeles basin, a classic example of conditions on a large scale for the buildup and nondispersal of polluted air.

of air layers beneath it. Unfortunately for Los Angeles, now the most notorious smog area in the world, the temperature of its surface air has also been lowered by passage over cool ocean water, and therefore it is slow to rise. The result, as pictured in Figure 13-9, is an *inversion* of the normal decrease in temperature with altitude. Inversion—warm air above, cool air below—confines pollutants near ground level by holding a blanket of warm air over the basin. More than 75,000 tons of fuel of all kinds is burned in Los Angeles daily. Mixed 2000:1 in air, the minimum dilution considered adequate for dispersal, the product gases would occupy 400 cubic miles. Under characteristic inversion, the volume available in the basin, however, is only 300 cubic miles. The result is a smog attack.

Initial identification of components in Los Angeles smog was made by California Institute of Technology biochemist A. Haagen-Smit. First he drew large volumes of air through paper filters to trap air-borne particles. Then he washed the filters with solvent and separated the dissolved substances by *liquid-solid chromatography* (*LSC*), a technique in which solution was allowed to trickle through a tube packed with an inert material such as powdered aluminum oxide. Dissolved components were adsorbed as bands different distances down the tube and later washed free for individual recovery. Chemical analysis proved oxygen-containing organic substances were in the air. Unable to identify specific compounds, Haagen-Smit sought to synthesize the mixture. Connecting in his mind organic compounds with automobiles and oil refineries, he tried to air-oxidize gasoline vapor in a flask. When the first experiments failed, he added ozone, a stronger oxidizer, and a haze resembling smog began to form. Exposed to sunlight it formed more rapidly, giving oxygenated products. Atmospheric reactions promoted by sunlight (*photochemical reactions*) were thus linked to smog.

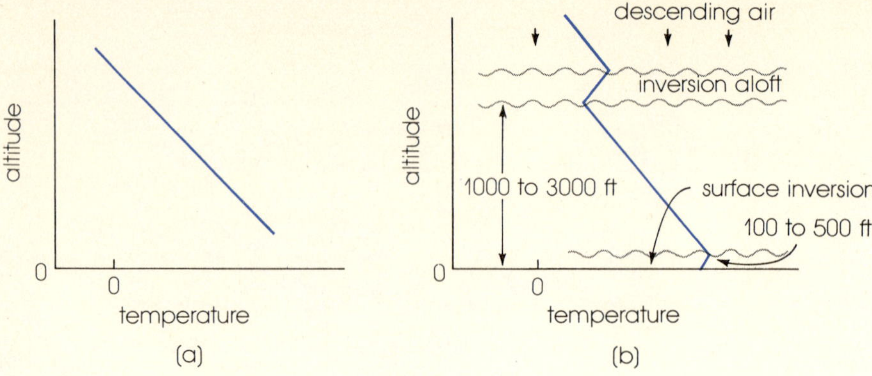

Figure 13-9 Lapse rate diagrams and inversion. (a) Normal cooling of the air with increasing altitude. (b) Abnormal conditions, with two kinds of inversion. Surface or radiation inversion, due only to night cooling of the air, is quickly dispersed by morning sun. Inversion aloft (subsidence inversion) is longer lasting. It prevents air circulation and limits air volume below it in which pollutants can be diluted. (c) Subsidence inversion at only 300 feet traps smog. (Photograph Courtesy Los Angeles Air Pollution Control District.)

Photochemical principles

Photochemistry is correlated by two principles. First, a reacting substance must absorb light energy. (If the energy is in the spectrum's visible region, the absorbing substance must appear colored to the eye.) Second, because light is quantized, each molecule activated by light must absorb at least 1 light quantum (1 photon).

In Chapter 3 we learned that when radiation wavelength is known, its energy can be calculated (Eq. 3-1). We now go a step further and say that if bond strengths in molecules are known (Table 13-2), we can calculate which bonds may break when radiation of a given wavelength is absorbed, and thereby predict the odd-electron fragments called *free radicals* that

Table 13-2. Bond energies[a]

Bond	Bond energy kcal/mole	Bond	Bond energy kcal/mole
C—C	83.1	O—H	110.6
C=C	147	S—H	81.1
C≡C	194	C—N	69.7
O—O	33.2	C—O	84.0
C—H	98.8	N—O	41.4
H—H	104.2	N—N	38.4

[a] Average energies for polyatomic molecules. Values are obtained by experiments that measure energies required to separate atoms.

might appear. Both ultraviolet and visible light contain enough energy to break single covalent bonds.

Example 13-2 The shortest radiation reaching the earth in quantity has a wavelength of 2900 Å (2.90×10^{-5} cm). If absorbed by a molecule containing a nitrogen–oxygen single bond, can it break the bond?

Solution Equation 3-1 is

$$E = h\nu = \frac{hc}{\lambda}$$

where E is energy, h is Planck's constant, ν is frequency, c is the speed of light, and λ is wavelength. Substituting with the proper units gives

$$E = \frac{(6.63 \times 10^{-27} \text{ erg sec})(3.00 \times 10^{10} \text{ cm/sec})}{2.90 \times 10^{-5} \text{ cm}} = 6.86 \times 10^{-12} \text{ erg/photon}$$

This energy applies to one bond. A mole (Avogadro's number) of bonds requires a mole of photons (called 1 einstein). To convert to kilocalories we need to know that 2.39×10^{-11} kcal = 1 erg. By factor-unit computation,

$$? \text{ kcal/mole} = \frac{6.86 \times 10^{-12} \text{ erg}}{\text{bond}} \left(\frac{6.02 \times 10^{23} \text{ bonds}}{\text{mole}} \right) \left(\frac{2.39 \times 10^{-11} \text{ kcal}}{\text{erg}} \right)$$

$$= 98.7 \text{ kcal/mole}$$

The energy of the radiation is more than twice as great as the N—O bond energy given in Table 13-2. We conclude, therefore, that 2900 Å (ultraviolet) light can break the bond.

Reactions producing photochemical smog

A partial analysis of typical smoggy air is given in Table 13-3. Research has shown that the components essential to smog production are hydrocarbons and nitrogen oxides.

In cities the main source of hydrocarbons and nitrogen oxides is the automobile. More than half the 130 million tons of man-made pollutants released yearly in the United States come from transportation. For example,

Table 13-3. Typical concentrations in photochemical smog

Component	Concentration near ground level ppm by volume
Aldehydes, RCHO	0.6
Carbon dioxide, CO_2	340
Carbon monoxide, CO	40
Hydrocarbons, RH	5
Ozone, O_3	0.4
Nitrogen oxides, NO_x	0.2
Sulfur dioxide, SO_2	0.2
Water vapor, H_2O	2000

a standard-size car traveling 12,000 miles per year on leaded gasoline exhausts the following pound quantities of pollutants: carbon monoxide, 300; nitrogen oxides, 90; lead, 3.3; halogens, 3; sulfur dioxide, 4; plus smaller quantities of other substances. In addition, it loses 220 pounds of hydrocarbons from its crankcase and by evaporation.

The important primary photochemical process in smog is the reaction of nitrogen dioxide with light. As evidenced by its reddish-brown color, NO_2 absorbs in the visible region. Light energy breaks nitrogen–oxygen bonds, giving nitric oxide and atomic oxygen:

$$NO_2 + h\nu = NO + O \tag{13-6}$$

Oxygen atoms then trigger a series of interdependent reactions, including those yielding ozone. Reaction complexity can only be hinted at in the following simplified summary.

Atomic oxygen reacts with certain hydrocarbons to give oxygenated odd-electron species called *acyl radicals*. It also reacts with molecular oxygen yielding ozone which reoxidizes nitric oxide:

$$O + R'H \longrightarrow R-\overset{\overset{\displaystyle O*}{\|}}{C}\cdot$$

$$O + O_2 \overset{M}{=} O_3$$

$$O_3 + NO = NO_2 + O_2 \tag{13-7}$$

Acyl radicals react with oxygen to give peroxyacyl radicals:

$$R-\overset{\overset{\displaystyle O}{\|}}{C}\cdot + O_2 = R-\overset{\overset{\displaystyle O}{\|}}{C}-O-O\cdot$$

Peroxyacyl radicals can then react with hydrocarbons to give several classes of oxygenated compounds including ketones $R-\overset{\overset{\displaystyle O}{\|}}{C}-R'$ and aldehydes $R-\overset{\overset{\displaystyle O}{\|}}{C}-H$. Peroxyacyl radicals can also react with nitric oxide to give oxyacyl radicals plus nitrogen dioxide, and with oxygen to yield more oxy-

---

* —R is a chain composed of covalently bonded carbon and hydrogen atoms such as, $H_3C-CH_2-CH_2-CH_2-$; R′ designates a group similar to R. Organic compounds are discussed in Chapter 20.

acyl radicals plus ozone:

$$R-\overset{\overset{O}{\|}}{C}-O-O\cdot + RH \longrightarrow R-\overset{\overset{O}{\|}}{C}-R' + R-\overset{\overset{O}{\|}}{C}-H,\ etc.$$

$$R-\overset{\overset{O}{\|}}{C}-O-O\cdot + NO = R-\overset{\overset{O}{\|}}{C}-O\cdot + NO_2$$

$$R-\overset{\overset{O}{\|}}{C}-O-O\cdot + O_2 = R-\overset{\overset{O}{\|}}{C}-O\cdot + O_3 \tag{13-8}$$

Finally, oxygenated organic species may react with nitrogen dioxide, forming peroxyacyl nitrates (PAN), compounds that cause eye irritation and crop damage even at parts per billion (ppb) concentration levels:

$$R-CO_x\cdot + NO_2 \longrightarrow R-\overset{\overset{O}{\|}}{C}-O-O-NO_2 \tag{13-9}$$

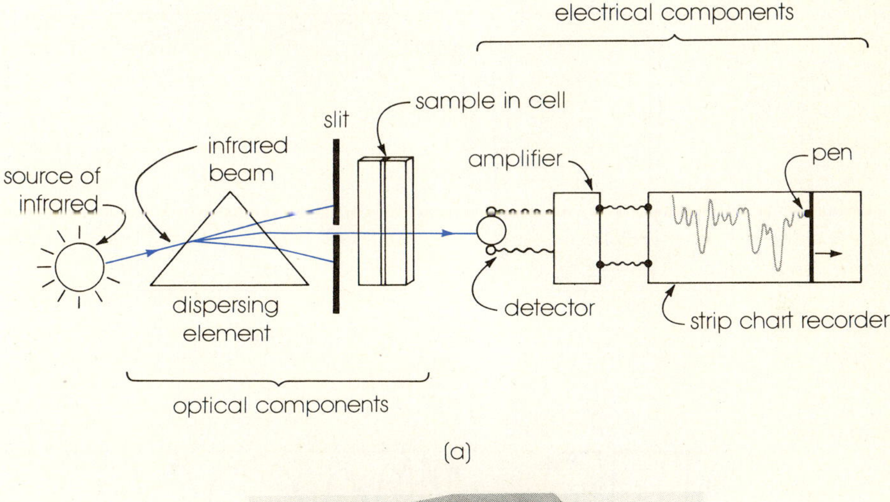

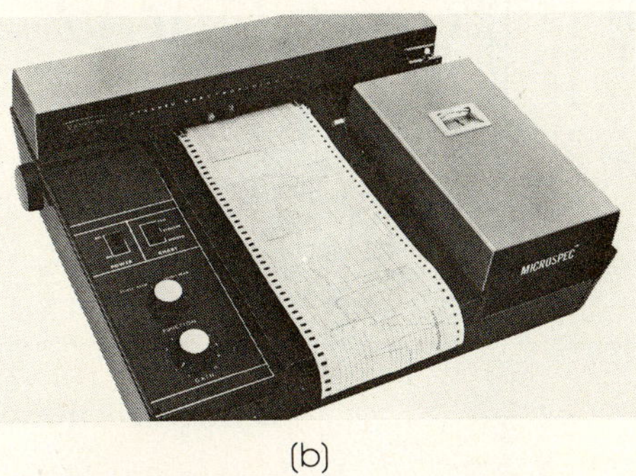

(b)

Figure 13-10 Infrared spectrometry. (a) Schematic drawing of the essentials of a single-beam instrument. Infrared energy is heat energy. The energy source may be an electrically heated rod of silicon carbide, the detector a sensitive thermocouple (see Figure 14-1). Lens, cells, and prisms are cut from sodium chloride crystals because glass and quartz absorb infrared. Sodium chloride optics permit operation in the 1 to 16 micron wavelength range (1 $\mu = 10^{-4}$ cm). The complexity of infrared spectra makes an automatic recorder necessary. (b) A modern instrument. Sample is held in the cell with the small white cap at the upper right. (Courtesy Beckman Instruments, Inc.)

Studying air pollution; infrared spectrophotometry

Reaction schemes like the foregoing are deduced with the aid of infrared analytical equipment (Figure 13-10). Two spectra, patterns of energy absorption produced by samples irradiated by infrared in a spectrometer, are shown in Figure 13-11.

Molecules have internal energy related to their (1) electron distribution about the nuclei (*electronic energy*), (2) vibration of atoms or groups of atoms about equilibrium positions (*vibrational energy*), (3) rotations about the molecular axes (*rotational energy*), and (4) whole-molecule movement from place to place (*translational energy*). Subjected to a range of energies, molecules absorb those energies corresponding to allowed energy transitions and transmit the remainder. Absorption of infrared energy in the wavelength segment of 2 to 100 microns (2×10^{-4} cm to 10^{-2} cm) causes molecular vibration. Typical modes of vibration in a small molecule are shown in Figure 13-12. Almost all compounds absorb across the infrared region. Thus even simple molecules such as CO_2 and H_2O give complex and unique spectra. By matching known with unknown spectra, chemists

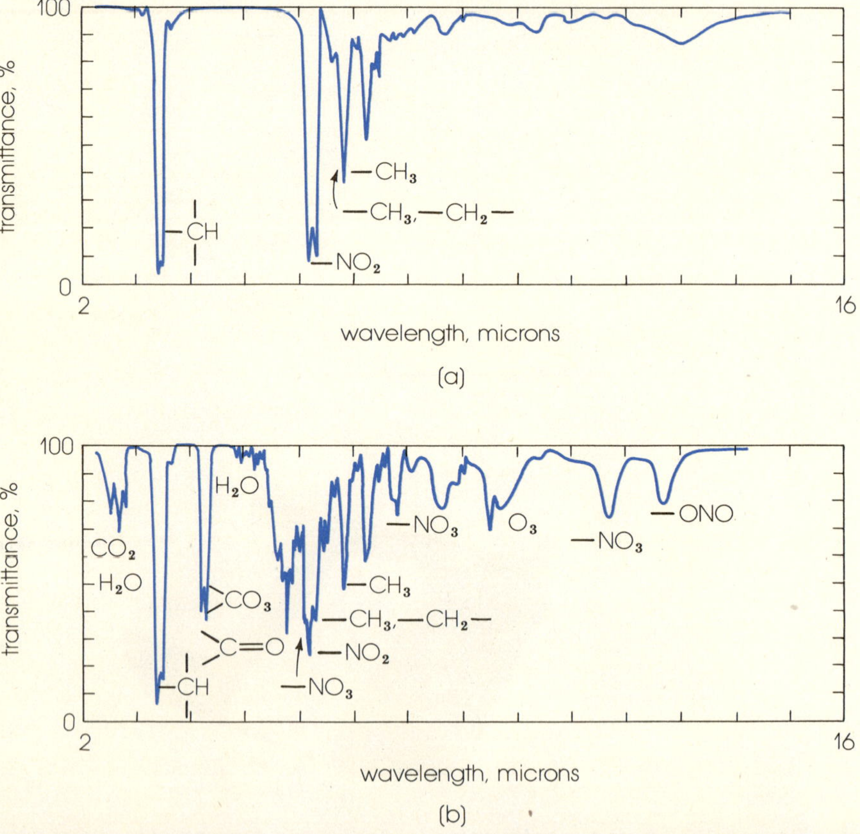

Figure 13-11 Infrared spectra (redrawn). (a) One of the C_8H_{18} hydrocarbons found in gasoline, mixed with NO_2 and O_2, before irradiation with 2900 Å ultraviolet light. Groups of atoms responsible for infrared absorption are noted. (b) After irradiation. Some of the new absorption peaks are identified by comparison to known substances. The spectrum resembles that of liquids isolated in cold traps from photochemical smog.

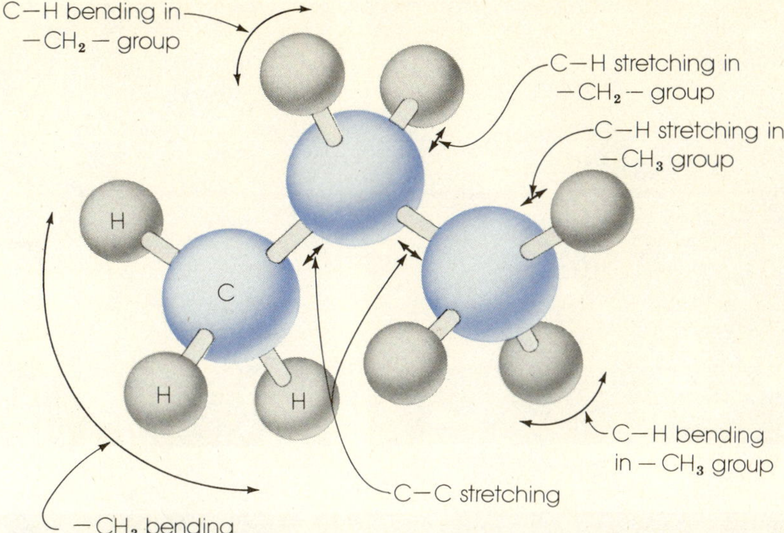

Figure 13-12 Some vibrational stretching and bending modes, illustrated with a ball-and-stick model of propane (CH_3—CH_2—CH_3), which are produced by absorption of infrared energy. Absorption peaks corresponding to each movement are found in other molecules having the same structural characteristics.

identify molecules as well as characteristic parts of molecules (for example, a $C{=}O$ bond).

13-4. THE CHEMISTRY OF INDIVIDUAL ATMOSPHERIC CONTAMINANTS

Particles

The air contains a great variety of tiny solid and liquid particles collectively called *aerosols*. The greatest mass and largest number fall in the 0.1 to 10 micron (μ) range of diameters.

Particles come from many sources. Analysis of dust collected over several eastern American cities shows the presence of silica and compounds of 20 metals, including calcium, sodium, aluminum, and iron in abundance ratios suggesting their sources are coal ash and basaltic rock (see Table 5-1). Other particles found at various locations contain salt from ocean spray, volatile oils from plants ("smoke" in the Smoky Mountains), pollen from plants, ash from volcanoes, minerals from mining operations, asbestos from vehicle brake linings, rubber from rubber tires, lead from gasoline, and mixtures that are condensates of smaller aerosols. Air over a dirty city may contain a billion particles per cubic foot, leading to a monthly dustfall of 500 tons per square mile. In a year's time the average city dweller breathes in 1.5 pounds of aerosols. Larger particles are caught in the mucous of the nose and throat; smaller ones are carried into the lungs.

Aerosols can be counted by a cloud chamber technique (see Figure 2-17) in which cloud density is related to the number and size of particles on which water has condensed. Surface features of particles are studied with the aid of the electron microscope (Figure 13-13). In any microscope, magnification bears an inverse relationship to the wavelength employed to "illuminate" an object, and the limit of resolution in an object's image is

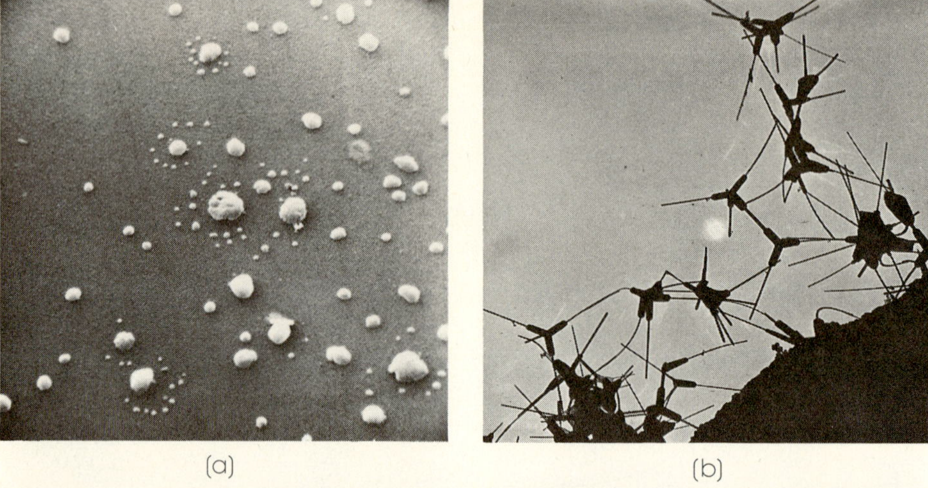

Figure 13-13 Electron microscopy. (a) Airborne particles (×5250) from multiple-source pollution. Sample was taken by airplane between 1000 and 2000 feet over southwestern Pennsylvania. The droplets may be sulfuric acid. (b) Zinc oxide particles (×3000). Sunlight sensitizes a ZnO surface so that gas reactions take place on it, complicating air pollution problems. (c) A scanning electron microscope. Left, controls; right, microscope, with vacuum provisions. (Photographs; a, courtesy Meteorology Research, Inc., from work performed pursuant to Contract CPA 22-69-20 with the National Air Pollution Control Administration, Department of Health, Education and Welfare, photograph by J. Devaney, JPL; b, courtesy E.S.C. Bowler, Electron Microscope Laboratory, University of California, Los Angeles, School of Engineering; c, courtesy K Square Corporation, Pittsburgh, Pa.)

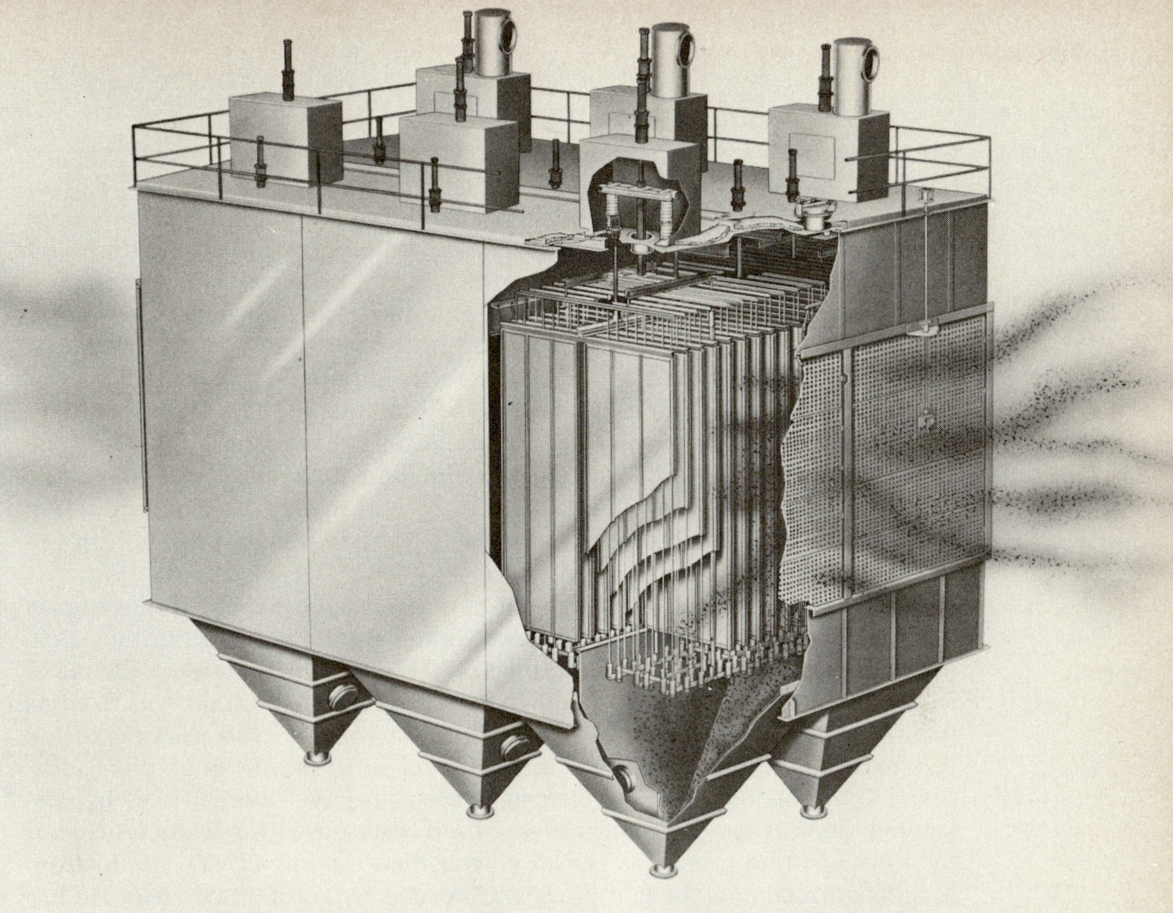

Figure 13-14 An electrostatic precipitator. Gas containing suspended particles enters at the right and exits, cleaned, at the left. Particles are charged near the wires and collected on the plates. (Courtesy Western Precipitation Division of Joy Manufacturing Company.)

somewhat less than the wavelength used. With visible light, detail is lost in an object having a diameter of less than about 3000 Å. With electron-beam illumination, objects only a few angstroms in diameter (thus individual, larger molecules) are distinguishable.

Aerosols may affect the weather. Hot air rising from a city carries a heavy particulate loading. As it cools its particles act as condensation nuclei that can make rain and snow downwind. An indication that aerosols are increasing worldwide comes from the observation that since 1940 average global air temperature has dropped 0.3°C. The decrease is blamed on particles which, by making the air more opaque, limit the amount of sunlight reaching the earth.

A device used by industry to curb particle emissions is the *electrostatic precipitator* (Figure 13-14). It contains closely spaced alternate rows of vertical collecting plates and fine wires. When high-voltage direct current is delivered to the wires, there is created about them a phenomenon called *corona discharge* that ionizes gases. As charge is transferred to particles in the gas stream, the particles attach themselves to the grounded plates while the cleaned gas continues on. The plates are later shaken mechanically to deposit collected material in bins below.

Carbon dioxide

Carbon dioxide in the air is part of the earth's *carbon cycle*. The carbon cycle is a natural chain of chemical events through which CO_2 is converted to carbon compounds in the biosphere (collectively, all living matter), and the carbon compounds break down to CO_2 again. Eventually all carbon goes through the atmosphere. Geochemists give the following material balance. The oceans, which contain 110,000 billion tons of CO_2 (partly as bicarbonate ion, HCO_3^-), and the atmosphere, containing 2730 billion tons of CO_2, yearly intertrade an estimated 200 billion tons of CO_2. Likewise, there is a yearly balance between the entrance of 60 billion tons to CO_2 into the atmosphere from the biosphere's respiration and slow decay and the uptake of 60 billion tons in photosynthesis (Eq. 13-2). Finally, there is production of 0.1 billion ton of CO_2 by volcanism offset by removal of 0.1 billion ton of CO_2 by the weathering of rocks.

Man influences the carbon cycle through fuel burning and farming. In 1970 estimated world production of coal and lignite was 3.30 billion tons, crude oil 2.30 billion tons, natural gas 0.70 billion ton. Assuming the materials to be 85 percent carbon and that 90 percent of the total was used as fuel, it is easy to show almost 18 billion tons of CO_2 was released in the production of energy. From recently cleared soil there is exuded an additional 2 billion tons of CO_2. Nature is absorbing the overbalances by dissolving them in seawater and causing their reaction with silicate rocks, but the rates of these processes are far slower than the rate of CO_2 production. Man has mined coal for 800 years and burned half of it since 1940. He has pumped oil for 100 years and burned half of it in the last 15 years.

To date man has taken 130 billion tons of fossil carbon from the earth. Its combustion has resulted in a net usage of 0.07 percent of the atmosphere's store of O_2. We may well ask, as carbon fuel consumption continues, *will we eventually run out of oxygen?* The answer is no, even if oxygen renewal due to photosynthesis were to stop. Total mass of the atmosphere is 5.73×10^{15} tons (Example 13-1). About 23 percent by weight is O_2, or 1.33×10^{15} tons. Dividing O_2 mass by the earth's surface area (5.48×10^{15} ft^2) gives available oxygen over every square foot of 0.242 ton or 484 pounds. Geochemists estimate the planet's total available carbon (living and dead, on the surface, underground, and in ocean sediments) amounts to 0.492 pound per square foot. To burn carbon completely, $C + O_2 = CO_2$, 32 pounds of O_2 (1 lb-mole) is required for each 12 pounds of carbon (1 lb-mole). To burn the carbon equivalent to each square foot of planetary surface would require $0.492 \times 32/12$, or 1.31 pounds of O_2. That is, however, but $100 \times 1.31/484$, or 0.27 percent of the O_2 available, equivalent to a reduction in atmospheric O_2 from its present volume percent of 20.948 to 20.678. This calculation is not intended to infer that air pollution is insignificant or that oxygen lack in a localized area cannot occur. It does mean that molecular oxygen is apparently one resource we will never use up.

Carbon dioxide and water vapor are strong absorbers of infrared energy. Any addition of them to the air enhances its *greenhouse effect*. Like a window, the atmosphere transmits visible light waves from the sun but retains infrared waves that the earth radiates. The atmosphere does so by absorbing the heat energy in the vibrational and rotational energy of its molecules, as represented in Figure 13-12. In the last hundred years 360 billion tons of CO_2 were released to the atmosphere by fuel combustion, increasing the CO_2 concentration by 13 percent. In response, average air temperature gradually rose 1°C and glaciers receded (see Problem 11). If continued, a weight equal to 70 percent of the amount of CO_2 now present in the air will be released during the next hundred years. A consequence could be melting of the polar ice fields, where 2 percent of the planet's total water is imprisoned. Oceans would rise 200 feet with flooding on every continent.

Countering these dire possibilities is the effect of particles on air clarity and the surety that as temperature increases, evaporation increases, leading to greater cloud cover. Cloud and ice cover now reflect 36 percent of the sunlight reaching the earth. For a rise of 1 percent in reflectivity, mean surface temperature is lowered 1°C. The data suggest that changes in water vapor and carbon dioxide concentrations produced profound weather alterations over geologic time and will do so again in the future. Increasing the earth's reflectivity to 41 percent would lower the mean temperature 4°C and bring on a new ice age.

Carbon monoxide

Excluding CO_2, the quantity of CO in city air exceeds that of all other pollutants combined. Prior to 1972 it was generally assumed man's activity, and not nature, was the prime source of carbon monoxide. Then chemists found contributions from the two sources could be estimated because they yield CO containing distinctively different mixtures of carbon and oxygen isotopes. Global air analysis with a mass spectrometer has since led to the estimate that natural sources put about 3.5 billion tons of CO into the air each year, which is more than 90 percent of the total addition. A major source of the carbon is methane, CH_4, which comes from decaying organic matter in rice paddies, tropical forests, and swamps. Oxidation of methane in the troposphere (Figure 13-7) is believed to be carried out by hydroxyl radicals created by the photolysis of water vapor:

$$CH_4 \xrightarrow{HO\cdot} H_2O + H_3C\cdot \xrightarrow{HO\cdot} H_2O + CO \xrightarrow{HO\cdot} H_2O + CO_2 \qquad (13\text{-}10)$$

Other natural sources of CO include forest fires and certain biological processes occurring in the ocean. Production of CO is balanced by its oxidation to CO_2 in the air (Eq. 13-10), and its absorption in some soils via biological action that is not yet understood. Carbon monoxide is relatively inert. Its residence time in the troposphere is estimated to be one to two months.

Carbon monoxide is a colorless, odorless, poisonous gas. At low, continuous levels existing in canyonlike city streets it can dull brain activity and have other harmful effects. The function of hemoglobin (Hb) in the blood is to carry oxygen from the lungs. Carbon monoxide, with 200 times more affinity for hemoglobin, ties up hemoglobin as carboxyhemoglobin (COHb), causing oxygen starvation in cells. Atherogenic symptoms (degeneration of blood vessels) and the deposition of cholesterol (a waxy substance that can restrict flow in blood vessels) have also been correlated with COHb concentration.

Sulfur dioxide

About 75 percent of the sulfur dioxide in the air probably originates as H_2S from volcanoes and the action of bacteria on organic matter. Contribution of the latter can be assessed from the different $^{34}S:^{32}S$ isotope ratio in its gas. The remaining SO_2 comes directly from burning sulfur-containing fuels. High sulfur coal (1.5 to 3.5 percent sulfur) comprises 80 percent of all coal mined. Sulfur dioxide and SO_3 aerosols form with atmospheric moisture (Figure 13-13a) and return to the earth by precipitation. Rain falling downwind from coal-burning communities is acidic. Eventual neutralization of SO_3 gives sulfates.

In the United States some 35 million tons of SO_2 is emitted to the atmosphere annually, and the amount is increasing 5 to 10 percent per year. One approach to removing SO_2 from smokestack gases is by scrubbing with alkaline materials (for example, $CaO + SO_2 = CaSO_3$). Ironically, removal of SO_2 from the air allows levels of strong oxidants such as O_3 to rise because oxidants are otherwise used up oxidizing SO_2.

Nitrogen oxides

Average air concentrations in parts per million of nitrogen oxides are N_2O, 0.35; NO_2, 0.01; NO, a trace. The bulk of the nitrogen oxides seems to be made naturally (lightning, soil bacteria) but not all sources are known. Man contributes NO from combustion processes in which atmospheric N_2 is oxidized directly. In the atmosphere NO is oxidized to NO_2. In city air NO_2 concentration varies directly with automobile traffic volume.

Nitrogen dioxide affects biological systems. If it attacks such protein molecules as cell enzymes (represented with saw-tooth lines in the following equations), the enzymes lose their vital catalytic activity. Thus sulfhydryl groups may be oxidized to disulfides, linking molecules together:

$$2 \text{ (—SH)} \xrightarrow[H_2O]{NO_2} \text{(—S—S—)} + H_2O + N_2 \qquad (13\text{-}11)$$

And amino groups may be removed completely and replaced by hydroxy groups:

$$\text{(—NH}_2) \xrightarrow[H_2O]{NO_2} \text{(—OH)} + N_2 \qquad (13\text{-}12)$$

As can be surmised from Table 13-4 data, the problem of cleaning automobile exhaust of nitrogen oxides must be considered in relation to curbing all emissions. The job is not easy because changes in operating conditions to lower one pollutant can act to raise another. Leaner gasoline–air mixtures and higher combustion temperature, for example, decrease hydrocarbon and CO emissions but increase NO_x. Ideally a catalyst will be found that simultaneously causes chemical reduction of nitrogen oxides to nitrogen as carbon monoxide is being oxidized to carbon dioxide:

$$CO + NO_x \longrightarrow CO_2 + N_2 \qquad (13\text{-}13)$$

Lead

Recent research indicates that 3 percent by weight of all solids suspended over American cities may be lead-containing aerosols. Primary source is

Table 13-4. Regulating automobile emissions

Emission	Levels before controls g/mi	Recommended 1970 levels g/mi	Federal proposed limit for 1976 g/mi
Carbon monoxide	80	23	11
Hydrocarbons	11	2.2	0.5
Nitrogen oxides	4.0	3	0.9
Particulates	1	0.3	0.1

leaded gasoline, as proved by mass spectrometer studies that have related lead isotope ratios to specific brands of gasoline and to the mines from which the lead came. The long history of lead in the environment has been traced through corings taken from ocean sediments and Greenland ice masses. Lead mining started about 2500 B.C. In 1750, the start of the Industrial Revolution, a significant increase in lead pollution began. Since 1940, when lead additives became common in gasoline, the trend of lead contamination has climbed sharply.

In 1970 about 20 percent of the nation's lead consumption went into gasoline. Three-quarters of that, some 270,000 tons, was exhausted to the atmosphere as lead bromochloride, $PbBrCl$, and other compounds. Although the figure is now declining, average big city air still contains 3 micrograms (μg) of lead per cubic meter, about 1000 times the concentration that can be attributed to natural causes. Evidence of lead inhalation by city dwellers is present in their blood, where levels between 15 and 40 micrograms per 100 milliliters have been found. Long-term effects of lead in the body are not yet known.

Before lead mining began, water contained only 0.015 ppb of lead (0.015 μg/liter). Today average rainfall over the United States contains 36 ppb of lead. Fortunately most of the metal seems to get filtered out by natural processes before it reaches public water supplies. The U.S. Public Health Service rejects drinking water containing more than 50 ppb of lead.

In 1970 an estimated \$3 billion equipment changeover began in the petroleum industry to make a lead-free 91–octane gasoline equivalent to former regular-grade fuel. The approach has been to produce more high-octane hydrocarbons from crude oil for blending with low-octane hydrocarbons. With the new gasoline, a relatively simple ($\sim$ \$100) and effective catalytic muffler is expected to be usable on automobiles to oxidize hydrocarbon emissions. The same unit, typically containing 5 pounds of mineral oxides thinly coated with platinum, does not work with leaded gasoline because lead poisons its surface (see Figures 17-4, 17-5).

13-5. AIR CONTAMINATION AND SOLID WASTES

How should solid waste be managed?

A city is a giant dynamic system that takes in air, water, fuels, metals, and other raw materials and pours out useful products, heat, and solid and liquid wastes. Americans are champion waste makers. With less than 6 percent of the world's people we are each year using 40 percent of all resources being consumed. In his lifetime the average American uses 26 million tons of water, 21,000 gallons of gasoline, 10,000 pounds of meat,

28,000 pounds of milk, 9000 pounds of wheat, and large quantities of coal, paper, steel, glass, and other commodities. Abundance inevitably leads to waste. Our yearly trash collection in cities totals 210 million tons (sample contents: 50 billion cans, 30 billion bottles), equivalent to 6 pounds per capita per day. Its disposal, costing $5 billion, furnishes employment for 350,000 men working with 100,000 trucks.

There is no cheap, good way to get rid of trash. Dumped off the coast, it pollutes the ocean; piled up, it is an unsightly breeding grounds for rats; buried, it may contaminate surrounding water supplies when rainwater leaches the dump site; burned, it fouls the air. Nevertheless, as city landfill area gets scarce, incineration becomes the disposal method of choice. The cost of incineration is expected one day to rival the cost of public education.

Depending on size, a trash disposal plant such as the scheme labeled "New" in Figure 13-15 might cost $50 million. Wastes would be burned at 1200°C. Gases would be cleaned in an electrostatic precipitator. Heat generated would be used to make steam for sale and for turning machinery to make electric power. The power would heat an electric furnace capable of fusing worthless ash into a hard slag usable in construction.

Nonincineration schemes for city trash handling include recycling: sorting out metals and glass for remelting, shredding and fermenting garbage to make fertilizer, pyrolyzing rubber to make carbon black, and transforming paper into oil or alcohol.

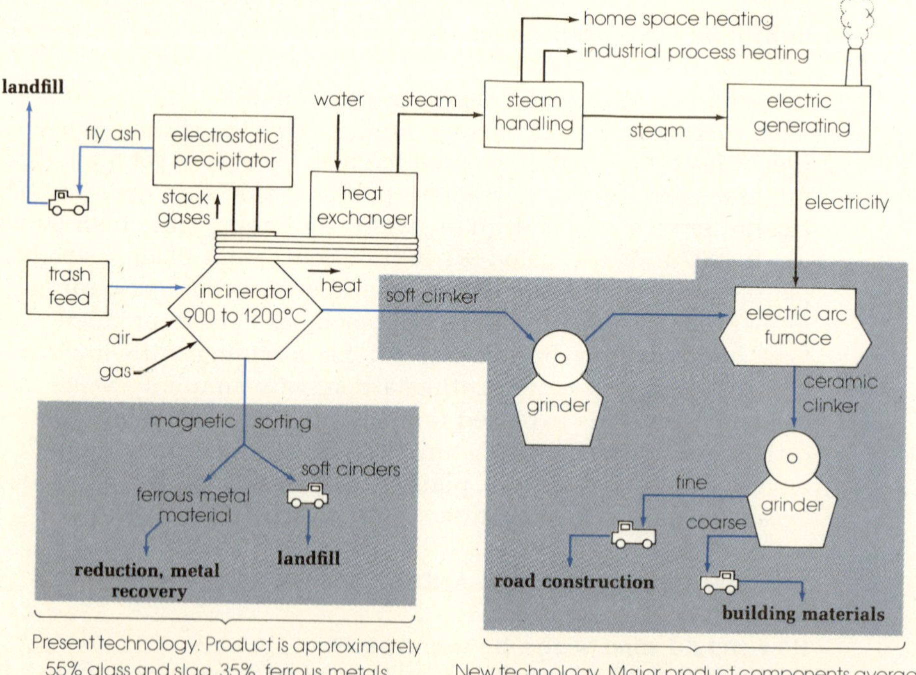

Figure 13-15 A plan for municipal trash disposal by burning. Reduction in volume is 90 percent. Trash yields about half as much heat as the same weight of coal, based on typical northeastern city dump contents: 49 percent paper, 20 percent food scraps, 10 percent metal, 10 percent glass, 3 percent yard trimmings, 2 percent plastics, 2 percent wood, cloth, and rubber, and 4 percent inert material such as ashes.

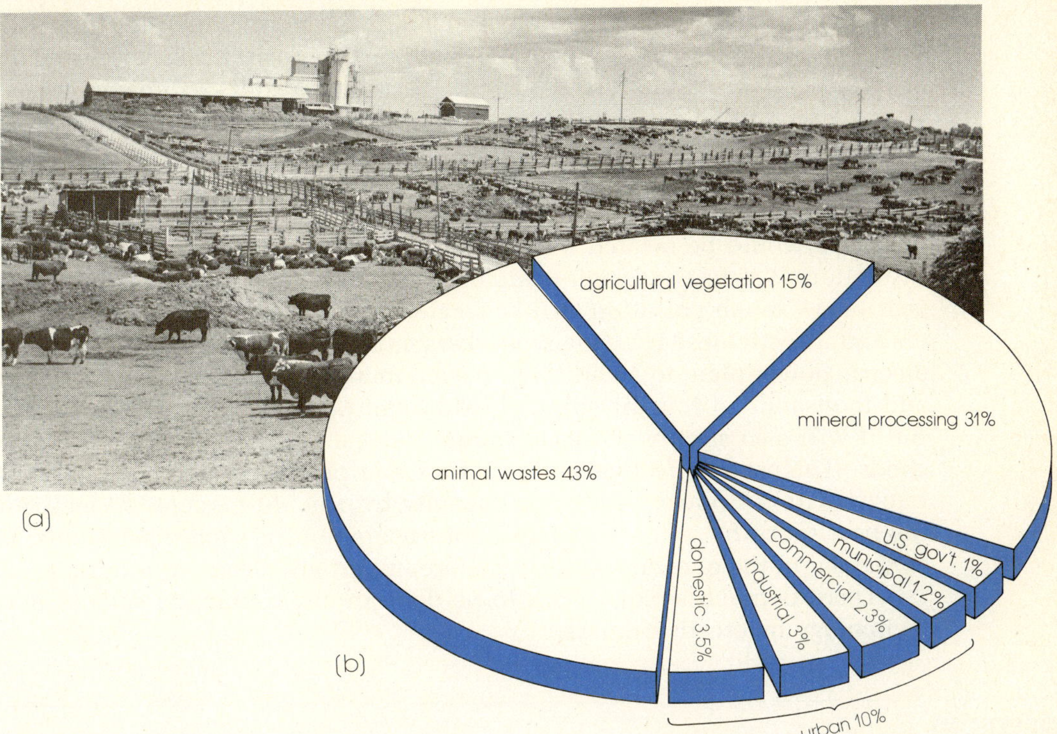

(a)

(b)

Figure 13-16 Agricultural wastes. (a) A cattle feed lot near Omaha, Nebraska. The largest such lot, feeding 80,000 head of cattle, generates a mass of daily excrement equivalent to that from a city of 1 million people. Approximately 115 million cattle, 60 million hogs, and 400 million chickens, and 110 million turkeys are being raised in the United States. (b) Solid wastes in the United States—approximate percentages by weight. Although urban waste receives most attention (compacted, a year's collection would build a wall 18 feet thick and four stories high stretching from Los Angeles to New York City), it is only one part of the overall waste problem. (Courtesy U.S. Department of Agriculture.)

An even larger waste disposal problem than the one described above is that created by the nation's agricultural activities (Figure 13-16). Happily much of the agricultural waste is cellulose–derived and therefore renewable.

The U.S. Bureau of Mines has recently demonstrated two experimental schemes for converting cellulosic wastes. In the *pyrolysis* method, dry manure is heated for 6 hours at 900°C to yield gas, oil, and carbon. In the *hydrogenation* process, manure is heated for 20 minutes at up to 340 atm pressure with H_2, CO, and H_2O. From 1 ton of dry cellulosic waste comes 3 barrels of low-sulfur fuel oil (95 percent yield):

$$(CH_2O)_x + CO + H_2 \longrightarrow (CH_2)_x + H_2O$$

A third, privately developed process, converts cellulosic waste to sugars, protein, and methane gas via a week of fermentation. The solid product is suitable as an animal feed supplement.

Tradeoffs

Each scheme for cleansing the environment involves economic considerations (somebody must pay the bill) and broad scientific knowledge (somebody must understand the problem). One of the most important things to know is the basic environmental principle that everything is related to everything else. Any effort at change, therefore, results in change somewhere else, so that one must accept "tradeoffs" rather than expect simple and perfect solutions. If, for example, we decide that production of synthetic fibers should be halted because it requires too much electric power and synthetics are hard to dispose of, we then have to determine what new problems would be created if we go back to producing more natural fiber, say cotton. Is irrigation water available? (A cotton crop requires 25 inches a year.) How much pollution will be created by applying the needed fertilizers and insecticides? How many workers must be relocated? Or consider an electric power plant to be built in an economically depressed region. It will add thousands of tons of sulfur dioxide and particles per year to the air. But it will also supply power to industries that employ people and create goods. To install effective antipollution equipment on a total fixed-cost budget will cut power-producing capacity by, say, 40 percent. At least for the near term the painful tradeoff, not uncommon in such discussions, is contained in the question, Which is more important, clean air or more jobs?

There is much serious work to be done in environmental science and technology by creative minds.

QUESTIONS

1. Upper atmosphere A rocket-mounted mass spectrometer returns the positive ion spectrum of Figure 13-17 from an altitude of 55 mi. Tentatively identify the ions. Explain.

2. Spectrum The infrared absorption spectrum of CO_2 is shown in Figure 13-18. How does it help explain the greenhouse effect?

3. Ozone Atmospheric ozone may be estimated by bubbling a known air volume through an iodide solution and comparing the resulting yellowish color of triiodide ion with the colors of a set of previously prepared color standards:

$$O_3 + 3I^- + H_2O = I_3^- + 2OH^- + O_2$$

Why does ozone analysis give a quick check on smog conditions?

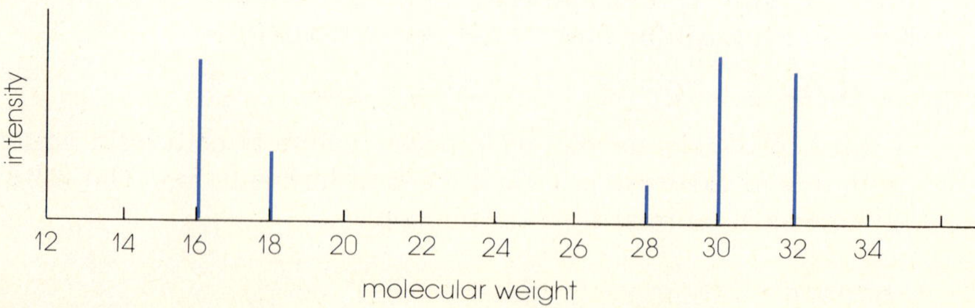

Figure 13-17

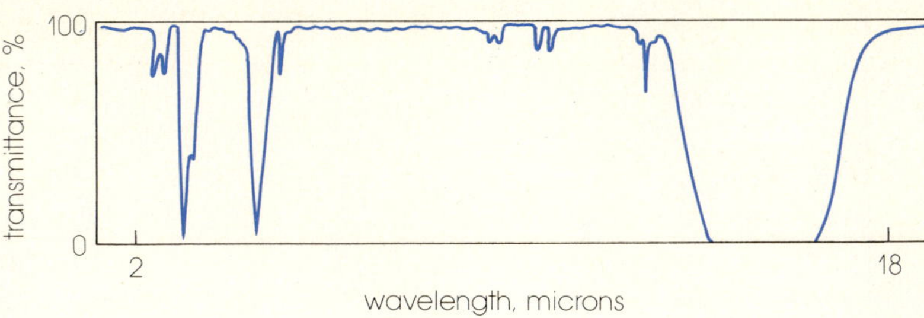

Figure 13-18

4. Pollution **(a)** A *primary* pollutant remains in the atmosphere in the form in which it was emitted. A *secondary* pollutant is produced in the atmosphere. List several air pollutants of each kind. **(b)** Jet airplane traffic between Chicago and Detroit is heavy. Precipitation has increased along this corridor. Give an explanation. **(c)** There are more chickens in the United States than people living here. What pollution problems does this cause?

5. Lead Vapors of leaded gasoline passed over a 0.15 percent solution of iodine in mineral oil at 20°C are sent into a chamber of cold air. Within 90 sec, ice crystals have formed in the cold chamber. What implications do you see in the experiment?

6. Carbon monoxide **(a)** A cycle involving CO and CH_4 operates in nature. Explain. **(b)** There is a *diurnal* variation of the CO concentration at the ocean's surface. What does diurnal mean? How do you interpret this experimental fact?

7. Climate **(a)** The growing season in England has been shortened by about two weeks since 1950. Give a possible explanation. **(b)** How might global weather be affected as developing nations become industrialized? Explain.

8. Tradeoffs Cellulosic trash might be packed deep underground and vaporized with a nuclear explosion (cost ~ $5 million). This would release carbon monoxide and hydrogen (including some tritium) which might be recovered and used in the synthesis of useful chemicals such as ammonia and methyl alcohol. List advantages and disadvantages to the scheme.

9. Tradeoffs You are a scientific adviser appointed to help a city government examine technical problems. The following question arises: Should the city build a complete trash incineration system (Figure 13-15) or should it contract with the railroad to haul trash 150 mi for dumping into an abandoned coal mine? List some of the economic and scientific data you would like to have to answer the questions. What tradeoffs are involved?

10. Air analysis A continuous air analyzer in a city gives the results shown in Figure 13-19. What interpretations can you make?

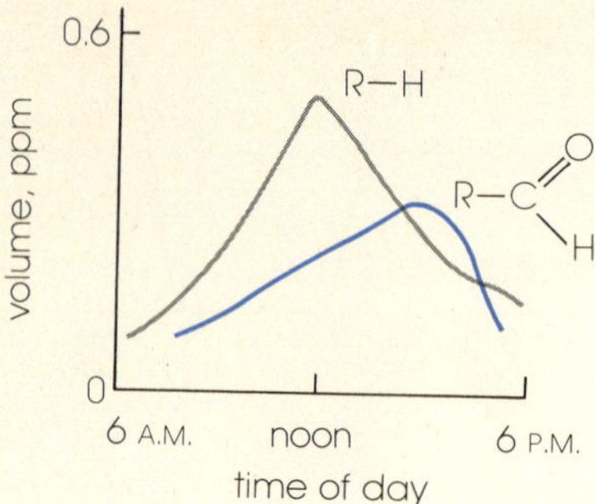

Figure 13-19

PROBLEMS

11. Carbon dioxide **(a)** Construct a graph as begun in Figure 13-20 by plotting, in different colors, year vs. CO_2, and year vs. °F (as given in Table 13-5). Extrapolate the curves to estimate Icelandic conditions in 1990. Write in the expected values. **(b)** List several factors that are making these forecasts uncertain.

12. Atmospheric temperature Using data from Figure 13-7 plot a graph of atmospheric temperature (°C on abscissa) vs. altitude in miles. Add horizontal lines across the sheet to denote sphere boundaries and label the spheres.

13. Lead Earthworms dug up near a highway are found to contain a maximum of 330 ppm of lead. It is known that a lead level of 200 ppm will kill an adult mallard duck. How many grams of 330-ppm-Pb worms would presumably be a lethal meal for a 2000-g mallard?

14. Photochemical reaction Is it possible for light of 4200 Å wavelength to break a carbon–carbon single covalent bond? Show calculations.

15. Automobiles Roads are being built in the United States at the rate of 90 million ft/yr. Automobiles are being built at the rate of 11.5 million/yr. **(a)** Measure the length of an "average" automobile. **(b)** If the new roads are two lanes wide, and the new cars are lined up on them two abreast and bumper to bumper, what percent of the total new road length will be occupied?

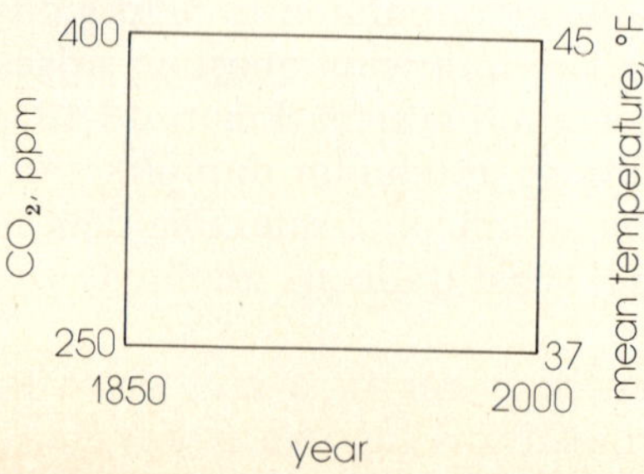

Figure 13-20

Table 13-5. Variations in Icelandic temperature, and atmospheric CO_2

Year	Atmospheric CO_2 ppm	Mean Iceland temperature °F	Year	Atmospheric CO_2 ppm	Mean Iceland temperature °F
1870	285	37.8	1920	330	38.4
1880	290	37.9	1930	308	38.7
1900	287	38.2	1940	315	39.3

16. Gasoline In 1971 gas stations in the United States pumped 96.34 billion gal of gasoline. If 105 million vehicles obtained an average mileage of 13.6 mi/gal, how many gallons a year did each vehicle burn and how many miles did it travel?

17. Automobile pollution Consider the following data. An experimental automobile travels 12 mi for each gallon of leaded gasoline used. In 1 yr it is driven 12,000 mi. Each gallon of gasoline contains 2.0 g of lead. Gasoline density is 5.8 lb/gal. Engine inefficiency is such that 9.0 percent of the gasoline is burned to carbon monoxide:

$$2C_8H_{18} + 17O_2 = 16CO + 18H_2O$$

Calculate the pounds per year of **(a)** gasoline used; **(b)** lead emitted; **(c)** gasoline burned to CO; **(d)** CO emitted.

18. Dust A student in a small town weighs a beaker, half fills it with distilled water, puts a screen over it, then places it in a protected area to collect airborne dust. After a month she evaporates the water and weighs beaker and residue. The residue weighs 20 mg. The cross-sectional area of the beaker mouth is 40 cm². Calculate the yearly dustfall in grams per square meter.

19. Particles Living room air may contain 2 million dust particles per cubic foot. Assume a person at rest in the room taking 18 breaths a minute, each of 0.30 liter volume. **(a)** In 24 hr how many particles enter the lungs? **(b)** If the average particle weighs 10^{-13} g, what milligram weight of dust will be inhaled under these conditions in a year?

20. Sulfur dioxide A certain power plant, per hour, burns 60 tons of coal containing 2 percent sulfur by weight. If all sulfur is converted to SO_2 and released to the air where each pound-mole of SO_2 occupies 400 ft³, how many cubic feet of SO_2 are produced each 24 hr?

21. Carbon monoxide One hundred milliliters of cigarette smoke is bubbled through a palladium chloride solution. Carbon monoxide in the smoke reacts as follows:

$$CO + PdCl_2 + H_2O = CO_2 + Pd\downarrow + 2HCl$$

(a) If the residue of palladium metal weighs 10.6 mg, what milligram weight of CO was in the sample? **(b)** If the smoke sample weighed 140 mg, what percent by weight CO was in it?

22. Recycling Six million automobiles are junked each year in the United States. Eighty-five percent are reclaimed but the remainder pile up. **(a)** If a junk car is worth $60, what is the dollar value of a year's worth of unreclaimed cars? **(b)** If each car contains 1.1 tons of iron, how many tons of iron are not reclaimed? **(c)** Estimate or otherwise determine the weight in pounds of an "average" discarded automobile tire. Explain the origin of your figure. Assume 180 million tires are discarded in the United States annually. If on pyrolysis 2 lb of tires yields 1 lb of carbon black, how many pounds of carbon black (for use in new tires) are potentially available from this source?

23. Pollution cost A certain community of 300,000 people produces 5.5 lb of trash per person per day. Cost of trash collection and disposal is $1.20/100 lb. How many dollars must be gathered through taxes or fees to pay the trash disposal bill for a year?

24. Incineration Polyvinyl chloride (PVC) polymer (a unit of which is shown below), used as plastic packaging, creates an incineration problem because it releases HCl which corrodes equipment:

$$2 \; \text{-(}CH_2\text{—}CHCl\text{)-} \; + 5O_2 = 4CO_2 + 2H_2O + 2HCl$$

If one year's domestic production of PVC, 180 million lb, is burned as shown, how many pounds of hydrogen chloride gas can theoretically be recovered with proper technology?

25. Incineration Municipal incinerator fly ash may contain 2 to 9 oz of silver and 0.02 to 0.05 oz of gold per ton. Assume a total of 40 million tons of trash is burned per year in the nation's 200 largest municipal incinerators, and that each 20 tons of trash yields 1.0 ton of fly ash collectable by electrostatic precipitation. Using the minimum figures above, what is the total dollar value of the silver and the gold content at prices of $2.50 and $90 per ounce, respectively?

14 HOW DO GASES BEHAVE?

14-1. Measurement of gas properties
1. Heat and temperature are related.
2. Readings on different temperature scales are interconvertible.
3. Various properties are used to measure temperature.
4. Pressure can be measured using various principles.
5. Gas density can be determined in several ways.

14-2. The gas laws
1. At constant temperature, gas volume and pressure are inversely proportional.
2. The law relating gas volume and temperature is, in effect, a definition of temperature.
3. There is a lowest temperature in nature.
4. At constant pressure, gas volume and Kelvin temperature are directly proportional.
5. The variables P, V, and T can be combined into one law.

6. Standard conditions are defined for gas calculations.
7. The general perfect-gas equation combines the laws of Boyle, Charles, and Avogadro.
   a. The perfect gas obeys the perfect-gas equation exactly.
   b. The units of R must fit the problem being solved.
8. Gas density varies with temperature and pressure.
9. The total pressure of a mixture of gases is related to the partial pressures of the component gases.
10. Evaporation of a liquid creates a vapor pressure above the liquid.
11. Humidity and dew point are related to air temperature and water vapor pressure.
12. Diffusion (or effusion) rate is inversely proportional to the square root of gas density.
13. Gas molecular weights can be found from diffusion (or effusion) rate, gas density, or Avogadro's law.

14-3. The kinetic-molecular theory
1. Brownian movement is evidence of molecular impacts.
2. The kinetic-molecular theory qualitatively explains the gas laws.

14-4. Statistical mechanics
1. Statistical mechanics attempts to deduce macroscopic properties of substances through microscopic theory.
2. Kinetic energy is the energy of motion.
3. The distribution of molecular velocities in a gas is predicted by the Maxwell-Boltzmann law.
4. Molecular average velocity, collision frequency, and mean free path can be calculated.

14-5. Real gases and nonideality
1. Real gases may behave ideally or nonideally.
2. The effectiveness of van der Waals attraction depends on the closeness of molecules.
3. The Joule-Thompson effect reflects nonideal gas behavior.

Gases have no bounding surface; thus they expand indefinitely and completely fill any container. Compared to the density of a solid or liquid, the density of a gas is normally small. Gas density is markedly sensitive to changes in temperature and pressure, and the effects of the changes are nearly independent of the gas used. A few laws, derived originally from simple experiments and later from a theory of gas structure, interrelate gas temperature, pressure, volume, and mass. The theory, called the *kinetic-molecular theory,* is a powerful aid in understanding other phenomena as well.

14-1. MEASUREMENT OF GAS PROPERTIES

Temperature

Heat is a form of energy that flows spontaneously from a hotter body to a colder body. The heat energy obtainable from an object or a chemical reaction depends on the amount of material involved. Flame from a camp fire and a forest fire, for instance, may be at the same temperature, but the latter emits much more heat.

Temperature is the level at which heat energy exists. Hotness (the presence of heat) and coldness (the absence of heat) are determined with a thermometer and expressed in degrees.

Temperature scales are established from temperature standards that are reproducible natural constants. G. Fahrenheit, a German instrument maker who invented the mercury-in-glass *thermometer* (1714), devised a scale for his instrument by calling the temperature of the human body 100 degrees and the lowest temperature attainable with a salt–ice mixture zero degrees. Today the Fahrenheit scale is based on the freezing point (32°F) and boiling point (212°F) of water. In 1742 Swedish astronomer A. Celsius proposed the centigrade (C, centi–, 100) scale, establishing the freezing point of water as zero degrees and the boiling point of water as 100 degrees centigrade. Scientific work uses the centigrade, also called the *Celsius* scale. On the Celsius scale the freezing and boiling points of water differ by 100 degrees. On the Fahrenheit scale the difference is $212 - 32$ degrees or 180 degrees. The ratio of intervals, 100:180 degrees, is the ratio of 5:9. At -40 degrees the two scales coincide. These numbers are used in the interconversion of Celsius and Fahrenheit readings as follows:

$$9\,(°C + 40) = 5\,(°F + 40) \tag{14-1}$$

Use of Eq. 14-1 is illustrated in Example 14-1.

> **Example 14-1** Normal body temperature is 98.6°F. What is it in Celsius degrees?
>
> **Solution** Substitute into Eq. 14-1:
>
> $$9(°C + 40) = 5(98.6 + 40)$$
>
> $$9(°C) + 360 = 493 + 200$$
>
> $$°C = \frac{693 - 360}{9} = 37.0°C$$

Any temperature-dependent property of matter may in principle be used to measure temperature. A few examples are listed in Table 14-1. Of these the greatest resolution is attained with the quartz thermometer, 0.0002°C. The thermocouple thermometer is illustrated in Figure 14-1.

Ordinary thermometry fails when only a very limited amount of the medium being measured contacts the thermometer. Thirty-five miles above the earth, for example, a thermometer absorbs heat from the sun but does not record the temperature of the few gas particles present. Their temperature is obtained indirectly by setting off rocket-carried explosives and recording the time required for the sound to reach the earth. Temperature in absolute or Kelvin degrees (see Figure 14-7) is proportional to the square of the speed of sound in a gas.

Pressure

Pressure is force per unit area. *Standard pressure* is the pressure of the air at sea level, 14.7 pounds per square inch (abbreviated psi). This is also called 1 atmosphere (atm) of pressure. Measured with a *barometer* (Figure 13-3), *standard pressure* is equivalent to a column of mercury 29.9 inches or 760 mm high. The pressure equivalent to 1 mm of mercury is called 1 torr. Standard pressure in several different units is

$$14.7\ psi = 1\ atm = 29.9\ in. = 760\ mm = 760\ torr \tag{14-2}$$

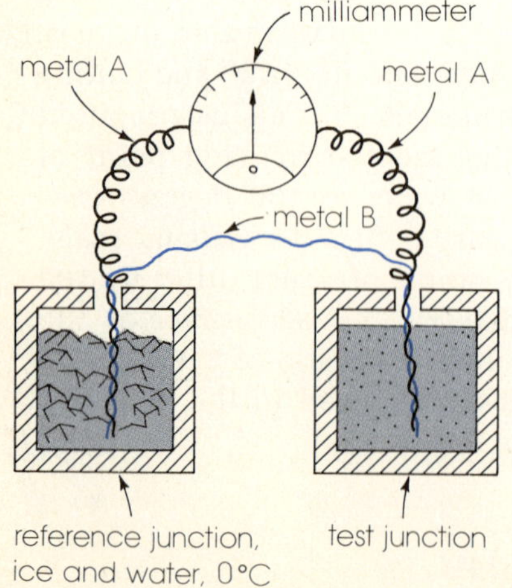

Figure 14-1 A thermocouple, based on the Seebeck (thermoelectric) effect discovered in 1821 (see Problem 17). Wires of two different metals joined at both ends and maintained at different temperatures set up a voltage difference in the circuit that depends on the temperature difference between the junctions. Joining several thermocouples produces a thermopile. Astronomers use thermpiles to measure the temperatures of stars.

Table 14-1. Measuring temperature

Principle	Thermometer	Useful range °C
Change in liquid volume	Mercury in glass	−30 to 650
Change in gas pressure	Fixed gas volume	Down to −272, using helium
Change in electric resistance	Platinum wire	−260 to 800
Change in crystal oscillation	Oscillating quartz crystal	−50 to 300
Development of electric potential	Thermocouple; two dissimilar metals	Up to 2200, using tungsten and rhenium

Some pressure-measuring devices are listed in Table 14-2 and illustrated in Figure 14-2.

Density

In the laboratory the density of a gas is obtained by simply weighing a known volume of the gas. For example, if a 500-ml flask weighs 100.00 g evacuated and 100.90 g filled with carbon dioxide at laboratory temperature and pressure, the density of CO_2 under these conditions is 0.90 g/0.500 liter = 1.80 g/liter.

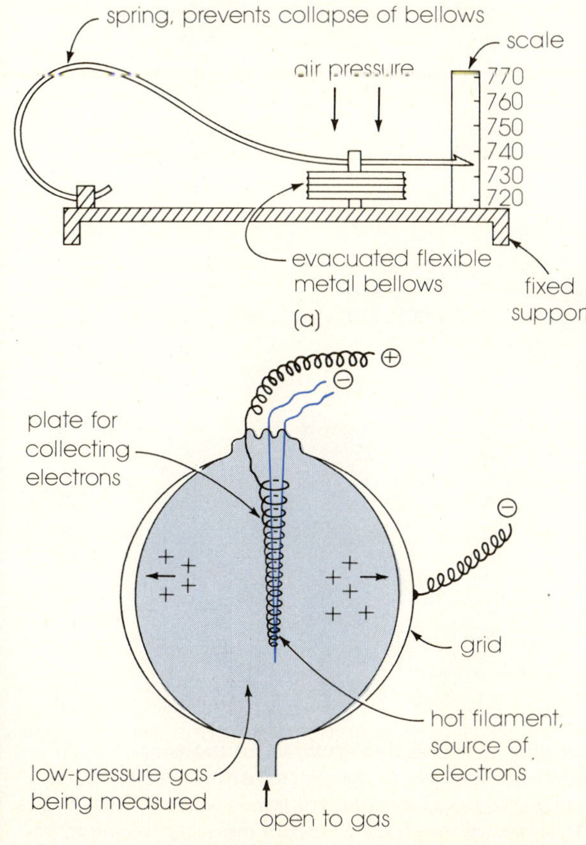

Figure 14-2 Pressure indicating instruments. (a) Principle of the aneroid (fluidless) barometer. It is independent of gravity. Changes in atmospheric pressure cause flexure of the bellows which moves the pointer across the scale. (b) An ionization gauge, useful in very low-pressure (high vacuum) environments like the upper air. The heated filament produces electrons that are accelerated toward the positive plate. Accelerating voltage is sufficient to give electrons the energy needed to produce positive ions upon collision with gas molecules. As positive ions strike the negative grid, they produce an electric current proportional to' their number, hence to gas pressure.

Table 14-2. Measuring pressure

Principle	Instrument	Useful range torr
Support of a mercury column	Mercury barometer	1 to 10^3
Flexing of a thin evacuated metal bellows	Aneroid barometer	1 to 10^3
Uncoiling of a hollow metal tube containing gas being measured	Bourdon tube	Up to 5×10^5
Conversion of gas molecules to ions which cause electric current flow	Ionization gauge	Down to 10^{-10}

In atmospheric research, density may be determined indirectly using the rate of fall of a balloon or rate of change in time a satellite takes to complete an earth orbit. In either case the object is decelerated in proportion to air drag, which is directly related to air density.

14-2. THE GAS LAWS

Boyle's law

The manner in which gas volume depends on pressure was systematically investigated first by Robert Boyle in England in 1662. He found that *the volume of a gas at constant temperature is inversely proportional to the*

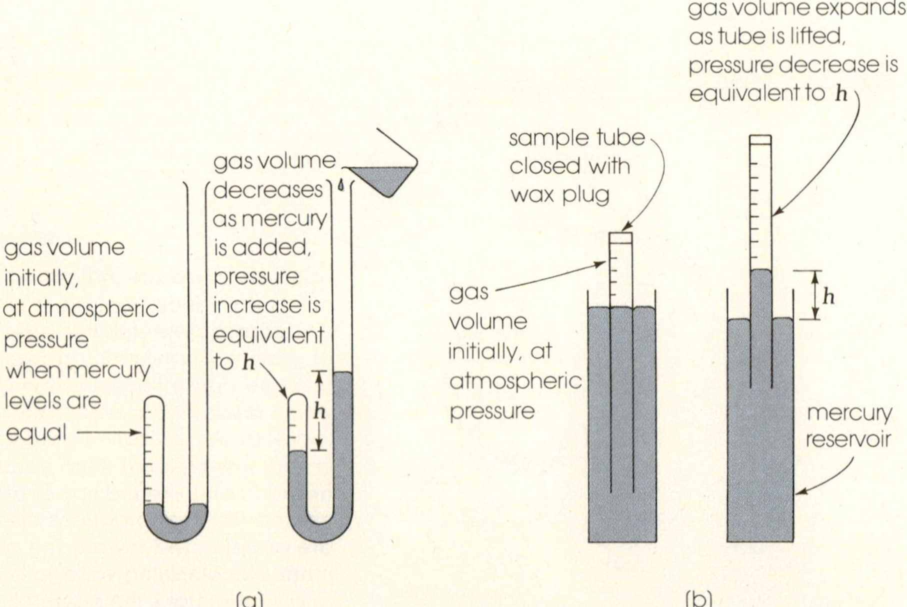

Figure 14-3 Boyle's apparatus for studying gas pressure–volume relationships. (a) A volume of gas is confined at atmospheric pressure in the closed end of the tube. As mercury is added, the pressure is increased and the gas volume decreases. (b) A gas sample is sealed off at atmospheric pressure in a tube above a mercury reservoir. As the tube is raised, the pressure on the gas is decreased and its volume increases.

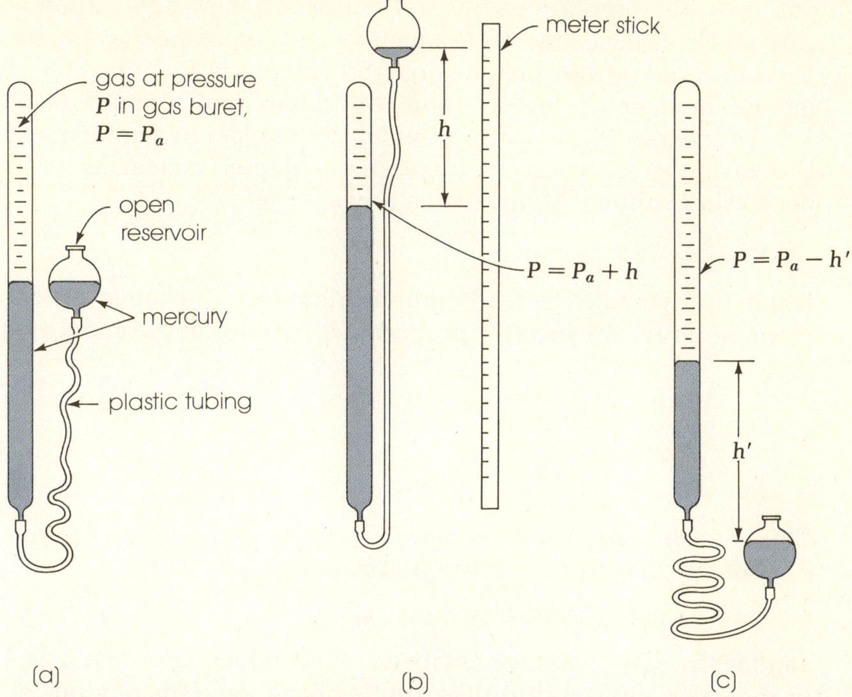

Figure 14-4 A laboratory apparatus for demonstrating Boyle's law at pressures near atmospheric. The reservoir is raised (b) or lowered (c) to raise or lower the pressure on the gas. Temperature is constant. Atmospheric pressure is designated P_a.

pressure on the gas. The statement is known as *Boyle's law.* For pressures slightly in excess of atmospheric pressure Boyle used the device of Figure 14-3a, first confining a gas sample with mercury, then compressing the gas to a new volume by adding more mercury. For measurements at slightly less than atmospheric pressure he used the apparatus of Figure 14-3b, confining a gas sample over mercury, then raising the sample tube to, in effect, lower the mercury level and cause the gas volume to stretch out.

The apparatus of Figure 14-4, which serves the function of both pieces in Figure 14-3, can be used to demonstrate Boyle's law. In position (a) pressure P of the gas sample in the graduated tube (buret) is the same as the atmospheric (barometric) pressure P_a because the mercury levels are equal and the reservoir is open to the air. In position (b) the reservoir has been

Table 14-3. Gas volume variation with pressure (constant temperature)

Trial	Atmospheric pressure torr	Difference in mercury levels torr	Gas pressure torr	Gas volume liter
1	749	0	749	0.0328
2	749	−117	632	0.0384
3	749	−198	551	0.0445
4	749	+290	1039	0.0235
5	749	+500	1249	0.0195

raised, so the pressure on the gas is P_a *plus* the height difference in mercury levels as measured on the meter stick. In position (c) the reservoir has been lowered so the pressure on the sample is P_a *minus* the height difference in mercury levels. Table 14-3 gives typical experimental results.

An inverse relationship is evident in the last two columns of the table: Decreasing pressure gives increasing volume; increasing pressure gives decreasing volume. Mathematically we write

$$V \propto \frac{1}{P}$$

where the symbol $\propto$ is read "proportional to." To change a proportionality to an equality, we insert a proportionality constant k and an equals sign:

$$V = k\frac{1}{P}$$

or

$$PV = k \tag{14-3}$$

This says the product of pressure times gas volume is a constant.

For the first pair of values in Table 14-3,

$$P_1V_1 = (749)(0.0328) = 24.6 \text{ torr liter}$$

and for the other pairs we obtain the PV products 24.3, 24.5, 24.4, and 24.4 in the same units. Within the limits of the experiment's range and precision, the results support Eq. 14-3.

Because PV is a constant, we can equate initial (1) and final (2) PV conditions for a gas system. This gives an alternate mathematical form of the law:

$$P_1V_1 = P_2V_2 \tag{14-4}$$

At low pressures, all gases follow Boyle's law exactly. At high pressures, deviations appear, as explained with Figure 14-13.

Example 14-2 One thousand cubic feet of oxygen gas at room temperature and pressure is to be compressed to a volume of 175 ft³. What pressure is needed?

Solution We will make several assumptions: **(a)** Boyle's law holds over the pressure range; **(b)** temperature remains constant; **(c)** room pressure is 740 torr. Initial conditions are 1000 ft³ and 740 torr. Final conditions are 175 ft³ and P torr. Substituting into Eq. 14-4 gives

$$(740)(1000) = (P)(175)$$

from which

$$P = 4.23 \times 10^3 \text{ torr}$$

The problem of Example 14-2 can be solved in a less stereotyped way. We know from previous discussion (and practical experience) that a gas contracts under pressure (as with the down stroke of a bicycle pump) and expands when pressure is relieved (as air issuing from a punctured tire). According to Boyle's law, to decrease the volume from 1000 to 175 cubic feet requires a pressure increase in the same ratio: 1000:175. We use the ratio as a factor in a factor-unit setup of the solution:

$$? \; torr = 740 \; torr \left(\frac{1000 \; ft^3}{175 \; ft^3} \right) = 4.23 \times 10^3 \; torr$$

The law of Charles and Gay-Lussac

The relationship between gas volume and temperature when pressure is kept constant was investigated in the year 1800 in France by J. Charles and Gay-Lussac, and in England by Dalton. Their experiments took this form: They warmed volumes of various gases from 0° to 100°C and determined the fractional change in volume for the 100°C temperature span. Their results approximated today's accepted value, 0.3663.

Thus when a gas is heated at constant pressure from 0° to 100°C, it increases in volume 36.63 percent. If heated from 0° to 1°C, it increases 0.3663 percent in volume. The corresponding fractional increase 0.003663 is called the *coefficient of volume change*. It is equivalent to the fraction 1/273. This means that the volume of a gas at 0°C increases 1/273 of its volume for each increase of 1°C. And the volume decreases 1/273 of its volume at 0°C for each decrease of 1°C. These facts were originally summarized by the statement

$$V = V_0 + 0.003663 V_0 \, \Delta t \tag{14-5}$$

where V_0 is gas volume at 0°C, Δt is temperature change in degrees centigrade, and V is the final volume.

Equation 14-5 is not an especially useful statement of the temperature–volume law. Derivation of a better equation can be obtained using the apparatus of Figure 14-5. In position (a), with the stopcock open, the mercury reservoir is held at such a height that some convenient volume of air (or other gas) enters the buret. The stopcock is closed, and temperature read on the thermometer. A liquid or gas at constant temperature is then circulated through the jacket to cool or heat the entrapped air. When the temperature is uniform throughout the system, mercury levels are readjusted to equalize them, and the temperature and volume read again. Typical laboratory data are given in Table 14-4.

The obvious increase of volume with temperature proves to be a straight-line relationship (Figure 14-6), and the results can be interpreted as a *definition of temperature: At constant pressure, temperature is a property that varies in a linear manner with the volume of a gas.* Downward extrapolation of the plot intersects the temperature axis at about −273°C. At that temperature gas volume presumably would be zero (although all gases condense to liquids or solids before −273°C), indicating that there is a lowest conceivable temperature in nature. This temperature, −273.15°C, is called *absolute zero,* a point at which molecules have minimum energy. Very low temperatures are attained by demagnetizing paramagnetic substances previously cooled in liquid helium. Lowest temperature attained thus far is only 0.000001°C above absolute zero. The laws of thermodynamics forbid actual attainment of absolute zero.

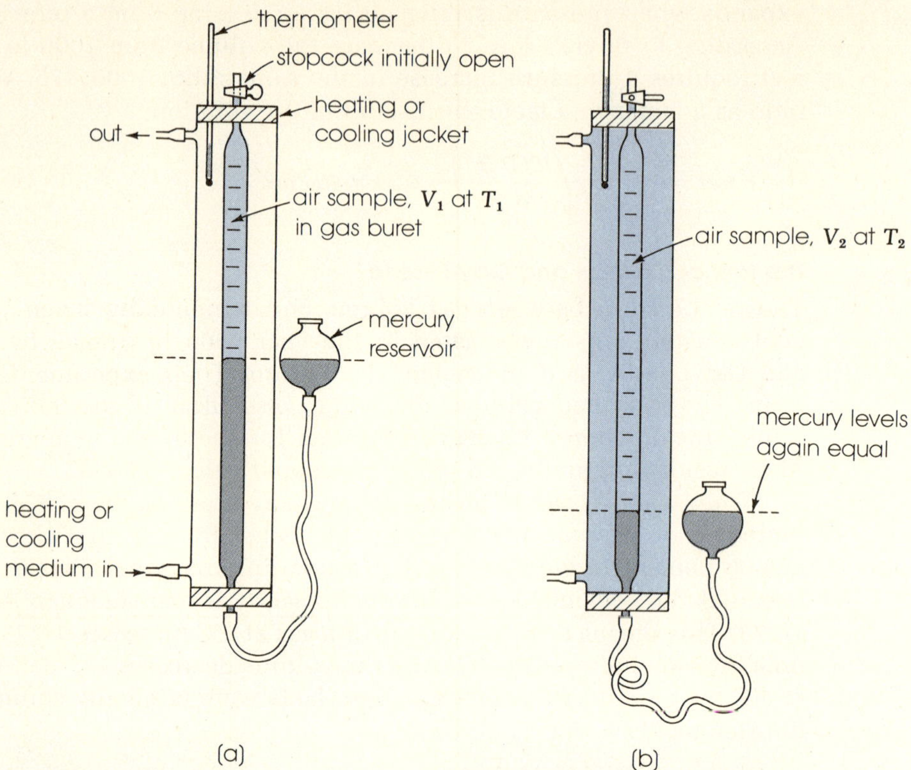

Figure 14-5 An apparatus for demonstrating the dependence of gas volume on temperature. The temperature in (b) is greater than the temperature in (a), making V_2 greater than V_1. Pressure on the confined gas is maintained at barometric pressure by adjusting the height of the reservoir to keep the mercury levels equal.

The straight-line graph might tempt one to write $V \propto t$ (volume is directly proportional to Celsius temperature), followed by $V = kt$, from which V/t should be constant. A second look at the table data, however, will show that the relationship is linear but not a direct proportion (doubling t does not double V), and thus V/t is not constant. What is needed is a new temperature scale to which gas volume is related directly.

The new scale was conceived by Kelvin in 1854. Today it is called the *International System Kelvin scale*. The unit is the Celsius degree which is called a *Kelvin*. Absolute zero is 0°K and the triple point of water (Appendix Table A-1) is considered to be 273.16°K. The triple point is a natural

Table 14-4. Gas volume variation with temperature (constant pressure)		
Trial	Temperature °C	Volume ml
1	−33.0	20.4
2	21.0	25.0
3	58.0	28.2
4	80.0	30.0
5	160	36.8

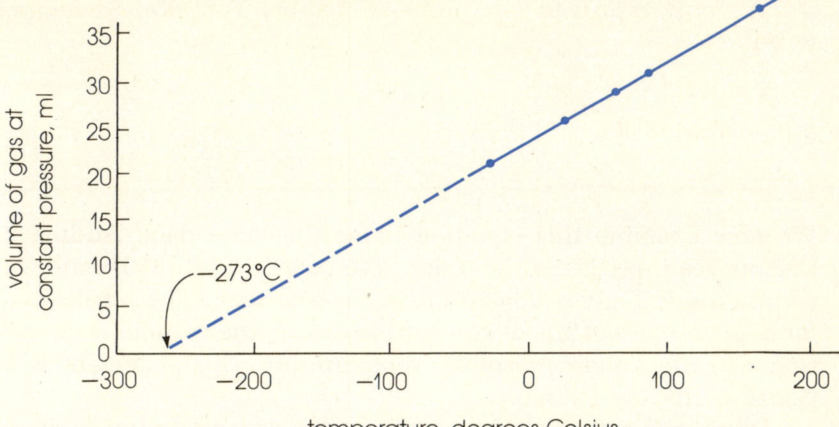

Figure 14-6 Dependence of gas volume on temperature, a plot of Table 14-4 data. Extrapolation locates absolute zero.

defining point 0.01°K above the freezing point of water at a pressure of 0.0060 atm. To convert a Celsius or centigrade reading to a Kelvin reading, we add 273.15. For most purposes this is rounded off at 273:

$$°K = °C + 273 \tag{14-6}$$

A comparison of the Kelvin, Celsius, and Fahrenheit temperature scales is made in Figure 14-7.

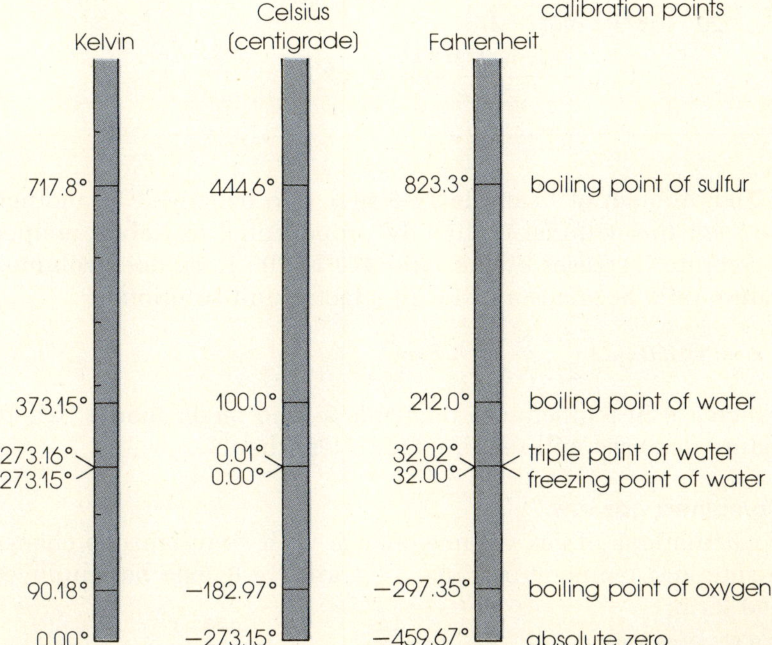

Figure 14-7 The three commonly used temperature scales, named for their inventors. The first two use the centigrade degree (also called the Celsius degree or Kelvin). The Fahrenheit degree is five-ninths as large as the centigrade degree. Some of the reference points used in thermometer calibration are indicated.

We now return to our problem. Letting T be Kelvin temperature we write

$$V \propto T$$

followed by

$$V = kT \quad \text{or} \quad \frac{V}{T} = k \tag{14-7}$$

We now examine this equation using the table data. Adding 273 to the Celsius readings gives, in order, 240°, 294°, 331°, 353°, and 433°K. And computing V/T gives 0.0850 ml/°K in each case. The results indicate that *for a given mass of gas at constant pressure, the volume is directly proportional to the Kelvin (absolute) temperature.* This statement is the *law of Charles and Gay-Lussac.*

When two sets of volume-temperature data are to be compared, a more convenient form of the law is

$$\frac{V_1}{T_1} = \frac{V_2}{T_2} \tag{14-8}$$

Its use is illustrated in the next example.

Example 14-3 A gas sample at 20.0°C has a volume of 100 ml. What will the volume be at 40°C?

Solution We will assume constant pressure. Substituting into Eq. 14-8 gives

$$\frac{100 \text{ ml}}{20° + 273} = \frac{V_2}{40° + 273}$$

$$V_2 = \frac{(100 \text{ ml})(313°K)}{293°K} = 107 \text{ ml}$$

The problem of Example 14-3 may also be solved in another way. We have seen that volume is directly proportional to Kelvin temperature. As temperature increases by the ratio 313°K:293°K, so does volume. The temperature ratio becomes a factor in a factor-unit solution:

$$? \text{ ml} = 100 \text{ ml} \left(\frac{313°K}{293°K} \right) = 107 \text{ ml}$$

An increase in temperature demands a ratio larger than 1, whereas a temperature decrease will require a ratio less than 1.

A combined gas law

For calculations of gas volume change with simultaneous changes in both pressure and temperature, Eqs. 14-4 and 14-8 may be employed in combined form:

$$\frac{P_1 V_1}{T_1} = \frac{P_2 V_2}{T_2} \tag{14-9}$$

Problems solvable by Eq. 14-9 can also be solved by the factor-unit approach, as illustrated in Example 14-4.

Example 14-4 A balloon is partially inflated with helium at ground level. Its volume is 200 ft³, temperature is 20°C, pressure is 740 torr. What will the volume be at a 19-mi altitude where temperature is −40°C and pressure 10 torr?

Solution Original volume V_1 will be multiplied by temperature and pressure factors to give the final volume V_2,

$$V_2 = V_1 (T \text{ factor})(P \text{ factor}) \tag{14-10}$$

Initial temperature is $20° + 273 = 293°K$. Final temperature is $-40° + 273 = 233°C$. From Charles' law we know a decrease in Kelvin temperature means a proportionate decrease in volume. The T factor will therefore be less than one, 233°K/293°K.

Initial pressure is 740 torr; final pressure is 10 torr. According to Boyle's law, a decrease in pressure gives a proportionate increase in volume. The P factor will thus be greater than one, 740 torr/10 torr. Inspection of the two factors shows the latter will have the greater effect, and V_2 will be larger than V_1 by about 60 times:

$$V_2 = 200 \text{ ft}^3 \left(\frac{233°K}{293°K}\right)\left(\frac{740 \text{ torr}}{10 \text{ torr}}\right) = 11.8 \times 10^3 \text{ ft}^3$$

Standard conditions

By convention standard temperature and pressure (STP) conditions are defined for reference in gas calculations. When no other conditions are given, standard conditions are often assumed. *Standard temperature is 273°K (0°C). Standard pressure is 1 standard atmosphere.* (Equivalent pressures are given in Eq. 14-2.)

The general perfect-gas equation

The combined gas law equation, $P_1 V_1 / T_1 = P_2 V_2 / T_2$, says that PV/T is a constant for a given quantity of gas. That constant is called the *universal gas constant,* or simply the *gas constant,* and is designated R. Letting the quantity of gas be n moles, we write

$$PV = nRT \tag{14-11}$$

Equation 14-11 is called the *general perfect-gas equation,* after the theoretical concept of a "perfect" gas whose behavior the equation describes exactly. Under normal conditions the equation describes the behavior of real gases quite well also and is the most compact way we have of interrelating all the variables.

When the volumes of 1 g-molecular weight of various gases are measured at STP, they are found to average 22.4 liters (roughly 0.8 ft³). Chemists assume the same is true for any gas, and call *22.4 liters the molar volume of a gas at standard conditions.* At STP, 22.4 liters is the volume of 2.02 g of H_2, 28.8 g of air, 32.0 g of O_2, etc.

Example 14-5 Calculate a value for the gas constant R.

Solution The numerical value of R depends on the units of the other terms in the perfect-gas equation. Because no units are given, we are free to select a set from values we know. From prior discussion we know that 1 mole of gas occupies 22.4 liters at STP. Standard temperature is 273°K, standard pressure is 1 atm. Therefore

$$R = \frac{PV}{nT} = \frac{(1\ atm)(22.4\ liters)}{(1\ mole)(273°K)} = 0.0820\ atm\ liter/mole\ °K$$

Had we chosen to use 760 torr instead of 1 atm for P, the answer would have been 62.3 torr liters/mole °K. There are thus several values for R, and we must employ the one appropriate to the units used in the problem.

Example 14-6 A 50-liter boiler containing 100 g of liquid water is heated to 250°C. What is the internal pressure?

Solution We assume that all the water will be in the vapor state. The molecular weight of H_2O is 18.0. The number of moles of water present is $100/18.0 = 5.56$. Rearranging, then substituting into Eq. 14-11 with the value of R calculated above, we obtain the pressure in atmospheres,

$$P = \frac{nRT}{V} = \frac{(5.56)(0.0820)(250 + 273)}{50} = 4.8\ atm$$

Calculation of gas density

The density of a gas at standard temperature and pressure can be calculated by dividing its gram-molecular weight by the gram-molecular volume. For example, at STP, we know 32.0 g of oxygen occupies 22.4 liters. Density is 32.0/22.4, or 1.43 g/liter.

The density of a mixture of gases at STP can be found by first computing an effective gram-molecular weight for the mixture and dividing that number by 22.4 liters.

Example 14-7 Find the approximate density of air at STP.

Solution By volume, air is approximately 79 percent N_2 and 21 percent O_2 (Table 13-1). Taking these percentages of the gram-molecular weights of N_2 and O_2 and adding gives the gram weight of 22.4 liters of air. Division then gives the density:

$$(0.79)(28\ g) + (0.21)(32\ g) = 28.8\ g/22.4\ liters\ of\ air$$

$$D = \frac{28.8\ g}{22.4\ liters} = 1.3\ g/liter$$

Dalton's law

John Dalton, who had a lifelong interest in meteorology, began a series of measurements on water vapor in the air in 1799. In one experiment he added water vapor to a confined volume of dry air and noted an increase in total pressure equal to the gas pressure of the water vapor added. He correctly concluded that *the total pressure of a dilute, nonreacting mixture of gases is equal to the sum of the partial pressures of the individual gases.* (The partial pressure of a gas is the pressure it would exert if alone in the volume occupied by the mixture.) This is a statement of *Dalton's law.* Mathematically,

$$P_{\text{total}} = P_1 + P_2 + P_3 + \cdots + P_n \qquad (14\text{-}12)$$

The terms on the right side are partial pressures of gas-mixture components 1 through n. Use of the law is illustrated in Example 14-8.

Vapor pressure

A volatile liquid standing in an open container gradually evaporates into the air. If the container is closed, however, and pumped free of air, evaporation is limited to the space over the liquid. A dynamic equilibrium soon becomes established there between molecules of evaporating liquid and molecules of condensing vapor. The pressure above a liquid due only to its own vapor at equilibrium is called its *vapor pressure.* The temperature at which vapor pressure equals 760 torr is called the *normal boiling point* of the liquid. The way vapor pressure changes with temperature is shown in Figure 14-8. Water vapor pressure data are given in Table 14-5.

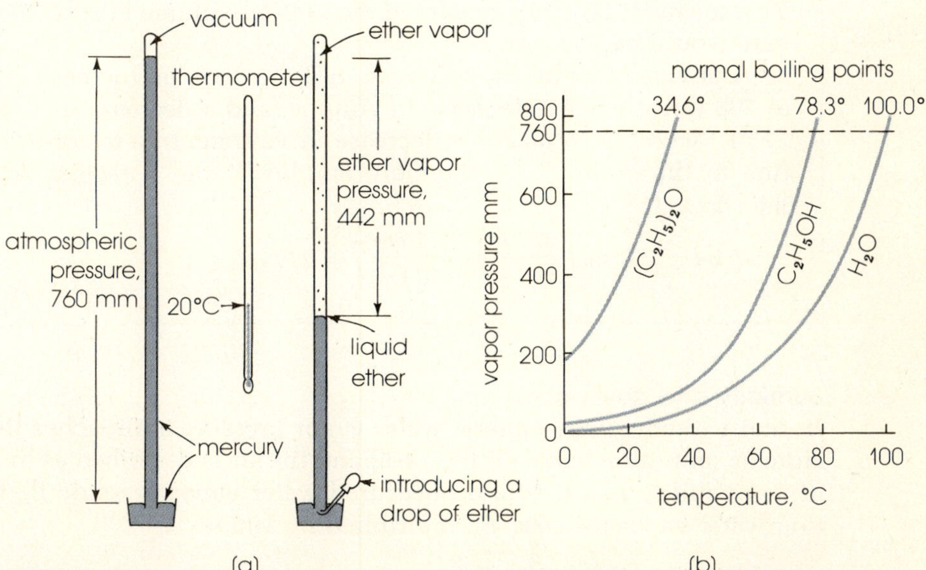

Figure 14-8 Vapor pressures of liquids. (a) A volatile liquid introduced under a barometer tube rises to the top and evaporates in the vacuum there. As equilibrium is established between excess liquid and vapor, the vapor forces the mercury column down in direct measure of its pressure. At 20°C the vapor pressure of ether is 442 mm. (b) Vapor pressure plots of diethyl ether, ethyl alcohol and water.

Table 14-5. Pressure and weight of water vapor at various temperatures

Temperature °C	Temperature °F	Water vapor pressure torr	Water vapor in saturated air g/m³
−10 (ice)	14	2.1	2.16
0	32	4.6	4.85
10	50	9.2	9.40
20	68	17.5	17.3
30	86	31.8	30.4
40	104	55.3	51.1
50	122	92.5	
60	140	149	
70	158	234	
80	176	355	
90	194	526	
100	212	760	

Example 14-8 Two hundred milliliters of hydrogen gas is collected by water displacement (Figure 13-4) at a barometric pressure of 750 torr and a temperature of 20°C. What is the volume of (dry) H_2 at STP?

Solution Having bubbled through water, the H_2 is assumed to be saturated with water vapor. In Table 14-5 we note that at 20°C the water vapor pressure contribution to the system's total pressure is about 18 torr. From Dalton's law,

$$750 \ torr = P_{H_2} + P_{H_2O \ vapor} = P_{H_2} + 18 \ torr$$

$$P_{H_2} = 732 \ torr$$

This means if H_2 alone occupied the 200-ml volume at 20°C, the pressure would be 732 torr.

Conversion of the H_2 volume to STP involves an increase in P (732 to 760 torr), hence a decrease in volume, and a decrease in T (293° to 273°K) which also means a decrease in volume. In a factor-unit solution to the problem we therefore employ T and P factors both less than 1:

$$? \ ml = 200 \ ml \left(\frac{273°K}{293°K}\right)\left(\frac{732 \ torr}{760 \ torr}\right) = 179 \ ml$$

Humidity and dew point

In moist weather atmospheric water vapor pressure approaches its equilibrium or saturation value (Table 14-5) and the air is described as humid. *Relative humidity* is the ratio of measured water vapor pressure P_m to saturation water vapor pressure P_s, multiplied by 100:

$$relative \ humidity = \frac{P_m}{P_s} \times 100 \qquad (14\text{-}13)$$

We know from experience that warm air can hold more water vapor than cold air. Thus if air saturated at one temperature is raised in temperature, it becomes less than saturated. Conversely, when unsaturated air is cooled, it moves toward a saturated condition. The temperature at which water (in some cases, ice) deposits on a solid surface is called the *dew point*. By definition saturated air is at its dew-point temperature.

Example 14-9 On a certain day when the temperature is 86°F, the atmospheric water vapor pressure is found to be 17.5 torr. What is the relative humidity? The dew point?

Solution In Table 14-5 we read that saturated air at 86°F has a water vapor pressure of 31.8 torr. The saturation or humidity ratio is therefore 17.5/31.8. The relative humidity is

$$\frac{17.5}{31.8} \times 100 = 55.0 \text{ percent}$$

The dew point is that temperature at which the observed vapor pressure equals the saturation or table-value pressure. Reading the table again, we see that not until the temperature falls to 68°F will this occur.

The dew point is 68°F.

Graham's law

Thomas Graham, a Scotchman, was one of the first physical chemists—a scientist whose research on such subjects as heats of chemical reaction, viscosity of liquids, and transport of gases combined principles from physics and chemistry. In experimentation done around 1830 he investigated how gases *diffuse* through one another. He found that diffusion rate was inversely proportional to the square root of gas density; that is,

$$r \propto \frac{1}{\sqrt{D}}$$

from which, with insertion of a proportionality constant k, he wrote

$$r = \frac{k}{\sqrt{D}} \quad \text{or} \quad r\sqrt{D} = k \tag{14-14}$$

The experimental data of Table 14-6 show how Eq. 14-14 may be tested.

Graham's law is often stated with reference to the comparison of two gases: *Under the same conditions the diffusion rates of two different gases*

Table 14-6. Diffusion of gases at 1 atm pressure and 25°C

Gas	Diffusion rate r ml/min	Density D g/liter	$\sqrt{D}$	$r\sqrt{D}$
H_2	4.90	0.0818	0.286	1.40
N_2	1.30	1.15	1.07	1.39
CO_2	1.00	1.80	1.34	1.34
Cl_2	0.790	2.90	1.70	1.34

are inversely proportional to the square roots of their densities or of their molecular weights. For gases 1 and 2,

$$\frac{r_1}{r_2} = \frac{\sqrt{D_2}}{\sqrt{D_1}} = \frac{\sqrt{M_2}}{\sqrt{M_1}} \tag{14-15}$$

Equation 14-15 also applies to *effusion.* The rates at which gases effuse (escape) through a tiny hole into an evacuated space are inversely proportional to the square roots of their densities or molecular weights. (An effusion apparatus is illustrated with Problem 34.)

Equation 14-15 can be used to find gas molecular weights. It is also used to calculate the efficiency of separations by diffusion. As mentioned in Chapter 4, uranium isotopes were isolated by diffusion during World War II. Separation of uranium-235 from uranium-238 was accomplished by preparing the gaseous hexafluorides (molecular weights 349 and 352, respectively) and pumping them through barriers specially made to contain holes of the proper size. With molecular weights nearly the same, the ratio of diffusion rates (the enrichment factor) was close to unity and separation required passage of the gas mixture through several thousand diffusion chambers:

$$\frac{\text{rate of } ^{235}UF_6}{\text{rate of } ^{238}UF_6} = \frac{\sqrt{352}}{\sqrt{349}} = 1.0043$$

Experimental gas molecular weights

Chemists determine molecular weights as a means of identifying substances. Our brief acquaintance with gas principles already makes three experimental methods available:

1. Diffusion or effusion rates of a known and an unknown gas are measured. Equation 14-15 is then used to find the molecular weight of the unknown.
2. The weight of a volume of gas (in effect its density) is determined at STP. Because a gram-molecular weight of any gas occupies 22.4 liters, a simple proportion leading to the weight of 22.4 liters of the

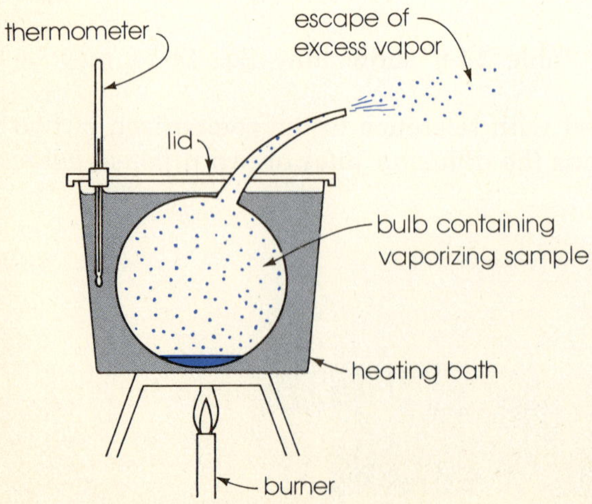

Figure 14-9 Molecular weights of volatile liquids from vapor density; the method of J. Dumas (France 1827). After the bulb is weighed full of air at room temperature, excess sample liquid is drawn into it. The bulb is then heated in the bath until excess vapor ceases to come from the tip. The temperature is read on the thermometer. The tip is sealed and the bulb cooled to room temperature and weighed. The tip is broken off under water, allowing water to enter and fill the bulb. The bulb is weighed full of water to obtain its volume (1 g of $H_2O = 1$ ml). The laboratory barometer is read. Vapor molecular weight is calculated.

gas gives its molecular weight (Example 12-5). The molecular weight of a volatile liquid may be determined in the same way by measuring the density of its vapor (see Example 14-11). An early vapor density device is pictured in Figure 14-9.

3. Equal volumes of a known and an unknown gas at the same temperature and pressure are weighed. According to Avogadro's principle (Section 12-2) the two samples contain equal numbers of molecules. The weight ratio is therefore the same as the ratio of their molecular weights. The molecular weight of the unknown is found by multiplying the experimental ratio by the molecular weight of the known (see Example 14-10).

Example 14-10 Exhaust gas from a gasoline engine is cooled and collected in a flask. Flask plus exhaust gas weigh 148.05 g. At the same temperature and pressure the flask filled with nitrogen gas weighs 147.48 g. The flask evacuated by vacuum pump weighs 145.35 g. Calculate the molecular weight of the engine exhaust.

Solution The weight of exhaust gas is $148.05 - 145.35 = 2.70$ g. The weight of nitrogen gas is $147.48 - 145.35 = 2.13$ g. The known molecular weight of N_2 is 28.0. Moleculer weight M of the exhaust gas is larger by the factor that compares the weights of equal volumes:

$$M = 28.0 \left(\frac{2.70 \ g}{2.13 \ g} \right) = 35.5$$

Example 14-11 Fuel taken from a racing car is a yellowish oily liquid boiling at about 100°C. By the Dumas method (Figure 14-9) 0.250 liter of its vapor at 127°C and 740 torr weighs 0.453 g. What is the molecular weight?

Solution We will solve the problem two ways. In the first method we will find the STP volume of the vapor and from it calculate the weight of 1 gram-molecular volume.

Using either Eq. 14-9 or 14-10,

$$V_2 = 0.250 \ liter \left(\frac{740}{760} \right) \left(\frac{273}{400} \right) = 0.166 \ liter$$

The *weight* of vapor is, of course, not affected by any volume change, hence at STP 0.166 liter of vapor weighs 0.453 g. The gram weight of 22.4 liters at STP is the gram-molecular weight:

$$? \ g = 22.4 \ liters \left(\frac{0.453 \ g}{0.166 \ liter} \right) = 61.1 \ g$$

The same result can be obtained in one step by using a modified form of the perfect-gas equation (Eq. 14-11). For n, the moles of gas, we will substitute m, the weight of gas, divided by M, its molecular weight:

$$PV = \left(\frac{m}{M}\right) RT \tag{14-16}$$

In the units of the problem R has the value 62.3 torr liters/mole °K (see the end of Example 14-5). Substitution gives

$$M = \frac{(0.453)(62.3)(400)}{(740)(0.250)} = 61.1$$

(A chemical analysis shows the compound is nitromethane, CH_3NO_2.)

14-3. THE KINETIC-MOLECULAR THEORY

Brown, Bernoulli, and heat

Today we are accustomed to thinking of gases as composed of molecules in rapid motion, yet experimental support of the idea has been available only since the mid-nineteenth century. Modern gas theory began with physicist R. Hooke, a contemporary of Boyle, who suggested that moving particles in a gas exerted pressure by striking the walls of their container. (Boyle thought of gas particles as tiny springs coiling and uncoiling with changes in pressure.) Evidence of molecular movement was reported in 1828 when R. Brown, a Scottish botanist, observed under a microscope that pollen grains suspended in a drop of water were maintained in continuous and erratic motion. In other demonstrations he showed that any suspended microscopic solid (smoke in air, for example) had a restless movement that resisted the settling force of gravity. Brownian movement was correctly explained as due to bombardment of the observable particles by the ever-moving invisible molecules of the suspending medium (Figure 14-10). Increasing temperature increased the movement.

In 1738 Swiss physicist D. Bernoulli derived an equation to describe a model gas composed of submicroscopic spheres in motion. As pictured in

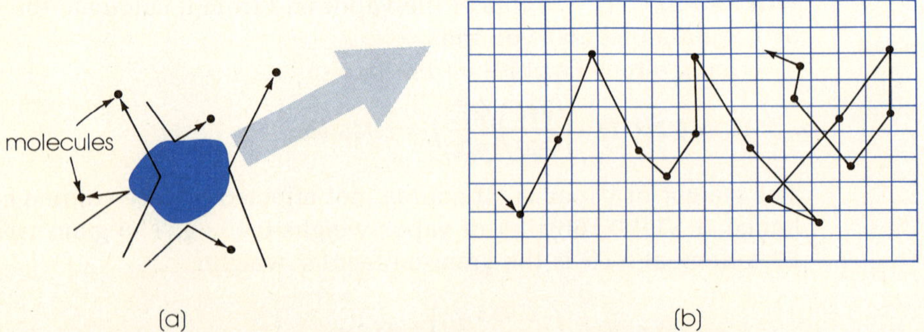

Figure 14-10 Brownian movement. (a) A microscopic-sized particle moves when struck by molecules suspending the particle. The smaller the particle, the more erratic its movement as it acquires the characteristics of molecular motion. The particle is shown moving to the right because more molecules at one instant happen to have hit it on the left. (b) Movement of the particle as seen using side lighting under a microscope. Position in noted on a reference grid at regular time intervals.

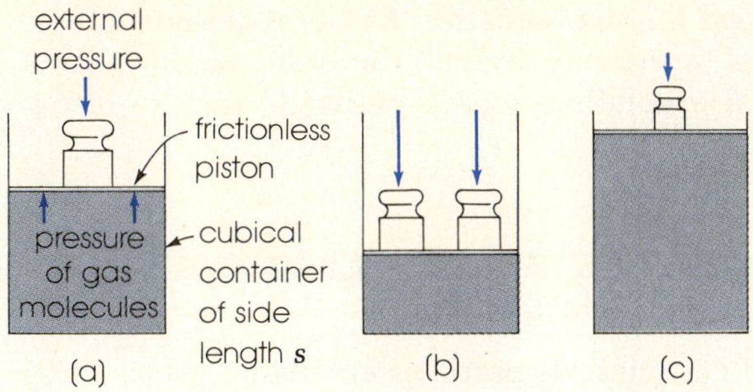

Figure 14-11 Bernoulli's visualization of gas behavior. Each fixing of piston position results in a new equilibrium condition in which pressure of the confined gas created by its molecules bombarding the piston balances the applied external pressure.

Figure 14-11 (a), he imagined the gas confined in a cubical box by a piston, its particles in chaotic motion and exerting a pressure on the piston just equal to the external pressure. As external pressure is increased   (b), gas volume decreases until enough molecules strike the piston is unit time to again balance the applied pressure; decreasing external pressure (c), causes the gas to expand to a new volume in which fewer particles must strike the piston in unit time to balance the applied pressure. The equation Bernoulli deduced was

$$P = \frac{Nmv^2}{3s^3}$$

where P is gas pressure, N is the number of confined gas particles, m is the mass of one particle, v is particle velocity, and s is the length of a side of the cubical container; s^3 is a volume. Replacing s^3 with V (volume) gave the most important equation available for more than a century in the study of gas dynamics:

$$PV = \frac{Nmv^2}{3} \tag{14-17}$$

Bernoulli knew that a more complete explanation of gas behavior would include terms related to heat. He was, however, unable to put the idea into mathematical form because the nature of heat had not yet been explained. Then, and for the next hundred years, heat was considered to be a physical fluid that some experimenter would eventually capture. The prevailing view came from an adaptation that Dalton made of Lavoisier's *caloric* concept. Lavoisier thought that gases expanded upon addition of caloric, an invisible material whose particles repelled each other, and contracted as caloric was taken away. Dalton described "caloric" as expansible heat stuff holding embedded gas particles.

Our present concept of heat began to take shape when all attempts to weigh heat failed and when the significance of the production of heat by friction, and friction by motion, was realized. In the 1840s English physicist J. Joule connected heat and energy by determining the amount of work required to produce a unit of heat. He did so in several independent ways, the most accurate of which related the heating of a fluid to the mechanical energy of a paddle wheel that stirred the fluid. His experiments ended the caloric theory and introduced a new theory. Besides Joule and Kelvin, its developers were the giants of the science called *statistical mechanics* (Section 14-4), the mathematical physicists R. Clausius (Germany), J. Maxwell (England), and L. Boltzmann (Austria). They added temperature to the equations and further developed the mathematics. Earlier experimentalists had made observations on real gases and derived the basic gas laws from them. Now theorists advanced a detailed energy–molecule picture of nature.

The kinetic model and the gas laws

The kinetic model of a gas upon whose idealized mechanics gas behavior would be described was given four characteristics:

1. The gas is composed of particles. Its particles are many, yet so small that the volume they occupy is normally negligible compared to the volume of the gas.
2. The particles have insignificant attraction for one another.
3. The particles are in constant, rapid, random motion.
4. The particles collide with one another and with the container walls without losing average kinetic energy.

The model proved to be an excellent one—easy to understand, reliable in helping relate theory to experimental facts, and capable of providing, as outlined below, new refinement of the gas laws.

1. Boyle's law A gas is readily compressed because its molecules occupy relatively little space. A gas expands readily under decreased pressure because, unlike the strong forces in liquids and solids, its intermolecular forces are very weak. When a gas is compressed so its volume is, say, cut in half, its molecules strike the walls twice as often. That doubles the pressure in the container and just balances the external compressive force.

Bernoulli gave Boyle's law a quantitative interpretation with his equation, $PV = Nmv^2/3$. For a known quantity of gas (N particles, each of mass m), the right side is a constant if particle velocity is constant—meaning $PV = k$, which is Boyle's law. Actually, not all molecules in a gas have the same velocity but Maxwell, who worked with a distribution of particle velocities, obtained a result having the same mathematical form:

$$PV = \frac{Nm\bar{v}^2}{3} \tag{14-18}$$

where $\bar{v}^2$ is an averaged value of v^2. Equation 14-18 is called the *kinetic-theory equation*.

2. The law of Charles and Gay-Lussac Boyle's law holds only at constant temperature. As expected then, Eq. 14-18 shows that $\bar{v}^2$ is directly proportional to absolute temperature T. It follows that Eq. 14-18 applies to the law of Charles and Gay-Lussac as well as to Boyle's law, and it can be written

$$PV = kT$$

where k is a constant incorporating all the other terms. With P held fixed, the equation can be arranged to read $V/T = k$, which is the volume–temperature law (Eq. 14-7).

Temperature was thus not included in the original postulates of the kinetic theory because it was not recognized as it is today to be a measure of particle agitation. With the development of statistical mechanics, however, scientists discovered that *the absolute temperature of a gas is directly proportional to the average kinetic energy of its molecules.*

3. Dalton's law The kinetic theory assumes negligible cohesive forces between molecules. In a mixture of gases each gas consequently acts as if it were alone in the container and exerts a pressure proportional to the number (moles) of its particles present, their mass, and their velocity. The total pressure of the mixture is the sum of the individual (partial) pressures. This is a restatement of Dalton's law.

4. Graham's law Molecules of the ideal gas are taken to be point masses—that is, particles of insignificant size. It is obvious, however, that the molecules of two different real gases have different masses and, at a given temperature and pressure, have different average velocities. It follows that when gases are compared by their ease of escape through a small opening into a vacuum, escape rate bears an inverse relationship to molecular weight.

The relationship can be deduced from $PV = N m \bar{v}^2/3$, which may be rewritten $m\bar{v}^2 = k$ for the situation in which gas quantity, pressure, and volume are kept constant. Solving for $\bar{v}$ gives

$$\bar{v} = \frac{k'}{\sqrt{m}} \tag{14-19}$$

This says that the average molecular velocity is inversely proportional to the square root of a molecule's mass. The equation is like Eq. 14-14, which is Graham's law.

14-4. STATISTICAL MECHANICS

Distribution of molecular velocities

The law governing the distribution of molecular velocities as derived by Maxwell and improved by Boltzmann is the cornerstone of statistical mechanics. Statistical mechanics has been eminently successful in explaining the properties of matter at large by examining the energy distribution in its molecular parts.

The energy possessed by a body because of its motion is called *kinetic energy* E_t defined by the equation

$$E_t = \tfrac{1}{2} m v^2 \tag{14-20}$$

where m is body mass and v is its velocity. Maxwell assumed that average kinetic energy of molecules increased with temperature, and that at the same temperature all gases had the same average kinetic energy per molecule. Thus molecular mass and velocity were inversely related to keep $m\bar{v}^2/2$ constant. He also assumed that as a result of *elastic collisions*—that is, those in which energy was transferred without loss from one molecule to another—a distribution of molecular velocities became established in any gas.

The manner in which the total energy of a gas is shared among its molecules is illustrated in Figure 14-12; points for plotting such curves can be calculated from the equation given there. At any temperature a small fraction of the molecules travels comparatively slowly and a small fraction

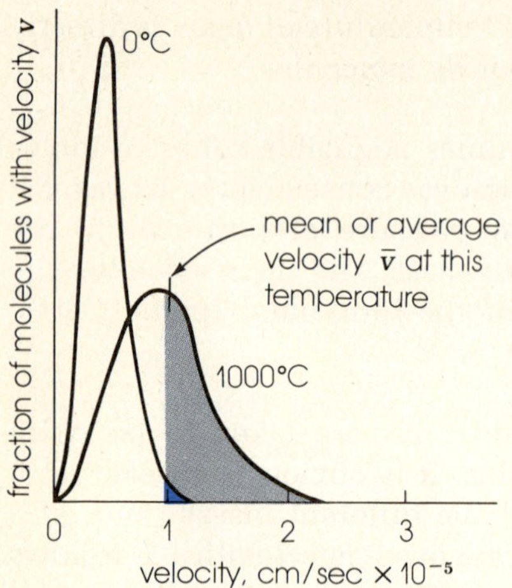

Figure 14-12 Distribution of molecular velocities for nitrogen gas at two different temperatures, as calculated from the Maxwell-Boltzmann distribution equation,

$$\frac{\Delta N}{N} = 4\pi \left(\frac{m}{2\pi kT}\right)^{3/2} e^{-mv^2/2kT} v^2 \Delta v$$

where $\Delta N/N$ is the fraction of molecules having velocities between v and Δv, m is the mass of one molecule, k is Boltzmann's constant (1.38×10^{-16} erg/molecule °K), T is absolute temperature, and e is 2.72, the base of natural logarithms.
As indicated by the shaded areas, the fraction of molecules with high velocities (say above 1×10^5 cm/sec) increases markedly with temperature.

comparatively rapidly, but most travel near the average velocity. The higher the temperature, the more flattened the distribution curve and the further its maximum moves to the right. This means that both the fraction of molecules having higher velocities and those with average velocity increase with increasing temperature.

Fifty years after the distribution equation was introduced it was confirmed experimentally by Stern. Using a molecular beam apparatus (similar in concept to the one pictured in Figure 3-10, but without the magnet) Stern allowed a hot beam of metal atoms to pass through a slit and fall under the normal influence of gravity onto a measuring device. As fall varied with velocity of the atoms (slower atoms fell farther), a distribution of distances was obtained. Plotting fall distance against signal intensity gave a curve like those of Figure 14-12.

Average velocity

The mathematics supporting the kinetic theory permit calculation of molecular velocities which scientists average in various ways. For approximate purposes the averages may be considered to be the same, and we will use only one. We will call it the *average velocity* ($\bar{v}$) and say it is equal to the sum of all the molecular velocities divided by the number of velocities considered. At absolute temperature T, the average velocity in centimeters per second of molecules having molecular weight M is given by

$$\bar{v} = \sqrt{\frac{8RT}{\pi M}}^{\,*} = 1.46 \times 10^4 \sqrt{\frac{T}{M}} \tag{14-21}$$

Example 14-12 Calculate the average velocity of O_2 molecules at STP.

Solution The molecular weight of oxygen is 32.0 and standard absolute temperature is 273°K. Using Eq. 14-21,

$$\bar{v} = 1.46 \times 10^4 \sqrt{\frac{273}{32}} = 4.26 \times 10^4 \ cm/sec$$

That is equivalent to a velocity of 1400 ft/sec or 960 mi/hr.

---

* R, the gas constant, has the value 8.31×10^7 ergs/mole °K.

Collision frequency

Now that we know molecular velocities are high, we may inquire how often molecules collide. Statistical mechanics furnishes an equation to give the answer:

$$Z = \frac{\sqrt{2}}{2}\pi n^2 \sigma^2 \bar{v} = 2.22 n^2 \sigma^2 \bar{v} \tag{14-22}$$

where Z is the total number of collisions in 1 cm³ of gas in 1 second, n is the number of molecules in that volume, σ (sigma) is a molecule's *collision diameter* (length of cross section), and $\bar{v}$ is again the average molecular velocity in centimeters per second.

Example 14-13 Given the collision diameter of an O_2 molecule, 3.40×10^{-8} cm, find the total number of collisions in 1 cm³ of oxygen at standard conditions.

Solution At STP Avogardro's number of molecules occupy 1 molar volume of gas (22.4 liters). In 1 cm³ are 6.02×10^{23} molecules/22,400 cm³ $= 2.69 \times 10^{19}$ molecules. With such a large number moving in a confined space at the average velocity calculated in Example 14-12, we can anticipate a very large number of collisions per second. Substituting into Eq. 14-22 we find

$$Z = (2.22)(2.69 \times 10^{19})^2 (3.40 \times 10^{-8})^2 (4.26 \times 10^4)$$

$$= 7.91 \times 10^{28} \text{ collisions/cm}^3 \text{ sec}$$

A somewhat more readily grasped number is obtained by dividing the collisions per second by the number of molecules present, giving the collisions experienced in 1 sec by just one molecule:

$$\frac{7.91 \times 10^{28}}{2.69 \times 10^{19}} = 2.94 \times 10^9 \text{ collisions/molecule sec}$$

The number is still very large.

Mean free path

The last calculation demonstrates that except at low pressures, where molecular population is drastically reduced, a gas molecule cannot travel far without colliding with another molecule. The average distance between collisions is called the *mean free path*, λ (lambda). Mean free path (in centimeters) can be calculated from still another equation of statistical mechanics:

$$\lambda = \frac{1}{\sqrt{2}\pi n \sigma^2} = \frac{0.226}{n \sigma^2} \tag{14-23}$$

where n and σ have the same meaning as in Eq. 14-22.

Maxwell used the concept of mean free path to explain how a gas conducted heat. From it he also predicted that the viscosity of a gas would be dependent on temperature but independent of pressure. Although doubted at first by critics (they claimed high pressure must increase the resistance of a gas to flow), experiment proved Maxwell right.

Example 14-14   Calculate the mean free path of O_2 at STP.

Solution   We will use Eq. 14-23 and the values of n and σ from Example 14-13:

$$\lambda = \frac{0.226}{(2.69 \times 10^{19})(3.40 \times 10^{-8})^2} = 7.27 \times 10^{-6} \ cm$$

For a better feel of the answer we can divide the mean free path by the collision diameter of a molecule and obtain the number of diameters a molecule moves in a straight line on average before hitting another molecule. The result is

$$\frac{7.27 \times 10^{-6}}{3.40 \times 10^{-8}} = 214 \ diameters/collision$$

Escape of the atmosphere

An interesting application of statistical mechanics can be made in discussing possible escape to outer space of the earth's atmosphere. Species with the best chance of leaving the earth are light exosphere ions having high speed and low collision frequency. At a height of 300 miles where the kinetic temperature may be 1500°K and the pressure 10^{-8} torr (Figure 13-7), only a few tens of millions of particles (ions mostly) exist per cubic centimeter compared to more than 10^{19} particles at STP. Under these conditions the average velocity of an oxygen ion O^+, for example, is about 3100 miles per hour and its mean free path 35 miles. Moving further from the earth it undergoes collisions considerably less frequently. At a 600-mile altitude its mean free path is 300 miles.

When we ask what particle velocity is needed for escape from the planet's gravitational field, the force of gravity must be included in the calculation. From Newton's law of universal gravitation we get the equation that relates escape velocity to the mass of the body from which escape is to be made:

$$v_e = \sqrt{\frac{2G_0M}{r}} \tag{14-24}$$

G_0 is the gravitational constant (6.67×10^{-8} dyne cm²/g²), M is the mass of the earth (5.97×10^{27} g), r is the earth's radius (6.37×10^{8} cm), and v_e is escape velocity whose units are centimeters per second. Substitution of these numbers into Eq. 14-24 gives an escape velocity of 1.12×10^{6} cm/sec (2.51×10^{4} mi/hr).

The influence of gravity decreases as the square of the distance from the earth's center. At a 300-mile altitude the force is 0.865 times the surface force. That is still sufficient to keep captive the ions that possess low to average velocity. Those that through collisions attain velocities several times the average, as some must to satisfy the Maxwell-Boltzmann distribution law, can be lost if traveling in a direction away from the earth.

Is there danger that the atmosphere will soon stream away? The answer is no. According to calculations made by English astrophysicist J. Jeans a planetary atmosphere will remain captive if the surface escape velocity is greater than five times the average velocity of atmospheric gas molecules. Comparing v_e just calculated with v found in Example 14-12, we see the value of the ratio $v_e/\bar{v}$ is approximately 26 for the earth. Even especially energetic particles in very high altitudes, which may have free paths measuring many miles, will generally curve back from space, drawn in by gravity.

14-5. REAL GASES AND NONIDEALITY

Van der Waals forces

In developing the kinetic theory scientists used the concept of an ideal gas that obeys the gas laws perfectly. Real gases, however, deviate from ideality, especially at lower temperatures and higher pressures, an indication that cohesive intermolecular forces and the volume occupied by molecules themselves are factors to be considered.

Attraction that comes into play between molecules when free path is decreased by increased pressure is *van der Waals attraction*, the same effect that holds molecular crystals together (Section 10-2). The normally weak attraction between the positive nuclei in one molecule for the negative electrons of the next molecule becomes influential when the molecules are nearly in contact. When gases are liquefied, it is this force that overcomes nuclei–nuclei and electron–electron repulsions, and conversely it is the same force that is overcome when a liquid is converted to a gas.

Nonideal behavior in real gases can be detected by plotting gas-law data gathered over a range of experimental conditions and noting deviation from straight-line expectations. For example, at fixed temperature we expect pressure times volume to be a constant. A plot of *PV* vs. *P* for any real gas it not, however, the anticipated straight line over a wide pressure range, as Figure 14-13 shows. Nitrogen and oxygen are seen initially exhibiting negative deviations from Boyle's law. Their *PV* products are less than ideal because van der Waals attraction reduces the theoretical volumes. At higher pressures N_2 and O_2 molecules occupy enough volume themselves to prevent ideal compression, giving a larger *PV* product than expected. By contrast hydrogen shows positive deviation at all pressures. This means

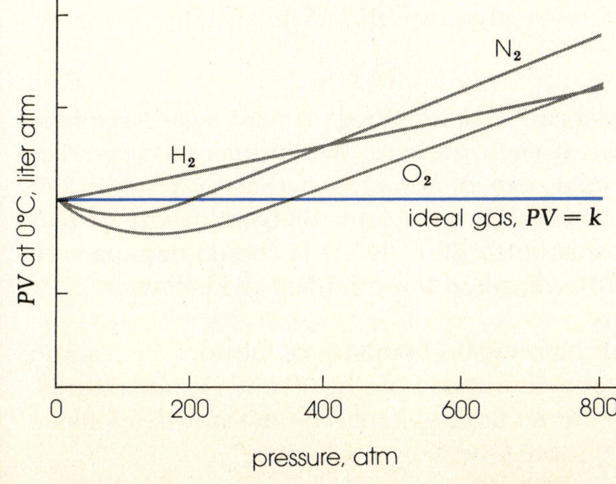

Figure 14-13 The PV product of several gases. All real gases depart from ideality at higher pressures. If Boyle's law were followed, PV for each gas would lie on the ideal line.

that although the H_2 molecules are closer together at higher pressures, the attraction between them remains small. Hydrogen's behavior is not unexpected; at 0°C it is almost 253°C above its normal boiling point.

Nonideality in real gases due to attractive forces is also demonstrated by the *Joule-Thompson effect*. An ideal gas with no intermolecular attractions should expand from a compressed state into a vacuum with no change in temperature because the gas does no work that would take kinetic energy from its molecules. The temperature of most real gases is lowered on expansion, however, an effect taken advantage of in designing refrigeration equipment. The work done is against a gas's van der Waals forces, and energy is subtracted from the gas's kinetic energy. Because absolute temperature is directly proportional to kinetic energy, gas temperature falls.

QUESTIONS

1. STP Why is it necessary in work with gases to specify a standard temperature and pressure?

2. Temperature **(a)** Why do scientists believe there is a lowest temperature? **(b)** The interior temperatures of very hot stars may be 100 billion degrees. Do you think that is the highest temperature nature allows?

3. Pressure On a particular day a mercury barometer stands 30 in. high. A barometer tube filled with another liquid stands 300 in. high. **(a)** How are the densities of the two liquids related? Explain. **(b)** Give one advantage and one disadvantage of a barometer filled with water.

4. Temperature The length of a solid is affected by temperature changes (application: bridges are built with expansion joints). Suggest how one could measure temperature in the laboratory using this principle.

5. Density **(a)** Knowing Boyle's law we can deduce that gas density is directly proportional to pressue. Explain. **(b)** Knowing Charles' law we can deduce that gas density is inversely proportional to absolute temperature. Explain.

6. Ideal gas **(a)** What is an ideal gas? **(b)** In an experiment a gas sample is subjected to various pressures. At each pressure the volume is measured. How, from pressure–volume data, can one decide if the gas is or is not behaving ideally? **(c)** When the valve on a tank of compressed carbon dioxide is opened, gas that rushes forth into the air is cooled enough so a snow of Dry Ice (solid CO_2) forms. Explain the nonideal gas behavior.

7. Heat, temperature (library) In several chemistry or physics books and also in a dictionary find and copy definitions of "heat" and "temperature." Then define the two terms in your own words. Include the words *intensive* and *extensive* in your definition (see Question 4, Chapter 1).

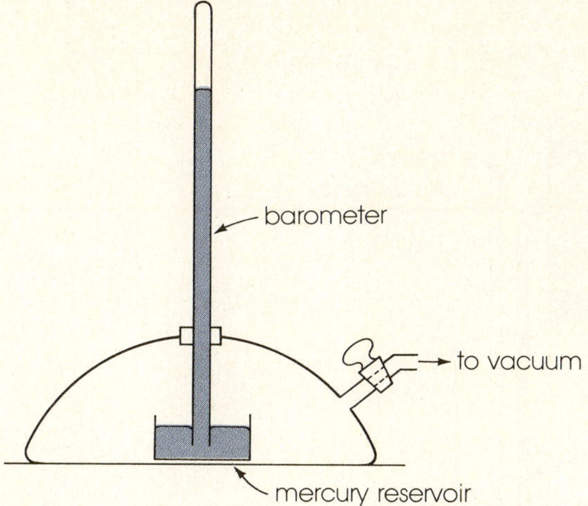

Figure 14-14

8. Terminology Define, illustrate, or otherwise explain: effusion, dew point, relative humidity, vapor pressure, boiling point.

9. Terminology As in Question 8: Brownian movement, mean free path, collision diameter, van der Waals attraction, average gas molecule velocity.

10. Barometer At room conditions a barometer is sealed into a bell jar and the outlet connected to a vacuum pump (Figure 14-14). Explain the action of the barometer as air is pumped out of the jar.

11. Kinetic theory Study Figure 14-15. Use the kinetic-molecular theory to explain what happens when stopcock S is opened.

12. Pressure Study Figure 14-16. When the apparatus is operating in the laboratory, a barometer there reads 750 mm. What is height h in millimeters? Explain.

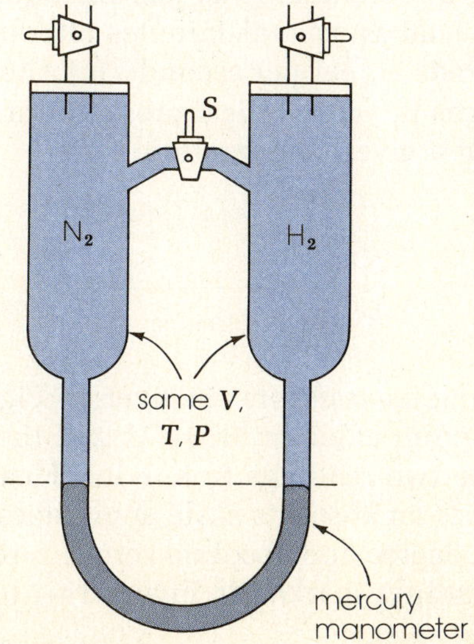

Figure 14-15

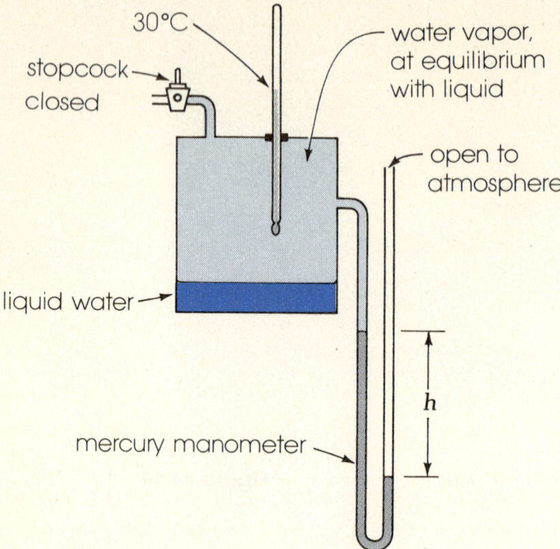

Figure 14-16

13. Molar volume **(a)** With reference to gases, define the term "molar volume." **(b)** Is the following equation correct? Explain.

gas molecular weight = molar volume × gas density

(c) At STP 1 molar volume of ideal gas is 22.414 liters. At STP the gram-molecular volumes of hydrogen and carbon dioxide are 22.430 and 22.263 liters, respectively. Explain.

14. Vapor pressure At 80°C the equilibrium vapor pressures of two pure organic liquids, aniline ($C_6H_5NH_2$) and 1-bromobutane (C_4H_9Br), are 18 and 370 torr, respectively. **(a)** Which liquid has the greater intermolecular attractive forces? **(b)** Which liquid has the higher boiling point? Explain.

15. Kinetic-molecular theory Use the kinetic-molecular theory to explain the following: **(a)** A gas changes to a liquid when cooled sufficiently. **(b)** Drops of perfume put in a corner of a room disappear and the odor eventually permeates the room. The effect takes several minutes although room temperature molecules travel hundreds of feet per second. **(c)** The molecular weight of gas A is twice that of gas B, yet identical pressures are exerted by a mole of each in 1-liter flasks at a given temperature.

PROBLEMS

16. Temperature **(a)** Several *defining fixed points* are given in Figure 14-7. Two higher points are the freezing point of silver, 960.8°C, and the freezing point of gold, 1063.0°C. Convert the two readings to Kelvin (°K) and Fahrenheit (°F). **(b)** The *Rankine scale* is an absolute scale sometimes used by engineers. Its unit is the Fahrenheit degree and absolute zero is zero degrees Rankine. Convert the Fahrenheit readings marked in Figure 14-7 to degrees Rankine (°R).

Table 14-7. Calibration of a thermocouple

Defining substance	Temperature °C	Milliammeter reading
Cs, melting	29	62
H_2O, boiling	100	152
Sn, melting	232	253
$FeCl_3$, boiling	315	300
Zn, melting	420	337

17. Temperature A thermocouple (Figure 14-1) with ice and water in the reference junction is calibrated with defining points as shown in Table 14-7. **(a)** Plot a calibration curve (temperature on abscissa). **(b)** If the meter reads 225 mA (milliamperes) when the test junction is in contact with a gas, what is the gas temperature? Indicate it on your plot.

18. Air pressure **(a)** Plot the data in Table 14-8. Mark each point with the name of the mountain. **(b)** Using Figure 14-8b calculate the boiling point of water on each mountain peak. Put the figures after the names on the graph. **(c)** Atop Mt. Ranier in Washington water boils at 87°C. What is its altitude?

19. Boyle's law From Table 14-3 data **(a)** plot P (abscissa) versus V, and **(b)** P versus $1/V$; **(c)** explain the shape of each plot.

20. Volume–temperature **(a)** A plastic bag of air having a volume V is sealed at 27°C and put in a Dry Ice chest where the temperature is -40°C. What is the final volume, assuming constant pressure? **(b)** A balloon of helium has a volume of 100 liters at STP. To what temperature must it be heated, at constant pressure, to increase the volume 25 percent?

21. Lifting power A typical balloon for cosmic ray research in the stratosphere is made of very thin Mylar plastic, weighs only 50 lb and has a volume of 3 million liters. What weight can it lift if 10 percent filled at ground level with helium at STP? (Hint: The difference in weight between 1 liter of helium and 1 liter of air, at the same conditions, is the "lifting power" of 1 liter of helium.)

Table 14-8. Altitude-pressure relationships

Location	Altitude ft	Atmospheric pressure at peak torr
Sea level	0	760
Black Mountain, Kentucky	4,140	660
Clingman's Dome, Tennessee	6,642	600
Mt. Alverstone, Alaska	14,500	450
Mt. McKinley, Alaska	20,320	365
Mt. Everest, Nepal	29,200	240

22. Molecular weight A sample of pure oxygen is passed through a silent electric discharge several times and some is converted to ozone: $3O_2 = 2O_3$. At STP 1.00 liter of the mixture weighs 1.57 g. What is the effective molecular weight of the mixture?

23. Density Bacteria buried with trash in sanitary landfills initially produce methane and carbon dioxide in approximately equal volumes. **(a)** What is the effective molecular weight of this gas mixture at STP? **(b)** What is its density at STP?

24. Density A 1.00-liter flask filled at STP with oxygen weighs 131.43 g. Filled under the same conditions with an unknown gas it weighs 132.00 g. Find **(a)** the weight of the flask evacuated; **(b)** the density of the unknown; **(c)** the molecular weight of the unknown.

25. Density **(a)** Calculate the density of atomic hydrogen at STP. **(b)** Convert that density (D_1) to a density (D_2) at the conditions of outer space, which we will assume to have only atomic hydrogen present at 3°K and a pressure of 3.7×10^{-19} atm. Employing a factor-unit approach,

$$D_2 = D_1 \,(T \, factor)\,(P \, factor) \tag{14-25}$$

Use reasoning similar to that given in Example 14-4 to decide whether each factor should be smaller or greater than one.

26. Reaction yield One pound-mole (a molecular weight measured in pounds) of any gas occupies 359 ft³ at STP. A "smog-free" automobile engine designed to run on methyl alcohol burns 9.70 gal (64 lb) of alcohol in a test:

$$CH_3OH + O_2 \longrightarrow CO_2 + H_2O$$

(a) Balance the equation. **(b)** How many cubic feet of CO_2 are produced at STP?

27. Gas constant Calculate a value for R, the gas constant, in these units: ml torr/mole °K.

28. Combined law A sample of argon is collected over water at 25°C and 730 torr. If the gas is dried by passing it over anhydrous $CaCl_2$, then confined at STP, what is its final volume if initial volume was 100 ml?

29. Combined law In the Van Slyke method for determining CO_2 in the blood, a measured blood volume is shaken with acid over mercury and the released CO_2 collected in a small calibrated tube at known temperature and pressure. Using 0.80 ml of blood, 1.00 ml of CO_2 is collected at 20°C and 330 torr. Find **(a)** the CO_2 volume at STP, and **(b)** the STP volume percent of CO_2 in the blood sample.

30. Vapor pressure Estimate the normal boiling point of mercury from the data of Table 14-9.

Table 14-9. Vapor pressure of mercury

Temperature °C	Vapor pressure torr	Temperature °C	Vapor pressure torr
200	17	320	376
250	74	340	558
290	198	350	673

31. Gas volume One gram of liquid water at 0°C and 1 atm is converted to steam at 100°C and 1 atm. **(a)** Calculate the steam volume. **(b)** Assuming the volume of the H_2O molecules themselves to be 1.00 ml (why?), what percentage of the steam is "empty" space?

32. Air analysis Air is 0.03 percent CO_2 by volume. Calculate the milligrams of CO_2 in 1000 ml of air at STP.

33. Gas mixture A tank has a pressure of 20 atm. It contains a mixture that is 65 percent H_2, 25 percent He, and 10 percent N_2 by volume. **(a)** What is the partial pressure of each gas? **(b)** If the helium could be removed, what would tank pressure be?

34. Effusion From a large volume of contaminated air a chemist isolates a nitrogen oxide. To get its molecular weight he pumps the gas into the sample tube of a *Bunsen effusiometer* (Figure 14-17), displacing mercury into the reservoir. Next he closes the stopcock. After attaching a pinhole outlet to a vacuum line, he opens the stopcock and records the time of rise

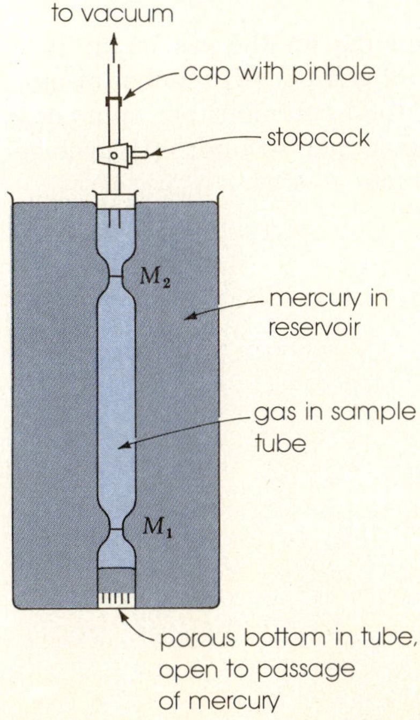

Figure 14-17

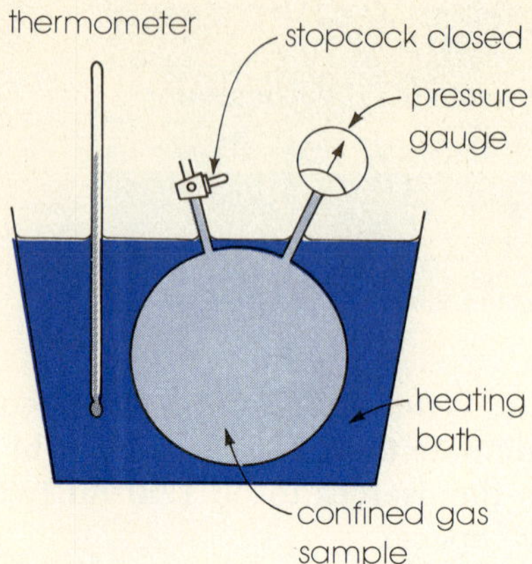

Figure 14-18

of mercury between reference marks M_1 and M_2 as the gas effuses out of the pinhole. He repeats the procedure with hydrogen gas, whose molecular weight he knows. If 25 ml of H_2 requires 30 sec, and 25 ml of the nitrogen oxide 140 sec to effuse, what is the molecular weight of the latter? Which nitrogen oxide is it? (Give its formula.)

35. Law derivation The following data are gathered with the apparatus of Figure 14-18: At 300°, 350°, and 525°K, gas pressure readings are 1.20, 1.40, and 2.10 atm, respectively. Offhand one might conclude pressure is directly proportional to absolute temperature: $P \propto T$. If true, then (as in the derivation of Eq. 14-3) $P = kT$. and $P/T = k$, a constant. **(a)** Verify the last statement. **(b)** What is P when T is 640°K? **(c)** In the manner in which Eq. 14-4 was derived, convert $P/T = k$ into an equation for equating two sets of P/T conditions. (The result is called *Amonton's law*.)

36. Gas properties Find the following for the gas in the flask of Figure 14-19: **(a)** pressure; **(b)** moles; **(c)** grams; **(d)** number of molecules; **(e)** density; **(f)** mean free path; **(g)** total collisions per cubic centimeter per second; **(h)** collisions per molecule per second; **(i)** average molecular velocity; **(j)** average molecular kinetic energy.

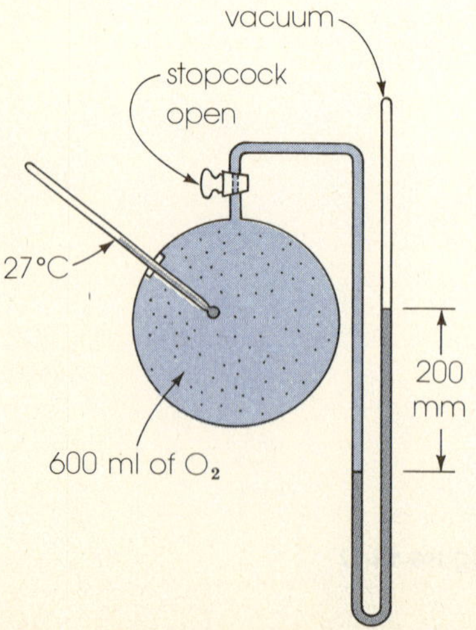

Figure 14-19

Table 14-10. Velocity of sound in gases

Gas	Molecular weight	Velocity of sound in the gas mi/hr at STP
H_2	2.02	2830
CH_4	16.0	1000
O_2	32.0	710
CO_2	44.0	600
air	28.8	?

37. Law derivation **(a)** As in Problem 35 derive a "law" from the data of Table 14-10 (that is, write a proportionality, rewrite it as an equation with a proportionality constant, test it to see if k is constant). **(b)** Use your law to predict the velocity of sound in air.

38. Water vapor Table 14-5 data are used by air conditioning engineers to calculate how much moisture can theoretically be removed from air during cooling. **(a)** Plot column 1 vs. 4, temperature on the abscissa. Extrapolate the curve to about 55°C. **(b)** By calculation, using the perfect-gas equation, show that air saturated at 20°C does indeed contain 17.3 g of water vapor per cubic meter. **(c)** If 10^5 m³ of saturated air is cooled from 23.0° to 13.0°C, how many liters of liquid water are recovered? **(d)** If the 13.0°C air (condensed water removed) is reheated to 23.0°C, what is its percent relative humidity? **(e)** Calculate a value for the last column of the table at 50°C and compare to the value you are able to read from your graph.

39. Mean free path Calculate the mean free path of a helium atom under the conditions of interstellar space if the total population of matter is only 1 atom/cm³. The collision diameter of helium is 2.18×10^{-8} cm. Express the result in miles (1 mi $= 1.61 \times 10^5$ cm).

40. Escape velocity The radius of Mars is 3.39×10^8 cm. Its mass is 6.58×10^{26} g. Calculate the escape velocity in centimeters per second and compare to the escape velocity of the earth. What implications do you see regarding the Martian atmosphere?

WHAT ARE WATER AND SOLUTIONS?

Experimentation with the use of warm-water discharge from an electric generating station. Water used to cool the plant (rear) goes through a lagoon (foreground) before flowing back into Long Island Sound. The growth of oysters in protected trays is accelerated by one year. (Courtesy Long Island Lighting Company at Northport, New York.)

15

WHAT IS THE CHEMISTRY OF OUR WATER ENVIRONMENT?

THE IMPORTANT CONCEPTS

15-1. The ocean

1. Natural waters vary greatly in mineral content.
2. Ocean salinity is apparently regulated by a geochemical cycle encompassing volcanism, weathering, and mantle movement.
3. Nitrogen and phosphorus cycles are related to ocean productivity of food.
4. Only a few minerals are obtained commercially from the ocean.
   a. Proposed mining of manganese nodules points up problems of ocean exploitation.

15-2. Desalination processes

1. Largest desalination plants today use distillation.
2. Electrodialysis and reverse osmosis plants use membranes.

3. The water-ice change in freeze processing requires less energy
   than the water–steam change in distillation.
   a. Energy requirements are calculated knowing water mass,
      temperatures, and latent heats.

15-3. Water pollution; some problems
1. Water demand will soon exceed the minimum dependable
   supply.
2. Eight classes of water pollutants are identified.
   a. Electric plants produce waste heat.
   b. Excess nutrients cause eutrophication.
   c. Iron pyrite causes acidic mine drainage.
   d. Chlorinated hydrocarbons persist in the environment.
   e. Mercury and cadmium are mobilized in the environment.
   f. Crude oil is common on the ocean.

15-4. Monitoring water and water pollution
1. Water quality is determined by a few routine tests.
2. Aerial spectral scanning is a wide-scale monitoring technique.

15-5. Municipal wastewater treatment
1. Wastewater has a solids loading and an oxygen demand.
   a. Oxygen requirement is expressed as a biochemical oxygen
      demand value.
2. Primary and secondary sewage treatments will furnish main service for the foreseeable future.

Earth is unique for harboring life and a large quantity of liquid water. Eighty percent of all animals live on or in the water which covers 71 percent of the planet (Figure 15-1). So far as is known life without water is impossible—most living things are 60 to 65 percent water. Somewhat more than half the water in a human being is found in his cells, the remainder in extracellular medium. Extracellular fluid is a 0.9-percent salt solution. Resembling a 1:4 dilution of seawater in ion content, it suggests life began long ago in saline seas and continues to carry its aqueous medium along.

Many scientists, engineers, and technicians are working today in water chemistry and related fields. Their activities include evaluating the ocean as a source of food and minerals; using heating, freezing, and membrane technologies to obtain fresh water from salt water; identifying water pollutants and origins of pollution; and improving methods of treating wastewater to minimize its degrading influence on the environment.

15-1. THE OCEAN

Geochemistry of seawater

As described in Chapter 5 all available water is believed to have come from the earth's interior. Perhaps as much as 90 percent of it was expelled when the core formed (see Figure 5-12), with the remainder outgassing through volcanism. The water supply circulates globally; it evaporates into the

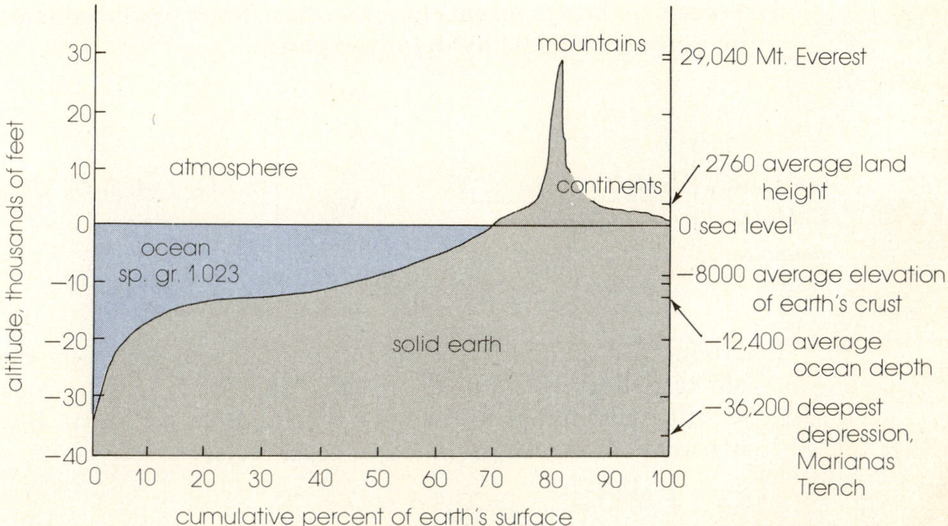

Figure 15-1 Schematic profile of the earth's surface, plotted to show the cumulative percent having an elevation above the ocean's deepest part.

Table 15-1. Major dissolved species in natural waters

Substance	Ocean ppm	Lower Mississippi River ppm
Cl^-	19,000	10
Na^+	10,600	14
SO_4^{2-}	2,700	40
Mg^{2+}	1,300	12
Ca^{2+}	400	40
K^+	380	2
HCO_3^-	142	100
SiO_2	12	11
NO_3^-	1	1

atmosphere, condenses, falls as rain, and returns as rivers flowing down to the sea.

Seawater and a typical fresh water are compared in Table 15-1. Despite the wide differences in concentrations and component ratios in these two samples (Na:Ca is 76 times larger in seawater, for example), seawater is apparently not undergoing much change as rivers mix with it. Analysis of sediments indicates that the *salinity* or salt content of seawater has remained stable since the oceans were formed about 170 million years ago. This steady state condition is explained by an evaportaion–condensation balance, and by assuming that as the world's rivers dump in their yearly load of 4 billion tons of salts, an equal weight of salts precipitates on the ocean floor. By proposing events that take place over time periods ranging up to millions of years, geochemists take into account movement of material not only through water but through the atmosphere and earth as well. The cycle shown in Figure 15-2 illustrates the grand scale on which geochemistry operates. Its main processes are as follows:

1. Volcanism brings magma to the surface. Rock produced is weathered by rain made acidic by dissolved gases:

$$\left. \begin{array}{l} NaAlSi_3O_8 \\ CaMgSi_2O_8 \\ KAlSi_3O_8 \end{array} \right\} + \underbrace{H_2O + CO_2 + NO_x}_{rainwater} \longrightarrow \left. \begin{array}{l} Na^+ \\ K^+ \\ Ca^{2+} \\ Mg^{2+} \\ HCO_3^- \\ NO_3^- \end{array} \right\} + \underset{silica}{SiO_2} + \underset{clay}{Al_2Si_2O_5(OH)_4} \qquad (15\text{-}1)$$

granite, basalt, feldspar   *soluble ions*

2. Products of weathering, some in solution, some as suspended solids, are carried by rivers to the ocean. Clay reacts with various ions in solution and settles to the bottom with other sediment. Bicarbonate is used both in the formation of coral reefs,

$$Ca^{2+} + 2HCO_3^- = CaCO_3\downarrow + H_2O + CO_2$$

and the maintenance of the acid–base balance essential to aquatic life,

$$CO_2 + H_2O \rightleftharpoons H_2CO_3 \rightleftharpoons H^+ + HCO_3^- \qquad (15\text{-}2)$$

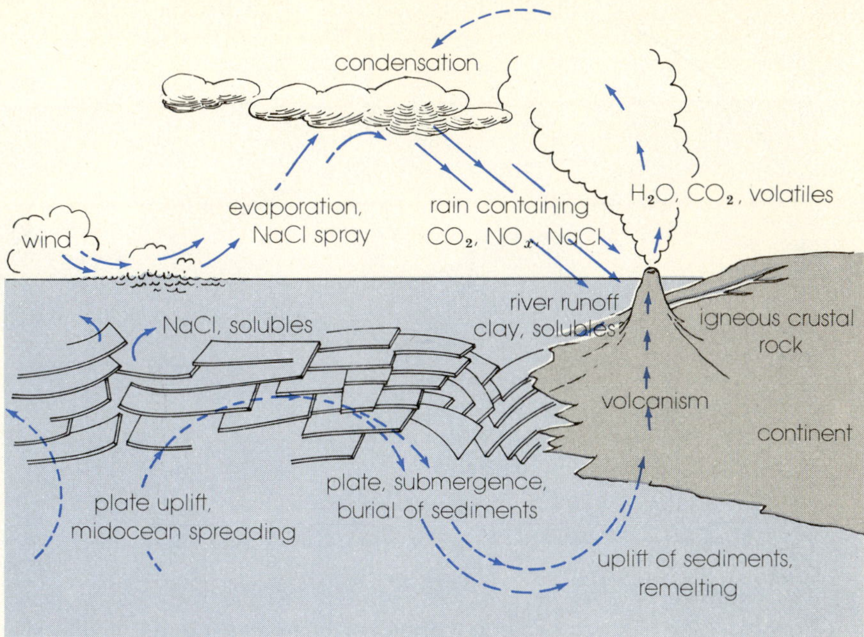

Figure 15-2 A model for the geochemical cycle that maintains the ion balance in the ocean. Large-scale convective flow that seems to be occurring in the mantle causes sea-floor movement. Geological stresses come from temperature differences in the earth.

3. Clay and sediment become buried. Movement in the mantle, originating in midocean and spreading large blocks or plates of the crust toward deep trenches, submerges the sediments under continents. Heat and pressure melt the material. Magma and volatile materials are pushed upward and the cycle begins again.

Food from the ocean

As growth of human population races against food supply and an insufficiency of animal protein looms as a dietary problem to millions, it is logical to ask if the sea is not a vast food source that is being insufficiently exploited. Answers can be derived from a chemical understanding of the cyclical movement of nutrients with respect to marine life. Other than carbon, the important nutrients are nitrogen and phosphorus.

The nitrogen cycle is modeled in Figure 15-3. It begins with fixation of N_2 in the atmosphere by electrical storms and man's burning of fuels, followed by entry of nitrate into the ocean. Nitrate that reaches the bottom is reduced by denitrifying bacteria to N_2 which returns to the atmosphere. Nitrate in the water becomes biochemically fixed as protein in algae and bacteria. The sequence of increasingly higher organisms through which nutrients and energy pass is called a *food web* or *chain*. As animals eat the algae and then are eaten by other animals, protein nitrogen moves through the web. Nitrogen returns to the water as animal waste products and as ammonia from dead organisms. Oxidation of ammonia by oxygen produces nitrate ion which again is used by microorganisms as a nutrient. Two-thirds of the ocean's dissolved nitrogen (6.3×10^{11} tons) occurs as nitrate

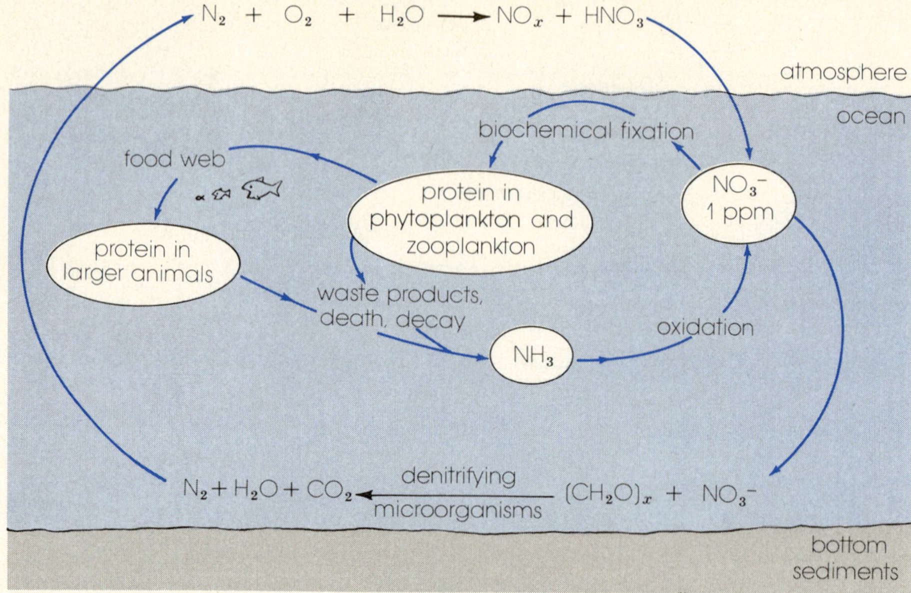

Figure 15-3 The nitrogen cycle as it operates through the ocean. Phytoplankton and zooplankton are, respectively, microscopic floating plants and animals found especially in surface waters.

and comprises a substantial part of all fixed nitrogen on the planet. Sediments contain little nitrogen. It is assumed therefore that losses balance gains and that nitrogen in seawater is constant.

The phosphorus cycle (modeled in Figure 15-4) begins with runoff from land where dissolving rocks and agricultural operations are mainly responsible for adding some 50 million tons of phosphate yearly to the

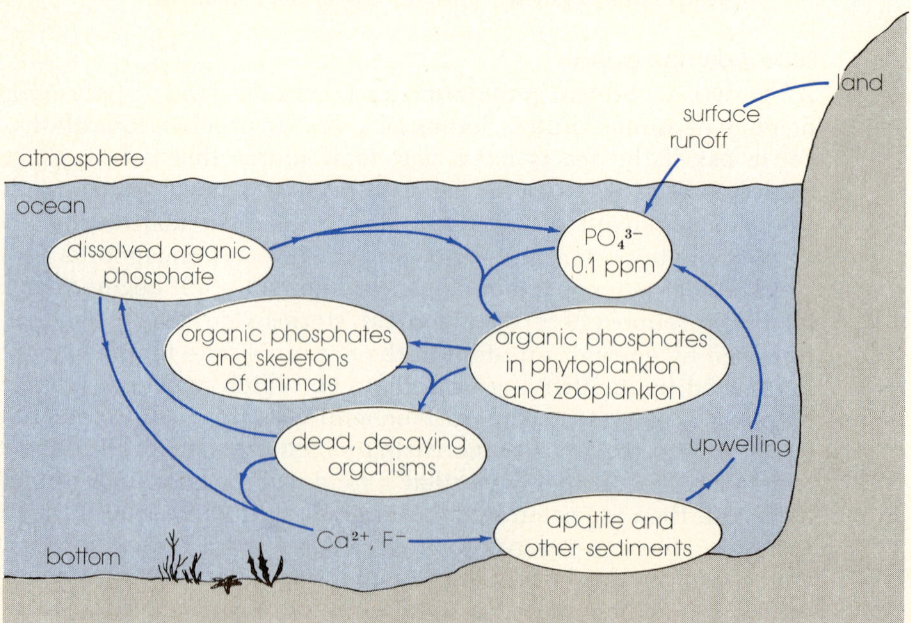

Figure 15-4 The phosphorus cycle that operates through the ocean.

ocean's huge store of 335 billion tons. Phosphate not taken up in the food web precipitates as the mineral apatite, $Ca_5(PO_4)_3F$. It is the solubility of apatite that controls phosphate availability. Bacteria and plants incorporate phosphate into vital organic compounds. Animals also utilize phosphorus to build calcium phosphate skeletons. When organisms die, their phosphate returns to the water or falls to the bottom. As upwelling currents bring phosphate back to the sunlight zones, the cycle starts again.

Fertile regions where nutrients are abundant represent only about 0.2 percent of the ocean's entire surface. From chemical determination of nutrient concentrations and census counts of organisms there, scientists estimate the ocean produces 500 billion tons of wet phytoplankton a year. From that green mass comes perhaps 250 million tons of marine animals—fish, crustaceans, mollusks. If 50 percent were harvested (today's harvest is 30 percent), the total catch would be only 2 percent of the world's food production—and half that would be rough varieties used presently only as farm animal feed supplement. Thus although the ocean covers most of the earth, the total productivity of plants on land exceeds that in the ocean by perhaps 50 percent. The main reason is that food in the ocean is much more dilute. To get a greater yield will require massive marine cultivation, and techniques for that have not yet been worked out.

Minerals from the ocean

Some minerals are obtained in quantity from the ocean today and a few others will be commercialized tomorrow, but both lists are surprisingly brief.

Out of seawater itself come fresh water by distillation, sodium chloride by solar evaporation, bromine by chlorine displacement, and magnesium metal by electrolysis of magnesium chloride prepared from precipitated magnesium hydroxide. From the *continental shelf* (the projection of continents which by international agreement is terminated at the 650-foot depth line) come petroleum, natural gas, and sulfur. The shelf area of the United States is equal to 36 percent of the combined dry land area of the 50 states. It has yielded $4 billion worth of raw materials in the last 20 years and is believed to hold much more. Its salt domes and phosphate deposits are untapped, as are gold, platinum, and tin placers (gravels) lying off the coast of Alaska.

Out beyond continental slopes, where exploration has largely been limited to instrument readings, calcium phosphate and manganese nodules are the only mineable deposits found thus far. The latter are intriguing for the geochemical and geopolitical questions they raise.

Manganese dioxide nodules accrete apparently self-catalytically near where sea-floor spreading brings up fresh material from the mantle. They grow about 1 inch in thickness every 25,000 years. An average nodule may be golf-ball size and contain 24 percent manganese, 14 percent iron, and smaller percentages of nickel, copper, cobalt, zinc, chromium, vanadium, tungsten, and lead. The central Pacific Ocean basin is estimated to contain a trillion tons of the material. At an indicated growth rate of 10 million tons per year, accretion over that area alone would satisfy world needs for manganese, nickel, and cobalt. Perplexing questions follow. Who "owns" the ocean and what should be done with profits from it? If mining were to strip the sea floor and stir up sediments, how would marine life be affected? How would ocean mining upset established economic patterns? Presently African mines provide the world with much of its cobalt, chromium, manganese, copper, and phosphate rock. In so doing they supply the African nations with most of their foreign exchange. Cheap sea minerals might bankrupt them. On the other hand the United States buys large quantities of metals. We import 97 percent of the nickel we require, 95 percent of the manganese, and 88 percent of the cobalt. For us ocean metals may soon become a necessity.

15-2. DESALINATION PROCESSES

In this century population in the United States has more than doubled. With accompanying increases in industrialization and urbanization the demand for fresh water has increased 10-fold and is increasing today at an estimated rate of 30,000 gallons a minute. Water shortages are not uncommon in cities, and everywhere water quality is going down. Even for modest growth to continue we, and all industrial nations, will not only need to conserve this essential resource but we will be turning to the ocean and other saline sources for more.

Distillation

Distillation has been known for 2000 years (Archimedes described it) as a means of obtaining fresh water from the sea. As seawater is heated, dissolved gases and water vapor escape, leaving behind a brine of nonvolatile salts. Condensation of the vapor on a cool surface yields fresh water.

The most rudimentary means for distillation is a greenhouse arrangement in which solar energy that comes through the glass heats water. Water vapor condensing on the glass is caught in troughs. Although the energy is free, it is low level and variable with weather and time of year (see Problem 11).

For large-scale distillation, *multistage flash evaporation* (Figure 15-5) has been developed most extensively. Seawater under pressure is heated to about 100°C and sprayed into a hot chamber maintained at a somewhat lower pressure. Steam comes off suddenly (flashes), rises in the chamber, condenses, and gives up heat to tubes that are bringing raw seawater into the process. Under the tube bundle, distillate drips into a tray for transport to storage. The remaining brine, now cooler, is sprayed into the next chamber, which is held at a lower pressure than the first. Flashing occurs again, brine temperature drops, vapor condenses on the tubes and transfers more heat to incoming seawater. The concentrated brine is conducted to the next chamber, held at a still lower pressure, and flashing is continued. In cooling from 100° to 60°C seawater gives up 7.1 percent of its weight as distillate.

Efficient heat exchange is the characteristic of multistage distillation plants. If seawater is brought in at 20°C, heated to 100°C, flashed in the chambers, and eventually discharged at 30°C a practical minimum of heat is wasted with the returned brine.

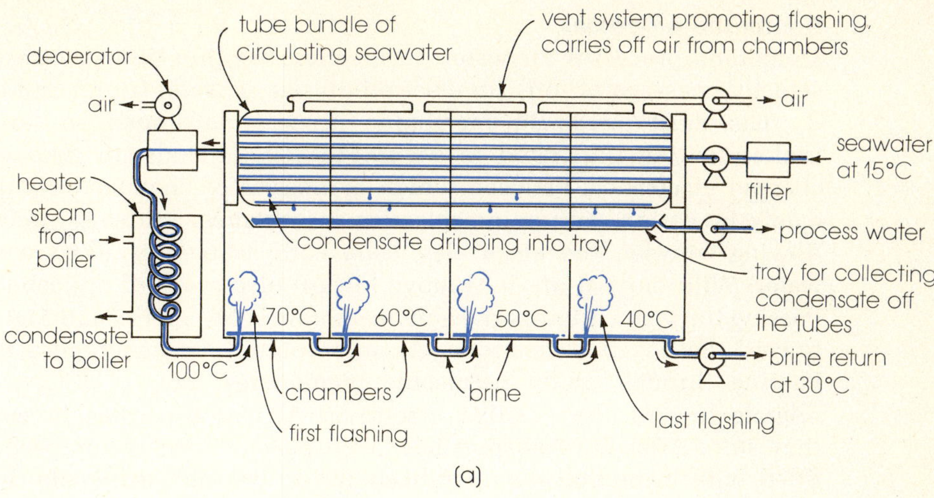

Figure 15-5 Multiple-effect flash evaporation. (a) Schematic drawing of the process. (b) One of the world's largest seawater distilling plants, at Shuaiba, Kuwait; each unit produces 1.2 million gallons of drinking water per day. A major advantage of flash evaporation is that scale (calcium sulfate and calcium carbonate, particularily) is minimized because water evaporates from the surface of warm brine and not from a hot metal surface as in ordinary distillation. Solids remaining suspended in the brine are returned to the ocean. (Courtesy Weir Westgarth, Limited, Scotland.)

Membrane processes

Membrane processes for water purification use thin-film surfaces to allow selective passage of ions (in *electrodialysis*) or water (in *reverse osmosis*).

The most advanced membrane process developed so far is electrodialysis. As illustrated in Figure 15-6 it depends on two oppositely charged electrodes, between which is a series of chambers divided alternately by cation- and anion-selective membranes. With no direct current flowing between the electrodes, each compartment is filled with saline water. With current on, ions move toward electrodes of opposite polarity. Passing through cation-permeable membranes, Na^+, Ca^{2+}, and Mg^{2+} migrate toward the negative electrode. At anion-permeable membranes they stop. Passing through anion–permeable membranes, SO_4^{2-}, HCO_3^-, and Cl^- migrate toward the positive electrode. At cation-permeable membranes they stop. After ion transport has taken place, alternate compartments are filled with brine and demineralized water. *Potable* (drinkable) water and salt water are drawn off through separate pipes.

Another membrane method is the reverse osmosis (R.O.) process. As will be explained in Chapter 16 *osmosis* is the spontaneous transport of solvent from a dilute solution to a more concentrated solution through a semipermeable membrane. When pure water and salt water are on opposite sides of a membrane that is permeable only to water, water molecules come through and dilute the salt solution (Figure 15-7a). The pressure required on the salt side to prevent mass transfer is called the *osmotic pressure*. Osmosis is reversed by putting a pressure in excess of the osmotic pressure on the salt solution side. Water then diffuses through the membrane in the op-

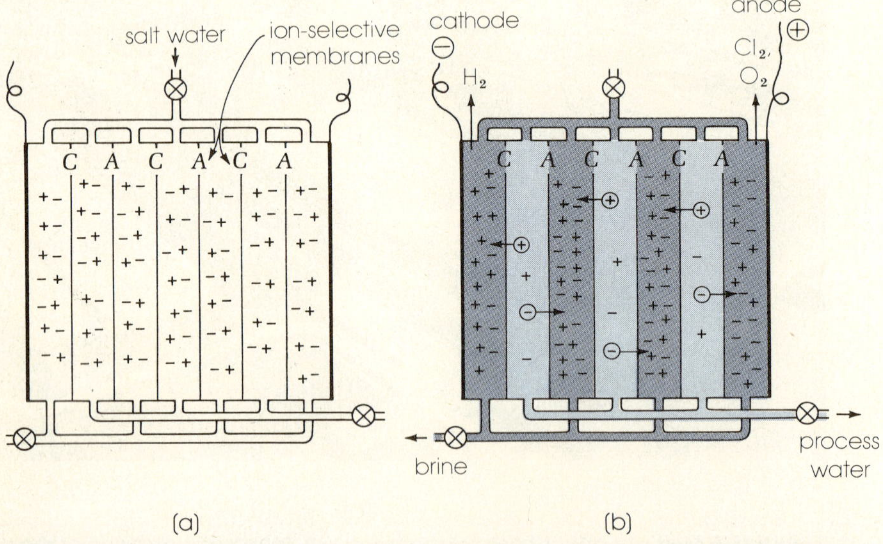

Figure 15-6 Principles of electrodialysis. (a) A bank of cells separated by cation-permeable (C) and anion-permeable (A) membranes, with no electric current flow. (b) When direct current is passed through the cell bank, cations and anions migrate in opposite directions. Due to the alternate arrangement of membranes, ions can only pass through one barrier before being stopped by the next. After a time, alternate compartments contain concentrated brine and demineralized water. Membranes may be 0.1 to 1 mm thick and spaced 1 mm apart. The direct current causes some electrolysis to occur, giving the gases shown at anode and cathode.

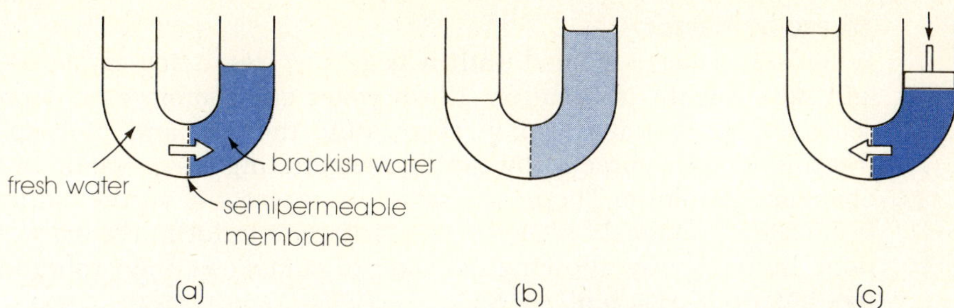

Figure 15-7 Principles of osmosis (Gk. osmos, push). (a) Osmosis. Water moves spontaneously through the semipermeable membrane into the solution side. (b) Water has diluted the more concentrated solution. (c) Reverse osmosis. Pressure applied to the solution side forces water from the solution.

posite direction, increasing the volume of fresh water and concentrating the brine (Figure 15-7c).

Membrane processes at present do not compete with distillation in seawater conversion; for as salinity increases, so does the membrane power requirement. Electrodialysis and reverse osmosis can be used on mildly mineralized or brackish waters, and such plants are operational (Figure 15-8). Membrane methods commonly reduce salt content by 90 percent, yielding, for instance, potable water containing 400 ppm dissolved solids from feed having 4000 ppm dissolved solids. Distillation reduces salt content 100 percent.

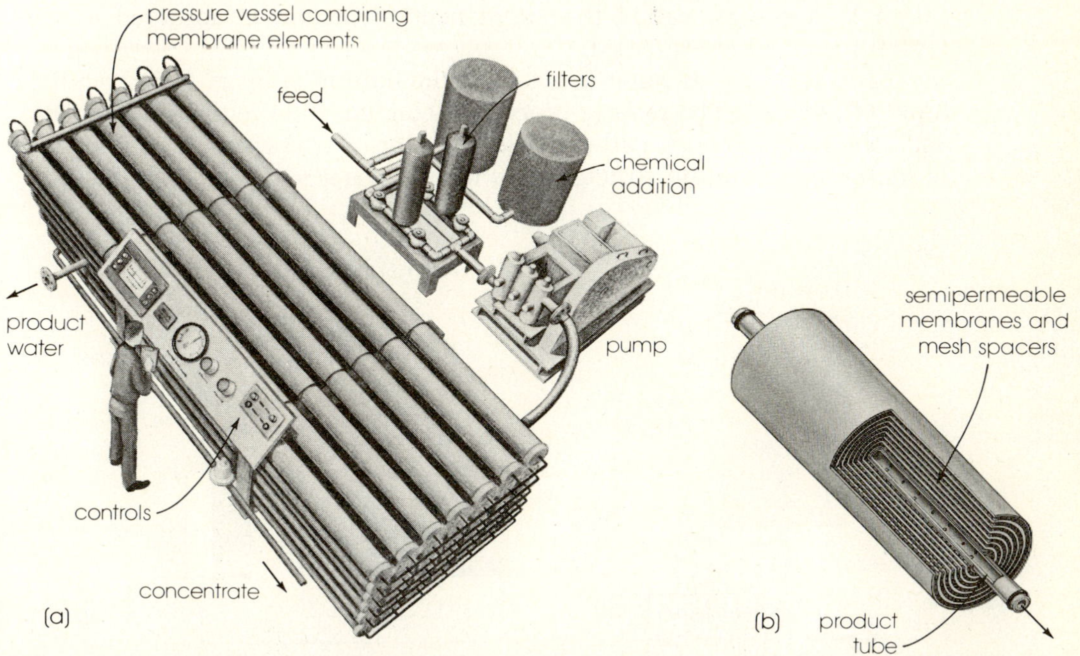

Figure 15-8 A 250,000-gallon per day reverse osmosis installation. (a) A standard arrangement of equipment. Pump pressure overcomes osmotic pressure and reverse osmosis takes place. (b) One of the membrane elements. Membranes are in a spiral arrangement supported by alternating layers of wire mesh. Demineralized water (typically 70 to 80 percent of the feed volume) passes through the membrane and exits from the central tube. (Courtesy Gulf General Atomic.)

Freeze processes

When salt water is cooled until it begins to freeze, ice crystals are formed and salts remain in solution. Fresh water can therefore be obtained from seawater by partial freezing, segregating the ice, and melting it. When freezing is done directly by rapidly evaporating water vapor in a vacuum chamber, the Joule-Thompson effect (Chapter 14) carries sufficient heat away from previously cooled seawater for ice to form. Freezing can also be done indirectly by allowing seawater to contact an inert refrigerant liquid like butane (C_4H_{10}, b.p., $-0.5°C$).

In the *vacuum freezing-vapor compression method* (Figure 15-9) seawater is degassed, cooled, and pumped into a crystallizer chamber held at a pressure of 3 to 4 torr. Flash evaporation cools the water, freezing half of it as an ice–brine slush or slurry. The slurry is pumped into a washer–melter where brine is washed from the ice crystals. The ice is then pushed upward into a central melter column while brine flows out through screens. As vapor exits from the crystallizer, it is compressed and introduced into the top of the melter column. Condensing on the washed ice it melts the ice and adds to the total product volume. Fresh water flows out of the melter.

Latent heats of fusion and vaporization

An important advantage of freeze processing is that the water–ice transformation involves less energy than the water–steam transformation.

To convert 1 g of ice to water at its melting point of 0°C requires the input of 79.7 cal. (The reverse process of freezing requires removal of 79.7 cal/g.) This energy, called the *latent heat of fusion,* is absorbed as ice-crystal forces are diminished.

To convert 1 g of water to steam at the boiling point of 100°C requires input of 540 cal. (The reverse process of condensation requires removal of 540 cal/g.) This energy, called the *latent heat of vaporization,* is absorbed in loosening the liquid structure and freeing water vapor.

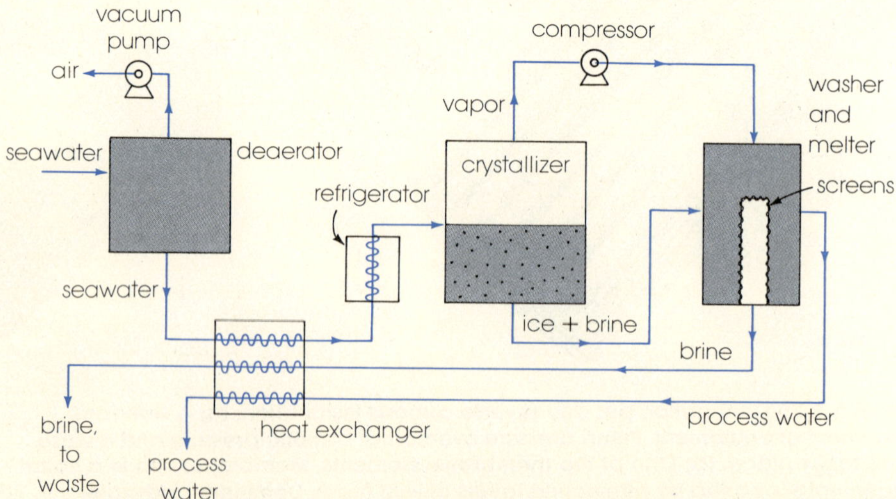

Figure 15-9 Direct freezing of seawater with vapor compression. Main disadvantage of the process is that the small ice crystals produced cannot be washed clean of brine.

For every substance known the solid-liquid transition requires less heat than the liquid-gas transition—in the case of water, about one-seventh as much. Use of latent heats is illustrated in the following example.

Example 15-1 Two hundred grams of steam at 100°C is to be converted to water at 0°C. If all the heat liberated is used to melt ice at 0°C to liquid water at 0°C, how many grams of ice can be melted?

Solution Condensation of the steam releases 540 cal/g at 100°C. For 200 g,

$$(5.40 \times 10^2)(2 \times 10^2) = 10.8 \times 10^4 \, cal$$

Cooling 100°C water to 0°C water releases 1 cal/g °C (for the specific heat of liquid water, see Chapter 12). This amounts to

$$(1 \, cal/g \, °C)(2 \times 10^2 \, g)(1 \times 10^2 \, °C) = 2 \times 10^4 \, cal$$

Total heat released is the sum, 12.8×10^4 cal.

To melt 1.00 g of ice at 0°C requires 79.7 cal. With 12.8×10^4 cal available for melting, the weight of ice that can be melted is

$$\frac{12.8 \times 10^4 \, cal}{7.97 \times 10 \, cal/g} = 1.61 \times 10^3 \, g$$

15-3. WATER POLLUTION; SOME PROBLEMS

Water use patterns

Modern community life is possible only when fresh water is abundantly available and a system operates for disposing of liquid and solid wastes. Water must be found, transported, stored, purified, distributed, used, recollected, treated, and returned to nature. A torrent of water (a billion gallons per day or more in the biggest metropolitan areas) flows through every city and in every instance exits with its quality degraded.

Rainfall over the United States averages 30 inches per year. *Daily* rainfall of 4300 billion gallons generates a runoff in waterways of 1200 billion gallons per day. About half this, 580 billion gallons, is classified as the minimum dependable supply we have to use. Daily water withdrawals today total 300 billion gallons, of which 130 billion gallons is used up (evaporated, etc.) and 170 billion gallons is sent into receiving waters. Returns are 14 percent of total supply and 29 percent of minimum supply. When in 30 years returns exceed the minimum supply, "once-through" water will have become a luxury of the past. Cleanup for immediate reuse will then be mandatory because water the city discards one day will be the only water there is to use the next.

Industry takes 50 percent of all water used (mostly for cooling), agriculture takes 33 percent, municipalities take 7 percent. The last figure is given perspective as follows. Seven percent of total daily use of 300 billion gallons is 21 billion gallons. Urban dwellers in the United States now number about 173 million. Average daily use per person in municipalities is therefore $21 \times 10^9/173 \times 10^6$ or 120 gallons. About half the volume represents sewage flow.

Classes of pollutants

In the early 1900s the chief problem with public water supply (since solved by filtration and chlorination) was its infestation of bacteria that caused typhoid fever, an often fatal disease. Today the problems are those of sheer volume of water demand and the great variety of contaminants that water contains.

The U.S. Public Health Service has cited eight general classes of water pollutants:

1. *Industrial heat,* a nonchemical pollutant from steel making, petroleum refining, and other heavy industry requiring much cooling water.
2. *Organic chemicals,* including synthetic compounds such as pesticides.
3. *Other chemicals and minerals* originating in chemical, metallurgical, and mining operations.
4. *Plant nutrients,* primarily nitrogen and phosphorus from agricultural land and domestic sewage.
5. *Oxygen-demanding wastes,* principally organic matter from sewage plants and industries; the wastes require aerobic bacteria for decomposition.
6. *Infectious agents* or bacteria and viruses common in sewage and animal wastes.
7. *Radioactive elements* originating in mining, atom bomb testing, radioisotope research, and the processing of nuclear fuel.
8. *Sediments* from eroding land, mining and agricultural areas, and urban developments.

Some details will follow.

The problem of waste heat

Discharge of large quantities of heat into the environment is called *thermal pollution.* When bodies of water become heated, the solubility of oxygen decreases. Oxygen-dependent organisms may then die or reproduce abnormally.

One of the producers of a great amount of waste heat is the electric generating industry. This is most unfortunate because energy can be considered the ultimate resource upon which all manner of transformations depend. To meet demands for energy in the United States, steam-electric generating capacity is being doubled every 10 years. If that growth is to continue into 1995, the industry estimates 100 nuclear and 400 conventional power stations will have to be built. Their daily cooling water requirements alone will be 400 to 500 billion gallons, a volume equivalent to the entire Mississippi River system. Even with likely improvements in technology, fossil-fueled stations will still dump 40 to 50 percent of their heat on the environment and nuclear stations may waste 70 percent. Large installations being most efficient, their construction will further intensify both population concentration and thermal pollution. In the Boston–Washington area, for example, waste heat is expected to approach 65 cal/cm² per day, a figure which is 50 percent of the sun's radiation incident there during the winter. Obviously heat discharge is going to aggravate water problems. It may even reach such levels that local climate will be affected.

Engineering greater efficiency into new plants is one approach to the problem. Another is installation of evaporative cooling towers and cooling ponds to keep heat out of natural waters. More imaginative answers are found in proposals to locate around a central power station activities that can utilize its waste heat. Homes, food-growing greenhouses, and ponds (see the figure on page 359) could be warmed, industries run, sewage and solid waste incinerated, and salt water desalted cheaply enough for use in agriculture (see Figure 4-22). Warm water that continued to be discharged might be used to keep winter waterways open for shipping and to extend the growing season by heating agricultural land via underground pipes.

The problem of eutrophication

When water becomes oversupplied with nutrients, algae and other plant life proliferate. The end result is a depletion of dissolved oxygen due to dying plant material and aerobic bacteria, and the onset of anaerobic conditions that kill fish and give the water an unpleasant taste and odor. Taken together the effects are called *eutrophication*.

The principal nutrient elements found in fresh water (as in seawater) are nitrogen, phosphorus, and carbon. One source of soluble nitrogen is the nitrogen oxides in the air (Figure 15-3). Man-made NO_x alone represents as much nitrogen as is disposed of in domestic sewage, and the airborne quantity may double in the next 20 years. Another major nitrogen source is all aspects of agriculture, including animal feed lots (Figure 13-16). Reducing these sources of pollution is difficult and would embody cost-cleanliness tradeoffs as discussed in Chapter 13. To limit meaningfully the use of synthetic fertilizers, for instance, would have the effect of raising prices of food and natural fiber. Not to fertilize chemically at all would mean the earth could feed far fewer people than it now does.

Phosphorus enters the water system from agricultural activities and municipal and industrial effluents. Phosphorus seems easier to control than nitrogen. Algae, however, apparently store and reuse phosphorus in an almost catalytic manner so that concentrations as small as 0.01 ppm in water can exert appreciable effects on cell growth.

During the year 1970 1.4 billion pounds of phosphorus was estimated to have entered the freshwater system of the United States. About a third came from detergents containing sodium tripolyphosphate (STP, $Na_5P_3O_{10}$) as a water-softening agent (it forms soluble complexes with the Ca^{2+} and Mg^{2+} of hard water). In answer to laws several states passed to limit detergent phosphate, soap manufacturers produced new formulations, substituting for STP either Na_2SiO_3 and Na_2CO_3, or NTA (sodium salt of nitrilotriacetic acid, $N(CH_2CO_2Na)_3$). The first type in general proved too alkaline. Preliminary experiments on the latter indicated it might induce cancer and defective births in laboratory animals, and it was withdrawn from American markets pending further research. At present the best way to handle phosphate from cities is to remove it at the sewage disposal plant as described in Section 15-5.

The problem of acidic mine drainage

Acidic mine drainage (AMD) is the name given 8 million tons of sulfuric acid that seep out yearly from coal and copper mines and pollute 10,000 miles of American streams.

Acidic mine drainage begins in the seams of both active and abandoned mines with the moist air oxidation of iron pyrite (FeS_2) to sulfate. Ferrous ions formed at the same time are oxidized to ferric ions. The normally slow oxidation is catalyzed by microorganisms. Ferric ions oxidize more pyrite, giving more ferrous and sulfate ions. Ferric ions also react with water to give hydrogen ions and a precipitate of hydrated ferric oxide. Water polluted with acidic mine drainage therefore is a dilute sulfuric acid solution carrying iron:

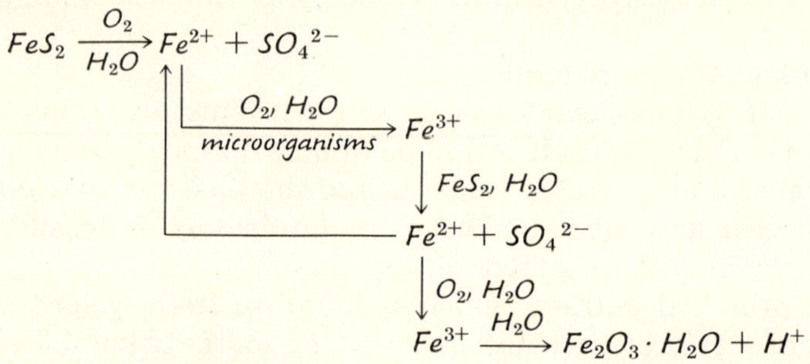

$$FeS_2 \xrightarrow[H_2O]{O_2} Fe^{2+} + SO_4^{2-}$$

$$\xrightarrow[microorganisms]{O_2,\ H_2O} Fe^{3+}$$

$$\xrightarrow{FeS_2,\ H_2O} Fe^{2+} + SO_4^{2-}$$

$$\xrightarrow{O_2,\ H_2O} Fe^{3+} \xrightarrow{H_2O} Fe_2O_3 \cdot H_2O + H^+ \tag{15-3}$$

The *reaction mechanism* suggests solutions to the problem. One is to stop the first reaction step by completely eliminating oxygen, a difficult approach because mines usually contain cracks that open them to air. Nevertheless, several depleted mines have been sealed with an atmosphere of nitrogen, and acid from them has diminished. A second solution is to stop catalysis in the second reaction step by eliminating microorganisms. Experiments with bacteriocides are under way.

The problem of chlorinated hydrocarbons

Man-made chlorinated hydrocarbons, synthesized in large quantity since 1950, have become dispersed throughout the world. Some are long lived and toxic to a variety of species.

Most familiar is DDT (abbreviation for dichlorodiphenyltrichloro-ethane),

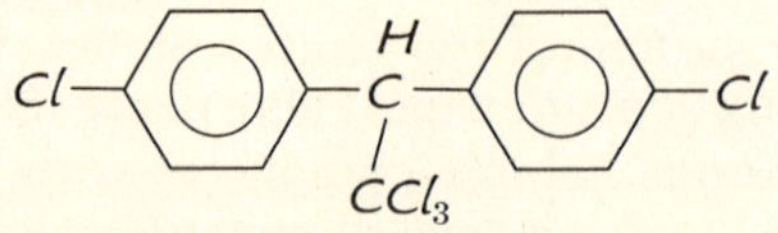

first synthesized in 1874 and rediscovered as an insecticide in 1939 by Paul Muller (Switzerland 1900–   , Nobel prize in medicine or physiology 1948). Largest user has been the World Health Organization (WHO). In 1969 WHO stated DDT has been

The main agent in eradicating malaria in countries whose populations total 550 million people, of having saved 5 million lives and prevented 100 million illnesses in the first eight years of its use, of having recently reduced the annual malaria death rate in India from 750,000 to 15,000 and of having served at least 2 billion people in the world without causing the loss of a single life by poisoning from DDT alone.

The case against DDT can be summarized by the characteristics that have caused it to be called "the universal contaminant." It is unusually persistent (half-life in the environment measured in years), highly mobile (residues travel by air and water), and soluble in fat (it moves up the food web, amplifying in fatty tissues). In ocean water DDT concentration is about 0.05 ppb. In phytoplankton it is 0.02 ppm, in zooplankton 0.04 ppm, in fish 0.5 to 20 ppm, in fish-eating birds up to 70 ppm, and in fatty portions of the average American 12 ppm.

Health authorities have set 7 ppm as the limit of DDT residues in most foods. Although DDT at low levels may be harmless in man, there is evidence it has noticeable effects on other animals. In pelicans and some other birds concentrations as low as 1 to 3 ppm apparently cause the liver to produce an overabundance of the enzymes that destroy the sex hormone having responsibility for transferring calcium stored in bone hollows to the synthesis of egg shells. As is the case of birds with Newcastle's disease and birds disturbed by intruders during nesting, the result is thin-shelled eggs that break when the parent sits to hatch them, and thus diminution of the species.

The trend today is away from "hard" broadly effective pesticides. Under development are more specific agents, including insect viruses and bacteria, hormones that prevent insects from maturing or egg laying, otherwise harmless insects that prey on insect pests, and chemical sex attractants whose odor lures insects to traps.

A DDT-like contaminant also detected throughout the world now in rain, birds, fish, mammals, and milk is mixtures of polychlorinated biphenyls (PCBs). Some of the fifty PCBs identified thus far apparently come from DDT, which degrades environmentally into a variety of compounds. The direct source is commercial production that involves chlorination of the hydrocarbon biphenyl:

$$\text{biphenyl} \xrightarrow{Cl_2} \xrightarrow{Cl_2} \text{, etc.}$$

Polychlorinated biphenyls are nonflammable fluids used in industrial heat transfer, hydraulics, and insulation, and as additives in plastics, adhesives, and paints. Production in the United States to date has been 0.4 million tons, of which perhaps half has wound up in the environment through volatilization and water dispersal of waste materials that contained PCBs. Like DDT, PCBs are persistent, mobile, water insoluble, able to affect the growth rate of some microorganisms and the egg-laying ability of birds (including chickens), and capable of moving up the food chain. Data are lacking on low-level long-term effects, but use of the compounds is being curtailed. The probable presence of PCBs in virgin paper pulp (from which they can transfer to food via paper packaging) and in the fatty tissues of most people is a reminder of how readily seemingly stable chemicals, although not deliberately spread about, can soon appear almost everywhere.

The problem of metals

Toxic trace metals are an insidious environmental hazard. Some metals are essential for human health in small amounts (Fe, Zn, Mn, Co, Cu, Cr, Mo), a few are plainly poisonous, and others may cause complex physiological changes almost too subtle to detect. We will discuss two examples, mercury and cadmium.

In 1970 mercury was discovered in Great Lakes fish. A search implicated nearby electrochemical plants using mercury cells (see Figure 18-7). It was then recalled that in 1969 some people in Japan had died from eating mercury-polluted fish, and research began in several laboratories to find how mercury was being removed from sediments in which it was commonly thought to be inert. Chemists found mercury is mobilized through chelation with organic molecules in the water, and by way of anaerobic organisms that produce such compounds as dimethyl mercury, $(CH_3)_2Hg$. Mercury then becomes concentrated in the fatty portions of fish directly or through the food web.

Although localized mercury contamination may be traced to industrial sources (American industry has consumed 82,000 tons of mercury since 1900), nature is the major distributor of the metal. This is not surprising because mercury is a volatile element (vapor pressure, 0.3 torr at 100°C). Recent research at the U.S. Department of Agriculture indicates a natural cycle operates in which about 100 tons of mercury per day is released world-wide by plants and soils and then returned to earth by way of dustfall. Plants take mercury from the soil, by some mechanism reduce it to the metal, and release the metal to the air. Average soil in the United States contains 71 ppb of mercury. Seawater contains a total of 25 million tons of mercury or 2500 times as much as present world annual production. Mercury can be considered ubiquitous in at least trace amounts, therefore, and living things have developed a certain tolerance for it.

Cadmium is potentially more dangerous than mercury because in the body one of its actions appears to be the replacement of zinc, which is vital to the digestion of fats. As undigested material accumulates in the circulatory system, the result may be hypertension or high blood pressure, and heart disease. Some 23 million Americans suffer some hypertension, and cadmium may be the cause in many cases. One source of the metal is the galvanizing in water pipes, where cadmium occurs as an impurity with zinc and leaches out slowly. Food sources include shell fish (which concentrate cadmium by a factor of about 1200 times), and highly refined rice, flour, and sugar that during processing lose zinc while being enriched in cadmium. Cadmium in air comes from steel mill smoke stacks and the combustion of leaded gasoline and plastics. Tobacco smoke also contains the metal. Cadmium intake (1.5 μg) from smoking one pack of cigarettes is at least 1.5 times that from the average daily diet.

The problem of crude oil

The greatest threat to marine life today is crude oil. An estimated 1.5 million tons of oil is introduced into the ocean each year through natural seepage, offshore drilling, accidents, and ocean shipping (half the world's annual shipping load, 750 million tons, is oil). In addition, up to 4.5 million tons of oil may be washing into the ocean via waterways from disposal sites and leaks on land. Oil covers many square miles of ocean. By one estimate floating residues are already equal in weight to that of phytoplankton near the surface, and the oil film may be extensive enough to alter light reflectivity of the whole planet. Concentrated in the film are metals and chlorinated hydrocarbons. From there these substances become dispersed into the atmosphere. The ocean surface is a major source of atmospheric particles.

Most petroleum compounds are hydrocarbons (see Figure 21-1). As lighter hydrocarbons evaporate from seawater, heavier ones concentrate. The more active compounds become oxidized by oxygen. Some are utilized as food by bacteria. Others poison bacteria. Most seem to become adsorbed on particles that sink to the bottom where microorganisms gradually break them down. Hydrocarbons are fat soluble and may circulate for a time through the environment. As one example 3,4-benzopyrene ($C_{20}H_{12}$), a hydrocarbon known to induce cancer in humans, has been identified in plankton and in sediments located beneath ocean shipping lanes. Cancerous fish have been found there as well as in coastal waters that receive sewage and industrial effluents. Oil pollution must be minimized. We would live precariously with a poisoned ocean.

15-4. MONITORING WATER AND WATER POLLUTION

Modern analytical methods

Chemical analysis of water and wastewater is the basis for judging water quality and making decisions about water use and treatment.

A water's general character is quickly established by measuring its acidity–alkalinity with a pH meter (see Figure 17-7), electric conductance (an indication of total ion concentration), and total dissolved solids (evaporation to dryness of a known volume). For more information individual components are then determined. Sodium, potassium, calcium, and magnesium may be run by flame photometry (see Figure 6-4). Silica, boron, phosphate, nitrate, nitrite, ammonia, fluoride, manganese, iron, trace metals (copper, lead, zinc, mercury, etc.), and some other substances are usually determined colorimetrically—that is, by the intensity of color they develop with color reagents (see Figure 22-16a). Chloride is determined by precipitation with a standard silver nitrate solution. Bicarbonate is found by titration (see Figure 16-10) with a standard hydrochloric acid solution. Sulfate may be determined gravimetrically by precipitation with barium chloride.

In instrumental analytical methods the trend is toward automation. An example of automated equipment is shown in Figure 15-10.

To monitor pollution on a wide scale, *aerial spectral photography* is being used, including that practiced from spacecraft. The analyst first deter-

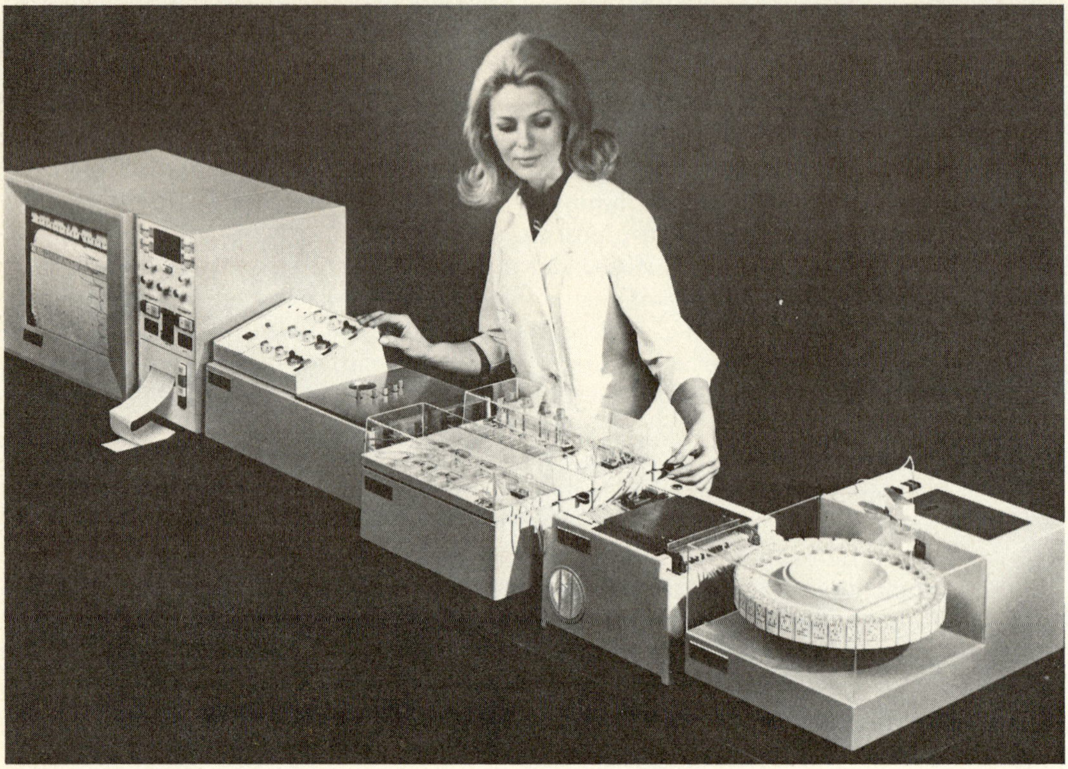

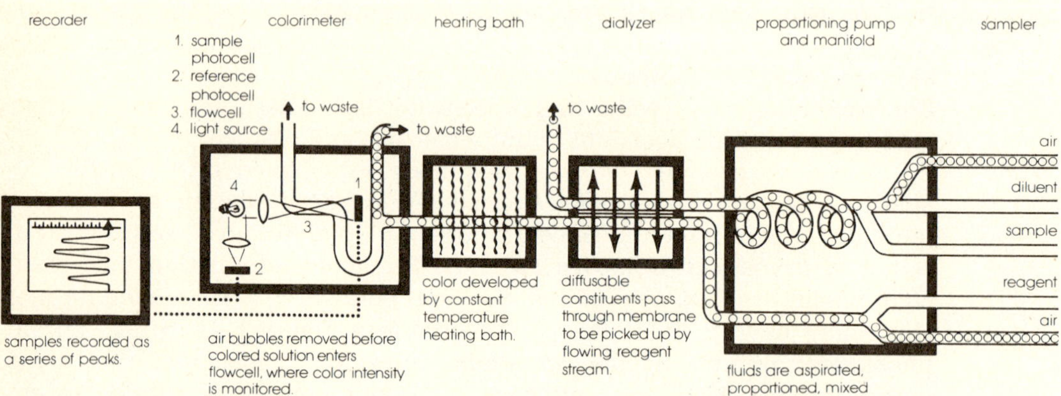

Figure 15-10 Automated analysis. (a) A series of interconnected modules that perform stepwise wet-chemical (solution) analyses. Operating from right to left the units include a sample holder, reagent dispenser and pump, dialyzer (mixer), heater, colorimeter, and printer–recorder of analytical results. (b) Schematic view, simplified. Air bubbles are used to segment the flowing streams of samples and the standard solutions.

To consider one example, PO_4^{3-} is analyzed as follows. Ammonium molybdate reagent is added to the sample, giving ammonium phosphomolybdate, $(NH_4)_3P(Mo_3O_{10})_4$. That is reduced with stannous chloride to molybdenum blue (see Group VIB, Chapter 8). The intensity of the blue color, which is proportional to phosphate concentration, is measured by the colorimeter and recorded. (Courtesy Technicon instruments Corporation, Tarrytown, New York.)

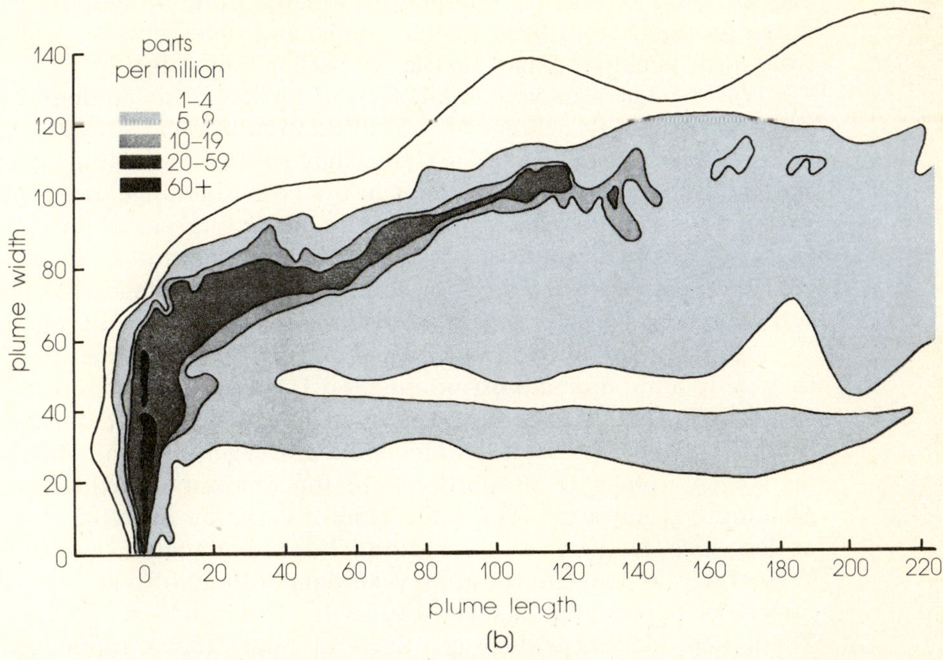

Figure 15-11 Aerial surveillance of water-borne pollution. (a) Photograph from 910 feet of a buoyant plume of fibrous pollutant (left) dumpted into the Niagara River from an industrial plant near Buffalo, New York. Particles 10^{-4} inches in diameter are being released at a concentration of about 100 ppm. The camera is equipped with a filter to record especially light of 6500 Å wavelength, which the pollutant reflects best. (b) Interpretation of the photo (both axes in feet). (Courtesy K.R. Piech, Cornell Aeronautical Laboratory.)

mines by experiment the wavelength of light best reflected by the pollutant. He next equips a camera with a spectral filter to allow only that wavelength to reach the film. He calibrates the system by photographing painted panels of known reflectance. He then photographs the polluted area and mathematically translates the densities of his pictures into concentration units of pollutant (Figure 15-11).

In principle each pollutant has a different spectral reflectance which can be interpreted both quantitatively and qualitatively. A quantitative measure of algae, for example, can be made as a gauge of eutrophication. To photograph dissolved (molecular) pollutants requires definition of very narrow wavelength regions. This technique, probably requiring the beam of a laser to give light of adequate intensity, has not yet been made practical.

15-5. MUNICIPAL WASTEWATER TREATMENT

Characterizing sewage

Municipal wastewater is a mixture of domestic sewage and industrial wastes. Manufacturing may account for two-thirds of total volume. Dissolved inorganic ions are mainly those already present in natural fresh waters (Table 15-1), with concentrations somewhat increased. Dissolved and suspended organic compounds (many unidentified) include breakdown products from fecal matter, foods, and detergents, as well as wastes from meat packing, paper, textile, and other industries.

Wastewater is grossly characterized by its *solids content* and *oxygen demand*. About 14 percent of the solids in municipal sewage are particles with diameters larger than 0.01 cm. They settle on standing and are called *settleable solids*. Smaller particles in the colloidal range (see Section 16-4), which do not settle, comprise 6 percent of the total solids and include bacteria and viruses. Some 70 percent of the solids are of molecular size and are called *dissolved solids*. After treatment in a two-stage sewage plant, the solids remaining are typically 94 percent of the dissolved type.

Oxygen demand refers to the need of oxygen for oxidation of material, especially ammonia and organic matter. Under specific aerobic conditions, the oxygen that sewage microorganisms require to oxidize a wastewater sample is called the sample's *biochemical oxygen demand* (BOD). The rate of oxygen uptake is proportional to the amount of oxidizable material remaining at any time. The same kind of equation describes the oxidation as describes the decay rate of radionuclides (explanation with Figure 5-13). When the rate constant for an oxygen-demanding pollutant has been determined by experiment, its half-life in the environment can be estimated. Theoretical biochemical oxygen demand values (usual units: milligrams of oxygen required per liter of water) are calculated by simple stoichiometry. Consider a water polluted with glucose sugar. The equation for its complete oxidation is

$$C_6H_{12}O_6 + 6O_2 = 6CO_2 + 6H_2O$$

The molecular weights of glucose and oxygen are 180 and 32, respectively. If 1 liter of water contains 180 mg of glucose (or its oxygen demand equivalent), the biochemical oxygen demand is $32 \times 6 = 192$ mg O_2/liter.

Total oxidizable sewage components, including those not readily attacked by bacteria, can be determined by injecting the water sample into a furnace equipped to combust all carbon to carbon dioxide. Carbon dioxide is then measured quantitatively by its infrared absorbance (Figure 13-10).

Primary and secondary treatments

Eighty-five percent of industrial wastewater volume comes from the primary metals, chemical, paper, and petroleum industries. Ninety percent of the biochemical oxygen demand comes from the chemical, paper, and food industries. The magnitude of the problem of disposing of these wastes is indicated by the data in Table 15-2.

The processes that have been developed over more than 100 years to handle wastewater and sewage are called *primary* and *secondary processes*. Simple, inexpensive (a few cents per thousand gallons treated), and reasonably effective, they will continue to furnish the main cleanup service for the foreseeable future.

Primary treatment is a crude process that removes grit and other settleable solids, grease, and some suspended solids by screening, skimming, flocculation, and sedimentation. The collected solids, called *primary sludge*, may then be allowed to digest in a tank before being disposed of, perhaps in a sanitary landfill. The aqueous portion is aerated for several hours or sent to an open-air oxidation pond. Pond bacteria digest oxidizable matter and release nutrients. Algae use the nutrients and by photosynthesis release oxygen for the bacteria. Periodically algae growth is skimmed off for disposal. The fluid is chlorinated for disinfection and odor control and released to receiving waters.

Secondary treatment consists of primary treatment followed by a more effective biochemical process. In municipal sewage plants the most widely used secondary process is the *activated sludge treatment*. After incoming sewage is screened, it is run into a large tank and inoculated with about 25 percent its volume of activated sludge (see below). For 4 to 8 hours air is bubbled through and the mixture stirred mechanically while aerobic bacteria oxidize most of the dissolved organic matter. The mixture is then transferred to a settling tank where an "activated sludge" of microorganisms and other suspended material settles out. Water is siphoned off, chlorinated and disposed of in a river, dumped on a spreading area for seepage into groundwater, or perhaps sprayed over a golf course—the average course in summer can soak up 100,000 gallons per day. A portion of the sludge is returned to the aeration tank. The remainder is usually run into closed digestion tanks. A digestion tank may operate at 35°C and hold a batch of sludge for a month. During digestion anaerobic organisms liquefy most of the solids and produce gas. The gas, about 70 percent methane, is burned to

Table 15-2. Estimated daily quantities of pretreated waste in the United States (1972)

Source	Wastewater billion gal/day	BOD million lb/day	Settleable and suspended solids million lb/day
All manufacturing	47	77	63
Domestic sewage	21	24	28

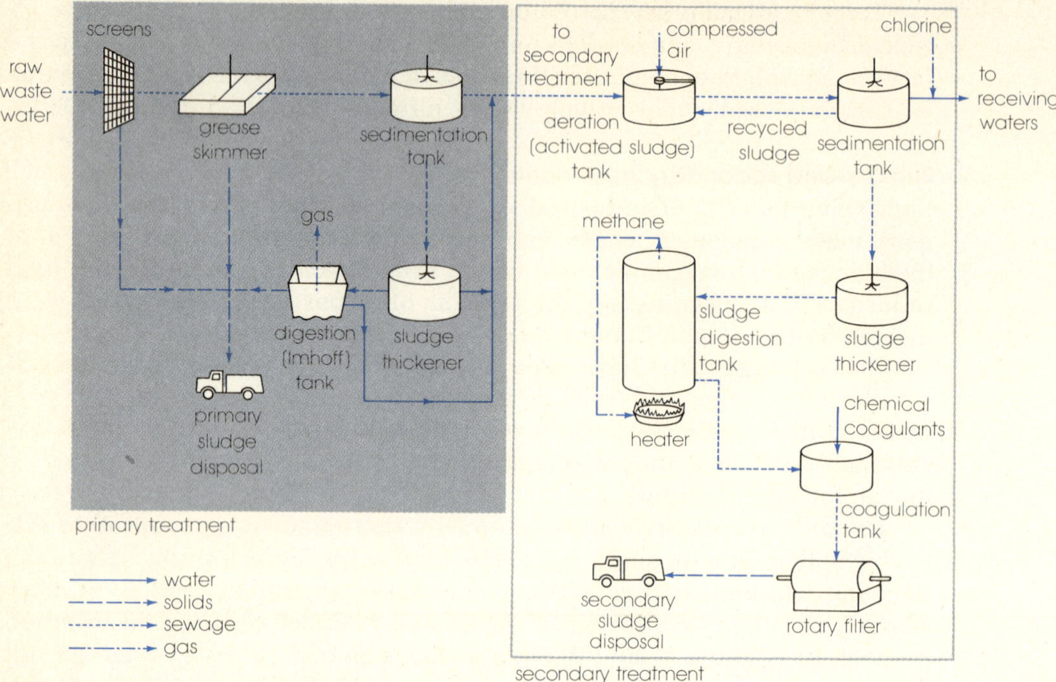

Figure 15-12 Schematic view of primary and secondary sewage treatments. Per 1000 gal of average domestic sewage, primary treatment costs 3 to 5 cents, secondary treatment 5 to 20 cents. If secondary processing is not used chlorine is added at the line marked "to secondary treatment" and the effluent is dumped. In place of the activated sludge tank some installations employ a trickling filter over which wastewater is sprayed and in which aerobic organisms degrade oxidizables.

heat the digester and run the plant's machinery. In round numbers one cubic foot of gas is produced per day per person on the system. Digested sludge is chemically coagulated and vacuum filtered. Sludge may be dumped in a landfill or in the ocean, or it may be incinerated, dried, and sold as a low-grade fertilizer for nonedible plants (toxic metals may be concentrated in it). An efficient combination of primary and secondary treatments (Figure 15-12) typically reduces suspended solids 90 percent, biochemical oxygen demand 90 percent, total nitrogen 50 percent, total phosphorus 30 percent.

New sewage treatments

The growing reality of polluted water supplies has hastened improvements in existing processes and the invention of advanced methods for wastewater treatment.

One development has been the addition of oxygen to compressed air fed to activated sludge tanks. Increase in the rate of biochemical oxidation is emphatic enough so sewage oxidation may become a major use of industrial oxygen. Another development has been the addition of chemicals at some step in the processing to precipitate phosphate so it can be removed by filtration (see Problem 22). Still another improvement is the use of charcoal as a molecular adsorption medium. Charcoal activated by heating with steam can have a surface area of 125 acres per pound (a large installation may contain 0.2 million pounds). Research since 1970 has demonstrated that molecules in domestic raw waste fall into two molecular weight classes: those above 1500 (46 percent of total mass) and those below 400. By making the mixture alkaline and heating, the former are broken into the smaller sizes that are more effectively removed by charcoal filtration. The filtrate is chlorinated and released to public waters. The charcoal is burned free of acquired impurities and recycled. The success of adsorption methods indicates physiochemical processes will sometime replace biological steps and eliminate sludge handling which often represents 50 percent of a sewage plant's operational expense.

Obviously any additional sewage treatment to that being used now will cost money. To bring every sewage treatment facility in the United States to acceptable standards for effluent quality will cost at least $30 billion, or $600 per family. Facilities for special treatments will cost still more. It will be money well spent.

QUESTIONS

1. The ocean The rate at which sediments form on the ocean bottom can be determined by carbon-14 dating. Explain.

2. DDT substitutes (library) Here are three experimental approaches to finding substitutes for DDT. **(a)** Introduce *juvenile hormones* that prevent the maturing of insect larvae. **(b)** Breed and release natural insect enemies. **(c)** Lure insect pests into traps with *pheromones*. From library references report on one approach. Cite references.

3. Desalination Both distillation and freeze plants for seawater conversion are engineered to conserve heat. Explain.

4. Boats There are 8 million private boats on the waterways of the United States. **(a)** What kinds of pollution do they create? **(b)** If you were writing boating regulations into law, what sorts of antipollution rules would you favor?

5. Cooling towers Huge towers may evaporate thousands of gallons of water per minute in dissipating waste heat at an electric power generating station. How might the water vapor affect local weather in summer? in winter?

6. Biochemical oxygen demand (BOD) In the tomato canning industry wastewater with a BOD of thousands of pounds per million gallons is common. **(a)** What does that mean? **(b)** How is BOD treated to minimize its pollution?

7. Acidic mine drainage (AMD) Addition of thiocyanate to ferric solutions gives an intensely red-colored complex ion, $Fe^{3+} + SCN^- = FeSCN^{2+}$. The reaction may be used to indicate that a stream is polluted with AMD. Explain.

Table 15-3. Pollutants in rainwater

Rainfall in.	Soluble substances ton/mi²	Insoluble substances ton/mi²
10	7.5	18
25	18	47
35	25	66

8. Thermal effects In 1969 the Federal Water Quality Administration (FWQA) ruled that the temperature of wastewater entering the Great Lakes be limited to 1°F above the temperature of ambient (circulating) lake water. What is the purpose of the rule? What effects can the dumping of hot water have?

9. Rain Rain washes substances out of the air, adding them to the pollution burden in public waters. Data like that given in Table 15-3 are obtained by analyzing rainfall over large cities. **(a)** How would you analyze a rainwater sample for total insolubles? **(b)** Name several specific soluble and insoluble components of rainwater.

10. Natural steam (library) Geothermal steam like that harnessed at The Geysers in northern California (Figure 15-13) is being used to turn electric generators. Obvious advantages over other electrical plants are the relative environmental cleanliness and the free heat. A disadvantage is that other gases and highly mineralized water may accompany the steam. From library reference to at least one geology or environmental science text, gather some facts that allow you to discuss the matter in more detail.

PROBLEMS

11. Solar still In Phoenix, Arizona, the average heat received on a horizontal surface per day is 522 cal/cm². **(a)** How much heat is received per

Figure 15-13 (Courtesy Thermal Power Company.)

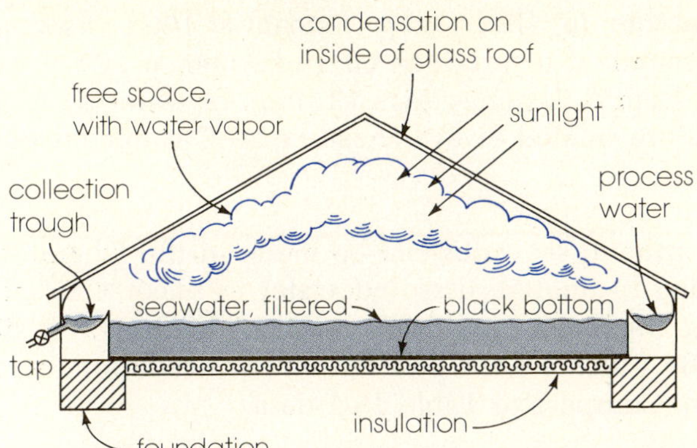

Figure 15-14

square foot of surface? **(b)** An experimental solar still (Figure 15-14) has an active area of 50 ft². If it operates at 50 percent efficiency how many gallons (1 gal = 3780 ml) of 30°C water can be evaporated in a day? (Consider water to be heated from 30° to 100°C then converted to steam at 100°C.)

12. Ocean resources The concentration of magnesium in seawater is 1300 ppm. **(a)** Calculate the pounds of magnesium in 1 million gal of seawater (1 gal = 8.5 lb). **(b)** In the Dow process 500,000 lb of magnesium is obtained from each 50 million gal of seawater handled. Calculate the percent of efficiency of the process.

13. Ocean volume **(a)** Total ocean volume is 320 million mi³. In each cubic mile is a great wealth of minerals. (Of course, a cubic mile of ocean is a great deal of water.) Suppose a chemical plant is built to process seawater at a rate of 1 mi³/yr. How many gallons must be pumped through it per minute (1 ft³ = 7.48 gal)? **(b)** Total ocean mass is 1.56 × 10¹⁸ ton. Runoff water from all continents is about 3.55 = 10¹³ ton/yr. At that rate of inflow, and assuming no losses, in how many years would today's quantity of ocean water have accumulated? Why is this figure not the age of the ocean?

14. Sulfur cycle The sulfur cycle that operates in nature is important to environmental scientists. From the following description of its steps construct a diagram like Figures 15-3 and 15-4. **(a)** Hydrogen sulfide comes from decaying animal and vegetable matter, from anaerobic bacterial reduction of sulfites, and from volcanoes and hot springs. **(b)** Hydrogen sulfide is oxidized to elemental sulfur by sulfur bacteria. **(c)** Hydrogen sulfide is also oxidized to sulfur dioxide and sulfites by aerobic bacteria. The reverse (reduction) process is accomplished by anaerboic bacteria. **(d)** Elemental sulfur is converted to sulfur dioxide and sulfites by combustions (for example, the burning of coal). **(e)** Sulfur is oxidized to sulfur trioxide and sulfates by man's manufacture of sulfuric acid and by aerobic bacteria. **(f)** Sulfur dioxide and sulfites are also oxidized (by air) to sulfur trioxide and sulfates. The reverse (reduction) process is accomplished by anaerobic bacteria. **(g)** Sulfates are used by plants as a nutrient, giving plant organic sulfur. **(h)** Animals eat plants, giving animal organic sulfur. **(i)** The excrement of animals and the death and decay of animals and plants give hydrogen sulfide [step (a), above], closing the cycle.

15. Heat of vaporization (a) Given 60 g of steam at 100°C, how much heat (calories) must be removed to produce 60 g of water at 20°C? (b) A man working hard on a humid day may lose 13 liters of water by evaporation. How many calories are carried away? (Assume 30°C water raised to 100°C then vaporized.)

16. Seawater A synthetic seawater can be made in the laboratory by dissolving the following in enough distilled water to make 1 kg of solution: 26.9 g NaCl, 3.7 g $MgCl_2$, 2.1 g $MgSO_4$, 1.3 g $CaSO_4$, 0.7 g KCl, 0.2 g $Ca(HCO_3)_2$, 0.1 g KBr. Calculate the concentration of each *ion*, expressing it in parts per million. Compare to Table 15-1 data.

17. Sediments Sediments build up slowly on the ocean floor. (a) Assuming a sedimentation rate of 5 mm/1000 yr, calculate the kilometer thickness that would be observed today if the ocean floor were as old as the continental crust, 3.5 billion yr. (b) Seismic studies and corings (the removal of cylinderical samples from the bottom) show sediments range from depths of 1 to 800 m. How do you interpret that fact, based on part (a)?

18. Trace elements In 1919 Haber investigated the possibility of recovering gold from seawater and using it to pay off Germany's war debt. His analysis showed 0.004 mg of gold present per metric ton (2200 lb) of seawater. At a recent price of $3.20 per gram, how many metric tons of seawater would theoretically have to be processed to recover $1 billion worth of gold?

19. DDT (library) In the past 20 years about 160 million lb of DDT has been sprayed in India. In a library reference find the land area of India in square miles. If spraying covered 20 percent of the country, how many milligrams of DDT were applied, on an average, to each square meter in the spray area (2.59×10^6 m² $= 1$ mi², 1 lb $= 4.54 \times 10^5$ mg)?

20. Water demand Daily usage and public water supply in the United States have increased this century as shown in Table 15-4. (a) In separate colors plot the two data sets (years on abscissa). Connect points of each set with straight lines. (b) List several ways in which the problem predicted for the 1980s may be met.

Table 15-4. Water usage and supply in the United States

Year	Water usage billion gal/year	Water system supply billion gal/year
1900	40	90
1920	75	125
1940	125	265
1950	175	300
1960	225	340
1970	300	425
1980	500 (est.)	550 (est.)

21. Solid waste A certain town of 25,000 persons generates a sewage flow of 140 gal per person per day. Sedimentation analysis shows the wastewater carries 6 ft³ of grit per million gallons. Bulk density of wet grit is 170 lb/ft³. How many tons of the material must the city truck out and dispose of each year?

22. Phosphate removal Phosphate is to be recovered from sewage effluent by adding magnesium oxide and utilizing ammonium ion naturally present:

$$5H_2O + H_2PO_4^- + MgO + NH_4^+ = Mg(NH_4)PO_4 \cdot 6H_2O\downarrow$$

How many pounds of magnesium ammonium phosphate hexahydrate can in theory be produced per week by a small city plant processing daily 5 million gal (41.7 million lb) of wastewater containing 30 ppm $H_2PO_4^-$?

16

WHAT ARE LIQUIDS AND SOLUTIONS?

THE IMPORTANT CONCEPTS

16-1. Liquids

1. Properties of liquids lie between those of gases and solids.
2. A hole model explains liquid viscosity, compressibility, and diffusion.
3. A kinetic-molecular model explains liquid density, vapor pressure, surface tension, and capillarity.

16-2. Solutions and solubility

1. A solution is a molecular dispersion of solute in solvent; seven two-component types are recognized.
2. Mechanisms of dissolving are known.
    a. A useful generalization is "Like dissolves like."
3. Solutions may be described as dilute, concentrated, saturated, or supersaturated.

4. The force between ions in solution is related to the solvent's dielectric constant.
5. Solution processes may be exothermic or endothermic.
6. Le Châtelier's principle guides prediction of equilibrium shifts when system conditions are changed.

16-3. Molal solutions and colligative properties

1. Properties of ideal solutions are predictable.
2. Colligative properties are proportional to the number of molecules (or moles) in a unit quantity of solvent.
    a. Changes in vapor pressure, freezing point, boiling point, and osmotic pressure are colligative properties.
3. Raoult's law relates vapor pressure of a solvent to its mole fraction in solution.
4. Simple mathematical statements connect solution molality with the colligative properties.

16-4. Colloids

1. Colloidal particles fall within a certain size range.
2. Colloids have a large specific surface and high adsorptive ability.

16-5. Electrolyte solutions

1. Arrhenius proposed ions to account for the behavior of electrolyte solutions.
    a. Solutes can be classified by electric conductance of their solutions.
    b. Electrolytes exhibit abnormal colligative effects.

16-6. Acid–base theories

1. The Arrhenius theory deals with H^+ and OH^-.
2. The Lowry-Brønsted theory focuses on proton donor–acceptor processes.
    a. Acids and bases are related as conjugate pairs.
    b. Acid–base strengths can be compared.
    c. Solvents are describable as leveling and nonleveling.
3. The Lewis theory describes electron-sharing processes.
    a. Acid–base reaction leads to a coordinate–covalent bond.

16-7. Volumetric analysis

1. Molar solutions may be used in chemical analysis.
2. Volumetric analyses involve titrations.
3. Hydrolysis, a reaction with solvent water, affects acid–base indicators.

Liquids are composed of molecules that have freedom of movement although remaining close to one another. The characteristics of liquids are intermediate between those of solids and gases (Table 16-1). Water, for example, is like ice in having a definite density and being difficult to compress. Like steam, however, water flows and lacks characteristic shape.

Solutions are mixtures that contain substances at a molecular level of dispersion. Most chemical processes, including natural ones, occur in solution. Water is the most abundant liquid and water mixtures are the most common kind of solution.

16-1. LIQUIDS

Properties of liquids

The structure of a solid can be described by its intermolecular forces (see Table 10-1). The gaseous condition can be described by its kinetic activity (see Eq. 14-17, for example). A liquid, however, is composed of molecules lying close enough together for the forces between them to be appreciable, yet the molecules possess enough kinetic energy for them to remain mobile until cooled to the freezing point. Thus far no theory describes the liquid condition exactly.

Liquid theory began in the 1870s with van der Waals, who extended his kinetic-molecular study to the liquid state which he treated as a gas in a condensed condition. Modified gas equations continued to be used to describe liquids until the 1930s when Debye and others took X ray diffraction pictures of liquids. The patterns obtained were similar to those of amorphous solids such as glass, and proved that liquids have some structural organization.

The liquid model we will use is pictured in Figure 16-1. We will consider a liquid to possess the local order of an imperfect close-packed lattice, although lacking regularity at longer range. Disorder in the structure involves about 3 percent "free volume" or "holes." As molecules and holes shift about, ordered cells disappear and reform.

The cell or hole theory is useful in describing the *viscosity, compressibility,* and *diffusion* of liquids.

A liquid has less resistance to flow—that is, lower viscosity—as temperature is raised. The effect is explained as an increase in molecular energy with temperature, which causes hole volume to increase. As intermolecular forces are decreased, the liquid's fluidity increases. Flow takes place when molecules and holes slide by one another.

Liquids can be compressed only a few percent. Compression is pictured as reducing hole volume to give a close-packed structure. Further

Table 16-1.   Three states of matter compared

State	Shape	Diffusion	Compressibility	Density g/ml at STP	Thermal expansion %/°C
Solid	Fixed	Slight	Difficult	1 to 20	Small, 0.01
Liquid	Fit container	Slow	Difficult	0.5 to 2[a]	Small, 0.1
Gas	Fill container	Rapid	Easy	1×10^{-4} to 50×10^{-4}	Medium, 0.3 (at constant pressure)

[a] Excluding mercury.

compression is resisted very powerfully because atomic electron clouds are then in contact.

The diffusion or spontaneous mixing of two liquids in contact is visualized as a movement of molecules facilitated by the presence of holes. When a molecule moves, it leaves a hole. As another molecule comes in to occupy that position, it creates a new hole. Diffusion is slow due to many molecular collisions.

In discussing such other properties of liquids as *density, vapor pressure, surface tension,* and *capillarity,* we borrow concepts from the kinetic-molecular theory of gases.

Change in liquid density with temperature is explained as a change in the average kinetic energy of molecules which causes them to occupy effectively more or less volume. A heated liquid expands. Expansion leaves a given volume (say 1 ml) with less mass and a lower density. As a liquid is cooled, it contracts. A given volume then has greater mass and greater density.

As the temperature of a liquid in an open container is raised, energy exchange gives some of its molecules at the surface enough energy to escape. Rising vapor exerts a pressure against the atmosphere and the liquid evaporates. As more heat is added, vapor pressure increases until it equals the atmospheric (barometric) pressure; at that temperature the liquid boils.

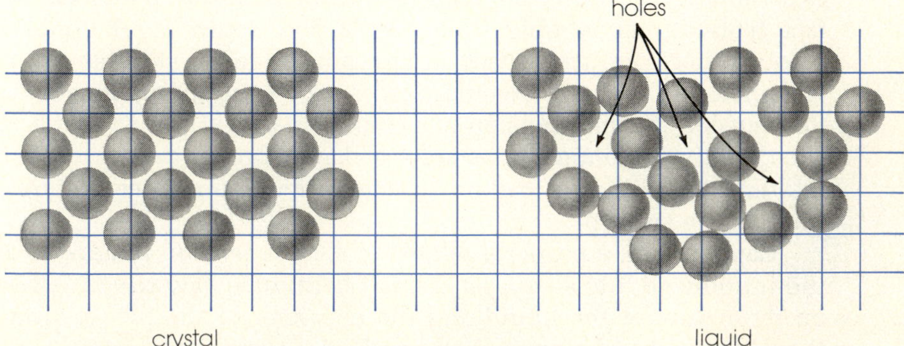

Figure 16-1 Order in crystals and liquids. Regions of order in the liquid extend only short distances before being interrupted by holes. In general the normal solid state of a substance is more dense than its liquid state (water is an exception.) Liquid behavior is presently best approximated by computer application of equations of motion to a random distribution of closely packed spheres.

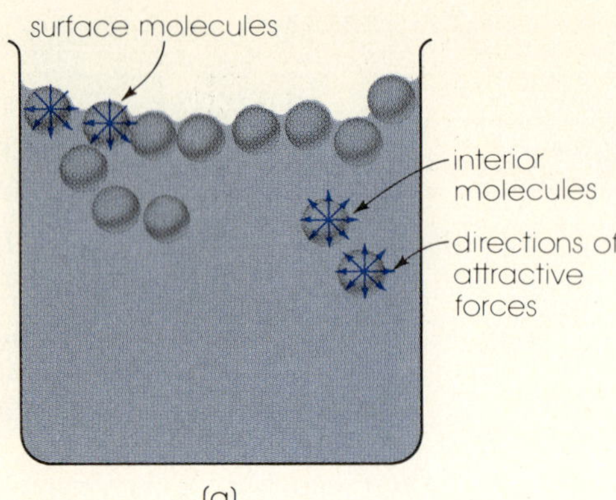

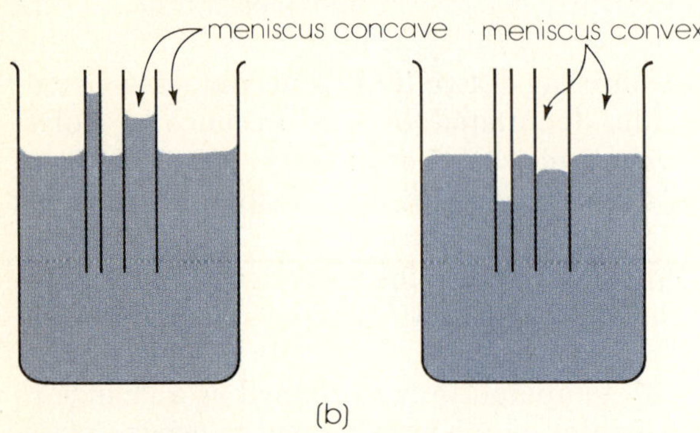

Figure 16-2 Surface tension and capillarity. (a) The attractive force on surface molecules is downward, giving a liquid surface the behavior of being under a stress. The force acting at right angles to the surface that maintains surface integrity is called surface tension. (b) Left, capillary rise of a liquid such as water in tubes of a material like glass, which it wets. Right, capillary depression of a liquid such as mercury in tubes of glass or other material, which it does not wet.

Some of the heat added prior to the boiling point raises the liquid's temperature. Heat added at the boiling point, however, is carried away entirely as the energy of escaping molecules. At the boiling point the average kinetic energy of molecules remaining in the liquid does not change and thus the temperature of boiling is constant. As shown in the vapor pressure curves of Figure 14-8b, standard pressure for defining normal boiling points is 760 torr. Normal boiling points, tabulated in reference books, are one means of identifying liquids and assessing their purity.

A molecule surrounded by others in the bulk of a liquid (Figure 16-2a) experiences attraction from all directions. By contrast a molecule at the surface feels attraction only from beside and below it. A liquid is drawn inward, therefore, its surface is minimized, and a skinlike exterior is established due to intermolecular forces. The contracting force is called *surface tension*. Surface tension is a characteristic of liquids that accounts for such familiar phenomena as spherical rain drops (they are misshapen by fall) and the ability of certain insects to move on water without breaking through the surface.

Capillarity is a property related to surface tension (Figure 16-2b). A liquid introduced into a capillary (small-bore) tube may rise or fall depending on the nature of the liquid and the character of the tube's surface. When there is attraction between liquid and solid surface, the liquid is said to wet the surface. The attraction pulls liquid upward and the liquid's *meniscus* (Gk. *meniskos*, crescent) or top surface is observed to be concave. When liquid and solid surfaces repel one another, the liquid's internal attractive forces predominate. The liquid is then depressed at the tube walls and its meniscus is convex. Capillary action contributes to the mechanical weathering of rock and development of soil when intergranular spaces have capillary dimensions through which water can climb. Capillarity also accounts for the rise of water from the roots of plants, a matter of 300 feet or more in the tallest trees.

16-2. SOLUTIONS AND SOLUBILITY

Some definitions and solution types

A solution is a homogeneous material whose composition can be varied. It is usually obvious that a solution is composed of a *solvent* or dissolving medium that is in major amount, and one or more dissolved species called *solutes* that are in lesser amounts. When the quantity of solute is quite small compared to the quantity of solvent, the solution is described as *dilute*. When the solute is in appreciable quantity, the solution is called *concentrated*.

Solutions are prepared in the laboratory so chemical reactions can be conducted with them. Quantities of solute are then easily measured by taking the required volumes of solution. Reaction rate can be varied readily by dilution with solvent, and reaction heat can be controlled in the same way. When reactions are carried out in solution, the products may be readily collected by such elementary physical means as distillation and filtration.

The three states of matter — solid, liquid, and gas — can be combined in nine different ways to make two-component mixtures. Liquids and solids cannot be said to dissolve in gases, however, so that solutions actually are limited to seven of the nine possible combinations (Table 16-2). Least important are gas-in-solid and liquid-in-solid solutions. Most important are the three types in which a liquid is the solvent.

Why does the solute dissolve?

A solute is expected to dissolve in a solvent when (1) a reversible reaction occurs that leads to solvated species, (2) a more profound and less readily reversed chemical reaction takes place that changes the solute into a completely new species, or (3) a physical dispersion results. Polar solvents tend to dissolve polar or ionic solutes, whereas nonpolar solvents tend to dis-

Table 16-2. Solution types

Solute	Solvent	Example
Gas	Gas	Air; oxygen, etc., in nitrogen
	Liquid	Magma; steam in molten rock
	Solid	Hydrogen in platinum
Liquid	Gas	
	Liquid	Wine; alcohol in water
	Solid	Dental amalgam; mercury in silver
Solid	Gas	
	Liquid	Seawater; salts in water
	Solid	Brass; zinc in copper (an alloy)

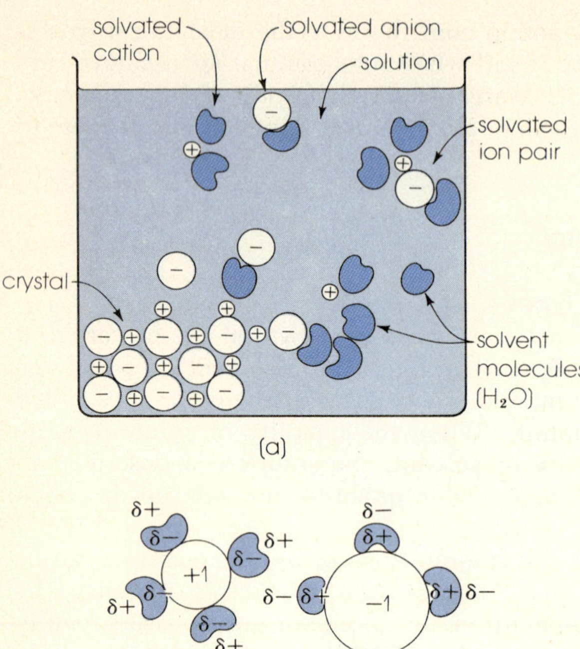

Figure 16-3 Dissolution of a salt in water. (a) Water dipoles separate the crystal lattice by coming between ions and moving with them into the body of the solvent to create a solution. Stirring aids the process. Dilute solutions favor separate, hydrated ions. In more concentrated solutions ion pairs are found held together loosely by coulomb attraction. (b) Water dipoles surround ions with a sphere of opposite charge, reducing the attraction between ions.

solve nonpolar solutes. *Like dissolves like* is the simplest generality chemists use to correlate solubility and chemical constitution. No general rule is available for predicting how soluble a solute will be in a given solvent.

An example of the first of these three mechanisms is the dissolution of a salt such as NaBr in water (Figure 16-3). We know from previous discussion (Section 9-2) that water molecules are polar. Their somewhat positive hydrogen atoms are attracted to negative bromide ions whereas the somewhat negative oxygen atoms are attracted to positive sodium ions. As water molecules surround individual ions, they lower interionic attractions and lift ions from the surface of the solid. Due to natural molecular movement, the hydrated ions travel slowly through the solution and finally become uniformly distributed. If temperature is kept constant and more solute is present than will dissolve, an *equilibrium* becomes established wherein the rate at which ions are removed from the crystal surface is equal to the rate at which they are redeposited. The solution is then said to be *saturated*. Before equilibrium is established, the solution is unsaturated at the given temperature.

With certain compounds a *supersaturated solution* can be readily prepared. Such a solution is more concentrated than a saturated solution. By carefully cooling a saturated aqueous solution of sodium acetate, for example, a supersaturated solution is obtained. Not being at equilibrium the solution is unstable. Addition of a crystal of solid sodium acetate will "seed" crystal formation and excess solute will begin precipitating. It will continue to precipitate until equilibrium between dissolved and undissolved solute is established. The solution will then be saturated.

A measure of a solvent's ability to lower the force acting between ions in solution is given by its *dielectric constant*. The dielectric constant for water at 25°C is about 80, one of the highest values known. It means the work needed to separate two ions is $\frac{1}{80}$ as much in water as it is when the two ions are the same distance apart in the vapor state. As a result, ionic compounds are water soluble. The effect is illustrated in Figure 16-3b. Alignment of polar solvent molecules around salt ions partially neutralizes the ionic charges and shields the ions from one another. Dielectric constants of polar solvents (HCN, H_2SO_4, etc.) are high; those of nonpolar solvents (CCl_4, C_6H_6, etc.) are low. Other generalizations regarding the solubilities of ionic compounds in water have been given in Chapter 11 under the heading *Metathesis*.

An example of the second mechanism is the dissolving of calcium metal in water:

$$Ca + H_2O \longrightarrow Ca^{2+} \text{(hydrated)} + 2OH^- \text{(hydrated)} + H_2\uparrow$$

The result of the solute–solvent chemical reaction is a solution containing new species. As they are formed, the solvent solvates them.

An example of the third mechanism is the dissolving of a nonreactive gas such as helium in water. Helium atoms are simply dispersed homogeneously to give a solution without any interaction with the solvent. As would be expected such solutions readily lose solute.

Temperature and solubility; Le Châtelier's principle

Temperature affects solubility. When solution temperature is raised, the water solubility of gases decreases and the solubility of liquids and solids usually increases. The effect of temperature on the solubility of some salts is shown in Figure 16-4.

When a salt dissolves in water, an energy change takes place. Energy is absorbed as crystal bonds are broken, and energy is simultaneously re-

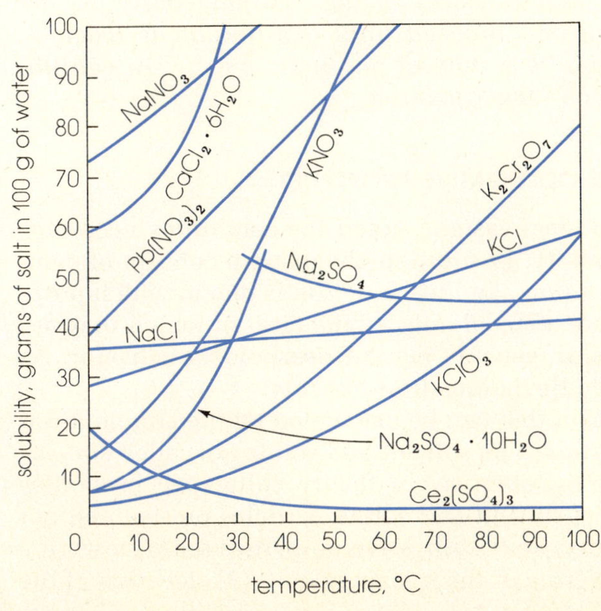

Figure 16-4 Solubility curves for some soluble salts in water. Temperature change can increase or decrease solubility. Data are collected by evaporating weighed samples of saturated solutions at given temperatures then weighing the salt residue (see Problem 17). When solubility exceeds approximately 1 g/100 g of water, the salt is called "soluble." When the solubility is less than approximately 0.1 g/100 g of water, the salt is called "insoluble." Salts classed as "slightly soluble" have solubilities between these values.

leased as ions are hydrated. For instance,

$$I.\ K^+NO_3^- = K^+ + NO_3^- - heat$$
$$II.\ K^+ + NO_3^- + 2H_2O = K^+H_2O + NO_3^-H_2O + heat \tag{16-1}$$

If the energy absorption in reaction I is larger than the energy release in II (in this example it is), then the overall solution process is *endothermic* (heat of solution is negative). In those cases,

$$salt + water = solution - heat \tag{16-2}$$

When KNO_3 is dissolved in water, the temperature of the mixture continues to drop until the solution is saturated. If heat is applied to the saturated solution, solubility is increased and more salt can be dissolved.

If the energy term of reaction II is larger than that of I, the overall solution process is *exothermic* (heat of solution is positive). Then

$$salt + water = solution + heat \tag{16-3}$$

$Ce_2(SO_4)_3$ and Na_2SO_4 (Figure 16-4) are examples of salts that heat up a solution as they dissolve. When additional external heat is applied to their saturated solutions, solubility is decreased and solute begins to precipitate.

The generalizations above can be deduced from a thermodynamic principle discovered in 1888 by French chemist Henri Le Châtelier. A statement of *Le Châtelier's principle follows: If conditions are changed in a system at equilibrium, the equilibrium will shift in the direction that tends to restore the original conditions.*

Consider an endothermic solution process at equilibrium. If heat is added, the system can restore equilibrium by absorbing heat. In the system described by Eq. 16-2 this means the reaction will proceed further to the right and salt solubility will increase. If heat is added to an exothermic process at equilibrium, equilibrium is again restored by heat absorbance. For the process in Eq. 16-3 this is accomplished when the reaction proceeds further to the left and solute comes out of solution.

Systems both chemical and physical change spontaneously in the direction toward equilibrium and proceed until equilibrium is reached. The true equilibrium condition is a state of minimum energy. At equilibrium spontaneous change is no longer possible.

16-3. MOLAL SOLUTIONS AND COLLIGATIVE PROPERTIES

In Chapter 14 we defined an ideal gas and noted the conditions necessary for real gases to approach ideality. In solution chemistry a concept of comparable utility is the *ideal solution*. An ideal solution is one in which properties of solute and solvent are affected only by the ratio in which they are mixed. Very dilute solutions, in general, exhibit nearly ideal behavior. As concentration increases, so do deviations from ideality.

The properties of a solution that can be calculated by the simple arithmetic averaging of the properties of its solvent and solute are called *colligative* (L. *colligatus*, to collect) *properties*. In theory colligative properties depend only on the number of solvent and solute particles present and not at all on their chemical identity. Four colligative properties are (1) lowering of the vapor pressure, (2) lowering of the freezing point, (3) elevation of the boiling point, and (4) change in osmotic pressure. The first effect is the most important one.

Molality

A molal (m) solution is prepared by weight. As a weight concentration, molality does not change with temperature. (A solution prepared by volume does change concentration because volume (or density) varies with temperature.)

The molality of a solution is defined as the number of moles of solute in 1 kg (1000 g) of solvent. A 1 m aqueous solution contains 1 mole or 1 gram-formula weight of solute dissolved in 1 kg of water.

Example 16-1 Analysis of a sample of honey shows it is 80 percent simple sugars ($C_6H_{12}O_6$) and 20 percent water by weight. Calculate its molality.

Solution Although sugar is in the larger amount, we will consider water the solvent. The weight of sugar dissolved in 1 kg of H_2O is

$$? \text{ g sugar} = 1000 \text{ g } H_2O \left(\frac{80 \text{ g sugar}}{20 \text{ g } H_2O} \right) = 4000 \text{ g}$$

The molecular weight of the sugar is $(6 \times 12) + (12 \times 1) + (6 \times 16) = 180$. The moles of sugar in 4000 g is

$$\frac{4000 \text{ g}}{180 \text{ g/mole}} = 22.2 \text{ moles}$$

Because 22.2 moles of sugar would be dissolved in 1 kg of solvent, by definition the solution is 22.2 molal (an unusually concentrated solution).

Raoult's law

In the 1880s French chemist F. Raoult carried out a series of experiments to find how the physical properties of solvents varied with solute concentration. He found that when a solution was prepared from a nonvolatile solute such as sugar and a volatile solvent such as water, the vapor pressure due to the solvent was reduced. Furthermore, in a dilute solution the solution vapor pressure was directly proportional to the fraction of solvent molecules in it.

Given, for example, a solution consisting of $\frac{1}{10}$ solute molecules and $\frac{9}{10}$ solvent molecules, Raoult could predict that its vapor pressure would be $\frac{9}{10}$ the vapor pressure of pure solvent at the same temperature. These fractions are called *mole fractions* (symbol, X). The mole fraction of the solvent X_1 is

$$X_1 = \frac{\text{moles of solvent}}{\text{moles of solvent} + \text{moles of solute}} \tag{16-4}$$

A statement of *Raoult's law* is as follows: *The vapor pressure of a volatile solvent in a dilute solution is proportional to its mole fraction.* Mathematically,

$$P = P_0 X_1 \qquad\qquad (16\text{-}5)$$

where P is the vapor pressure of the solution and P_0 the vapor pressure of pure solvent (at the same temperature).

Vapor pressure data can be obtained by using a device like the one in Figure 14-8. When vapor pressure curves of solvent and solution of a nonvolatile solute are plotted, results resemble the lines of Figure 16-5a. At each temperature, in accordance with Raoult's law, solution vapor pressure is less than the vapor pressure of pure solvent. When vapor pressure data are available at freezing and boiling temperatures, plots like Figure 16-5b can be made. At the lower end it can be seen that the freezing point of the solution is less than the freezing point of pure solvent. At the upper end the solution boiling point is greater than the boiling point of pure solvent.

A simple explanation for the lower vapor pressure of the solution is that its exposed surface consists of both solvent and solute molecules, whereas at the surface of pure solvent only solvent molecules exist. If, for instance, all molecules in a solution are the same size but $\frac{1}{10}$ of the surface is nonvolatile solute molecules and $\frac{9}{10}$ solvent molecules, then $\frac{9}{10}$ as many solvent molecules are required under equilibrium conditions in the vapor above the solution to balance the number leaving the surface. The equilibrium vapor pressure of the solution is therefore ideally $\frac{9}{10}$ that of pure solvent.

Change in freezing point due to dissolved solute is explained by assuming that the solute interferes with the ordered alignment of solvent molecules required for crystal formation at the freezing point. Change in boiling

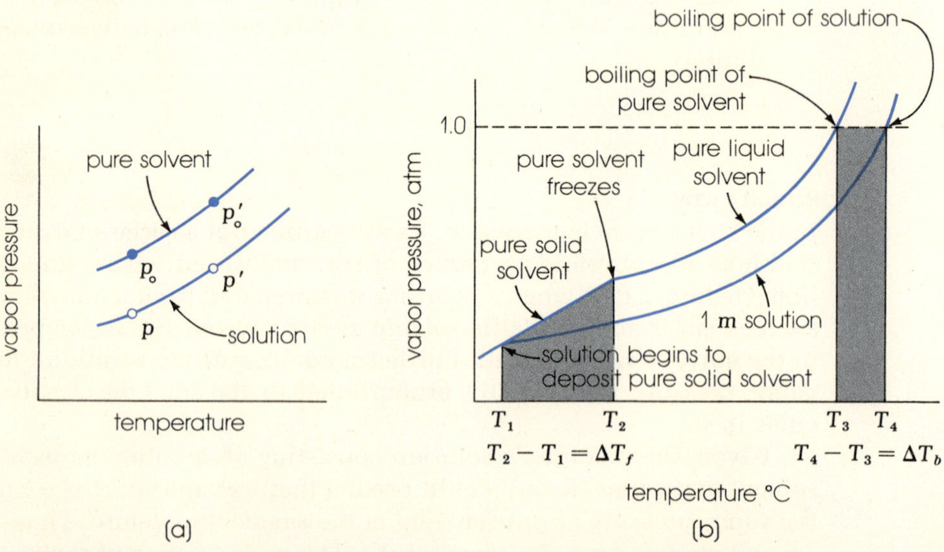

Figure 16-5 Vapor pressure curves. (a) Curves such as those established by water and a water solution of a nonvolatile solute. At all points solution vapor pressure is lower. (b) Vapor pressure curves (not to scale) extended to show at the extremities how the solvent's freezing point has been depressed and its boiling point raised. Because the lower curve is that of a 1 m solution, ΔT_f and ΔT_b are, respectively, the molal freezing point and molal boiling point constants. The values are characteristic of each solvent. For any solvent ΔT_f is always larger than ΔT_b.

point due to dissolved nonvolatile solute is explained, as above, by its lowering of the vapor pressure. Boiling does not take place (in an open container) until vapor pressure equals atmospheric pressure. With fewer solvent molecules at the solution's surface, a higher temperature is needed to attain that pressure.

Example 16-2   A solution is made by dissolving 60 g of urea, $(NH_2)_2CO$ (mol. wt. 60) in 180 g of water (mol. wt. 18). What is the vapor pressure at 100°C? Will the solution boil at 100°C if the atmospheric pressure is 760 torr? (Urea is nonvolatile.)

Solution   We will use Eq. 16-4 to find the mole fraction of water, then Eq. 16-5 to calculate the solution vapor pressure. The moles of solute and solvent are, respectively,

$$\frac{60\ g\ (NH_2)_2CO}{60\ g/mole} = 1.0\ mole \qquad \frac{180\ g\ H_2O}{18\ g/mole} = 10\ moles$$

The mole fraction of water is $\frac{10}{11}$ or,

$$\frac{10\ moles}{1.0\ mole + 10\ moles} = 0.91$$

At 100°C (its normal boiling point) the vapor pressure of water is 760 torr (see Figure 14-8b). The vapor pressure of the solution is

$$P = (0.91)(760\ torr) = 692\ torr$$

The solution does not boil at 100°C but at some temperature higher. At 100°C its vapor pressure is lower than atmospheric pressure.

Freezing point of a solution

It is common knowledge that a solution freezes at a lower temperature than does the pure solvent. Thus seawater begins freezing about 2°C below the freezing point of pure water, and magma begins to freeze hundreds of degrees below the freezing point of its solvent, pure silica. In 1883 Raoult discovered that various nonionic solutes at the same concentration in a given solvent caused the same degree of freezing-point lowering. Other experiments with similar, dilute solutions showed the effect to be directly proportional to molal concentration. Mathematically,

$$\Delta T_f = K_f m \tag{16-6}$$

where ΔT_f is the change in or lowering of the solvent's normal freezing point, K_f is the molal freezing-point constant of the solvent (also called its *cryoscopic constant*), and m is the molality; K_f values are determined from experiments that give data like those in Figure 16-5. The value for water is 1.86. The units are degrees Celsius kilograms of solvent per mole of solute. It follows from Eq. 16-6 that the freezing-point depression in Celsius degrees of a water solution having molality m is 1.86 m. Water freezes at 0°C. Therefore, the solution freezing point in degrees Celsius is $0 - 1.86\ m$, or

$-1.86\ m$. These relationships allow one to predict the freezing point of a solution given its molality, or to calculate the molecular weight of the solute given the solution's freezing point and the weights of solute and solvent it contains.

Example 16-3 An "antifreeze" is prepared for automobile radiator use by dissolving 3000 g of methyl alcohol, CH_3OH (mol. wt. 32.0), in 3000 g of water. At what temperature will the solution begin to deposit ice (when will the solvent, water, start freezing)?

Solution The moles of alcohol are

$$\frac{3000\ g}{32.0\ g/mole} = 93.8\ moles$$

The solution contains 3 kg of water. In 1 kg of water there is $93.8/3 = 31.3$ moles of alcohol. The solution is therefore 31.3 molal. A 1 m solution shows a freezing point depression of 1.86°C. A 31.3 m solution should show a depression of

$$(31.3)(1.86) = 58.2°C$$

In theory the solution will not begin freezing before a temperature of $-58.2°C$ is reached.

Example 16-4 A solution prepared by dissolving 1.00 g of a nonionic organic compound in 50 g of benzene has a freezing point of 4.00°C. The freezing point of pure benzene is 5.49°C and its K_f value is 5.08. Calculate the molecular weight of the solute.

Solution The freezing point of benzene has been lowered $5.49 - 4.00 = 1.49°C$. This is ΔT_f; K_f is given in the statement of the problem. Solution molality can be found by rearranging Eq. 16-6 as follows:

$$\frac{\Delta T_f}{K_f} = m = \frac{1.49}{5.08} = 0.293$$

If 1000 g of benzene instead of 50 g had been used to prepare the solution, it would have contained $1000/50 = 20$ times as much solute, or $(20)(1.00\ g) = 20$ g. Because the solution is 0.293 molal, it contains 0.293 mole of solute per 1000 g of benzene. Therefore 20 g of solute is also 0.293 mole of solute. The weight of 1 mole of solute is

$$?\ g = 1\ mole\left(\frac{20\ g}{0.293\ mole}\right) = 68\ g$$

Solute molecular weight is 68.

Boiling point of a solution

In a dilute solution made by dissolving a nonvolatile nonelectrolyte in a volatile solvent, the elevation of the solvent's normal boiling point is ideally directly proportional to the molality. Mathematically,

$$\Delta T_b = K_b m \tag{16-7}$$

where ΔT_b is the boiling-point elevation, K_b is the solvent's molal boiling-point constant (also called its *ebullioscopic constant*), and m is the molality; K_b values, like K_f values, are obtained by experiment (Figure 16-5.) For water, K_b is 0.52. The units are degrees Celsius kilograms of solvent per mole of solute. This means a 1-molal solution boils 0.52°C above the boiling point of pure water under the same conditions of pressure. At 760 torr a 1 m aqueous solution boils at $100.00 + 0.52 = 100.52°C$.

Boiling-point data can be used to determine molecular weights, as illustrated in the next example.

Example 16-5 A solution prepared by dissolving 69 g of fructose, a sugar, in 1 kg of water begins boiling at 100.20°C (760 torr). Calculate the molecular weight of fructose.

Solution The normal boiling point of water is 100.00°C. The boiling point rise is $100.20 - 100.00 = 0.20°C$. Because the solution is made from 1 kg of water, we can solve directly for solution molality by rearrangement of Eq. 16-7:

$$\frac{\Delta T_b}{K_b} = m = \frac{0.20}{0.52} = 0.385$$

Sixty-nine grams of solute is 0.385 mole. The weight of 1 mole is

$$?\,g = 1\ mole \left(\frac{69\ g}{0.385\ mole} \right) = 179\ g$$

The molecular weight of fructose is approximately 179.

Osmotic pressure

We are already acquainted with the phenomena of osmotic pressure (see Figure 15-7). In dilute solution osmotic pressure (symbol, π) is a function of absolute temperature T, the volume V of solution used, and the number of moles n_2 of solute in solution. The relationship, discovered in 1887 by van't Hoff, is such that the solute acts as if it were a confined gas describable by the perfect-gas law (Eq. 14-11):

$$\pi V = n_2 RT \tag{16-8}$$

It is convenient to choose a value for R, the universal gas constant, that will give π in atmospheres (because osmotic pressures can be large), and to define solution concentration as a molality m. For water solutions the relationship becomes

$$\pi = 0.082 m T \tag{16-9}$$

The osmotic pressure of a 1 m solution at 0°C (273°K) is

$$\pi = (0.082)(1)(273) = 22.4\ atm$$

When osmotic pressure is known, the molecular weight of the solute

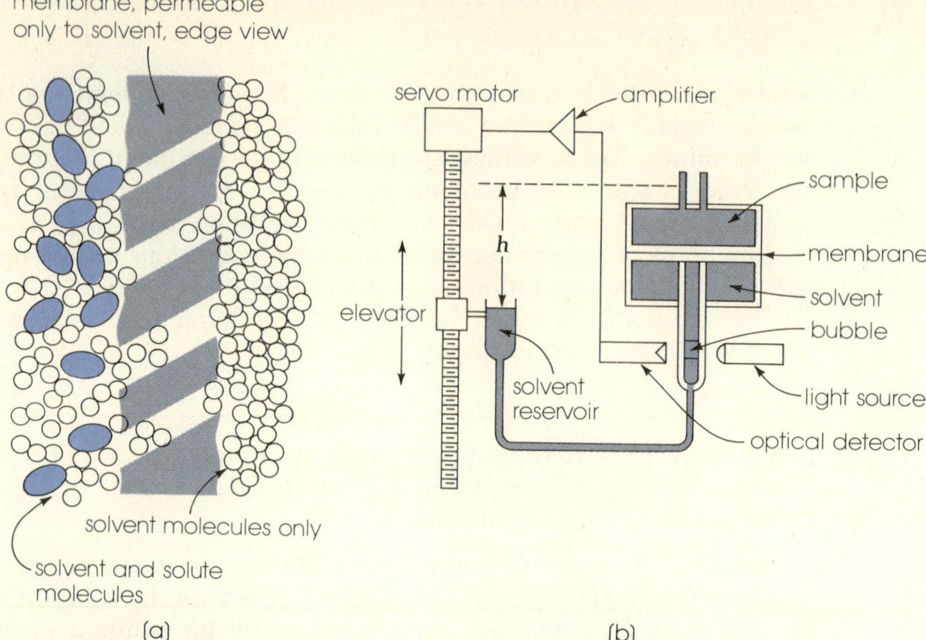

Figure 16-6 Osmosis and osmometry. (a) A mechanistic explanation of osmosis. Solvent-membrane collisions take place more often on the solvent side, resulting in a net transfer of solvent molecules into the solution side. (b) Schematic view of a membrane osmometer. Sample solution is introduced above the semipermeable membrane and pure solvent below the membrane. The tendency for solvent molecules to pass through the membrane is opposed automatically by a mechanism that adjusts the height of the solvent reservoir (and thus hydrostatic head h on the solvent side) until solvent flow is just prevented. Solvent flow is followed via the air bubble that rises and falls. Bubble motion, monitored by the optical detector, is translated into elevator adjustment of solvent reservoir and read on a counting device. At equilibrium the counter reading is equivalent to the solution's osmotic pressure. Molecular weights between 10^4 and 10^6 can be determined with the instrument. (Courtesy Hewlett-Packard, F & M Scientific Division.)

can be found. The method is especially useful for obtaining the molecular weights of large molecules in dilute solutions as illustrated in Example 16-6. The principles of osmotic pressure are studied today by engineers in designing reverse osmosis units for desalting water, and by biologists interested in fluid transport through membranes of living cells. An *osmometer* is illustrated in Figure 16-6b.

Example 16-6 A protein is partially broken down. A solution is prepared by dissolving 0.055 g of the products in 100 ml of water. At 27°C the measured osmotic pressure is 1.5 torr. What is the average molecular weight of molecules in the sample?

Solution We first convert the osmotic pressure to a pressure in atmospheres:

$$? \ atm = 1.5 \ torr \left(\frac{1 \ atm}{760 \ torr} \right) = 1.98 \times 10^{-3} \ atm$$

Next we rearrange Eq. 16-9 to solve for molality:

$$m = \frac{\pi}{0.082T} = \frac{1.98 \times 10^{-3}}{(0.082)(27 + 273)} = 8.02 \times 10^{-5}$$

Now 0.055 g of sample in 100 g of water is equivalent to 0.55 g of sample per 1000 g of water. Therefore 0.55 g is 8.02×10^{-5} mole. The weight of 1 mole is

$$? \, g = 1 \, mole \left(\frac{0.55 \, g}{8.02 \times 10^{-5} \, mole} \right) = 6858 \, g$$

By definition the solute's average molecular weight is 6.9×10^3.

Osmosis can be explained on the basis of the kinetic model shown in Figure 16-6a. On the pure solvent side of a semipermeable membrane more solvent molecules are available to pass through the membrane than on the solution side where some of the molecules are those of solute. A net transfer therefore takes place from solvent side to solution side as a consequence of the previously stated principle that all systems change spontaneously in the direction that decreases their capacity for change; that is, they go toward the equilibrium condition. In osmosis equilibrium means solvent is transferred in a direction that tends to make the concentrations on both sides of the membrane equal. If concentrations are equal, so are the vapor pressures.

16-4. COLLOIDS

Characteristics of colloidal systems

In the 1860s Graham (Graham's law, Eq. 14-14) discovered that certain substances such as glue and natural gums diffused unusually slowly in water and could not pass through a (semipermeable) animal bladder membrane as did salts. In experimenting with membrane separations he coined the words *colloid* (gluelike), *crystalloid* (saltlike), and *dialysis* (separation).

Unlike a true solution, which consists of dissolved molecules or ions, a colloidal system is a stable two-phase dispersion. In general it is made up of fine particles called the *dispersed phase*, held in a medium called the *dispersing phase*. The dimensions of colloidal particles are between about 10^{-4} and 10^{-7} cm (10^4 to 10 Å). Particles larger than that settle out of suspension under the influence of gravity; particles smaller may form true solutions and not display colloidal behavior.

Colloid chemistry is the study of phenomena related to small particles, aggregates, bubbles, drops, active surfaces, thin filaments, and films. (In the latter cases only one dimension of the dispersed phase is in the colloid range.) A classification of systems is given in Table 16-3 (compare with Table 16-2).

The dispersed phase of a colloid system may consist of (1) aggregations of molecules or very small particles, or (2) single giant molecules (*macromolecules* or *polymers*). Examples of the first kind are clay suspended in river water (a sol), dust and water droplets suspended in air (aerosols), and soaps and detergents that form stringy clusters called *micelles* of perhaps a

Table 16-3. Colloid system types

Dispersed phase	Dispersing phase	General name	Example
Gas	Gas		
	Liquid	Foam	Whipped cream
	Solid	(None)	Lava
Liquid	Gas	Liquid aerosol	Fog
	Liquid	Emulsion	Milk
	Solid	Gel (sometimes)	Jelly
Solid	Gas	Solid aerosol	Smoke
	Liquid	Sol	Carbon ink
	Solid	Solid sol	Rock

hundred molecules each. Examples of the second kind include proteins (see Figure 23-13b) and synthetic high polymers (see Figure 21-2). Macro-molecular weights begin in the thousands and range into the millions.

Colloid systems are studied by using (1) the electron microscope (particles too small for ordinary microscopy), (2) Brownian movement (see Figure 14-10), (3) osmosis (colloids exhibit only small colligative effects because the molecular solute to solvent ratio is small), (4) the Tyndall effect (a beam of light penetrating a colloidal dispersion is visible due to scattering by the particles), and (5) the ultracentrifuge (particles are brought down from suspension if a sedimenting force well in excess of the earth's gravity is exerted—see Figure 23-15).

The most unique property of colloids is their enormous total surface area compared to coarse-grained matter of equal mass. This characteristic can be illustrated by imagining a centimeter cube of material divided into colloid-size cubes 10^{-7} cm on a side. In place of 1 cube with a surface area of 6 cm^2 there are now 10^{21} cubes with 10 millionfold the surface area—60 million cm^2, or about 1.5 acres.

Surfaces have energy, or an ability to do work, as evidenced in adsorption. Capacity for adsorption and surface area are approximately proportional. Activated charcoal is a most effective adsorbant of small organic molecules from gas or fluid streams because its spongy structure gives it a very high ratio of surface to weight. The Cottrell precipitator (see Figure 13-14) works because colloidal particles can be made to assume a surface charge that is later neutralized. Colloidal particles in solution can adsorb ions from solution. If they adsorb ions of only one charge as a first or *primary layer*, they acquire the charge sign of those ions and become mutually repelling. Repulsion by coulomb force is mainly responsible for the stability of sols. Sols are *flocculated* for removal by filtration by heating or by adding salts whose ions, as they are absorbed, counteract the charge of the primary layer.

16-5. ELECTROLYTE SOLUTIONS

The Arrhenius ionization theory

Solutes that give conducting solutions are called *electrolytes*; solutes that do not give conducting solutions are *nonelectrolytes*. Two hundred years ago Cavendish showed aqueous salt solutions conducted an electric cur-

Figure 16-7 Svante Arrhenius (Sweden 1859–1927, Nobel prize in chemistry 1903). The ionization theory published by Arrhenius as his doctoral thesis was regarded with suspicion at first because most chemists were unable to conceive of an ion with properties entirely different than those of the element(s) from which it came. Arrhenius later reinforced the theory when he pointed out that the unusually large colligative effects noted in salt solutions could be explained by assuming salts were composed of ions. Thus an ionic substance could give a multiple of the expected number of particles in solution. (Nobel Foundation.)

rent. Fifty years later Faraday suggested that electric current caused salt molecules to break up into charged particles. In 1884 Arrhenius (Figure 16-7) published the theory that *ions* were formed when electrolytes dissolved in water, and electric current was then carried by the ions. He visualized that if current was applied to two electrodes standing in a salt solution, the ions would move toward electrodes of opposite charge.

The concept of ions, received skeptically at first, remains the basis of electrolyte theory today.

Solution conductance can be studied with the device pictured in Figure 16-8. Based on degree of conductance in water solution, three classes of solutes can be distinguished.

1. Nonelectrolytes Sugars, alcohols, and other compounds yielding only neutral molecules are nonelectrolytes. Conductance is zero.

2. Weak electrolytes A few heavy metal salts, weak acids, and weak bases (Section 11-3) are weak electrolytes. Solutions of weak electrolytes conduct slightly because they contain relatively few ions. We will symbolize weak ionization with the shorter of two arrows in equations. In the case of the weak electrolyte acetic acid, for example,

$$HC_2H_3O_2 \rightleftharpoons H^+ + C_2H_3O_2^-$$

As dilution of a weak electrolyte is increased, so is its percent of dissociation. In 0.1, 0.01, and 0.001 m solutions acetic acid dissociates 1.3, 4.2, and 14 percent, respectively. At very high dilution weak electrolyte dissociation approaches 100 percent.

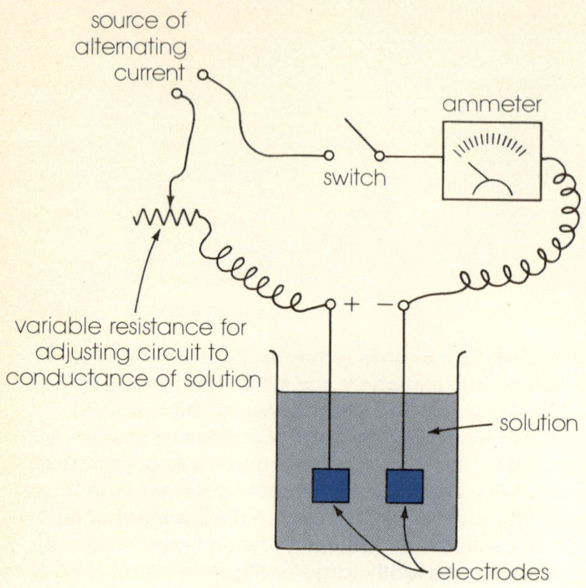

Figure 16-8 A simple device for getting relative conductances of solutions. With the switch closed, current flows through the circuit as indicated on the meter. Charge is carried through the solution by ions that migrate toward electrodes of opposite sign.

3. Strong electrolytes Most salts, strong acids, and strong bases are strong electrolytes. Electrolytes may originate as molecular substances such as hydrogen iodide which water causes to ionize,

$$HI = H^+ + I^-$$

or ionic substances such as KBr whose lattice of ions water molecules take apart.

Electrolytes and colligative properties

The effect on colligative properties of an electrolyte in solution is much larger than the effect of a nonelectrolyte at the same apparent molality. In quite dilute solutions salts like KBr and $CuSO_4$ give double, and salts like CaI_2 and K_2SO_4 give triple the expected effects. The explanation is that an electrolyte solution prepared to be, say, 0.01 m is actually n(0.01 m), where n is the number of ions per salt "molecule." Thus 0.01 m KBr is 0.01 m in K^+ and 0.01 m in Br^-, or 0.02 m in solute particles. Colligative properties depend on the number of dissolved particles. Therefore, the effects will be twice those found in a 0.01 m nonelectrolyte solution.

16-6. ACID–BASE THEORIES

Classical acid–base ideas

Acids and alkalis (later called *bases*) are mentioned in manuscripts that predate the alchemical era. Boyle described acids as solutions that tasted sour, changed the color of vegetable extracts used as "indicators," and lost their character in reacting with alkalis. Alkalis he described as solutions that tasted bitter, felt slippery, changed the color of indicators, and lost their character in reacting with acids. Lavoisier attempted to characterize acids chemically. He concluded that all acids contained oxygen as the "acidifying principle." A few decades later, however, Davy showed that hydrochloric acid contained only hydrogen and chlorine. Through the nineteenth century chemists thought of acids as hydrogen carriers.

From his studies on electrolytes Arrhenius devised an acid–base theory which, though narrow, is still useful. In water solution he assumed acid properties were caused by the hydrogen ion. Observed strengths of various acids were therefore due to differing degrees of dissociation that released different amounts of H^+. He defined a strong acid as one completely dissociated in dilute solution, and a weak acid as one slightly dissociated.

Alkali or base properties Arrhenius ascribed to the hydroxyl ion. Strong bases he said were highly dissociated; weak ones, weakly dissociated:

$$KOH \longrightarrow K^+ + OH^-$$
$$NH_4OH \rightleftharpoons NH_4^+ + OH^-$$

Neutralization of a strong acid by a strong base was then describable as reaction between their characteristic ions:

$$H^+ + OH^- = H_2O$$

The Lowry-Brønsted theory

In 1923 the Arrhenius theory was extended by T. Lowry (England) and by J. Brønsted and N. Bjerrum (Denmark). They proposed that an acid be defined as a substance capable of donating a proton, H^+.

From the definitions it followed that for each acid there was a related base:

$$acid \rightleftharpoons proton + base$$
$$HCl \longrightarrow H^+ + Cl^-$$
$$HCN \rightleftharpoons H^+ + CN^- \tag{16-10}$$

The related pair, called a *conjugate acid–base pair*, differed only by a proton.

The bare proton is an unusual species, being 10^5 times smaller than any other ion (10^{-13} cm instead of 10^{-8} cm in diameter). Because of the extreme charge concentration, most protons in aqueous solution are attached to water molecules. Formation of the hydrated proton, called a *hydronium ion,* is indicated as follows:

$$H_2O + H^+ \rightleftharpoons H_3O^+$$

It is always correct to write H_3O^+ for hydrogen ion in water solution, although chemists sometimes shorten the notation to H^+.

Use of the hydronium ion in equations helps interpret Lowry-Brønsted acid–base theory by showing that dissociation can be thought of as a *protolysis* or proton transfer. Transfer of a proton from an acid to a base yields another acid and another base. Proton transfer between the latter gives back the original acid and base:

$$acid\ 1 + base\ 2 \rightleftharpoons acid\ 2 + base\ 1$$
$$HI + H_2O \rightleftharpoons H_3O^+ + I^-$$
$$HBr + NH_3 \rightleftharpoons NH_4^+ + Br^-$$
$$H_3O^+ + OH^- \rightleftharpoons H_2O + H_2O \tag{16-11}$$

Acid 1 and base 1 comprise one conjugate pair; acid 2 and base 2, another. Again, the partners of each pair differ only by a proton.

A substance able to act as either a proton donor (acid) or acceptor (base) in Lowry-Brønsted theory is termed *amphiprotic* or *ampholytic* (Gk. *ampho*, both). In the presence of a stronger base than itself it acts as an acid, yielding a weaker base. With a stronger acid than itself it acts as a base, yielding a weaker acid. An example is bicarbonate ion:

$$HCO_3^- + OH^- \rightleftharpoons H_2O + CO_3^{2-}$$
$$\text{acid} \qquad \text{strong base} \qquad\qquad \text{weaker base}$$

$$HCO_3^- + H_3O^+ \rightleftharpoons H_2O + H_2CO_3$$
$$\text{base} \qquad \text{strong acid} \qquad\qquad \text{weaker acid}$$

Acid strength is measured by ability for proton donation; *base strength*, by proton acceptance. Perchloric acid is one of the strongest acids in water solution; its protolytic reaction is virtually complete:

$$HClO_4 + H_2O \longrightarrow H_3O^+ + ClO_4^- \tag{16-12}$$
$$\text{acid 1} \quad \text{base 2} \qquad \text{acid 2} \quad \text{base 1}$$

Both acids 1 and 2 have ability to transfer a proton, whereas bases 1 and 2 may accept a proton. Because the equilibrium point lies far to the right (forward reaction proceeds much further than the backward reaction), and we know both $HClO_4$ and H_3O^+ are strong, we can conclude that ClO_4^- is a weaker base than H_2O. Protolysis reactions proceed in the direction of the weaker acid or the weaker base.

In water all acids commonly classed as strong appear equally strong because of equal ability in transferring protons to water. By its ready proton acceptance water is said to exert a *leveling effect* on the acids. The strongest acid that can exist in appreciable quantity in any solvent is the solvent's conjugate acid. The conjugate acid of the base H_2O is H_3O^+; in water solution all strong acids are leveled to its strength.

To measure the relative proton-donating abilities of strong acids requires a solvent having less proton-accepting power than water, meaning a weaker base than water. Such a solvent is glacial (anhydrous) acetic acid. Although itself an acid, it shows amphiprotic character in the presence of a much stronger acid by accepting the latter's proton:

$$HClO_4 + HC_2H_3O_2 \rightleftharpoons H_2C_2H_3O_2^+ + ClO_4^- \tag{16-13}$$

In the nonleveling solvent acetic acid, the order of strength for common strong acids is

$$HClO_4 > HI > HBr > H_2SO_4 > HCl > HNO_3$$

Relative strengths of weak acids are found by electric conductance measurements on their water solutions. "Weak" means limited proton transfer. Consequently, water is a *differentiating* solvent for weak acids. The weaker the acid, the fewer the ions formed and the lower the conductance. Weak acids and bases are considered quantitatively in Chapter 17.

The Lewis theory

The Lowry-Brønsted theory expanded the meaning of the terms "acid" and "base." It correlated many more reactions than did the Arrhenius theory and prompted acid–base studies in solvents other than water. It was limited, however, to interpreting only proton-transfer processes.

A more general acid–base theory, which also appeared in 1923, was authored by Lewis (see Figure 9-2). He proposed that an *acid* be defined as any substance that could accept a pair of electrons, and a *base* be defined as any substance that could donate a pair of electrons. Acid–base reaction then led to formation of a coordinate covalently bonded product:

acid + *base* $\longrightarrow$ *coordinate covalent bond*

$$H^+ + \; \overset{H}{\underset{H}{>}}O\!:\; = \; \overset{H}{\underset{H}{>}}O\!\nearrow^{H^+}$$

$$:\!\overset{..}{C}l\!-\!\underset{:\underset{..}{C}l:}{\overset{:\overset{..}{C}l:}{\underset{|}{\overset{|}{B}}}} \;+\; :\!\overset{..}{C}l:^- \;=\; :\!\overset{..}{C}l\!-\!\underset{:\overset{..}{C}l:}{\overset{:\overset{..}{C}l:}{\underset{|}{\overset{|}{B}}}}\!\leftarrow\!\overset{..}{C}l:^-$$

$$Pb^{2+} + :\!\overset{..}{\underset{..}{S}}:^{2-} = Pb\!\leftarrow\!\overset{..}{\underset{..}{S}}: \tag{16-14}$$

The Lewis theory encompasses both the Arrhenius and Lowry-Brønsted theories. It also extends acid–base concepts to nonprotolytic reactions. The Lowry-Brønsted theory becomes a special case of the Lewis theory in which (1) an acid is a substance that gives a proton (although by Lewis definition the proton itself is the acid), and (2) a base is a donor of an electron pair to a proton. Potentially all positive ions and any substance with an incomplete electron octet are Lewis acids, whereas all substances with unshared electron pairs are potentially Lewis bases. Chemists use both theories. In aqueous solution chemistry Lowry-Brønsted theory may be easiest to apply. In other cases Lewis theory is often the better interpreter of reactions.

16-7. VOLUMETRIC ANALYSIS

Molar solutions

A *volumetric analysis* is a chemical analysis in which the amount of some substance (called the *analate* or *titrand*) is determined from the volume of standard solution (the *titrant*) used to react with it. A *standard solution* is one whose concentration is accurately known. Volumetric analysis began about 1850 as a means to attain quality control in chemical manufacture and is actively practiced in laboratories everywhere today.

Solution concentration is often expressed as *molarity*. A solution containing 1 gram-molecular weight (1 mole) of solute per liter of solution is a 1 molar (1 M) solution.*

Solution molarity can be defined by the statement,

$$molarity = \frac{moles\ of\ solute}{liters\ of\ solution} \tag{16-15}$$

* One gram-mole of substance can also be called 1 gram-formula weight of substance, and so a 1-molar solution can be called a 1-formal (1 F) solution. We will not distinguish between the two definitions.

Letting molarity be M and liters of solution be V, Eq. 16-15 can be rearranged to

$$moles\ of\ solute = M\ (molarity) \times V\ (liters) \tag{16-16}$$

The weight of solute in solution is

$$grams\ of\ solute = M \times V\ (liters) \times \frac{grams\ of\ solute}{mole\ of\ solute} \tag{16-17}$$

In most laboratory analyses volumes are measured in milliliters instead of liters. Calculations therefore are made using millimoles instead of moles (1 mole = 1000 mmoles) and milligrams instead of grams. Like the three equations above,

$$molarity = \frac{mmoles\ of\ solute}{milliliters\ of\ solution} \tag{16-18}$$

$$millimoles\ of\ solute = M\ (molarity) \times V\ (milliliters) \tag{16-19}$$

$$milligrams\ of\ solute = M \times V\ (milliliters) \times \frac{milligrams\ of\ solute}{millimole\ of\ solute} \tag{16-20}$$

Preparing molar solutions

Chemists prepare standard solutions by (1) dissolving a weighed amount of solute in solvent and diluting to a known volume, or (2) diluting a previously prepared standard solution to known volume. In the first method solute is weighed on an analytical balance (Figure 16-9), transferred to a *volumetric flask* (Figure 16-10a), and solvent is added to the flask's cali-

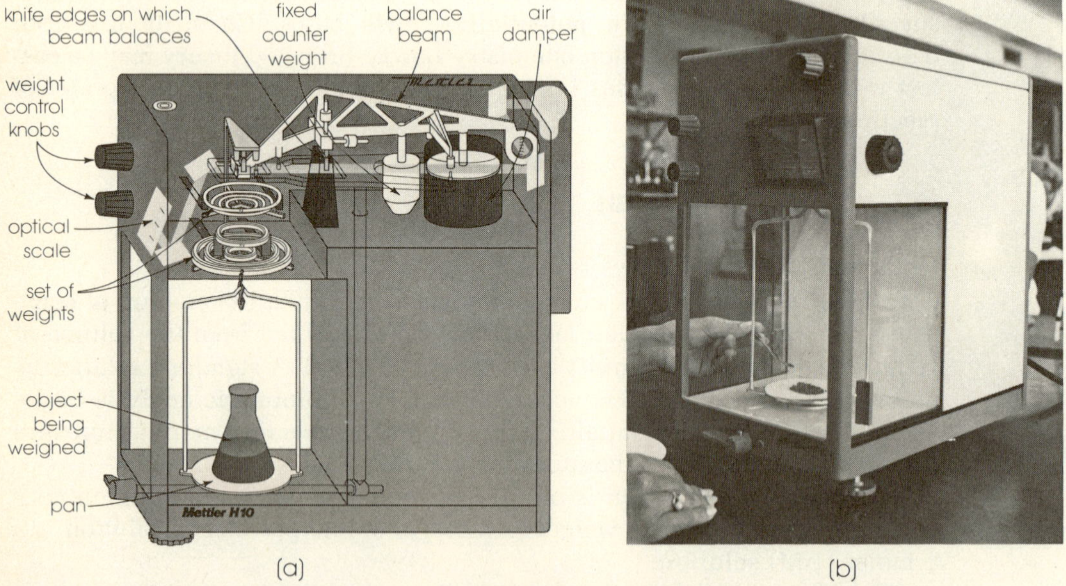

Figure 16-9 A modern single-pan balance. (a) Schematic diagram. The object to be weighed is put on the pan. Weights are then removed by levers attached to the weight control knobs in an amount equal to the mass of the object. The air damper brings the beam to rest quickly. (b) The balance in use. (Courtesy Mettler Instrument Corporation).

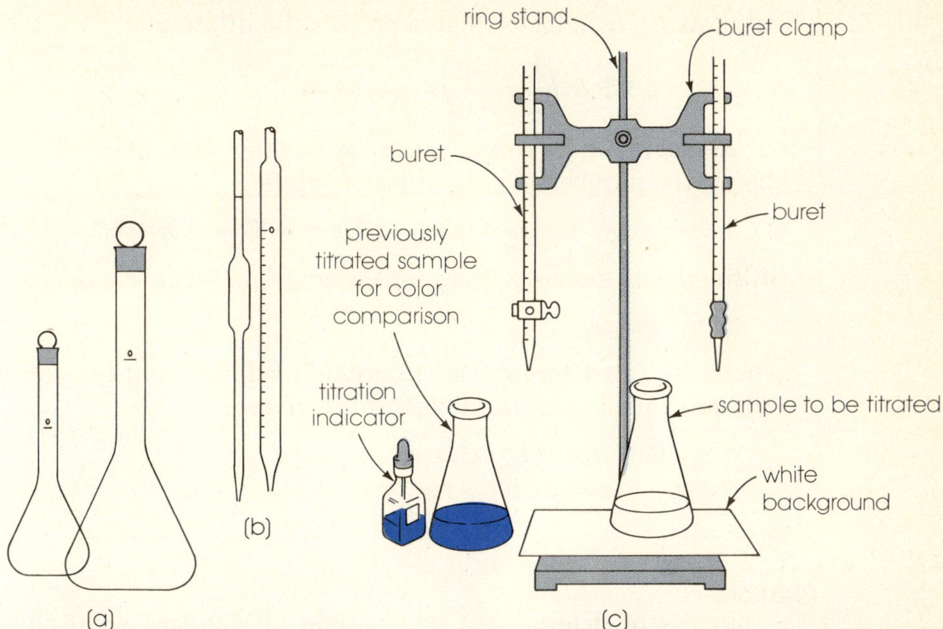

Figure 16-10 Equipment for volumetric analysis. (a) Volumetric flasks. The long narrow neck allows accurate addition of solvent to the calibration level. (b) Pipets. Left, a fixed volume pipet, right, a graduated pipet. Solution is sucked up by mouth or rubber bulb to the mark, then allowed to drain. (c) Setup for titrations. Burets are long, graduated glass tubes of uniform bore; 50-ml burets readable to 0.01 ml are most common.

bration mark. In the second method a pipet (Figure 16-10b) is used to withdraw accurately a known volume of standard solution that is then drained into the volumetric flask and diluted with solvent to the mark.

Example 16-7 How can 500 ml of 0.200 M $AgNO_3$ be prepared, starting with solid $AgNO_3$?

Solution A 0.200 M solution by definition contains 0.200 mole of solute per liter. In 500 ml, or $\frac{1}{2}$ liter, there will be 0.100 mole of solute. The weight of 1 mole of silver nitrate is $108 + 14 + (3 \times 16) = 170$ g. One-tenth mole is 17.0 g. To make the desired solution we weigh out 17.0 g of $AgNO_3$, transfer it to a 500-ml volumetric flask, add distilled water, swirl to dissolve the solute, then fill with water to the calibration mark, stopper, and mix well by shaking.

Example 16-8 A 50.0-ml sample of 0.300 M HCl is transferred by pipet to a 250-ml volumetric flask and diluted with water to the mark. What is the new molarity?

Solution We can solve the problem by simply multiplying the original molarity by a factor that expresses the dilution:

$$? \, M = 0.300 \, M \left(\frac{50.0 \ ml}{250 \ ml} \right) = 0.0600 \, M$$

Alternatively we can begin by saying that dilution does not change the number of millimoles of solute:

millimoles of solute in original solution = millimoles of solute in final solution

Millimoles is molarity times milliliters (Eq. 16-18). Therefore

$$M_1 V_1 = M_2 V_2 \tag{16-21}$$

where the sub 1 terms refer to original solution and the sub 2 terms refer to diluted solution. Substitution gives

$$(0.300)(50.0) = (M_2)(250)$$
$$M_2 = 0.0600 \ molar$$

Titrations

The process of determining the volume of standard solution that reacts with a quantity of titrand is called *titration* (Figure 16-10c). In theory the chemist adds standard solution from a buret until the *equivalence point* is reached; that is, until an amount of titrant has been introduced that is exactly equivalent chemically to the titrand originally present. In practice the chemist estimates the equivalence point as best he can by securing an *end point* marked by a chemical indicator added to the mixture or by the use of suitable instrumentation (for example, the pH meter in Figure 17-7). The difference between equivalence point and end point is called *titration error*.

In early acid–base work indicators were colored extracts from plants such as litmus. Today they are mostly synthetic organic dyes. When the analytical reaction is complete and the standard solution first appears in slight excess, it reacts with the indicator. As the indicator's molecular structure is altered, it marks the end point by changing color. The volume of titrant used is read on the buret and the necessary calculations are made.

Besides acid–base titrations (Example 16-9), there are precipitation titrations (analate is precipitated, Example 16-10), complexometric titrations (analate is complexed), and oxidation–reduction titrations (analate is oxidized or reduced).

Example 16-9 Vinegar, a product of the fermentation of natural mixtures such as grape juice, is a water solution of acetic acid. A vinegar is analyzed as follows: A 10.0-ml sample is diluted with distilled water and a few drops of phenolphthalein indicator are added. Titration requires 33.2 ml of 0.250 M NaOH. Calculate the molarity of the vinegar as an acid, and the grams of acetic acid present per liter.

Solution The balanced molecular equation for the neutralization is

$$HC_2H_3O_2 + NaOH = NaC_2H_3O_2 + H_2O$$

The number of millimoles (Eq. 16-19) of NaOH is

$$millimoles = (33.2\ ml)(0.250\ M) = 8.30$$

The reaction equation shows a 1:1 acid–base molecular ratio. Therefore 8.30 mmole of acetic acid was in the 10.0-ml sample. The molarity (Eq. 16-18) of vinegar as an acetic solution is

$$M = \frac{8.30\ mmole}{10.0\ ml} = 0.830$$

The gram-molecular weight of acetic acid is 60.0 g. A 1 M solution contains 60.0 g of acetic acid per liter. A 0.830 M solution contains

$$(60.0\ g/mole)(0.830\ mole/liter) = 49.8\ g/liter$$

Example 16-10 An oil field brine similar to seawater is known to contain 19.0 g of chloride ion per liter. If a 5.00-ml sample is taken for analysis, what volume of 0.135 M $AgNO_3$ will be needed to precipitate all the chloride? (The indicator is K_2CrO_4. The end point is signaled by the appearance of red Ag_2CrO_4 in the white precipitate of AgCl.)

Solution The balanced ionic equation is

$$Ag^+ + Cl^- = AgCl\downarrow$$

The weight of 1 mole of Cl^- is 35.5 g. The brine contains

$$\frac{19.0\ g/liter}{35.5\ g/mole} = 0.535\ mole/liter$$

By definition (Eq. 16-15) the chloride concentration is 0.535 M. The number of millimoles (Eq. 16-19) of chloride in the sample is

$$(5.00\ ml)(0.535\ M) = 1.68\ mmole$$

The silver:chloride reaction ratio is 1:1. Thus 1.68 mmole of $AgNO_3$ is required to precipitate 1.68 mmole of chloride. The volume of $AgNO_3$ required (Eq. 16-19) is

$$ml = \frac{1.68\ mmole}{0.135\ M} = \frac{1.68\ mmole}{0.135\ mmole/ml} = 12.4$$

Alternatively we can solve the problem by adapting the dilution formula (Eq. 16-21), $M_1V_1 = M_2V_2$. In 1:1 reacting ratio cases the terms on one side will refer to titrant; on the other side to titrand. Substitution gives

$$(0.535\ M)(5.00\ ml) = (0.135\ M)V_2$$
$$V_2 = 12.4\ ml$$

Hydrolysis

When a titration involves a strong acid and strong base, it is carried out until an indicator indicates that the solution is *neutral* (solution contains the same number of H^+ and OH^-). When a weak acid is titrated with a strong base, or a weak base is titrated with a strong acid, however, the end point is not the neutral point because one of the products formed undergoes *hydrolysis* (Gk. *hydro*, water + *lysis*, loosening).

An ion is said to hydrolyze when it reacts with water to give a weak acid and hydroxyl ion, or a weak base and hydrogen ion. As a chemical reaction, hydrolysis therefore can be thought of as the opposite of neutralization. In Lowry-Brønsted terms hydrolysis is a proton-transfer process with water as a reactant. Anions like CN^- that are the conjugate bases of weak acids (HCN) undergo hydrolysis to give basic solutions. Cations like NH_4^+ that are the conjugate acids of weak bases (NH_4OH) undergo hydrolysis to give acidic solutions.

Consider acetic acid titrated with sodium hydroxide,

$$HC_2H_3O_2 + Na^+ + OH^- = Na^+ + C_2H_3O_2^- + H_2O$$

Sodium ion is inert. Acetate ion, however, is a weak base and hydrolyzes:

$$H_2O + C_2H_3O_2^- \rightleftharpoons HC_2H_3O_2 + OH^-$$

acid 1   *base 2*   *acid 2*   *base 1*

Although the hydrolysis proceeds less than 0.01 percent in 0.1 M solution, it is sufficient to produce enough OH^- to make the solution alkaline and to affect an indicator. An indicator changing color in somewhat basic solution is therefore used in titrating a weak acid with a strong base.

For comparison consider titrating ammonium hydroxide with hydrochloric acid:

$$NH_4OH + H^+ + Cl^- = NH_4^+ + Cl^- + H_2O$$

In this case chloride ion is inert. But ammonium ion is a weak acid and hydrolyzes:

$$NH_4^+ + H_2O \rightleftharpoons H_3O^+ + NH_3$$

acid 1   *base 2*   *acid 2*   *base 1*

Again hydrolysis is slight, but enough H_3O^+ is produced to affect indicators. An indicator changing color under somewhat acidic conditions must be employed in weak base–strong acid titrations.

In a strong acid–strong base titration like

$$H^+ + Br^- + K^+ + OH^- = K^+ + Br^- + H_2O$$

there is no hydrolysis. Neither K^+ nor Br^- reacts with water because neither can lead to a weak base or a weak acid. In strong acid–strong base titrations an indicator changing color at or near the neutral point is used.

QUESTIONS

1. Viscosity   Viscosity is a measure of the forces within a material that cause its resistance to flow. Consider the following liquids, listed in order of increasing viscosity number at 20°C: methyl alcohol (CH_3OH), 0.59; water (HOH), 1.00; n-propyl alcohol (C_3H_7OH), 2.26. Why do you think water is more viscous than methyl alcohol? Why is n-propyl alcohol more

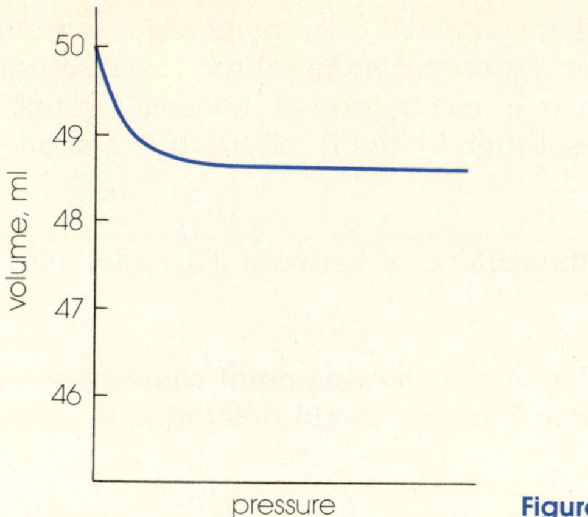

Figure 16-11

viscous than either of the other two liquids? (If possible examine ball-and-stick or other models of the molecules.)

2. Liquids   When water is drained from a glass tube, a film of water remains behind. When mercury is drained from a glass tube, it drains completely (saving the necessity of drying the equipment). Explain.

3. Solubility   Refer to Figure 16-4. **(a)** What is observed when a KCl solution saturated at 100°C is cooled to 10°C? **(b)** When potassium dichromate is dissolved in water, is heat absorbed or released? Explain. **(c)** Would heating tend to dissolve more or less potassium dichromate? Use Le Châtelier's principle to explain.

4. Liquid structure   **(a)** The result of subjecting 50 ml of liquid to high pressure is shown in Figure 16-11. Explain, using the hole theory of liquid structure. **(b)** When most solids are melted, the volume increases by a few percent. Explain.

5. Raoult's law   Discuss the information contained in Figure 16-12. What is the significance of each marked temperature?

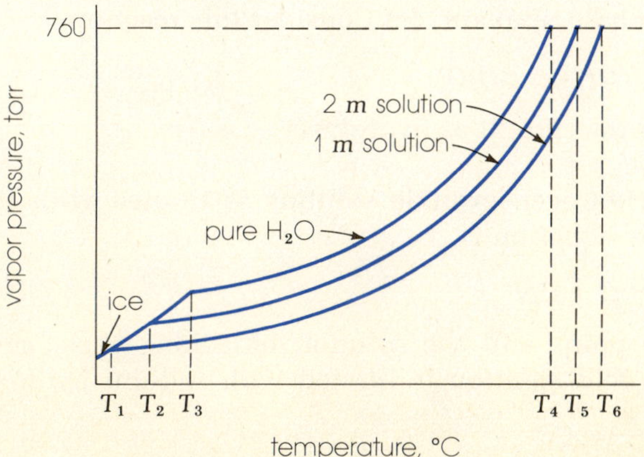

Figure 16-12

6. Solution types **(a)** A beaker contains a solution of sugar in water at room temperature. How might you test to find if the solution is unsaturated, saturated, or supersaturated? If it is electrolytic or nonelectrolytic? **(b)** How could you test a 0.1 M acid solution to find if the solute is a strong acid or a weak acid?

7. Concentration Explain the difference between a 1-molal and a 1-molar solution.

8. Colligative properties The molal freezing-point constant of pure acetic acid is 3.9°C kg/mole. Explain how you could determine the value experimentally.

9. Vapor pressure An open beaker of concentrated sugar solution and an open beaker of dilute sugar solution are placed in a container and the container is sealed. Describe what happens over a period of time, and why.

10. Membrane behavior When red blood cells are placed in distilled water, they *hemolyze* (swell). When placed in a concentrated salt solution, they *crenate* (shrink). Explain.

11. Membrane process A dilute aqueous solution of animal protein glue and salt is put in a sealed semipermeable plastic bag. The bag is placed in a beaker of distilled water. Periodically beaker water is removed and replaced with fresh distilled water. After several days, what remains in the bag? Explain.

12. Concentration Plants that resist moisture loss in dry climates and plants that resist freezing in winter weather have relatively high solute concentrations in their cell fluid. Explain.

13. Ionization theory Arrhenius used the observation that electrolyte solutions show "abnormal" colligative properties to support his ionization theory. Explain.

14. Acid–base theory **(a)** All Lowry-Brønsted acid–base reactions are proton-transfer reactions. Explain. **(b)** Water can act as both a Lowry-Brønsted acid and a base. Explain. **(c)** Consider this reaction:

$$Al^{3+} + 6H_2O = Al(OH)_3 + 3H_3O^+$$

Use Lewis theory to interpret it as hydrolysis.

15. Titrations **(a)** Hydrogen cyanide solution is titrated with potassium hydroxide. The ionic equation is

$$HCN + OH^- = H_2O + CN^-$$

At the equivalence point will the solution be acidic, basic, or neutral? Explain. **(b)** Nitric acid solution is titrated with sodium hydroxide. The molecular equation is

$$HNO_3 + NaOH = NaNO_3 + H_2O$$

At the equivalence point is the solution acidic, basic, or neutral? Explain.

16. Definitions Define and illustrate **(a)** surface tension; **(b)** colligative property; **(c)** osmosis; **(d)** amphiprotic behavior; **(e)** leveling effect; **(f)** equivalence point; **(g)** end point.

PROBLEMS

17. Solubility A saturated solution of KBr in water is prepared at 80°C, then evaporated. From the following data calculate the solubility of KBr at 80°C. Express it as g KBr/100 g H_2O.

beaker + saturated solution	*128.25 g*
beaker empty, dry	*78.25*
beaker + KBr, dry, after evaporation	*100.45*

18. Freezing point The carabid beetle, native to Alaska, is able to live through winters as cold as $-40°C$ because in the fall its system manufactures glycerol, $C_3H_8O_3$ (mol. wt. 92), a soluble nonelectrolyte that becomes mixed with its body fluid. **(a)** Assuming glycerol to be the only solute, what is its molality? **(b)** How many grams of it are present per gram of fluid?

19. Freezing point When a sample of human blood is cooled, it begins freezing at $-0.56°C$. **(a)** What is its effective molal concentration? **(b)** How many grams of the sugar glucose ($C_6H_{12}O_6$, mol. wt. 180) would have to be dissolved in 1 kg of water to give a solution of the same molality?

20. Osmotic pressure A biochemist wants to measure the osmotic pressure in living cells. On a plate at 27°C he puts a drop of solution containing 1.84 g of glycerol (Problem 18) in 100 g of H_2O. Then he adds some green algae cells and observes them under the measuring eyepiece of a microscope. He notes that the average cell radius of 10^{-5} cm does not change on standing. What is the osmotic pressure in the cells?

21. Colloids Colloids are characterized by their large (and chemically active) surface area. **(a)** Calculate the surface area of a cube of material 1.00 cm on a side. **(b)** If that cube is sectioned into colloid-size cubes 10^{-5} cm on a side, how many small cubes would lie along one edge of the original cube? **(c)** How many small cubes would be exposed on one face of the original cube? **(d)** What is the total number of small cubes in the original cube? **(e)** What is their total surface area in square feet (929 cm$^2 = 1$ ft^2)?

22. Concentration Solution concentration is sometimes expressed as a weight percent:

$$\text{percent by weight solute} = \frac{\text{weight of solute}}{\text{weight of solution}} \times 100 \qquad (16\text{-}22)$$

At 90°C a saturated solution of cane sugar contains 448 g of sugar in 100 g of water. What is the percent by weight sugar in the mixture?

23. Acid–base Write equations illustrating Lowry-Brønsted behavior of HNO_3 (a strong acid) and HNO_2 (a weak acid) in water. Label the conjugate acid–base pairs. Use arrows of different lengths to indicate equilibrium displacement.

24. Acid–base The ion $H_2PO_4^-$ is an amphiprotic substance. Write an equation showing how it acts like a base in the presence of hydronium ion. Write another equation showing how it acts like an acid in the presence of hydroxyl ion. Use arrows as in the previous problem.

25. Molarity How many **(a)** grams of HCl (mol. wt. 36.5) are in 0.200 liter of 0.200 M solution? **(b)** millimoles of NaOH are in 150 ml of 0.350 M solution? **(c)** milliliters of 2.40 M NaCl must be diluted to 500 ml to prepare 0.170 M solution? **(d)** millimoles of KBr are in 238 mg?

26. Osmotic pressure A solution contains 1.0 g of hemoglobin (the red coloring matter of blood) dissolved in 1 kg of water. At 0°C the osmotic pressure of the solution is 2.5 torr. What is the molecular weight of hemoglobin?

27. Freezing point A student who wants to determine the temperature one winter night puts a glass of beer outside on his window ledge. After a time only a very thin crust of ice has appeared on the surface. Assuming 100 g of beer contains just 3.7 g of ethyl alcohol C_2H_5OH (we will ignore the presence of a few grams of carbohydrate and protein), calculate the approximate freezing point of the beer (and the night temperature).

28. Solubility Use the data of Table 16-4 to prepare a graph like Figure 16-4. Use it to answer the following: **(a)** How many grams of ammonium nitrate are needed to saturate 400 g of water at 30°C? **(b)** What happens if a saturated solution at 90°C and containing 100 g of H_2O is cooled to 5°C? **(c)** How much water is needed to just dissolve 1 kg of ammonium nitrate at 50°C?

29. Titration The sulfate ion in a 100-ml sample of seawater requires 25.3 ml of 0.105 M barium chloride solution to precipitate it all as barium sulfate. **(a)** Write the molecular and net ionic equations. **(b)** Calculate the sulfate molarity of seawater. **(c)** How many millimoles of sulfate ion are in

Table 16-4.   Solubility of NH_4NO_3

Temperature °C	Solubility of NH_4NO_3 g/100 g H_2O
0	118
20	192
40	297
60	421
80	580
100	871

the sample? **(d)** How many grams of sulfate ion (ionic wt. 96.0) are in 1 liter of seawater?

30. Titration A 100-ml sample of tap water is titrated for its bicarbonate content and found to require 15.0 ml of 0.0630 M HCl:

$$HCO_3{}^- + H^+ = H_2O + CO_2$$

(a) What is the bicarbonate molarity of the water? **(b)** How many milli-moles of bicarbonate ion are in the sample? **(c)** How many grams of bicarbonate ion (ionic wt. 61.0) are in 1 liter of tap water?

THE IMPORTANT CONCEPTS

17-1. Reaction rates and chemical equilibrium

1. The dynamic equilibrium state is attained from either the forward or reverse reaction direction.
2. At equilibrium, forward and reverse rates are equal and concentrations do not change.
3. Reaction rates usually increase with temperature.
4. A reaction may proceed via an activated complex; the complex may be a resonance hybrid.
5. A catalyst provides a reaction path having a lower activation energy.
6. Catalysts are described as homogeneous or heterogeneous.
   a. Reactions with heterogeneous catalysts probably involve chemisorption–desorption at active surface sites.
   b. Inhibitors poison active sites.

7. The rate of a reaction is related to component concentrations, as stated by the mass action law.
8. A K expression can be written by inspection of the balanced chemical equation.
    a. The magnitude of K tells how far a reaction can proceed but not how fast the reaction goes.
    b. K may have units.
9. Changes in concentration, pressure, or temperature will shift the equilibrium according to Le Châtelier's principle.
10. A catalyst does not change the equilibrium position.

17-2. Ionic equilibria
1. Ionization constant expressions are written for weak acids and weak bases, but not for strong electrolytes.
2. A K expression is written for water.
3. The pH scale expresses H^+ concentration.
    a. pH, $[H^+]$, pOH, and $[OH^-]$ are interrelated in H_2O solutions.
4. The pH is measured with indicators or the pH meter.
    a. The pH meter uses calomel and glass electrodes.
5. A buffer is a mixture with capacity to regulate pH.
6. Equilibrium principles are applicable to "insoluble" compounds.
7. Ionic equilibria are shifted by the common ion effect.
8. Precipitation can be predicted by comparing ion product and solubility product constant.

Rates at which chemical reactions occur vary widely. Ions in solutin generally react immediately because they are mobile and their coulomb forces act in all directions. Reactions between covalently bonded molecules in solution are usually slower because bonds must first be broken and certain orientations between reacting species may be needed to produce the products. Some reactions are practically instantaneous, as in a natural gas explosion, but others, like those in smog (Section 13-3), are slower. Reactions in solids, such as changes during geological processes, are very slow. Reaction rates are affected by the nature and concentration of the reactants, temperature, and catalysts. The last three can be regulated by the experimenter.

Many reactions are *reversible*; as reactants give products, products react to give back reactants. When the rate of the forward reaction equals the rate of the reverse reaction, the system is said to be at *equilibrium*. At the equilibrium point both reactions continue without any noticeable overall change. As changes in experimental conditions are imposed on an equilibrium system, the equilibrium shifts to counteract the changes. A knowledge of this cause and effect allows chemists to control reaction direction and speed.

Equilibrium is concisely defined by an *equilibrium constant expression* that can be written by inspection of the system's balanced chemical equation. The *equilibrium constant K* is determined experimentally. Its magnitude indicates in which direction the equilibrium lies, although it does not tell how fast equilibrium is attained; K varies with temperature but is independent of concentrations and pressures of the reacting substances. Important calculations on equilibrium systems, including systems in nature, can be made with the aid of table values of equilibrium constants.

17-1. REACTION RATES AND CHEMICAL EQUILIBRIUM

The equilibrium condition

About 1890 German chemist M. Bodenstein made one of the early detailed studies on equilibrium in a gas-phase reaction. In one series of tubes he sealed hydrogen and iodine to study the reaction

$$H_2 + I_2 \longrightarrow 2HI$$

and in another series of tubes he sealed hydrogen iodide to study the reverse reaction,

$$2HI \longrightarrow H_2 + I_2$$

The data gave him information on the reversible reaction

$$H_2 + I_2 \rightleftharpoons 2HI \tag{17-1}$$

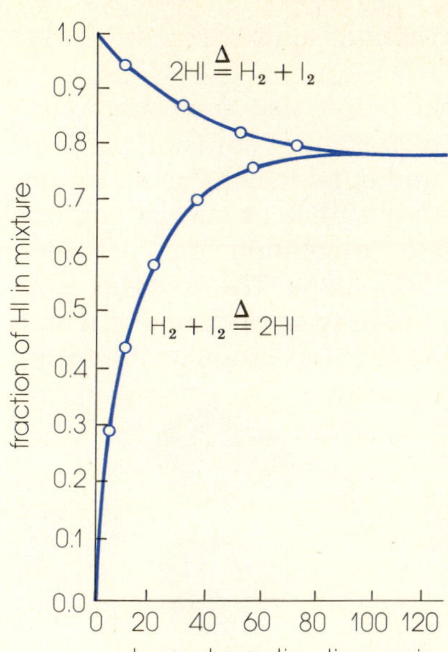

Figure 17-1 Study of the system $H_2 + I_2 \rightleftharpoons 2HI$ at 718°K (445°C). Equilibrium is established starting with either pure reactants or pure product.

Heating the samples in a chamber maintained at 445°C, he periodically removed a tube, chilled it rapidly to halt the reaction, then chemically analyzed the contents. Results are shown in Figure 17-1. The lower curve represents the synthesis of HI; the upper curve, the decomposition of HI. In either case the steady state called *equilibrium* was reached after about 90 minutes in which 78.6 percent of the material was in the form of HI, 21.4 percent in the form of $H_2 + I_2$.

Systems like the one above proceed spontaneously toward an equilibrium point where the proportions of the components no longer change. Reactants initially react rapidly because collisions between their many molecules are numerous. Product molecules, initially scarce, can react only slowly. As reactants are consumed, however, the rate of the forward reaction slows down. As products increase in concentration, the rate of the reverse reaction accelerates.

Equilibrium is characterized by being attainable from either a forward or backward direction and by being dynamic or moving. Concentrations become constant at equilibrium not because the reactions quit, but because opposing reaction rates become equal. The dynamic character of the hydrogen iodide system can be demonstrated by preparing an equilibrium mixture from HI, H_2, and radioactive I_2. After a time, measurements will show some of the radioactive iodine is chemically combined in HI. If equilibrium were a static or stationary condition, all radioactivity would remain in the I_2.

Temperature effects and the activated complex

In Figure 14-12 we saw that a rise in temperature increases the average kinetic energy of molecules and hence collision frequency. We can anticipate, therefore, that reaction rates generally increase with temperature. It was first pointed out by Arrhenius that a temperature rise can also significantly increase the number of molecules having the energy necessary to overcome repulsive forces and react. Some reactions approximately double in rate for each 10°C rise in temperature.

Modern theory deemphasizes the visualization that molecules containing a certain amount of energy react at the instant of collision, in favor of the concept of a change of orbital shapes and bond lengths as molecules approach one another (see Figure 9-9). With a sufficient energy content, and at the proper distance of approach and orientation, an "activated complex" or "transition state" is thought to take shape. The complex is an intermediate presumably capable of uncoupling in two ways, one of which gives back the original reactants, the other the reaction products. Picturing the hydrogen iodide system as

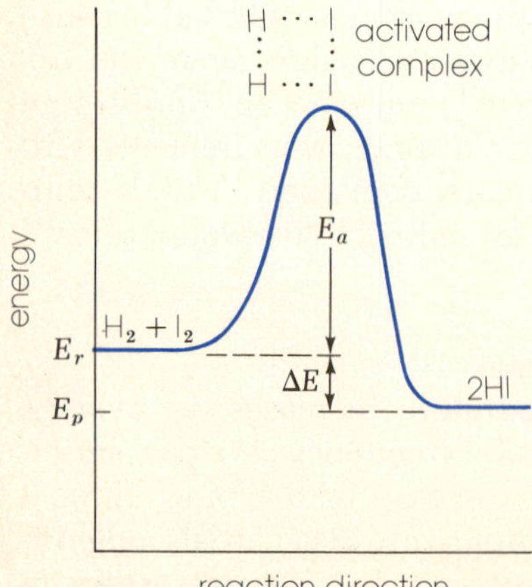

reactants   complex   products

$$(17\text{-}2)$$

we can add a bit of detail by assuming the complex is a resonance hybrid possessing a certain energy content and geometry:

$$\left\{ \begin{matrix} H & I \\ | & | \\ H & I \end{matrix} \quad \begin{matrix} H\cdots I \\ \vdots & \vdots \\ H\cdots I \end{matrix} \quad \begin{matrix} H-I \\ \\ H-I \end{matrix} \right\} \tag{17-3}$$

The lifetime of activated complexes is usually extremely short, in certain cases on the order of 10^{-12} second. In that interval one set of bonds breaks and another set forms.

Figure 17-2 indicates the energy position of the complex in the hydrogen–iodine system. It can be approached from either direction by energy absorption which loosens bonds. As this always endothermic reaction proceeds over the "hump," the system goes to a lower-energy state with energy release and new bonds become established. Energy absorbed by molecules during complex formation is called the *activation energy* E_a.

Figure 17-2 Schematic representation of the energy relationships in the $H_2 + I_2 \rightleftharpoons$ 2HI system. At 718°K activation energy E_a is about 40 kcal and ΔE is −4 kcal/mole. The "reaction direction" coordinate can be interpreted as reaction time. The interval pictured may be only 10^{-10} sec.

The difference between the energy of a system's products p and reactants r is the energy of the reaction ΔE:

$$E_p - E_r = \Delta E$$

We have already encountered this relationship (Eq. 12-10). When $E_r < E_p$, ΔE is positive and the reaction is endothermic. When $E_r > E_p$, ΔE is negative and the reaction is exothermic. The synthesis of HI is an example of the latter case.

Effect of a catalyst on reaction rate

A substance that lowers the energy hump upon which the activated complex is perched is called a *catalyst*. The word was invented by Berzelius and later defined by Ostwald, "A catalyst is a substance that changes the velocity of a chemical reaction without itself appearing in the end products." A catalyst increases reaction rate by changing the reaction mechanism — that is, by providing a pathway for the reaction that has lower energy than the uncatalyzed route (Figure 17-3). With reduced energy requirements, more molecular collisions become effective per unit of time.

An *inhibitor*, also called a *negative catalyst* or poison, is a substance that decreases reaction rate. It may do so by preventing the formation or breakdown of the activated complex, thereby forcing the reaction to take a pathway of higher energy. Tetraethyl lead, for example, probably prevents very fast explosive reactions or "knock" in the internal combustion engine by reacting with intermediate compounds formed in the burning of gasoline. In biological processes it is not uncommon for inhibitors to be products of reactions. Such reactions are self-regulating.

Catalysts distributed uniformly in a reacting system, like ions in solution or one gas in another, are called *homogeneous*. In *homogeneous catalysis* reaction rate depends on catalyst concentration. An example is the mixture of nitrogen oxides used to promote the oxidation of sulfur dioxide in the lead-chamber process for making sulfuric acid. After participating in

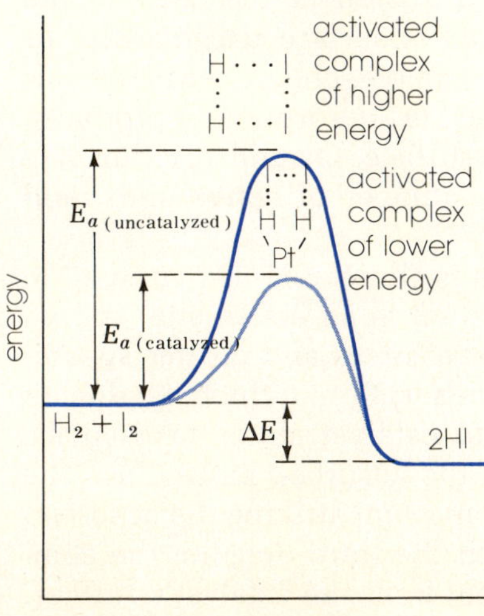

Figure 17-3 Effect of a catalyst. The catalyst (platinum) provides a route of lower energy for both the forward and reverse reactions.

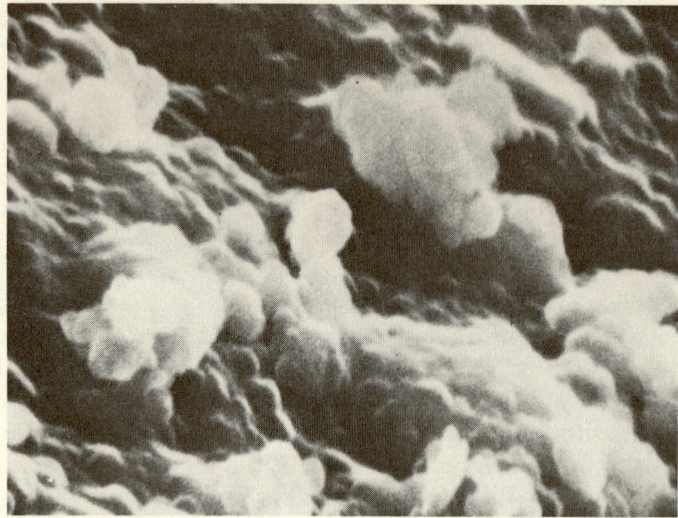

Figure 17-4 Electron photomicrograph (×55,000) of a catalyst used in reforming the configuration of petroleum hydrocarbons into high-octane gasoline. The material consists of aluminum oxide impregnated with 0.5 percent platinum and 1 percent chloride (the latter is not visible). The uneven surface provides a very large area for adsorption. Some catalysts used for automobile exhaust purification are similar in appearance. (Courtesy Universal Oil Products Company and Scientific American 225 (6), 47 (1971).)

the synthesis of an intermediate compound, nitrosulfuric acid, NO and NO_2 are released in the next step for reuse:

$$NO + NO_2 + O_2 + 2SO_2 + H_2O = 2NOHSO_4$$

$$2NOHSO_4 + H_2O = NO + NO_2 + 2H_2SO_4$$

Catalysts that perform as a separate phase are called *heterogeneous*. Most common are solids that accelerate gas or liquid reactions (Figure 17-4). *Heterogeneous catalysis* involves (1) diffusion of reactants to the catalyst surface, (2) *chemisorption* of reactants at surface irregularities or electronically active sites, (3) surface reactions among chemisorbed species that include formation of the activated complex, (4) desorption of products, and (5) diffusion of products away from the surface. Overall reaction rate varies with the catalyst's total surface area, number of active sites, and speed of the relatively slow diffusion steps.

A mechanism for heterogeneous catalysis is illustrated in Figure 17-5. The first diagram depicts movement of a reactant to an active site, and its chemisorption. Chemisorption involves valence forces and greater specificity and process heat than does physical adsorption (which is due to weaker van der Waals forces). Attachment to catalyst atoms takes place with a weakening and stretching of bonds in the adsorbed molecules. The second diagram shows molecules of the other reactant striking the adsorbed species and forming the activated complex. In the third diagram the complex decomposes and products desorb (release) from the catalyst's surface. If A_2 and B_2 are reactants, and their direct combination $A_2 + B_2 = 2AB$ is the uncatalyzed reaction with high activation energy, the catalyst will direct reaction along an alternate route with lower activation energy. For instance,

$$A_2 + \textit{catalyst} \rightarrow \textit{catalyst} \overset{A}{\underset{A}{\diagup}} \xrightarrow{B_2} \textit{catalyst} \overset{A}{\underset{A}{\diamondsuit}} B_2 \rightarrow \textit{catalyst} + 2AB \uparrow$$

$$\textit{adsorption} \qquad\qquad \textit{complex} \qquad\qquad\qquad \textit{desorption}$$

If a catalyst's active sites become occupied by a *poison*, they are deactivated. Arsenic, carbon monoxide, and volatile sulfur compounds poison many catalysts, including some in living organisms. The introduction in 1970 of catalytic exhaust systems on automobiles to oxidize combustible residues was made possible by the production of gasoline that did not contain lead, another catalyst poison.

One-sixth of all manufactured materials (about $100 billion worth) in

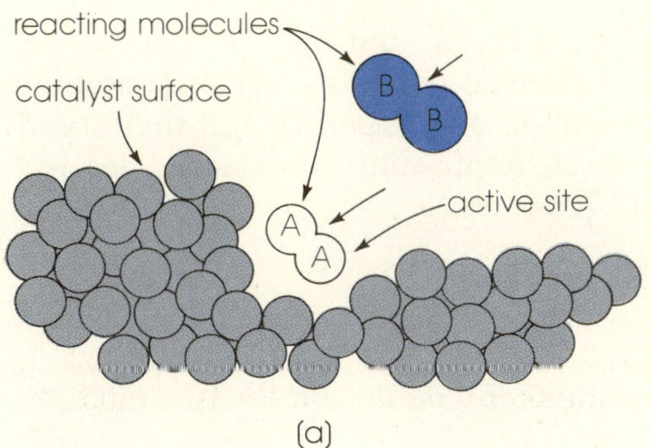

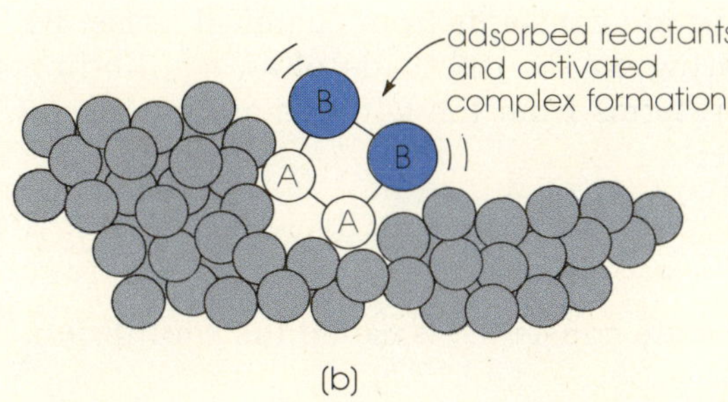

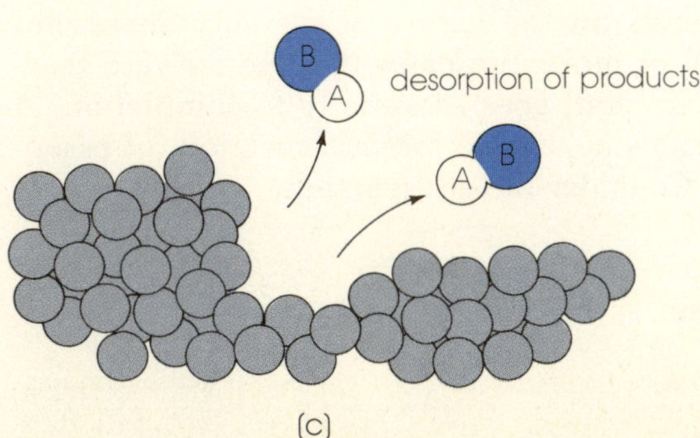

Figure 17-5 A mechanism for heterogeneous catalysis.

the United States depends on the use of catalysts. Gasoline, rubber, acids, and plastics are examples of products mass produced by catalytic processes.

Effect of concentration on reaction rate; the law of chemical equilibrium

The relationship between reaction rate and concentration was worked out during 1864–1879 by mathematician C. Guldberg and chemist P. Waage, Norwegians and brothers-in-law. Called the *law of chemical equilibrium* or *law of mass action*, it states that reaction rate is proportional to the product of the molecular concentrations of the reactants, each concentration term raised to a power equal to its coefficient in the reaction equation. They deduced the law from published experimental data. It was later derived theoretically.

We will derive a mathematical form of the law starting with the general reaction

$$aA + bB = cC + dD$$

where A, B, C, and D are chemical substances, and a, b, c, and d are, respectively, their coefficients in the balanced equation. We will assume that A and B react directly in a single step to give C and D, and that C and D react in a single step to give A and B. Representing molar (moles per liter) concentrations with the symbol [ ], we write

$$rate\ of\ forward\ reaction = k[A]^a[B]^b$$

$$rate\ of\ reverse\ reaction = k'[C]^c[D]^d$$

where k and k' are proportionality constants. By definition the two rates are equal at equilibrium:

$$k[A]^a[B]^b = k'[C]^c[D]^d \qquad (17\text{-}4)$$

We will rewrite Eq. 17-4 to separate constants from chemical terms. By convention chemical terms appearing on the right side of an equilibrium equation are put in the numerator, terms from the left side are put in the denominator:

$$\frac{k}{k'} = K = \frac{[C]^c[D]^d}{[A]^a[B]^b} \qquad (17\text{-}5)$$

where K, a combination of the two rate constants, is called the *equilibrium constant*.

The value of K for a given reaction is determined by experiment (Example 17-1). Its magnitude depends on the nature of the substances and temperature. A large K value, 10^5 or more, indicates that the forward reaction (as the reaction equation is written) goes essentially to completion. A small K value, 10^{-5} or less, indicates negligible forward reaction, or essentially complete reverse reaction. Consider the gas reaction,

$$H_2 + Cl_2 = 2HCl$$

for which the equilibrium constant expression is

$$K = \frac{[HCl]^2}{[H_2][Cl_2]} \qquad (17\text{-}6)$$

At 1200°C the experimental value of K is 2.5×10^4. This means substantial conversion to hydrogen chloride takes place. For the reverse reaction, $2HCl = H_2 + Cl_2$ at the same temperature, the value of K is $1/2.5 \times 10^4$ or 4×10^{-5}. The tendency for HCl to decompose at 1200°C is slight.

Example 17-1 Hydrogen and iodine are introduced into a 1-liter flask and held at 400°C. After equilibrium is established, the flask is cooled quickly to maintain the equilibrium mixture, then analyzed. The molar concentrations of the components are found to be H_2, 2.00×10^{-3}; I_2, 1.76×10^{-3}; HI, 15.0×10^{-3}. Calculate K for the reaction $H_2 + I_2 = 2HI$.

Solution The form of the equilibrium expression for HI synthesis is like Eq. 17-6:

$$K = \frac{[HI]^2}{[H_2][I_2]}$$

Substitution of the equilibrium concentrations gives K at 400°C:

$$K = \frac{(15.0 \times 10^{-3})^2}{(2.00 \times 10^{-3})(1.76 \times 10^{-3})} = \frac{(15.0)^2}{(2.00)(1.76)} = 63.9$$

In the example above the units moles per liter attached to each term cancel. In another case they may not. For instance given the gas reaction

$$CO + 2H_2 = CH_3OH$$

for which the equilibrium constant expression is

$$K = \frac{[CH_3OH]}{[CO][H_2]^2}$$

K will have the units $1/(moles/liter)^2$.

Shifting the equilibrium position

It is of great advantage, especially in industrial processes, to know how to make changes that will cause incomplete reactions to give a larger yield of products. A change in composition of the equilibrium mixture with a change in experimental conditions is called a *shift of the equilibrium*. Equilibrium shifts can be accomplished by changing concentrations, pressures, or temperature. Qualitative predictions about the effects are made by applying Le Châtelier's principle that an equilibrium system reacts to nullify stresses.

Consider once again the equilibrium,

$$H_2 + I_2 \rightleftharpoons 2HI + 4 \; kcal$$

If we add H_2 or I_2, Le Châtelier's principle tells us that the system will respond by a shift to the right, the direction in which excess reactant can be used up. When equilibrium is restored, at the same temperature, the ratio $[HI]^2/[H_2][I_2]$ will again be equal to K. A shift to the right can also be caused by withdrawing some HI because in that direction the system produces HI and compensates for the loss.

If heat is added to a system at equilibrium, Le Châtelier's principle dictates that the equilibrium position will shift in the direction of heat absorption. In the equation above an increase in temperature causes a right-to-left shift and a decrease in the value of K because in that direction the equation reads, "heat plus hydrogen iodide yields iodine plus hydrogen (heat is absorbed)." Conversely, when the reaction is cooled (if not cooled so low that the molecules no longer have adequate reaction energy), more HI is produced (heat is given off) and the value of K increases ($K = 369$ at 100°C; compare to the result in Example 17-1.)

An increase in temperature favors an endothermic reaction. A decrease in temperature favors an exothermic reaction.

When components in a system are gases, the equilibrium may be shifted by changes in pressure. If pressure on the system as a whole is increased, Le Châtelier's principle predicts a change to decrease the pressure—that is, a shift in the direction in which fewer molecules exist. If pressure is decreased, the equilibrium will shift to increase the pressure and the reaction goes in the direction in which more molecules exist. In the reaction $H_2 + I_2 \rightleftharpoons 2HI$ a change in total pressure has no effect because the same number of molecules exist on both sides. In the gas reaction $2H_2 + CO \rightleftharpoons CH_3OH$, however, an increase in total pressure causes system condensation and a shift of the equilibrium toward the right. A decrease in pressure will favor system expansion and a shift of the equilibrium to the left.

Insertion of a catalyst into an equilibrium system does not affect the equilibrium position. As shown in Figure 17-3 a catalyst lowers the energy barrier; but it lowers it to the same level for both the forward and backward reactions, accelerating them equally. The result is no change in the ratio of equilibrium concentrations.

An application of Le Châtelier's principle

During the nineteenth century the population of western Europe tripled. By 1900 governments faced with grain scarcities due to limitations in natural fertilizer supply called on chemists to realize the dream of converting atmospheric nitrogen to a form usable by all green plants.

The answer was provided by Fritz Haber (Germany 1868–1934, Nobel prize in chemistry 1918) who developed the catalytic synthesis of ammonia from its elements:

$$3H_2 + N_2 = 2NH_3 + 22 \ kcal$$

From Le Châtelier's principle Haber reasoned that the following would favor ammonia production: high pressure (condensation of the system from 4 volumes of gas to 2), lowest temperature at which the synthesis would proceed at reasonable speed (facilitating the exothermic reaction), and addition of H_2 and N_2 as they were consumed and removal of NH_3 as it was formed (process uses up reactants, makes more product). The conditions he finally used were a 3:1 H_2 to N_2 feed mixture, 200 atm, and 500° to 600°C. To attain equilibrium more quickly he employed an osmium–uranium catalyst. The process was commercialized by a group working under C.

Bosch, a chemical engineer. Bosch supervised studies on the metallurgy of special steels for the high-pressure equipment, development of a cheap, rugged catalyst (iron), and the preparation of the chemical raw materials by the (1) *producer gas* and (2) *water gas* reactions:

$$\text{(1) } air + coke \rightarrow CO + N_2$$

$$\text{(2) } steam + coke \rightarrow CO + H_2$$

The Haber-Bosch process has been important in world history and economics. Without it World War I would have been much shorter, for it supplied Germany with most of her needs for fertilizers and explosives based on nitric acid (see Question 6). Today ammonia fertilizer plants operate throughout the world with the science and technology worked out 70 years ago from the rules of chemical equilibrium.

17-2. IONIC EQUILIBRIA

Dissociation of weak acids and bases

The principles discussed above which govern gas reaction equilibria also apply to equilibria in aqueous solution. Equilibrium expressions for the latter employ molar concentrations of dissolved molecules and ions. Solvent concentration is considered as unchanging and in effect incorporated in the equilibrium constant.

When a weak acid or weak base is dissolved in water, an equilibrium is established between its dissociating molecules and recombining ions. For example, the dissociation and equilibrium equations for the weak acid HF are

$$HF = H^+ + F^-$$

$$K_a = \frac{[H^+][F^-]}{[HF]} = 6.7 \times 10^{-4} \qquad (17\text{-}7)^*$$

For the weak base ammonia the corresponding equations are

$$NH_3 + H_2O = (NH_4OH) = NH_4^+ + OH^-$$

$$K_b = \frac{[NH_4^+][OH^-]}{[NH_4OH]} = 1.8 \times 10^{-5} \qquad (17\text{-}8)$$

K_a and K_b are called *ionization* or *dissociation constants* (for acid and base, respectively). They are special cases of the general equilibrium constant K; K_a values for some weak acids are listed in Table 17-1.

Acids and bases are classed as *very weak* (dissociation very slight in, say, 0.1 M solution) when K *is* less than about 10^{-8}, *weak* when K is between 10^{-2} and 10^{-8}, *moderately strong* when K is near 1, and *very strong* (dissociation approximately 100 percent in 0.1 M solution) when K is 10 or greater. Values of K_a and K_b are obtained through measurements on solution acidity, colligative properties, and electric conductance.

---

* The equations are written in simplest form for convenience. To incorporate the role of water in the Lowry-Brønsted sense we write

$$HF + H_2O = H_3O^+ + F^-$$

$$K = \frac{[H_3O^+][F^-]}{[HF][H_2O]} \qquad K_a = \frac{[H_3O^+][F^-]}{[HF]}$$

Table 17-1.   Ionization constants of some weak acids, 25°C

Acid	Dissociation	K_a
Acetic	$HC_2H_3O_2 = H^+ + C_2H_3O_2^-$	1.8×10^{-5}
Boric	$H_3BO_3 = H^+ + H_2BO_3^-$	5.8×10^{-10}
Carbonic	$H_2CO_3 = H^+ + HCO_3^-$	4.5×10^{-7a}
Hydrocyanic	$HCN = H^+ + CN^-$	4.0×10^{-10}
Nitrous	$HNO_2 = H^+ + NO_2^-$	4.5×10^{-4}
Phosphoric	$H_3PO_4 = H^+ + H_2PO_4^-$	7.5×10^{-3a}

[a] Value given is for the dissociation of the first proton, as indicated. Each proton dissociation thereafter must be done against the attraction of a negative ion, hence succeeding ionization constants are smaller; H_3BO_3 only dissociates one H^+.

Example 17-2 A 0.10 M solution of cyanic acid, HCNO is 4.5 percent dissociated according to electric conductance measurement. Calculate K_a.

Solution The acid was prepared to be 0.10 M, but at equilibrium 4.5 percent is dissociated into H^+ and CNO^-. The concentration of each ion is therefore 4.5 percent of 0.10 M. And undissociated acid has a concentration of $100 - 4.5$ percent $= 95.5$ percent of 0.10 M. For "bookkeeping" purposes we correlate the figures with the dissociation equation:

$$\underset{\substack{(0.10)(0.955) \\ = 0.096\ M}}{HCNO} \underset{95.5\ percent}{\overset{4.5\ percent}{\rightleftharpoons}} \underset{\substack{(0.10)(0.045) \\ = 0.0045\ M}}{H^+} + \underset{\substack{(0.10)(0.045) \\ = 0.0045\ M}}{CNO^-}$$

We then find K_a by substitution into the equilibrium equation,

$$K_a = \frac{[H^+][CNO^-]}{[HCNO]} = \frac{(4.5 \times 10^{-3})^2}{9.6 \times 10^{-2}} = 2.1 \times 10^{-4}$$

Cyanic acid is about as strong as nitrous acid (Table 17-1).

Dissociation of water; the water constant

Water dissociates* slightly. Its equilibrium equations are

$$H_2O = H^+ + OH^-$$

$$K = \frac{[H^+][OH^-]}{[H_2O]}$$

The concentration of water is essentially constant. The product of K and $[H_2O]$ is therefore another constant designated K_w; K_w is called the *water constant* or the *ion product for water*:

* The dissociation of water can also be treated as a self-protonation to again make evident the part solvent plays:

$$2H_2O = H_3O^+ + OH^-$$

$$K = \frac{[H_3O^+][OH^-]}{[H_2O]^2} \qquad K_w = [H_3O^+][OH^-]$$

$$K_w = [H^+][OH^-]$$

At 25°C electrical conductance measurement of pure water proves that on average only 1 in every 555 million H_2O molecules is dissociated. This corresponds to the presence of 1.00×10^{-7} mole/liter of H^+ and of OH^-. Substitution into the previous equation gives

$$(1.00 \times 10^{-7})(1.00 \times 10^{-7}) = 1.00 \times 10^{-14}$$

or

$$K_w = [H^+][OH^-] = 1.00 \times 10^{-14} \tag{17-9}$$

The p scale; pH

Given either $[H^+]$ or $[OH^-]$ for a water solution, one can calculate the other by use of Eq. 17-9 (we will assume 25°C throughout). If, for instance, a small amount of acid is added to pure water, we know from Le Châtelier's principle the water equilibrium will shift to the left until the ion product $[H^+][OH^-]$ is again equal to K_w. This requires that some H^+ be used up by combination with an equal quantity of OH^-. If the final hydrogen ion concentration is, say, 10^{-6} M, or 10 times larger than originally, then the equilibrium hydroxyl ion concentration must be 10^{-8} M, or 10 times smaller than originally, because in every case the ion product must be 10^{-14}.

To describe a solution as neutral, acidic, or basic, chemists give a value for the hydrogen ion concentration. At room temperature an aqueous solution is *neutral* when $[H^+] = 10^{-7}$ M, *acidic* when $[H^+] > 10^{-7}$ M, and *basic* when $[H^+] < 10^{-7}$ M.

A handy method of expressing the typically small and cumbersome hydrogen ion values found in dilute solutions was introduced in 1909 by Danish biochemist S. Sørensen. Called the *pH scale* (p for "power" or exponent), it expresses the molar hydrogen ion concentration as the logarithm (to the base 10) of its reciprocal, which is equivalent to its negative logarithm (see Appendix C). The reciprocal of a small number is a large number, and the logarithm of the large number is a positive number of convenient size:

$$pH = \log \frac{1}{[H^+]} = -\log [H^+] \tag{17-10}$$

The reciprocal relationship means that as acidity increases (basicity decreases), pH decreases. And as acidity decreases (basicity increases), pH increases. Being logarithmic, the pH scale is structured in intervals representing powers of ten. A solution of pH 2 (10^{-2} M H^+), for instance, is 10 times stronger as an acid than a solution of pH 3 (10^{-3} M H^+), and 1000 times stronger than one of pH 5 (10^{-5} M H^+).

Example 17-3 Calculate the pH of 0.001 M HCl.

Solution HCl is a strong acid and completely dissociated:

$$HCl \xrightarrow{\text{100 percent}} \underset{\substack{(10^{-3})(1) \\ = 10^{-3} \text{ M}}}{H^+} + \underset{\substack{(10^{-3})(1) \\ = 10^{-3} \text{ M}}}{Cl^-}$$

From Eq. 17-10 the pH is

$$pH = \log\left(\frac{1}{10^{-3}}\right) = \log 10^3 = 3$$

When $[H^+]$ and $[OH^-]$ are not exactly negative integral powers of 10, a log table (Table 17-2) or the slide rule is used in pH calculations.

Example 17-4 A sample of sauerkraut juice has a hydrogen ion concentration of 3.0×10^{-4} M. Calculate its pH.

Solution From Eq. 17-10,

$$pH = \log \frac{1}{3 \times 10^{-4}}$$

To simplify we first divide the exponential term:

$$\log\left(\frac{1}{3 \times 10^{-4}}\right) = \log\left(\frac{10^4}{3}\right)$$

Now $\log (X/Y) = \log X - \log Y$ (Appendix C). Therefore

$$\log\left(\frac{10^4}{3}\right) = \log 10^4 - \log 3$$

Log 10^4 is 4; log 3 (Table 17-2 or the slide rule) is 0.48. So

$$pH = \log 10^4 - \log 3 = 4 - 0.48 = 3.5$$

Use of the p scale is not restricted to the hydrogen ion. Hydroxyl ion concentration can be expressed as a pOH, where pOH is defined like pH:

$$pOH = \log \frac{1}{[OH^-]} = -\log [OH^-]$$

For certain calculations it is convenient to express equilibrium constants using p notation. For the water constant K_w

$$pK_w = \log \frac{1}{K_w} = -\log K_w$$

Table 17-2. Logarithms

Number	Log
1.0	0.00
2.0	0.30
3.0	0.48
4.0	0.60
5.0	0.70
6.0	0.78
7.0	0.84
8.0	0.90
9.0	0.95

Table 17-3. Acidity-alkalinity relationships; aqueous solutions at 25°C

$[H^+]$	$[OH^-]$	pH	pOH	Familiar examples
10^0	10^{-14}	0	14	1 M HCl
		1	13	
				Gastric fluid
10^{-2}	10^{-12}	2	12	
				Lemon juice
		3	11	Carbonated beverages
10^{-4}	10^{-10}	4	10	
				Tomatoes
		5	9	Tea
				Urine
10^{-6}	10^{-8}	6	8	
				Milk
		7	7	Pure water
				Blood
10^{-8}	10^{-6}	8	6	
				Baking soda
		9	5	
				Household cleanser
10^{-10}	10^{-4}	10	4	
		11	3	
				Household ammonia
10^{-12}	10^{-2}	12	2	
				0.1 M Na_3PO_4
		13	1	
10^{-14}	10^0	14	0	1 M NaOH

This leads to the use of p notation for entire equilibrium equations. The water equilibrium $[H^+][OH^-] = K_w = 1.00 \times 10^{-14}$ can be written as follows:

$$pH + pOH = pK_w = 14.00 \tag{17-11}$$

Equation 17-11 states that the sum of pH and pOH for any water solution is 14.00. Here is a helpful relationship, because if one p value is known, the other can be found by simply subtracting the known value from 14.00. The pH numbers of solutions are generally between 0 (1 M H^+) and 14 (10^{-14} M H^+) as indicated in Table 17-3.

To summarize:

1. When pH < pOH, $[H^+]$ > $[OH^-]$, and the solution is acidic.
2. When pH = pOH, $[H^+]$ = $[OH^-]$, and the solution is neutral.
3. When pH > pOH, $[H^+]$ < $[OH^-]$, and the solution is basic.

Measuring pH

Finding the concentration of an acid or base is one of the most frequently made laboratory measurements. In approximate work chemists assess pH with acid–base indicators like those used to locate end points in titrations. Different indicators change color at different pH values. The range over which an indicator changes color completely is about 2 pH units. By using

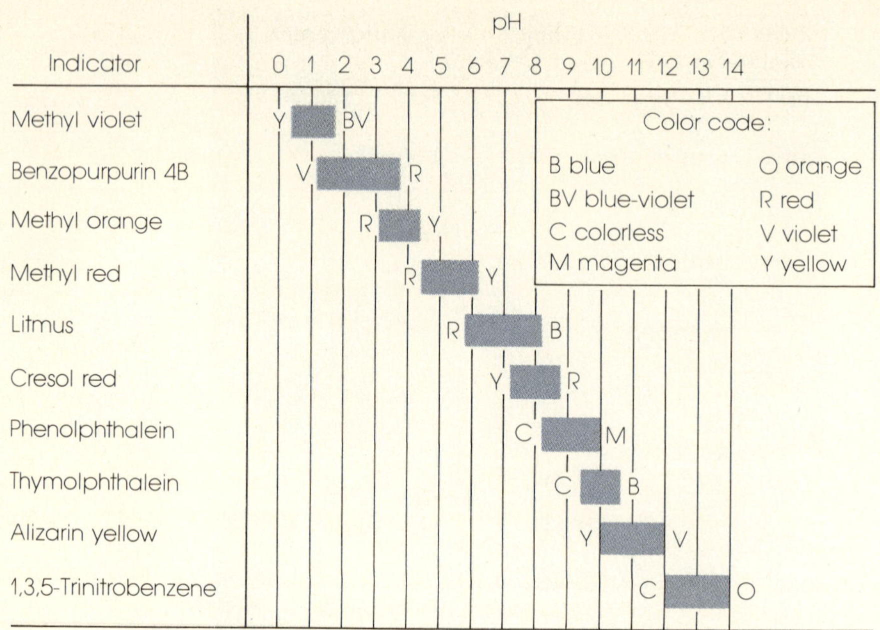

Figure 17-6 A chart of representative acid–base indicators, with pH range and color change of each.

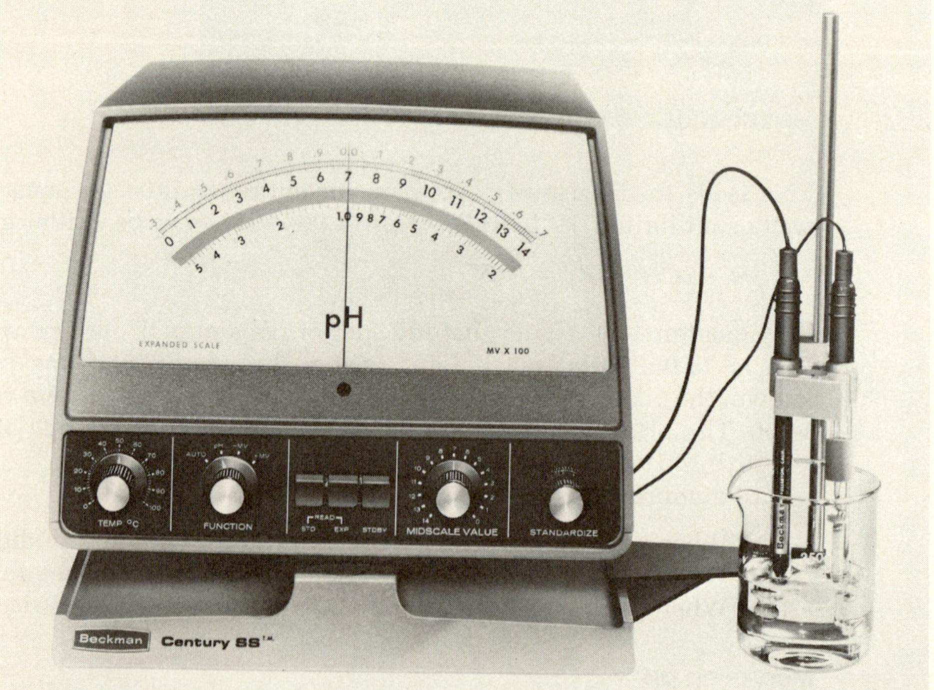

Figure 17-7 A pH meter. The solution whose pH is being measured is in the beaker. The electrode on the right is the calomel electrode, the other is the glass electrode. The principle of the latter was discovered in 1906 by M. Cremer, a German biologist, and developed by Haber (Courtesy Beckman Instrument Company.)

several indicators and comparing the colors developed in reference solutions of known pH, the pH of an "unknown" solution can be estimated. (See Figure 17-6 and Question 12.)

For more precise determination of H⁺ concentration a pH *meter* is used (Figure 17-7). This instrument employs two electrodes, a *calomel reference electrode* containing Hg, Hg_2Cl_2 (calomel), KCl and KCl solution, and a working or indicating *glass electrode* containing Ag, AgCl, and 0.1 M HCl. The glass electrode, having a thin-walled end of special glass through which hydrogen ions can move, is pH responsive. When the electrodes are placed in the solution to be measured, a small voltage develops between them depending on solution pH. The voltage, measured with a sensitive electronic voltmeter connected to the electrodes, is read in pH units.

Buffer solutions

A solution of a weak acid and its salt, or a weak base and its salt is called a *buffer solution*. Buffer solutions have the useful property of maintaining their pH during dilution and despite additions of small amounts of strong acid or strong base. Human blood is an example of a natural buffer. Various substances, including dissolved CO_2 ($CO_2 + H_2O \rightleftharpoons H_2CO_3$), cause it to be buffered at about pH 7.4 through the acid–salt system H_2CO_3–HCO_3^-. If the pH of the blood drops below 7.0 or rises above 7.8 due to disease, its oxygen transport ability is impaired. Human life depends, therefore, on buffering action that maintains the physiological pH.

One component of a buffer must be capable of combining with added base, and the other with added acid, so that either kind of addition is neutralized as a pH disturbance. In the blood the neutralizing reactions are

$$H_2CO_3 + OH^- = HCO_3^- + H_2O$$
acid

$$HCO_3^- + H^+ = H_2CO_3 \tag{17-12}$$
base

If H_2CO_3 is used in a reaction, more is formed by combination of CO_2 and H_2O. If HCO_3^- is used, more is formed by the dissociation of H_2CO_3. Dissociating weakly, neither carbonic acid nor bicarbonate ion can alter the hydrogen ion concentration much. Shifts in Eq. 17-12, therefore, leave solution pH practically unchanged.

The pH of a buffer solution can be calculated using the equilibrium expression of the system's weak acid or weak base as illustrated below.

Example 17-5 A sample of blood is found to be 1.24×10^{-2} M in HCO_3^- and 1.10×10^{-3} M in H_2CO_3 (dissolved CO_2). Calculate its pH.

Solution From Table 17-1 we get the equilibrium expressions we need:

$$H_2CO_3 = H^+ + HCO_3^-$$

$$K_a = \frac{[H^+][HCO_3^-]}{[H_2CO_3]} = 4.5 \times 10^{-7}$$

Substitution of the known quantities gives H^+ concentration, then pH:

$$4.5 \times 10^{-7} = \frac{[H^+](1.24 \times 10^{-2})}{1.10 \times 10^{-3}}$$

$$[H^+] = 4.0 \times 10^{-8}$$

$$pH = log\left(\frac{1}{4.0 \times 10^{-8}}\right) = log\ 10^8 - log\ 4.0 = 7.4$$

Solubility product

When an excess quantity of a slightly soluble ionic compound is stirred in water, a solution saturated in the compound's ions is formed and the remaining solid settles to the bottom. A dynamic equilibrium then exists as ions are simultaneously deposited on and dissolved from the solid's surface. (Figure 16-3 helps visualize the process.)

Consider silver bromide. Its slight solubilization is described by

$$AgBr = Ag^+ + Br^-$$

and

$$K = \frac{[Ag^+][Br^-]}{[AgBr]}$$

The concentration of solids does not change with the quantity present (mass per unit volume is density, a constant). Any solid term is assigned a value of 1, therefore, and in effect is dropped from the equilibrium law expression. The equilibrium constant related to solubility is called the *solubility product constant*, abbreviated K_{sp}. For silver bromide,

$$K_{sp} = [Ag^+][Br^-] = 5.2 \times 10^{-13} \tag{17-13}$$

The expression $[Ag^+][Br^-]$ is called the *ion product*. As before, the square brackets mean molar concentrations.

Some K_{sp} values are given in Table 17-4. Note that coefficients in the balanced equations again appear as powers in the equilibrium expressions. K_{sp} increases with temperature for those compounds whose solubility increases with temperature. All K_{sp} values are small because the solubility product principle is valid only for saturated solutions that contain relatively few ions (meaning the solute is sparingly soluble).

Table 17-4. Solubility product constants, 25 °C

Compound	Equilibrium expression	K_{sp}
Barium sulfate	$K_{sp} = [Ba^{2+}][SO_4^{2-}]$	1.0×10^{-10}
Calcium sulfate	$K_{sp} = [Ca^{2+}][SO_4^{2-}]$	6.1×10^{-5}
Cupric sulfide	$K_{sp} = [Cu^{2+}][S^{2-}]$	8.5×10^{-45}
Ferric hydroxide	$K_{sp} = [Fe^{3+}][OH^-]^3$	1.5×10^{-36}
Lead bromide	$K_{sp} = [Pb^{2+}][Br^-]^2$	3.9×10^{-5}
Magnesium hydroxide	$K_{sp} = [Mg^{2+}][OH^-]^2$	1.8×10^{-11}
Silver chloride	$K_{sp} = [Ag^+][Cl^-]$	1.8×10^{-10}
Silver chromate	$K_{sp} = [Ag^+]^2[CrO_4^{2-}]$	1.1×10^{-12}

Any method that permits experimental determination of solubility gives data for calculating a solubility product constant. For somewhat soluble compounds such as calcium sulfate a known volume of saturated solution may be evaporated and the residue weighed (see Example 17-6). For less soluble compounds such techniques as colorimetry, electric conductance, and counting the radioactivity of an evaporation residue are used to find solubility.

Example 17-6 A student is asked to determine experimentally the solubility product constant of calcium carbonate. He shakes an excess of pure, solid $CaCO_3$ with pure water at room temperature and after the solution is saturated, filters it and evaporates 1.0 liter. The residue of $CaCO_3$ weighs 6.9 mg. What is the K_{sp}?

Solution For substitution into the K_{sp} expression it is necessary to have the molar concentrations of calcium and carbonate ions. The experimental solubility of $CaCO_3$ is 6.9 mg/liter or 6.9×10^{-3} g/liter. The molecular weight of $CaCO_3$ is $40 + 12 + 48 = 100$. The molar solubility is

$$\frac{6.9 \times 10^{-3} \ g/liter}{10^2 \ g/mole} = 6.9 \times 10^{-5} \ mole/liter$$

For every mole of $CaCO_3$ that dissolves, 1 mole of Ca^{2+} and 1 mole of CO_3^{2-} go into solution. Therefore in the saturated solution 6.9×10^{-5} mole/liter of each ion exists. By definition,

$$[Ca^{2+}][CO_3^{2-}] = (6.9 \times 10^{-5})^2 = K_{sp} = 4.8 \times 10^{-9}$$

Common ion effect and precipitation

We can predict by Le Châtelier's principle that if either $Ca(NO_3)_2$ or Na_2CO_3, both soluble salts, is added to a saturated solution of $CaCO_3$, the solubility of the calcium carbonate will be decreased and some will precipitate. That is, the equilibrium

$$CaCO_3 \rightleftharpoons Ca^{2+} + CO_3^{2-}$$

will be displaced toward the left as the system acts to use up the added substance and reestablishes the condition in which the ion product $[Ca^{2+}][CO_3^{2-}]$ has the K_{sp} value. The effect of added solute that furnishes one of the ions already in an equilibrium system is called the *common ion effect*. It is illustrated below.

Example 17-7 Calcium is precipitated as $CaCO_3$ from a sample of hard water by addition of an excess of carbonate ion (from sodium carbonate). When equilibrium is attained, the carbonate concentration in solution is 0.01 M. What is the solubility of calcium carbonate in the mixture?

Solution At equilibrium the following must be true:

$$[Ca^{2+}][CO_3^{2-}] = K_{sp} = 4.8 \times 10^{-9}$$

With the carbonate concentration increased over its value in saturated calcium carbonate (Example 17-6), the calcium concentration must be decreased correspondingly:

$$[Ca^{2+}](0.01) = 4.8 \times 10^{-9}$$

$$[Ca^{2+}] = 4.8 \times 10^{-7} \; mole/liter$$

This value is also the solubility of calcium carbonate because the moles of $CaCO_3$ in solution must be the same as the moles of calcium ion in solution. By the common ion effect the added carbonate has decreased the solubility of $CaCO_3$ about 140-fold from that calculated previously.

When will a precipitate form?

The table value of the solubility product constant can be used to determine when a substance will precipitate from a mixture of its ions. If the experimental ionic concentrations when substituted into the K_{sp} expression give an ion product (I.P.) larger than the K_{sp}, the concentrations are in excess of saturation values and a precipitate forms. If the experimental ion product is less than the K_{sp}, the solution is less than saturated and no precipitation is possible. If the ion product equals the K_{sp}, the solution is saturated.

The three conditions are summarized as follows:

1. When I.P. $> K_{sp}$, precipitation occurs until I.P. $= K_{sp}$.
2. When I.P. $< K_{sp}$, solution is unsaturated and precipitate cannot form.
3. When I.P. $= K_{sp}$, solution is saturated and at equilibrium.

Example 17-8 Seawater contains 1300 mg of magnesium (at. wt. 24.3) per liter. If 1 liter of seawater is mixed with 1 liter of 0.020 M NaOH, will $Mg(OH)_2$ precipitate?

Solution We will first calculate the molar concentration of each ion in the mixture and compare the ion product to the K_{sp}. We will assume none of the other ions normally present in seawater has any influence on the result. The magnesium concentration in seawater is

$$\frac{1.3 \; g/liter}{24.3 \; g/mole} = 0.0535 \; M$$

When the two solutions are mixed, each dilutes the other. At the moment of mixing, therefore, $[Mg^{2+}]$ is 0.027 M, $[OH^-]$ is 0.010 M (NaOH, a strong base, is 100 percent dissociated). The equilibrium expressions are

$$Mg(OH)_2 \rightleftharpoons Mg^{2+} + 2OH^-$$

$$K_{sp} = [Mg^{2+}][OH^-]^2$$

The ion product is

$$I.P. = (2.7 \times 10^{-2})(1.0 \times 10^{-2})^2 = 2.7 \times 10^{-6}$$

The ion product is larger than the K_{sp} value (Table 17-4); $Mg(OH)_2$ therefore precipitates. It continues to precipitate until the solution is just saturated with $Mg(OH)_2$. At that point I.P. $= K_{sp}$ and the system has reached equilibrium.

QUESTIONS

1. Definitions Define, explain, and illustrate **(a)** reversible reaction; **(b)** equilibrium state; **(c)** activated complex; **(d)** catalyst; **(e)** Le Châtelier's principle; **(f)** weak acid; **(g)** water constant; **(h)** solubility product constant; **(i)** ion product; **(j)** common ion effect.

2. Catalyst Solid cooking fats are produced by causing hydrogen gas under pressure to react with a vegetable oil in the presence of a catalyst of finely divided nickel metal. **(a)** Why is the catalyst in powder form rather than in, say, a thin sheet? **(b)** If the oil is not first purified to free it of sulfur compounds, the catalyst works for only a short time before its action stops. What might be the explanation?

3. Equilibrium expression **(a)** What is the general form of equilibrium constant (K) expressions? **(b)** In applying such expressions why are concentrations expressed in moles per liter rather than in, say, grams per liter?

4. Reaction rate If the rate of beat of an insect's wings is controlled by a chemical reaction, would you expect insects to be more active in cold or in hot weather? Explain.

5. Equilibrium In the early 1900s an endothermic process was demonstrated by Norwegians Birkeland and Eyde for fixing nitrogen by passing air through a high-temperature electric arc. It promised to be a rival of the Haber-Bosch process if cheap electric power was available:

$$N_2 + O_2 = 2NO - 43.2 \; kcal$$

According to Le Châtelier's principle how could NO yield be increased? Explain.

6. Equilibrium In 1902 Wilhelm Ostwald (Germany 1853–1932, Nobel prize in chemistry 1909) worked out a catalytic method for oxidizing ammonia (and thus for preparing nitric acid):

$$I. \; 4NH_3 + 5O_2 \overset{Pt}{=} 4NO + 6H_2O + heat \quad (all \; gases)$$

$$II. \; 2NO + O_2 = 2NO_2 + heat \quad (all \; gases)$$

$$III. \; 6NO_2 + 2H_2O = 4HNO_3 + 2NO + heat$$

(a) Reaction I is run at 550°C. What is the effect on the equilibrium of raising the temperature? of increasing the total pressure? of adding oxygen? **(b)** Answer the same questions for II. **(c)** In I what is the effect on reaction rate if the catalyst is removed? **(d)** Should cooling or heating be provided to promote III? Explain.

7. Equilibrium position The following equilibrium operates in nature during the metamorphosis of dolomitic limestones:

$$CaMg(CO_3)_2 \rightleftharpoons CaCO_3 + MgO + CO_2 - heat$$

$$\text{dolomite} \qquad \text{calcite} \quad \text{periclase}$$

(a) If at some time in geologic history the partial pressure of CO_2 in the atmosphere was greater than it is today, how would the limestone equilibrium have been affected? Explain. **(b)** Which compound(s) would you expect to find if a limestone zone was subjected to volcanic heat? Explain.

8. Equilibrium constants **(a)** In a certain reaction, $A \rightleftharpoons B$, there is found four times as much A as B at equilibrium. What is the numerical value of K? (Solve by inspection.) **(b)** In one experiment the reaction $A + B \rightleftharpoons 2C$ comes to equilibrium with equal quantities of A and B and six times as much C as either of them. What is the value of K?

9. Effect of pressure In an all-gas reaction when the sum of the coefficients of reactants equals the sum of the coefficients of products, a change in total system pressure (at constant temperature) does not affect the equilibrium position. Explain.

10. Indicators Given a 0.1 M solution of an acid, HX, explain how the use of indicators would enable you to determine whether HX is weak or strong.

11. pH **(a)** Which has the higher pH, 0.1 M acetic acid or 0.1 M nitrous acid (Table 17-1)? Explain. **(b)** When one is asked the pH of 10^{-2} M HNO_3 he can immediately give the correct answer. Asked for the pH of 10^{-2} M HNO_2, however, he must make a calculation involving several steps. Explain.

12. pH A sample of milk of magnesia, a remedy for excess stomach acidity, is divided among several test tubes. Cresol red is red in one, thymolphthalein is colorless in another. **(a)** In what pH range is the sample? **(b)** What indicator should be tried next to determine the pH more closely? Explain.

13. Common ion The common ion effect was illustrated in connection with precipitation (Example 17-7). The effect also operates in acid and base solutions. Consider a solution of the weak acid HF (Eq. 17-7). What effect will the addition of NaF (a soluble salt) have on the acid's dissociation? Will the pH of the mixture be higher, lower, or the same as the pH of acid solution alone? Explain.

14. Basic buffer The pH of 0.1 M ammonium hydroxide is measured with a pH meter and found to be 11.1. Some ammonium chloride is dissolved in it. Is the pH now the same, higher, or lower than before? (See Eq. 17-8.) Explain.

15. Solubility **(a)** Use the K_{sp} data of Table 17-4 to tell which has the greater (molar) solubility: barium sulfate or calcium sulfate. Explain. **(b)** Why are K_{sp} values calculated from data on saturated solutions only? **(c)** Why are K_{sp} values small? **(d)** How could a radioactive tracer be used to prove that a dynamic equilibrium exists between a slightly soluble salt and its saturated solution?

PROBLEMS

16. Equilibrium constant Write K expressions for the following high-temperature gas reactions:

(a) $4HCl + O_2 = 2Cl_2 + 2H_2O$
(b) $CO_2 + H_2 = H_2O + CO$
(c) $CS_2 + 4H_2 = 2H_2S + CH_4$

17. Equilibrium constant **(a)** In one synthesis of HI from $H_2 + I_2$ at 426°C these equilibrium concentrations (moles per liter) were found: H_2, 0.020; I_2, 0.021; HI, 0.153. Calculate K. **(b)** In another similar experiment these equilibrium concentrations (moles per liter) were found: H_2, 0.040; I_2, 0.050. What was the concentration of HI? (Use K from above.)

18. Equilibrium constant **(a)** Write the equilibrium constant (K) expression for the reaction $2SO_2 + O_2 = 2SO_3$. **(b)** At 515°K in one experiment the following were molar concentrations at equilibrium: SO_2, 6.00×10^{-5}; O_2, 0.308×10^{-3}; SO_3, 1.20×10^{-3}. Calculate K.

19. Ionization A new weak acid HA discovered in research is 1 percent ionized in 0.1 M solution. **(a)** In the manner of Example 17-2 write an equation for the ionization and include calculations of concentrations of species in solution. **(b)** What is the solution pH? **(c)** What color would methyl red indicator give in the solution? **(d)** Write an expression for K_a and calculate its numerical value from part (a). **(e)** Is HA a stronger or weaker acid than acetic acid? Explain.

20. Ionization constant A 0.1 M solution of hydrazoic acid gives a reddish-violet color with benzopurpurin 4B (Figure 17-6):

$HN_3 = H^+ + N_3^-$

(a) What is the pH (nearest whole number)? **(b)** What is the corresponding molar concentration of H^+? **(c)** At equilibrium what are the approximate molar concentrations of N_3^- and HN_3? **(d)** Write the equilibrium constant expression. **(e)** Using the values from (b) and (c) and the expression from (d) calculate the approximate value of K_a.

21. pH What is the pH of **(a)** 10^{-2} M HCl? **(b)** 10^{-2} M KOH? **(c)** 4×10^{-3} M HCl? **(d)** 7×10^{-4} M KOH? **(e)** Black coffee whose $[H^+] = 10^{-5}$ M?

22. Buffers A buffer solution has the ability to use up either added acid or base. One of the buffer mixtures contained in blood is $H_2PO_4^- - HPO_4^{2-}$.

Write one chemical equation showing how the system uses up H^+ added to it, another how it uses up added OH^-.

23. Buffer An example of a basic buffer is a mixture of ammonium hydroxide and ammonium nitrate. Show with one equation how the system uses up H^+ added to it, another how it uses up added OH^-.

24. Buffer A buffer solution is 0.10 M each in hydrocyanic acid and sodium cyanide. **(a)** Write the equilibrium (K_a) expression for HCN. Use it and Table 17-1 data to calculate $[H^+]$ in the buffer solution. **(b)** Calculate the corresponding pH.

25. Solubility product When 1.0 liter of saturated $PbSO_4$ (mol. wt. 303) solution is evaporated, a residue of the salt weighing 41.7 mg is recovered. **(a)** Calculate the salt's solubility in moles per liter. **(b)** What is the molarity of each of its two ions in solution? **(c)** Calculate the K_{sp}.

26. Solubility product A solution saturated at room temperature with the slightly soluble salt $MnCO_3$ is analyzed and found to contain 0.50 mg of manganous ion per liter. **(a)** What is the molar concentration of Mn^{2+}? of CO_3^{2-}? **(b)** Calculate the K_{sp}.

27. Ion product **(a)** Equal volumes of 0.002 M silver nitrate and 0.002 M sodium chloride are mixed. Will a precipate of silver chloride form? **(b)** Equal volumes of 0.020 M lead nitrate and 0.020 M sodium bromide are mixed. Will a precipitate of lead bromide form? Show all calculations.

28. Solubility **(a)** Given the K_{sp} of $PbBr_2$ (Table 17-4), calculate its molar solubility. Begin by letting x be the molar solubility of lead bromide. Then

$$PbBr_2 = Pb^{2+} + 2Br^-$$
$$x\ M \qquad 2x\ M$$

(b) Convert the lead molarity to milligrams of Pb^{2+} per liter. **(c)** A public health regulation states that drinking water should contain no more than 0.1 mg of lead per liter. By how many times is the safe limit exceeded in saturated $PbBr_2$?

SEVEN

HOW ARE CHEMISTRY AND ELECTRICITY RELATED?

Largest single-circuit mercury cell chlorine plant in the United States (at Lake Charles, Louisiana). Approximately the size of a football field, the unit consumes enough electric power to serve 175,000 homes. It produces 600 tons of Cl_2 and 650 tons of $NaOH$ a day. (Courtesy PPG industries.)

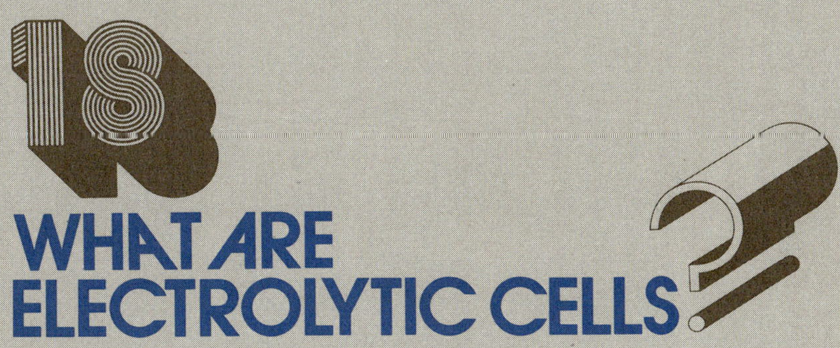

WHAT ARE ELECTROLYTIC CELLS?

THE IMPORTANT CONCEPTS

18-1. The beginnings of electrochemistry
1. Current electricity can be produced by bimetallic systems.

18-2. Electrode reactions
1. Electrolysis current is carried by ions.
2. Oxidation and reduction reactions take place at electrode surfaces during electrolysis.
   a. Oxidation and reduction are defined as electron loss and gain.
   b. Electrodes are designated anode and cathode for their respective oxidizing and reducing abilities.
3. A few rules allow prediction of products of electrolysis.

18-3. Quantitative relationships in electrolysis

1. Faraday's laws concern quantities of reactant and electricity, and the reactant's equivalent weight.
   a. An equivalent is definable as N electrons, 1 faraday, or 9.65×10^4 coulombs.
2. Pumping of electricity and water can be compared.
   a. A series of simple formulas permits calculation of electrical quantity, flow rate, potential, power, and work.

18-4. Applications of electrolytic processes

1. Electrolytic industries turn out useful products and consume large amounts of electric power.
   a. Electrolytic processing includes metal production and refining, metal plating and coating, metal forming, and the manufacture of chemicals.

Matter is electrical in nature and some of its most important particles—electrons, protons, and ions—carry electric charge. When an electric potential is applied between electrodes standing in a molten salt or an ionic solution, ions migrate to electrodes of charge opposite their own and chemical reactions take place. Quantitative aspects of this process of electrolysis are important in analytical chemistry and the chemical industry.

18-1. THE BEGINNINGS OF ELECTROCHEMISTRY

Static electricity is electric charge at rest. *Current electricity* is electric charge in motion. Current electricity was discovered near the end of the eighteenth century by two Italians, biologist L. Galvani and physicist A. Volta. Galvani, after noting that static electricity caused the muscles of dead animals to twitch, was surprised to find that dissected frog legs also twitched when suspended from copper hooks on an iron bar. The phenomenon was investigated by Volta who concluded that Galvani's "animal electricity" was not due to some mysterious residual force in animal nerves but to a "metallic electricity" produced by the contact of two dissimilar metals. Using an electrometer (see Figure 2-5) to measure the relative activities of metals, he established an electrochemical series. He determined that touching samples of zinc and silver together left them oppositely charged; he then proceeded to show that when the two metals were separated by blotting paper soaked in brine and connected with a metal clip, they gave a small but continuous electric current (Figure 18-1).

Chemistry benefited immediately from Volta's research because his battery made *electrolysis* possible, the process by which chemical changes are

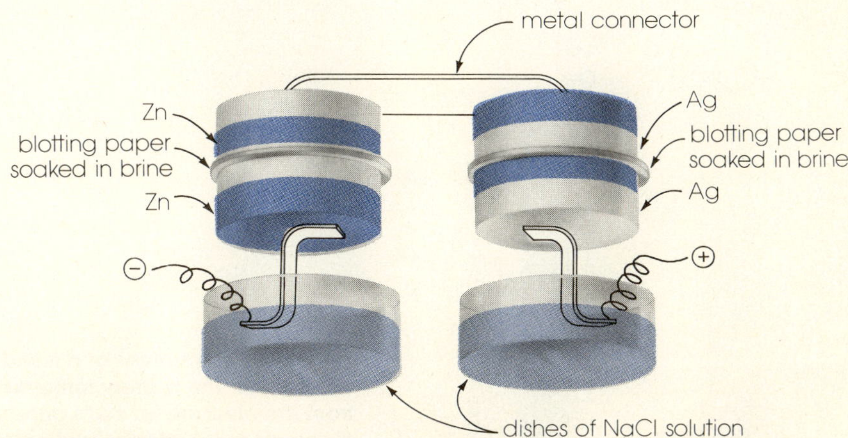

Figure 18-1 Volta's bimetallic "pile," the earliest battery and first source of steady electric current.

caused by electric current. The English chemist Humphry Davy demonstrated its usefulness in chemical analysis by showing that hydrogen, metals, and alkaline substances gathered at the negative electrode during electrolysis, and oxygen and acidic substances appeared at the positive electrode. By electrolyzing molten hydroxides Davy was the first to isolate the elements sodium, potassium, boron, magnesium, calcium, strontium, and barium (see Figure 1-4).

18-2. ELECTRODE REACTIONS

Electrolysis of a molten electrolyte

In an electrical conductor, current is carried by free electrons. In a molten salt or conducting solution, current is carried by the movement of ions. Ionic substances, their melts, and water solutions are all called *electrolytes*. Electric current can be caused to enter and leave an electrolyte by means of solid (usually metal) connectors called *electrodes*. A pair of electrodes, electrolyte, and container comprise an *electrochemical cell*.

Basically there are two kinds of electrochemical cells. In a *voltaic* (also called *galvanic*) *cell* chemical reactions take place that produce electrical energy. In an *electrolytic cell* chemical changes take place only when electrical energy is supplied from the outside. A voltaic cell (or a series of them called a *battery*) can be thought of as an electron pump. Connected to an electrolytic cell it causes reactions to occur in the latter. Using Le Châtelier's principle we can correctly predict that the reaction which most readily gives up electrons will take place at the electrode from which the pump removes electrons, while the reaction that most readily accepts electrons will occur at the electrode to which the pump supplies electrons.

Figure 18-2 illustrates electrolysis of a molten salt. When the circuit is

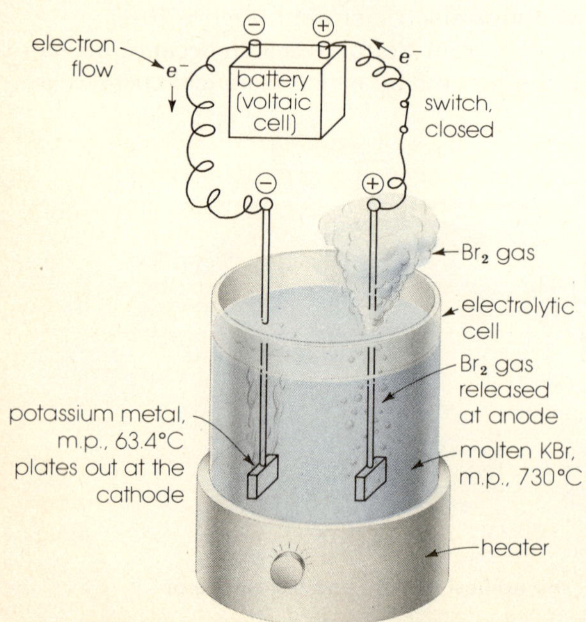

Figure 18-2 Electrolysis of a fused (molten) electrolyte. The battery removes electrons from the electrolytic cell's anode, sends electrons to the electrolytic cell's cathode. Platinum electrodes are usually used when chemically inert surfaces are required.

completed by closing the switch, electrons begin to flow. The electrode connected to the battery's positive pole becomes positive or electron deficient, and attracts negative bromide ions. As ions contact the electrode, electrons transfer from bromide ions to electrode metal, leaving bromine atoms on the metal surface. The atoms then combine into molecules:

$$2Br^- = 2Br + 2e^-$$
$$2Br = Br_2$$

An increase in valence number or oxidation state due to electron loss is called *oxidation*. Bromide ions are oxidized. The electrode at which oxidation occurs is named the *anode*. In an electrolytic cell the anode is positive.

The electrode connected to the battery's negative pole becomes negative or electron rich, and attracts positive potassium ions. At its surface electrons are transferred to K^+, giving a deposit of potassium metal:

$$K^+ + e^- = K$$

A decrease in valence number or oxidation state due to electron gain is called *reduction*. The electrode at which reduction takes place is named the *cathode*. The cathode is negative in an electrolytic cell.

The overall cell reaction is the sum of the two electrode reactions. For balance we write two electrons in each:

$$\begin{array}{ll} \text{at anode:} & 2Br^- \longrightarrow Br_2 + 2e^- \\ \text{at cathode:} & 2K^+ + 2e^- \longrightarrow 2K \\ \hline & 2K^+ + 2Br^- = 2K + Br_2 \end{array} \qquad (18\text{-}1)$$

Electrolysis of aqueous solutions

Electrolysis of a water solution is complicated by the possibility of electrode reactions by both water and solute. If chemically active electrodes are used, they add further complications. The following generalizations are useful in predicting what reactions take place during electrolysis of aqueous solutions at metal electrodes.

1. When the anode is more active than platinum or gold, it can be oxidized (see the Activity Series, Table 11-5). Oxidation may give an adherent oxide ($Pb \rightarrow PbO_2$) or the metal may go into solution as

$$Fe = Fe^{2+} + 2e^-$$

2. When the anode is inert like platinum and the solute anion is not oxidizable (examples: F^-, NO_3^-, SO_4^{2-}, PO_4^{3-}), electrons may be removed from water:

$$2H_2O = 4H^+ + O_2 + 4e^- \qquad (18\text{-}2)$$

3. When the cation is a metal more active than manganese (meaning that the cation is more difficult to reduce than Mn^{2+} as noted in Column P, Table 11-5), electrons may be given to water:

$$2H_2O + 2e^- = H_2 + 2OH^- \qquad (18\text{-}3)$$

4. When the cation is a metal below manganese in the activity series,

electrons are given to the cation which then plates out on the cathode:

$$Sn^{2+} + 2e^- = Sn$$

5. When the anion is more readily oxidized than water [examples: Br^-, I^-, $Fe(CN)_6^{4-}$] and the anode is an inactive metal, the anion is oxidized:

$$Fe(CN)_6^{4-} = Fe(CN)_6^{3-} + e^-$$

Example 18-1 An aqueous solution of calcium iodide is electrolyzed at platinum electrodes. What reactions take place?

Solution From Table 11-5, Column P, we do not expect Ca^{2+} to be reduced in aqueous solution. If current is to flow, therefore, water must accept electrons. At the cathode the reaction of Eq. 18-3 takes place.

At the anode iodide is oxidized:

$$2I^- = I_2 + 2e^-$$

The overall reaction is

$$Ca^{2+} + 2I^- + 2H_2O = Ca^{2+} + H_2\uparrow + 2OH^- + I_2\downarrow$$

The net result is the release of hydrogen gas at the cathode, accumulation of a dark purple deposit of iodine around the anode, and a rise in solution pH due to an increasing concentration of hydroxyl ions. Calcium ion remains unaffected.

Example 18-2 An aqueous solution of nickel nitrate is electrolyzed using platinum electrodes. What reactions take place?

Solution The positive anode attracts nitrate ions but cannot oxidize them (NO_3^- does not yield $NO_3 + e^-$), and the electrode itself is inert. Therefore electrons are removed from water, giving the reaction of Eq. 18-2.

Dipositive nickel is reducible to metal in aqueous solution (Table 11-5). At the cathode it accepts electrons:

$$Ni^{2+} + 2e^- = Ni$$

The overall reaction, balanced for a gain and a loss of four electrons, is

$$2Ni^{2+} + 4NO_3^- + 2H_2O = 2Ni\downarrow + 4H^+ + 4NO_3^- + O_2\uparrow$$

As electrolysis proceeds, the cathode increases in size due to a plating of nickel metal. At the anode oxygen gas bubbles are released. The green color of nickel ion gradually fades and the pH drops as hydrogen ions increase in number. When all nickel is plated, a solution of nitric acid remains.

18-3. QUANTITATIVE RELATIONSHIPS IN ELECTROLYSIS

Electrical units and Faraday's laws

Faraday (Figure 18-3) discovered that when a single reaction takes place at an electrode during electrolysis, the following are true:

1. The weight of substance produced is directly proportional to the quantity of electricity passed through the cell.
2. The weights of different substances produced by the same quantity of electricity are directly proportional to their equivalent weights.

These relationships, known as *Faraday's laws*, are independent of electrolyte concentration and temperature.

The quantity of substance that is oxidized by losing 1 mole (Avogadro's number, N) of electrons, or that is reduced by gaining 1 mole of electrons is called 1 gram-equivalent or 1 *equivalent* (abbreviated 1 "eq") of that substance. The equivalent weight of an electrolysis reactant is found by dividing its formula weight by the electron change per formula weight in the balanced equation that describes its electrode reaction. In the electrolysis of fused KBr,

$$2K^+ + 2e^- = 2K$$
$$2Br^- = Br_2 + 2e^-$$

passage of N electrons through the melt will yield $2K/2$ (or $K/1$) = 39.1 g of K at the cathode, and $Br_2/2 = 79.9$ g of Br_2 at the anode.

As found by Millikan the charge on the electron is -1.60×10^{-19} coulomb. Every ion carries the electronic charge or some multiple of it. The charge of N electrons is thus $(-1.60 \times 10^{19})(6.02 \times 10^{23}) = -9.65 \times 10^4$ coulomb. In most calculations the negative sign is dropped, and this quan-

Figure 18-3 Michael Faraday (England 1791–1867). Although essentially nonmathematical and self-taught, Faraday was one of the greatest experimental physicists. He discovered the laws of electrolysis (1834) and was the first to demonstrate the workings of the electric motor, generator, transformer, and dynamo, four machines indispensable today. The most far reaching of his scientific theories was the concept of fields of force. His rise from a lowly position as Davy's assistant to one of world eminence prompted a critic to remark that Davy's greatest discovery was Faraday. (Edgar Fahs Smith Memorial Collection, University of Pennsylvania.)

tity of positive electricity is called *one faraday*. By definition,

$$1 \text{ equivalent of electricity} = 1 \text{ mole of electrons} = 1 \text{ faraday} \tag{18-4}$$

And by experiment,

$$1 \text{ faraday} = 9.65 \times 10^4 \text{ coulombs} \tag{18-5}$$

The transfer of electricity through a conductor by means of a battery or electric generator can be thought of as analogous to moving water through a pipe with a mechanical pump. Quantity is measured in *coulombs* of electricity and in cubic feet (or gallons) of water. Rate of flow (also called *current*), defined as the quantity passing a point per unit of time, is measured electrically in coulombs per second and in water flow in cubic feet per second. One coulomb per second is called an *ampere* (amp). The number of coulombs of electricity passing through a conductor is found by multiplying current by time:

$$coulombs = amperes \times seconds \tag{18-6}$$

There are N electrons in 1 faraday. Consequently,

$$6.02 \times 10^{23} \text{ electrons}/9.65 \times 10^4 \text{ coulombs} = 6.02 \times 10^{23} \text{ electrons}/9.65 \times 10^4 \text{ amp-sec}$$
$$= 6.24 \times 10^{18} \text{ electrons/amp-sec}$$

(This is approximately the current carried by a household electric light bulb.)

A conductor offers resistance to the flow of electricity just as a pipe resists fluid flow. The unit of electrical resistance is the *ohm*. One ohm is the resistance at 0°C of a uniform column of mercury 106.3 cm long and containing 14.45 g of mercury.

In the same manner that flow rate in a pipe depends on the fluid pressure difference from one end to the other (expressed, say, in pounds per square inch), flow rate between two points in a wire depends on their difference in electrical pressure. Electrical pressure, also called electric *potential* or *electromotive force* (emf), is measured in *volts* (V). About 1820 German physicist G. Ohm found that the current (*I*) that flows through a metallic conductor is directly proportional to the potential (*E*) and inversely proportional to the resistance (*R*). The relationship is called *Ohm's law*. It is often written

$$E = IR \tag{18-7}$$

or

$$volts = amperes \times ohms$$

The *volt* is thus defined as the potential or pressure needed to maintain a current of 1 ampere when the resistance is 1 ohm.

Power, the rate at which work is done or energy is supplied in the pumping of water, can be expressed as horsepower and calculated by multiplying the quantity of water moved per second by the water pressure. Similarly, electric power (*P*) is the product of current and potential. The practical unit is the *watt* (W) (after J. Watt, Scottish inventor of steam power equipment) or the kilowatt (1000 W = 1 kW). One watt is the power developed by 1 ampere flowing through a potential difference of 1 volt.

Algebraically,

$$P = EI \tag{18-8}$$

or

$$watts = volts \times amperes$$

Energy or *work* required in pumping water is calculated by multiplying the power by the time interval t during which power is expended (and expressed, for example, in horsepower-hours). One watt-second is a unit of electrical energy or work; it is equal to 1 *joule* (J). One joule is 10^7 ergs. Previous definitions then make 1 joule equivalent to 1 volt-coulomb:

$$1 J = 1 \ W\text{-}sec = 1 \ V\text{-}coulomb = 10^7 \ ergs$$

When larger quantities of energy are expended, a more practical unit is the watt-hour (Wh):

$$1 \ Wh = 3600 \ W\text{-}sec = 3600 \ J$$

Defining energy or work W by

$$W = EIt = Pt \tag{18-9}$$

we can write

$$watt\text{-}hours = volts \times amperes \times hours = watts \times hours$$

Electrolytic cell calculations

Calculations dealing with quantities involved in electrolytic processes are an extension of the stoichiometry developed in Chapter 12. The weights of substances oxidized and reduced depend only on the quantity of electricity (coulombs) passed through the cell. One equivalent weight of substance reacts per mole of electrons or per faraday. Power requirements and costs are important in industrial computations.

Example 18-3 What weight of cadmium will be deposited from a $CdCl_2$ solution by passage of a current of 1.50 amp for 30.0 min?

Solution Each cadmium ion requires two electrons for reduction:

$$Cd^{2+} + 2e^- = Cd$$

One gram-atomic weight (112.4 g) of cadmium therefore requires 2 N electrons; N electrons will reduce half as much, which is 1 gram-equivalent weight: $112.4/2 = 56.2$ g. By definition, N electrons is 9.65×10^4 coulombs or 1 faraday.

The number of coulombs passing through the cell is computed from the current and the time in seconds by Eq. 18-6:

$$(1.50 \ amp)(30.0 \ min)(60 \ sec/min) = \ 2700 \ coulombs$$

The weight of cadmium is found by

$$? \ g \ Cd = 2.70 \times 10^3 \ coulombs \left(\frac{5.62 \times 10 \ g \ Cd}{9.65 \times 10^4 \ coulombs} \right) = 1.57 \ g$$

Example 18-4 What STP volume of Cl_2 is produced at the anode during the process described in Example 18-3?

Solution The anode reaction is

$$2Cl^- = Cl_2 + 2e^-$$

This means 2 faradays will displace 1 gram-*molecular* volume (22.4 liters) of Cl_2 at STP. And 1 faraday, or 9.65×10^4 coulombs, will displace 1 gram-*equivalent* volume or $22.4/2 = 11.2$ liters.

In Example 18-3 we calculated that 2.70×10^3 coulombs of electricity participated in the cathode reaction. Because the same quantity of electricity always flows into one electrode as out of the other electrode, 2.70×10^3 coulombs must also be responsible for reaction at the anode. Therefore,

$$? \text{ liter } Cl_2 = 2.70 \times 10^3 \text{ coulombs} \left(\frac{1.12 \times 10 \text{ liters}}{9.65 \times 10^4 \text{ coulombs}} \right) = 0.313 \text{ liter}$$

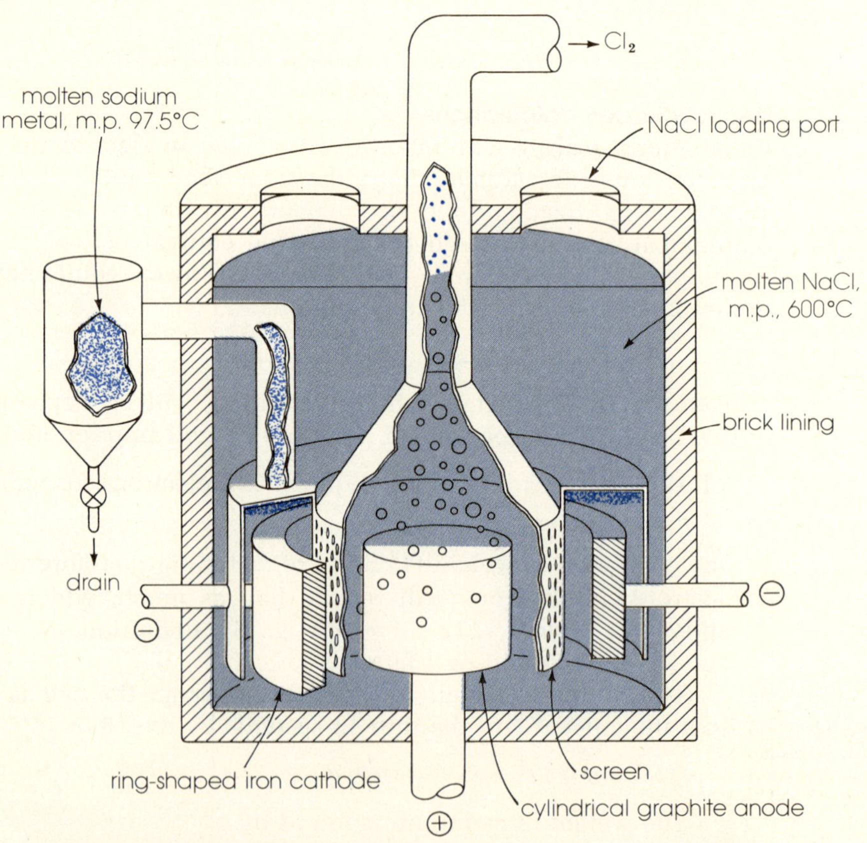

Figure 18-4 A cell for industrial production of sodium and chlorine. Some CaCl$_2$ is mixed with the NaCl to lower the melting point. The cathode is separated from the salt mixture by a screen that allows entry of salt while retaining molten sodium.

> **Example 18-5** Consider a Downes cell (Figure 18-4) operating at a potential of 7.00 V and a current of 26,000 amp. Calculate its power requirement, the energy expended in 8.00 hr of operation, and the electrical cost if power is purchased for 2.74 cents/kWh.
>
> **Solution** Power is the product of potential and current (Eq. 18-8):
>
> $$P = (7.00 \text{ V})(2.6 \times 10^4 \text{ amp}) = 1.82 \times 10^5 \text{ W}$$
>
> Energy is the product of power and time (Eq. 18-9):
>
> $$W = (1.82 \times 10^5 \text{ W})(8.0 \text{ hr}) = 1.46 \times 10^6 \text{ Wh} = 1.46 \times 10^3 \text{ kWh}$$
>
> Cost is the product of energy used and price per energy unit:
>
> $$cost = (1.46 \times 10^3 \text{ kWh})(2.74 \text{ cents/kWh}) = 4.00 \times 10^3 \text{ cents} = \$40$$

18-4. APPLICATIONS OF ELECTROLYTIC PROCESSES

Plants employing electrolysis to manufacture substances, deposit coatings, and mill metal surfaces comprise the *electrolytic industries*. Together with industries that use high-temperature electric furnaces to make alloys, graphite, calcium cyanamid, abrasives, and other materials, these industries are concentrated around hydroelectric centers like Niagara Falls. They consume about 13 percent of all electric power generated in the United States. Their functions can be classified as follows:

1. Electrolysis in fused salts
   (a) Electrometallurgy
2. Electrolysis in aqueous solutions
   (a) Electrometallurgy
   (b) Electrodeposition: electroplating, electrocoating, electroforming
   (c) Electrochemical machining
   (d) Electroproduction of chemicals

Electrometallurgy

The direct use of electricity in chemically extracting and refining metals is called *electrometallurgy*. Extraction from ores usually requires two steps: first, a purely chemical procedure in which a salt of the metal is prepared from purified mineral; then electrodeposition of the metal from fused salt or aqueous solution.

Most of the alkali and alkaline earth metals are produced by the electrolysis of fused chlorides. When the chloride is not naturally occurring, it is prepared chemically. Lithium chloride, for example, is prepared from its mineral spodumene by

$$LiAlSi_2O_6 \xrightarrow[1100°C]{K_2SO_4} Li_2SO_4 \xrightarrow{Na_2CO_3} Li_2CO_3 \xrightarrow{HCl} LiCl \tag{18-10}$$

Fused salt cells like that in Figure 18-4 operate at high currents where resistance heating keeps the electrolyte and product metal molten.

Aluminum, all of which is produced electrolytically (Figure 18-5),

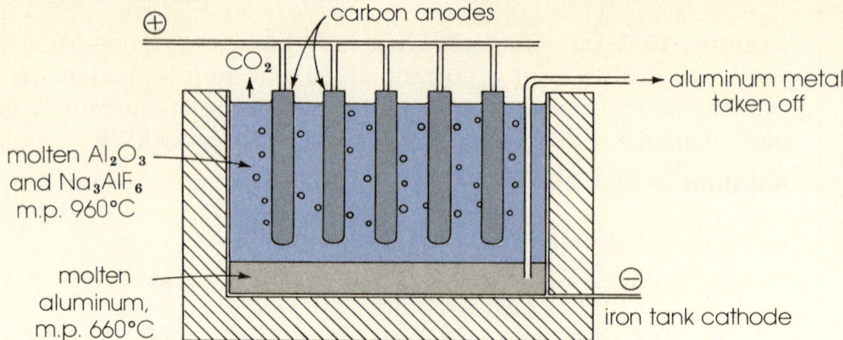

Figure 18-5 A cell for industrial production of aluminum.

at anode: $C + 2O^{2-} = CO_2 + 4e^-$
at cathode: $Al^{3+} + 3e^- = Al$

Typical cell conditions: temperature, 1000°C; emf, 4.5 V; current, 120,000 amp.

uses about 45 percent of all electric power required by the electrolytic industries. The electrolyte is a fused mixture of aluminum oxide, aluminum fluoride, and sodium fluoroaluminate. The oxide is prepared from bauxite ore by alkaline extraction, followed by acidification of aluminate ion and dewatering the hydroxide:

$$impure\ Al_2O_3 \xrightarrow[filter]{OH^-} Al(OH)_4^- + residue$$
$$\Big\downarrow \begin{array}{c} CO_2, \\ filter \end{array}$$
$$Al(OH)_3 \xrightarrow{1000°C} Al_2O_3 \qquad\qquad (18\text{-}11)$$

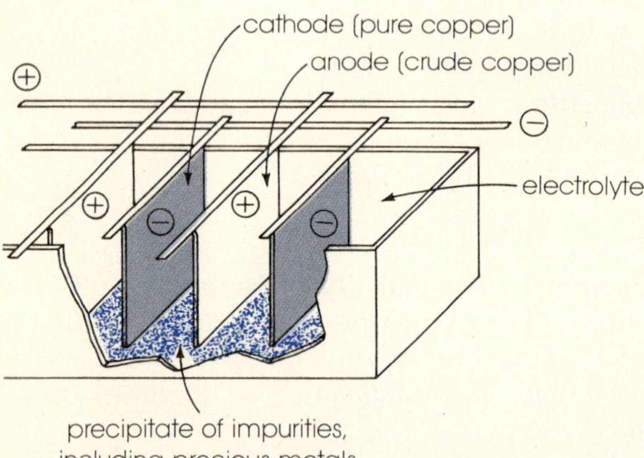

Figure 18-6 Electrorefining of copper. Anodes weighing about 500 pounds are cast out of crude copper obtained from ore or scrap. Cathodes are initially thin sheets of pure copper. A number of electrodes and cells are electrically interconnected. As electrolysis proceeds anodes dissolve while cathodes thicken:

at anode: $Cu = Cu^{2+} + 2e^-$
at cathode: $Cu^{2+} + 2e^- = Cu$

Active impurities such as zinc and iron go into solution whereas other impurities—silver, antimony, gold, platinum—drop to the bottom where they are recovered. Typical operating conditions: temperature, 60°C; emf, 0.2 V; current density, 20 amp/ft².

Most metals are obtainable by electrolysis of their fused salts. Commercial production, however, depends on costs relative to reduction by chemical means. The rare earths and tantalum are examples of less familiar metals produced economically by electrometallurgy.

The *electrochemical refining* of a metal begins by casting impure plates of the metal which have been obtained by reduction of ore. The plates are made the anodes in a cell where cathodes are pure metal and electrolyte is an aqueous solution of one of the metal's salts. Electrolysis dissolves the anode and plates out pure metal at the cathode. Cadmium, cobalt, lead, manganese, nickel, tin, silver, and copper (Figure 18-6) are refined this way.

Electroplating

The electrochemical deposition of metal for undercoating, corrosion protection, and surface decoration is called *electroplating*. The object to be plated is made the cathode. The anode is often made of the plating metal. The electrolyte may be a simple salt in acid solution ($CuSO_4 + H_2SO_4$), or a complex salt in base solution ($NaCu(CN)_2 + Na_2CO_3$). A circuit for electroplating is illustrated in Figure 18-8.

Nonconducting materials such as plastics can be plated by first etching with acid and chemically depositing an undercoating upon which a second metal is electroplated. The undercoating can be chemically deposited by a process called *electroless plating* in which the electrolyte solution contains a reducing agent such as sodium hypophosphite that spontaneously reduces metal ions:

$$Ni^{2+} + H_2PO_2^- + H_2O = Ni + H_2PO_3^- + 2H^+ \tag{18-12}$$

It is sometimes possible to codeposit two metals from solution to produce an alloy plate such as lead–tin (solder plate). In solder plating the anode is a lead–tin alloy bar, the electrolyte is a mixture of lead and tin fluoroborates and fluoroboric acid, HBF_4.

Electroforming

Articles made by relatively thick electrodeposition on a removable mold are described as *electroformed*. Phonograph record manufacture uses electroforming. The original recording is cut with a needle on an aluminum disk thinly coated with plastic. After chemically depositing a very thin film of silver in the grooves, nickel is electroformed on top. Peeled off, the nickel disk is a master negative. The "master" is coated to prevent adhesion, and positive "mothers" electroformed on top of it in nickel. From the mothers negative chromium "stampers" are electroformed. Stampers mounted two in a press are used to hot press records from plastic sheets.

Electrolytic production of chemicals

A considerable industry operates for the electrolytic production of chemicals. A major portion is devoted to making chlorine and sodium hydroxide from sodium chloride (see figure on page 447). A single such cell is shown in Figure 18-7. Its novel feature is the mercury cathode. Because sodium is far above hydrogen in the chemical activity series (Table 11-5), one would ex-

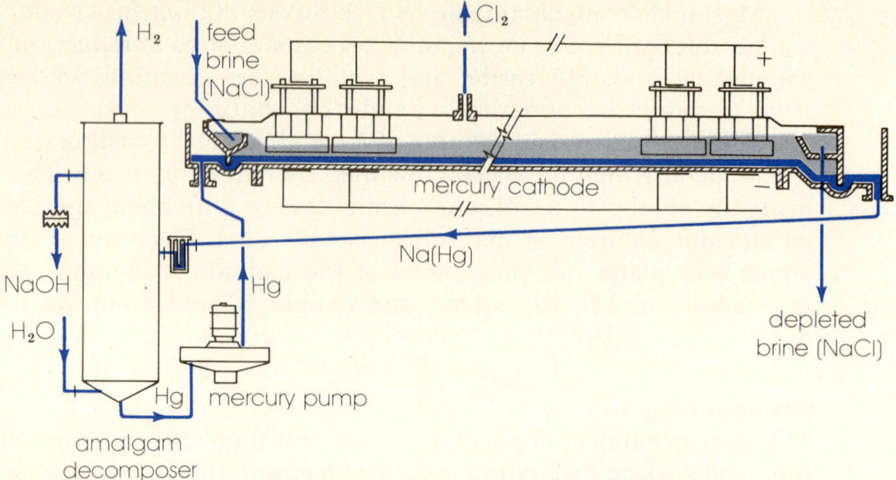

Figure 18-7 The De Nora mercury cell for producing electrolytic chlorine and sodium hydroxide. All chlorine used in the United States for water purification, bleaching, and sanitation is made electrolytically. Upper right, the primary cell. Brine is electrolyzed between graphite plate anodes and a flowing mercury cathode, giving chlorine gas and an amalgam of sodium. Lower left, the secondary cell. Amalgam is decomposed with electric current and fresh water to give sodium hydroxide solution and hydrogen gas. Depleted brine is then resaturated and recirculated. By-product hydrogen may be used in the hydrogenation of edible oils and the synthesis of ammonia and hydrochloric acid. Primary and secondary cell reactions, respectively, are

$$2NaCl + xHg \xrightarrow{\text{electrol.}} 2Na(Hg)_x + Cl_2$$

$$2Na(Hg)_x + 2H_2O \xrightarrow{\text{electrol.}} 2NaOH + H_2 + 2xHg$$

(Courtesy PPG Industries.)

pect hydrogen from water to be reduced rather than sodium ions. Metal surfaces vary greatly, however, in their ability to catalyze the reaction $2H = H_2$, and a somewhat greater voltage than theoretical (a so-called *overvoltage*) is needed to make electrolysis proceed with the evolution of hydrogen gas. On a mercury surface the overvoltage is so great at high-current density that the reduction product is not hydrogen but sodium metal, which amalgamates with the mercury. The amalgam is decomposed with water to give hydrogen and aqueous sodium hydroxide.

When a cell for the electrolysis of sodium chloride is designed to allow mixing of chlorine and sodium hydroxide, sodium chlorate and hydrogen are the products. Intermediate sodium hypochlorite, formed by the reaction

$$2NaOH + Cl_2 = NaClO + NaCl + H_2O$$

is oxidized:

$$NaCl + 3H_2O \xrightarrow{\text{electrol.}} NaClO_3 + 3H_2$$

Electrolysis of sodium chlorate solution gives sodium perchlorate:

$$NaClO_3 + H_2O \xrightarrow{\text{electrol.}} NaClO_4 + H_2$$

Other inorganic chemicals produced by electrolysis include MnO_2 from

$MnSO_4$ solution; $KMnO_4$ from K_2MnO_4 solution; F_2 from fused $KF \cdot 2HF$; NaOH and H_2SO_4 from Na_2SO_4 solution; H_2, D_2 (heavy hydrogen), and O_2 from H_2O; and hydrogen peroxide from sulfuric acid:

$$2H_2SO_4 \xrightarrow{electrol.} H_2S_2O_8 + H_2$$

$$H_2S_2O_8 + 2H_2O \xrightarrow{distil} 2H_2SO_4 + H_2O_2 \uparrow \tag{18-13}$$

Except in laboratory experiments, few organic chemicals are produced electrolytically. Limitations are imposed by the low solubility of most carbon compounds in electrolyte solutions and the tendency for reactive by-products to form.

QUESTIONS

1. Battery A battery's terminals are unmarked. Describe an experiment that would enable you to determine their polarity.

2. Electrolysis Criticize the following. In the electrolysis of aqueous potassium chloride solution at platinum electrodes it is a fact that hydrogen is released at the cathode and the solution becomes alkaline. This may be explained as the plating at the cathode of potassium metal followed by its immediate reaction with water:

$$2K + 2H_2O = 2KOH + H_2$$

3. Equivalents What is the electrochemical meaning of the equivalent weight of an element?

4. Electroless plating A kind of electroless plating is illustrated when a strip of tin metal is placed in a solution of silver nitrate. Explain.

5. Electrical units Supply the missing units; explain:

$$\frac{96,500}{6.02 \times 10^{23}} = 1.60 \times 10^{-19} \; coulomb$$

6. Definitions What is **(a)** electrolysis; **(b)** the faraday; **(c)** the coulomb; **(d)** the ampere; **(e)** the ohm?

7. Definitions What is the difference between **(a)** an anode and a cathode; **(b)** static and current electricity; **(c)** the current carrier in a wire and in a conducting solution; **(d)** power and work, in electrical terms?

8. Electrolysis A sunken ship is to be lifted from the ocean bottom. Plastic bags, containing seawater and equipped with an arrangement of inert internal and external electrodes are attached to the ship. Electrolysis current is applied to the electrodes to fill the bags with hydrogen gas. Is the internal electrode the anode or cathode? Explain. What are the products at the other electrode?

PROBLEMS

9. Electrolysis Write equations like those of Eq.18-1 for the electrolysis of the following at platinum electrodes: **(a)** fused $CaCl_2$; **(b)** aqueous $MgBr_2$; **(c)** aqueous Na_3PO_4.

10. Units Summarize the analogies given in the chapter between the pumping of water and of electricity by preparing a table like Table 18-1 and inserting the appropiate terms or units in each space.

11. Electrical units **(a)** Show that 1 faraday is equivalent to 1 amp flowing for 26.8 hr. **(b)** If a 0.250-amp current flows through a circuit at a potential of 1.10 V, what is the circuit's resistance in ohms?

12. Electrical units In a circuit 10^{19} electrons pass a reference point each second. **(a)** How much current (amperes) is carried? **(b)** A current of 0.50 amp flows through a wire under a potential of 6.0 V for 30 min. Calculate the watt-seconds, volt-coulombs, and ergs of energy expended.

13. Electrical units In the circuit of Figure 18-8, B is a battery and E an electrolytic cell at whose cathode silver is being plated. Ammeter A reads 0.400 amp, voltmeter V reads 0.250 V. **(a)** What is the resistance of cell E in ohms? **(b)** If current flows for 20.0 min how many faradays are transported? **(c)** What is the power in watts? **(d)** What is the energy input in joules?

14. Faraday's laws In Figure 18-9 battery B causes electrolysis at platinum electrodes in three electrolytic cells connected in *series arrangement* (all current passes in turn through each cell). E_1 contains Sb^{3+}, E_2 Sn^{2+}, E_3 Au^+. Assuming 100 percent efficiency in the plating of each metal, what weights are deposited on the three cathodes by the passage of 0.100 faraday?

15. Faraday's laws When 1836 coulombs is passed through a KI cell 2.414 g of I_2 is deposited. From these data calculate a value of the faraday to four significant figures.

16. Electrolytic chlorine In 1973 about 1.7×10^5 tons of Cl_2 was used in the United States to disinfect municipal drinking water. **(a)** How many day's production is that for the plant pictured on page 447? **(b)** If 10^{13} gal of water was treated, express the concentration of chlorine used in parts per million parts of water (1 gal $= 8.33$ lb). **(c)** Water at the kitchen tap may

Table 18-1. Comparing the pumping of water and electricity				
	Quantity	Flow rate	Pressure	Power	Energy
Water					
Electricity					

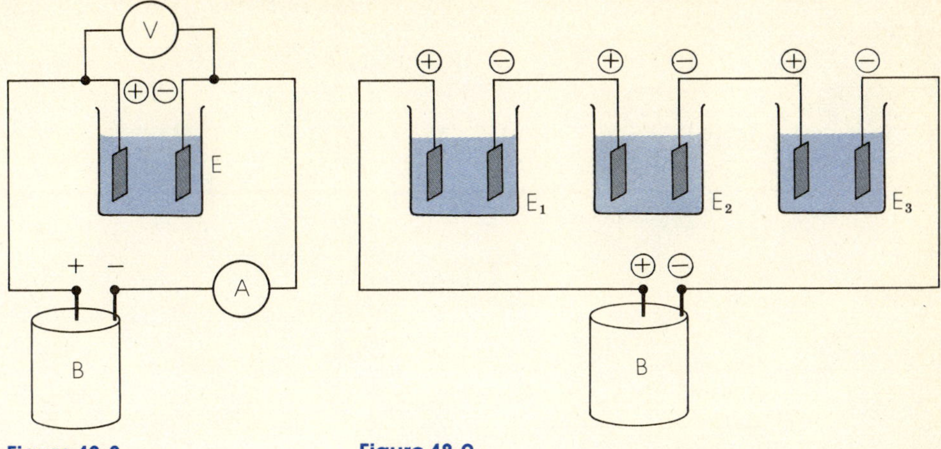

Figure 18-8   **Figure 18-9**

contain less than 1 ppm Cl_2. Explain the difference between this value and your answer in (b).

17. Industrial electrochemistry A typical cell for the electrolysis of molten $BeCl_2$ at 400°C uses a potential of 9.0 V and a current of 520 amp. Calculate the **(a)** power requirement in watts; **(b)** electrical energy expended in 24 hr of operation in kilowatt-hours; **(c)** dollar costs in (b) if power is purchased at 2.17 cents/kWh.

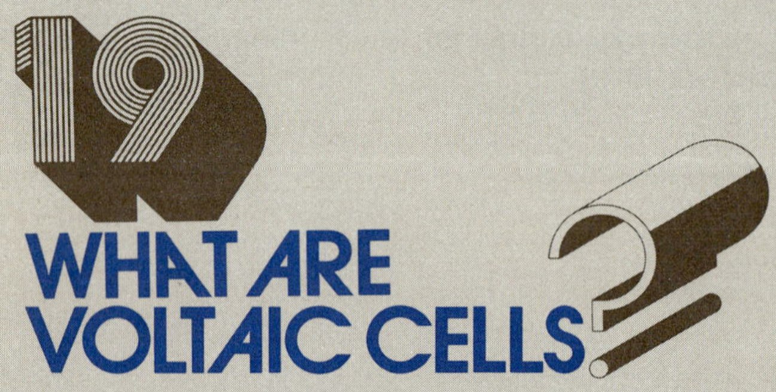

WHAT ARE VOLTAIC CELLS?

THE IMPORTANT CONCEPTS

19-1. Electric current from chemical reactions

1. Two half-cells properly combined make a voltaic cell.
   a. Electron and ion movement can be traced through electrodes, external circuit, solution, and salt bridge.
   b. Electrons are repelled at the negative anode and attracted at the positive cathode.
2. A cycle of charge transfer operates when a voltaic cell is connected to drive reactions in an electrolytic cell.

19-2. Standard electrode potentials

1. Oxidation occurs in one half-cell of a voltaic cell, reduction in the other half-cell.
2. Standard conditions and a standard half-cell are references for calculations.
3. The emf of a whole-cell comprising the hydrogen half-cell and another half-cell at standard state is $E°$ for the latter.

  a. Any half-cell reaction is at least theoretically reversible.

  b. Standard half-reaction equations listed in order of decreasing oxidation ease are a table of oxidation potentials.

 4. A few rules guide the combination of two half-cell equations to give the spontaneous whole-cell redox equation and E°_{cell}.

19-3. Practical voltaic cells

 1. Primary and secondary cells differ in chargeability.

 2. Electric vehicles may become practical for urban use.

  a. Electric power has several advantages over gasoline power.

  b. Energy-density and power-density requirements demand invention of high-performance batteries.

19-4. Future strategies for electric power

 1. An economy might be based on hydrogen.

 2. A fuel cell produces electricity directly from combustion.

 3. Capture of the diffuse energy of sunlight may become practical.

 4. A silicon solar cell converts light to electricity through its semiconductor properties.

19-5. Bioelectricity

 1. Electric potentials in living cells are common.

19-6. Corrosion

 1. Various voltaic mechanisms can be identified in corrosion.

  a. Metal corrosion is controlled by surface treatment or application of a counter current.

19-7. Writing redox equations

 1. In a redox reaction the oxidizer accepts electrons and is reduced; the reducer gives electrons and is oxidized.

 2. The half-reaction method is a way to balance redox reactions.

When a piece of zinc metal is dropped into a solution of copper sulfate, a spontaneous reaction occurs; zinc dissolves, copper metal appears, and heat is released:

$$Zn + Cu^{2+} = Zn^{2+} + Cu + heat$$

According to definitions given in Section 18-2 zinc atoms have been oxidized through electron loss and cupric ions reduced by electron gain.

The chemical energy of oxidation-reduction reactions can be converted to useful electrical energy by an arrangement that separates the reactants, diverts the electron flow over an outside conductor, and allows for transport of ions through the solution to balance concentration changes that accompany the process. In Chapter 18 we called such a system a voltaic or galvanic cell. In this chapter we examine voltaic cells, both theoretical and practical, and the reactions that make them go.

19-1. ELECTRIC CURRENT FROM CHEMICAL REACTIONS

The Daniell cell

The copper–zinc cell was introduced in 1836 by English chemist J. Daniell as an improvement on Volta's battery (Figure 18-1). In the Daniell cell of Figure 19-1 reactants are shown in separate beakers. Each beaker is described as a *half-cell*. Each half-cell consists of an electrode and an electrolyte solution. The electrodes are connected with wires through a meter that indicates current flow (an ammeter) or potential difference (a voltmeter). The solutions are connected through an inverted U-tube called a *salt bridge* which contains an aqueous gel of an electrolyte whose ions are free to migrate but which do not become oxidized or reduced. The cycle of operations, with its balance of electrons and charges, proceeds as follows:

1. A zinc atom goes into solution as a zinc ion, leaving behind its two valence electrons on the surface of the zinc electrode.
2. The bipositive charge of the new zinc ion is compensated for in solution by two chloride ions entering the zinc half-cell from the large reservoir of chloride ions held by the salt bridge.
3. The two electrons are pumped into the wire. At the opposite end two electrons appear on the surface of the copper electrode. They reduce a copper ion to a copper atom which plates out.
4. The disappearance of a copper ion leaves an excess sulfate ion in the copper half-cell solution. Its charge is neutralized by two unipositive potassium ions that diffuse toward it from the bridge.

Using the conventions introduced in Chapter 18, we again define the *anode* as the electrode where oxidation takes place and electrons leave the

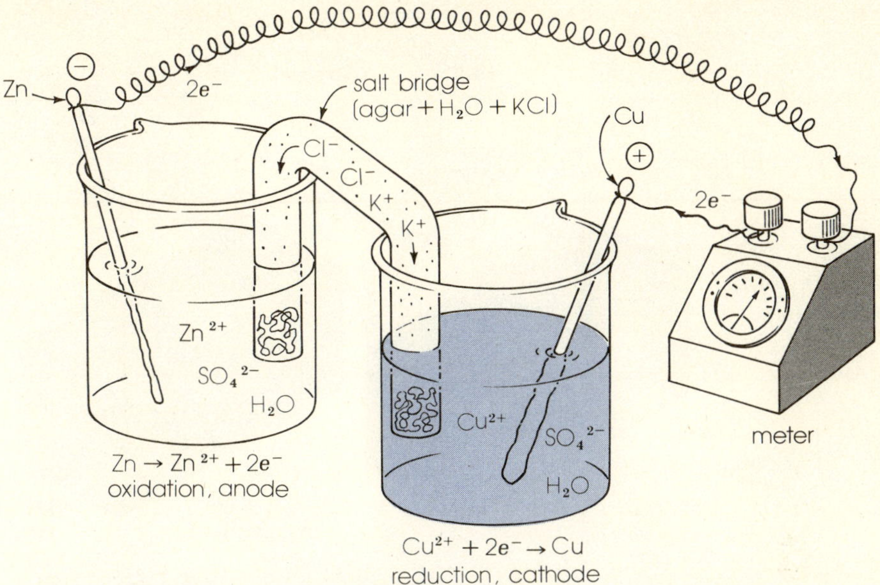

Figure 19-1 The Daniell cell. Charge is carried through the external circuit (wires, meter) by electrons and through the solutions by ions.

cell, and the *cathode* as the electrode where reduction takes place and electrons enter the cell. In the Daniell cell zinc metal is the anode, and copper metal is the cathode. Free electrons are more numerous at zinc; therefore, we call the anode negative and the cathode, where free electrons are less numerous, positive. Electrons are repelled at the negative anode and attracted to the positive cathode. If we define electric current as a flow of electrons, then current flows over the wire from anode to cathode in a galvanic cell.

The force causing electrons to move from one electrode to the other is the potential or electromotive force (emf) of the whole-cell. The emf depends on the ion-concentration ratio in the half-cells and the temperature. When both zinc and copper ions are 1 molar and the temperature is 25°C, the emf of the Daniell cell is 1.1 volts.

We can now clarify the difference between a voltaic cell and an electrolytic (electrolysis) cell.

Figure 19-2 shows a cell of each type interconnected. The spontaneous oxidation–reduction reaction in the voltaic cell drives an electrolysis reaction in the otherwise inert electrolytic cell and dictates the direction of electron flow in the circuit and the polarity of all electrodes. As a magnesium atom is oxidized to a magnesium ion in the voltaic cell, two electrons are sent into the wire. At the chemically inert platinum electrode at the wire's opposite end, two electrons appear. They reduce two hydrogen ions to hydrogen atoms which combine as a hydrogen molecule. Magnesium is an anode because oxidation occurs there; it is negative because it is a source of electrons. Platinum is a cathode because reduction occurs there; it is negative for being connected to negative magnesium. At the voltaic cell's tin electrode a stannous ion picks up two electrons and is reduced. The demand for two electrons is transmitted across the wire to the other platinum electrode where two iodide ions give up two electrons and become an iodine mole-

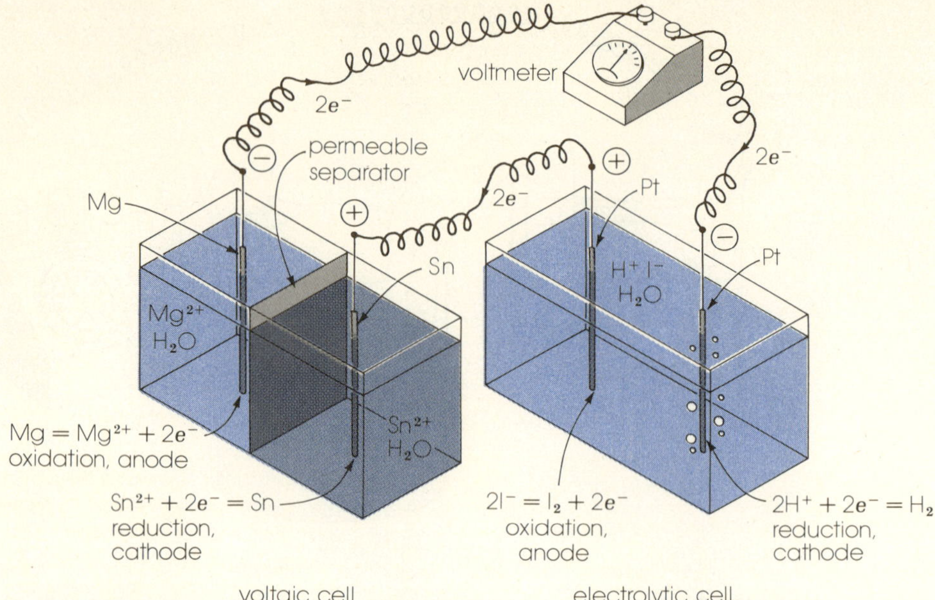

Figure 19-2 Electrode names, signs, and reactions in a circuit consisting of a voltaic and an electrolytic cell.

cule. Tin is a cathode because reduction takes place there; it is positive because it must be opposite in charge to magnesium, and it attracts electrons. The connected platinum electrode is positive because tin is positive, and that platinum electrode is an anode because oxidation occurs there.

19-2. STANDARD ELECTRODE POTENTIALS

The standard hydrogen electrode

A chemically active metal like zinc goes into solution as ions, leaving electrons behind. As ion concentration increases, the process tends to reverse itself so that in theory an equilibrium state is reached. For divalent metal M, equilibrium is described as $M = M^{2+} + 2e^-$.

To discuss the reaction at a single electrode chemists write a partial or half-cell reaction equation. When two electrodes are combined as in Figure 19-1 to make a (whole) cell, one partial equation describes a reduction, the other an oxidation. The sum is the overall reaction equation.

The creation and separation of charge at an active metal electrode cause an electrical pressure or potential to develop between the metal and its ions in solution. No method has been devised for measuring the absolute value of a single-electrode potential. It is possible, however, to get definite numerical values of such potentials by defining a set of standard conditions and comparing all electrode reactions to an electrode reaction whose potential has an assigned value. *Standard conditions* chosen are 1 molar concentrations for dissolved substances, 1 atmosphere pressure for gases, and a temperature of 25°C. A half-cell potential at standard state is designated E°.

The reference electrode is the hydrogen electrode (Figure 19-3). By general agreement its standard state potential is zero:

$$H_2 = 2H^+ + 2e^- \qquad E° = 0.00 \ V$$

(19-1)

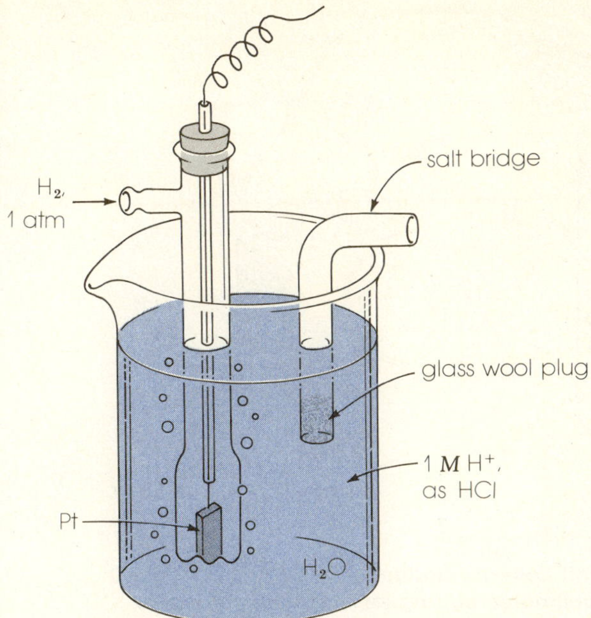

Figure 19-3 A standard hydrogen electrode. Hydrogen gas at 1 atm pressure bubbles over a surface of platinum coated with platinum black (finely divided platinum) which is also in contact with 1 M H⁺.

Because the hydrogen electrode potential is zero, the standard potential of any other half-cell is the voltage developed when, at standard state conditions, that half-cell is combined with the hydrogen electrode. The combination illustrated by Figure 19-4, for example, shows how $E°$ (0.44 V) for the iron electrode can be established. Iron dissolves spontaneously and electrons flow from the iron electrode to the hydrogen electrode where they reduce hydrogen ions. Reactions are

$$\begin{aligned}
\textit{at anode, oxidation:} \quad & Fe \longrightarrow Fe^{2+} + 2e^- \\
\textit{at cathode, reduction:} \quad & H_2 \longleftarrow 2H^+ + 2e^- \\
\hline
& 2H^+ + Fe = Fe^{2+} + H_2
\end{aligned} \qquad (19\text{-}2)$$

If instead of an iron half-cell, a copper half-cell (copper metal electrode dipping into 1 M Cu^{2+}) is connected to the hydrogen electrode at standard state, the measured potential is 0.34 volt. Cupric ions spontaneously plate out on the copper electrode and electrons flow from the hydrogen to the copper side. The reactions are

$$\begin{aligned}
\textit{at anode:} \quad & H_2 \longrightarrow 2H^+ + 2e^- \\
\textit{at cathode:} \quad & Cu \longleftarrow Cu^{2+} + 2e^- \\
\hline
& Cu^{2+} + H_2 = 2H^+ + Cu
\end{aligned}$$

A half-reaction like $Fe = Fe^{2+} + 2e^-$, which has a greater tendency to proceed as an oxidation than does the half-reaction oxidation of hydrogen, $H_2 \rightarrow 2H^+ + 2e^-$, will be *assigned a positive $E°$ value.* That means its reductant (Fe) is a better electron donor than is H_2. We list it above hydrogen in a table of half-reactions, which are all written as oxidations. A half-reaction like $Cu = Cu^{2+} + 2e^-$, which has a greater tendency to proceed as a reduction than does the half-reaction reduction of hydrogen ion, $H_2 \leftarrow 2H^+ + 2e^-$, is *assigned a negative $E°$ value.* Its oxidant (Cu^{2+}) is a better electron acceptor than is H^+. In our table we list it below hydrogen.

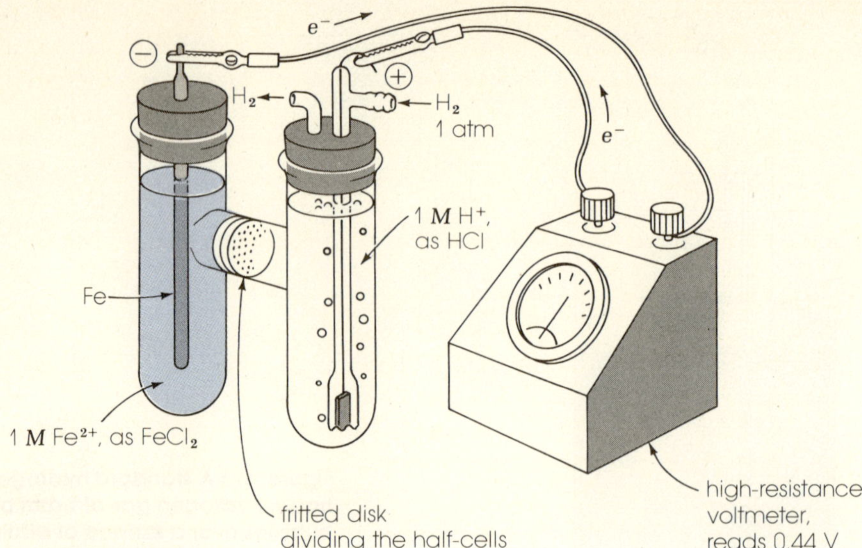

Figure 19-4 Combination of two half-cells at standard state. Because E° = 0.00 V for the hydrogen half-cell, the voltmeter reading for the whole cell is also the standard state potential for the other half-cell. A salt bridge is not needed in this experimental arrangement because ions can travel through the permeable disk. For each Fe²⁺ produced, 2H⁺ are converted to H₂ and 2Cl⁻ migrate from the hydrogen side to the iron side.

Any half-reaction we will discuss is at least theoretically reversible. Reversal of an oxidation half-reaction gives a reduction half-reaction. By the law of energy conservation the absolute value of $E°$ remains the same, but its sign is reversed in calculations. The magnitude of $E°$ is a measure of the tendency for the half-reaction to proceed as written. For the reactions described above the order is

$$Fe = Fe^{2+} + 2e^- \qquad E° = 0.44\ V$$
$$H_2 = 2H^+ + 2e^- \qquad E° = 0.00\ V$$
$$Cu = Cu^{2+} + 2e^- \qquad E° = -0.34\ V$$

Table 19-1. Standard oxidation potentials, 25 °C

Half-reaction	*E° volts*	*Half-reaction*	*E° volts*
$Li = Li^+ + e^-$	3.05	$Ag = Ag^+ + e^-$	−0.80
$Mg = Mg^{2+} + 2e^-$	2.37	$HNO_2 + H_2O = NO_3^- + 3H^+ + 2e^-$	−0.94
$Al = Al^{3+} + 3e^-$	1.66	$NO + 2H_2O = NO_3^- + 4H^+ + 3e^-$	−0.96
$Zn = Zn^{2+} + 2e^-$	0.76	$2Br^- = Br_2 + 2e^-$	−1.06
$Fe = Fe^{2+} + 2e^-$	0.44	$2H_2O = O_2 + 4H^+ + 4e^-$	−1.23
$Sn = Sn^{2+} + 2e^-$	0.14	$2Cr^{3+} + 7H_2O = Cr_2O_7^{2-} + 14H^+ + 6e^-$	−1.33
$H_2 = 2H^+ + 2e^-$	0.00	$2Cl^- = Cl_2 + 2e^-$	−1.36
$H_2S = S + 2H^+ + 2e^-$	−0.14	$Pb^{2+} + 2H_2O = PbO_2 + 4H^+ + 2e^-$	−1.46
$Sn^{2+} = Sn^{4+} + 2e^-$	−0.15	$Mn^{2+} + 4H_2O = MnO_4^- + 8H^+ + 5e^-$	−1.51
$Cu = Cu^{2+} + 2e^-$	−0.34	$O_2 + H_2O = O_3 + 2H^+ + 2e^-$	−2.07
$2I^- = I_2 + 2e^-$	−0.53	$2HF = F_2 + 2H^+ + 2e^-$	−3.06
$Fe^{2+} = Fe^{3+} + e^-$	−0.77		

When standard electrode potentials are determined by the method of Figure 19-4 and tabulated from most positive to most negative, the chemical activity series of Table 11-5 is reproduced quantitatively and other half-reactions added to it. A representative listing is given in Table 19-1. With all half-reactions written as oxidations, that is,

$$reduced\ form = oxidized\ form + electron\,(s) \tag{19-3}$$

the compilation is called a *table of oxidation potentials*.

Use of standard oxidation potentials

Data from Table 19-1 are used by chemists to (1) deduce the feasibility of a redox process, (2) write the chemical equation for the process, and (3) calculate the theoretical whole-cell potential E°_{cell} developed by a combination of any two half-cells. These objectives are attained as follows:

1. When two half-reactions are to be combined, they are first copied, in their order of appearance, from Table 19-1. Because the half-reactions are all given as oxidations, one of the two will be reversed to provide a reduction. The half-reaction with the higher oxidation potential (E° value) will proceed as an oxidation (left to right, as given in the table). The half-reaction with the lower oxidation potential will proceed as a reduction (its table direction reversed). Thus the half-reaction standing higher in the table will go as written; the lower half-reaction will be reversed.

2. If the number of electrons is not the same in the two half-reactions, they are balanced by inserting coefficients. The multiplication does not alter the E° values, as E° pertains only to half-reaction driving force and not to reactant quantities.

3. The equation with the lower potential is subtracted with its E° value from the equation with the higher potential and its E° value. The result is the overall or whole-cell reaction (with a cancellation of electrons), and the overall or whole-cell potential E°_{cell}. The sign of E°_{cell} will be positive. A positive E°_{cell} signifies that the reaction is spontaneous as written; that is, it proceeds from left to right with the release of energy.

Example 19-1 Use Table 19-1 to find the spontaneous reaction in the Daniell cell and its standard state potential.

Solution The half reactions, in table order, are

$$Zn = Zn^{2+} + 2e^- \qquad E^\circ = 0.76\ V$$

$$Cu = Cu^{2+} + 2e^- \qquad E^\circ = -0.34\ V$$

The number of electrons is the same in each, so no balancing is required. The lower half-reaction will be subtracted from the upper half-reaction. To subtract we reverse both the direction of the lower half-reaction and the sign of its E° value, then add. The result is the overall spontaneous reaction (reading left to right) and its standard state voltage:

$$Zn \longrightarrow Zn^{2+} + 2e^- \qquad E° = 0.76\ V$$
$$\underline{Cu \longleftarrow Cu^{2+} + 2e^- \qquad E° = 0.34\ V}$$
$$Zn + Cu^{2+} = Zn^{2+} + Cu \qquad E° = 1.10\ V$$

Example 19-2 A company wants to make electronic circuits by photographically printing them on thin copper sheets then removing excess copper with a mild oxidizing agent. Is ferric chloride solution a feasible oxidizer?

Solution From Table 19-1 the two half-reactions related to the problem, in listed order, are

$$Cu\ \ = Cu^{2+} + 2e^- \qquad E° = -0.34\ V$$

$$Fe^{2+} = Fe^{3+} + e^- \qquad E° = -0.77\ V$$

The copper half-reaction has the higher oxidation potential and therefore goes as written, that is, as an oxidation. The ferrous-ferric half-reaction must proceed in the opposite direction with ferric ion being reduced. We arrive at the overall spontaneous reaction by first balancing the electrons (multiplying the iron half-reaction, but not its $E°$ value, by 2), then subtracting the lower from the upper half-reaction. As in Example 19-1 subtraction is equivalent to reversing both the direction of the lower half-reaction and the sign of its $E°$ value, and adding:

$$Cu \longrightarrow Cu^{2+} + 2e^- \qquad E°\ \ = -0.34\ V$$
$$\underline{2Fe^{2+} \longleftarrow 2Fe^{3+} + 2e^- \qquad E° = 0.77\ V}$$
$$Cu + 2Fe^{3+} = Cu^{2+} + 2Fe^{2+} \qquad E°_{cell} = 0.43\ V$$

Ferric ions will indeed oxidize (and solubilize) copper metal.

19-3. PRACTICAL VOLTAIC CELLS

Primary cells and storage cells

A *primary cell* is an electric cell that cannot be recharged. The most familiar example is the dry cell, of which more than a billion are manufactured yearly in the United States. In the dry cell pictured in Figure 19-5 the anode is the zinc container, the cathode a rod of carbon and manganese dioxide. The electrolyte is a moist paste of $ZnCl_2$–NH_4Cl. The potential developed is 1.5 volts:

at anode: $\qquad\qquad\qquad\qquad Zn \longrightarrow Zn^{2+} + 2e^- \qquad\qquad E° = 0.76\ V$

at cathode: $2MnO(OH) + 2NH_3 \longleftarrow 2MnO_2 + 2NH_4^+ + 2e^- \qquad E° = 0.76\ V$

The cell is nonrechargeable because if the direction of electron flow is reversed by connection to, say, a more powerful cell, Mn(III) is not reoxidized to Mn(IV).

Storage cells are electric cells that can be recharged. An everyday ex-

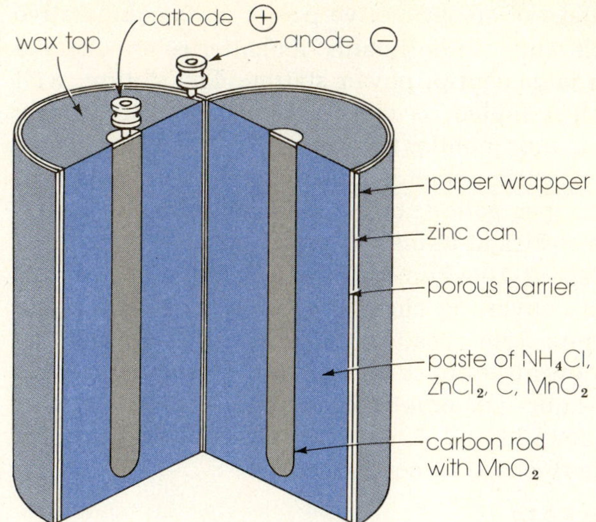

Figure 19-5 The common dry cell. It was invented in 1867 in France by G. Le-clanché. Cell potential gradually falls due to increasing internal resistance caused by precipitation of Zn(OH)₂. The effect is partially offset by reaction of Zn(OH)₂ with NH₃ to form the soluble complex ion Zn(NH₃)₄²⁺ which diffuses away from the anode.

ample is the lead–acid automobile battery (Figure 19-6) The electrodes are thin lead grids arranged alternately. Anodes are filled with spongy metallic lead, cathodes with lead dioxide. The electrolyte is sulfuric acid. The acid's concentration (initially 38 percent H_2SO_4 by weight) decreases during discharge as lead sulfate forms on the electrodes. Battery condition is checked by measuring the acid's density with a hydrometer. Each cell develops a potential of about 2 volts,

$$at\ anode: \qquad Pb + SO_4^{2-} \longrightarrow PbSO_4 + 2e^- \qquad\qquad E° = 0.36\ V$$
$$at\ cathode: \quad PbSO_4 + 2H_2O \longleftarrow PbO_2 + SO_4^{2-} + 4H^+ + 2e^- \qquad E° = 1.69\ V$$

The battery is recharged by reversing the electron flow. Each of the half-reactions then goes in the direction opposite that indicated above.

The battery-powered car

For the foreseeable future the internal combustion engine seems to be the only practical means for moving land vehicles over long distances. Commuting and service vehicles, however, may someday be electric.

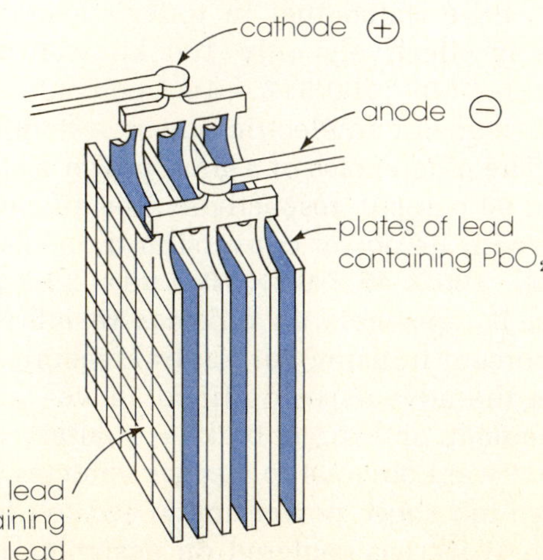

Figure 19-6 The lead storage cell. Plates are separated by porous plastic sheets to prevent short circuiting if they buckle. A 12-V battery containing six cells of 29 plates each may deliver 500 amp of starting current. Plates stand in approximately 5 M sulfuric acid.

The most obvious advantage of electromotive power is its comparative cleanliness. The vehicle itself emits no pollutants. Its batteries are charged by electricity produced by a large central power station. The station, well controlled and perhaps nuclear fueled, is cleaner than the many poorly maintained gasoline engines in automobiles that it would, in effect, replace.

The second advantage of electric power is lower cost. Fifteen gallons of gasoline, tax-free at 40 cents per gallon, costs $6.00 and carries an automobile about 200 miles. In electrical terms the performance is equivalent to an energy generation of about 100 kilowatt-hours (kWh). A battery is about 70 percent efficient in converting chemical energy to electrical energy. To get 100 kilowatt-hours from a battery consequently requires that it contain $100/0.70 = 143$ kilowatt-hours of energy. An average household today may use 700 kilowatt-hours per month. Charging a vehicle's battery would be in excess of that amount and go at a rate of, say 2.45 cents per kilowatt-hour. Complete battery charging cost would be

$$143 \ kWh \times \$0.0245/kWh = \$3.50$$

Expected savings per 2000 miles is $\$6.00 - \$3.50 = \$2.50$ or, over a year's driving of 12,000 miles,

$$\$2.50 \times 12,000/200 = \$150$$

Vehicle use of 50 miles a day of urban driving is equivalent to perhaps 25 kilowatt-hours. The battery should be charged every day of use. Household electricity is 30 amperes at 220 volts. Charging from an outlet in the garage between 10 P.M. and 6 A.M. easily stores the daily requirement:

$$220 \ V \times 30 \ amp \times 8 \ hr = 52.8 \ kWh$$

Most cities have excess generating capacity at night, so new electric plants might not be needed to maintain electric cars.

A third advantage of electric power is its thermal efficiency. The heat of combustion of gasoline is 2.97×10^4 kilocalories per gallon. One kilocalorie is equivalent to 1.16×10^{-3} kilowatt-hour. When 15 gallons of gasoline is used as fuel,

$$15 \ gal \times 2.97 \times 10^4 \ kcal/gal \times 1.16 \times 10^{-3} \ kWh/kcal = 517 \ kWh$$

of energy is produced. But we have stated that in today's average automobile 15 gallons of gasoline is effectively only 100 kilowatt-hours of energy. The efficiency is thus only $100 \times 100/517 = 19.3$ percent.

To demonstrate that the efficiency of an electric power system is superior, we begin with these data: The efficiencies of a good power station and its transmission lines are 45 and 90 percent, respectively; the efficiencies of a battery and an electric motor with its power conversion train are 70 and 90 percent, respectively. Overall, $100 \times 45 \times 90 \times 70 \times 90 = 25.5$ percent. This is still low. Nevertheless, it represents an increase in efficiency of $100 \times (25.5 - 19.3)/19.3 = 32$ percent in using our natural resources.

Although electric cars have the advantages outlined above — plus simplicity, quietness, smooth movement, and direct braking — battery research has not solved the problems that were obvious in electric vehicles 50 years ago: their cost is too high, range too short, power and speed too low, and weight of batteries is excessive. Work has centered on designing batteries

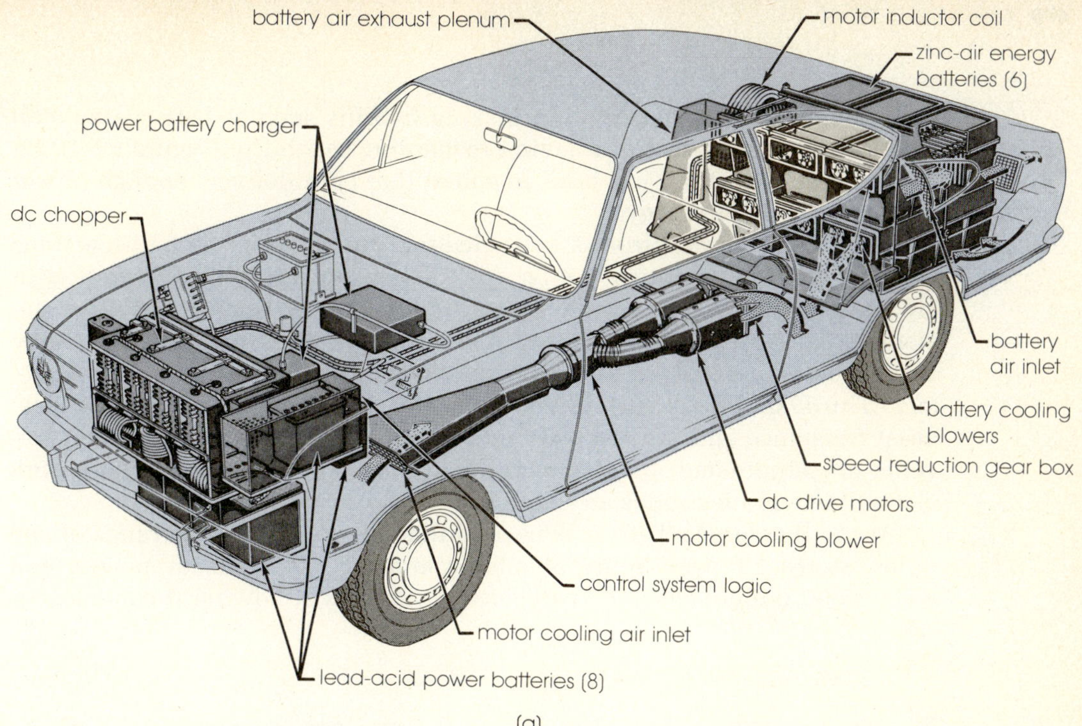

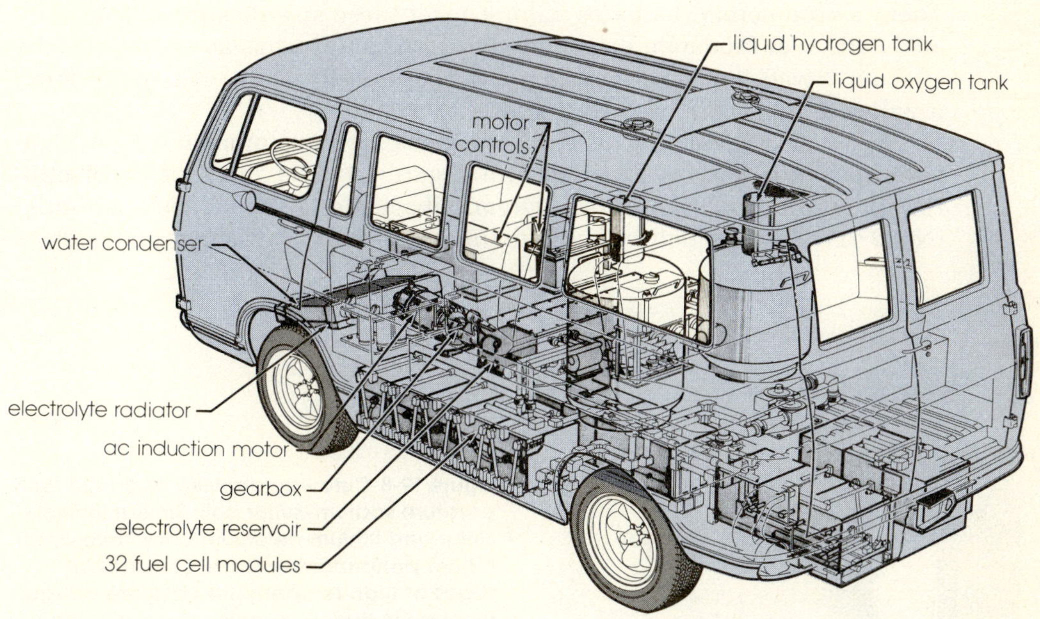

Figure 19-7 Experimental electrical vehicles. (a) A passenger car containing zinc-air batteries in the rear for high storage of energy (adequate vehicle range) and lead–acid batteries in the front for high power (adequate vehicle performance). The zinc–air batteries are rechargeable only by replacing all 300 zinc plates after each 150 miles of travel, and thus are too costly. (b) A vehicle powered by an oxygen–hydrogen fuel cell array. Besides high cost its disadvantage is the hazard of carrying liquified O_2 and H_2 and 45 gallons of potassium hydroxide electrolyte. (Courtesy General Motors Research Laboratories.)

with greater energy density, and making hybrids such as battery–fuel cell (Section 19-4) combinations having recharging capability (Figure 19-7). So far no practical battery has been invented that both delivers enough power and stores enough energy.

To get a large emf from a cell, electrochemists examine half-reactions whose oxidation potentials are far apart. They consider first elements with low electronegativity (periodic table Groups IA, IIA) for anodes, and elements with high electronegativity (Groups VIA, VIIA) for cathodes. They next consider elements of low equivalent weight to make a light-weight battery. Third, they look for systems that will have low internal resistance. Least resistance and highest electrode reaction rates are attained when solvents are eliminated and temperature is raised to the point where both electrolyte and electrodes are molten.

We will assume a battery is needed that gives a small car a range of 200 miles, stores 10^5 watt-hours of energy and 5×10^4 watts of power, and weighs 500 pounds. Its *energy density*, a measure of how far it can take the car, will be

$$10^5 \ Wh/500 \ lb = 200 \ Wh/lb$$

Its *power density*, a measure of how fast the car can accelerate, will be

$$5 \times 10^4 \ W/500 \ lb = 100 \ W/lb$$

Today's commercial batteries cannot meet these specifications. The lead–acid battery at moderate discharge rate, for example, has an energy density of about 7 watt-hours per pound at a power density of 38 watts per pound; it will take our vehicle only 12 miles.

A cell of a general type with prospects for development into a high-performance battery is illustrated in Figure 19-8. In place of an electrolyte it employs a semiporous ceramiclike material called *beta alumina* ($Na_2O \cdot 11Al_2O_3$) which has conducting properties. Sulfur, an insulator, is

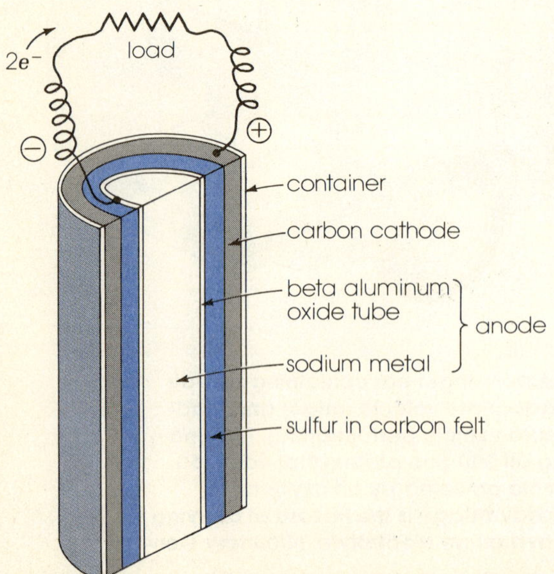

Figure 19-8 Cutaway drawing of a high temperature sodium–sulfur cell. Similar lithium–sulfur and lithium–selenium cells have even higher performance. General disadvantages of high-temperature cells are danger inherent in their very active ingredients, corrosive action on cell construction materials, and tendency of reactants to diffuse, causing self-discharge. Approximate reactions are

$$2Na + 3S \underset{charge}{\overset{discharge}{\rightleftharpoons}} Na_2S_3$$

(Courtesy E. J. Cairns and H. Shimotake, Argonne National Laboratory, Argonne, Illinois).

contained in a conducting pad of carbon which is the cathode. At about 300°C, sodium loses electrons and is oxidized to Na^+. Sodium ions travel through the alumina cup to react in the pad with sulfur, forming various sodium polysulfides. Electrons are collected by the cathode and used to reduce sulfur. The cell develops about 2.6 volts.

In theory a 400-pound sodium–sulfur battery could store a total of 54 kilowatt-hours and would have, at reasonable loading, a specific energy of 135 watt-hours per pound. Assuming installation in a 2000-pound car, which has an energy requirement under urban driving conditions of 0.27 kilowatt-hours per mile, the battery would give a range of

54 kWh/0.27 kWh/mi = 200 mi

19-4. FUTURE STRATEGIES FOR ELECTRIC POWER

Hydrogen economics; fuel cells

The growing need for electric power is well recognized. Electricity has its drawbacks, however. They include (1) the air and heat pollution attending its production; (2) the seeming impossibility of its direct storage (so production must match consumption); (3) the need for elaborate systems of production, distribution, and transmission; and (4) economic noncompetitiveness when energy is demanded in the form of heat — as most energy is. An alternative is an economy based on hydrogen.

As envisioned by electrochemist J. Bockris, atomic reactors held on platforms over ocean water deep enough for easy cooling would be used to electrolyze water to hydrogen and oxygen. The gases would be piped like natural gas to distribution stations and stored as liquids at low temperatures. Hydrogen could be burned directly with air in home furnaces and in conventional automobile engines. Hydrogen would also be used to reduce ores and to make chemicals such as ammonia. Reconversion of H_2 to electricity would be done in on-site fuel cells in the home and in industrial plants. In all uses the only combustion by-product would be water. If in A.D. 2000 the average American home uses 10 times as much electricity as it does today, fuel-cell water could supply all household needs for drinkable water.

A *fuel cell* is an electrochemical cell that continuously converts the chemical energy of a fuel and an oxidizer directly to electrical energy without any preliminary production of heat. The principle of causing a combustion to produce electric potential was discovered in 1839 but the first practical cell was not built until 1959. Operating ideally a fuel cell may be two to four times more efficient than a motor-driven generator in producing electrical energy.

Fuel-cell construction can be generalized as follows:

anode,fuel | electrolyte | oxidizer,cathode

where the commas indicate intimate contact between different phases, and the vertical lines indicate phase boundaries. Fuels may be liquids such as methyl alcohol and gasoline, or compressed gases such as hydrogen and methane. Oxidizers are pure oxygen or air. Electrolytes may be aqueous solutions of strong bases (KOH) or acids (H_2SO_4) in low-temperature cells

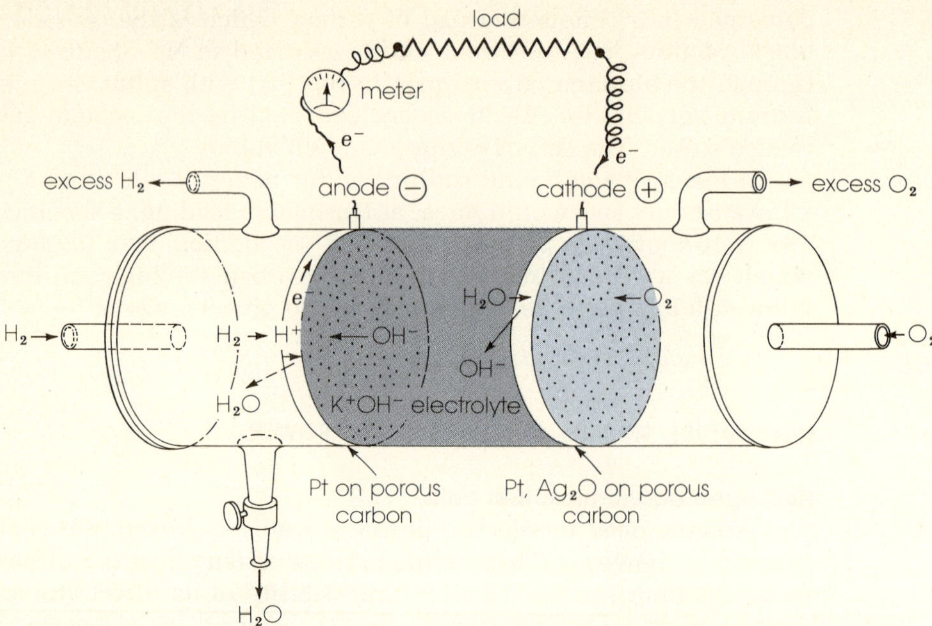

Figure 19-9 A hydrogen–oxygen fuel cell. It operates below 100°C. This type cell is also illustrated in Figure 19-7b.

($\sim 100°C$), or molten salt mixtures like $Na_2CO_3 + Li_2CO_3$ in high-temperature cells ($\sim 400°C$). Electrodes are porous to allow reactant–electrolyte contact at a metal surface. Electrodes may also need to possess catalytic activity as demanded by the reactants.

The most practical fuel cell to date is the type pictured in Figure 19-9:

$$Pt,H_2 \quad | \quad KOH \quad | \quad O_2,Ag_2O,Pt$$

Hydrogen gas diffuses through the platinized anode. Catalyzed by platinum, hydrogen molecules become hydrogen ions as electrons are given up to the metal and move to the external circuit. At the platinum surface the hydrogen ions react with hydroxyl ions from the aqueous KOH electrolyte to yield water. At the opposite electrode of platinum and silver oxide, oxygen, water, and electrons from the external circuit come together to make hydroxyl ions. The reactions are

$$\begin{aligned}
&at\ anode: &&2H_2 + 4OH^- \longrightarrow 4H_2O + 4e^- \\
&at\ cathode: &&4OH^- \longleftarrow 2H_2O + O_2 + 4e^-
\end{aligned}$$

$$(19\text{-}4)$$

Sunlight economics: photovoltaic cells

Another scheme for generating electric power in the future with no dependence on fossil fuels and with minimal environmental damage involves the use of sunlight.

One method of capturing solar energy is through specially coated surfaces which, by being transparent to visible light but opaque to infrared reemission, take advantage of the greenhouse effect (Chapter 13). Attaining temperatures of 500°C or more, the panels would transfer heat via liquid sodium pumped through them to a reservoir of liquid salts whose store of heat would be used to run conventional steam boilers, turbines, and electric generators. Calculations indicate that in sunny regions like the southwestern United States about 3 square miles of collecting surface and a 13 million gallon reservoir of cooling water would be equivalent to a 1000-megawatt generating plant, a size being built with atomic fueling today. High initial cost and questionable durability of the surface coating are present disadvantages to the plan.

Another approach is to convert solar energy directly to electricity through the use of *solar cells*. One proposal is to put a cell platform in stationary earth orbit (22,500 miles up) from where energy would be transmitted to the ground by microwave beam for reconversion into electricity. A 1000-megawatt electric plant would require input from a solar cell array having an area of 20 square miles.

A scheme for the direct production of electricity in private homes would feature a roof of solar cells that would store electricity by charging batteries. Ignoring the present high cost of the cell array the economics of the plan is as follows. For the southwestern United States we will estimate conservatively that the solar energy flux reaching the earth during middle daylight hours is 70 watts per square foot. Given roof panels measuring 2000 square feet and intercepting maximum light effectively for 4 hours per day, the energy available is

$$70 \; W/ft^2 \times 2000 \; ft^2 \times 4 \; hr/day \times 1 \; kW/1000 \; W = 560 \; kWh/day$$

Present efficiency of the solar cell is only 13 percent, however, so

$$560 \times 0.13 = 73 \; kWh/day \; of \; available \; energy$$

That would be enough for all household needs and the charging of the batteries of the family's electric car.

The practical solar cell today is made of silicon. The silicon atom crystallizes in the diamond lattice (Figure 6-7a) in which all four valence electrons are occupied in bonding. Normally a nonconductor, the structure becomes conducting when enough heat or light energy is supplied to it to raise the valence electrons into higher, vacant orbitals where electric current is carried. These properties make silicon an example of a *semiconductor*. In terms of the models of Figures 10-6 and 10-7 a semiconductor has a small forbidden gap between filled and unfilled electron levels. To promote an electron from a valence to a conduction zone in silicon requires about 1.1 electron volts, an energy within the capability of visible light. Maximum response is at 700 millimicrons (mμ), a prominent wavelength in sunlight.

The property of semiconduction is made permanent by "doping," that is, by controlled introduction of tiny amounts of other elements. Doping with a Group VA element such as arsenic, whose atoms contain five valence electrons, produces a crystal in which the "extra" electron is readily excitable because it is not needed in lattice bonding. Such a crystal is called an n (negative) crystal. Doping with a IIIA element such as boron, whose atoms have three valence electrons, gives a structure with "holes" into which electrons can move. This is a p (positive) crystal.

When a thin p crystal is formed on top of an n crystal a p–n junction diode is produced (Figure 19-10). Initially electrons move from the n side to the p side. In effect "holes" move in the opposite direction because the

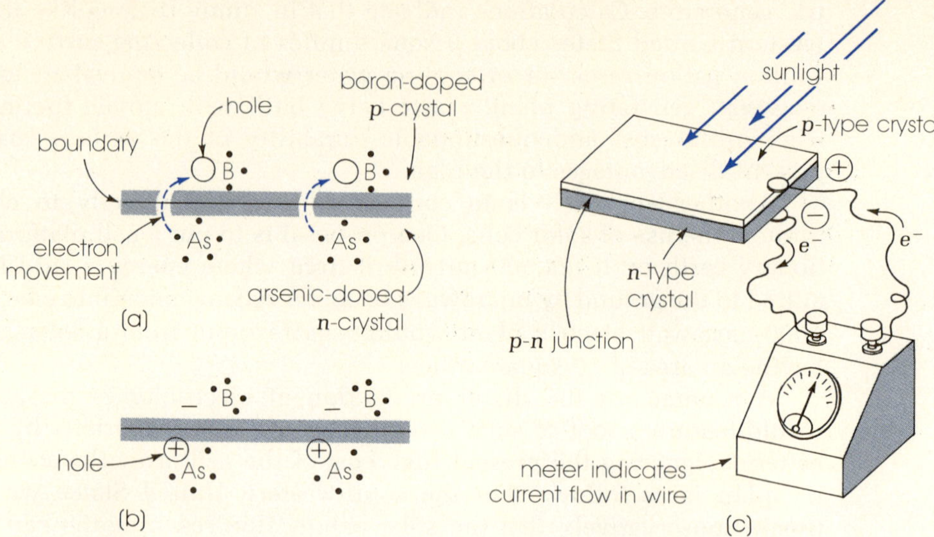

Figure 19-10 A solar cell. (a) When a thin layer of silicon doped with boron is put on a silicon crystal containing a trace of arsenic, electrons migrate from arsenic to valence "holes" around boron. (b) Migration makes the boron side negative and the arsenic side positive at the junction. (c) Sunlight striking the surface mobilizes electrons which move from the boron side to the arsenic side. The effect causes electrons to flow from the body of the cell through the meter and back into the cell's surface. With present technology a cell only 1 mm thick may develop 0.5 V at 0.03 amp/cm² of active area. An arrangement of many cells connected side by side is a solar battery.

departure of an electron leaves a hole. When equilibrium is established at the boundary, the p crystal has a negative layer of charge which prevents further influx of electrons, and an electric potential exists over the boundary.

In the dark a voltmeter connected across the cell reads zero. Exposed to sunlight, however, photon energy is absorbed in the boundary by electrons which jump to the conduction level. As the boundary equilibrium is upset, electrons move to the n side while holes move in the opposite direction. Electron pressure causes electrons to flow out of the negative n side through the wire and back into the positive p side where they recombine with holes. A collection of such cells is a solar battery.

19-5. BIOELECTRICITY

Electric fishes

Electric potentials caused by cellular ionic processes are found in both plants and animals. Potentials are especially noticeable in animal muscles and related nerve tissue. Largest bioelectric effects are found in special organs of fish like the electric eel (Figure 19-11). About 500 species of fish are equipped with electric organs which are used for attack, defense, and navigation.

Electric organs consist of thin cells called *electroplaques* or plates. Electroplaques have properties suggestive of both capacitors (see Figure

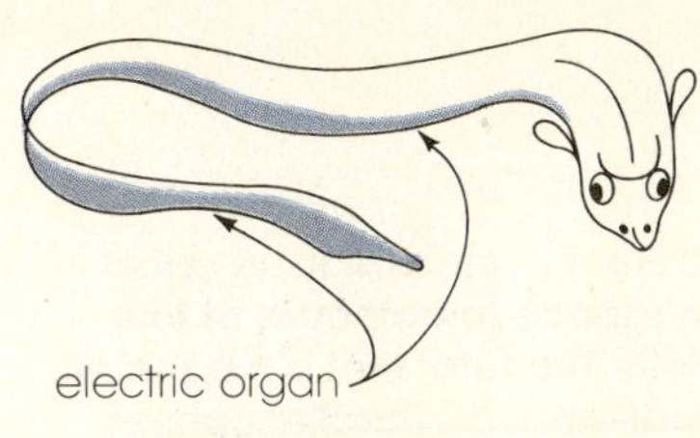

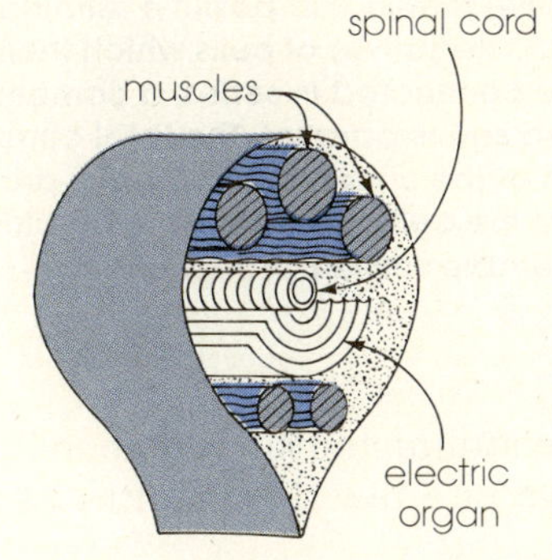

Figure 19-11 An electric eel and a partially dissected cross section of the tail.

9-4) and voltaic cells. Fluids inside and outside the electroplaques contain different solutes but have identical osmotic pressure. Outside, sodium and chloride ions predominate. Inside, potassium and organic ions prevail. The separation of Na^+ and K^+ is due to the semipermeability of the cell membrane which passes potassium ions preferentially. A potential of about 0.080 volt existing across the membrane is maintained by cellular energy that pumps out sodium ions as potassium ions come in. Under rest conditions the inner surface of the membrane is negative, and the outer surface is positive. One side of each electroplaque is innervated — connected to the nerve system that sends signals to it — and the other side is uninnervated. Signals start from the brain and on reaching a membrane alter its permeability, permitting ions and electric charge to flow across it. In the eel the polarity of the innervated side of each electroplaque reverses itself. Current then pulses, as represented in Figure 19-12, creating an electric field around the fish. A mechanism operates that regulates signal speed so that

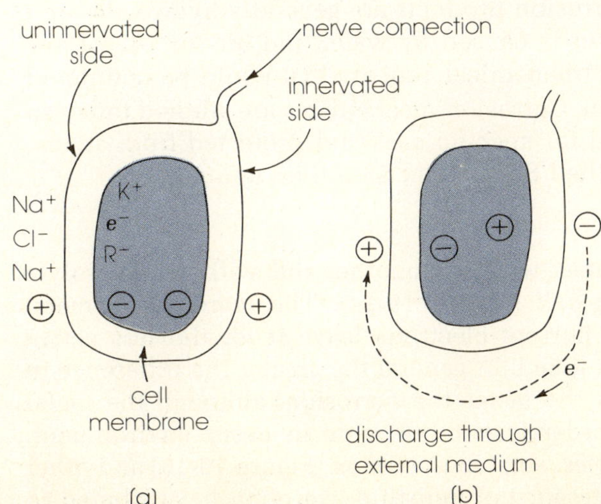

Figure 19-12 Electroplaques. (a) A resting electroplaque; R⁻ represents organic anions. (b) An electroplaque stimulated by a nerve impulse. Stimulation alters the permeability on the right side, changes the polarity across the membrane, lowers membrane resistance and allows current to flow.

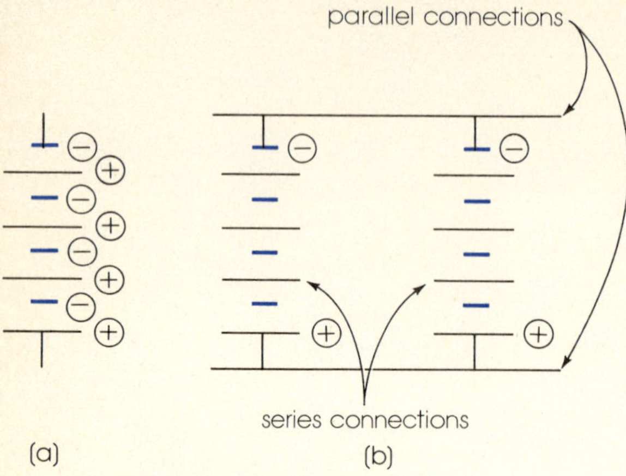

Figure 19-13 Electric circuits. (a) A series connection (anode to cathode) of four galvanic cells. The total emf is the sum of the individual emfs. (b) Parallel connection (positive terminal to positive terminal, negative to negative) of cells which themselves are connected in series, a combination called series-parallel. The total current is the sum of the currents individually produced by the cells. Nerve nets in electric fishes resemble series-parallel arrangements.

hundreds of thousands of electroplaques can be stimulated simultaneously. The organ then unloads its accumulated charge in a few thousandths of a second.

Electroplaques are connected in series (end-to-end), an arrangement that makes their voltages additive (Figure 19-13a). In a large eel they occur on both sides of the body as some 70 stacks containing up to 10,000 electroplaques each. Given a potential of 0.08 volt per electroplaque, maximum voltage on dry land is $0.08 \times 10^4 = 800$ volts.

Stacks are connected in parallel (side-by-side), an arrangement in which the current becomes the sum of the currents from the stacks (Figure 19-13b). An eel's current is low, about 1 ampere. In water its maximum power output is about

$$1\ amp \times 400\ V = 400\ W$$

That is enough to kill fish and to give a human a strong shock.

19-6. CORROSION

We will define *corrosion* (L. *corrosio*, eating away) as chemical attack on metal by its environment. Corrosion products are generally hydroxides and oxides because most corrosion is caused by water and air. In 1824 Davy proved corrosion was an electrochemical process that could be countered by reversing its galvanic action. Corrosion mechanisms are studied today so metals can be wisely selected for specific uses and protected from attack. Losses to corrosion in the United States total $6 billion annually.

Corrosion cells

Two dissimilar metals in contact with one another and with an electrolyte comprise a *voltaic corrosion cell* (Figure 19-14). The more active metal anode corrodes as corrosion current electrons leave it for the less active metal cathode which remains intact. In general the greater the difference in the two oxidation potentials, the faster the corrosion, although the metal activity series (Table 11-5) order is not the same in every environment.

Corrosion around impurities, scratches, strains (Figure 19-15) and other metal imperfections in the presence of moisture can often be attributed to

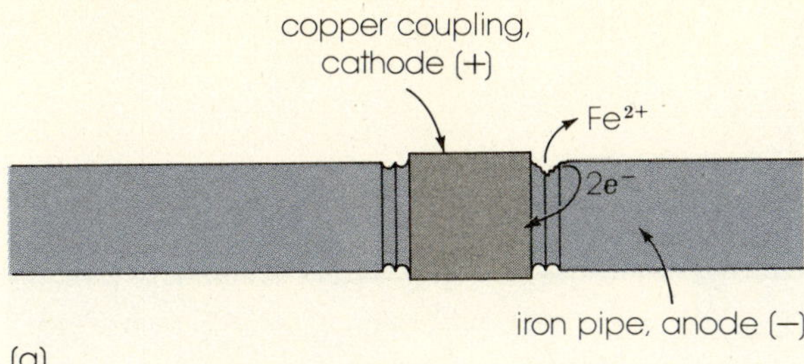

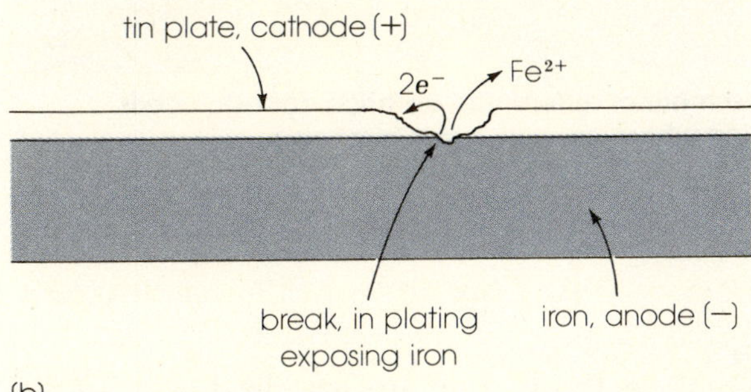

Figure 19-14 Voltaic corrosion cells. (a) Iron is more active than copper: iron dissolves, copper remains intact. The coupling should have been made of iron. (b) Most plating contains tiny holes. When, as here, the plating is less active than base metal, base metal corrodes, starting in the holes.

an oxygen corrosion cell. For instance,

$$\text{at anode:} \qquad 2Fe \longrightarrow 2Fe^{2+} + 4e^- \qquad\qquad (19\text{-}5)$$

$$\text{at cathode:} \quad 4OH^- \longleftarrow O_2 + 2H_2O + 4e^- \qquad\qquad (19\text{-}6)$$

followed by,

$$2Fe^{2+} + 4OH^- = 2Fe(OH)_2$$

$$4Fe(OH)_2 + O_2 = \underset{rust}{2Fe_2O_3 \cdot H_2O} + 2H_2O$$

Beneath the rust, corrosion continues.

A metal in contact with an electrolyte containing dissolved air at two different concentrations corrodes because a different potential exists in each region. Paradoxically, in a *differential aeration corrosion cell* the region of higher O_2 is cathodic, that of lower O_2 anodic. Consider a piece of iron standing in salt water. At any point some hydrogen ions exist so these reactions are possible:

$$Fe \longrightarrow Fe^{2+} + 2e^- \qquad and \qquad 2H \longleftarrow 2H^+ + 2e^-$$

Hydrogen clings to the iron and *polarizes* it; that is, hydrogen atoms tend to deactivate the surface and halt further reaction. Where air is relatively

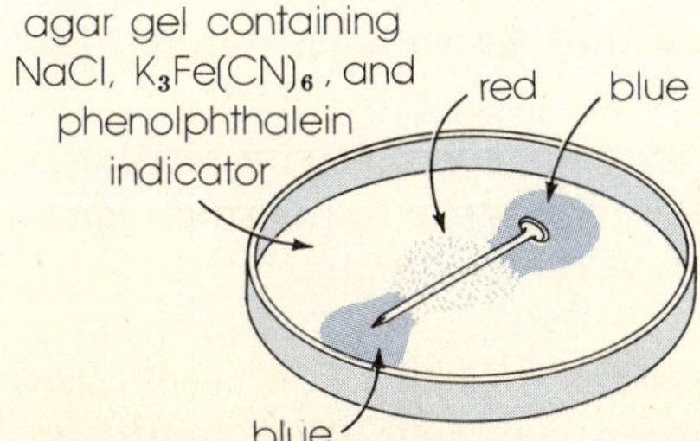

Figure 19-15 Demonstration of an oxygen corrosion (strain) cell. An ordinary iron nail is placed in a dish of gel containing the ingredients indicated. Nail head and point, having been strained during manufacture, are more active than the shaft. The ends of the nail, being anodic, liberate ferrous ions (Eq. 19-5) which combine with ferricyanide giving blue KFe[Fe(CN₆)]. The shank of the nail, being cathodic, causes hydroxyl ions to form. (Eq. 19-6) which turn the indicator red.

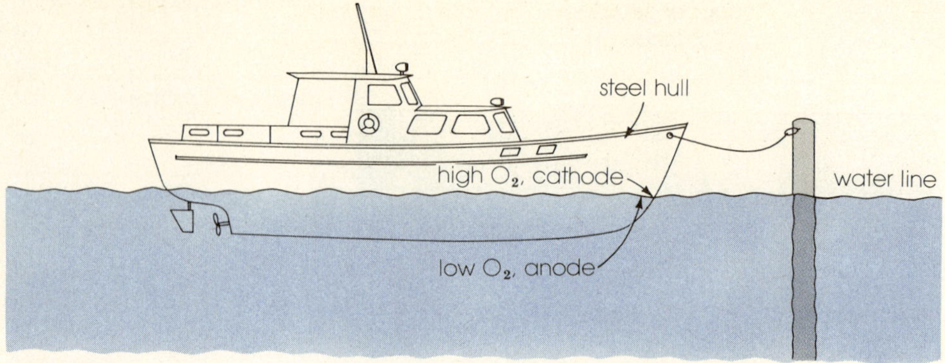

Figure 19-16 Conditions for establishment of differential aeration corrosion cells. Rusting occurs below the water line where oxygen is less available.

abundant, however, O_2 can react with the hydrogen layer and reactivate the surface by depolarizing it:

$$4H + O_2 = 2H_2O$$

Hydrogen atoms needed to replace those used up are produced as the region with more O_2 pulls electrons from the region of lower O_2. Corrosion occurs in the latter place (Figure 19-16).

Corrosion protection

Metal surfaces may be protected from corrosion by (1) coatings (electroplating, rubber lining, etc.), (2) *passivation* (treating with an oxidizer like HNO_3 which is thought to create a thin oxide film that lowers the surface's oxidation potential), (3) dehumidifying the air, (4) removing oxygen dissolved in the water, and (5) using *cathodic* protection. Cathodic protection is accomplished by attaching to the surface either a more active (sacrificial) metal that will corrode preferentially, or a source of direct current whose current flow is directed to oppose the flow of corrosion current (see Question 9).

19-7. WRITING REDOX EQUATIONS

Balancing equations; the half-reaction method

Redox is a contraction of the words "reduction" and "oxidation." In a redox reaction electron transfer between atoms occurs. The electron donor is called the *reducing agent,* and the electron acceptor is called the *oxidizing agent.* As the reducing agent gives up electrons, it is oxidized and its valence number becomes more positive (or less negative). As the oxidizing agent accepts electrons, it is reduced and its valence number becomes less positive (or more negative).

Some redox equations can be balanced by inspection. In the synthesis of a compound from its elements, for instance, we write the correct equation by merely knowing the valence numbers of the elements:

$$Mg + Cl_2 = MgCl_2$$

Typically, the metal is the reducing agent and the nonmetal is the oxidizing agent. The metal donates electrons to the nonmetal and is oxidized. The nonmetal accepts the electrons and is reduced.

The equation for displacement of a metal from solution by a more active metal such as

$$Al + 3Ag^+ = 3Ag + Al^{3+}$$

can also be written correctly on inspection by knowing valence numbers and balancing the total charge (in this case 3+) on both sides.

For more complex redox equations we need a systematic approach to balancing. The system we will use, called the *half-reaction method,* has already been informally introduced (Eq. 19-2). It is applied as follows:

1. Divide the equation to be balanced into two half-reactions, one for the oxidation, the other for the reduction. (Examples are given in Table 19-1.)
2. Balance, by itself, each half-reaction with respect to atoms and charges. One or more electrons will be included.
    To balance an incomplete half-reaction, it may be necessary to add H_2O, or H^+, or both, to either side if the reaction takes place in acid solution, and to add H_2O, or OH^-, or both, to either side if the reaction proceeds in base solution.
3. Balance the two half-reactions with one another by balancing the number of electrons donated in the oxidation with the number of electrons accepted in the reduction. This will mean multiplying each term in a half-reaction by the required integer.
4. Add the half-reactions to get the overall reaction equation. The electrons will cancel.

Example 19-3 In acid solution oxygen gas reacts with ferrous ions. The products are water and ferric ions. Use the half-reaction method to write the overall net ionic redox equation.

Solution We find the half-reactions in Table 19-1. We note there that the ferrous-ferric half-reaction has the larger potential. Following the convention established in Example 19-1, we write it as an oxidation (as given) and reverse the oxygen half-reaction. Ferrous ion will be spontaneously oxidized as oxygen gas is reduced:

$$Fe^{2+} \longrightarrow Fe^{3+} + e^-$$
$$2H_2O \longleftarrow O_2 + 4H^+ + 4e^-$$

To obtain electron balance we multiply the first half-reaction by 4. Addition gives the overall net ionic equation (electrons cancel):

$$4Fe^{2+} + O_2 + 4H^+ = 4Fe^{3+} + 2H_2O$$

Example 19-4 Complete and balance these half-reactions (acid solution):

$$Ti^{2+} \longrightarrow Ti^{3+}$$
$$BiO^+ \longleftarrow Bi + H_2O$$

Solution The first is easy; addition of one electron to the right side balances the charges (2+ on each side):

$$Ti^{2+} = Ti^{3+} + e^-$$

In the second half-reaction we note hydrogen appears only on the right side. With acid solution specified, we may add $2H^+$ to the left side. That furnishes the needed hydrogen and balances the atoms:

$$BiO^+ + 2H^+ \longrightarrow Bi + H_2O$$

The left side now has a total charge of 3+ and the right side has zero charge. We balance charge with electrons. Addition of three electrons to the left gives the balanced equation:

$$BiO^+ + 2H^+ + 3e^- = Bi + H_2O$$

Example 19-5 Balance the following half-reactions (acid solution). Then combine them to get a balanced redox equation in which antimonyl ion is oxidized by manganese dioxide:

$$SbO^+ \longrightarrow Sb_2O_5 + H^+$$
$$MnO_2 \longrightarrow Mn^{2+} + H_2O$$

Solution The first half-reaction has hydrogen only on the right side. It also has two antimony atoms on the right but only one on the left. We will first furnish hydrogen atoms to the left side by the addition of water and multiply SbO^+ by 2 for a balance of the antimony atoms:

$$2SbO^+ + H_2O \longrightarrow Sb_2O_5 + H^+$$

To balance the five oxygen atoms of the right side, we need three H_2O molecules on the left. The six hydrogen atoms introduced we balance by multiplying H^+ on the right side by 6:

$$2SbO^+ + 3H_2O \longrightarrow Sb_2O_5 + 6H^+$$

Atoms are now balanced but charges are not. We add four electrons to the right side to give a total charge of 2+ on both sides:

$$2SbO^+ + 3H_2O = Sb_2O_5 + 6H^+ + 4e^-$$

The second half-reaction requires multiplication of H_2O by 2 to balance the two oxygen atoms on the left side. Four H^+ must then be added on the left to balance the four hydrogen atoms on the right:

$$MnO_2 + 4H^+ \longrightarrow Mn^{2+} + 2H_2O$$

The atoms are now balanced. Charge balance is attained by adding two electrons to the left, giving a net 2+ charge to both sides:

$$MnO_2 + 4H^+ + 2e^- = Mn^{2+} + 2H_2O$$

The two balanced half-reactions are mutually balanced by multiplying the latter by 2, equalizing electron loss and gain between them. Addition gives

$$2SbO^+ + 3H_2O + 2MnO_2 + 8H^+ + 4e^- =$$
$$Sb_2O_5 + 6H^+ + 4e^- + 2Mn^{2+} + 4H_2O$$

Cancellation of electrons and elimination of duplication in water molecules and hydrogen ions gives the overall net ionic redox equation:

$$2SbO^+ + 2MnO_2 + 2H^+ = Sb_2O_5 + 2Mn^{2+} + H_2O$$

QUESTIONS

1. Cells **(a)** Distinguish between an electrolysis cell and a voltaic cell. **(b)** Is the following statement true? In every half-cell of a voltaic cell there is an element in two different oxidation states. Explain.

2. Laboratory cells What is a salt bridge? If omitted from the setup of Figure 19-1, what would be the result? Explain.

3. Standard potentials **(a)** What is an oxidizing agent? a reducing agent? **(b)** Suppose you are given a strip of cadmium metal and some cadmium chloride. Explain how you could experimentally determine that $E°$ for $Cd = Cd^{2+} + 2e^-$ is 0.40 V.

4. Oxidation potentials Refer to Table 19-1. **(a)** At standard state which is the stronger oxidizing agent: nitrate ion or ozone? **(b)** Which is the stronger reducer: stannous ion or ferrous ion? Explain.

5. Circuits When a light bulb at home burns out, other lights and electrical outlets continue functioning normally. Does that indicate the wiring is in series or in parallel? Explain.

6. Reaction direction **(a)** With reference to Table 19-1 what are standard state conditions? **(b)** Given a (whole) reaction at standard state such as

$$Cl_2 + Sn^{2+} = 2Cl^- + Sn^{4+}$$

how can one tell in which direction it will go spontaneously? Explain. In which direction does this reaction go?

7. Bioelectricity Electric fishes in the sea have high amperage–low voltage systems, whereas river-dwelling electric fishes have high voltage–low amperage systems. The two types are thus adapted for their environments. Explain.

8. Metal surface Nitric acid, a strong oxidizer, can be shipped in alumi-

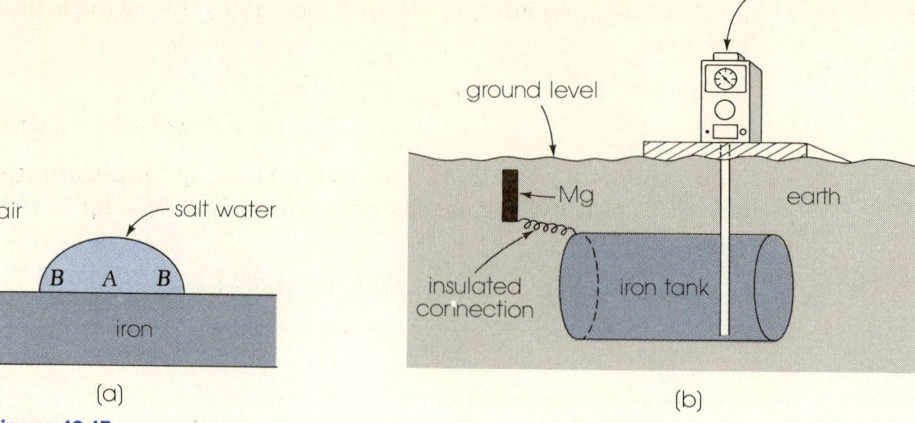

Figure 19-17

num tank cars. Hydrochloric acid put into the same tanks dissolves the aluminum. Explain.

9. Corrosion **(a)** Consider Figure 19-17a. Which area, A or B, will be anodic and corrode? Explain. **(b)** Consider Figure 19-17b. Explain how the magnesium bar prevents corrosion of the tank.

10. Corrosion **(a)** Four million pounds per year of Cu_2O is incorporated into marine paints in the United States to kill barnacles and other growths on ship hulls. Its disadvantage is this: When a break occurs on a painted iron surface, iron corrosion starts at that point. Explain. **(b)** An iron nail is pulled from an old fence. The head of the nail, which had been exposed to the weather, is relatively uncorroded compared to the end which was in the wood. Explain.

PROBLEMS

11. Practical cells Half-reactions for the dry cell and lead storage cell are given in the chapter. Derive the whole-cell equations and values of E^o_{cell}.

12. Aluminum (laboratory) Weigh an empty aluminum beverage can to the nearest gram. **(a)** How many cans may be made from 1 ton of aluminum? **(b)** If present production in the United States is 10 million such cans per year, how many tons of aluminum is used for this purpose? **(c)** If 1.7×10^4 kWh of electricity is needed to reduce 5 tons of bauxite ore to 1 ton of aluminum metal, how many kilowatt-hours correspond to your answer in (b)? **(d)** Assuming that the average household used 8.4×10^3 kWh/yr, how many households, on a yearly electric power demand basis, are equivalent to your answer in (c)?

13. Fuel cells A fuel cell that can operate with propane (C_3H_8) fuel, oxygen as the oxidizer, porous platinized electrodes, and an electrolyte of 3 M sulfuric acid at 85°C uses these reactions:

at anode: propane + water ⟶ carbon dioxide + hydrogen ions + electrons
at cathode: water ⟵ oxygen + hydrogen ions + electrons

(a) Translate the word "equations" into two chemical equations and balance them mutually. **(b)** Draw a schematic picture of the cell and label it. **(c)** A similar cell uses methyl alcohol (CH_3OH) instead of propane for fuel. Write anode and cathode equations and balance them mutually.

14. Balancing equations For each of the following, balance the half-reactions in the manner of Example 19-4: **(a)** $Br_2 + H_2O \rightarrow BrO_3^- + H^+$; **(b)** $H_2SeO_3 + H_2O \rightarrow SeO_4^{2-} + H^+$.

15. Balancing equations As in problem 14: **(a)** $Se + H_2O \rightarrow H_2SeO_3 + H^+$; **(b)** $V^{3+} + H_2O \rightarrow VO^{2+} + H^+$.

16. Redox equations Use Table 19-1 and the method of Example 19-5 to write complete net ionic equations: **(a)** Bromine will oxidize hydrogen sulfide; products are bromide and hydrogen ions and sulfur. **(b)** Chlorine will oxidize stannous ions; the only products are chloride and stannic ions. **(c)** In acid solution ozone will oxidize nitrous acid; products are oxygen and water, and nitrate and hydrogen ions.

17. Half-reactions Complete each one, as needed, and write balanced half-reaction equations: **(a)** $Ti^{2+} \rightarrow Ti^{4+}$; **(b)** $MnO_2 + H_2O \rightarrow MnO_4^- + H^+$; **(c)** $H_2O \rightarrow H_2O_2 + H^+$.

18. Practical cell The Edison storage cell employs 21 percent KOH as the electrolyte. Electrodes are steel plates, one coated with finely divided iron, the other with hydrated nickel(IV) oxide. Half-reactions and $E°$ values in base (B) solutions are

$$Fe + 2OH^- = Fe(OH)_2 + 2e^- \qquad E_B° = 0.88 \ V$$
$$Ni(OH)_2 + 2OH^- = NiO_2 + 2H_2O + 2e^- \qquad E_B° = -0.49 \ V$$

(a) Designate anode and cathode. **(b)** Write the overall cell reaction equation. **(c)** Calculate E_{cell}. **(d)** Calculate the voltage developed by three Edison cells in series.

19. Reaction feasibility As in Example 19-2, find $E_{cell}°$: **(a)** A chemist wants to devise a color test for ozone in the air. Is O_3 a strong enough oxidizer to convert iodide ion to (colored) iodine? **(b)** Copper oxide is treated with sulfuric acid, giving Cu^{2+} solution. Can copper metal be precipitated by adding tin metal?

20. Reaction feasibility As above: **(a)** A solution open to air contains acid and stannous ions. Can atmospheric oxygen cause oxidation to stannic ions (and give water as a by-product)? **(b)** An iron tank is painted with a primer containing finely divided metallic cadmium ($Cd = Cd^{2+} + 2e^-$, $E° = 0.40 \ V$). If a break occurs in the paint will cadmium at that point protect the iron from corroding?

21. Cell construction A cell is abbreviated as follows:

$$Sn \mid Sn^{2+}\,(1\ M) \parallel Mg^{2+}\,(1\ M) \mid Mg$$

(a) Draw a picture of it similar to Figure 19-1. Label anode and cathode. Show the direction of electron flow in the wire, and ion flow in the salt bridge. **(b)** Calculate E°_{cell}.

22. Battery A certain commercial 12-V automobile battery costs \$30. It weighs 22.2 lb empty and requires 7.8 lb of sulfuric acid solution to fill. At 80°F it has a peak (rated 100 percent) power output of 3000 W. At −20°F the power output is 1350 W. Calculate **(a)** current output; **(b)** power density in watts per pound (filled); **(c)** cost in dollars per pound; **(d)** percent efficiency at −20°F.

EIGHT

ORGANIC CHEMISTRY— WHAT ARE THE COMPOUNDS OF CARBON?

One of the major plants where synthetic organic chemicals are made (at Bishop, Texas). The plant produces pentaerythritol for paints and lubricants from acetaldehyde and formaldehyde:

$$H_3CC(=O)H + HC(=O)H \longrightarrow C(CH_2OH)_4.$$

(Courtesy Celanese Corporation of America.)

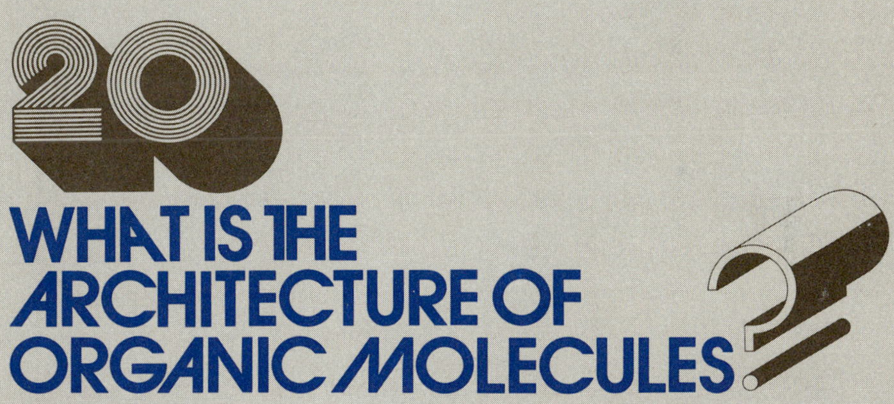

WHAT IS THE ARCHITECTURE OF ORGANIC MOLECULES?

THE IMPORTANT CONCEPTS

20-1 Establishing molecular formulas

1. Organic compounds may be analyzed via combustion.
2. Organic chemistry advanced through the overthrow of vitalism and advent of valence theory.
3. Graphical formulas helped explain addition and substitution reactions.
4. The Kekulé benzene model and Körner's method clarified isomerism and the character of aromatic compounds.

20-2 Stereochemistry

1. Organic molecules require viewing in three dimensions.
2. Asymmetric substances can exist in right- and left-handed forms that are mirror images of each other.

3. Asymmetric molecules are optically active; they can cause rotation of plane polarized light.
4. A polarimeter is used to measure optical rotation.
5. Several ways are known to resolve a dl (racemic) mixture.
6. Asymmetry is a consequence of carbon's tetrahedral structure.
7. Asymmetry can be established by a superposing test of models or by geometric examination of a model.
8. Attempts at total asymmetric synthesis have failed.
9. Geometrical isomerism is associated with nonrotation about the C=C bond.
10. C_5 rings and larger are puckered and strain-free.
11. Conformational analysis is a way of relating structure to properties.
12. Nuclear magnetic resonance (NMR) spectroscopy can give information on the structural environment and number of hydrogen atoms in a molecule.

20-3 Structure and atomic-orbital theory

1. Overlap of two s orbitals or an s and a p orbital gives a σ bond.
2. Lateral overlap of two p orbitals gives a π bond.
3. Orbital theory accounts for the geometry and bond multiplicity in alkanes, alkenes, and alkynes.
4. Benzene stability is correlated with π-electron delocalization.

The number of known chemical substances is about 6 million. Perhaps 95 percent are organic compounds, which means the chemical combinations of carbon exceed those of all other elements by a ratio of 20:1. The remarkable ability of carbon atoms to bond covalently with one another makes possible an almost infinite array of frameworks on which other atoms (H, O, N, S, particularily) may attach. Correlation of organic structural features has kept pace with synthesis through development of theory which today treats both a molecule's disposition in three dimensions and the quantum-mechanical description of its electrons.

20-1. ESTABLISHING MOLECULAR FORMULAS

Carbon–Hydrogen analysis

Organic chemistry began about 1800 in attempts to explain the "kingdoms of animals and vegetables." Animal chemists experimented with such things as body fluids and the technology of making soap from tallow. Plant chemists worked with dyes, sugars, oils, gums, and other materials extracted from plants. Laboratory synthesis of organic substances was not seriously attempted. With natural products coming from living matter, it was assumed a divine vital force, *vis vitalis,* was necessary for their creation.

Little progress was made in establishing the constitution of organic compounds until chemists found how to determine percentages of carbon and hydrogen through analysis by combustion. When reliable atomic weights and methods for determining molecular weights became available, the percentage data were used to calculate formulas. The method, developed first by Lavoisier, then by Dumas, Berzelius, and J. von Leibig, is illustrated below.

Example 20-1 Alcohols contain carbon, hydrogen, and oxygen. An alcohol formed from corn (by the action of yeast) and isolated by distillation is combusted in a stream of oxygen. From 0.500 g of alcohol come 0.585 g of water (absorbed in a tube of granular $CaCl_2$) and 0.957 g of carbon dioxide (absorbed in a tube of $NaOH + CaO$). In a separate analysis a sample of alcohol is vaporized and its vapor density found to be 2.07 g/liter at STP. Find the percentage elemental composition, the simplest (empirical) formula, and the molecular formula of the alcohol.

Solution The weights of hydrogen and carbon in 0.500 g of alcohol are found as follows:

$$? \text{ g H} = 0.585 \text{ g } H_2O \left(\frac{(2)(1.01 \text{ g H})}{18.0 \text{ g } H_2O} \right) = 0.0657 \text{ g}$$

$$? \text{ g C} = 0.957 \text{ g } CO_2 \left(\frac{12.0 \text{ g C}}{44.0 \text{ g } CO_2} \right) = 0.261 \text{ g}$$

The percentages of hydrogen and carbon in the sample are

$$\frac{0.0657 \times 100}{0.500} = 13.1 \text{ percent H}$$

$$\frac{0.261 \times 100}{0.500} = 52.2 \text{ percent C}$$

Oxygen was not analyzed. Its percentage is found by difference:

$$100 - (13.1 + 52.2) = 34.7 \text{ percent O}$$

The next step is to establish the *atom ratio* of the three elements. For convenience we will assume having 100 g of compound. By dividing the weight of each element in 100 g by its gram-atomic weight, we obtain the moles or gram-atoms of the elements. The gram-atom ratio is the same as the atom ratio. To reduce the ratio to that of small whole numbers, we divide each value by the smallest,

$$\frac{13.1 \text{ g H}}{1.01 \text{ g/g-atom}} = 13.0 \text{ g-atom}; \qquad \frac{13.0}{2.17} = 6.0$$

$$\frac{52.2 \text{ g C}}{12.0 \text{ g/g-atom}} = 4.35 \text{ g-atom}; \qquad \frac{4.35}{2.17} = 2.0$$

$$\frac{34.0 \text{ g O}}{16.0 \text{ g/g-atom}} = 2.17 \text{ g-atom}; \qquad \frac{2.17}{2.17} = 1.0$$

The empirical formula is C_2H_6O.

The gram-molecular weight of the vapor is the weight of 22.4 liters at STP:

$$? \text{ g} = 22.4 \text{ liters} \left(\frac{2.07 \text{ g}}{1 \text{ liter}} \right) = 46.4 \text{ g}$$

The empirical weight, calculated by adding up the atomic weights in the formula C_2H_6O, is

$$2 \times 12.0 + 6 \times 1.0 + 1 \times 16.0 = 46.0$$

The experimental molecular weight is also about 46. The accurate molecular weight is therefore 46.0, and the molecular formula is the same as the empirical formula. (The compound is ethyl alcohol, written C_2H_5OH.)

Early theories about carbon compounds

Chemical analysis convinced chemists that organic compounds had definite formulas. The formulas, however, suggested no classification scheme. Lack of system was blamed on nature's vital force until the German chemist

F. Wöhler made the chance discovery in 1828 that ammonium cyanate, a salt regarded as inorganic, could be converted by heat into crystalline urea, a compound classed as organic and previously available only from evaporation of urine:

$$NH_4CNO \overset{\Delta}{=} CO(NH_2)_2$$

As workers compiled organic chemical facts, they voiced ideas for correlating them. One, called the *radical theory*, came from observations by Dumas, Wöhler, and Liebig. It stated that related compounds like wood alcohol (CH_3OH), dimethyl ether (CH_3OCH_3), and methyl chloride (CH_3Cl) contained a common "radical" (here the methyl group) that could be a basis for classification. Another idea, called the *theory of four types*, was set forth by French chemist C. Gerhart. He proposed that organic compounds be thought of as radicals substituted for hydrogen atoms in four simple inorganic molecules: H_2, H_2O, HCl, and NH_3. Thus ethane (C_2H_5H) was "hydrogen type"; ethyl alcohol (C_2H_5OH) was "water type"; ethyl chloride (C_2H_5Cl) was "hydrogen chloride type"; and ethylamine ($C_2H_5NH_2$) was "ammonia type."

A more concrete contribution was made by Hermann Kolbe (Germany 1818–1884), a student of Wöhler's, who showed how chemical formulas could be deduced through stepwise synthesis. His preparation of acetic acid by two routes (Figure 20-1) was the first synthesis since Wöhler's urea synthesis to yield an organic compound from sources called *inorganic*. It persuaded chemists that carbon had a constant power of combination. It also was the decisive blow that ended "vitalism" as a working philosophy and replaced it with "mechanism"—the view that organic chemistry, like the rest of physical science, was governed by testable natural laws.

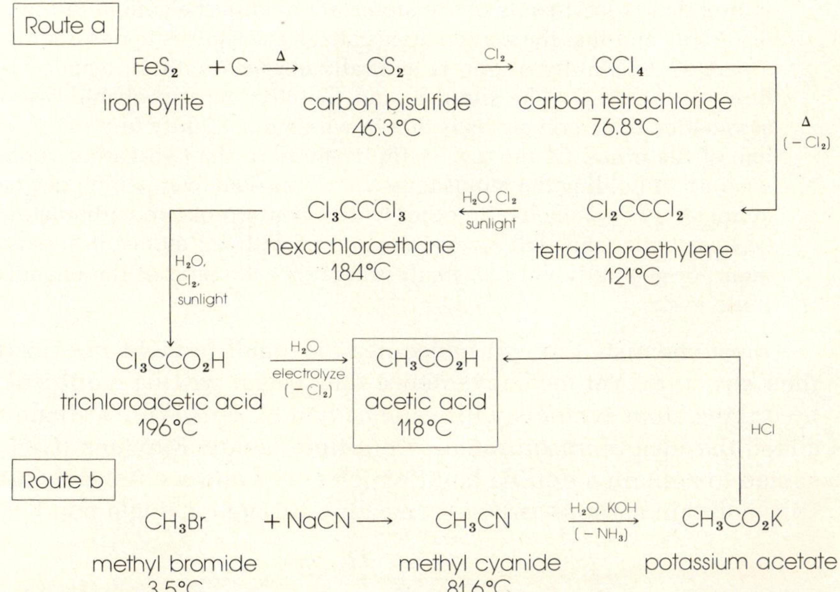

Figure 20-1 Kolbe's syntheses of acetic acid (with boiling points and today's formulas). This work led to the start of industrial organic chemistry by showing that commercially valuable carbon compounds could be built up from cheap raw materials.

Despite slow progress in synthesis, organic chemists were the first scientists to use the powerful valence theory. Its originator was the English chemist E. Frankland. From analysis of *organometallic* compounds like diethyl zinc $(C_2H_5)_2Zn$, he concluded that metals combined with only a certain number of radicals and perhaps every element could be assigned one or more combining numbers or "valences" expressing its saturation in chemical combinations.

In 1858 A. Couper (Scotland) and Kekulé (Figure 20-4), convinced of carbon's quadrivalency, began drawing structural pictures of organic compounds. They represented methane, for example, as

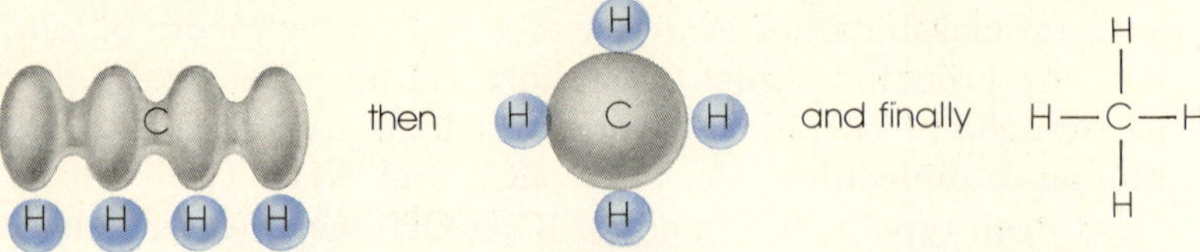

For compounds with more than one carbon atom Kekulé assumed carbon–carbon connections. He had compounds like

$$H-\underset{\underset{H}{|}}{\overset{\overset{H}{|}}{C}}-\underset{\underset{H}{|}}{\overset{\overset{H}{|}}{C}}-H \quad and \quad Cl-\underset{\underset{Cl}{|}}{\overset{\overset{Cl}{|}}{C}}-\underset{\underset{Cl}{|}}{\overset{\overset{Cl}{|}}{C}}-Cl$$

$$ethane \qquad\qquad hexachloroethane$$

in mind when with extraordinary insight for the time he wrote the following, which is still acceptable today:

In the cases of substances which contain several atoms of carbon, it must be admitted that at least some of the atoms are held in the compound by the affinity of carbon, and that the carbon atoms attach themselves to one another, whereby a part of the affinity of one is naturally engaged with an equal part of the affinity of the other. The simplest and therefore most probable case of such an association of carbon atoms is that in which one affinity unit of one is bound by one of the other. Of the 2×4 affinity units of the two carbon atoms, two are used up in holding the atoms together; six remain over, which can be bound by atoms of other elements. In other words, a group of two carbon atoms, C_2, will be hexatomic, and will form a compound with six atoms of a monatomic element, or generally with so many atoms that the sum of the chemical units of these is six.

Most chemists thought molecules too small to yield information about their structure. Yet for convenience they began writing *graphical formulas* by linking atom symbols with valence bonds. To Kekule's single bond was added the idea of *unsaturated or multiple bonds*. Ethylene (C_2H_4) was assumed to contain a *double bond* which could *add* a chlorine molecule and still maintain carbon–carbon connection through a single bond:

$$\underset{H}{\overset{H}{\diagdown}}C\!=\!C\underset{H}{\overset{H}{\diagup}} \;+\; Cl-Cl = Cl-\underset{\underset{H}{|}}{\overset{\overset{H}{|}}{C}}-\underset{\underset{H}{|}}{\overset{\overset{H}{|}}{C}}-Cl \qquad\qquad (20\text{-}1)$$

Acetylene (C_2H_2) was assigned a *triple bond* to account for its ability to add two chlorine molecules yet keep its two-carbon unit:

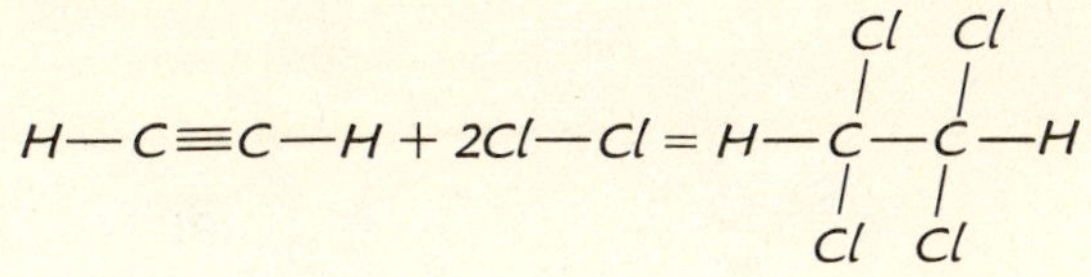

$$H-C\equiv C-H + 2Cl-Cl = H-\underset{\underset{Cl}{|}}{\overset{\overset{Cl}{|}}{C}}-\underset{\underset{Cl}{|}}{\overset{\overset{Cl}{|}}{C}}-H \qquad (20\text{-}2)$$

Benzene and isomerism

Chemists concluded from reactions like those above that *hydrocarbons* should be considered the parent compounds from which other compounds might be derived. They divided hydrocarbons into two classes: *aliphatic* ("fatty") and *aromatic* ("fragrant"). The former, like C_2H_6, with relatively high hydrogen-to-carbon ratios, seemed simple enough chemically. The latter, typified by benzene (C_6H_6) with its relatively low hydrogen-to-carbon ratio, remained to be explained. Then Kekulé, in an inspiration he described as coming to him during a day dream, perceived that if the benzene molecule were a hexagon of carbon atoms held together by alternate double and single bonds, the arrangement would use up three valency units of each carbon, leaving one free for hydrogen attachment:

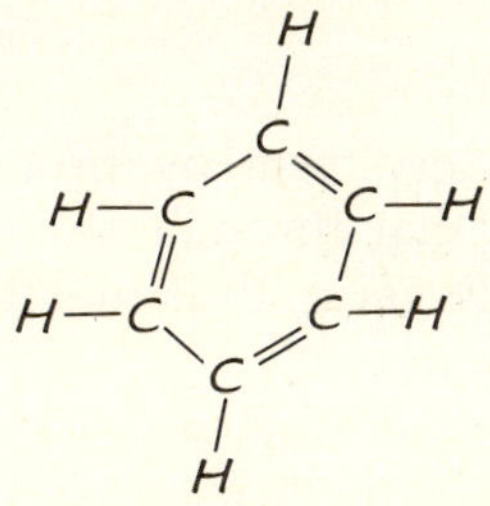

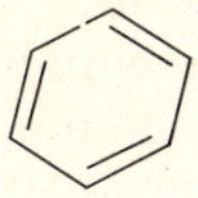

benzene

Using this structure he correctly predicted that only one monosubstituted benzene with a given substituent would be found, because the symmetry of a regular hexagon made replacement of one hydrogen the same as replacement of any other:

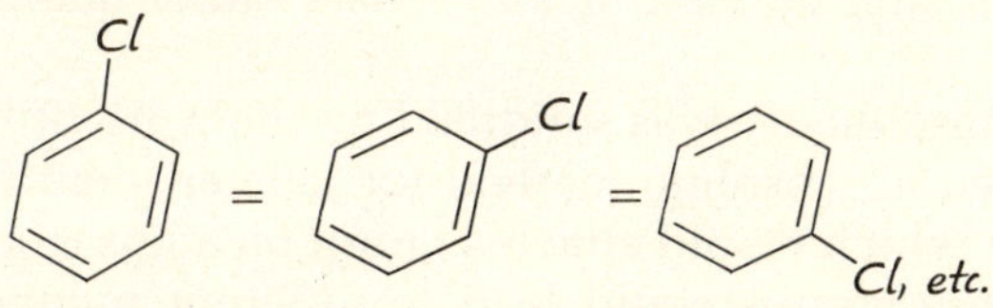

chlorobenzene

Kekulé used his model convincingly to discuss the possibilities of isomers in benzene chemistry—that is, compounds with the same molecular formula but different structure. He stated that only three disubstituted benzenes (such as $C_6H_4Br_2$) could be prepared, because only three different dispositions of two substituents about the ring were possible. Numbering the positions 1 through 6 he designated the isomers 1,4- (also called *para*), 1,2- (*ortho*), and 1,3- (*meta*). Bromination of bromobenzene with powdered iron catalyst was known to give three isomers and he visualized the

process as

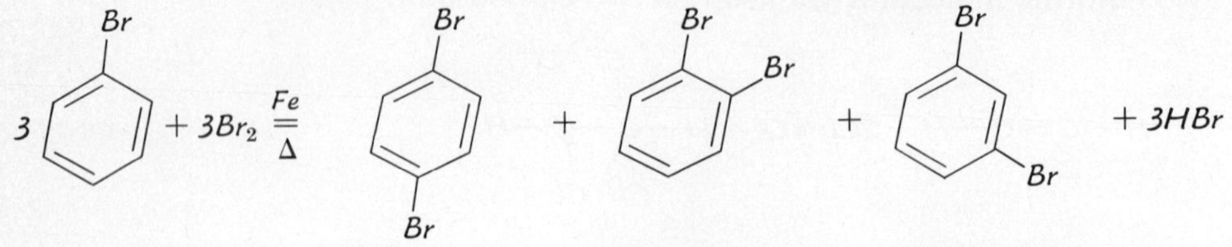

$$3 \text{[benzene]} + 3Br_2 \xrightarrow[\Delta]{Fe} \text{[1,4-dibromobenzene]} + \text{[1,2-dibromobenzene]} + \text{[1,3-dibromobenzene]} + 3HBr \qquad (20\text{-}3)$$

1,4-dibromobenzene   1,2-dibromobenzene   1,3-dibromobenzene

84 percent   12 percent   4 percent yield

Kekulé's contemporaries raised two objections. First, as an unsaturated compound with three ethylenelike double bonds, benzene should readily *add* three molecules of bromine (Eq. 20-1), not undergo *substitution of* bromine for hydrogen. Second, the proposed bond arrangement should lead to more than three dibromobenzenes as exemplified by the following molecules, one of which has a single bond and the other a double bond between the carbon atoms holding the halogen atoms:

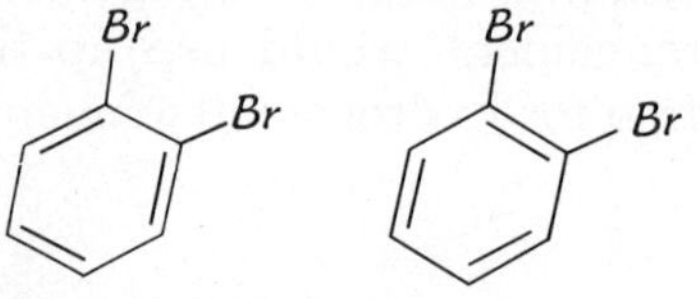

Kekulé gave an ingenious answer to the criticism by proposing an oscillating structure in which he imagined that through atomic vibrations single bonds became double bonds and vice versa, thus making all carbon-carbon bonds in the aromatic ring equivalent:

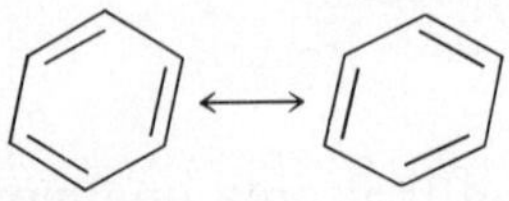

The real molecule, he said, would have a structure between these extremes and possess special stability due to its unusual symmetry. Except for the dynamics, his model is in remarkable accord with the theory of resonance developed 60 years later by Pauling and others out of quantum mechanics (Figure 20-14c).

Kekulé's structure theory was solidified by the work of his assistant W. Körner who devised an absolute method for isomer identification. Körner saw that a reaction which would attach a group on a hexagonal ring already holding one or more groups could lead to different numbers of isomeric products depending on the positions of the original substituents. To illustrate he prepared the three possible dichlorobenzenes, then heated each with a mixture of nitric and sulfuric acids. The reaction introduced a nitro group ($-NO_2$) into each molecule and gave mixtures of dichloronitrobenzenes (in various yields) which he separated by repeated crystallizations. As predicted by Kekulé the *para* dichloride yielded one, the *ortho* dichloride two, and the *meta* dichloride three isomers. He differentiated the products by their melting points:

(20-4)

20-2. STEREOCHEMISTRY

Optical activity

The structural formulas written in the mid-nineteenth century pictured flat molecules. As more compounds became known, the properties of some could not be explained by two-dimensional drawings and chemists concluded that they needed to write stereorepresentations — formulas in three dimensions. Means for examining the *stereochemistry* of molecules were already available in optical science.

Light behaves as if composed of many electromagnetic waves vibrating in every plane perpendicular to its direction of travel. In 1670 physicists M. Bartholinus and C. Huygens discovered, however, that when a beam of ordinary light was passed through a crystal of calcite ($CaCO_3$), it split into two beams, each of which vibrated in only a single direction. They described the beams as being *plane polarized*. In experiments in the early 1800s French physicists E. Malus, D. Arago, and J. Biot passed polarized light through equal-thickness plates cut from certain quartz crystals and found that the light beam was rotated to the right by one and to the left by another. The first they named a *d* or *dextrorotatory* (L. *dexter*, right) crystal, the second an *l* or *levorotatory* (L. *laevus*, left) crystal. Their results coincided with the observation by French mineralogist R. Haüy that two kinds of natural quartz existed, the crystal faces of one arranged as the *mirror image* of the faces of the other (Figure 20-2). Biot then discovered that some pure liquids as well as some compounds dissolved in water were also "optically active"; that is, they rotated a beam of plane polarized light. He surmised that optical activity was associated with the molecules themselves. In 1840 he built the first polarimeter (Figure 20-3) for investigating optical activity.

Rotation caused by an optically active substance depends on the thickness of the substance that the beam of plane polarized light traverses, the wavelength of light used, and the temperature. When the substance is in solution, rotation depends on solution concentration (see Problem 14) and may also depend on the solvent. Rotation occurs because the electric field of the light interacts with the asymmetric electric field of the asymmetric molecules comprising the sample.

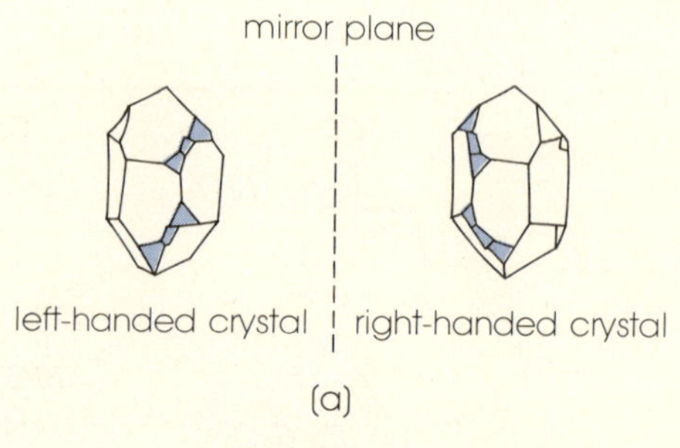

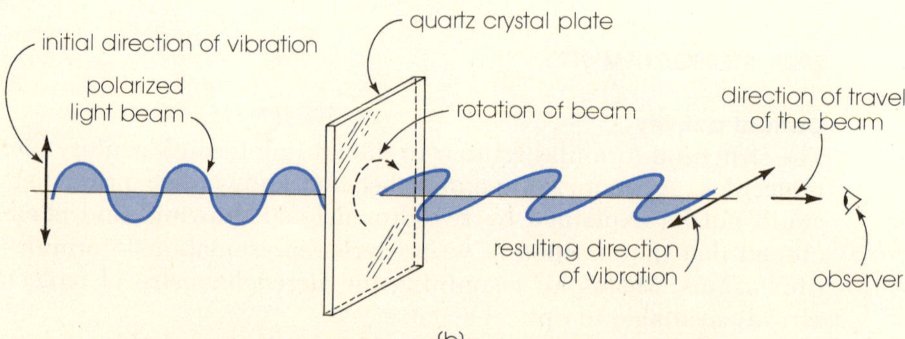

Figure 20-2 Asymmetric crystals. (a) The two crystal-face orientations of natural quartz, SiO_2. The forms bear a mirror or right- and left-hand relationship to one another. (b) A polarized light beam, vibrating in the plane of the paper, is rotated to the right (clockwise) with respect to the head-on observer as it passes through a dextrorotatory quartz crystal plate.

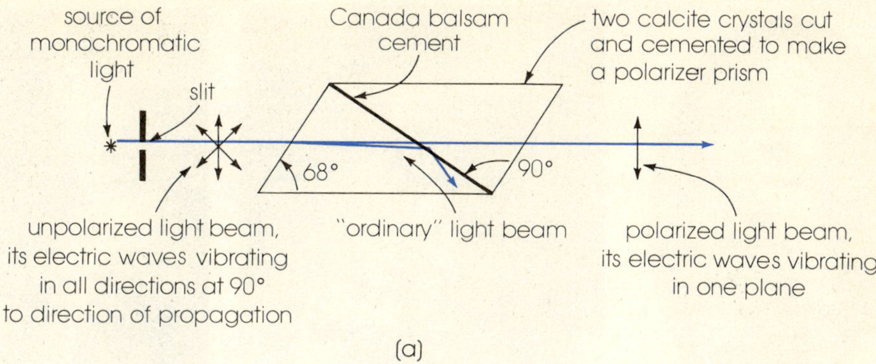

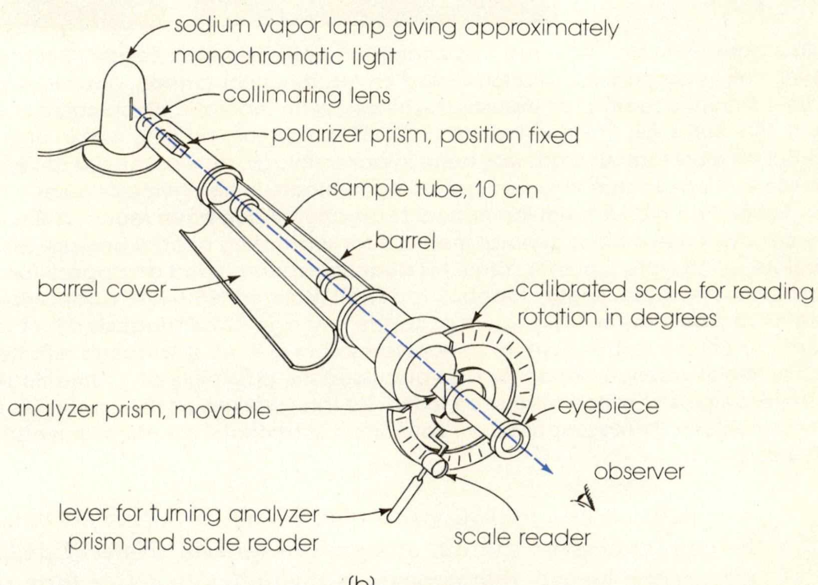

Figure 20-3 Polarimetry. (a) Producing plane polarized light. The polarizer prism's refractive properties produce plane polarized light with a known vibration direction while shunting out and absorbing the remainder of the initial beam. (Light may also be polarized by the use of Polaroid, a plastic film containing oriented crystals of a compound that transmit light vibrating in only one direction.) Partially plane polarized light is common in nature, being found in light from the blue sky and in light reflected from water. Its polarization is not detected by the unaided human eye but many insects and crustaceans apparently orient themselves to it. (b) A polarimeter. Approximately single-wavelength (monochromatic) light is plane polarized by the polarizer. The beam rotates on passing through the sample. The observer turns the analyzer to bring the polarized beam back to its original orientation, then reads the rotation on the scale.

Work of Pasteur

Pasteur (Figure 20-4), called the "father of microbiology" for his germ theory of disease, was also a pioneer stereochemist. For his very first research he chose to study crystals of potassium hydrogen tartrate taken from wine vat residues, seeking to learn why in water solutions one kind of crystal was dextrorotatory whereas a second kind, with the same chemical

Figure 20-4 Pioneers in structural organic chemistry. (Left) Frederich Kekulé (Germany 1829–1896). Kekulé originated notation based on quadrivalent carbon, and his ideas on benzene structure facilitated industrial syntheses with aromatic chemicals derived from coal. (Center) Louis Pasteur (France 1822–1895). Pasteur's training was in physical science but his most famous work was done in bacteriology. He was the first to separate a mixture of dextro and levo isomers, being fortunate in his choice of compounds—fewer than a dozen similar manual separations have been reported since. Later he discovered two other general methods for separating optical isomers. His last words before dying were inquiries about his students' research and an appeal for them to continue working. (Right) Jacobus van't Hoff (Holland 1852–1911, Nobel Prize in chemistry 1901). Van't Hoff was a student of Kekulé's. At age 22 he suggested that the molecular asymmetry responsible for optical isomerism resided in carbon's tetrahedral distribution of valence bonds. He also discussed the possibility of geometrical isomerism, examples of which were soon found. He later worked on the properties of electrolyte solutions. (Photographs Edgar Fahs Smith Memorial Collection, University of Pennsylvania.)

composition, was optically inactive. To obtain larger crystals he converted the compounds to sodium ammonium tartrate ($NaNH_4C_4H_4O_6$). Under the microscope he saw that crystals of the optically active form were all "right handed" whereas the inactive form consisted of an equal number of right- and left-handed crystals. He described how he proceeded with the latter,

> I carefully separated the crystals which were hemihedral* to the right from those which were hemihedral to the left, and examined their solutions separately in the polarizing apparatus. I then saw with no less surprise than pleasure that the crystals hemihedral to the right deviated the plane of polarization to the right, and those hemihedral to the left deviated it to the left; and when I took an equal weight of each of the two kinds of crystals, the mixed solution was indifferent toward the light in consequence of the neutralization of the two equal and opposite deviations.

Theorizing later, he asked,

> Are the atoms of the dextro acid grouped on the sprials of a right-handed helix, or situated at the corners of an irregular tetrahedron, or have they some other asymmetric grouping? We cannot answer these questions. But it cannot be a subject of doubt that there exists an asymmetric arrangement having a non-superposable image. It is not less certain that the atoms of the levo acid possess precisely the inverse asymmetric arrangement.

* A hemihedrally asymmetric crystal has only half the faces needed for symmetry.

Pasteur found that the *d* and *l* forms of a given compound had similar properties, with two important exceptions: (1) They rotated a beam of plane polarized light in opposite directions, and (2) they differed in their reactivities with other optically active compounds or with natural biologic systems having spatial orientation preferences. He demonstrated the latter by allowing *penicillium glaucum*, a mold, to grow in a nonoptically active nutrient mixture containing *dl*-tartaric acid. As the *d* form was consumed more rapidly than the *l* form, the mixture became levorotatory.

Pasteur remained interested in asymmetry throughout his life, wondering what asymmetric forces in nature might have originally induced chemical asymmetry, and how asymmetry continued to be preserved in the molecules of living systems. (We will return to these questions in the chapters on biochemistry.)

The tetrahedral carbon atom

Interpretation of Pasteur's stereochemical work was made in 1874 by van't Hoff (Figure 20-4). Starting with the simpler case of *d*- and *l*-lactic acid [$CH_3CH(OH)CO_2H$] he showed that a tetrahedral disposition of carbon's four valences could account for the acid's existence in two forms (Figure 20-5). In his words,

> In the case where the four affinities of an atom of carbon are saturated by four different univalent groups, two and only two different tetrahedra can be obtained, of which one is the mirror image of the other. . . . Every carbon compound which in solution impresses a deviation on the plane of polarization possesses (such) an asymmetric carbon atom.

Van't Hoff's conclusion is still sound. A carbon atom with four different groups attached is called an *asymmetric center*. It is the essential structural feature of most optically active compounds. We can test for asymmetry by making models of the molecule and its mirror image and seeing if one is superposable on the other (Figure 20-5d,e). If it is not superposable, the molecules constitute an optically active (*dl*) pair. Alternately, we can make a geometrical examination of the molecule's structural formula. If it does not have a point, plane, or line of symmetry, it is asymmetric; it and its mirror image will be an optically active pair. Each of the two isomers is variously called an *optical antipode, enantiomer,* or *enantiomorph* (Gk. *enantios* + *morph,* opposite form).

Attempts to synthesize optically active compounds from totally optically inactive starting materials have all failed. For example, in the bromination of propanoic acid, which produces an asymmetric center (*),

$$H_3C-\overset{\overset{\displaystyle H}{|}}{\underset{\underset{\displaystyle H}{|}}{C}}-CO_2H + 2Br_2 = H_3C-\overset{\overset{\displaystyle Br}{|}}{\underset{\underset{\displaystyle H}{|}}{\overset{*}{C}}}-CO_2H + H_3C-\overset{\overset{\displaystyle H}{|}}{\underset{\underset{\displaystyle Br}{|}}{\overset{*}{C}}}-CO_2H + 2HBr \quad (20\text{-}5)$$

d acid   l acid

bromine has equal opportunity of substituting for either of two hydrogen atoms identically situated on the middle carbon. As a consequence the product is an optically inactive mixture containing equal numbers of *d* and *l* molecules. Such a mixture is called a *racemic* (L. *racemus,* grape) *mixture* or a *racemate.*

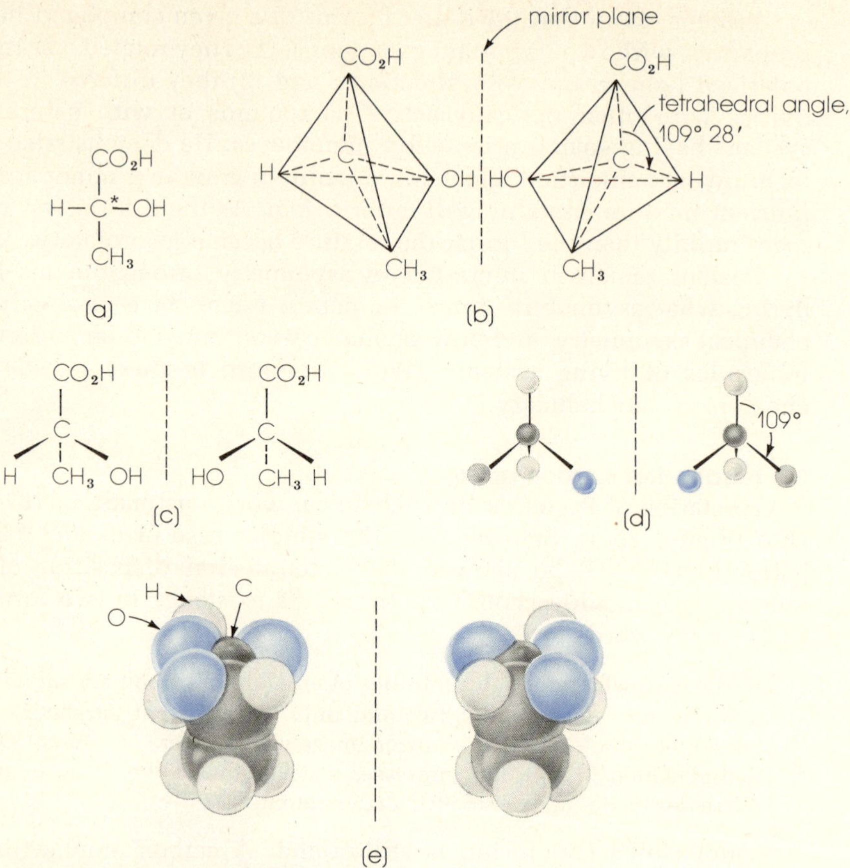

Figure 20-5 The stereochemistry of lactic acid. (a) If the molecule were flat, as this formula implies, lactic acid could only exist in one form because the plane in which the structure lies would be a plane of symmetry (and its mirror image would be superposable). (b) Van't Hoff's tetrahedra. The two molecules are mirror images, and nonsuperposable. They are a dl pair of optical isomers. (c) Projection formulas. The middle carbon, the carbon of the —CO_2H group and the bond connecting them lie in the plane of the paper. Behind the paper is —CH_3, in front are —H and —OH. (d) Ball-and-stick models. (e) Scale models. The scaleup from actual size shown here is about 10^8 times.

For Pasteur's two tartrates van't Hoff pictured the molecules as each composed of two tetrahedra, the bond between central carbon atoms being a union of the tetrahedra at one corner. As shown in Figure 20-6c, one structure is the mirror image of the other and not superposable with it. He also predicted another kind of tartaric acid, later synthesized by Pasteur. A model of it, called *meso* tartaric acid, is shown in Figure 20-6d. This time the two mirror-image structures are not different. If one is inverted, it will be seen to be identical to the other and therefore superposable. The reason for this is indicated in Figure 20-6b. Although the molecule contains two asymmetric carbons, the carbons are bonded to the same four groups, making possible an arrangement in which the upper and lower halves are mirror images of each other. The result is a plane of symmetry through the molecule and optical inactivity.

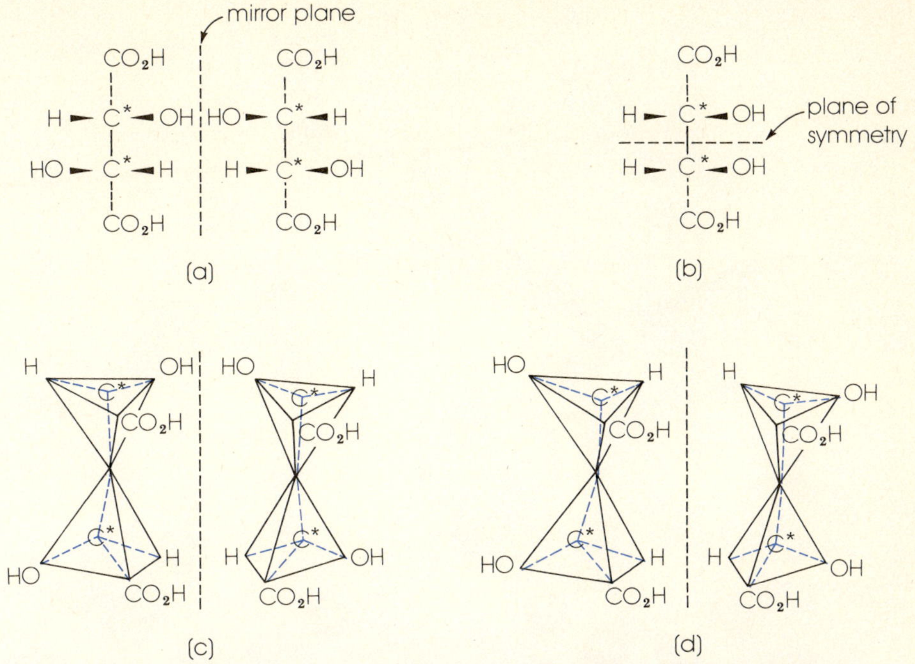

Figure 20-6 The stereochemistry of tartaric acid. (a) d, also called (+), and l, also called (−), tartaric acids. The structures are nonsuperposable mirror images of one another. (b) Meso tartaric acid. It is optically inactive because the molecule has a plane of symmetry. (c) d-tartaric and l-tartaric acids as van't Hoff conceived them. The tetrahedra emphasize the spatial arrangement around the two central carbon atoms. (d) Meso-tartaric acid. The mirror image forms are superposable by rotating one of them through 180 degrees.

Geometrical isomerism

In the same publication in which he related the tetrahedral carbon atom to optical activity, van't Hoff pointed out the possibility of another kind of isomerism which he thought should be common in compounds with carbon–carbon double bonds. If a single carbon–carbon bond could be represented as a freely *rotating* connection of tetrahedra at a common corner, then a double carbon–carbon bond would be a *nonrotating* connection of tetrahedra along a common edge. With two different groups on each of the doubly bonded carbons, two arrangements would be possible. As pictured in Figure 20-7 the structures comprise a pair of *cis* and *trans* forms, as defined previously (Section 10-1); van't Hoff called them an example of *geometrical isomerism*.

The first example of geometric isomers had already been described in the chemical literature but was not recognized as such until van't Hoff proposed his theory. The compounds were fumaric and maleic acids (Figure 20-8a). Van't Hoff assigned the *cis* configuration to maleic acid because of its notable ease of intramolecular dehydration — that is, the loss of water (at 100°C) between its two acid groups and formation of an anhydride (Figure 20-8b). Heating fumaric acid he got no reaction until 275°C, whereupon it rearranged to maleic acid and dehydrated to maleic anhydride.

A *cis* compound is not the mirror image of the corresponding *trans* compound and neither is it optically active unless other features make the

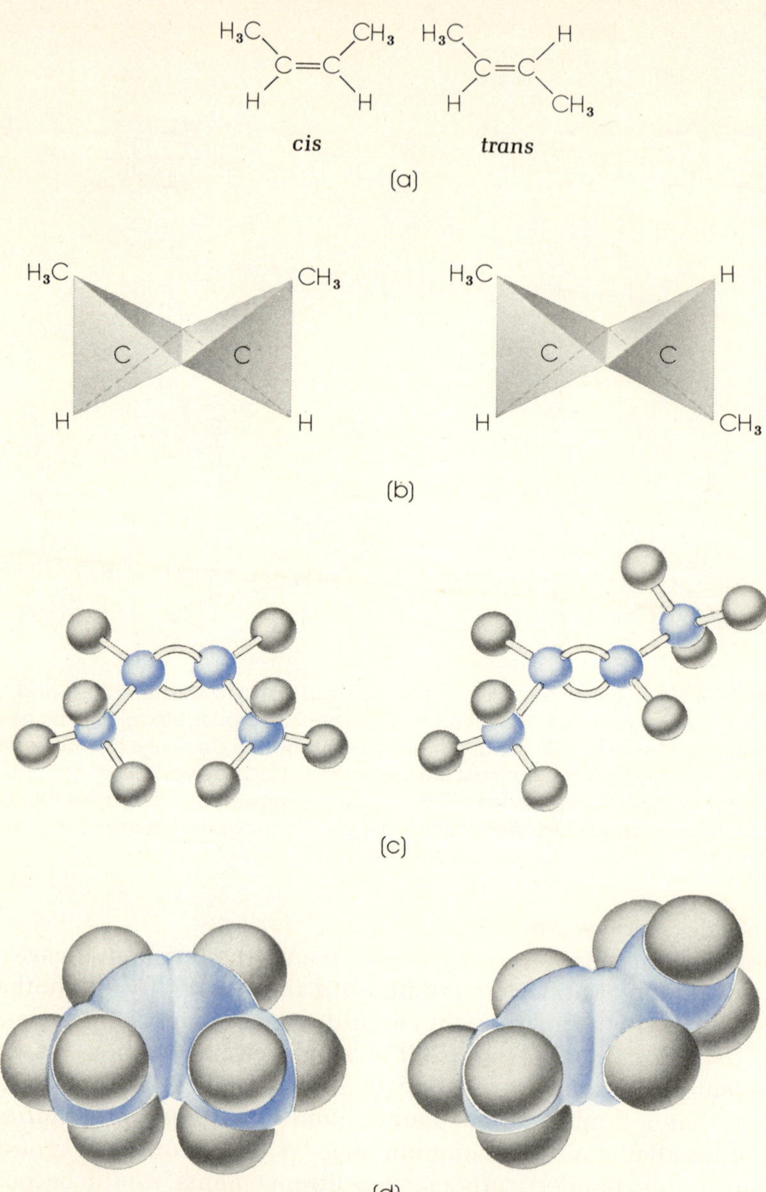

Figure 20-7 Geometric isomers in 2-butene. (a) Either the —CH₃'s or the —H's may be considered the reference groups. The two carbon atoms attached by the double bond and the four atoms attached to them all lie in the same plane. (b) Van't Hoff's visualization of the molecules. The tetrahedra are joined along one edge to represent a double bond with restricted rotation. (c) Ball-and-stick models of the same molecules. (d) Scale models of the same molecules.

molecules asymmetric. The properties of two geometric isomers may differ widely. Maleic acid melts at 130°C and has a solubility of 79 g in 100 g of water at 25°C. The corresponding data for fumaric acid are 270°C and 0.7 g.

Alicyclic compounds and conformational analysis

An *alicyclic* substance is a ring compound having the properties of aliphatic compounds (in contrast to aromatic ring compounds with ben-

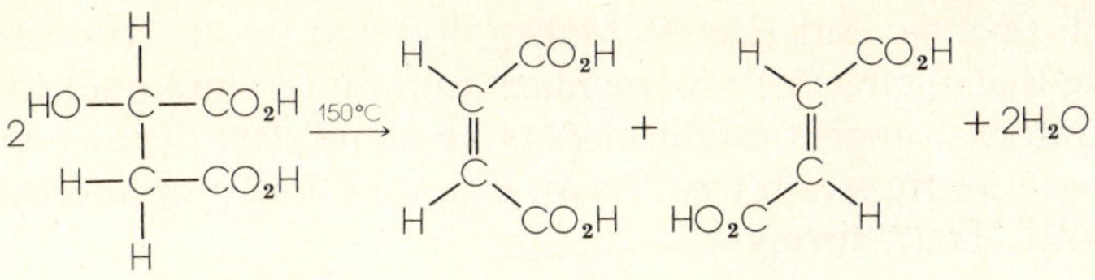

(a)

maleic acid,
4%

fumaric acid,
90%

$+ 2H_2O$

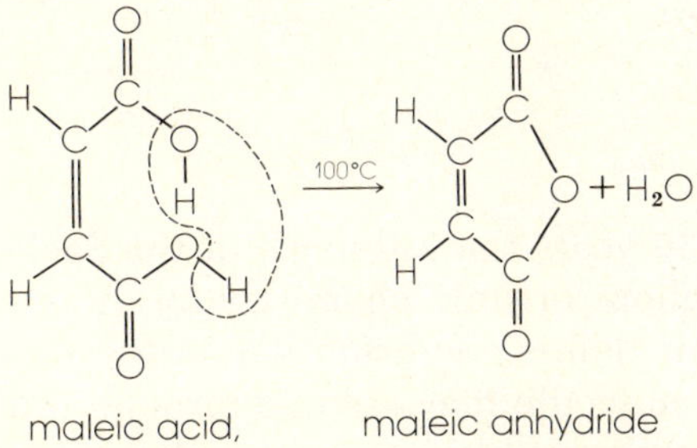

maleic acid,
cis

$100°C$

maleic anhydride

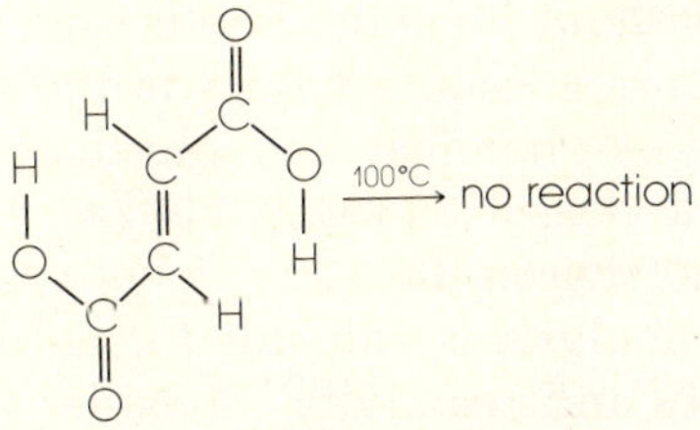

$+ H_2O$

fumaric acid,
(b)   trans

Figure 20-8 Geometric isomerism in maleic and fumaric acids. (a) Preparation from l-malic acid, a constituent of apple juice. (b) The cis isomer, with the acid groups adjacent, readily loses water to form an anhydride. The trans isomer cannot dehydrate within the same molecule because its acid groups are too far from one another and the double bond prevents rotation that would bring them together.

zenelike properties). In 1885 German chemist A. von Baeyer used the tetrahedral carbon concept to explain the overwhelming abundance of five- and six-membered aliphatic rings in natural products compared to rings of other sizes, and the ease of their laboratory preparation. Looking at the internal angles in geometric figures representing flat alicyclics of different sizes, he pointed out that those with five or six carbon atoms came closest to the 109-degree angles of the tetrahedral carbon atom. The difference between the calculated angles and 109 degrees he considered to be a measure of *ring strain* and thus molecular instability.

A few alicyclic molecules, assumed flat, with their internal angles are

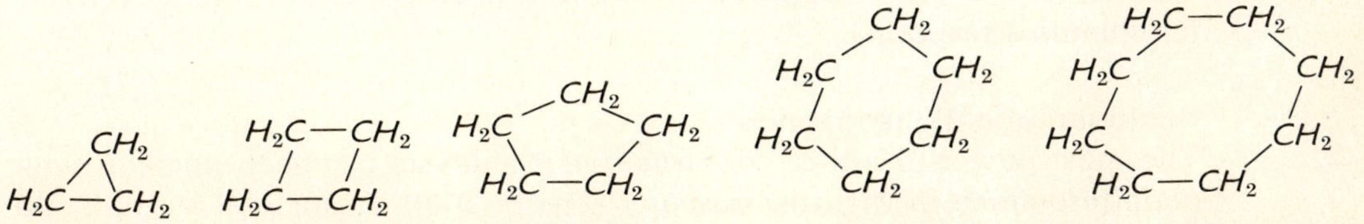

cyclopropane
60°

cyclobutane
90°

cyclopentane
108°

cyclohexane
120°

cyclooctane
135°

Ulrich Sachse, and later E. Mohr, modified the Baeyer strain theory by proposing that if rings of six members and more were *puckered* instead of flat, the internal angles might always be about 109 degrees and the structures consequently strain free. As an example they pictured cyclohexane in "chair" and "boat" forms:

chair   *boat*

In the past 30 years the influence of the Sachse-Mohr theory has been extended throughout organic chemistry by the methods of conformational analysis. We will define *conformation* as the preferred three-dimensional arrangements a molecule may assume through rotation about single bonds. The readily interconvertible forms are called *conformers* or *conformational isomers* to distinguish them from the isomers discussed earlier that cannot be interconverted without breaking bonds and putting the molecules back together again. Conformational analysis is the correlation of a molecule's conformations with its chemical and physical properties. Leaders in the field have been Odd Hassel (Norway 1897–   ) and Derek Barton (England 1918–   ) who shared the 1969 Nobel prize in chemistry.

Conformational analysis is valuable for telling how substitution affects a molecule's stability and reactivity. Consider an —OH group substituted for a hydrogen atom on cyclohexane, giving cyclohexanol (an alcohol). Two of the conformers of cyclohexanol are chair forms:

$$\text{(20-6)}$$

The six bonds that are perpendicular to the ring's "plane" (three directed upward and, alternately, three directed downward) are called *axial*. The other six bonds stick out roughly in the plane of the ring and are called *equatorial*. As a general rule substituents have the most room in equatorial positions, and that conformation in which they are equatorially situated is favored by its lower energy. The equilibrium of Eq. 20-6 lies toward the right, therefore, as the flipping movement of the ring, made possible by rotation about single bonds, converts each axial bond to an equatorial bond, and vice versa. The —OH, in an equatorial position, is also more available for chemical reactions.

Nuclear magnetic resonance

The most powerful aid to conformational analysis has been *nuclear magnetic resonance (NMR) spectroscopy* (Figure 20-9). A nucleus with an odd number of protons, or neutrons, or both (^1_1H, $^{14}_7N$, etc.) acts as if the nuclear charge is spinning around an axis and creating a tiny magnetic field. If such a nucleus is placed in a strong magnetic field it will absorb radio-frequency (RF) energy at specific frequencies and assume orientations that are its allowed quantum states. Reemitted energies, recorded as a series of peaks, make up the spectrum of natural frequencies at which the nucleus resonates.

The hydrogen nucleus has been most studied with NMR. It has only two allowed orientations or spin states. They are described as being aligned with (lower energy) or against (higher energy) the applied magnetic field. Transition from a lower to a higher state requires absorption of energy. The frequency at which an absorption peak appears is related to the structural environment of the hydrogen atom, and is evidence of the number of different structural features in a molecule involving hydrogen atoms. The area under a peak is proportional to the number of hydrogen atoms responsible for the absorption. Given the frequency ranges of structural features of known compounds for comparison, a chemist may identify an unknown compound from its NMR spectrum alone.

When ethane (H_3C—CH_3) is the NMR sample, for instance, only a single peak is obtained because all six hydrogens are equivalent. When propane (H_3C—CH_2—CH_3) is the sample, the spectrum consists of two peaks in the height ratio of 6:2, indicative of six hydrogens of methyl group type (—CH_3) and two of methylene group type (—CH_2—). With cyclohexane (p. 512) we might expect a spectrum consisting of a peak for axial hydrogen atoms and an equal peak for equatorial hydrogen atoms. Such is indeed observed at low temperatures (Figure 20-9c). At higher temperatures only one peak is obtained. This indicates that the molecule is then experiencing conformational changes so fast that hydrogen atoms in only one kind of "average" structural setting can be found.

Conformational analysis has greatly aided our understanding of the complex organic molecules essential to life processes in which structure and function are so intimately related. Cyclohexanelike rings are one of the structural features that have been looked at closely. They occur, for example, in most carbohydrates (Eqs. 23-11 and 23-12) and in steroids (Fig. 23-11).

20-3. STRUCTURE AND ATOMIC-ORBITAL THEORY

The use of atomic-orbital theory in describing covalent bonding was introduced in Section 9-3. Solutions of the Schrödinger wave equation give probabilities of finding the electron within a volume unit called an *atomic orbital*. A covalent bond is formed with a decrease in the system's energy when two atomic orbitals, each occupied by a single electron, "overlap." A coordinate covalent bond is formed when overlap occurs between an empty orbital and an orbital filled with two electrons. We will consider the overlap of two atomic orbitals to result in the bonding electron pair occupying a stable orbital in each of the two bonded atoms. When carbon is the atom bonded, we can assume that it uses hybrid bond orbitals. Depending on how s and p electron functions are mixed, various spatial arrangements are possible.

test tube with sample
magnet pole
S
magnet pole
N
RF receiver
and amplifier
RF transmitter
oscilloscope,
signal display
(a)

(b)

signal
field strength

signal
field strength

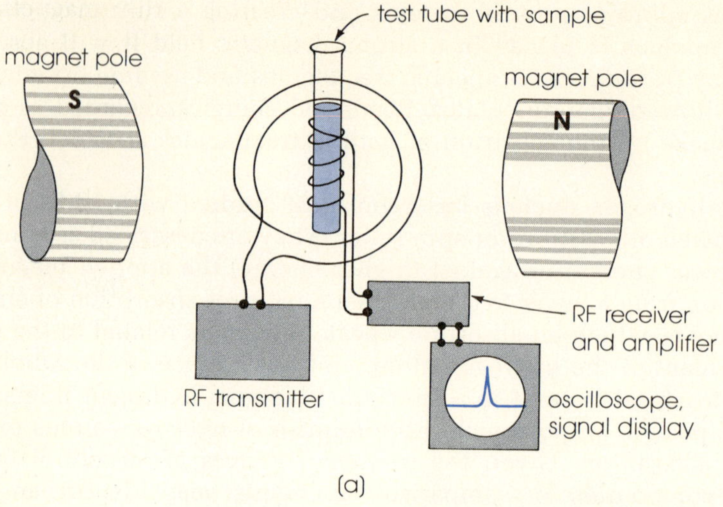

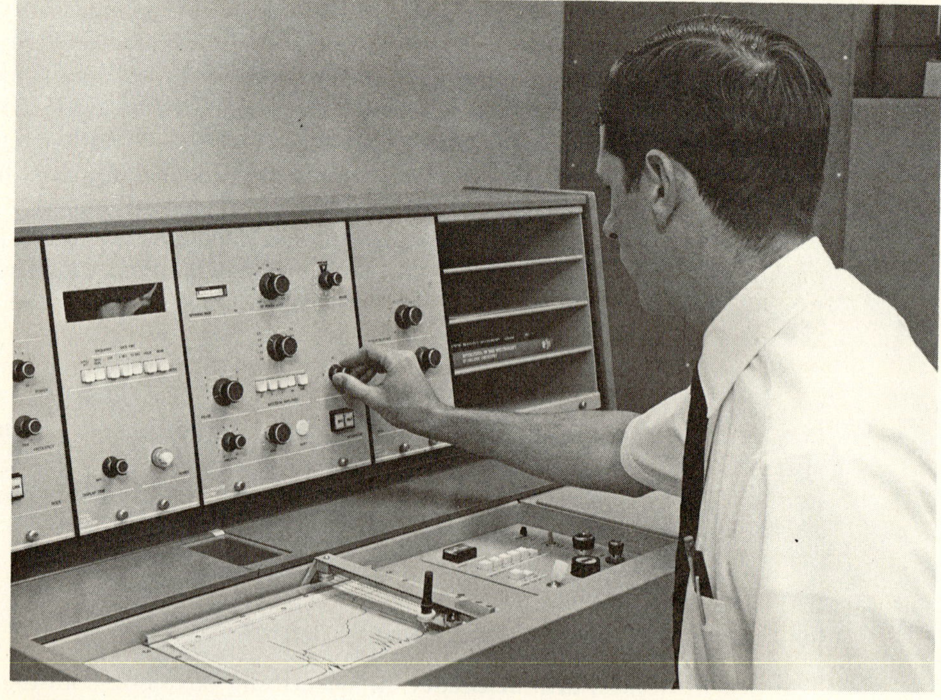

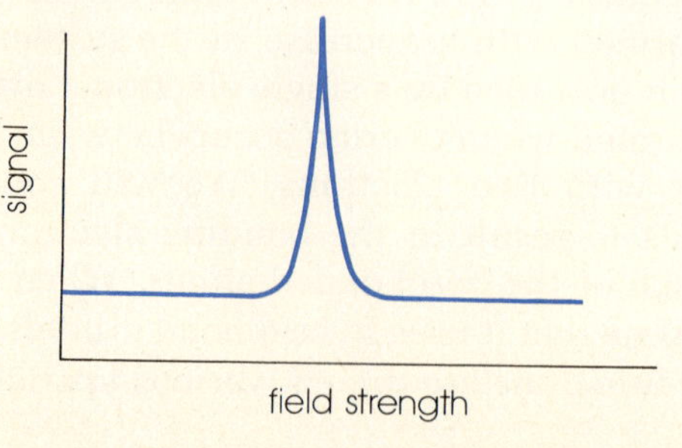

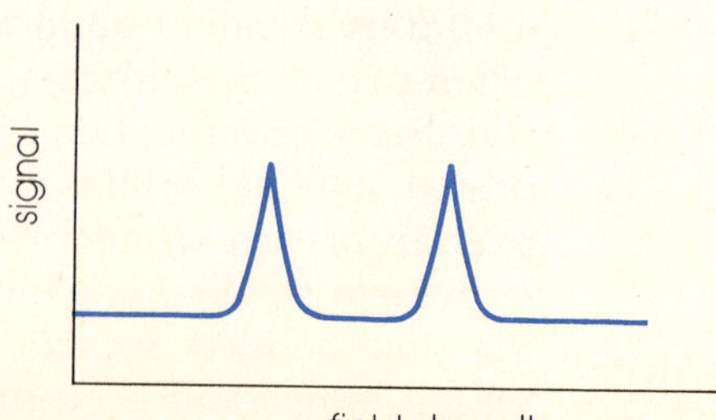

(c)

Figure 20-9 Nuclear magnetic resonance spectroscopy. The principles were worked out by Felix Bloch (Switzerland and United States 1905–   ) and Edward Purcell (United States 1912–   ) who shared the Nobel prize in physics in 1952. (a) A simplified schematic diagram of an NMR spectrometer. It consists of a powerful electromagnet, a radio-frequency transmitter (oscillator) connected to a coil that transmits energy to the sample, a radio-frequency receiver and amplifier connected to a second coil perpendicular to the first, a sample container, and a readout system. If magnet strength is held constant while the radio frequency is varied, the sample absorbs energies characteristic of the nuclei of its atoms. (b) A modern instrument. The recorder has traced out a typical spectrum. (c) Left, the spectrum of cyclohexane at room temperature. Right, the same at −70°C. Axial and equatorial hydrogen atoms are differentiated.

Research since 1969 has shown NMR methods may be useful in discriminating between malignant tumors and normal tissues. The former are characterized by an increased motional freedom of cell water detectable by NMR. (Photograph courtesy Varian Associates, Palo Alto, California.)

sp^3 hybridization in carbon

Carbon atom orbitals are designated $1s^2 2s^2 2p^2$, or

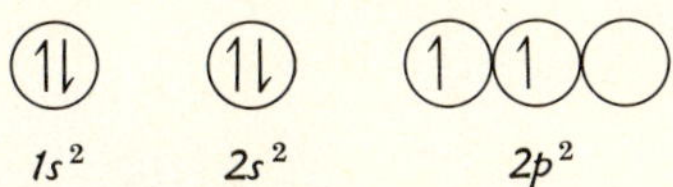

Accordingly, the atom might be expected to form one coordinate covalent bond ($2s^2$) and two covalent bonds ($2p^2$). Instead it forms four equivalent covalent bonds (in CH_4, for example). Quantum mechanicists account for this by mathematically combining the functions that describe the $2s$ and $2p$ electrons into an expression solvable four different ways. The solutions indicate that strongest bonding (lowest energy) will involve four new orbitals, called *hybrid orbitals* and designated sp^3, containing one electron each. Hybridization involves promotion of a $2s$ electron to the $2p$ level, giving

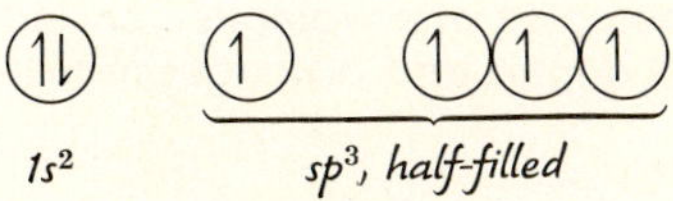

In methane four carbon–hydrogen covalent bonds are formed by overlap of the sp^3 orbitals with the singly occupied s orbital of four hydrogen atoms; the carbon atom then has filled orbitals:

When a bond is formed between two atoms by the overlap of two s orbitals or one s and one p orbital, the bond is called a *sigma* (σ) *bond*. In CH_4 all four bonds are σ bonds.

We have learned that hybridized bonds have a symmetrical spatial placement and maximum separation (Table 9-2); sp^3 hybrid bonds are tetrahedrally oriented (Figures 20-10, 20-11).

The carbon–carbon single bond

Ethane (H_3C—CH_3) is like methane. Both its carbon atoms are assumed to use sp^3 orbitals in bonding (Figure 20-11). The bond between the carbon

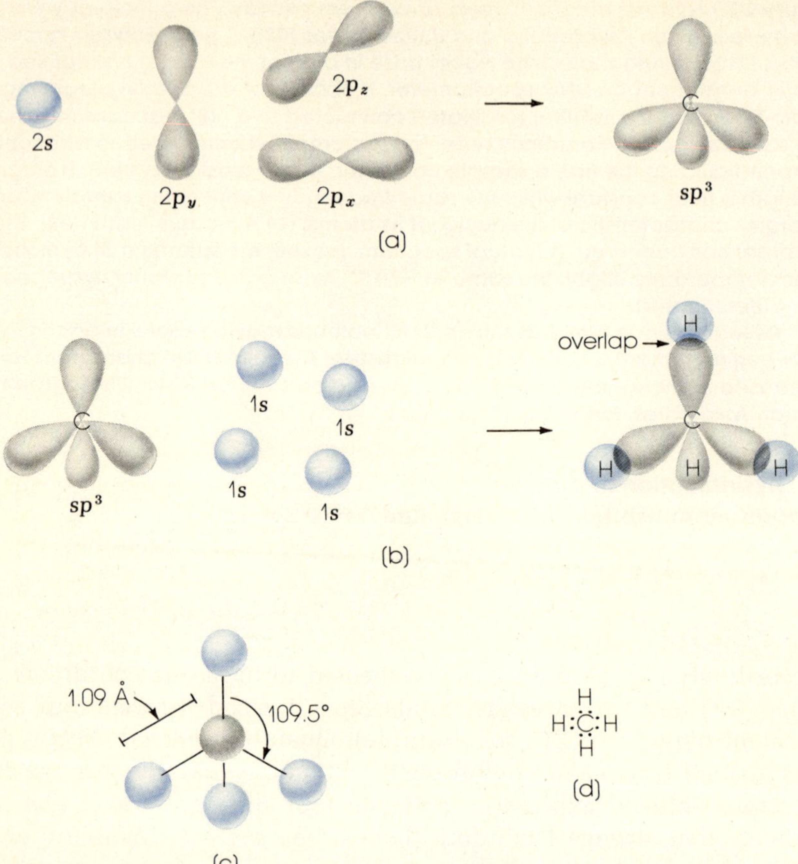

Figure 20-10 Methane. (a) Combination of the two 2p electrons and the two 2s electrons of a carbon atom give four identical sp³ hybrid orbitals with tetrahedral disposition. (b) Overlap of the hybrid orbitals with the 1s orbitals of four hydrogen atoms give a molecule of covalently bonded CH₄. (c) A ball-and-stick model with bond angles and distances marked. (d) The Lewis formula.

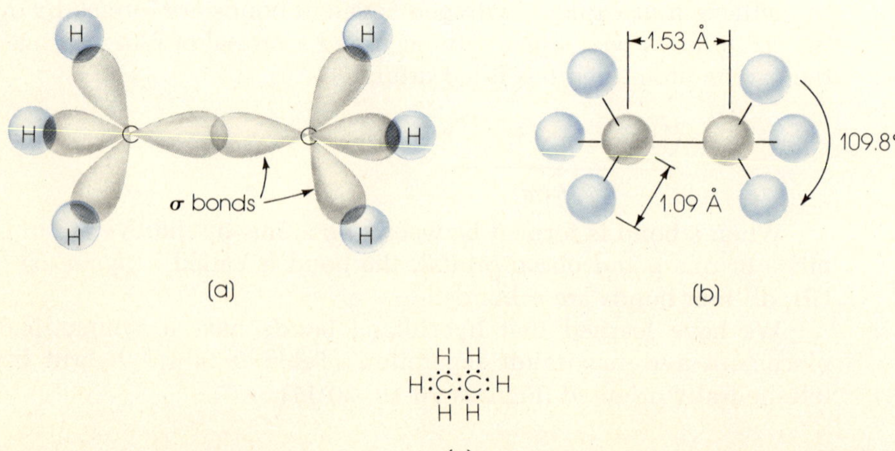

Figure 20-11 Ethane. (a) An atomic-orbital representation. (b) A ball-and-stick model. Indicated are the bonding angles, and C—C and C—H distances. (c) The Lewis formula.

atoms is formed by overlap of one sp^3 orbital from each carbon. Bonding is along the axis between the two nuclei, so that by definition it is a σ bond.

The carbon–carbon double bond

Each carbon in ethylene ($H_2C{=}CH_2$) is attached to three other atoms. According to orbital-bond theory, this means three equivalent carbon orbitals will be needed for bonding. The three are obtained through sp^2 hybridization in which one of the $2s$ electrons is promoted to a $2p$ level and three of the four atomic orbitals then make three sp^2 hybrid orbitals:

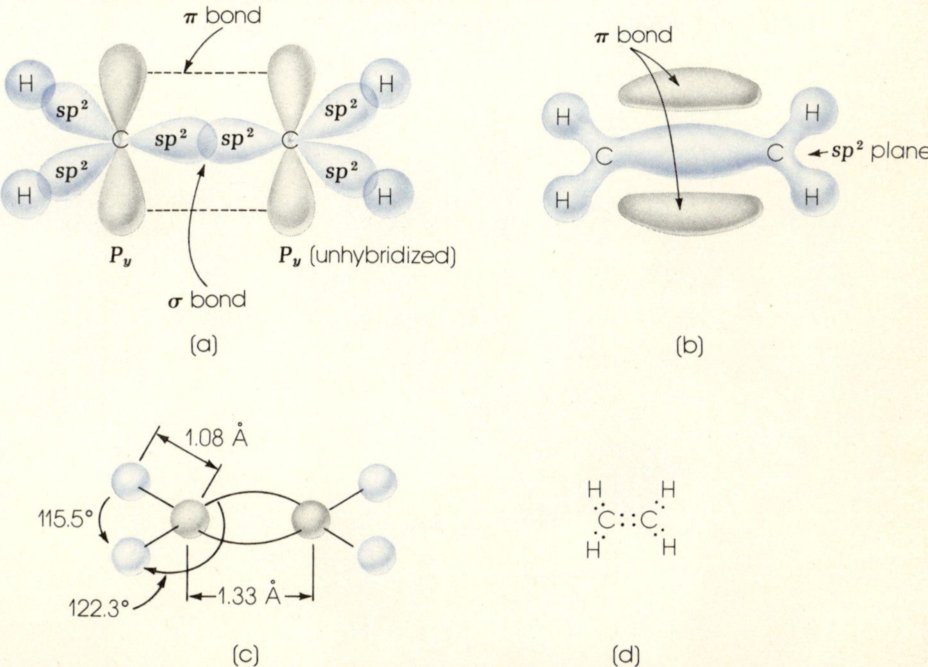

$$(20\text{-}7)$$

Sigma bonds are mainly responsible for the bonding in ethylene.

In sp^2 hybridization the three bonds ideally are mutually disposed at 120 degrees and lie in a plane (Table 9-2). The unhybridized p orbital is at right angles to the plane. In ethylene the p orbitals of the two carbon atoms are parallel to each other (Figure 20-12). Their overlap establishes a second two-electron bond between the carbons. Such an orbital, above and below the plane of the sp^2 hybrid structure, is called a *pi* (π) *orbital*. The bond is a π bond.

Rotation of the carbon atoms about the double bond is now restricted. Any twisting motion would require input of energy to lessen the orbital overlap.

Figure 20-12 Ethylene. (a) A model showing hybrid bonds and formation of the π bond. (b) A perspective model showing the π orbital above and below the plane in which all the atomic nuclei lie. (c) A ball-and-stick model, with experimentally determined bond distances and angles. A carbon-carbon double bond is shorter and stronger than a carbon-carbon single bond. (d) The Lewis formula.

The carbon–carbon triple bond

In acetylene (HC≡CH), each carbon is attached to two other atoms. The two equivalent orbitals that orbital theory requires are provided by sp hybridization:

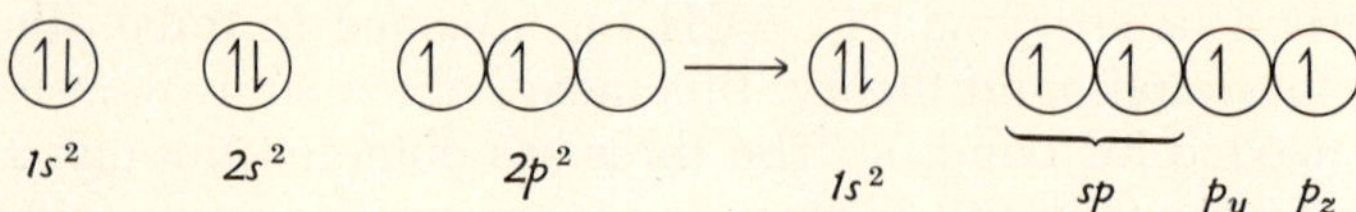

Each carbon uses its sp orbitals to form one σ bond with the other carbon and one σ bond with a hydrogen atom; sp hybridization is linear (Table 9-2), and measurements show that the four atoms of acetylene lie in a line. The unhybridized p orbitals are disposed at right angles both to the molecule's major axis and to one other. As they overlap laterally, two π bonds form (Figure 20-13). Together with the σ bond, they make a triple covalent bond. We already know that the triple bond is shorter and stronger than either a single or double bond (see Table 13-2).

Bonds in aromatic molecules

Molecules having structures similar to benzene are classed as *aromatic compounds*. In benzene (C_6H_6) each carbon atom is bonded to two other carbon atoms. Because measurements prove that the benzene ring (like ethylene) lies in one plane with 120-degree angles about each carbon, we expect sp^2 hybrid bonding. Each unhybridized p orbital projects at right angles above and below the plane of the ring. As they overlap, a donut-shaped π orbital is assumed to form (Figure 20-14). Each carbon–carbon bond, equivalent to three electrons, has a length intermediate between the lengths of single and double carbon–carbon bonds.

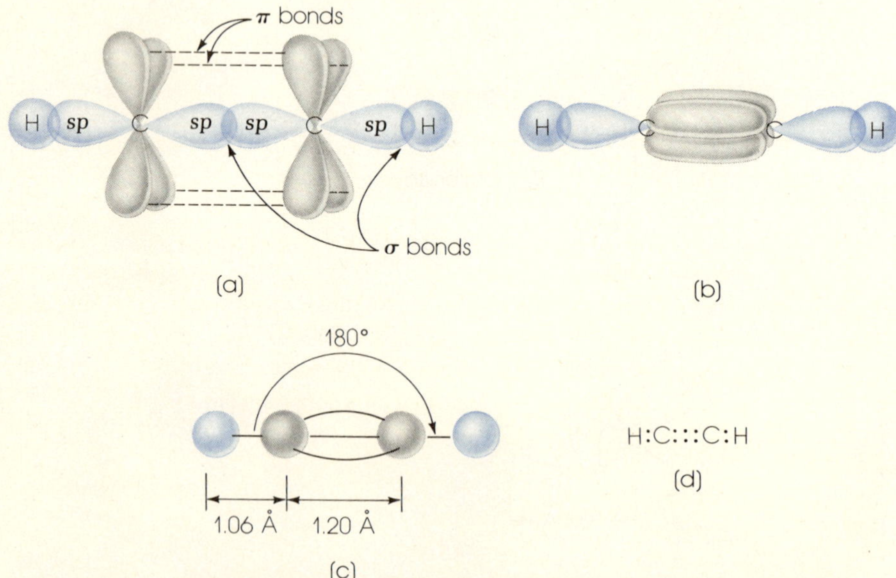

Figure 20-13 Acetylene. (a) A model showing the orientation of σ and π bonds. (b) A perspective view. The triple bond contains one σ and two π bonds. (c) A ball-and-stick model. Experimentally determined bond angles and distances are marked. (d) The Lewis formula.

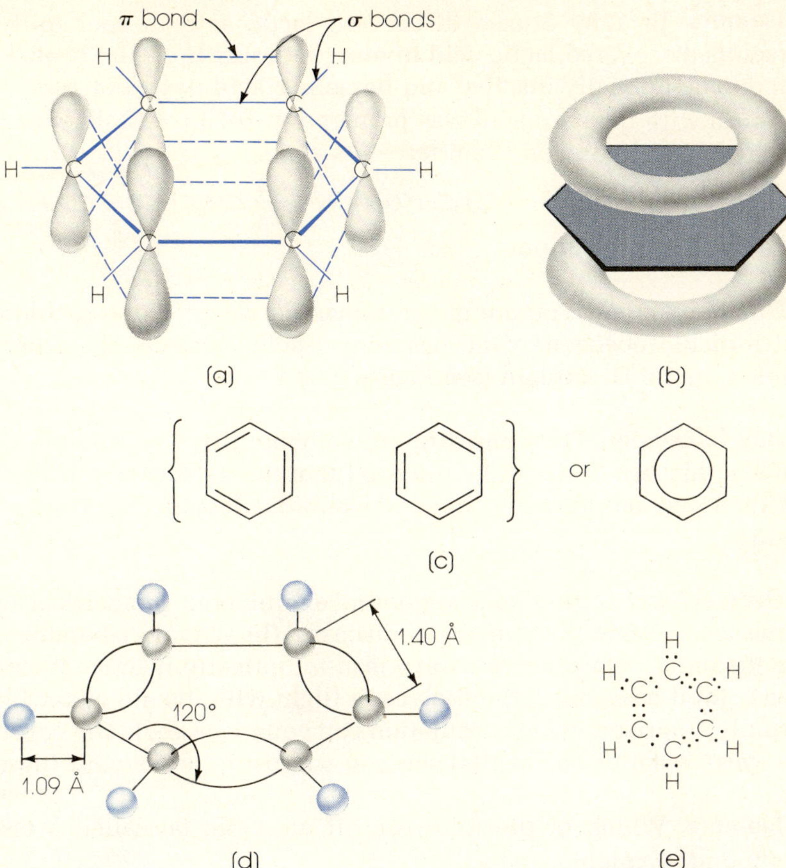

Figure 20-14 Benzene. (a) A model showing the ring carbon atoms and their unhybridized p orbitals at 90 degrees to the plane in which all the atoms lie. (b) Overlap of unhybridized p orbitals gives a large circular π orbital. The delocalization of electrons releases energy and stabilizes the molecule. (c) The resonance picture. The structure on the right is a hybrid structure made of the extreme Kekulé forms on the left. (d) A ball-and-stick model. Bond angles and distances, measured by electron diffraction, are marked. (e) A Lewis formula.

Electrons in extended π orbitals are said to be *delocalized*. Delocalization always leads to a decrease in energy. The energy difference between this state and one in which the same molecule would not have delocalized electrons is called the *delocalization* or *resonance energy*. Delocalization makes benzene much less susceptible toward addition reactions that typify nonaromatic unsaturated compounds like ethylene and acetylene. The latter lack comparable stablization.

QUESTIONS

1. Classification In 1841 French chemist C. Gerhart proposed that organic compounds be classified into families whose molecules contained the same number of carbon atoms. **(a)** What disadvantages do you see in his system? **(b)** Suggest another system.

2. Isomers In 1780 Sheele discovered lactic acid in sour milk. In 1807 Berzelius discovered lactic acid in muscle tissue. In 1849 Sheele's acid was found to be optically inactive and Berzelius' acid dextrorotatory. A product identical with Sheele's acid was prepared in 1873 by Wislicenus from acetaldehyde and hydrogen cyanide:

$$CH_3CHO + HCN \longrightarrow CH_3CH(OH)CN \xrightarrow[H^+]{H_2O} CH_3CH(OH)CO_2H$$

Interpret the observations.

3. Dipole moment You are given unmarked samples of o-dichlorobenzene and p-dichlorobenzene. One has zero dipole moment, the other a dipole moment of 2.5 D. Explain (see Figure 9-4).

4. Bond direction Only one difluoromethane (CH_2F_2) is known. How could you use this fact to discuss a proposal that the valence bonds of carbon are not disposed tetrahedrally but in pyramidal fashion as shown in Figure 20-15.

5. Optical activity **(a)** Can a compound exhibit both geometrical and optical isomerism? Define the terms, and discuss. **(b)** Why do laboratory reactions like the one in Question 2 always lead to optically inactive (racemic) mixtures? **(c)** If bees orient themselves in flight with the aid of sky light that is partially polarized, what speculation can you make about the constitution of bee eyes? What experiment(s) can you suggest to prove your ideas?

6. Isomers Which of the following (if any) can be called a *cis* or *trans* compound? Explain.

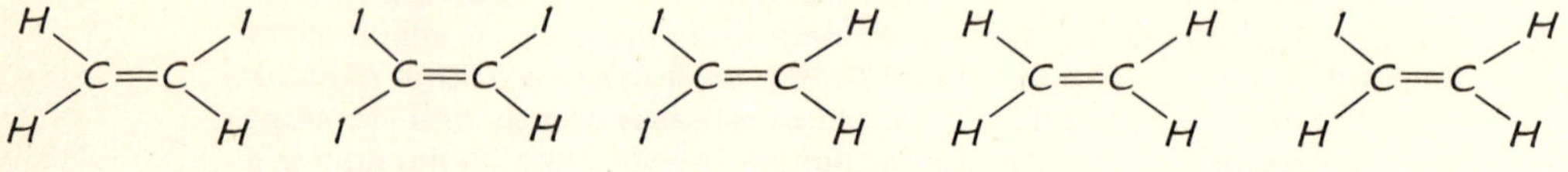

7. Isomers **(a)** How might a single chemical reaction distinguish between these diacids?

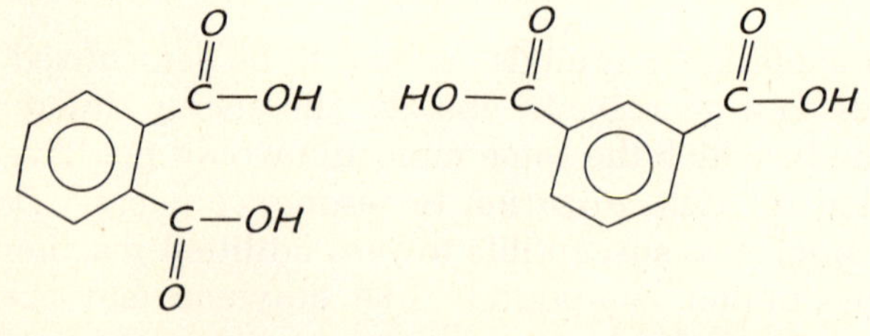

(b) 1,4-Hexadiyne ($HC{\equiv}C{-}CH_2{-}C{\equiv}C{-}CH_3$) is isomeric with benzene. Explain. **(c)** What alicyclic compound is isomeric with $CH_2{=}CH{-}CH_2{-}CH_3$?

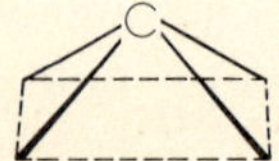

Figure 20-15

8. Definitions Define or otherwise explain **(a)** plane polarized light; **(b)** enantiomers; **(c)** Sachse-Mohr theory; **(d)** equatorial and axial positions; **(e)** extended π orbital.

9. Bonding (a) What are the differences between σ and π bonds? What are the similarities? Explain. **(b)** In molecules like ethylene, hybrid orbital theory offers an explanation for planarity, angles, and nonrotation (leading to *cis-trans* isomerism) about the carbon–carbon double bond. Explain.

10. Conformations The "eclipsed" conformation of dichloroethane can be represented as

The "staggered" conformation can be pictured as

(a) Explain the symbolism (called *Newman notation*) introduced in II and IV. **(b)** Which of the two conformers do you think is more stable? Explain.

---

11. Analysis A plastic used in packaging consists of very long chains of $-C_xH_y-$ units linked one after the other. If upon combustion 0.0462 g of plastic gives 0.145 g of CO_2 and 0.0594 g of H_2O, what are the minimum values of x and y?

12. Analysis An organic compound contains carbon, hydrogen, and sulfur. Its molecular weight is 216. Combustion of a 30.0-mg sample yields 73.3 mg of CO_2, 10.0 mg of H_2O, and 17.8 mg of SO_2. What are the empirical and molecular formulas?

13. Analysis Hippuric acid, isolated from the urine of the horse, cow, and camel, has a molecular weight of 179 and this elemental composition: 60.3 percent C, 5.0 percent H, 7.8 percent N, and the balance O. Calculate the molecular formula.

14. Polarimetry Specific rotation $[\alpha]$ of a substance in solution at a specified temperature and using light of a specified wavelength is given by the equation

$$[\alpha] = \frac{\alpha}{lc}$$

(20-8)

where α is the observed optical rotation in degrees, l the length of the sample tube in decimeters, and c the concentration in grams of solute per milliliter of solution. When 10 g of a sugar is dissolved to make 100 ml of solution, the observed rotation is $+14.5$ degrees in a 1-dm tube. Under the same conditions another solution of the sugar has a rotation of 5.2 degrees. **(a)** Calculate the specific rotation and the concentration of the second solution. **(b)** What would the specific rotation of the second solution be in a 2-dm tube?

15. Körner's method **(a)** Give structural formulas for all possible trimethyl-benzenes. **(b)** Show how monochlorination of remaining ring positions and Körner's principle can be used to distinguish among the trimethylbenzenes.

16. Isomers Write structures for all the isomers of **(a)** C_3H_5I; **(b)** C_4H_9I.

17. Isomers For each of the following that may exhibit optical activity draw the two isomers (as in Eq. 20-5): $CH_3CH_2CH_2I$, CH_3CHICH_3, CH_3CI_2,CH_3, CH_2IBr, CH_3CHICH_2I.

18. Isomers Draw the formula for a ring compound that is isomeric with CH_3—CH=CH—CH_2—CH_3; with CH_3—CH=CH—CH=CH—CH_3.

(a) Cyclopropane (C_3H_6) has rigid, planar ring structure. It has one conformation with three hydrogen atoms above and three below the ring's plane. Draw the structural formula. **(b)** When two hydrogen atoms are removed and two chlorine atoms put on the ring in their place, several different dichlorocyclopropane isomers are possible. Draw the structures of a *cis* form, a *trans* form, and another form. **(c)** Can any of these forms be optically active? Explain.

20. Bond types Summarize the carbon–carbon bonding discussed in the chapter by completing Table 20-1.

21. Conformations **(a)** Draw structural formulas for the boat and chair forms of cyclohexane. On each axial bond attach a letter A, on each equa-

Table 20-1. Bond types				
Bond	Hybridization	Geometry and angles about the carbons	Molecular orbital bond designation	Carbon-to-carbon distance
C—C				
C=C				
C≡C				

torial bond a letter E. Point out the difference between the two conformers in regard to the axial positions. **(b)** Use the same symbolism to picture cyclobutane. (The ring is bent; it can be visualized by folding a square piece of paper to bring two opposite corners together, creasing, then letting it fall open.) Alternate axial hydrogen atoms are above and below the ring.

22. Conformations n-Butane has the formula H_3C—CH_2—CH_2—CH_3. **(a)** Picture two of its conformers in the manner of Question 10, II and IV. **(b)** A third conformer is called "skew." Use the same symbolism to picture the structure that you think it has.

HOW DO ORGANIC COMPOUNDS REACT?

THE IMPORTANT CONCEPTS

21-1. Sources of organic raw materials
1. Distillation of coal and petroleum gives hydrocarbons and other compounds.

21-2. Nomenclature of aliphatic hydrocarbons
1. A few rules enable one to designate carbon number, multiple bonds, and branches in naming aliphatic hydrocarbons.

21-3. Reactions of alkanes
1. Alkanes burn, crack, isomerize, and halogenate by substitution.

21-4. Reactions of alkenes and alkynes
1. Multiple carbon–carbon bonds add hydrogen, halogen, sulfuric acid, and ozone.
2. Hydrogenation and bromination of C=C are stereospecific.

3. Identification of ozonolysis products establishes C=C positions in the reactant.

21-5. Reactions of aromatic hydrocarbons
1. Br^+, NO_2^+, and $^+CH_3$ are electrophilic reagents that attack aromatic rings.

21-6. Reactions of aliphatic hydrocarbon derivatives
1. A molecule is characterized by its functional group.
2. Alcohols may undergo dehydration, oxidation of —COH, and exchange of —OH by halogen.
3. An S_N1 reaction proceeds with loss of optical activity.
4. In an S_N2 reaction structure inversion occurs with retention of optical activity.
5. Reduction and cyanohydrin formation may occur at the carbonyl group.
6. Aliphatic acids are weakly ionized.
7. The mechanism of esterification is well known.
8. Amines are ammonia derivatives that can be alkylated, deaminated, and converted to salts.
9. Amides can be reduced and dehydrated.

21-7. Reactions of aromatic hydrocarbon derivatives
1. —NO_2 can be reduced, —R oxidized, and —NH_2 diazotized.
2. Ring substituents can be divided into two classes based on directive influence.
3. Resonance theory helps explain ortho-, para-, and meta-directing groups.

21-8. Synthetic polymers
1. Polymers are the major product of the organic chemical industry.
2. Natural rubber is a cis polymer of isoprene.
3. Monomers contain at least two reactive positions.

21-9. Commercializing a chemical process
1. Chemistry, engineering, and economics are interrelated in chemical manufacture.

Organic chemistry can be called the chemistry of hydrocarbons and their derivatives.

Organic reactions are characterized by their (1) slowness, because covalent rather than ionic bonds must be broken; and (2) production of mixtures, because several reaction paths of approximately equal energy (see Figure 17-3) are often available. Study of reactions is facilitated by a classification of organic compounds according to their reactive groups, and by research in recent decades that has outlined reaction mechanisms.

Synthetic organic products from chemical industry such as gasoline, rubber, and plastics have become essentials in modern society. The dozen most important industrial organic chemicals (mostly raw materials for syntheses) are listed in Table 21-1.

21-1. SOURCES OF ORGANIC RAW MATERIALS

Primary sources of most organic compounds are *coal* and *petroleum*, the concentrated remains of prehistoric life.

Coal is believed to have originated about 250 million years ago in the carboniferous era (see Figure 22-17) when vegetation was particularly lush. As dead plant matter underwent partial solubilization in water, microorganisms converted the cellulose to H_2O, CO_2, and CH_4. Heated and compressed while being submerged by earth movements, the remaining resins, containing 50 percent carbon, were changed successively to peat, brown coal, soft coal, and finally hard coal, which is 94 percent carbon.

When soft coal is heated without air, the following takes place:

$$\text{1 ton of soft coal} \xrightarrow[\text{distil}]{1150°C,} \begin{cases} 6000 \text{ ft}^3 \text{ of light gases} \\ 20 \text{ lb of coal oil} \\ 80 \text{ lb of coal tar} \\ 1600 \text{ lb of coke} \end{cases} \tag{21-1}$$

Table 21-1. Biggest volume organic chemicals — approximate 1972 production in the United States

Chemical	Production million tons	Chemical	Production million tons
Ethylene	9.0	Propylene	3.3
Benzene	4.0	Methyl alcohol	2.6
Ethylene dichloride	3.8	Ethyl benzene	2.5
Polyethylene	3.7	Styrene	2.3
Toluene	3.3	Formaldehyde	2.2
Urea, solution	3.3	Xylene	2.2

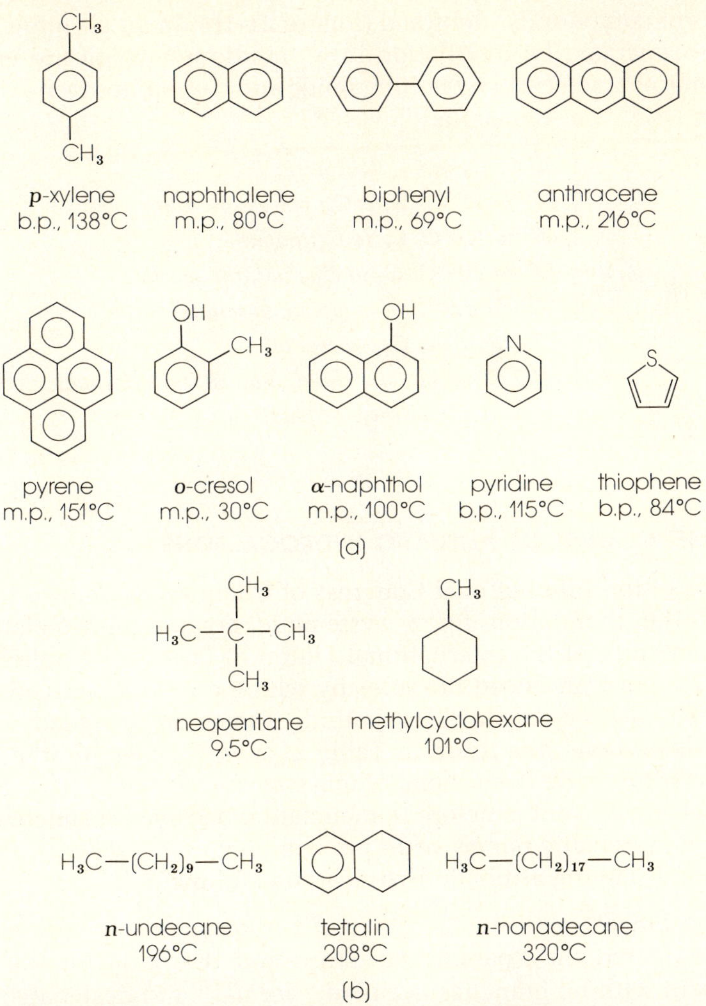

Figure 21-1 Organic raw materials. (a) Representative aromatic compounds from the destructive distillation of coal. Those with —OH groups are acidic and can be separated by dissolving in alkaline solution. Those containing nitrogen or sulfur are basic and will dissolve in acid solution. Hydrocarbons are not soluble in either solution. (b) Representative compounds from the distillation of crude oil, with their boiling points. (See Table 21-1 for others.) Each oil field gives a somewhat different mixture.

The oil and tar yield aromatic compounds (Figure 21-1a), of which benzene is the most important. In the process of coking 550 million tons of coal a year in the United States, 1 billion gallons of benzene is distilled.

Petroleum probably originated in sea shallows at least 15 million years ago. There tiny marine animals and plants sank and, under reducing conditions, putrified through the action of anaerobic bacteria. The first product possibly was *sapropel*, a dark mud found today on the bottom of the Black Sea and other places. Earth overfolding, heat, and pressure could have converted sapropel to petroleum.

Crude oil is primarily a mixture of hydrocarbons averaging 86 percent carbon by weight. The empirical formula is approximately CH_2. Aromatic substances are present in oil but aliphatics predominate among the hundreds of compounds identified (Figure 21-1b). In an oil refinery components are separated by distillation into "fractions" which are mixtures of compounds characterized by boiling range and *carbon number* — the number of carbon atoms per molecule:

$$\text{crude oil} \xrightarrow{\ distil\ } \begin{cases} 30°C,\ C_1\ to\ C_5;\ light\ gases \\ 20°\ to\ 60°C,\ C_5\ to\ C_6;\ petroleum\ ether \\ 60°\ to\ 100°C,\ C_6\ to\ C_7;\ naphtha \\ 38°\ to\ 204°C,\ C_5\ to\ C_{10};\ natural\ gasoline \\ 175°\ to\ 320°C,\ C_{10}\ to\ C_{15};\ kerosene \\ 275°C\ range,\ C_{11};\ diesel\ oil \\ 350°C\ range,\ long\ chain;\ lube\ oil,\ paraffin\ wax \\ residue,\ many\ rings;\ asphalt,\ sludge \end{cases} \tag{21-2}$$

21-2. NOMENCLATURE OF ALIPHATIC HYDROCARBONS

A meeting of the International Congress of Chemists at Geneva in 1892 established the foundation for a systematic organic nomenclature. Subsequent meetings of the International Union of Pure and Applied Chemistry (IUPAC) have produced the rules by which a unique derived name can be assigned to every organic compound. Names are assigned to *straight-chain* hydrocarbons like those in Table 21-2 and names of other aliphatic compounds are made from them. Nonsystematic or "common" nomenclature, which was invented before nomenclature became systematized, is still used when systematic names seem too complex.

Rules for naming aliphatic hydrocarbons follow.

1. The longest continuous chain of carbon atoms in the molecule is considered the "parent" structure and the basis for the name. A chain with no branches is called "normal" and designated n–.
2. The name ends in -*ane* when the chain contains only single bonds; -*ene* denotes a double bond, -*yne* a triple bond.
3. Chain carbon atoms are assigned consecutive numbers starting at the end nearest a multiple bond or nearest the most chain branching.

Table 21-2. Some straight-chain alkanes

Number of carbon atoms	IUPAC name	Molecular formula	Condensed structural formula	Boiling point °C
1	Methane	CH_4	CH_4	−162
2	Ethane	C_2H_6	CH_3CH_3	−89
3	n–Propane	C_3H_8	$CH_3CH_2CH_3$	−42
4	n–Butane	C_4H_{10}	$CH_3CH_2CH_2CH_3$	0
5	n–Pentane	C_5H_{12}	$CH_3CH_2CH_2CH_2CH_3$	36
6	n–Hexane	C_6H_{14}	$CH_3CH_2CH_2CH_2CH_2CH_3$	69
7	n–Heptane	C_7H_{16}	$CH_3CH_2CH_2CH_2CH_2CH_2CH_3$	98
8	n–Octane	C_8H_{18}	$CH_3CH_2CH_2CH_2CH_2CH_2CH_2CH_3$	126
9	n–Nonane	C_9H_{20}	$CH_3CH_2CH_2CH_2CH_2CH_2CH_2CH_2CH_3$	151
10	n–Decane	$C_{10}H_{22}$	$CH_3CH_2CH_2CH_2CH_2CH_2CH_2CH_2CH_2CH_3$	174

4. Names of branches formed from alkane hydrocarbons by removal of a hydrogen atom are derived by dropping the suffix *-ane* and adding *-yl*. (Such branches by themselves are called *alkyl groups*.)

5. The position of a multiple bond is indicated by inserting before the name of the parent structure the lower of the two numbers assigned to the carbons that hold the bond.

6. The position of a branch is indicated by inserting before its name the number of the chain carbon holding the branch.

7. A branched compound with a multiple bond is named in this order: the position number of each branch and the name of the branch, the position number of the multiple bond, the parent name, and the suffix describing the multiple bond. If several branches are present, they are named in alphabetical order. Like branches are numbered and named together; prefixes for them are *di-, tri-, tetra-*, etc.

8. Names and numbering of ring aliphatic compounds correspond to names and numbering of related chain aliphatics, with *cyclo-* used as a prefix.

9. Punctuation is as follows: Words are run together; commas separate numbers; dashes separate words and numbers.

Example 21-1 Name:

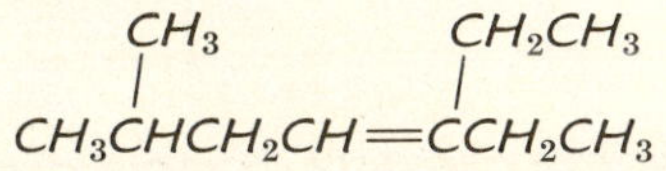

$$\begin{array}{ccc} CH_3 & & CH_2CH_3 \\ | & & | \\ CH_3CHCH_2CH & \!\!=\!\! & CCH_2CH_3 \end{array}$$

Solution To simplify we will write the carbon framework or skeleton (carbon atoms only) of the longest chain, then number it consecutively from the right, the end closest to the multiple bond:

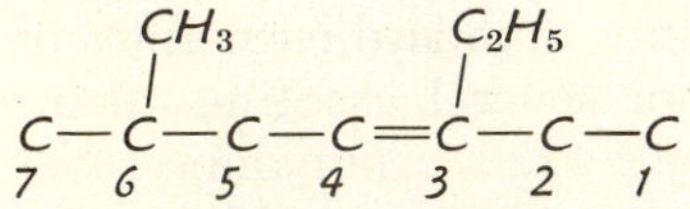

$$\begin{array}{ccccccc} & CH_3 & & & C_2H_5 & & \\ & | & & & | & & \\ C - & C - & C - & C = & C - & C - & C \\ 7 & 6 & 5 & 4 & 3 & 2 & 1 \end{array}$$

The double bond is between carbons 3 and 4. It will be designated by the lower number, 3. A branch is also on the 3-carbon. Derived from ethane, it is named *ethyl*. The branch on the 6-carbon, derived from methane, is called *methyl*. The parent structure with its seven carbon atoms is derived from heptane (Table 21-2); the double bond makes it a *heptene*.

Putting these stipulations in proper order, we arrive at the correct name: 3-ethyl-6-methyl-3-heptene.

Following are other examples of systematic naming:

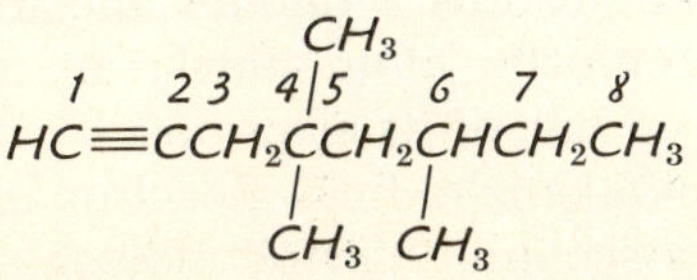

$$\overset{1}{HC}\equiv\overset{2}{C}\overset{3}{CH_2}\overset{4}{\underset{\underset{CH_3}{|}}{C}}\overset{5}{CH_2}\overset{6}{\underset{\underset{CH_3}{|}}{CH}}\overset{7}{CH}\overset{7}{CH_2}\overset{8}{CH_3}$$

4,4,6-trimethyl-1-octyne

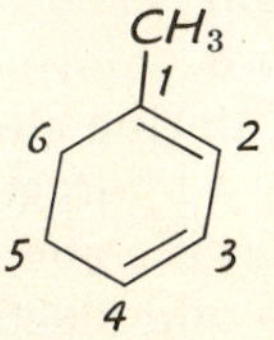

1-methyl-1,3-cyclohexadiene

21-3. REACTIONS OF ALKANES

Alkanes are all the compounds having the general formula C_nH_{2n+2}, where n is any integer. Alkanes are also called *paraffins* (L. *parum affinus*, little affinity). They are described as *saturated* to indicate the chemical inertness of their single covalent bonds. As already noted alkanes may have straight-chain, branched-chain, and cyclic structure.

Using butane as a representative alkane, four important alkane reactions are described as follows:

$$H_3CCH_2CH_2CH_3 \begin{cases} (1) \xrightarrow{O_2} CO_2 + H_2O + heat \\ \quad\quad complete\ combustion \\\\ (2) \xrightarrow[cat.]{press.} H_3CCH{=}CH_2 + CH_4 + C + H_2 \\ \quad\quad cracking \\\\ (3) \xrightarrow{cat.} H_3C\overset{\overset{\textstyle CH_3}{|}}{C}HCH_3 \\ \quad\quad isomerization \\\\ (4) \xrightarrow[h\nu]{Cl_2} ClCH_2CH_2CH_2CH_3 + H_3C\overset{\overset{\textstyle Cl}{|}}{C}HCH_2CH_3 + HCl \\ \quad\quad photochemical\ halogenation \end{cases}$$

$$(21\text{-}3)$$

1. All alkanes burn in air; in the forms of natural gas and gasoline they satisfy much of our energy demand. Standard enthalpies of complete combustion (kilocalories per mole) of several gasoline alkanes are n-hexane -929, n-heptane -1076, n-octane -1223, n-nonane -1370. The increase of 147 kcal/mole in each case corresponds to an additional $-CH_2-$ group.

2. Heating alkanes without air causes longer chains to break into smaller units, some of which have multiple bonds where hydrogen was lost. In oil refineries the use of pressure and catalysts gives control over the size of the molecules formed.

3. Mixtures of branched-chain isomers are produced from straight-chain alkanes through use of Lewis acid catalysts like BF_3 which initiate the reaction by removing a hydrogen atom with its electron pair. Branched-chain hydrocarbons have superior antiknock characteristics as motor fuels.

4. Alkanes can be chlorinated in the presence of light. The mechanism of chlorination is similar to that which leads to smog (Chapter 13). Reaction begins when a chlorine molecule absorbs a photon and the bond breaks, giving two chlorine atoms. As a chlorine atom attacks an alkane molecule, it takes away a hydrogen atom and one of its electrons, forming a hydrogen chloride molecule and leaving an alkane radical. Reaction between the radical, with its odd electron, and a second chlorine molecule gives a chlorinated alkane molecule and another chlorine atom. The chain reaction can be pictured as a *cycle*:

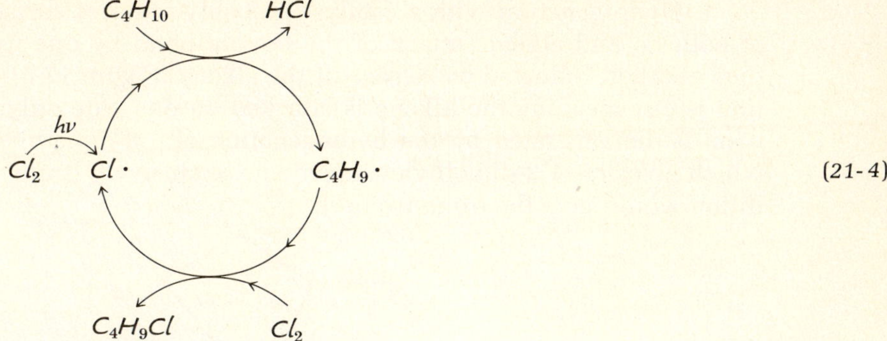

$$(21\text{-}4)$$

21-4. REACTIONS OF ALKENES AND ALKYNES

Alkenes are aliphatic hydrocarbons that contain at least one carbon–carbon double bond. Alkynes are aliphatic hydrocarbons that contain at least one carbon–carbon triple bond. Compounds of both classes are called *unsaturated* because with multiple bonds they contain fewer hydrogen atoms than the corresponding (saturated) alkanes. Strong acids and oxidizers attack the multiple bonds. Reaction is usually by *addition*,

$$C{=}C + X_2 = \overset{\displaystyle X}{\underset{\displaystyle |}{C}}{-}\overset{\displaystyle X}{\underset{\displaystyle |}{C}}$$

rather than by substitution. Reagents seeking sites like multiple bonds where electron density is high are said to be *electrophilic*. An unsaturated hydrocarbon is described as *nucleophilic* for its seeking out of substances that are electron deficient or are positively charged (like an atomic nucleus). Relatively rapid reaction at multiple carbon–carbon bonds is probably due to the ease with which weaker π bonds are broken and stronger σ bonds are formed.

Alkenes are also called *olefinic* (oil forming) hydrocarbons or *olefins*. 1-Butene, a typical olefin, undergoes addition reactions as follows:

$$(21\text{-}5)$$

1. Hydrogenation with a catalyst probably requires surface adsorption of both H_2 and alkene, attack of the double bond by one hydrogen atom then another, followed by release of the alkane (Figure 17-5). The H_2 addition is *cis*, meaning the alkene is attacked on one side only. This mechanism is demonstrated by the hydrogenation of 1,2-dimethylcyclopentene, which gives *cis*-1,2-dimethylcyclopentane exclusively (*trans* hydrogen addition would give the *trans* isomer):

2. Unlike halogenation of an alkane, halogenation of an alkene or alkyne is usually rapid, does not eliminate hydrogen atoms as HX, and does not require light. Used as a solution in inert CCl_4 solvent, Br_2 is red–brown. The fading of color as it reacts is a test for unsaturation.

Bromination takes place in two steps. The halogen attacks first one side of the double bond then the other. *Trans* addition can be demonstrated by the bromination of cyclopentene; the only product is *trans*-1,2-dibromocyclopentane:

Reactions like (1) and (2) above, which give only one of several possible isomeric products, are called *stereospecific reactions*.

3. Alkene hydration is accomplished by adding concentrated sulfuric acid, then hydrolyzing the alkyl hydrogen sulfate that is formed. The products are H_2SO_4 and an alcohol (2-butanol):

$$H_2C{=}CHCH_2CH_3 \xrightarrow{H_2SO_4} H_3C{-}\overset{\overset{\displaystyle OSO_3H}{\displaystyle |}}{C}HCH_2CH_3 \xrightarrow{H_2O} H_2SO_4 + H_3C\overset{\overset{\displaystyle OH}{\displaystyle |}}{C}HCH_2CH_3$$

Sulfuric acid adds as H^+ and $^-OSO_3H$. *The positive part of a reagent adds mainly to the double-bonded carbon atom having the greater number of hydrogens.* This rule was discovered by Russian chemist V. Markowni-koff in 1870. It is interpreted today as a two-step process, the first of which is an electrophilic attack (by the reagent's positive part) to yield the ion that is best stabilized by resonance:

$$C{=}C{-}C{-}C + H^+ \longrightarrow \begin{cases} \underset{\overset{|}{H}}{C}{-}\overset{+}{C}{-}C{-}C \quad \textit{more stable} \\[2em] \overset{+}{C}{-}\underset{\overset{|}{H}}{C}{-}C{-}C \quad \textit{less stable} \end{cases}$$

4. Ozone reacts with alkenes to give unstable ozonides. With acid and zinc, ozonides are cleaved into aldehydes or ketones or both (Table 21-3), whose identities are useful in assigning structure to the original hydro-carbon. Thus if a hexene reacts as follows,

$$C_6H_{12} \xrightarrow{O_3} \textit{ozonide} \xrightarrow[\text{Zn}]{HCl} \underset{\textit{an aldehyde}}{H_3CCH_2\overset{\overset{O}{\|}}{C}{-}H} + \underset{\textit{a ketone}}{O{=}C\overset{CH_3}{\underset{CH_3}{\diagup}}}$$

its structural formula must have been $H_3CCH_2CH{=}\overset{\overset{CH_3}{|}}{C}CH_3$. Ozone introduced oxygen where the double bond was.

Table 21-3. Hydrocarbon derivatives; functional groups[a]

Class of compounds	General formula	Class of compounds	General formula
Halide	$R{-}X,\ Ar{-}X$	Nitrile (cyanide)	$R{-}C{\equiv}N,\ Ar{-}C{\equiv}N$
Alcohol	$R{-}OH$	Carboxylic acid	$R{-}\overset{\overset{O}{\|}}{C}{-}OH,\ Ar{-}\overset{\overset{O}{\|}}{C}{-}OH$
Phenol	$Ar{-}OH$	Ester	$R{-}\overset{\overset{O}{\|}}{C}{-}O{-}R,\ Ar{-}\overset{\overset{O}{\|}}{C}{-}O{-}Ar$
Ether	$R{-}O{-}R,\ Ar{-}O{-}Ar$	Amine	$R{-}NH_2,\ Ar{-}NH_2$
Aldehyde	$R{-}\overset{\overset{O}{\|}}{C}{-}H,\ Ar{-}\overset{\overset{O}{\|}}{C}{-}H$	Amide	$R{-}\overset{\overset{O}{\|}}{C}{-}NH_2,\ Ar{-}\overset{\overset{O}{\|}}{C}{-}NH_2$
Ketone	$R{-}\overset{\overset{O}{\|}}{C}{-}R,\ Ar{-}\overset{\overset{O}{\|}}{C}{-}Ar$	Diazonium salt	$Ar{-}N{\equiv}N^+X^-$

[a] R— is an alkyl group, derived from an aliphatic hydrocarbon by removal of one hydrogen atom; Ar— is an aryl group, derived from an aromatic hydrocarbon by removal of one hydrogen atom. Mixed ethers, ketones, and esters are known: R—O—Ar, etc. Only primary alcohols and amines, and unsubstituted amides are shown.

21-5. REACTIONS OF AROMATIC HYDROCARBONS

The benzene ring is unsaturated, but its extended π-bond system makes it resistant to attack (see Figure 20-14). Powerful electron-seeking reagents created by Lewis acid catalysts will remove hydrogen atoms, however, so that substitution rather than addition characterizes benzene reactions:

$$(1) \quad \xrightarrow{Br_2, \; FeBr_3} \quad C_6H_5Br + HBr$$

halogenation

$$(2) \quad \xrightarrow{HNO_3, \; H_2SO_4} \quad C_6H_5NO_2 + H_2O$$

nitration

$$(3) \quad \xrightleftharpoons{ClCH_3, \; AlCl_3} \quad C_6H_5CH_3 + HCl$$

alkylation (21-6)

One mechanism is proposed for all three reactions. In bromination the electrophilic reagent is bromonium ion:

$$Br_2 + FeBr_3 = Br^+ + FeBr_4^-$$

in nitration it is nitronium ion:

$$HNO_3 + H_2SO_4 = NO_2^+ + HSO_4^- + H_2O$$

and in alkylation it is a positive alkyl ion (called an alkyl *carbonium* ion):

$$CH_3Cl + AlCl_3 = {}^+CH_3 + AlCl_4^-$$

We will consider alkylation briefly as representative of the three reactions. The carbonium ion is electrophilic. Attracted to the electron cloud of the benzene ring, it takes an electron pair from it and attaches itself to one of the ring carbon atoms:

$$C_6H_6 + {}^+CH_3 + AlCl_4^- = [C_6H_6(CH_3)]^+ + AlCl_4^-$$

Removal of the proton with a Lewis base reestablishes the stable aromatic ring system:

$$\text{(ring with } H, CH_3 \text{ and } + \text{) } + AlCl_4^- = \text{(ring with } CH_3 \text{) } + HCl + AlCl_3$$

Alkylation of the ring is reversible (nitration and halogenation are not). Toluene (methyl benzene) can also be alkylated further as the combination of $AlCl_3$ and CH_3Cl continues to substitute $—CH_3$ for ring hydrogen atoms.

21-6. REACTIONS OF ALIPHATIC HYDROCARBON DERIVATIVES

A compound formed by removing a hydrogen atom from a hydrocarbon and substituting another atom or group is called a *derivative* of that hydrocarbon (Table 21-3). The substituent is called a *functional group*. The important chemistry of the substituted molecule as a whole is often that of the functional group. Organic chemists correlate their subject by learning the characteristics of functional groups.

Alcohols

Alcohol names are derived from alkane names; in IUPAC nomenclature -e is dropped and -ol added. In common nomenclature *secondary* (sec-) and *tertiary* (tert-) are used to denote that the carbon attached to —OH is also attached to two other and to three other carbon atoms, respectively. *Iso*- means there is a methyl branch at the end of the chain.

$$\overset{4}{C}-\overset{3}{C}-\overset{2}{C}-\overset{1}{C}-OH \qquad \overset{\qquad OH}{\underset{4\quad3\quad2\quad1}{C-C-C-C}}$$

normal butyl alcohol secondary butyl alcohol
(1-butanol) (2-butanol)

$$\overset{C}{\underset{3\quad2\quad1}{C-C-C-OH}} \qquad \overset{OH}{\underset{3\qquad2\quad1}{C-C-C}} \atop C$$

isobutyl alcohol tertiary butyl alcohol
(2-methyl-1-propanol) (2-methyl-2-propanol)

The —OH group gives alcohols some waterlike character. Due to hydrogen bonding, alcohols have higher boiling points than the related hydrocarbons. Alcohols are very slightly acidic, giving not alkyl and hydroxyl ions as might be expected, but *alkoxyl* and hydrogen ions:

$$ROH \rightleftharpoons RO^- + H^+$$

Alcohols can be synthesized by alkene hydration, Eq. 21-5(3); carbonyl group reduction, Eq. 21-8(1); and fermentation. In fermentation (used to make alcoholic beverages), enzymes released from yeast cells catalyze the decomposition of sugars. Ethyl alcohol is one product:

$$C_6H_{12}O_6 \xrightarrow[H_2O]{yeast} 2CO_2 + 2H_3CCH_2OH$$

The reactions of 2-propanol illustrate alcohol reactions:

$$(1) \quad \xrightarrow[cat.]{\Delta} \quad H_3CCH{=}CH_2 + H_2O$$

intramolecular dehydration

$$(2) \quad \xrightarrow[cat.]{\Delta} \quad (CH_3)_2\overset{\overset{H}{|}}{C}{-}O{-}\overset{\overset{H}{|}}{C}(CH_3)_2 + H_2O$$

intermolecular dehydration

$$(3) \quad \underset{\rightleftharpoons}{\overset{HCl}{}} \quad H_3C\overset{\overset{Cl}{|}}{C}HCH_3 + OH^-$$

displacement

$$(4) \quad \xrightarrow{(O)} \quad H_3C\overset{\overset{O}{\|}}{C}CH_3 + H_2O$$

oxidation

(with the structure $H_3C\overset{\overset{OH}{|}}{C}HCH_3$ bracketed to reactions (1)–(4))

(21-7)

1. Dehydration within an alcohol molecule (loss of —OH and H from adjacent carbons) is accomplished by heating with concentrated sulfuric or phosphoric acids or passing the alcohol vapor through a hot tube packed with Al_2O_3. Tertiary alcohols are dehydrated most readily, primary alcohols least readily. (In a *primary* alcohol —OH is on a carbon atom that is attached to two other carbons.)

2. Elimination of a water molecule between two alcohol molecules leads to an ether, here diisopropyl ether. The *bimolecular dehydration* competes with reaction (1) above, so water loss may give both products.

The typical ether is a volatile liquid that is relatively chemically inert. Diethyl ether is used for anesthesia.

3. Displacement of a hydroxyl group by halide ion is most rapid with a tertiary alcohol, least rapid with a primary alcohol. Reasons for the reactivity order are discussed below.

Displacements involving tertiary compounds tend to go in two steps. The initial step is a slow dissociation, giving a carbonium ion:

$$R{-}\overset{\overset{R'}{|}}{\underset{\underset{R''}{|}}{C}}{-}OH = \overset{R\diagdown\diagup R'}{\underset{\underset{R''}{|}}{C^+}} + OH^-$$

The second step is a rapid combination with an electron-rich reagent (a nucleophile). If we assume the original reactant is asymmetric (—OH and three different R— groups attached to a carbon atom), the asymmetry tends largely to be lost because the carbonium ion is a free species that some of the time can be in a flat configuration with a plane of symmetry. Reacting on either "front" or "back" face with halide it gives a product that is a mixture of right- and left-handed molecules:

$$2\ \underset{\underset{R''}{|}}{R}\!\!\diagdown\!\!\underset{}{\overset{R'}{\diagup}}\!\!C^{+}\ +\ 2Cl^{-}\ =\ \underset{R''}{\overset{R}{Cl\cdots C\cdots R'}}\ +\ \underset{R''}{\overset{R}{R'\cdots C\cdots Cl}}$$

carbonium ion   nucleophile   dl pair

Reactions like this are called S_N1 *reactions* (substitution, nucleophilic, first order). First order means the reaction rate depends only on the concentration of the molecule converted, $RR'R''COH$.

Displacements on primary and secondary molecules tend to go in one step via an activated complex (Chapter 17). The approach of the nucleophile is from the molecule's back side or side opposite the group to be displaced. As a bond is formed with the approaching group and the bond broken with the leaving group, the molecule is *inverted* in configuration like an umbrella being turned inside out:

$$Cl^- \ + H\!\!\cdots\!\!\overset{R}{\underset{R'}{C}}\!\!-OH\ =\ \left[\overset{R}{\underset{H\quad R'}{Cl\cdots C\cdots OH}} \right]^{-}\ =\ Cl\!-\!\overset{R}{\underset{R'}{C}}\!\!\cdots H\ +\ OH^{-}$$

nucleophile   activated   inverted   leaving
                               complex   structure   group

If the inverted reactant was originally a right-handed molecule, the product is a left-handed molecule and optical activity is retained. Reactions of this type are called S_N2 *reactions* (substitution, nucleophilic, bimolecular). Bimolecular means the reaction rate depends on the concentration of both reactants.

The S_N1 reactions are favored in ionizing solvents like water with reactants that have large R— groups whose electrons can be partially dislocated to help stabilize the carbonium ion intermediate. The S_N2 reactions are favored in nonionizing solvents with reactants containing small R— groups that do not by physical bulkiness (called *steric hindrance*) block the backside attack.

4. In the laboratory alcohols may be oxidized with sodium dichromate in sulfuric acid solution; in industry the usual oxidizer is air with a catalyst. Under mild conditions tertiary alcohols are not oxidized. Secondary alcohols like 2-propanol are oxidized to ketones. Primary alcohols are oxidized to aldehydes:

$$RCH_2OH \ \xrightarrow{(O)}\ R\!-\!\overset{\displaystyle O}{\overset{\|}{C}}\!-\!H$$

Alcohols are used industrially as solvents and as reactants to make other oxygenated compounds.

Aldehydes and ketones

The characteristic group in aldehydes is $-\!\overset{\displaystyle O}{\overset{\|}{C}}\!-H$, whereas in ketones it is

$-\!\overset{}{C}\!-\!\overset{\displaystyle O}{\overset{\|}{C}}\!-\!\overset{}{C}\!-$. The common $-\!\overset{\displaystyle O}{\overset{\|}{C}}\!-$ unit is called a *carbonyl group*.

Aldehydes have common names derived from the acids to which they can be oxidized; thus formaldehyde → formic acid, acetaldehyde → acetic acid, etc. In IUPAC nomenclature aldehyde names come from the related hydrocarbons and the suffix *-al*:

$$
\underset{\substack{\textit{formaldehyde}\\ \textit{(methanal)}}}{HC\!-\!H} \qquad \underset{\substack{\textit{acetaldehyde}\\ \textit{(ethanal)}}}{H_3CC\!-\!H} \qquad \underset{\substack{\textit{butyraldehyde}\\ \textit{(butanal)}}}{H_3CCH_2CH_2C\!-\!H}
$$

Common names for ketones are made from the names of the two groups attached to the carbonyl carbon. Systematic ketone names come from related hydrocarbon names and end in *-one*:

$$
\underset{\substack{\textit{acetone}\\ \textit{(propanone)}}}{H_3CCCH_3} \qquad \underset{\substack{\textit{methyl ethyl ketone ("MEK")}\\ \textit{(butanone)}}}{H_3CCCH_2CH_3} \qquad \underset{\substack{\textit{methyl isobutyl ketone ("MIBK")}\\ \textit{(4-methyl-2-pentanone)}}}{H_3CCCH_2CHCH_3}
$$

Low-molecular-weight aldehydes and ketones are water soluble. Ketones are good solvents. Ketone odors are usually pleasant; aldehyde odors may be sharp.

The chemistry of aldehydes and ketones is that of the carbonyl group and the adjacent carbon atom (α carbon) with its hydrogen atoms. (The latter reactions are somewhat complex and will not be illustrated here.) Oxygen is more electronegative than carbon, so both σ and π electrons in the C=O bond are closer to the oxygen. The C=O bond is less reactive than the C=C bond; therefore some reagents that add readily to C=C (Br_2, HBr) give unstable products or do not add at all to C=O. Resonance stabilization accounts for both the superior strength and polar behavior of the carbonyl group:

$$
\left\{ \enspace \overset{}{\underset{}{C::\ddot{O}:}} \qquad \overset{+}{C}:\ddot{O}:^{-} \enspace \right\}
$$

Typical carbonyl reactions can be illustrated with propionaldehyde (propanal):

$$
H_3CCH_2C\!-\!H
\begin{cases}
(1) \quad \xrightarrow[\textit{press.}]{H_2,\ Pt} \ H_3CCH_2CH_2OH \\[4pt]
\hspace{3.2cm} \textit{reduction} \\[16pt]
(2) \quad \xrightarrow{(O)} \ H_3CCH_2C\!-\!OH \\[4pt]
\hspace{3.2cm} \textit{oxidation} \\[16pt]
(3) \quad \xrightarrow{HCN,\ OH^{-}} \ H_3CCH_2C\!-\!CN \\[4pt]
\hspace{3.2cm} \textit{cyanohydrin formation}
\end{cases}
$$

(21-8)

1. Aldehydes are reduced to alcohols with H_2 and a catalyst. (Reduction of $C{=}O$ to CH_2 can be done with zinc amalgam and acid.)

2. The ease with which aldehydes are oxidized to acids (or acid anions) is the basis of several aldehyde tests. *Fehling's test* uses an alkaline copper-tartrate solution (see Eq. 23-1). *Tollens' test* uses diammine silver ion ($AgNO_3$ in aqueous NH_3) as an oxidant; deposition of a silver mirror constitutes a positive reaction:

$$RCH + 2Ag(NH_3)_2{}^+ + 3OH^- = RCO^- + 2Ag\downarrow + 4NH_3 + 2H_2O$$

3. Cyanide ion adds to a $C{=}O$ bond (but not to $C{=}C$). The reaction is interpreted with reference to the $C{=}O$ resonance forms on page 538. The nucleophilic CN^- group donates an electron pair to the electrophilic carbon. Acidification of the resulting ion gives a *cyanohydrin*, a compound useful in further synthesis:

Carboxylic acids and esters

Carboxylic acids contain the *carboxyl group*, $-\overset{\displaystyle O}{\underset{\displaystyle \|}{C}}-O-H$, from which a proton can dissociate:

$$R-\overset{O}{\overset{\|}{C}}-OH \rightleftharpoons R-\overset{O}{\overset{\|}{C}}-O^- + H^+$$

The greater acid strength of the carboxyl's —OH ($K_a \simeq 10^{-5}$) compared to the —OH of an alcohol ($K_a \simeq 10^{-16}$) is attributable to resonance which stabilizes the acid anion:

Acids substituted on the α carbon by atoms more electronegative than hydrogen dissociate more readily because the substituent, by attracting electrons toward itself, loosens the O—H bond. Due to this electrical effect, called the *inductive effect,* fluoroacetic acid, for example, is about 200 times stronger than acetic acid.

Aliphatic monocarboxylic acids were named *fatty acids* in the early 1800s when chemists found that hydrolysis of fats and oils yielded the C_4 and higher acids. Certain unpleasant odors of animal origin (rancid butter, human perspiration) are due mainly to free fatty acids. Common names come from acid origins: formic acid (L. *formica*), ant; acetic acid (L. *acetum*), vinegar; caproic acid (L. *caper*), goat; IUPAC names are made from hydrocarbon names and the ending *-oic*:

$$
\underset{\text{formic acid}}{\underset{\text{(methanoic acid)}}{HC-O-H}}
\qquad
\underset{\text{isovaleric acid}}{\underset{\text{(3-methylbutanoic acid)}}{H_3CCHCH_2C-O-H}}
\qquad
\underset{\text{caproic acid}}{\underset{\text{(hexanoic acid)}}{H_3CCH_2CH_2CH_2CH_2C-O-H}}
$$

Esters contain the grouping $C-O-C$. Natural esters include animal and plant fats, oils, and waxes. Esters can be synthesized by combining acids with alcohols. Ester names are derived accordingly. Thus formic acid and methyl alcohol give methyl formate (methyl methanoate), $HC-O-CH_3$.

Acetic acid undergoes typical carboxylic acid reactions:

$$
H_3CC-OH
\begin{cases}
(1) \xrightarrow{\ NaOH\ } H_3CC-O^-Na^+ + H_2O \quad \text{neutralization} \\
(2) \xrightleftharpoons{\ H_3CCH_2-OH,\ H^+\ } H_3CC-O-CH_2CH_3 + H_2O \quad \text{esterification}
\end{cases}
\tag{21-9}
$$

1. Neutralization gives a salt, in this case sodium acetate. Sodium and potassium salts of long-chain fatty acids (see Table 23-1) are *soaps*. The reaction is reversed by adding a strong acid.

2. Elimination of water between an acid and an alcohol in the presence of a strong acid catalyst gives an ester. The reaction is reversible. One might assume that the water comes from combination of the acid's H^+ and alcohol's OH^-. The opposite is true, however, as demonstrated by esterifying with an alcohol containing isotopic oxygen. All the tracer element is found in the ester:

$$
H_3CC-(O-H+H)-{}^{18}O-C_2H_5 = H_3CC-{}^{18}O-C_2H_5 + H_2O
$$

The following mechanism is proposed to explain the result. First, H^+ attacks the electron-rich carbonyl oxygen giving a carbonium ion:

$$
H^+ + RC-O-H = R\overset{OH}{\underset{OH}{C^+}}
$$

Next the carbonium ion and alcohol interact via an electron pair on the alcohol oxygen to give an unstable intermediate. As the intermediate eliminates water, the catalyst is regenerated:

$$
\overset{OH}{\underset{OH}{RC^+}} + H-O-R \rightleftharpoons \left[\overset{OH}{\underset{OH}{RC}}-\overset{R}{\underset{H}{O^+}} \right] \rightleftharpoons RC-O-R + H_2O + H^+
$$

Amines and amides

Amines are ammonia derivatives in which one or more hydrogen atoms have been replaced by an alkyl or aryl group:

$$R-\ddot{N}\diagdown\,^{H}_{\,H} \qquad \,^{R}_{R}\diagup\ddot{N}H \qquad \,^{R}_{R}\diagup\ddot{N}\diagdown^{R}\;|\;R$$

primary amine   *secondary amine*   *tertiary amine*

Common names are made by adding "amine" to the name of the alkyl group attached to nitrogen: H_2NCH_3, methylamine; $HN(C_2H_5)_2$, diethylamine; etc. Naturally occurring amines have common names that give no structural information (see Questions 6 and 7). Systematic names are made from hydrocarbon names and the suffix "-amine." Example: $H_3CCH_2CH_2NH_2$, 1-propanamine.

Amines are weak bases. In aqueous solution the electron pair on the nitrogen atom accepts a proton:

$$R-\ddot{N}H_2 + H_2O \rightleftharpoons R-NH_3{}^+ + OH^-$$

Reactions of methylamine (methanamine) illustrate amine reactions generally:

$$H_3C-NH_2 \begin{cases} (1) \xrightarrow{HCl} H_3C-NH_3{}^+Cl^- \\ \qquad\qquad \text{salt formation} \\[1em] (2) \xrightarrow{H_3CNH_3{}^+Cl^-} H_3C-\underset{H}{N}-CH_3 + NH_4Cl \\ \qquad\qquad\qquad \text{alkylation} \\[1em] (3) \xrightarrow{HNO_2} H_3C-OH + N_2 + H_2O \\ \qquad\qquad \text{deamination} \end{cases}$$

$$(21\text{-}10)$$

    1. Amines react with acids to form salts. (The amine is regenerated by adding strong base). Most amines are liquids but their salts can be crystallized from solution.

    2. Primary amines give secondary amines when heated with an amine salt.

    3. Aliphatic primary amines are converted to alcohols by nitrous acid. Nitrogen, which is liberated quantitatively, is the basis of the *Van Slyke method* for amine nitrogen analysis. Each millimole of N_2 collected in a gas buret at STP is equivalent to 1 millimole of $-NH_2$.

Amides contain the grouping $-\overset{O}{\overset{\|}{C}}-NH_2$. As derivatives of carboxylic acids they are named from the acids; for example, $H\overset{O}{\underset{\|}{C}}-NH_2$, formamide, or methanamide.

Unlike amines, amides do not form salts with acids in water solution. Lack of basic character is explained by the electron-attracting ability of the carbonyl group which lowers the electronegativity of $-NH_2$ enough so the electron pair on nitrogen does not attach a proton. The effect is evident in the resonance hybrid:

$$\left\{ \begin{array}{cc} :\!O\!: & :\!\ddot{O}\!:^{-} \\ \| & | \\ R\ddot{C}\!-\!\ddot{N}H_2 & R\ddot{C}\!=\!\overset{+}{N}H_2 \end{array} \right\} \tag{21-11}$$

The —NH_2 group of an amide can be alkylated or deaminated like the amine —NH_2, Eqs. 21-10 (2 and 3). Acetamide (ethanamide) gives typical amide reactions:

$$
H_3C\!-\!\overset{\overset{\displaystyle O}{\|}}{C}\!-\!NH_2
\left\{
\begin{array}{l}
(1)\ \xrightarrow[\text{then } H_2O]{LiAlH_4,}\ H_3CCH_2\!-\!NH_2 + H_2O + LiOH + Al(OH)_3 \\
\qquad\qquad\qquad\qquad\qquad\quad reduction \\[2ex]
(2)\ \xrightarrow[\Delta]{P_4O_{10}}\ H_3CC\!\equiv\!N + H_3PO_4 \\
\qquad\qquad\qquad dehydration
\end{array}
\right.
\tag{21-12}
$$

1. Amides are reduced to amines with lithium aluminum hydride (or with H_2 and a catalyst).

2. Dehydration of an amide gives a nitrile (cyanide).

21-7. REACTIONS OF AROMATIC HYDROCARBON DERIVATIVES

Groups substituted on aromatic rings may react as they do when substituted on aliphatic chains; or their chemistry may be modified by electronic interaction with the ring's π cloud. For a sampling of reactions we will consider the reactivity of but three substituents: the nitro group, the alkyl group, and the amine group.

In general nitro groups substituted on either aliphatic or aromatic molecules are easily reduced. Aliphatic chains substituted on aromatic rings are more easily oxidized than the alkanes themselves. And aromatic amines are more easily converted to diazo compounds than are aliphatic amines:

$$(1)\quad \underset{NO_2}{\bigcirc} \xrightarrow{Fe,\ H^+} \underset{^+NH_3}{\bigcirc} \xrightarrow[(-H_2O)]{OH^-} \underset{NH_2}{\bigcirc}$$

reduction

$$(2)\quad \underset{CH_2CH_2CH_3}{\bigcirc} \xrightarrow{(O)} \underset{C\underset{OH}{\overset{O}{\diagup}}}{\bigcirc} + H_3C\overset{\overset{\displaystyle O}{\|}}{C}\!-\!OH$$

oxidation

$$(3)\quad \underset{NH_2}{\bigcirc} \xrightarrow[0°C]{H^+,\ HNO_2} \underset{^+N\equiv N}{\bigcirc} + H_2O \tag{21-13}$$

diazotization

1. Reduction of —NO$_2$ with a metal and acid gives an amine salt. Free amine is obtained by adding a strong base.

2. Oxidation of an alkylbenzene using aqueous sulfuric acid and a vigorous oxidizer like sodium dichromate oxidizes a side chain of any length at the carbon attached to the ring. Products shown above are benzoic and acetic acids.

3. When nitrous acid is added to an aliphatic amine, deamination occurs, Eq. 21-10 (3). An aromatic amine like aniline, however, with an ice-cold mixture of nitrous and hydrochloric acids gives a diazonium salt. Its stability can be explained as an energy lowering due to overlap of ring π electrons with N≡N π electrons.

Diazonium salts are valuable in synthesis. When boiled, for example, their solutions yield *phenols*:

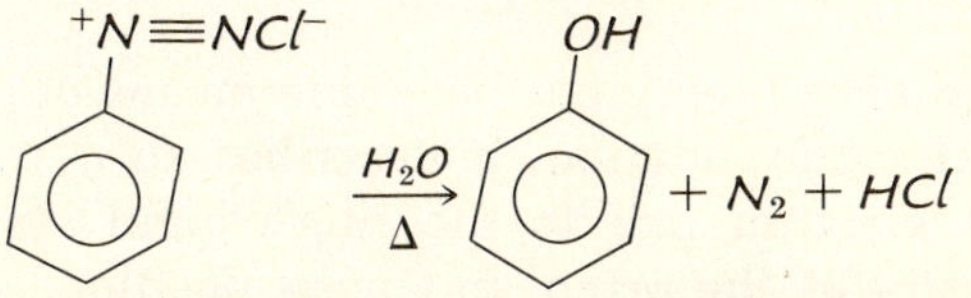

The product shown is phenol itself. Phenol also is the general name given all compounds having a hydroxyl group attached to an aromatic ring. In slightly alkaline solution diazonium salts undergo *coupling* with readily substituted aromatic compounds to yield colored *azo* compounds. Many dyes and indicators contain the azo linkage, —N—N :

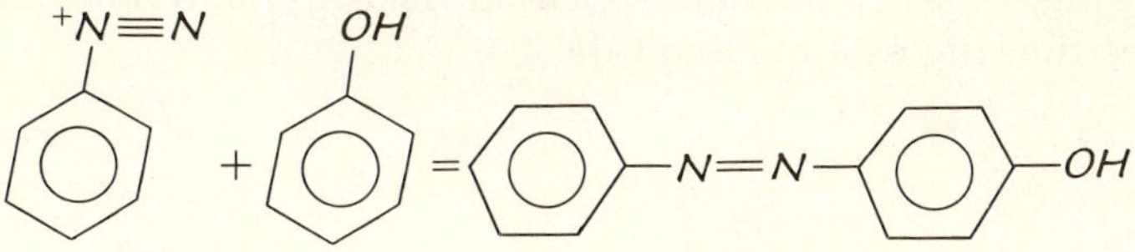

p-hydroxyazobenzene

Orienting influence of ring substituents

An atom or group substituted for a hydrogen atom on the benzene ring exerts effects manifested in both *orientation* or *directive influence,* and ease of reaction when a second substitution is made on the ring. Substituents are of two types. The first are called *ortho-para directors.* They promote attack at the *o* and *p* positions (see Eq. 20-4) and generally increase the ring's susceptibility to substitution. Strong *ortho-para* directors are —NH$_2$ and —OH; weaker are —CH$_3$ and the halogens —Cl and —Br. The second are called *meta directors.* They cause substitution to occur mainly at the *meta* positions while generally decreasing ring reactivity. Examples are —N—O,

—C≡N, —C—H, —C—O—H, and —C—NH$_2$.

Orientation rules were originally deduced from experiments and used empirically to guide syntheses. Thus to synthesize *p*-chloronitrobenzene a chemist knew he must first chlorinate benzene, then nitrate (—Cl an *o-p* director), then separate the isomers:

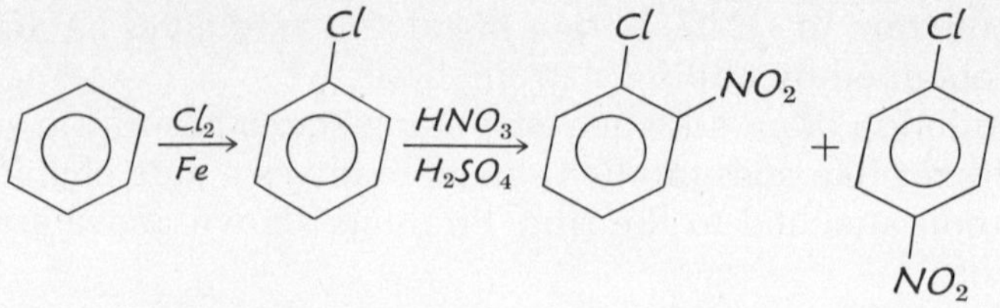

To make *m*-chloronitrobenzene the chemist first nitrated benzene then chlorinated (—NO_2 an *m*-director):

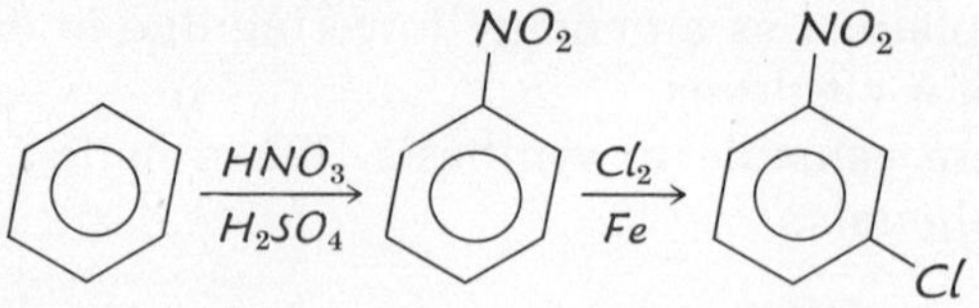

Since 1930 the resonance theory has been one means of giving the orientation rules a theoretical foundation. It describes an *o–p* director as a group that releases an electron pair to the ring's bond system where it increases electron density at the *ortho* and *para* positions. Consider the amino group, —$\ddot{N}H_2$. If its unused electron pair establishes a second bond between the nitrogen atom and the ring, then several plausible and approximately equal-energy forms are possible. As always in resonance theory, the structures written are different electronic arrangements derived by redistributing electron pairs in keeping with the octet rule. With an *o–p* director on the ring, negative charges appear at the *ortho* and *para* positions which enhance the ring as a nucleophile.

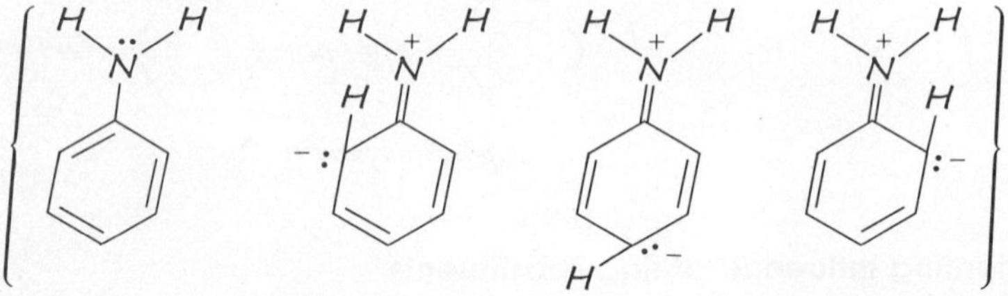

$$(21\text{-}14)$$

As an electrophilic reagent is attracted to one of those positions, the intermediate formed is stabilized because the positive charge remaining can be distributed about the ring:

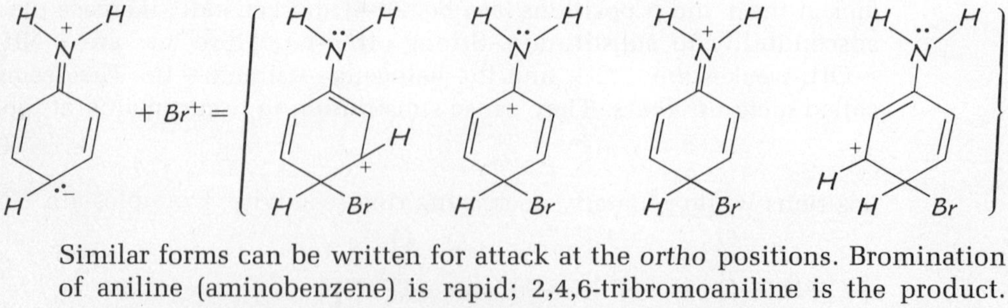

Similar forms can be written for attack at the *ortho* positions. Bromination of aniline (aminobenzene) is rapid; 2,4,6-tribromoaniline is the product.

A *meta*-orienting group according to resonance theory is an electron withdrawing group that *decreases* electron density at the *ortho* and *para* positions. Benzaldehyde, for instance, may be represented by the following resonance structures:

Withdrawal of an electron pair to the electronegative oxygen atom redistributes ring electrons so that positive charges appear at the *ortho* and *para* positions. Electrophiles attack the relatively unaffected *meta* positions.

21-8. SYNTHETIC POLYMERS

Terminology

A synthetic *polymer* (Gk. *poly* + *meros*, many parts) is a large molecule with a repeating structure. It may be synthesized by chemically combining many *monomer* (single part) units. When the units add to one another, as during synthesis of the familiar packaging material polyethylene (Eq. 21-15), the product is an *addition polymer*. When a small molecule like water is eliminated during polymerization, as in the synthesis of nylon (Figure 21-2), the product is a *condensation polymer*. When a single monomer is used, the product is a *homopolymer* (Gk. *homos*, same); when two different monomers are polymerized together, the product is a *copolymer*. Polymers that soften on heating are called *thermoplastic*; nonremeltable polymers are *thermosetting*. The names *resin* and *plastic* are often given to polymers used in molding applications. An *elastomer* is a rubberlike polymer.

Polymers are the major product of the synthetic organic chemical industry. In the United States polymers produced for packaging, automobiles, and furniture alone amount to 25 pounds per capita per year.

Rubber

Rubber was known to sixteenth century explorers of the Caribbean, who found that Indians obtained it by coagulating the milky sap (*latex*) of rubber trees with acidic fruit juices. In the nineteenth century rubber was deduced to be a mixture of long-chain polymeric alkenes (molecular weights ~ 60,000 to 360,000) having the empirical formula C_5H_8. Destructive distillation gave isoprene (2-methyl-1,3-butadiene) as the monomeric unit, a building block also found in pine tree gum and other natural products:

$$rubber \xrightarrow{\Delta} H_2C{=}\overset{\overset{\displaystyle CH_3}{|}}{C}{-}CH{=}CH_2$$

The technology for making synthetic rubber has been worked out since the 1930s and natural rubber has never regained its former importance.

To form a polymer, a molecule must possess at least two reactive positions at which it can attach two or more other molecules. An alkene such as ethene becomes polymerizable in the presence of oxide radicals produced by decomposition of an organic peroxide, $R{-}O{-}O{-}R = 2R{-}\dot{O}$. A radical attacking an alkene molecule causes an electron shift that opens the carbon–carbon double bond and produces a new radical:

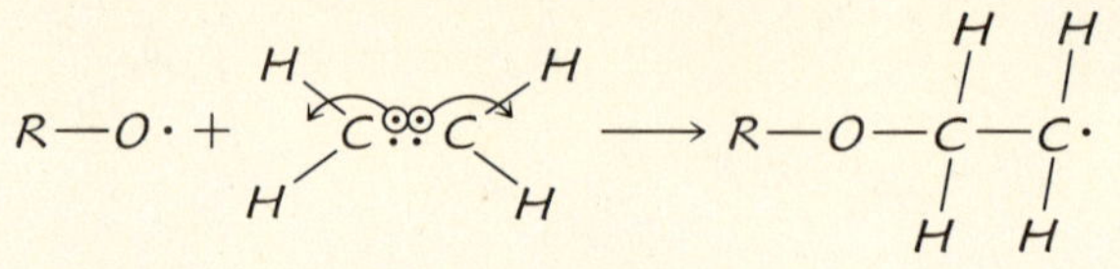

$$R\text{—}O\cdot + \quad C::C \quad \longrightarrow \quad R\text{—}O\text{—}C\text{—}C\cdot$$

A chain reaction begins as the new radical attacks another alkene molecule:

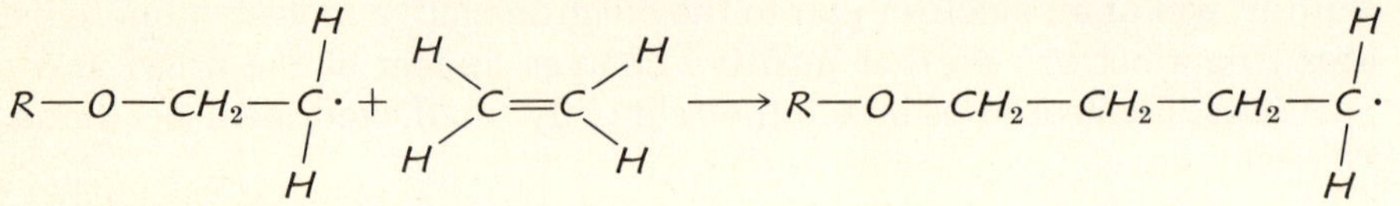

$$R\text{—}O\text{—}CH_2\text{—}C\cdot + \quad C{=}C \quad \longrightarrow \quad R\text{—}O\text{—}CH_2\text{—}CH_2\text{—}CH_2\text{—}C\cdot \qquad (21\text{-}15)$$

The chain continues building until terminated by reaction with a radical:

$$ROCH_2CH_2(CH_2CH_2)_xCH_2CH_2\cdot + \cdot OR = ROCH_2CH_2(CH_2CH_2)_xCH_2CH_2OR$$

Polymerizations like that above give a randomly oriented product. A product with more useful properties is obtained when polymerization is controlled to give a *stereoregular polymer*. Stereoregularity has been attained via catalysts introduced by Karl Ziegler (Germany 1898–1973) and Gulio Natta (Italy 1903–   ) who shared the Nobel Prize in chemistry in 1963. By a mechanism not yet completely understood their catalysts change the free radical route to an ionic route at lower temperature. A combination of $(C_2H_5)_3Al$ and $TiCl_4$, for example, with isoprene gives a quantitative yield of polymer which X ray diffraction shows has the same regular *cis* configuration and 9.1 Å repeat unit as natural rubber:

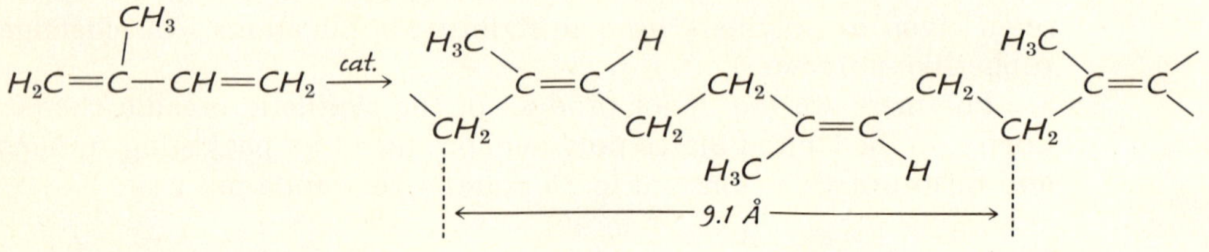

$$H_2C{=}C\text{—}CH{=}CH_2 \xrightarrow{\;cat.\;}$$

Polyesters

Polyesters may be synthesized by condensing a diacid with a dialcohol. The most important linear polyester is the product of reaction between terephthalic acid (from the oxidation of p-xylene, Figure 21-1) and ethylene glycol (from the hydrolysis of ethylene). It melts at about 270°C and can be made into fibers (*dacron*) or thin, tough films (*mylar*):

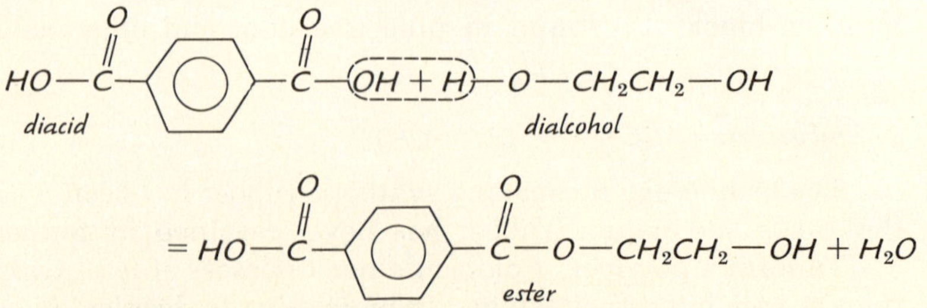

diacid   dialcohol

$$= HO\text{—}C\text{—}\bigcirc\text{—}C\text{—}O\text{—}CH_2CH_2\text{—}OH + H_2O \qquad (21\text{-}16)$$

ester

The ester product in this first step contains an acid group which can react with another ethylene glycol molecule. It also has an alcohol group which can react with another terephthalate molecule. Polyfunctionality is expressed by representing the polymer as a repeating unit,

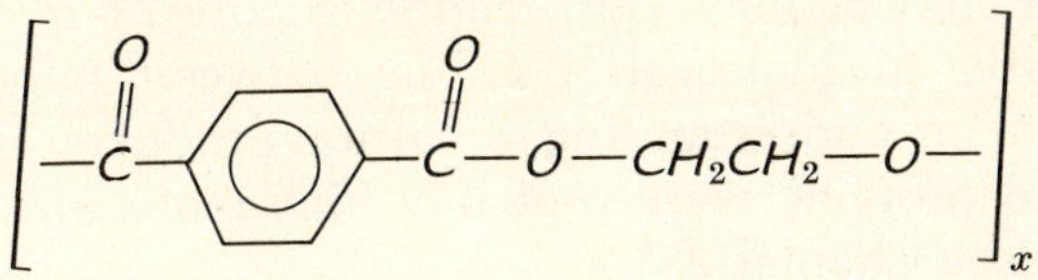

$$\left[-\overset{\overset{\textstyle O}{\|}}{C}-\!\!\bigcirc\!\!-\overset{\overset{\textstyle O}{\|}}{C}-O-CH_2CH_2-O-\right]_x$$

Polyamides

Reaction between a dicarboxylic acid and diamine leads to a polyamide in which the amide linkage, $-N-\overset{\overset{\textstyle O}{\|}}{C}-$, appears periodically. The best known

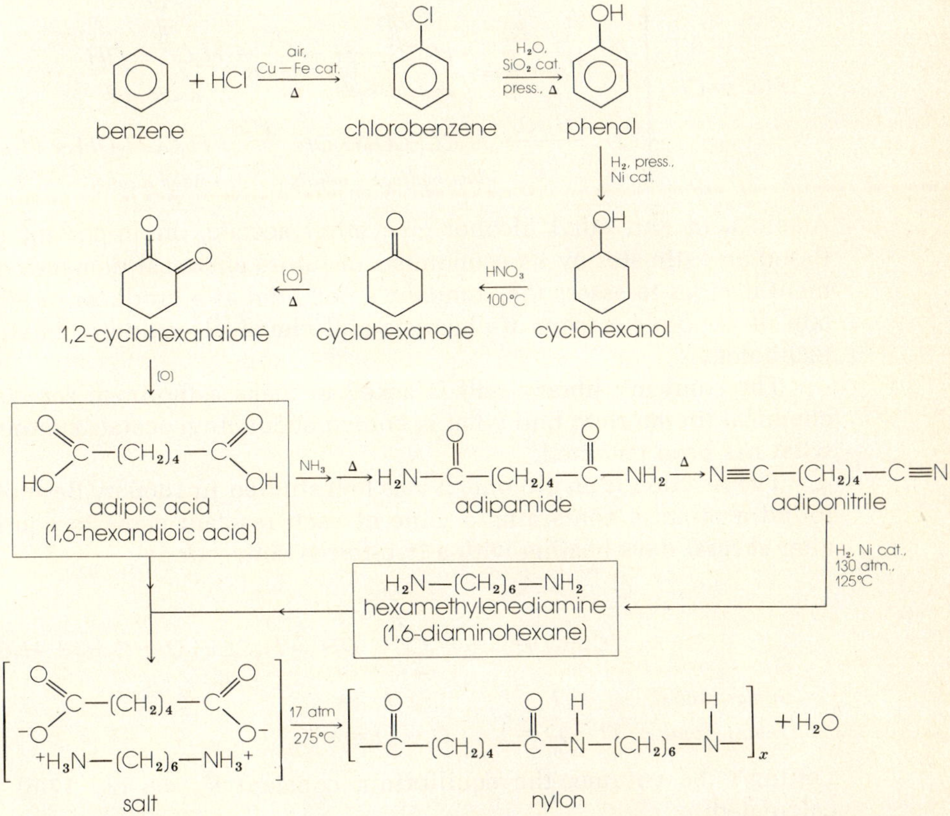

Figure 21-2 Synthesis of nylon. Benzene is chlorinated then hydrolyzed to phenol. The aromatic ring is hydrogenated giving an aliphatic alcohol. The alcohol is oxidized to a ketone, then to a diketone, and finally to a diacid as the ring opens. With ammonia the acid gives a diammonium salt which on heating loses water to yield a diamide. Heating the diamide gives a dinitrile whose triple bonds are hydrogenated to yield a diamine. Combination of diamine and diacid gives a salt which, when heated, undergoes condensation polymerization with the elimination of water. The polyamide chains of nylon have molecular weights between about 15,000 and 25,000.

synthetic polyamide is *nylon*, the first true synthetic fiber to be produced commercially. It was invented in 1934 as a substitute for silk by W. Carothers of the du Pont Chemical Corporation (Figure 21-2). Its high fiber strength is due partly to hydrogen bonding between polymer chains. Carothers' work led to an understanding of natural products. He stated that starch, cellulose, and proteins were probably high-molecular-weight condensation polymers (see Chapter 23).

21-9. COMMERCIALIZING A CHEMICAL PROCESS

Chemical industry translates chemical and physical processes from laboratory bench to manufacturing plant to produce chemicals society needs. In the following paragraphs a few features of such a development are discussed.

We will consider a company with expertise in ethylene technology. Two of its products are acetic acid and ethyl alcohol:

$$
H_2C{=}CH_2 \begin{cases} (1) \quad \xrightarrow[\text{Cu-Pd cat.}]{O_2} \; H_3C\overset{O}{\overset{\|}{C}}{-}H \xrightarrow[\text{Mn cat.}]{O_2} \; H_3C\overset{O}{\overset{\|}{C}}{-}OH \\[2pt] \qquad\qquad\qquad \text{acetaldehyde} \qquad\qquad \text{acetic acid} \\[6pt] (2) \quad \xrightarrow{H_2SO_4} \; H_3CCH_2OSO_2OH \xrightarrow{H_2O} \; H_3CCH_2OH + H_2SO_4 \\[2pt] \qquad\quad \text{ethyl hydrogen sulfate} \qquad\quad \text{ethyl alcohol} \end{cases}
$$

Acetic acid and ethyl alcohol give ethyl acetate, an important solvent. Based on estimates by its economists of future chemical demand, management decides to assess the company's potential as a producer of ethyl acetate. It appoints a team of chemists and chemical engineers to study the technology.

The company library staff is asked to make a thorough search in the chemical literature to find what is known about ethyl acetate synthesis and what has been patented.

The reaction is an old one. It was first studied in 1860 by Bertholet who found a mixture containing a mole of each reactant came to equilibrium after several days heating with a two-thirds conversion:

$$
C_2H_5OH \;+\; H_3C\overset{O}{\overset{\|}{C}}{-}OH \overset{\Delta}{\rightleftharpoons} H_3C\overset{O}{\overset{\|}{C}}{-}O{-}C_2H_5 + H_2O
$$

moles, initial:	1	1	0	0
moles, final:	$(1-\tfrac{2}{3})=\tfrac{1}{3}$	$(1-\tfrac{2}{3})=\tfrac{1}{3}$	$\tfrac{2}{3}$	$\tfrac{2}{3}$

Letting v be volume, the equilibrium constant K (see Eq. 17-5) can be calculated:

$$
K = \frac{[\text{ester}]\,[\text{water}]}{[\text{alcohol}]\,[\text{acid}]} = \frac{\left(\frac{\frac{2}{3}}{v}\right)\left(\frac{\frac{2}{3}}{v}\right)}{\left(\frac{\frac{1}{3}}{v}\right)\left(\frac{\frac{1}{3}}{v}\right)} = 4.0 \tag{21-17}
$$

Two things are unfavorable: The reaction is neither fast nor complete.

It is known that strong acid catalyzes esterification and this suggests a laboratory study. Strong acid catalysts are boiled with the reactants for

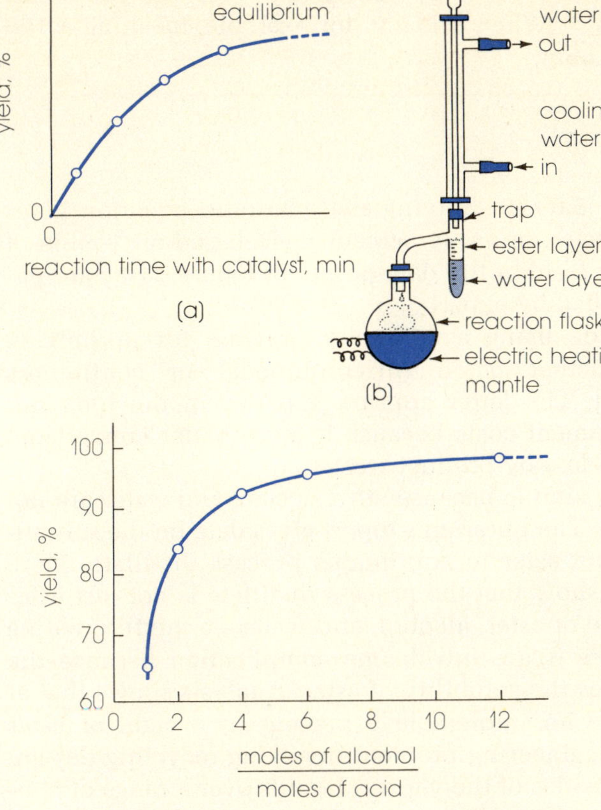

Figure 21-3 Studying the esterification reaction. (a) A time-yield curve (at the boiling temperature). Distillation will be carried out before the mixture reaches equilibrium in order to save time. Unused reactants will be returned to the reaction. (b) Laboratory setup for studying the effect of water removal on equilibrium. Condensate from the condenser drops into the trap, forming two layers. Water is thus removed from the reaction while ester overflows back into the heated flask. (c) A plot of alcohol:acid ratios versus yield. Yield is based on acid.

various periods of time and the reaction followed by titrating the remaining acetic acid with sodium hydroxide (see Figure 16-10). Sulfuric acid (b.p., 340°C) is chosen to be the catalyst because it will not distill with the ester during subsequent purification. Two percent by weight sulfuric acid is found to be optimum.

The catalyst hastens attainment of equilibrium within minutes (Figure 21-3a). To increase the conversion, alternatives controlled by the law of mass action are considered. In one series of experiments (Figure 21-3b) water is removed from the boiling mixture. In another series excess alcohol is used. As expected, both changes give increased product yield. The engineers will design equipment for combining the two operations and recovering unreacted alcohol.

Calculations are made to predict the effect of excess alcohol on conversion. In a 1-liter container 1 mole of acetic acid will be assumed mixed with some number, x, moles of alcohol. At equilibrium some number, y, moles of each product will have formed, and the original number of moles of each reactant minus y will remain:

$$C_2H_5OH + H_3CC\overset{\displaystyle O}{\overset{\|}{C}}{-}OH \overset{\Delta}{\rightleftharpoons} H_3CC\overset{\displaystyle O}{\overset{\|}{C}}{-}O{-}C_2H_5 + H_2O$$

moles, initial:	x	1.0	0	0
moles, final:	$(x - y)$	$(1.0 - y)$	y	y

By assigning arbitrary values to x, the values of y, the ester yield, can be calculated using Eq. 21-17. When x is 2.0, for example, meaning a 100 percent initial excess of alcohol,

$$K = 4.0 = \frac{(y)(y)}{(2.0 - y)(1.0 - y)}$$

from which, $3.0y^2 - 12y + 8.0 = 0$. Solving the quadratic equation gives $y = 0.84$ mole per liter of ester, or an 84 percent yield based on 1 mole of acetic acid used. In similar fashion the data points of Figure 21-3c are obtained. The results are verified by experiment.

Cost estimates are made using a computer program for optimizing operations to decide whether *batch* (discontinuous) or *continuous* processing should be used. The latter appears superior in the long run despite higher initial equipment costs because it gives better conversion, more uniform product, and less by-product waste.

Product recovery seems simple because ethyl acetate and water are not very soluble in one another. The librarian's report gives data on the solubility of ester in water, but not ester in continuous process distillate. More experiments are run. They show that the process distillate is not just ester and water, but an *azeotrope* of ester, alcohol, and water—a mixture with a constant boiling point. Here is an unwelcome complication because the presence of alcohol increases the solubility of ester. Analysis shows that at the distillation temperature an intolerable 9 percent by weight of ester formed is lost. This is an engineering problem involving recycling design. Studies are needed on viscosities of the various liquids over a range of temperatures. The laboratory conducts the studies. Pipe and pump particulars must be calculated from the data.

The engineers ask about the heat of the reaction. Is it exothermic or endothermic? For a given reactor, how much heating or cooling will be needed? What kinds of construction materials should be used to prevent product contamination yet keep building costs in hand?

The library report includes enthalpies of formation (see Chapter 12) in kilocalories per mole at 25°C;

$$\underset{-66.4}{C_2H_5OH\,(liq)} + \underset{-115.3}{H_3C\overset{\displaystyle O}{\overset{\|}{C}}{-}OH\,(liq)} = \underset{-110.7}{H_3C\overset{\displaystyle O}{\overset{\|}{C}}{-}O{-}C_2H_5\,(liq)} + \underset{-68.4}{H_2O\,(liq)}$$

The total on the left is -181.7, on the right -179.1. Product total minus reactant total is $+2.6$ kilocalories; this is the enthalpy of the reaction (Eq. 12-10). The reaction is slightly endothermic and will require heating.

Except for the catalyst all components of the system are volatile. The engineers decide that the reactant mixture can be sprayed directly into a hot tower where crude ester will be distilled off as it is formed. Rapid handling will save cost of a separate reaction vessel equipped with a distilling column.

A *pilot plant* producing 20 gallons of ester per day is built to study the concepts developed so far. Its operation suggests certain changes. It demon-

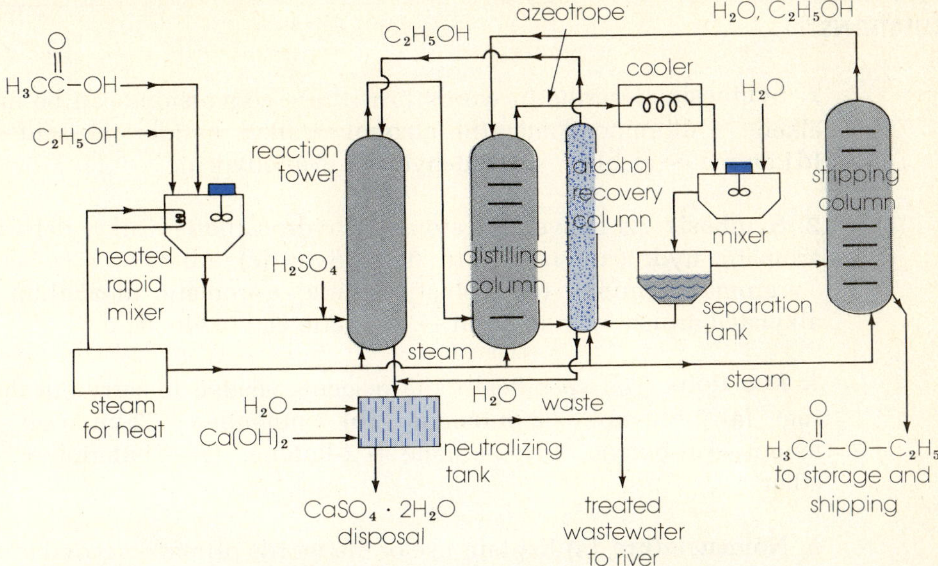

Figure 21-4 A flow chart for ethyl acetate production. Acid and excess alcohol are fed to the mixer (top, left). Heated and mixed, the solution, plus sulfuric acid catalyst as needed, is sent quickly to the reaction tower where acetic acid is mostly used up. At 81°C distillate is taken off the top (70 percent alcohol, 20 percent ester, 10 percent water) and sent into the distilling column operating at 71°C. From the top of the distilling column comes a three-component azeotrope that is 83 percent ester, 9 percent alcohol, and 8 percent water. It is cooled, mixed with water, and run into a separation tank. The water layer and water from the bottom of the distilling column go to an alcohol recovery distilling column which returns alcohol to the reaction tower and sends wastewater to the neutralizer. The top layer in the separator (93 percent ester, 5 percent water, 2 percent alcohol) goes to the stripping column where distillation sends water and alcohol, together with a little ester, back to the distilling column. The stripping column condensate, 99 percent ester, is sent to storage for drum loading and rail shipment. Wastewater is neutralized with lime, separated from calcium sulfate residue, and discharged into the river. (Production of ethyl acetate in the United States in 1973 was 240 million pounds.)

strates, for instance, that with efficient distillation equipment the wastewater will have low biochemical oxygen demand (Section 15-5). Some sulfuric acid will escape. It must be neutralized. A holding tank with automatic pH control will be needed.

When all questions are answered, the economics of the operation are calculated with computer assistance. Included are estimates on cost of land, materials, construction, equipment, insurance, utilities, operating personnel, maintenance, taxes. Life expectancy of both plant and its technology are discussed.

After considering the reports, the company's financial resources, and chemical market trends, management commits the company to commercializing the process. The plant, shown in Figure 21-4, having a production capacity of 1000 gallons per day (2.7 million pounds per year) is now constructed.

QUESTIONS

1. Synthesis Describe in words how these conversions can be made: **(a)** alkene → dibromoalkane; **(b)** alcohol → alkyl halide; **(c)** acid → amide; **(d)** amine → alcohol; **(e)** aldehyde → cyanohydrin.

2. Synthesis As above: **(a)** aromatic hydrocarbon → alkyl derivative; **(b)** aromatic hydrocarbon → nitro derivative; **(c)** aromatic nitro compound → aromatic amine; **(d)** aromatic amine → aromatic diazonium salt; **(e)** alkylated aromatic compound → aromatic carboxylic acid.

3. Reactions Tell specifically the reagents needed to carry out the following: **(a)** 2-butanol → 2-butanone; **(b)** 2-butanone → 2-butanol; **(c)** 2-butanone → n-butane; **(d)** 2-butane → 2-butene; **(e)** 2-butanol → di-2-butyl ether.

4. Nomenclature **(a)** Explain use of the words primary, secondary, and tertiary in describing these compounds:

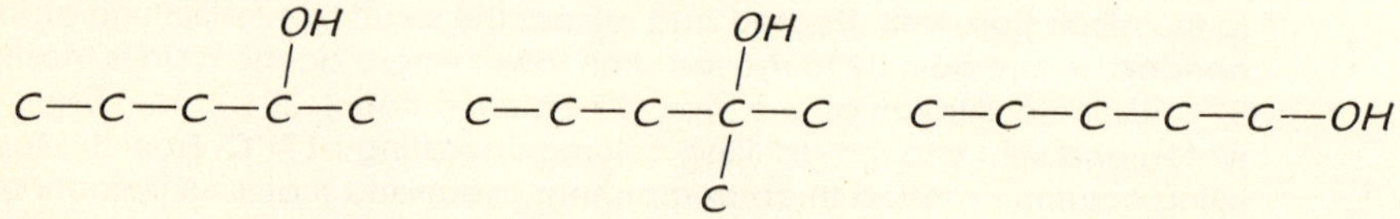

$$
\begin{array}{ccc}
& \mathrm{OH} & \\
& | & \\
\mathrm{C-C-C-C-C} & \mathrm{C-C-C-C-C} & \mathrm{C-C-C-C-C-OH} \\
& | & \\
& \mathrm{C} &
\end{array}
$$

(b) Give the three compounds systematic (IUPAC) names. **(c)** An organic chemist abbreviates four formulas this way: MeOH, *i*PrI, EtOEt, nBuCHO. Give the compounds systematic names.

5. Chemical tests How by *chemical* tests can one distinguish between **(a)** propanoic acid and 1-propanol; **(b)** 1-propanol and 2-propanol; **(c)** 1-propanol and propanal; **(d)** butane and butene; **(e)** methanol and water?

6. Identification Name the functional group(s) in each compound:

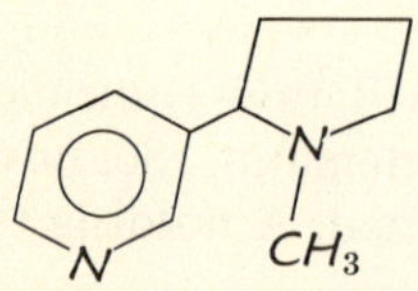

nicotine (from tobacco)

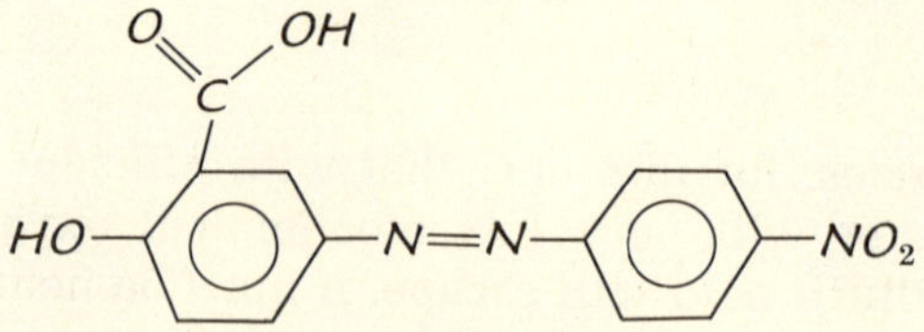

alizarin yellow R (a dye)

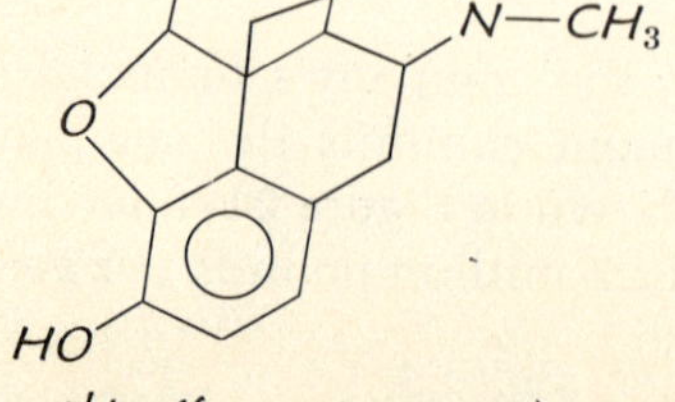

morphine (from opium poppy)

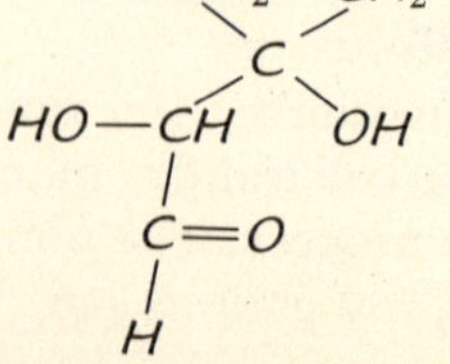

apiose (sugar from parsley)

7. Identification As before:

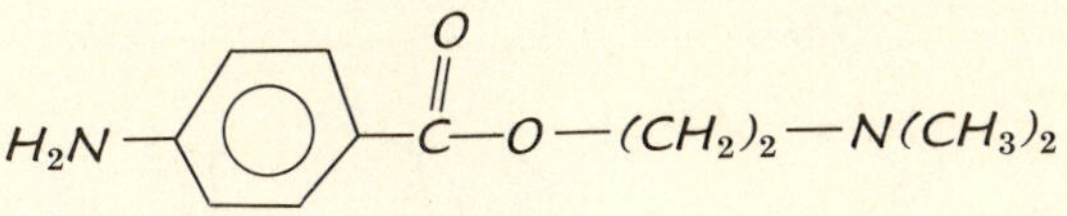

novacaine
(anesthetic)

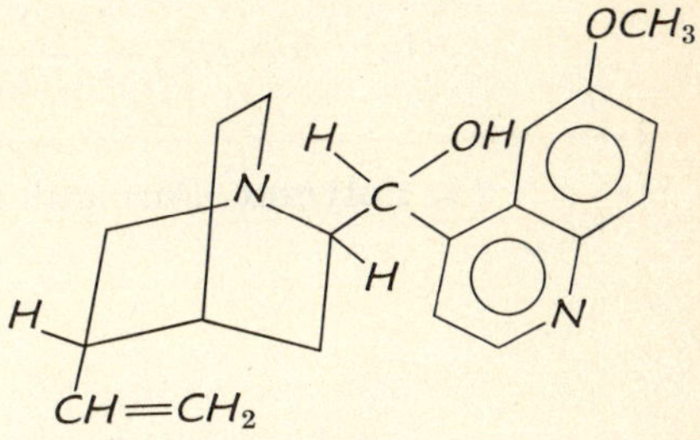

quinine
(malaria remedy)

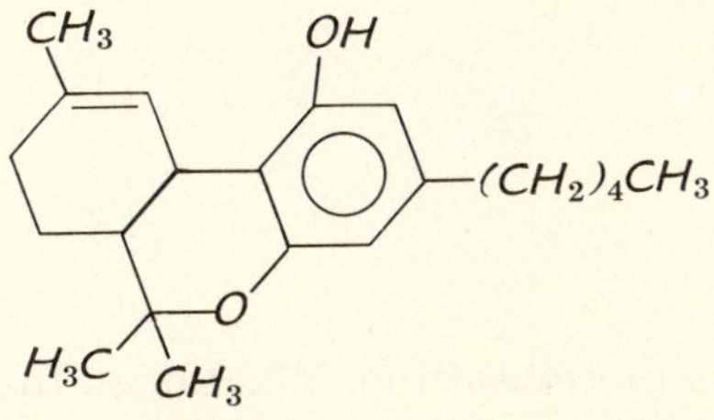

tetrahydrocannabinol
(from marihuana)

8. Definitions Explain the difference between **(a)** an alkane and an alkyl group; **(b)** aliphatic ring and aromatic ring; **(c)** amine and amide; **(d)** phenol and alcohol.

9. Definitions As above: **(a)** intermolecular and intramolecular dehydration of an alcohol; **(b)** deamination and diazotization of an amine; **(c)** nucleophile and electrophile; **(d)** esterification and neutralization of a fatty acid.

10. Definitions As above: **(a)** cracking and isomerization of an alkane; **(b)** ozonolysis and complete oxidation of an alkene; **(c)** a carbonyl group and a carboxyl group; **(d)** an aldehyde and a ketone.

11. Addition Which product predominates? Explain. Whose rule is illustrated?

$$2HBr + 2H_3CCH{=}CH_2 = H_3CCH_2CH_2Br + H_3CCHBrCH_3$$

12. Acid strength **(a)** Explain the inductive effect and how it relates to acid strength. **(b)** Arrange these acids in order of increasing strength and explain with the use of Table 9-1:

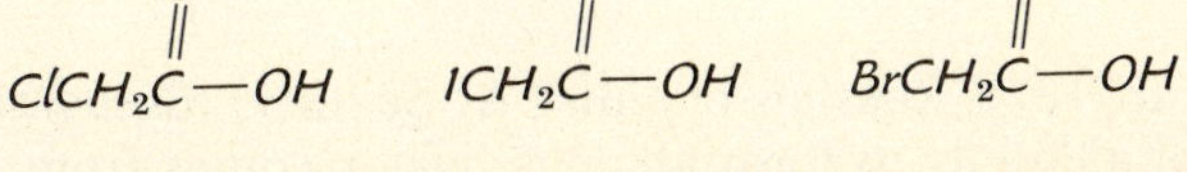

$$ClCH_2\overset{O}{\underset{||}{C}}{-}OH \qquad ICH_2\overset{O}{\underset{||}{C}}{-}OH \qquad BrCH_2\overset{O}{\underset{||}{C}}{-}OH$$

13. Orienting influence Name the main product(s) in each case (ϕ = the benzene ring):

(a) $\phi—NO_2 \xrightarrow[H_2SO_4]{HNO_3}$ (b) $\phi—\overset{\overset{\displaystyle O}{\|}}{C}—OH \xrightarrow[Fe]{Br_2}$ (c) $\phi—C_2H_5 \xrightarrow[AlCl_3]{ClCH_3}$

14. Polymer Formaldehyde and phenol react to give Bakelite resin:

$$HC(=O)—H + \text{(phenol, OH)} \xrightarrow{\Delta} \text{(Bakelite resin network)} \quad etc.$$

(a) Bakelite is thermosetting. What does that mean? **(b)** Is the above reaction an addition or condensation polymerization? Explain. **(c)** Is the product a copolymer or homopolymer? Explain.

15. Name reactions (library) In organic chemistry, reactions of general utility have been named for their discoverers. In an organic chemistry text find reference to the following "name reactions." Describe any two in general terms, giving types of reactants, conditions, and products. **(a)** Wurtz reaction; **(b)** Williamson ether synthesis; **(c)** Clemmensen reduction; **(d)** Beilstein test for halogen; **(e)** Cannizzaro reaction; **(f)** Gattermann-Koch aldehyde synthesis.

PROBLEMS

16. Combustion From the data given on enthalpies of combustion calculate $\Delta H°$ for the normal C_{11} and C_{15} alkanes.

17. Molecular weight A 0.21-g sample of an alkene adds 0.040 g of Br_2. Assuming one C=C bond per molecule, **(a)** What is the molecular weight of the alkene? **(b)** What is its molecular formula?

18. Homologous series A series of compounds describable by a general formula is called a *homologous series* (example: alkenes, C_nH_{2n}). **(a)** What is the general formula for monohydric alcohols (one —OH per molecule) derived from alkanes? **(b)** What is the general formula for aldehydes derived from all primary alcohols described in (a)?

19. Alcohols In 1875 Russian chemist A. Saytzeff discovered that in the dehydration of alcohols to form alkenes, water comes from —OH and the hydrogen atom from that adjacent carbon which is bonded to the fewest hydrogens. Write equations showing dehydration of **(a)** 2-pentanol;

(b) 2-methyl-3-pentanol; **(c)** 2,4-dimethyl-3-pentanol. **(d)** Name each alkene formed.

20. Alcohols A chemist has samples of 1-pentanol, 2-pentanol, and 2-methyl-2-butanol in unlabeled bottles. He adds HCl and $ZnCl_2$ to each and shakes. One gives an immediate reaction, forming the corresponding alkyl halide. Another gives the reaction after 5 minutes. The third does not react. How does this procedure (called the *Lucas test*) distinguish among the alcohols? Explain with equations.

21. Structure determination **(a)** Compound A is a 6-carbon alcohol. Dehydration gives 1-hexene. Write the structure and name of A. **(b)** Compound B is also a 6-carbon alcohol. Dehydration gives 2,3-dimethyl-2-butene. Write the structure and name of B.

22. Structure determination Compound A, C_4H_8O, and compound B, $C_5H_{10}O$, both give a positive Tollens' test. Upon hydrogenation with H_2 and Pt, A is converted to 1-butanol, B to 3-methyl-1-butanol. Write structural formulas and IUPAC names for A and B.

23. Esterification analogy When a mercaptan (a sulfur analog of an alcohol) reacts with a carboxylic acid, do you expect elimination of hydrogen sulfide or of water? Write the reaction equation using methyl mercaptan and propanoic acid. Explain.

24. Esterification Write equations showing ester formation between the following acids and alcohols. Name each ester. **(a)** Acetic acid + methyl alcohol. **(b)** Butanoic acid + ethanol.

25. Amines This is the Hofmann method (1849) for preparing amines: $RX + 2NH_3 = R—NH_2 + NH_4X$ (+ by-products; X is a halogen). Write equations showing reaction of the following alkyl halides with ammonia and name each amine formed: **(a)** iodoethane; **(b)** 1-bromobutane.

26. Amines Amines can be prepared by reducing nitro compounds: $R—NO_2 \xrightarrow{\text{(H)}} R—NH_2$. Show how the following amines can be prepared and give the systematic name of the nitro compound in each case: **(a)** 1-butylamine; **(b)** aniline; **(c)** p-methylaniline.

27. Amine structure A primary amine has a molecular weight of about 116. Subjected to the Van Slyke procedure 34.8 mg of amine gives 13.4 ml of N_2, collected dry at STP. How many —NH_2 groups are in the molecule?

28. Amides A general amide synthesis is

$$\underset{RC}{\overset{O}{\overset{\|}{}}}—O^-NH_4^+ \overset{\Delta}{=} \underset{RC}{\overset{O}{\overset{\|}{}}}—NH_2 + H_2O$$

Write equations showing how the ammonium salts of the following acids lead to amides, then give the systematic names of the amides: **(a)** formic acid; **(b)** caproic acid.

29. Synthesis Give equations showing how the following conversions can be made; add inorganic substances only, as needed: **(a)** 2-butene to 2-butanol; **(b)** 2-pentanol to pentane; **(c)** 2-butanol to 2,3-dibromobutane; **(d)** benzene to p-nitrotoluene; **(e)** toluene to 3-nitrobenzoic acid.

30. Polymerization process Show how reaction between oxalic acid and propylene glycol can lead to a polyester and indicate the reactive sites

in the first product:
$$HO-\overset{\overset{\displaystyle O}{\|}}{C}-\overset{\overset{\displaystyle O}{\|}}{C}-OH + HO-CH_2\overset{\overset{\displaystyle OH}{|}}{C}HCH_3 \rightarrow .$$

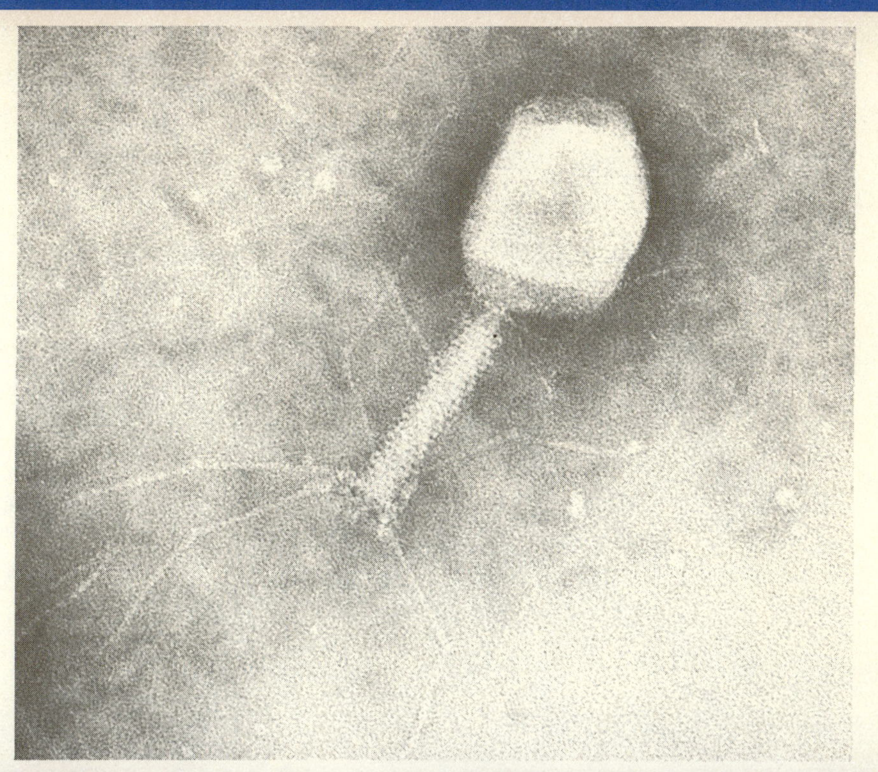

NINE

BIOCHEMISTRY— WHAT IS LIFE?

At the borderline of life; an electron micrograph of a virus particle (×320,000). All viruses are parasites that have some of the characteristics of living systems. This virus, called a bacteriophage, is capable of parasitizing certain bacterial cells and destroying them. The outer structure is made of proteins. The head contains a single molecule of DNA with a molecular weight of about 120 million. After the virus' six tail fibers contact a bacterium, the tail contracts and DNA is squeezed through it into the bacterial cell. The cell's metabolism then becomes diverted to producing virus DNA and protein, and within 20 minutes whole virus particles burst through the cell ready to infect other cells. At least one virus has been implicated in certain kinds of cancer in humans. (Courtesy of the Virus Laboratory, University of California, Berkeley.)

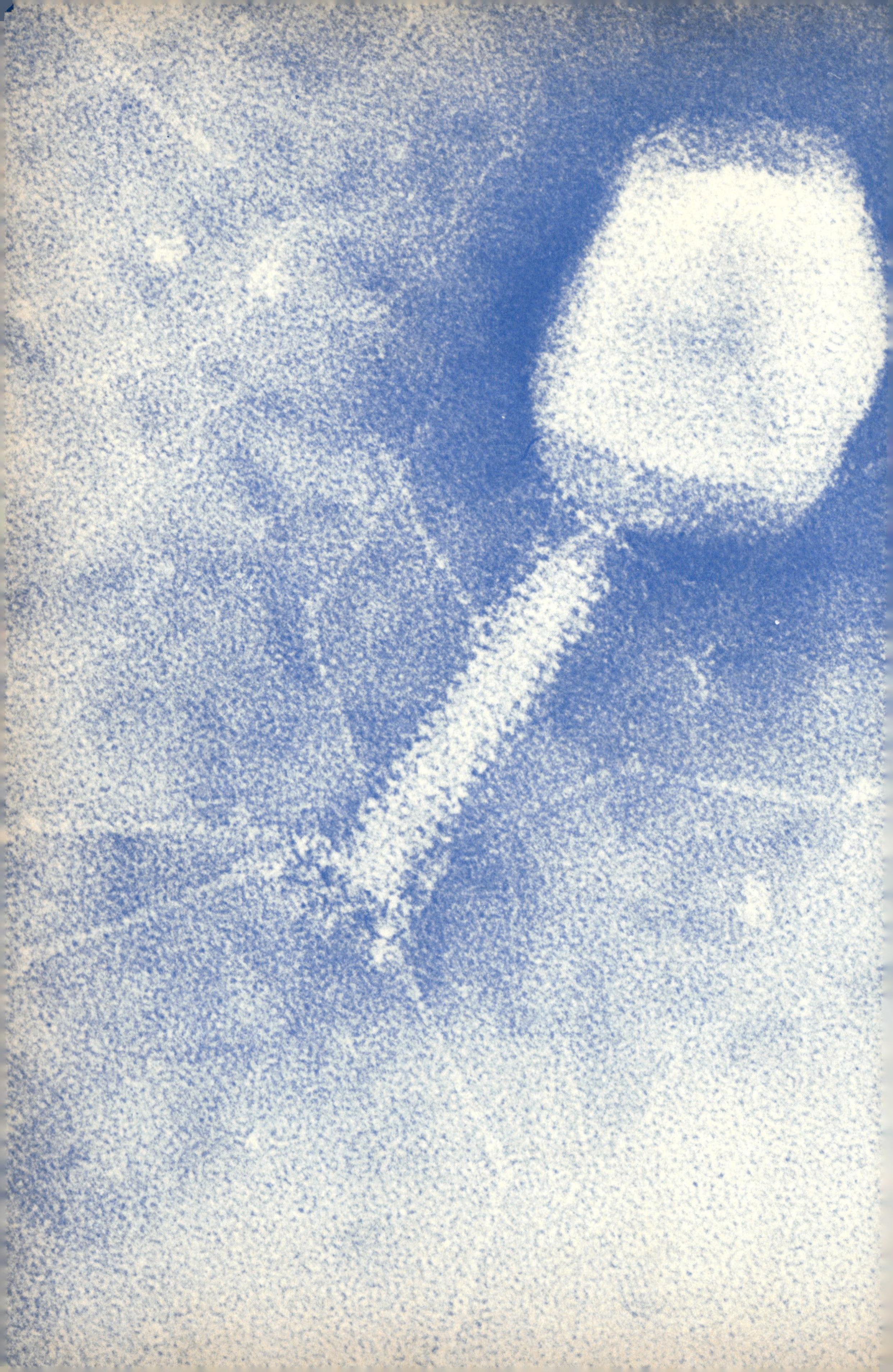

HOW DID LIFE BEGIN ?

THE IMPORTANT CONCEPTS

22-1. The Oparin-Haldane theory
1. The theory's basis is assumption of a reducing atmosphere first.
2. Some organic building units essential to life have been synthesized under simulated primitive earth conditions.

22-2. Biochemical evolution
1. Cell structure could have originated in amino acid polymers.
2. The origin of the asymmetry of natural compounds has not been established.

22-3. Biological evolution
1. First organisms were probably anaerobic heterotrophs that derived energy from fermentation.

2. Autotrophs arose when photosynthesis gave access to solar energy.
3. Two kinds of cells developed in nature, distinguished by their degree of internal structuring.

22-4 Photosynthesis; the photolysis reactions

1. Photosynthesis is the most consequential set of chemical reactions on earth.
   a. Two sequences operate: light capture, followed by reduction of CO_2.
2. Principal products of photolysis are NADPH, a reducing agent, and ATP, an energy transfer agent.
3. Chloroplasts contain the necessary compounds and structures for photosynthetic activity.
4. Electron (or hydrogen) reducing power can be traced from water through intermediate compounds to NADPH.
5. Photosynthetic membranes are intricate layer combinations of lipids, proteins, and pigments.

22-5. Photosynthesis; the reductive fixation reactions

1. Route of carbon in the fixation reactions was traced using $^{14}CO_2$ in vivo, and chromatography.
2. Photosynthetic carbon is rapidly incorporated into a variety of compounds, of which PGAl is particularly important.
3. From sunlight plant pigments absorb energy corresponding to their complementary colors.
4. Global efficiency of and total energy in photosynthesis can be estimated.

22-6. Change to an oxidizing environment

1. Photosynthesis could have changed the earth's atmosphere from a reducing to an oxidizing medium.
2. Absorbance of ultraviolet rays by atmospheric O_3 allowed life to emerge from the water.
3. Respiration gave life mobility and led to higher organisms than those fermentation can support.
4. A model of atmospheric O_2 increase can be correlated with findings in paleontology.

Biochemistry originated in the eighteenth century with Lavoisier's explanation of *respiration* as biological oxidation of food, and with experiments that showed living matter contained albuminous (protein), saccharine (carbohydrate), and oily (lipid) components not found in mineral sources. It gained status as a separate science called *physiological chemistry* in the latter part of the nineteenth century when organic chemistry was applied to problems of agriculture and medicine. Just as biology was unified by the cell theory of living matter, biochemistry found cohesion in discoveries which proved that (1) all cells have similar *metabolism*—that is, vital processes related to the intake of food and derivation of energy therefrom—and (2) complex cellular components are made from simpler molecules. In the twentieth century biochemistry has grown rapidly, stimulated by incorporation of ideas from physical and structural chemistry and development of powerful new analytical techniques such as electron microscopy and chromatography. Correlating the microstructure, function, and chemical reactions of cells, biochemistry today includes such areas as mechanisms for biological energy transfer, metabolic reaction sequences, reaction regulation by catalysts called *enzymes*, conformation of biomacromolecules, molecular basis of vision and memory, the behavior of viruses, and the composition and behavior of genetic material—the very root of life.

22-1. THE OPARIN-HALDANE THEORY

Spontaneous generation

Living systems are characterized by abilities to grow and reproduce, to convert nutrients to energy, to react to stimuli, and to alter their environmental adaptability through evolution. One way to understand life is to examine contemporary organisms. Another way is to begin at the beginning, to learn how life might have originated and developed.

Every religion and folklore describes a making of earth and a creation of man, an animal divinely touched with life. *Spontaneous generation*—earthworms springing from earth, weevils from grain—is also an ancient concept, and it was the working hypothesis well into the 1880s when scientists with microscopes were showing how mold appeared from "nowhere" to grow in nutrient solutions. Pasteur (Figure 20-4) proved, however, that growth did not begin in sterile solutions until unfiltered air carrying its normal load of invisible spores was admitted. His carefully conceived experiments doomed the spontaneous generation theory but left, as the only other scientific hypothesis, the idea that microscopic life called *cosmozoa* had long ago come from outer space to take root on earth. The cosmozoa theory evaded the question of the ultimate origin of life and, after

Figure 22-1 Aleksandr Oparin (USSR 1894–   ).
Beginning with the key assumption that the earth's
early atmosphere was reducing, Oparin proposed
that organic compounds of increasing complexity
formed first in the atmosphere, then in water, then
in minute enclosed systems where life began. He
has said that to understand the nature of life one
must also understand its origin and development.
(Courtesy Sovfoto.)

meteorites were examined for hitch-hiking germs (with negative results), it had few other tests to suggest.

The best scientific theory of *biopoesis* (origin of life) we have today was suggested in the 1920s by J. B. S. Haldane in England and by Oparin (Figure 22-1) in Russia. The theory's basic assumption is that simple molecules present on the barren early earth underwent a succession of chemical reactions to form organic compounds of increasing size and variation until the secrets of energy usage and self-replication were discovered. From then until the present *Darwinian evolution* (Charles Darwin, England, 1809–1882), based on "natural selection," or survival of the most fit to reproduce, is presumed to have operated and to account for the eventual emergence of human beings.

Synthesis in the atmosphere

Earlier theories assumed the earth's atmosphere has always contained oxygen because most known organisms respire, that is, use O_2 to oxidize their food. Oparin and Haldane, however, made the all-important assumption that little oxygen was initially present, that the earliest atmosphere instead contained hydrogen, methane, ammonia, water vapor, and hydrogen sulfide (see Figure 13-11). Energized by the earth's radioactivity, the fall of meteorites, violent electrical storms, and the sun's radiation, NO, N_2O, SO_2, SO_3, HCN, NCCN (cyanogen), $\cdot CH_3$, $:CH_2$, $\cdot NH_2$, $HO\cdot$, $NO\cdot$, $CN\cdot$, $H\cdot$, $O\cdot$, and other species could have formed and interacted to yield many different compounds.

Experiments to examine the possibility of *chemical evolution* were begun in the 1950s. Calvin (see Figure 22-12) showed that small aldehydes and carboxylic acids could be made without free oxygen if ionizing radiation was supplied to a mixture of carbon dioxide and water vapor. Urey (Figure 22-2), however, pointed out that if free hydrogen had been abundant in the early atmosphere, methane and not carbon dioxide would have been the stable primary carbon compound because at reasonable temperatures the

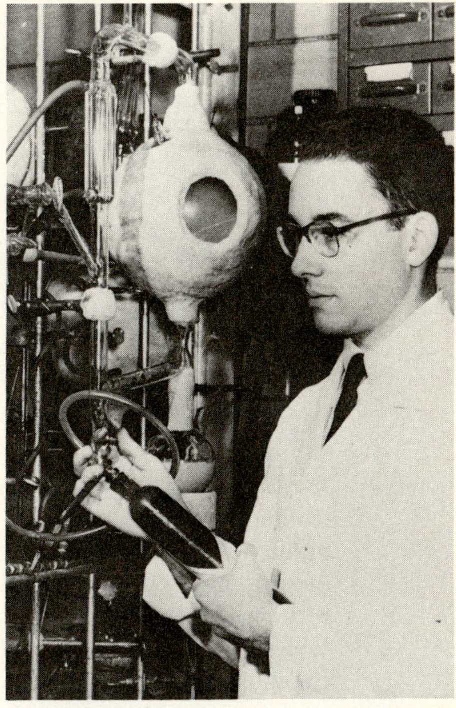

Figure 22-2 Scientists who contributed to biopoesis theory. (Left) Harold Urey (United States 1893– , Nobel prize in chemistry 1934). Urey discovered heavy hydrogen and worked on the atom bomb. Since World War II he has studied the chemical evolution of life and the solar system. (Right) Stanley Miller, a student of Urey's in 1953, with their apparatus for the synthesis of organic compounds under assumed primitive earth conditions. In a closed loop, sterile H_2O, CH_4, H_2, and NH_3 were boiled, passed as a gas through the flask (top, right) containing electrically sparking electrodes, cooled to condense the solution of chemical products formed, and siphoned back to the boiler. A week's cycling gave 19 products, including those in Table 22-1. When H_2S was a reactant, the product mixture contained some sulfur compounds. (Photographs: left, Courtesy University of California, San Diego; right, Compix, United Press International.)

equilibrium constant for the reaction $CO_2 + 4H_2 \rightleftharpoons 2H_2O + CH_4$ is large. With Miller, Urey demonstrated that electric sparking of a mixture of reducing gases gave a surprisingly diverse mixture of organic compounds which could be interpreted as precursors to larger molecules of biochemical interest (see Table 22-1). The following equations indicate how quickly complexity could increase in their mixture:

$$CH_4 + NH_3 = HCN + 3H_2$$
$$\text{hydrogen cyanide}$$

$$2CH_4 = C_2H_4 + 2H_2$$
$$\text{ethylene}$$

$$2HCN + C_2H_4 = NCCH_2CH_2CN + H_2$$
$$\text{succinodinitrile}$$

$$NCCH_2CH_2CN + 4H_2O = HO_2CCH_2CH_2CO_2H + 2NH_3 + H_2$$
$$\text{succinic acid} \tag{22-1}$$

Table 22-1.   Major products from the Urey-Miller experiment

Product (common name)	Formula	Percent yield (based on CH_4)
Formic acid	HCO_2H	7.80
Glycine	$H_2NCH_2CO_2H$	2.1
Glycolic acid	$HOCH_2CO_2H$	1.9
Alanine	$H_2NCH(CH_3)CO_2H$	1.1
Lactic acid	$CH_3CH(OH)CO_2H$	1.0
Acetic acid	CH_3CO_2H	0.51
β-Alanine	$H_2NCH_2CH_2CO_2H$	0.50
Propanoic acid	$CH_3CH_2CO_2H$	0.42
Succinic acid	$HO_2CCH_2CH_2CO_2H$	0.13

The theory of chemical evolution was strengthened in 1969 when astronomers discovered vast clouds of ammonia, water vapor, and formaldehyde in interstellar space (see Figure 5-7). Since then 20 other organic molecules have also been found in space, including hydrogen cyanide, a versatile intermediate for synthesis. One substance derivable in the laboratory from HCN is the *porphyrin* molecule, as noted later an essential pigment in both plants and animals:

Synthesis in the hydrosphere

Molecules formed in the atmosphere would have fallen with rain and been washed into warm, shallow lakes with minerals from land drainage. As concentrations increased, further synthesis could take place.

Complex as living things are, biochemists have found that only about 30 general kinds of building units are needed to account for such life essentials as proteins, enzymes, and nucleic acids, and the abilities to store and use energy. The units are two sugars (glucose and ribose; see Figure 23-3), lipids (see Eq. 23-18), diesters of phosphoric acid (Eq. 22-3), 20 amino acids (see Figure 23-12), and five nitrogen bases (adenine, cytosine, guanine, thymine, and uracil; see Table 24-1 and Eq. 24-2). Laboratory syntheses of some of these substances have been accomplished under assumed early earth conditions. Possible routes to increasingly complex molecules are indicated below.

1. *Sugars* have been formed by aldehyde reactions; further reaction can give higher polymers resembling starch and cellulose (Chapter 23). All

these compounds are called *carbohydrates* for the empirical formula CH_2O which fits most of them

$$6HC\overset{O}{\underset{H}{\diagdown}} \xrightarrow{hv} HC \left[\underset{H}{\overset{HO}{\underset{|}{C}}} \underset{OH}{\overset{H}{\underset{|}{C}}} \right]_4 C\overset{O}{\underset{H}{\diagdown}} \tag{22-2}$$

formaldehyde   *carbohydrate*

2. Lipids (fats and fatlike substances) are derivable from aliphatic acids and glycerol, an alcohol having three —OH groups. With phosphate ions, mixed carboxylate–phosphate esters may form. If a sugar then condenses with the phosphate group, the result is a typical phospholipid. Phospholipids are found almost universally in cells, particularily in cell membranes:

$$2RC\overset{O}{\underset{OH}{\diagdown}} + \begin{matrix} H_2C-OH \\ | \\ HC-OH \\ | \\ H_2C-OH \end{matrix} \xrightarrow[(-2H_2O)]{H^+} \begin{matrix} H_2C-O-\overset{\overset{O}{\|}}{C}-R \\ | \\ HC-O-\overset{\overset{O}{\|}}{C}-R \\ | \\ H_2C-O-H \end{matrix} \xrightarrow[(-H_2O)]{H_2PO_4^-}$$

fatty acid   *glycerol*   *diglyceride*

$$\begin{matrix} H_2C-O-\overset{\overset{O}{\|}}{C}-R \\ | \\ HC-O-\overset{\overset{O}{\|}}{C}-R \\ | \\ H_2C-O-\overset{\overset{O}{\|}}{\underset{\underset{O^-}{|}}{P}}-OH \end{matrix} \quad \overset{OH}{\underset{}{H_2C}}-(CHOH)_4-C\overset{O}{\underset{H}{\diagdown}} \xrightarrow[(-H_2O)]{}$$

mixed ester

$$\begin{matrix} H_2C-O-\overset{\overset{O}{\|}}{C}-R \\ | \\ HC-O-\overset{\overset{O}{\|}}{C}-R \\ | \\ H_2C-O-\overset{\overset{O}{\|}}{\underset{\underset{O^-}{|}}{P}}-O-CH_2-(CHOH)_4-C\overset{O}{\underset{H}{\diagdown}} \end{matrix}$$

phospholipid $\tag{22-3}$

3. *Alpha-amino acids* (like glycine and alanine, Table 22-1) can be condensed with heat to give chains resembling proteins (see Figure 23-18):

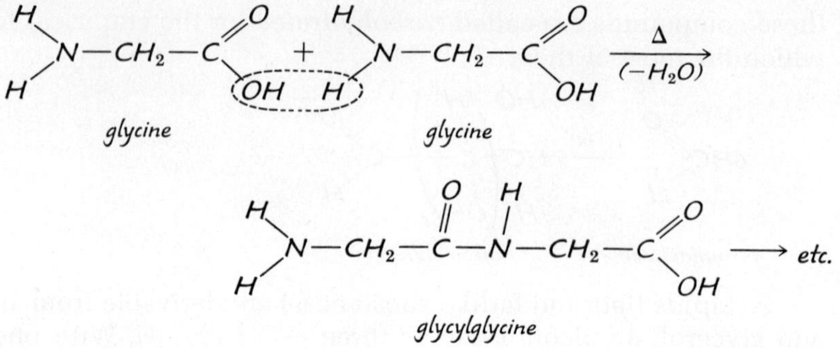

$$\text{glycine} + \text{glycine} \xrightarrow[(-H_2O)]{\Delta}$$

$$\text{glycylglycine} \longrightarrow \textit{etc.} \qquad (22\text{-}4)$$

In living matter today enzymes, most of which are proteins, come into being because enzymes already present guide their synthesis. The first enzymes, as well as the first cell enclosures, were probably proteins assembled as above.

4. *Cyclic nitrogen bases* (organic ring compounds containing nitrogen) have been synthesized from solutions of ammonium cyanide. In several steps,

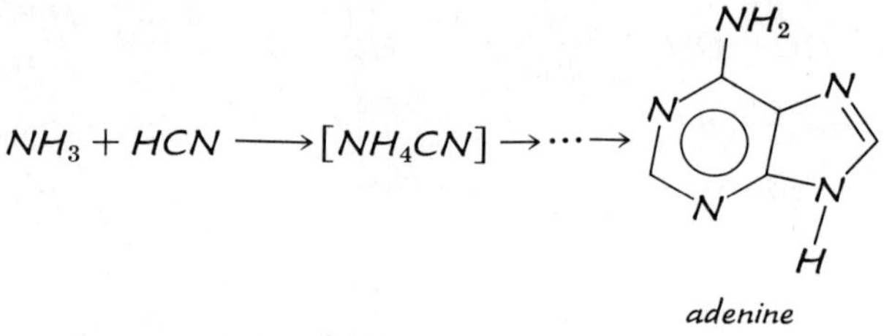

$$NH_3 + HCN \longrightarrow [NH_4CN] \rightarrow \cdots \rightarrow$$

adenine

5. *Nucleosides* (combinations of a 5-carbon sugar and a base) can be made by elimination of water:

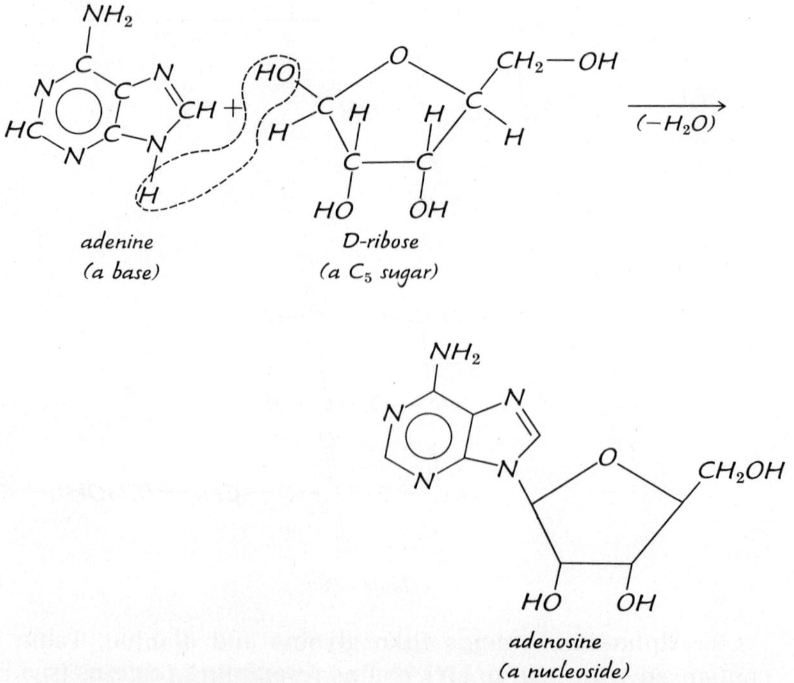

$$\text{adenine (a base)} + \text{D-ribose (a } C_5 \text{ sugar)} \xrightarrow{(-H_2O)}$$

$$\qquad (22\text{-}5)$$

adenosine
(a nucleoside)

Nucleotides (combinations of a nucleoside and an inorganic phosphate ion) can form with loss of another molecule of water:

adenosine

dihydrogen phosphate ion

adenylic acid (a nucleotide) (22-6)

Adenylic acid, also called *adenosine-5'-monophosphate* (abbreviated AMP), is closely related to adenosine triphosphate (ATP), which is the universal carrier of usable biochemical energy in cells (see Section 23-2).

6. Nucleic acids may have been synthesized abiogenically (without a living system) given ATP, enzymes, and nucleotides. Nucleic acids are polynucleotides of high molecular weight found in the nuclei of biological cells. One example is deoxyribonucleic acid (see DNA, Section 24-2). Under proper conditions DNA is self-replicating and the key to understanding the molecular biology of heredity. Representing a nucleotide unit like AMP as base–sugar–PO_4, nucleic acid structure is that of the following polymer:

$$\text{base} \qquad \left(\text{base} \qquad \right) \qquad \text{base}$$
$$-\text{sugar}-PO_4-\left(\text{sugar}-PO_4\right)_n-\text{sugar}-PO_4$$

Chemical evolution alone thus could have given carbohydrates, fats, proteins, nucleosides, nucleotides, perhaps even nucleic acids in a water solution that also contained inorganic substances. Now how did this mixture produce life?

22-2. BIOCHEMICAL EVOLUTION

The first cells

The simplest living thing is the biological cell. It is an enclosed chemical system that interacts with its surroundings through the cell wall mem-

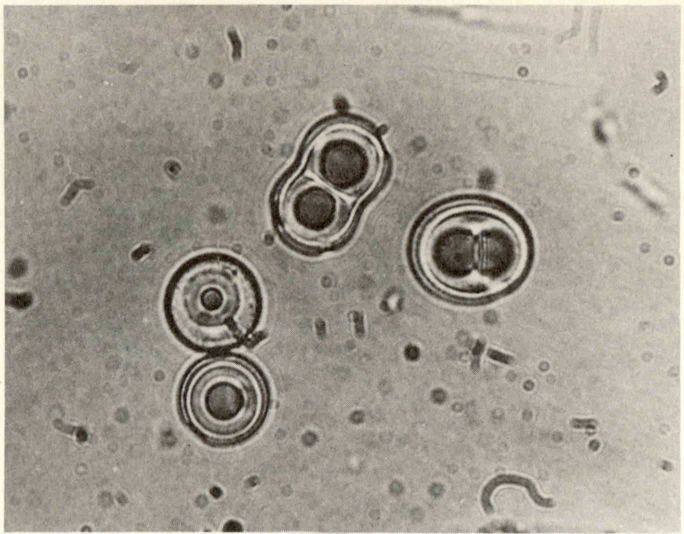

Figure 22-3 Protenoid microspheres. Their diameter is about 10^{-4} cm (or 1 μ). (Compare to cells in Figure 22-6.) (Courtesy Sidney Fox, Institute of Molecular Evolution, University of Miami, Florida.)

brane. One possible route to enclosed systems is from "thermal protenoid" polymers that can be made by heating dry amino acid mixtures. When treated with water, the proteinlike polymer yields cellular *microspheres*. As shown in Figure 22-3 some of the spheres are double walled like real cells. They also exhibit certain biocatalytic activity and selection in the molecules they allow to pass through their boundaries. Protenoids made in the presence of mixed polynucleotides even show a selective preference toward the latter, giving a hint of how the essential informational code, which as we will see later controls synthesis of proteins in modern cells, may have evolved.

Precells that incorporated (water-insoluble) hydrocarbons would have had protection against dissolution. Membranes could regulate the species that came in, and compounds inside would determine how they reacted. *Autocatalytic reactions* — that is, reactions catalyzed by their own reaction products — could have evolved and shown the way to cell replication. The best catalysts, some probably organometallic compounds, would be those promoting the fastest reactions and the reactions giving products most useful to the cells. Over long time periods almost every combination of available substances could have been tried.

Natural biopolymers (carbohydrates, proteins, nucleic acids) are universally optically active. Because syntheses (like Miller's) done without asymmetric starting materials give only equal quantities of right- and left-handed molecules (a racemic mixture), any life theory needs to explain how an asymmetric bias began. Several possible explanations based on natural processes are available, but the correct one(s) may never be established unequivocally. One theory is that once a single asymmetric compound formed, perhaps on the asymmetric catalytic surface of a *d*- or *l*-quartz crystal (see Figure 20-2a), it influenced subsequent syntheses to give asymmetric products. As the products were taken up in cells, they caused asymmetric syntheses that gave the stereoregular polymers needed for cellular stability.

As nutrients became depleted, "wrongs" probably disappeared and "rights" increased in mass and number, their subdivision perhaps forced by osmotic pressure or by mechanical action of sea waves. Stability of aggregates was greatest at a depth in water where the bond-breaking energy of solar radiation, unscreened for lack of atmospheric oxygen, could not reach. With no living organisms to prey on them in the warm sterile water and no destructive oxidation, precell buildup presumably continued. Then, in some manner, one or several kinds of precells made the momentous change from a physiochemical to a biological system by finding how to

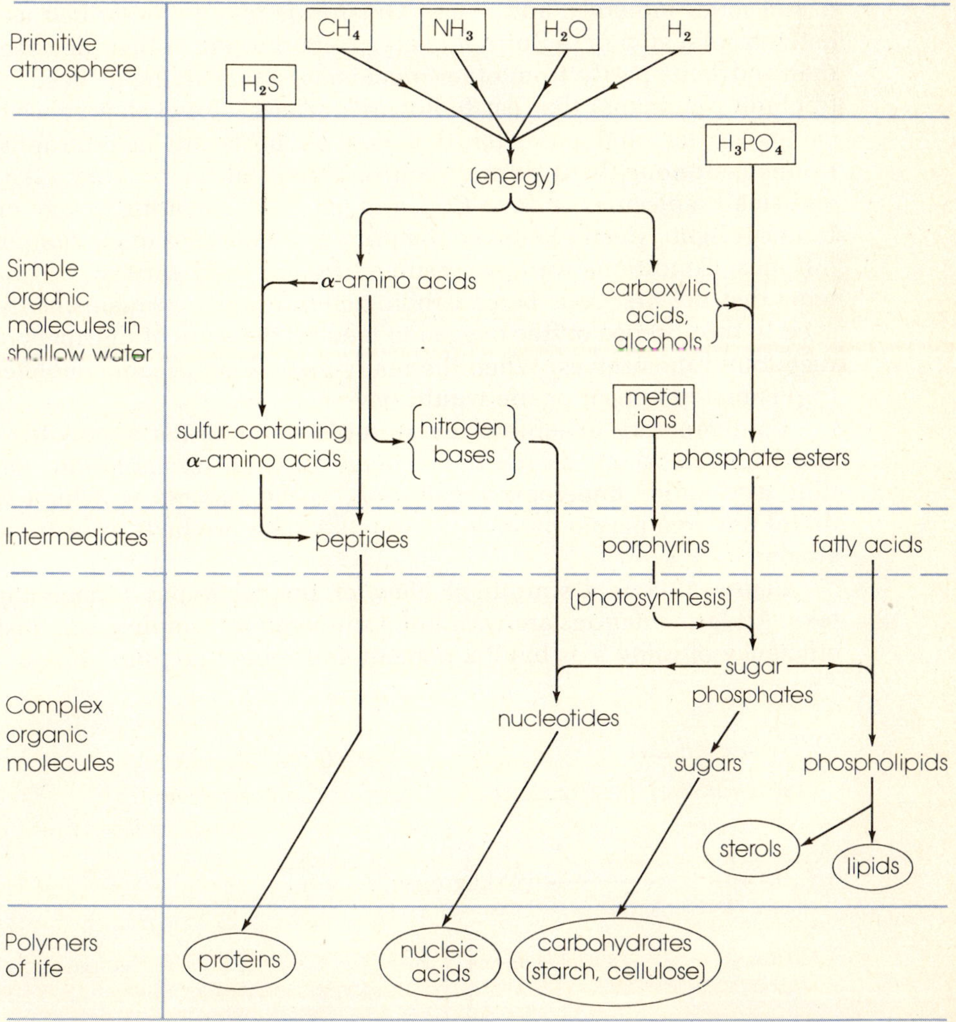

Figure 22-4 A scheme for chemical evolution. Lighter organic compounds (above the dashed lines) could have formed in the atmosphere, heavier ones in the ancient seas. Compounds below the dotted lines were eventually products of living things. (Adapted from J. D. Bernal, The Origin of Life on the Earth, Clark and Synge (Eds.) p. 52, Pergamon Press, New York, 1959.)

extract energy from their contents, synthesize their structural materials, and reproduce themselves through the unique self-replicating ability of the nucleic acids—and life had begun.

Establishing the essentials of chemical evolution (Figure 22-4) and forming the simplest of organisms required perhaps a billion years time and many trial and error experiments by nature, but the increasing complexity of products favored life. Even if the right combinations were highly improbable events, given enough trials they were bound to occur.

22-3. BIOLOGICAL EVOLUTION

The change in nutrition

Maintenance of any form of life requires *nutrition*, that is, acquisition and use of food, minerals, and water. Organisms may be described as feeding heterotrophically or autotrophically. *Heterotrophs* (other nourished) get their nutrients partly from other organisms, either living or dead. *Animals* (including humans) that eat biological material, *fungi* that subsist on decaying matter, and *parasites* that live on hosts are heterotrophs. *Autotrophs* (self-nourished) do not require biological foods. They take in ions and small molecules such as CO_2, and get their operating energy primarily from sunlight. Green plants are the principal examples of contemporary autotrophs. Conditions we have assumed for the early earth would have dictated that the first cells be anaerobic heterotrophs, or organisms that managed in the absence of free oxygen to feed off the soup of nutrients in which they found themselves. When the ready-made food became depleted, autotrophs had to appear or life would have expired.

Contemporary anaerobic heterotrophs include certain molds, yeasts, and bacteria, which derive their energy through *fermentation*. Fermentation, also called *anaerobic respiration*, is the process in which carbohydrates undergo partial breakdown to yield such products as lactic acid and ethanol.

Figure 22-5 is a simplified view of several routes fermentation may take. All fermentations are wasteful. Compared to complete combustion the efficiency of route V is but 2.2 percent and route I no more than 8 percent.

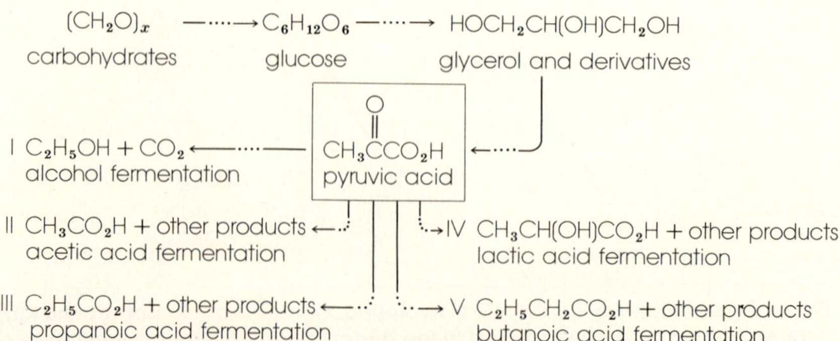

Figure 22-5 Fermentation (simplified). Pyruvic acid is a common intermediate. Glycolysis, a process equivalent to fermentation, is discussed in relation to its energy yield with Figure 23-4.

Probably the first anaerobes were even less efficient and required the diluting effect of surrounding water to keep from being poisoned by their own by-products. Doubtless their functions were slower also, because time had not yet permitted their finding the catalysts that make rapid and specific reaction sequences the rule of cells today. Their flicker of life may have been sustained by the energy from a single fermentive reaction.

As cells multiplied, using up the energy-rich environment, new organisms must have developed that could subsist on waste products of fermentation. The wastes they expelled represented degradation of one or more steps in structure and energy, and again new heterotrophs came into being which could feed on that mixture. In perhaps only a few thousand years carbon dioxide (as HCO_3^-) may have become the only readily available carbon compound in the sea. As the end product of the series of energy-degrading oxidations, it meant starvation to the organisms that had produced it. For life to hurdle this crisis a way had to be found to reverse the previous reactions, that is to obtain energy from somewhere, and to store it for use in products derived from the reduction of CO_2. The answer was the green pigment *chlorophyll,* an absorber of visible light. With it cells gained access to the practically limitless energy of the sun, and became capable of manufacturing their own food.

Photosynthesis is the process in green plants whereby solar energy is captured, used to energize chemical transformations, and stored as oxidizable chemical compounds. Photosynthesis made continuation of life possible and, as part of the carbon cycle (see CO_2, Section 13-4), has played a vital role in nature ever since. The carbon cycle can be summarized by the equation,

$$CO_2 + H_2O \underset{\substack{\text{respiration, fermentation} \\ \text{(energy derivation by cells)}}}{\overset{\substack{\text{photosynthesis in plants} \\ \text{(storage of solar energy)}}}{\rightleftarrows}} O_2 + \underset{\text{carbohydrate}}{(CH_2O)} \tag{22-7}$$

Development of two kinds of cells

As will be described later, nucleic acids, which are passed on at all levels of life from parent to offspring, make up molecular command units called *genes.* Genes determine structure and function of new cells. As sequences of subunits in nucleic acid polymers become slightly altered, the commands they give for chemical syntheses also become altered. Offspring, therefore, are not exact duplicates of parents. By some manner of change, perhaps the intermixing of free-floating nucleic acids followed by their capture in cells, there emerged in nature's scheme of things molecular programing for two kinds of cells.

The simpler, and therefore presumably the more primitive, types are called *prokaryote* cells. Modern prokaryotes are characterized by comparatively elementary and ill-organized internal parts, by genes not enclosed in a membrane sack, and by wide variations in cellular chemical processes. Today's representatives include bacteria (Figure 22-6a) which are generally heterotrophs, and blue-green algae which are photosynthetic autotrophs.

Cells of the second type are called *eukaryote* cells. Eukaryotes have relatively highly structured interiors, and carry out chemical reactions that

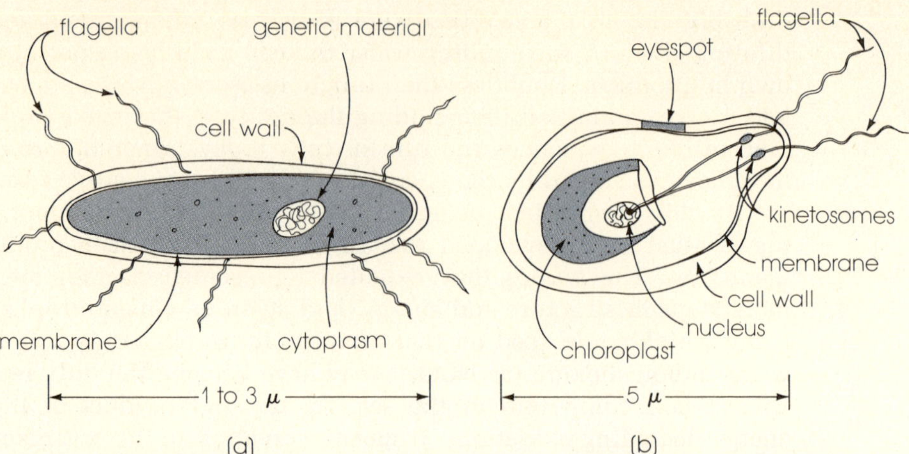

Figure 22-6 Modern unicellular organisms. (a) A (prokaryote) bacterial cell, typically the smallest cell known. It moves by action of the flagella. Cytoplasm is the term given biological material found between cell membrane and nucleus. All bacterial cell walls are made from two simple sugars and three or four of the same amino acids, indicating a common ancestry. (b) Chlamydomonas, a motile green alga (and a eukaryote). The eyespot is a structure that focuses light on a light-sensitive pigment. Connection to the kinetosomes seemingly causes the flagella to whip and move the organism toward the light where the chloroplast, which contains chlorophyll, can carry out photosynthesis.

are aerobic. Protein is combined with their nucleic acids to give filaments of genetic material called *chromosomes*. The chromosomes are membrane enclosed, forming a cell nucleus. Other distinct structures called *organelles*, which have special functions, are also present within the cell (Figure 22-6b). Judging from modern eukaryotes, ancient varieties must have been unusually adaptive to changing environmental conditions, swimming by means of motive organelles called *flagella*, exuding a protective coating to retain water during drought, and feeding either on available organic matter or making their own food by photosynthesis. Certain eukaryotes today can be changed in color from green to white by laboratory breeding techniques. Like typical plants the former are capable of photosynthesis. The latter, like animals, need to be supplied with nourishment. Both kinds exist in nature, intimating two lines of evolution from a single ancestor.

22-4. PHOTOSYNTHESIS; THE PHOTOLYSIS REACTIONS

Photosynthesis is the most massive and important sequence of chemical events operating on the earth. For every square foot of the planet's surface green plants produce 0.224 mole (0.0158 lb) of O_2 and use 0.224 mole (0.0217 lb) of CO_2 per year by way of the photosynthetic equation,

$$nCO_2 + nH_2O \xrightarrow{light} (CH_2O)_n + nO_2$$

Given the earth's surface area (5.48×10^{15} ft²), we can calculate that annually the total photosynthetic oxygen production is about 44 billion tons and carbon dioxide fixation is 60 billion tons. At these rates photosynthesis renews the equivalent of all the planet's atmospheric O_2 every 30,000 years and uses all the atmospheric CO_2 every 45.4 years. Bond-energy values (Table 13-2) tell us that photosynthesis causes the more stable oxygen–hydrogen bonds in water molecules to be broken and the less stable carbon–carbon, carbon–oxygen, and carbon–hydrogen bonds in carbohydrates to form. The latter bonds, with their higher energy content, represent stored sunlight that animals obtain by eating plants.

Experiments with labeled atoms (^{18}O) have shown that the oxygen taken into photosynthesized carbohydrates comes from CO_2, and that the oxygen expelled as by-product O_2 gas comes from water. This infers water becomes separated into its elements and that hydrogen atoms reduce CO_2:

$$\text{I.} \quad 2H_2O \xrightarrow{\text{light}} 4H + O_2\uparrow$$

$$\text{II.} \quad CO_2 + 4H \longrightarrow [CH_2O] + H_2O$$

Reaction I, called the *Hill reaction* after its discoverer, describes the *photolysis* of water. Though simple looking, it comprises a number of reactions whose mechanisms are less well understood than those of reaction II which describes the *reductive fixation* of carbon dioxide. In the following paragraphs we investigate I; in Section 22-5 we consider II, called the *Calvin cycle*.

Chloroplasts and the photolysis reaction

In green plants photosynthesis takes place in structures called *chloroplasts*. Chloroplasts contain arrangements of complex molecules like those in Figure 22-7 which absorb light and perform other functions. Hill demonstrated that if chloroplasts are separated from green leaves and put in a water solution containing electron acceptors like ferric salts ($Fe^{3+} + e^- = Fe^{2+}$), oxygen gas is liberated from the chloroplasts when the mixture is illuminated. When the light is turned off, the reaction stops. Thus in reaction I light is somehow trapped and its energy made available in chloroplasts for the breakdown of water.

Research to discover how the reducing power of electrons released in chloroplasts is used has proved that two general systems operate. Both incorporate a feature typical of biochemical processes: *cyclic systems of transport*. In photolysis the reducing power of electrons is transported from one substance to another by electron-rich compounds. Electron transport is equivalent to transport of hydrogen atoms because water containing hydrogen ions is always present, and $H^+ + e^- = H$.

Figure 22-8 shows one of the two systems of electron transport known to operate in photosynthesis. Parts of it are cyclic, although overall electron transport is noncyclic, being terminated in a compound used in other sequences. The scheme begins (at the lower left) with light striking a bundle of photoreceptive molecules which channel the energy to an associated chlorophyll molecule that will be the reaction center. As one of the chlorophyll's electrons is promoted to a higher quantum level and then returns to the original level, its energy is transferred through intermediates (not completely known) to water, giving hydrogen ions, electrons, and oxygen gas:

$$4H_2O \xmidarrow{h\nu} [4H + 4OH] = 4H^+ + 4e^- + 2H_2O + O_2 \tag{22-8}$$

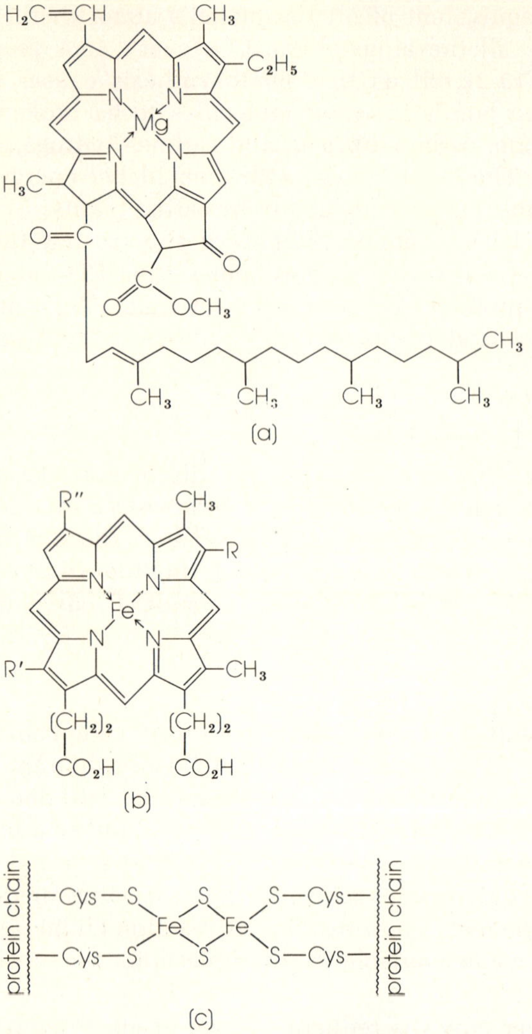

Figure 22-7 Chloroplast components used in photolysis. (a) Chlorophyll a, a magnesium porphyrin with a long side chain. (b) A cytochrome. The R's are different groups. (c) A ferredoxin (mol. wt. ~ 6000). It contains two, four, six, or more iron atoms in combination with sulfur, the amino acid cystine (Figure 23-12), and protein. It may have been one of the first proteins formed on the early earth. Ferredoxins and cytochromes are found in many biochemical reactions in which the redox ability of their iron atoms facilitates electron transport.

Oxygen escapes and the hydrogen ions are absorbed in chloroplast water.

Electrons from reaction centers are transferred to, and reduce, a compound identified thus far only as compound 550: oxidized $C_{550} + 2e^- \rightarrow$ reduced C_{550}. As reduced C_{550} gives up the acquired electrons, it becomes reoxidized. Electron acceptor this time is cytochrome b which gets the electrons via several steps not requiring input of light energy: oxidized cytochrome $b + 2e^- =$ reduced cytochrome b.

As cytochrome b is being reduced, part of the electron energy is drained off for the synthesis of the nucleotide adenosine triphosphate from adenosine diphosphate and inorganic phosphate ion: $ADP + P_i = ATP$. In

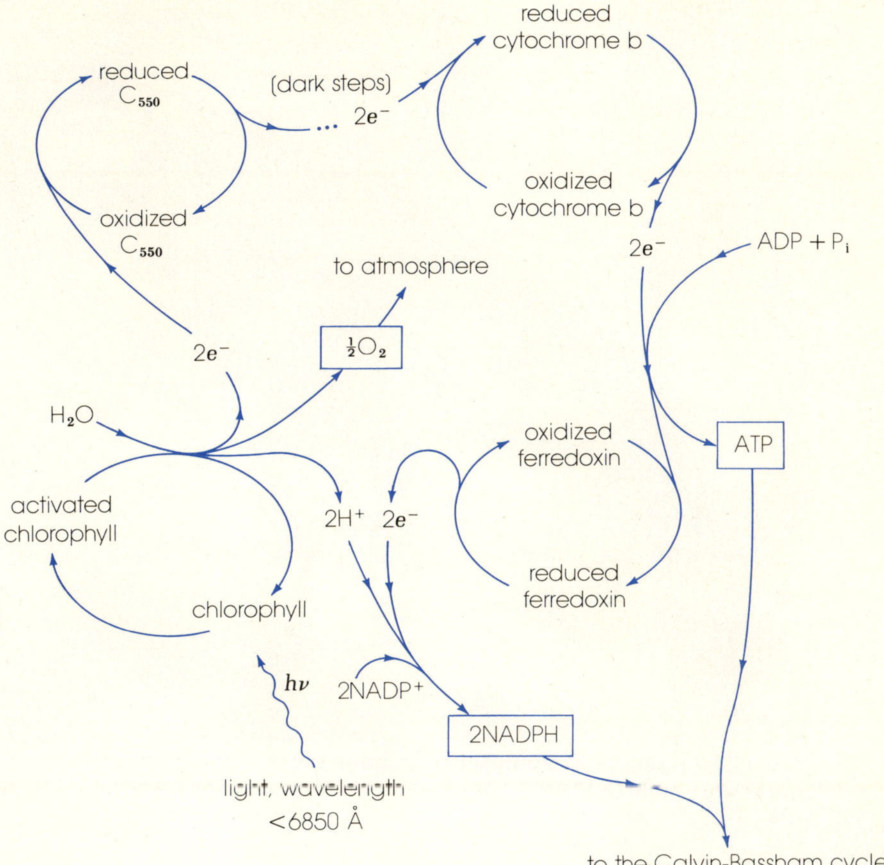

Figure 22-8 Noncyclic electron transport in the photolysis part of photosynthesis, a scheme clarified by the work of D. Arnon and others at the University of California, Berkeley. Light-dependent and independent processes alternate as electron reducing power is cycled. Net results of photolysis are

$$2H_2O + NADP^+ = NADPH + H^+ + H_2O + \tfrac{1}{2}O_2$$
$$2ADP + 2P_i = 2ATP$$

Chapter 23 we examine the role of ATP in providing cellular energy. Reduction of CO_2 requires ATP.

Excited by light energy, electrons are transferred from reduced cytochrome *b* to ferredoxin, reducing the latter: oxidized ferredoxin + $2e^- =$ reduced ferredoxin. Reduced ferredoxin transfers its electrons to protons in surrounding water. The hydrogen reducing power produced reduces a compound called *nicotinamide adenine dinucleotide phosphate*. The reaction is abbreviated as $NADP^+ + 2H = NADPH + H^+$. The compound NADPH is the carrier for hydrogen that reduces CO_2 in the Calvin cycle. This completes the photolysis reactions.

The site of photosynthesis

Cells and their features vary widely. We will therefore describe the structures responsible for photosynthesis as they might be found in a model cell

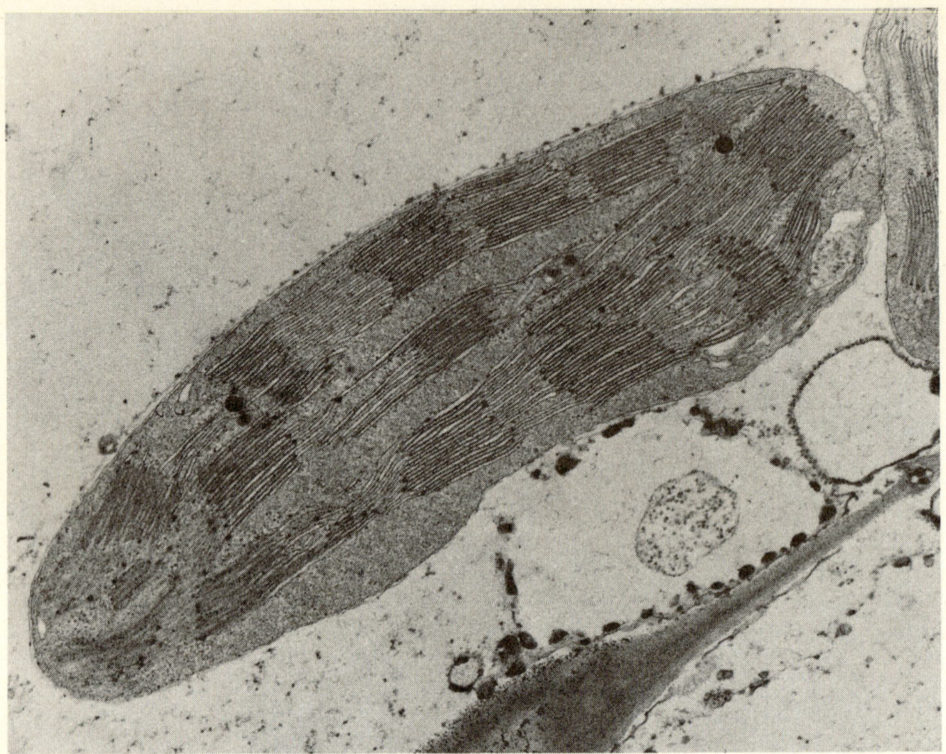

Figure 22-9 A single chloroplast from a bean plant. Lamellar structure and darker grana are visible (×39,000). (Courtesy W. W. Thomson, University of California, Riverside).

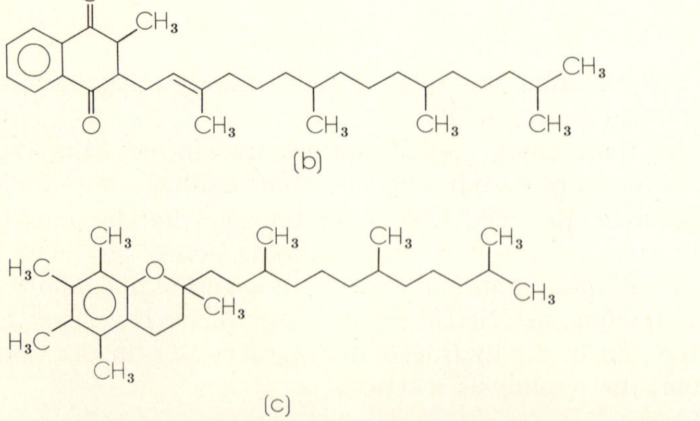

(a)

(b)

(c)

Figure 22-10 Some compounds found in quantasomes. (a) β-Carotene. Its conjugated structure makes it a strong absorber of visible light. (b) Vitamin K_1. (c) α-Tocopherol, one of the substances that comprise the mixture called vitamin E. β-Carotene and the side chains of the other two compounds are described as isoprenoid, derivatives of isoprene, $H_2C{=}CH{-}C(CH_3){=}CH_2$.

of higher plants (see Figure 22-11a). The cell's diameter is 70 microns (700,000 Å), so it and its larger parts will be readily observable under an ordinary microscope. Inside the cell are perhaps 25 chloroplasts, dark-green oblate organelles 5 microns (50,000 Å) long.

The structure and workings of chloroplasts have been examined chemically and with the electron microscope. Figure 22-9 is an electron micrograph of a cross section of a chloroplast. The internal organization is *lamellar* (layerlike), composed of roughly parallel membranes each 30 Å thick and extending lengthwise through the whole organelle. Folded and thickened sections occurring one on top of the other in the fashion of stacked

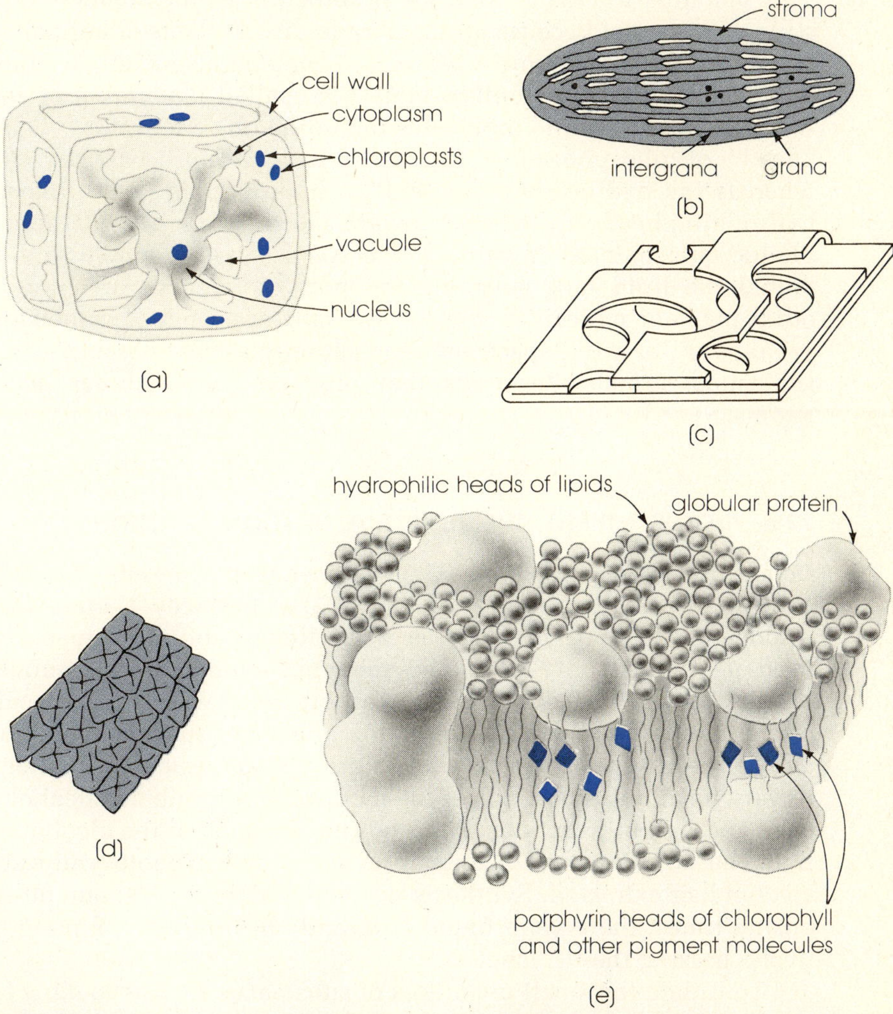

Figure 22-11 Schematic dissection of the photosynthetic apparatus. (a) A model cell of a higher plant. Each such cell contains numbers of chloroplasts and other organelles in the semifluid material called cytoplasm. (b) Chloroplasts have a membrane structure within which are stacks called grana. (c) A closer look, showing the relationship between grana and chloroplast membranes. (d) When membrane material is removed, neatly packed units called quantasomes are seen. (e) Cross section of a quantasome or photosynthetic membrane. (b, adapted from von Wettstein; c, adapted from Wehrmeyer; e, adapted from Green, Singer, and Lenard.)

coins are called *grana* (see Figure 22-11b,c). Photosynthetic pigments are concentrated in the grana.

When the surfaces of grana are removed, high magnification discloses a substructure resembling a tile roof (see Figure 22-11d). A thousand "tiles," called *quantasomes*, occupy a layer in a single granum disc. The quantasome, about 180 Å long and 100 Å thick, is thought to constitute one to four basic photosynthetic units. Molecules in a quantasome include lipids, enzymes, and proteins, each with a total molecular weight of 1 million, 230 chlorophyll molecules, and a variety of supplementary pigments like those of Figure 22-10.

The exact molecular arrangement in quantasomes is not yet known. This amounts to saying that the architecture of membranes in general is still controversial because all membranes, from plants or animals, have features in common. Figure 22-11e is a membrane model. It consists of a double row of round protein molecules called *globular proteins*, among which are various lipid molecules including phospholipids. The phosphate "head" of phospholipids (Eq. 22-3) is *hydrophilic* (water compatible) whereas the hydrocarbonlike "tail" is *hydrophobic* (water rejecting). The hydrophilic heads are pointed toward the membrane's outside, the hydrophobic tails pointed inward. In the special case of photosynthetic membranes the lipid tails surround the porphyrin heads of chlorophyll and other pigment molecules whose own tails appear to be embedded in the proteins. Filling the space between chloroplast membranes is a colorless gel called *stroma*. Reactions that produce carbohydrates occur in the stroma.

22-5. PHOTOSYNTHESIS; THE REDUCTIVE FIXATION REACTIONS

Details of reactions in which CO_2 is reduced were worked out in the 1950s by a group headed by Calvin (Figure 22-12) at Berkeley. Carbon-14 had been discovered there 10 years earlier by S. Ruben and M. Kamen, and Calvin used its radioactivity to follow the appearance of compounds as photosynthesis proceeded *in vivo*. His test organism was the single-celled green alga chlorella. Following illumination of a water suspension of chlorella he added, in the dark, a bicarbonate solution made from $^{14}CO_2$. After timed intervals, he allowed portions of the mixture to run into hot alcohol to kill the cells and halt all reactions. After boiling off most of the alcohol to concentrate the solution he separated the products of photosynthesis that the alcohol had extracted. Synthesis was rapid: after 5-, 30-, and 90-second exposure times, there were found 5, 8, and 15 different compounds, respectively, bearing the ^{14}C label.

To isolate the small quantities of substances he was looking for, Calvin used chromatography, one of the techniques to which biochemistry owes much of its spectacular recent advance. Separating mixtures by selective adsorption on filter paper (*paper chromatography*) or in a column containing inert packing material (*column chromatography*) were procedures known in the 1800s. They were refined around 1910 by Russian botanist M. Twsett. Twsett separated chlorophyll and other photosynthetic pigments as colored bands in a glass tube packed with calcium carbonate through

Figure 22-12 Melvin Calvin (United States, 1911– , Nobel prize in chemistry 1961). Besides research on photosynthesis Calvin has investigated organic geochemistry, identifying organic substances in ancient rocks to establish the trail of biological evolution. In the oldest rock yet found (3.1 eons) he identified hydrocarbons like the long side chain on the chlorophyll molecule (Figure 22-7a). If related, then some form of photosynthesis has been operative on earth since the earliest life. (Courtesy J. A. Bassham, Lawrence Radiation Laboratory, University of California, Berkeley.)

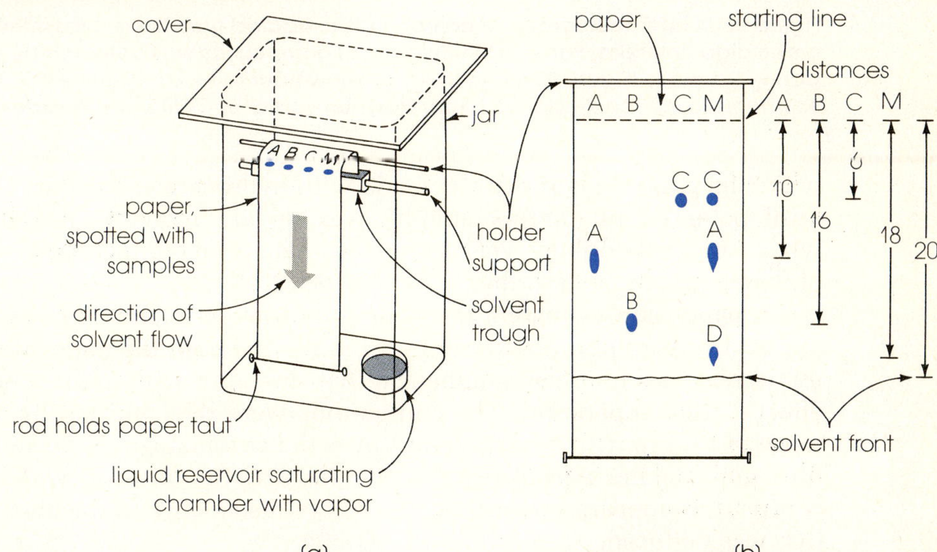

Figure 22-13 Paper chromatography. (a) Spots of sample mixture M, and known substances A, B, and C are applied to chromatography paper. The sheet is hung in a jar with the end nearest the spots in developing solvent and the jar is covered. Solvent moving by capillary action carries components different distances down the sheet depending on their solubility and other factors. (b) After the run. When the solvent has moved well toward the far end of the paper, the sheet is removed from the jar, the position of the solvent front marked, and the strip allowed to dry. The spots are located and R_f (retardation factor) values calculated,

$$R_f = \frac{\textit{distance of travel by component}}{\textit{distance of travel by solvent front}}$$

Solvent has moved 20 cm. Components A, B, C, and D have moved 10, 16, 6, and 18 cm, respectively; R_f values are therefore 0.5, 0.8, 0.3, and 0.9. Tentatively M contains A and C and a component not anticipated, D. The spots may be cut out, eluted with solvent, and analyzed further by infrared spectrometry and other techniques.

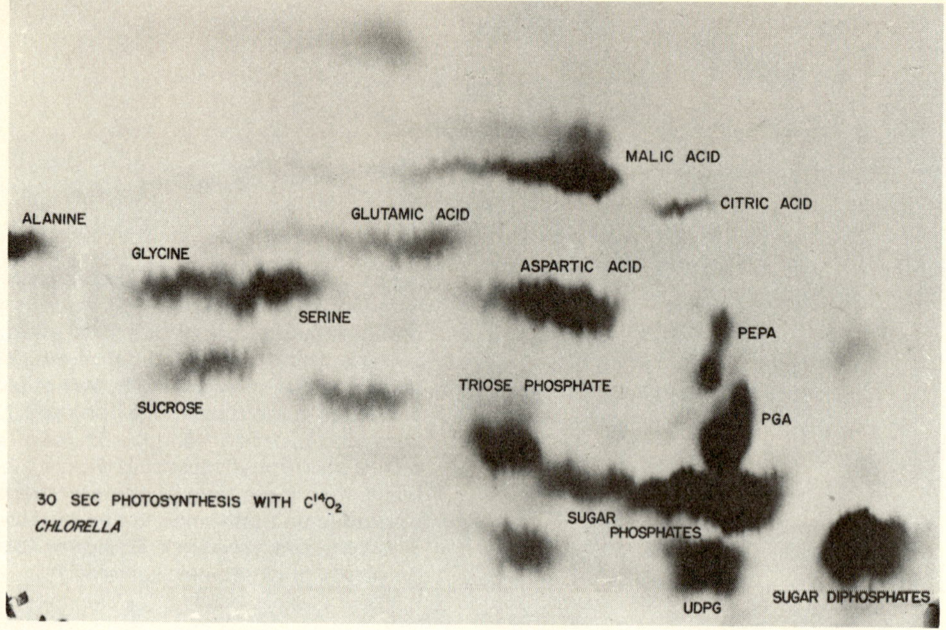

Figure 22-14 Radioautograph of compounds extracted from the photosynthesizing green alga chlorella. Partial identification of compounds was obtained by comparing R_f values (Figure 22-13) to those of known compounds. (Courtesy J. A. Bassham, Lawrence Radiation Laboratory, University of California, Berkeley.)

which he poured plant extracts made with hydrocarbon solvents. The powerful potential of chromatography was largely ignored until the 1940s when its methodology was improved. The general technique of paper chromatography is described with Figure 22-13.

A paper chromatogram from the Calvin experiments is shown in Figure 22-14. After chromatographing in one direction, the paper was turned 90 degrees and chromatographed in that direction with another solvent to effect further separation. The compounds were then located by placing a piece of film sensitive to ^{14}C β rays over the chromatogram, developing the film, and making a print from it. By studying the order in which labeled compounds appeared with time, the reaction sequence in the absorbance of CO_2 was deduced.

The series of reactions in the reduction of CO_2 is called the *Calvin-Bassham cycle*. Important parts of it are shown in Figure 22-15. Each turn of the cycle fixes one carbon atom (designated *C in the figure). Thus three turns are required to produce one C_3 phosphoglyceraldehyde (PGAl) molecule as illustrated, and the cycle must turn six times to yield one molecule of glucose (C_6). Each step requires catalysis by a specific enzyme. In all at least 11 enzymes are required.

Carbon fixation begins (Step 1a) with reaction between ribulose diphosphate (RDP) and carbon dioxide and water to give carboxylated RDP. This C_6 compound splits in two (Step 1b) giving phosphoglyceric acid (PGA), half of whose molecules contain one reference carbon *C. Conversion of PGA to glycerate diphosphate (GDP) requires introduction of energy and one phosphate ion per molecule. These ingredients are furnished (Step 2) by ATP from photolysis (Figure 22-8); GDP is then reduced (Step 3) by reaction with NADPH, a hydrogen carrier also produced in photolysis. By-products ADP, NADP$^+$, and P$_i$ (inorganic phosphate) are recycled to the photolysis reaction for regeneration.

Phosphoglyceraldehyde is the end product of carbon fixation to this point. In fact, PGAl can be considered the real product of photosynthesis

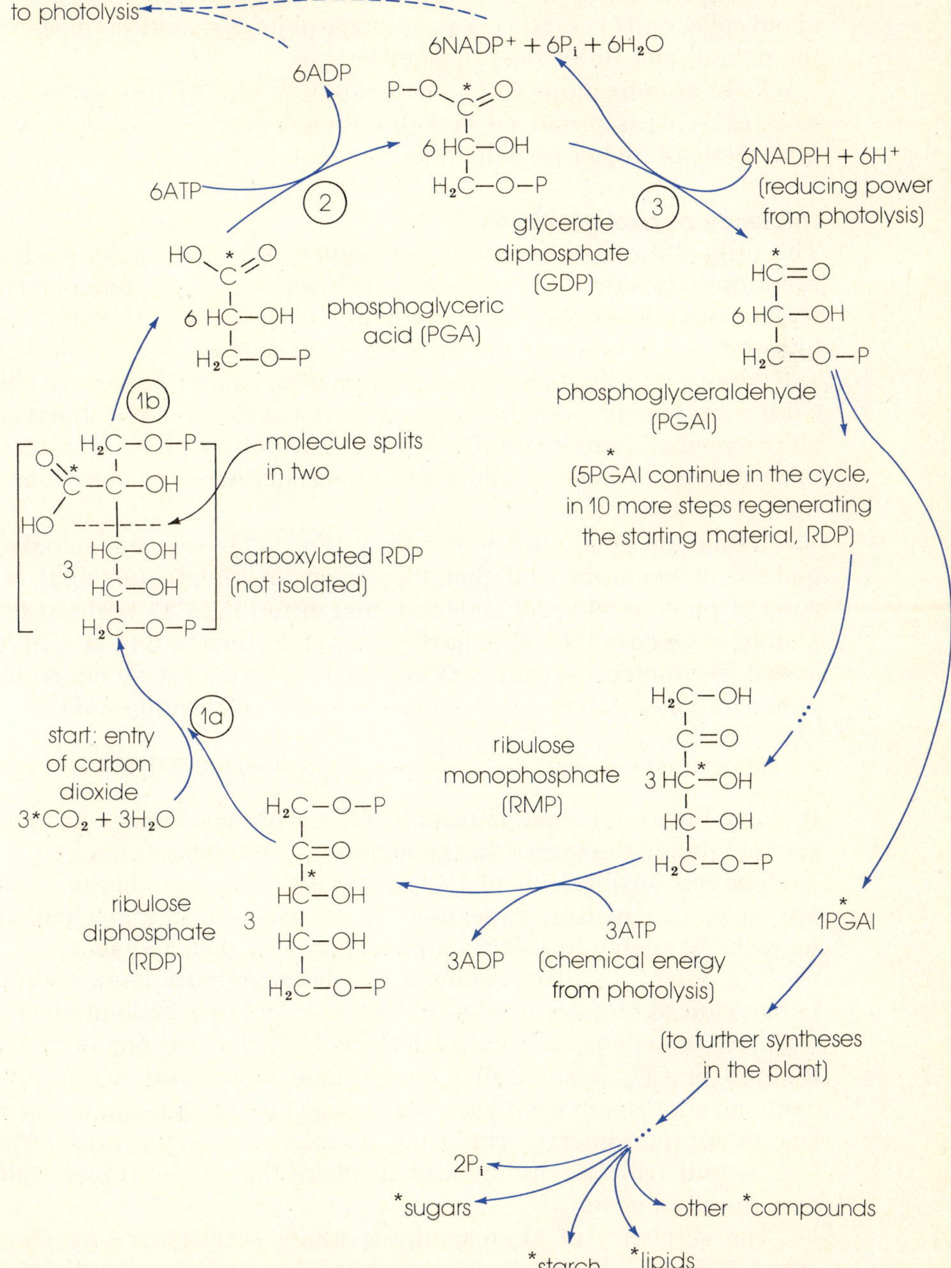

Figure 22-15 Reductive fixation of CO_2 in photosynthesis. The *C denotes the position in which ^{14}C is found experimentally if the cycle goes one full turn. The net reaction for three turns is

$$3CO_2 + 9ATP + 6NADPH + 6H^+ \longrightarrow$$
$$PGAl + 9ADP + 8P_i + 6NADP^+ + 6H_2O$$

rather than higher carbohydrates such as glucose. Five-sixths of the PGAl produced is retained in the cycle to make more RDP, which again attaches CO_2. The remaining PGAl goes to higher synthesis in the plant, including (1) production of sugars that can be used as an energy source in nonphotosynthesizing cells, (2) production of compounds needed in photosynthesis including chlorophyll, (3) manufacture of materials for new cells and repair of old cells, and (4) synthesis and storage of large quantities of starches (as in the potato) and lipids (as in peanuts).

Reduction fixation of CO_2 occurs only in photosynthesizing cells. However, as PGAl is produced and distributed it makes possible synthesis in other cells of higher carbohydrate molecules.

Efficiency of photosynthesis

The utilization of solar energy in photosynthesis may be studied with a spectrophotometer (Figure 22-16a). The way in which living green cells interact with visible light is shown in Figure 22-16b. Least transmittance (greatest absorption) occurs at wavelengths of 4350 Å (violet–blue light) and 6750 Å (red light). Substances appearing colored to the eye absorb their *color complement* from the range of wavelengths in the visible region of the electromagnetic spectrum. Chlorophyll, which is green, absorbs red light most strongly, whereas yellow and orange carotenes absorb violet and blue light (see Question 8).

Using the relationship $E = hc/\lambda$ (Eq. 3-1), we can calculate by the method of Example 13-2 that the energy of violet–blue light is 66 kcal/mole of photons (66 kcal/einstein), that of red light 42 kcal/einstein. When 1 mole of glucose (180 g) is burned in a calorimeter, 684 kcal of heat is released. By the conservation of energy law this must be equivalent to the photoenergy absorbed when 1 mole of glucose is synthesized:

$$6CO_2 + 12H_2O + xh\nu \xrightarrow{\text{chloroplasts}} C_6H_{12}O_6 + 6H_2O + 6O_2 - 684 \text{ kcal}$$

If violet–blue light alone is responsible for photosynthesis, then at 100 percent efficiency the term x in the equation is 684/66 or about 10. If red light is alone responsible, x is 684/42 or about 16. The number of photons actually needed, however, appears to be 48, meaning the mechanism of photosynthetic energy transfer is not as simple as sketched above.

The energy involved globally in photosynthesis is exceedingly large. Conversion of 6 moles of CO_2 ($6 \times 44 = 264$ g) to 1 mole of glucose (180 g) requires, as stated, an input of 684 kcal. That corresponds to 2.4 million kcal/ton of CO_2 fixed. Using an estimate of 60 billion tons of CO_2 undergoing conversion each year, we arrive at an annual capture of 1.4×10^{17} kcal of sunlight energy. The same amount of energy, in the form of gasoline, would serve the automotive needs of the entire United States for the next hundred years.

The efficiency of photosynthesis under ideal laboratory conditions is about 90 percent, based on a comparison of light quanta absorbed by chloroplasts to energy stored as synthesized carbohydrate. On the basis of total light striking a corn field, however, photosynthesis is notably inefficient. Due to reflection and other radiative losses, as well as the limited total area of chloroplasts exposed, less than 5 percent of the incident sunlight can be accounted for in carbohydrate product.

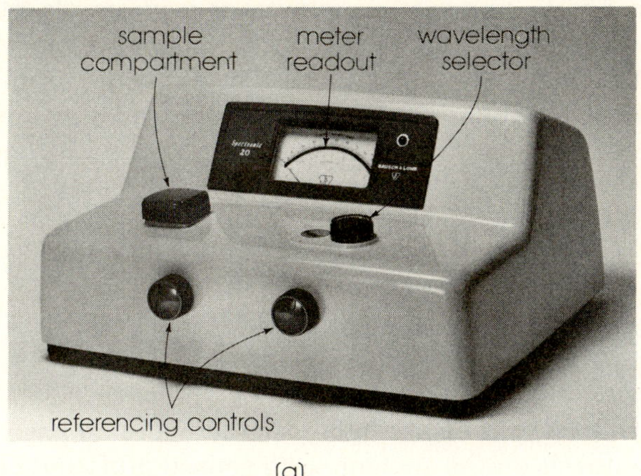

(a)

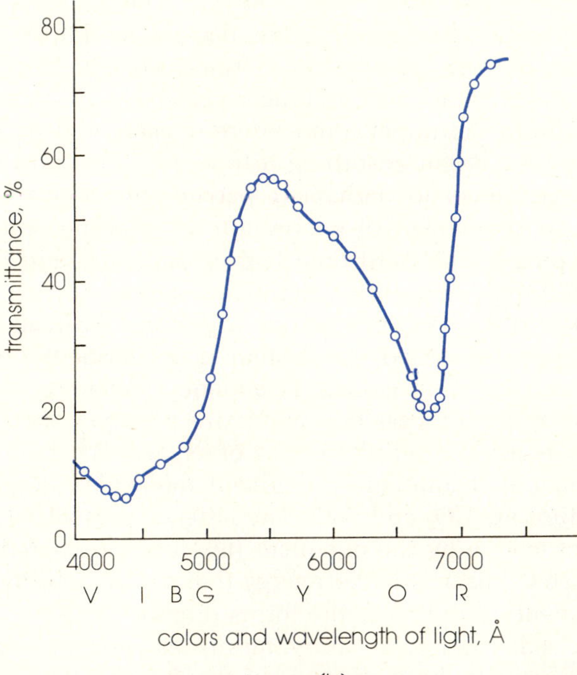

(b)

**Figure 22-16 Visible spectrophotometry. A spectrophometer for
the wavelength range 3400 to 9600 Å. The fundamental principle
of the instrument is that of Figure 13-10a. The wavelength is selected and the instrument set to read 100 percent transmittance
of light, using in the sample compartment a test tube containing
only solvent and reagents required in the particular method. That
tube is replaced with another containing the sample to be measured, and the meter read for percentage of transmittance of the
sample. (b) The transmission spectrum of a green alga suspended in water, plotted point by point by the procedure
above. (For color abbreviations see Figure 2-1.) (Courtesy Bausch
and Lomb, Rochester, N.Y.)**

One of the hopes of research on photosynthesis is to increase its practical yield—more sugar from sugar cane, for instance, instead of more leaves. Another goal is to prevent the growth of undesirable plants like weeds. Even an elementary understanding of photosynthesis suggests improvements in the practice of agriculture. For example, crop yield increases with CO_2 concentration. One way to increase CO_2 availability on still days is to stir the air. Another is to develop plants with rougher surfaces, giving them greater area. Another is to plant the crop to expose a maximum of green surface to the sun.

22-6. CHANGE TO AN OXIDIZING ENVIRONMENT

The rise of animals

Decomposition of water vapor by sunlight (see the figure on page 289) and the start of photosynthesis on the ancient earth caused profound changes in all life forms. Over geologic time as the two processes released free oxygen in large volume, reducing substances were oxidized (in the air, $H_2S \rightarrow SO_2 \rightarrow SO_4^{2-}$, in the water, $Fe^{2+} \rightarrow Fe^{3+} \rightarrow Fe_2O_3$, etc.), and the earth's entire air supply, originally reductive, became oxidative. With that reversal of conditions life was cut off forever from its origin. Thereafter the reactions of chemical evolution were no longer possible and life could only spring from other life. In the upper atmosphere oxygen, and its photodissociation product ozone, began absorbing ultraviolet light. Together with the rise of free oxygen level to perhaps 1 percent of today's level, this screening of lethal short-wavelength energy allowed marine organisms to emerge from the depths which until then had presumably held protective custody of all living things. Fermentive organisms starved as their food was oxidized. But in their place oxygen-adaptive organisms evolved and multiplied. At an oxygen level called the *Pasteur point* fermentation ceased being important and respiration gained ascendency. Respiration, the fundamental power-generating process that maintains all functions associated with life, essentially requires a cellular intake of oxygen. Whereas the products of fermentation are incompletely oxidized molecules (Figure 22-5), products of respiration are CO_2 and H_2O. The latter, representing a 20-fold greater derivation of energy by the organism (684 kcal compared to about 34 kcal/mole of glucose burned) is the energy that made mobility possible and led to development of all higher life forms (Figure 22-17).

Paleontologists, scientists who study *fossils* as indicators of life in previous geological periods, have discovered the fossil record of animals starts abruptly a half billion years ago in marine sediments, and that before then there is evidence only for algae, bacteria, and fungi. Their observation can be interpreted as dating the appearance in the seas of enough calcium ions to make the manufacture of protective shells possible, an essential condition for the emergence of complex multicelled animals.

About 100 million years later in the Silurian period fossil remains prove land plants, amphibians, and land animals were plentiful. As huge forests grew up in the Carboniferous period by photosynthesizing the accumulation of CO_2 which volcanoes had presumably released during 2 billion years of activity, the oxygen level may have reached today's level. Dinosaurs appeared. Then some 260 million years later during the Cre-

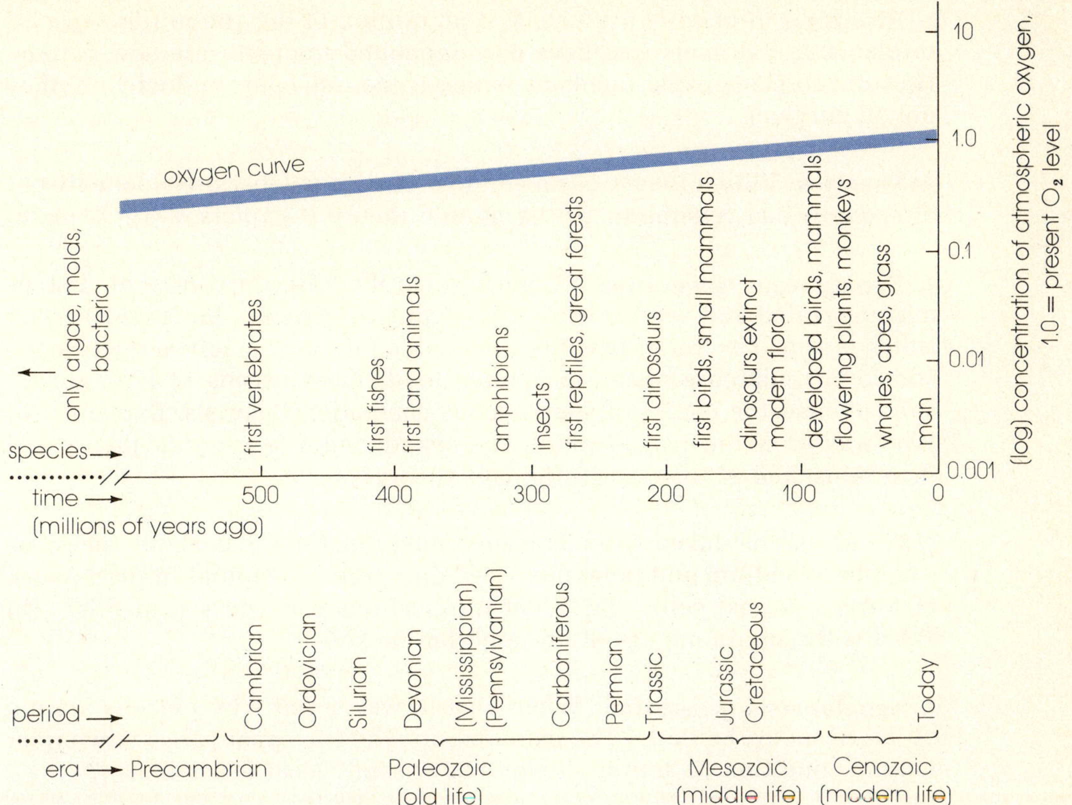

Figure 22-17 A model for the evolution of life with the rise of oxygen pressure. Oxygen from both photosynthesis and photodecomposition of water vapor is assumed here. (Modified to a higher O₂ level, from L. V. Berkner and L. C. Marshall, The Origin and Evolution of Atmospheres and Oceans, Brancazio and Cameron (Eds.), p. 120, Figure 6.15, Wiley, New York, 1964.)

taceous period the dinosaurs became extinct. Their demise may have been caused by a drop in atmospheric CO_2, reducing the greenhouse effect (Chapter 13) that had kept their environment warm. If an age of cold followed, forests died, were buried, and became beds of coal. Photosynthetic activity fell. Again carbon dioxide increased in the atmosphere from volcanism and another warming trend began.

Mammals have a history 70 million years long, apes 25 million, first humans perhaps 2 million, modern humans only a few tens of thousands of years. Though last to appear, human beings are the supreme animals. We have relatively long life and the superior intelligence that allows us to think abstractly, to use speech, to amass knowledge, and to regulate our environment. The next step may be finding how to control our evolution molecularily. We explore the possibilities in Chapter 24.

QUESTIONS

1. Evolution (library) **(a)** From the discussion in the chapter, what is chemical evolution? **(b)** What is meant by biological evolution? Consult at least one beginning biology text to formulate a brief answer for (b).

2. Evolution Biologists estimate that 98 million of the 100 million species of plants and animals that have ever inhabited the earth are now extinct. How did so many experiments by nature come into being and why did they not all survive?

3. Theories Which theory has a sounder basis, the Oparin-Haldane theory of spontaneous generation or the atomic theory (Chapters 1–4)? Explain.

4. Spontaneous generation A great scientific discovery may at first be misinterpreted and send scientists in wrong directions. **(a)** In 1670 when Dutch lens maker A. Leeuwenhoek assembled the first practical microscope and found microorganisms in spoiled food, these "seeds of life" greatly strengthened the claims of spontaneous generation theorists. Explain. **(b)** Why was spontaneous generation first abandoned as a scientific theory and then reinstated as the Oparin-Haldane theory?

5. Water Water taken from Precambrian formations has about the same calcium-to-sodium and potassium-to-sodium ratios as found in protoplasm of today's animal cells. **(a)** What implications can you see in this? **(b)** What is the minimum age of Precambrian rocks?

6. Spontaneous generation When Oparin was asked why precells are not observed today, he said, "The main reason why life cannot arise now under natural conditions is that it already has arisen." What did he mean?

7. Photosynthesis Various bacteria contain a compound called *luciferin* and a protein called *luciferase*. When molecular oxygen reacts with luciferin in the presence of luciferase, the result is a flash of light and the bacteria twinkle. It has been suggested that this mechanism for getting rid of unwanted oxygen was evolved at the time when the effects of photosynthesis on the atmosphere were becoming important. Explain.

8. Color Figure 22-18 is the color wheel of Munsell (1915) with wavelength ranges marked (Ångstroms). Colors directly opposite are complements of each other. Tell how the wheel aids interpretation of Figure 22-16.

9. Animals, plants Animals are described as motile heterotrophs, plants as immotile autotrophs. Explain.

10. Early reactions Rifle bullets are fired down a pipe containing hydrogen, water vapor, methane, and ammonia gases. When the gas mixture is shaken with water, traces of simple organic compounds are found in solution. **(a)** Explain what happened. **(b)** What "bullets" probably flew through the atmosphere of the very young earth with possibly similar results?

11. Labeled carbon Calvin ran Miller-type experiments using ultraviolet to irradiate mixtures of water, ammonia, hydrogen cyanide, and ^{14}C-labeled methane. What was the purpose of the labeled compound?

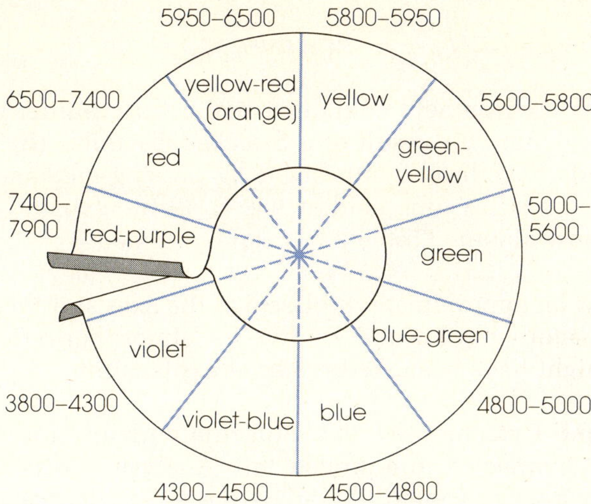

Figure 22-18

12. Life Why is life postulated to have started in the hydrosphere rather than on dry land or in the air?

13. Definitions Define or otherwise explain **(a)** paper chromatography; **(b)** R_f value; **(c)** eukaryote cell; **(d)** prokaryote cell.

14. Photosynthesis **(a)** Green plants absorb visible light (Figure 22-16) and use its energy for photosynthesis. Laboratory experiments show they also can absorb ultraviolet energy of wavelengths that, however, do not penetrate the atmosphere. What significance relative to evolution and atmospheric changes might be read into this? **(b)** The partial pressure of oxygen gas around chloroplasts regulates the rate of photosynthesis by deactivating enzymes involved in CO_2 fixation reactions. How could you prove oxygen's effect, aided by the technique illustrated in Figure 22-14? What might the experimental results look like?

PROBLEMS

15. Chemical evolution Formaldehyde and hydrogen cyanide are produced and used up in the Urey-Miller type of experiment (Table 22-1). Write balanced equations in the following sequence of reactions that shows a route to glycine via electric discharge in a mixture of gases assumed in the early atmosphere. **(a)** Methane plus water give formaldehyde plus hydrogen. **(b)** Formaldehyde plus hydrogen cyanide give hydroxymethylnitrile ($HOCH_2CN$). **(c)** Hydroxymethylnitrile plus ammonia give aminomethylnitrile plus water. **(d)** Aminomethylnitrile plus water give glycine plus ammonia.

16. Chemical evolution Alpha-amino acids can be made via the Strecker synthesis (1850):

$$RCHO \xrightarrow[NH_3]{HCN} RCH(NH_2)CN \xrightarrow[H^+]{2H_2O} RCH(NH_2)CO_2H$$

(a) If the early earth environment contained hydrogen cyanide, propanal, ammonia, and water, show the result of a Strecker reaction. **(b)** Fourteen different amino acids have been synthesized by passing methane, ammonia, and steam down a silica tube held at 950°C. Write a single, balanced equation showing how alanine (Table 22-1) can be formed.

17. CO$_2$ curve Draw an approximate duplicate of the axes and the O$_2$ curve of Figure 22-17. In another color add a curve for atmospheric CO$_2$ as you believe the trend might have been. Write a brief explanation.

18. Molecular weight Cytochrome c is a molecule with an iron-porphyrin head (Figure 22-7b) connected to a protein tail. Analysis shows the molecule is 0.45 percent iron by weight. Calculate the molecular weight of cytochrome c.

23

WHAT ARE CARBOHYDRATES, LIPIDS, AND PROTEINS?

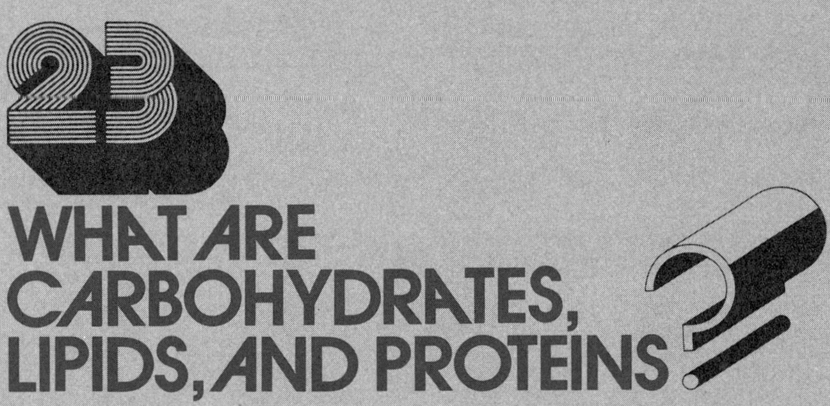

THE IMPORTANT CONCEPTS

23-1. Structure of carbohydrates

1. Carbohydrates may be divided into three classes.
2. Analysis proves glucose, a typical carbohydrate, has an aldehyde group, five hydroxyl groups, and six carbons.
3. A compound's optically active isomers are predictable from its number of asymmetric carbon atoms.
4. Fischer D and L designations are based on the two ways the glyceraldehyde structure can be written.
5. Expansion of a sugar chain by one carbon means creation of a new asymmetric center and a new dl pair of compounds.
6. Sugar acids may form inner esters called lactones; lactones are reducible to aldoses and ketoses.
7. A sugar diacid may be symmetric and optically inactive.

8. Mutarotation is explained by inner hemiacetal formation.
9. Glycosidic α and β bonds in oligosaccharides are selectively broken by enzymes.
10. Starch is polymaltose, cellulose is polycellobiose; they differ in the glycosidic bond.

23-2. Carbohydrate metabolism
1. Digestion in the mouth and stomach leads to absorption of molecules in the small intestine.
2. Glycolysis converts glucose to lactic acid and stores chemical energy in ATP.
3. The last C_2 fragment from food metabolism is oxidized in the Krebs cycle.
   a. Coenzyme A is the carrier of the C_2 fragment.
4. The reducing power of NADH and $FADH_2$, produced in glycolysis, is used to reduce O_2.
5. Made with energy from the respiratory chain, ATP furnishes its energy to cells.
   a. Two high-energy bonds are contained in ATP.
   b. The compound formed when ATP transfers a phosphate group to it is reactive.
6. Most energy-releasing reactions in respiration occur in mitochondria.

23-3. Lipids
1. Lipids include glycerides, lipoproteins, steroids, certain vitamins, and waxes.
2. Glyceride hydrolysis gives fatty acids and glycerol.
3. Glycerides are characterized via saponification and halogenation.
4. Gas chromatography is a powerful means for analyzing small quantities of volatile mixtures.
5. Phospholipids have properties of emulsifiers.
6. Fat-soluble vitamins are hydrocarbonlike.
7. Waxes are esters.
8. Lipids are oxidized in mitochondria and carried through the Krebs cycle by coenzyme A.
9. Coenzyme A also has a role in lipid synthesis.
   a. Isoprenoid units combine to give steroids.
10. Steroids are classified by carbon number; some have hormonal activity.

23-4. Proteins
1. Proteins are α-amino acid polymers classifiable as simple or conjugated.
2. Amino acids can be separated by distilling their esters.
3. Amino acids self-protonate to make zwitterions.
4. Ninhydrin is a color reagent for protein and amino acid analysis.
5. Most natural α-amino acids have L configuration.
6. Peptide nomenclature is based on amino acid names, and the N- and C-terminal ends.

7. Peptides can be synthesized by water loss between amino acids; unwanted reactions must be blocked.
8. Enzyme effectiveness has been explained by lock-and-key and orbital-steering mechanisms.
9. Sulfa drugs and heavy metals inhibit enzymes.
10. Dialysis, gel filtration, and electrophoresis are used to separate proteins.
11. Protein purity may be measured by electrophoresis, ultracentrifugation, and enzymatic activity.
12. Protein molecular weights are determinable from metal content and in other ways.
13. Primary structure can be determined by end-group analysis and reagents that cleave specific bonds.
14. Secondary structure is determined by X ray diffraction.
15. Hydrogen bonding is important in secondary structure.
16. Proteins are theoretically infinite in variation.
17. Amino acids from digested proteins are oxidized in the Krebs cycle or used in protein synthesis.

Biology did not develop a chemical–molecular basis until well into the twentieth century. Until then scientists were reluctant to believe that biological structures were built from large molecules (polymers), although as early as 1870 proteins had been crystallized (indicating an orderly arrangement of like molecules in living matter), and high molecular weights had been established (by osmotic pressure measurements) for starches and proteins.

Eventually chemists found that living things are primarily composed of four light elements—carbon, hydrogen, oxygen, and nitrogen—combined in three main classes of compounds—carbohydrates, lipids, and proteins. Carbohydrates, the major suppliers of energy that cells require, are the most abundant. Lipids are a high-energy foodstuff and a component of tissues and membranes. Proteins make up half the bulk of animal tissue. Most enzymes, hormones, and transfer agents of both oxidizing and reducing power are also proteins. The structure, function, biosynthesis, and reactions of carbohydrates, lipids, and proteins comprise a major part of biochemistry.

23-1. STRUCTURE OF CARBOHYDRATES

Classification and nomenclature

Around 1800 Proust, Gay-Lussac, and other French chemists found that destructive distillation of sugars, starch, and wood cellulose gave two products—charcoal and water. The results suggested a general formula, $C_x(H_2O)_y$, and a name, *hydrates de carbone*. The name *carbohydrates* is used today for that class of natural products, containing generally only carbon, hydrogen, and oxygen atoms, in which an —OH group or hydroxyl-derived group (aldehyde, ketone, ether) is attached to most of the carbon atoms.

When experiments proved that sucrose, the C_{12} sugar from sugar cane and sugar beets, could be hydrolyzed into two simpler C_6 sugars, glucose and fructose, and further that potato starch when hydrolyzed yielded only glucose, chemists began to recognize three classes of carbohydrates: (1) *monosaccharides*, three-, four-, five-, and six-carbon molecules; (2) *oligosaccharides*, containing several monosaccharide units (disaccharides, trisaccharides, etc.); and (3) *polysaccharides*, made up of many monosaccharide units and having molecular weights between about one thousand and one million.

Saccharide nomenclature was worked out. The ending *-ose*, signifying sugar, was added to word parts used to describe the main functional group and the chain length. A five-carbon sugar with an aldehyde function became an aldopentose; a six-carbon sugar with a ketone function, a ketohexose; etc.

Figure 23-1 Emil Fischer (Germany 1852–1919, Nobel prize in chemistry 1902), one of the greatest natural products chemists. Among Fischer's many accomplishments he (1) clarified sugar structures stereochemically; (2) determined the structures of nitrogen bases such as purine and certain derivatives,

purine   *uric acid*   *caffeine*

(3) separated amino acids from protein hydrolysates, synthesized peptides (protein precursors) from amino acids, and outlined protein problems in enzyme chemistry; and (4) discovered the class of sleep-inducing drugs called barbiturates. Following the deaths of three sons in World War I he committed suicide. (Edgar Fahs Smith Memorial Collection, University of Pennsylvania.)

Structure of glucose

The foundations of carbohydrate chemistry were established during the late 1800s by German chemists, especially Fischer (Figure 23-1) and his associates. At that time only a few saccharides had been isolated. Among the monosaccharides, the best characterized was glucose.

Up to 1880 it was known that hydrolysis of starch gave glucose, that the molecular weight of glucose was about 180, and thus that six "carbohydrate" units (CH_2O, formula weight 30) were in the simple sugar,

$$starch \xrightarrow{H_2O,\ H^+} n(CH_2O)_6$$

Glucose was also known as a "reducing sugar," meaning it contained a readily oxidizable function (the aldehyde group). One oxidizer used, basic cupric tartrate complex, gave an easily recognized yellow to red precipitate of cuprous oxide (*Fehling's test*). Representing the remainder of the molecule by R—,

$$R-C\overset{O}{\underset{H}{}} \xrightarrow[(OH^-)]{Cu^{2+},\ C_4H_4O_6^{2-}} R-C\overset{O}{\underset{O^-}{}} \cdot + Cu_2O\downarrow \qquad (23\text{-}1)$$

Glucose was further known to contain five alcohol groups because five acetyl chloride molecules were needed to esterify them:

$$\text{glucose} \xrightarrow[(-5H_2O)]{5H_3C-C\overset{O}{\underset{Cl}{}},\ OH^-} \text{glucose pentaacetate} + 5Cl^- \qquad (23\text{-}2)$$

glucose   *glucose pentaacetate*

Heinrich Kiliani showed that glucose could be assigned linear structure (as assumed in Eq. 23-2) by causing hydrogen cyanide to add to the aldehyde group, hydrolyzing the resulting nitrile to an acid, reducing all —OH groups to —H with hydriodic acid and identifying the final product as n-heptanoic acid, a linear molecule:

$$\text{(23-3)}$$

glucose   n-heptanoic
acid

The D sugars

In 1874 van't Hoff had announced the concept of the tetrahedral carbon atom. He had also discussed the rule that the

$$\textit{number of optically active isomers} = 2^n \qquad \text{(23-4)}$$

where n is the number of dissimilar asymmetric carbon atoms in a compound.

Noting the formula for glucose contained four asymmetric centers (*),

$$HO-\underset{H}{\overset{H}{C}}-\overset{H}{\underset{OH}{C^*}}-\overset{H}{\underset{OH}{C^*}}-\overset{H}{\underset{OH}{C^*}}-\overset{H}{\underset{OH}{C^*}}-\underset{H}{\overset{O}{C}}$$

Fischer realized that this sugar was but one of 2^4 or 16 isomeric aldohexoses theoretically capable of existence. During the next few years he was able to establish the structures of twelve of them, and the other four were proved later. Characterization by optical rotation (see Figure 20-3) was especially helpful.

For reference purposes Fischer related carbohydrate structures to the structure of one of the simplest carbohydrates, the aldotriose glyceraldehyde, $H_2C(OH)C^*H(OH)CHO$. Glyceraldehyde, with but one asymmetric carbon, he knew could exist in only the two forms shown in Figure 23-2.

Fischer assigned dextrorotatory (designated d or $+$) glyceraldehyde, the structure labeled D. With the carbonyl group written at the top, D meant that the —OH on the asymmetric carbon was attached on the molecule's right side; L meant the —OH group was on the left side. For convenience in writing he considered the structures pressed flat:

D-glyceraldehyde   L-glyceraldehyde

Because these formulas represented three-dimensional models, they could only be compared by rotating in the plane of the paper and not lifted from it. We continue to observe these conventions.

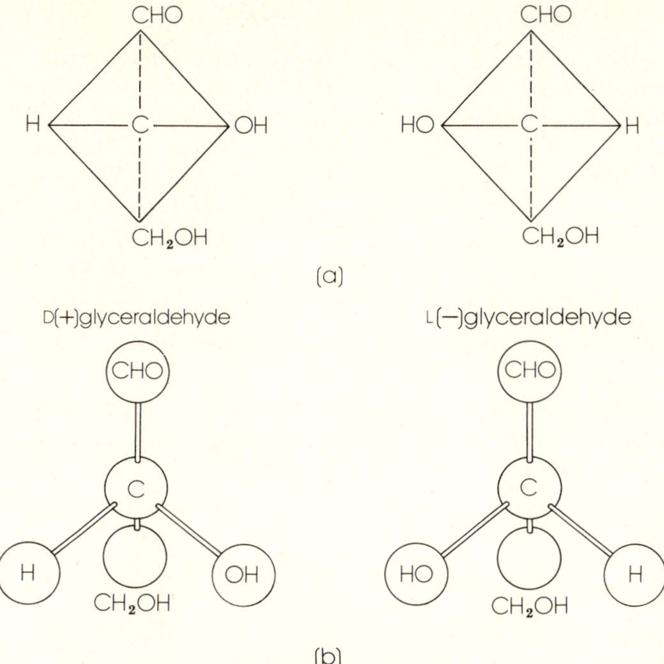

Figure 23-2 Fischer stereochemical conventions, illustrated with glyceraldehyde. (a) Tetrahedral models, (b) Ball-and-stick models equivalent to (a). In 1957 it was proved by X ray diffraction that the assumed configurations are the actual configurations.

Sugars whose structure was related to that of D-glyceraldehyde Fischer designated D-*series sugars;* those related to L-glyceraldehyde, L-*series sugars.* Glucose is a D-series sugar. The manner in which it and seven other aldohexoses, as well as six lower members of the D series, can be built up from D-glyceraldehyde is presented in Figure 23-3.

Syntheses indicated in the figure are carried out by lengthening the carbon chain one atom at a time. We will illustrate the classical method for saccharide chain expansion using D-erythrose as a starting compound.

1. Addition of HCN to the carbonyl group generates a pair of isomeric cyanides or nitriles. They differ only in the configuration at the newly generated optically active center (*):

$$
2\;
\begin{array}{c}
O\!\!=\!\!CH \\
| \\
HC\text{—}OH \\
| \\
HC\text{—}OH \\
| \\
CH_2OH
\end{array}
+\;2HCN \;=\;
\begin{array}{c}
CN \\
| * \\
HC\text{—}OH \\
| \\
HC\text{—}OH \\
| \\
HC\text{—}OH \\
| \\
CH_2OH
\end{array}
\;+\;
\begin{array}{c}
CN \\
| \\
HO\text{—}C\overset{*}{\text{—}}H \\
| \\
HC\text{—}OH \\
| \\
HC\text{—}OH \\
| \\
CH_2OH
\end{array}
\qquad (23\text{-}5)
$$

D(−)erythrose   isomeric nitriles

2. Hydrolysis of the nitriles gives a pair of corresponding isomeric acids:

$$\underbrace{\begin{array}{c} CN \\ | \\ HC^*\!-\!OH \\ | \\ HC\!-\!OH \\ | \\ HC\!-\!OH \\ | \\ CH_2OH \end{array} + \begin{array}{c} CN \\ | \\ HO\!-\!C^*\!H \\ | \\ HC\!-\!OH \\ | \\ HC\!-\!OH \\ | \\ CH_2OH \end{array}}_{\text{\textit{isomeric nitriles}}} + 4H_2O + 2H^+ \rightarrow \underbrace{\begin{array}{c} CO_2H \\ | \\ HC^*\!-\!OH \\ | \\ HC\!-\!OH \\ | \\ HC\!-\!OH \\ | \\ CH_2OH \end{array} + \begin{array}{c} CO_2H \\ | \\ HO\!-\!C^*\!H \\ | \\ HC\!-\!OH \\ | \\ HC\!-\!OH \\ | \\ CH_2OH \end{array}}_{\text{\textit{isomeric acids}}} \quad (23\text{-}6)$$

3. The two acids are not mirror images of one another. They (and their salts) therefore have different solubilities in water, and separation by crystallization is possible.

4. With strong acid the sugar acids are converted to their *lactones*. A lactone is an inner ester, characteristic of compounds that contain a carboxyl group and hydroxyl group so situated that when a water molecule is

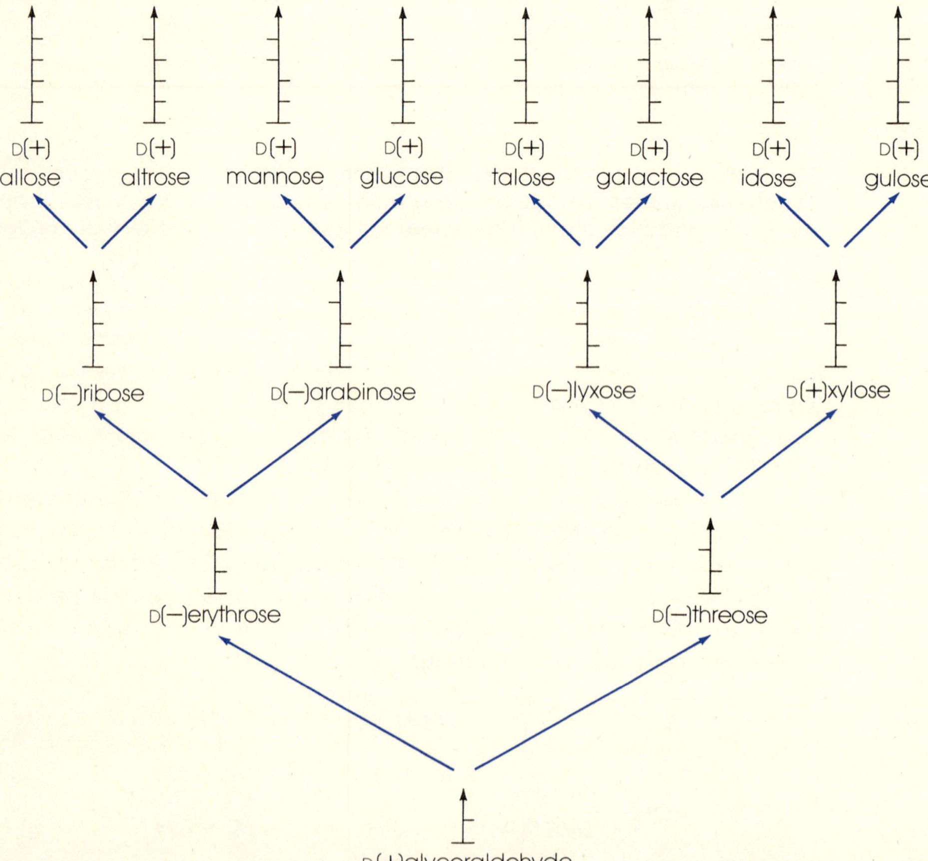

Figure 23-3 Fischer's structure assignments in the D series of sugars. A similar L series is derivable from L-glyceraldehyde, giving a total of 16 possible aldohexoses. All have been isolated or synthesized. Only three are found in nature (and they are the only three that yeasts are able to ferment): D(+)mannose, D(+)glucose, and D(+)galactose. Symbols used:

$$\uparrow \text{ is } \begin{array}{c} O \\ \| \\ C \\ | \end{array}\!\!\diagup H \quad , \perp \text{ is } CH_2OH, \vdash \text{ is } HC\!-\!OH$$

eliminated between them, a ring closes containing five atoms (γ lacetone) or six atoms (δ lactone):

$$
\begin{array}{ccc}
\text{CO}_2^- & \text{CHOH} & \text{CHOH} \\
\text{*CHOH} & \text{*CHOH} & \text{*CHOH} \\
\text{HC—OH} \xrightarrow{\text{H}^+} & \text{HC—OH} \xrightarrow{-\text{H}_2\text{O}} & \text{HC—OH} \\
\text{HC—OH} & \text{HC—OH} & \text{HC} \\
\text{CH}_2\text{OH} & \text{CH}_2\text{OH} & \text{CH}_2\text{OH} \\
\text{ion} & \text{acid} & \gamma\ \text{lactone}
\end{array}
\tag{23-7}
$$

5. The lactones are reduced with sodium amalgam. When mild acid conditions are used, the product is an aldose:

$$
\begin{array}{cc}
\text{*CHOH} & \text{*CHOH} \\
\text{HC—OH} \xrightarrow[\text{H}^+]{\text{Na–Hg}} & \text{HC—OH} \\
\text{HC} & \text{HC—OH} \\
\text{CH}_2\text{OH} & \text{CH}_2\text{OH} \\
\text{lactone} & \text{aldopentose}
\end{array}
\tag{23-8}
$$

The two aldopentoses derived in the five-step process from D-erythrose are

$$
\begin{array}{cc}
\text{HO—CH} & \text{HC—OH} \\
\text{HC—OH} & \text{HC—OH} \\
\text{HC—OH} & \text{HC—OH} \\
\text{CH}_2\text{OH} & \text{CH}_2\text{OH} \\
\text{D}(-)\text{arabinose} & \text{D}(-)\text{ribose}
\end{array}
$$

Now which compound is which? Fischer answered the question by first oxidizing with nitric acid to convert the compounds to dicarboxylic acids:

$$
\begin{array}{cccc}
\text{HO—CH} & \text{HO—CH} & \text{HC—OH} & \text{HC—OH} \\
\text{HC—OH} \xrightarrow[\Delta]{\text{HNO}_3} & \text{HC—OH} & \text{HC—OH} \xrightarrow[\Delta]{\text{HNO}_3} & \text{HC—OH} \\
\text{HC—OH} & \text{HC—OH} & \text{HC—OH} & \text{HC—OH} \\
\text{CH}_2\text{OH} & \text{CH}_2\text{OH} & \text{CH}_2\text{OH} & \text{CH}_2\text{OH} \\
\text{D}(-)\text{arabinose} & \text{asymmetric} & \text{D}(-)\text{ribose} & \text{plane of symmetry}
\end{array}
\tag{23-9}
$$

Then he examined the optical rotation of the products. The ribose-derived acid, its top and bottom halves mirror images of one another, was an optically inactive *meso* compound (like *meso*-tartaric acid, Figure 20-6). The arabinose-derived acid, being asymmetric, was optically active.

Ring structure of sugars

Observations by Fischer and others indicated that the linear or chain formula being used for glucose was inadequate: (1) The —CHO group gave only weak aldehyde reactions, (2) two different *hemiacetals* formed with methanol,

$$R-C\overset{O}{\underset{H}{\diagdown}} + H_3C-OH \underset{\longleftarrow}{\overset{HCl}{\rightleftharpoons}} R-\underset{H}{\overset{OH}{\underset{|}{\overset{|}{C}}}}-O-CH_3 \qquad (23\text{-}10)$$

aldose   *alcohol*   *hemiacetal*

and (3) two different optically active forms of glucose could be obtained by crystallization, one showing an initial optical rotation of $+113$ degrees, the other an initial rotation of $+19$ degrees, and both, on standing in solution, changing gradually to a value of $+52$ degrees, a change called *mutarotation*.

The experimental facts were best explained by assuming closure of a ring through intramolecular hemiacetal formation. A five-membered ring compound of this kind is called a *furanose*; a six-member ring compound, a *pyranose*, after the well-known oxocyclic compounds,

furan   *and*   *pyran*

As ring closure gives the aldehyde carbon an —OH group, that carbon becomes a new asymmetric center, which accounts for the two forms that mutarotate. These forms, designated α and β, differ only in the orientation of —H and —OH at the first carbon atom C-1. They are interconvertible through the open chain form that exists only in trace quantities. Picturing the rings in chair form (see Eq. 20-6), mutarotation is described as

α-D-glucopyranose   *chain form*

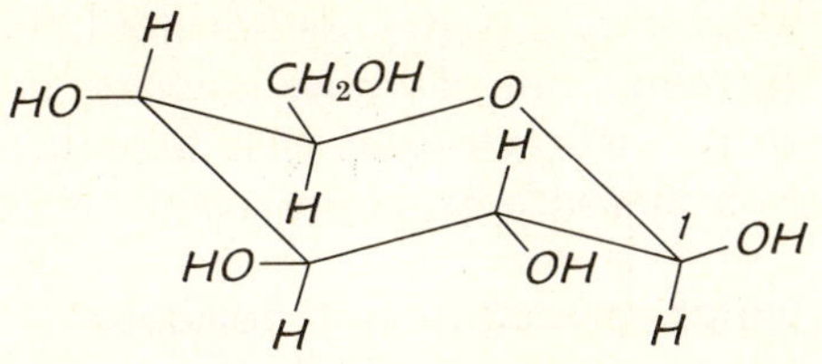

$$\beta\text{-D-glucopyranose} \tag{23-11}$$

Sucrose and glycosidic bonds

Oligosaccharides consist of cyclic monosaccharide units joined through oxygen atoms. Sucrose, for example, is a disaccharide combination of D-glucose and D-fructose, the latter a ketohexose. Hydrolysis gives the monosaccharides:

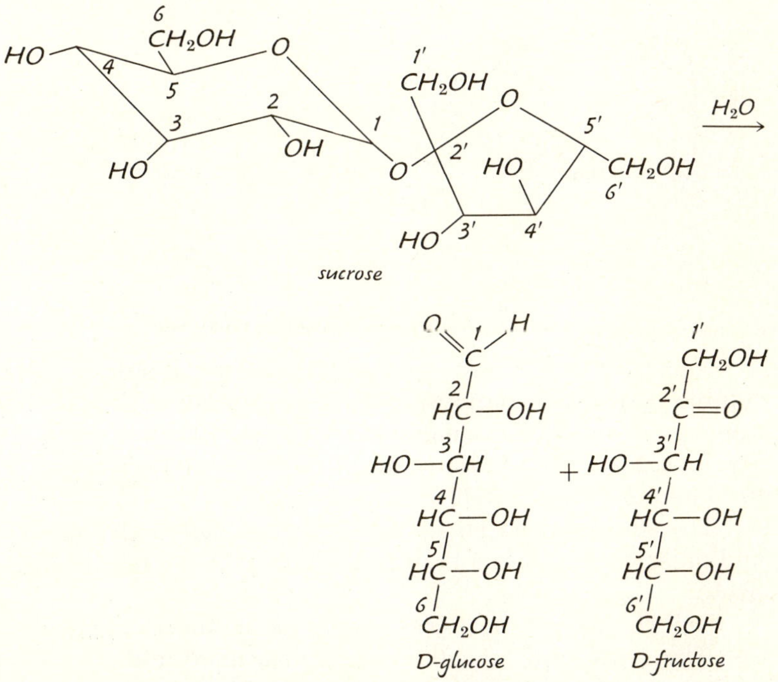

$$\text{sucrose} \xrightarrow{H_2O} \text{D-glucose} + \text{D-fructose} \tag{23-12}$$

The ring structure of fructose is explained by the formation of a *hemiketal* (analogous to the hemiacetal of glucose) in which the ketone carbon at C-2′ is joined to C-5′ through an oxygen atom. When a sugar hemiacetal or hemiketal —OH group reacts with an —OH from another molecule with loss of water, the product is called a *glycoside* and the oxygen atom bridging the two molecules is called a *glycosidic bond*. The glycosidic bond in sucrose is formed by elimination of H_2O between the acetal —OH of glucose (C-1) and the ketal —OH of fructose (C-2′). Sucrose thus has no potentially active carbonyl ($\diagup$C=O) function, shows no carbonyl reactions (such as oxidation by Fehling's reagent), and is classified as a *nonreducing* sugar.

The configuration about the glycosidic link is biochemically important because enzymes show selective ability in causing hydrolysis at that point. α-D-Glycosidase, an enzyme found in yeast, splits any glycosidic bond formed by α-D-glucose but not by β-D-glucose (Eq. 23-11). It splits sucrose, therefore, one of whose halves is best named α-D-glucopyranose. (The other half is β-fructofuranose.) Specificity of enzyme action and optical rotation measurements are used to assign configuration to glycosidic bonds.

Starch, glycogen, and cellulose

Most natural carbohydrates are high polymers which upon hydrolysis give monosaccharides as the principal product.

Plants store carbohydrate from photosynthesis in the form of starches. Starches are made up of *maltose units*, each consisting of two glucose units. Glycosidic linkages in starches are designated α, 1,4', meaning α orientation and a joining of C-1 of one glucose unit to C-4' of the next. Starch is not a reducing carbohydrate because only the few aldehyde groups available on terminal glucose units can be oxidized.

starch: the repeating maltose unit

Natural starches are separable into two fractions. The more water-soluble part, called *amylose*, generally consists of linear chains (as above) ranging in molecular weight from 10^4 to 10^6. The other fraction, called *amylopectin*, which is water insoluble, has a denser, branched-chain structure. Its molecular weight may reach 10^8.

The equivalent of starch in animals is called *glycogen*. Found in liver and muscle cells, it is a branched-chain polysaccharide resembling amylopectin.

Half the carbon in plants occurs as cellulose, molecular weight about 10^5. The repeat unit is *cellobiose*, a disaccharide that yields glucose upon hydrolysis. The glycosidic bonds are β,1,4':

cellulose: the repeating cellobiose unit

Examination with X ray diffraction and the electron microscope shows that cellulose chains occur naturally in bundles that have both amorphous and crystalline regions. The orderliness of the latter is a consequence of interchain hydrogen bonding brought about by the β-glycoside bonds whose orientation causes chains to line up parallel to one another.

Carnivorous animals (humans included) can digest starch but lack the enzymes necessary to hydrolyze the $\beta,1,4'$ bonds of cellulose. Cud-chewing animals and termites, however, are hosts to bacteria whose enzymes allow them to use cellulose for food.

23-2. CARBOHYDRATE METABOLISM

Digestion

The process by which large molecules of food are hydrolyzed to smaller molecules as they pass down the digestive tract before being absorbed in the animal body is called *digestion*. In the mouth food is reduced in size by chewing. Simultaneously it is mixed with saliva, a water solution of inorganic ions (mainly Na^+, Cl^-, HCO_3^-, pH $\sim$ 6.8) and organic substances that include α-amylase, the enzyme that begins the breakdown of starch. In the stomach food particles become mixed with gastric juice, a dilute solution of hydrochloric acid (pH $\sim$ 1.6) containing pepsin, an enzyme that starts the decomposition of proteins, and gastric lipase, an enzyme that causes hydrolysis of lipids.

As liquefied food passes into the duodenum and small intestine, it contacts acidic juices excreted by the intestinal walls and pancreas. This mixture contains enzymes for the hydrolysis of all nutrients. Included are the enzymes lactase, maltase, and sucrase for the disaccharides lactose (milk sugar), maltose (from starches), and sucrose. Digestion of lipids, which are water insoluble, is aided by emulsifying agents (described later) secreted from the gall bladder in a fluid called *bile*.

Digestion products, now small molecules, are absorbed through the walls of the small intestine. Undigested food, indigestible material (chiefly cellulose), bacteria, and the toxic substances they produce, as well as excess secretion products, pass into the large intestine, then to the anus for elimination as feces.

Glycolysis

Absorbed glucose is carried from intestinal walls mainly by the circulating blood system to the liver, an organ that regulates its distribution according to body demand. Some glucose is stored in the liver as glycogen while the remainder is passed to body tissues and into cells.

In cells glucose undergoes partial oxidation to lactic acid. This complex process, called *glycolysis* or the glycolytic pathway, proceeds in the absence of free oxygen and is similar to fermentation (Figure 22-5).

Some of the reactions of glycolysis are shown in Figure 23-4. (For simplification no enzymes are included.) The sequence begins (left, center) with the phosphorylation of glucose to glucose-6-phosphate by adenosine triphosphate, ATP. In two stages glucose-6-phosphate is converted to fructose-1,6-diphosphate by another molecule of ATP. Fructose-1,6-diphosphate is split into D-glyceraldehyde-3-phosphate and dihydroxyacetone phosphate, two triose phosphates that are isomeric and in equilibrium with each other. As the former is used up in the next series of steps, the latter undergoes rearrangement to make more of it. In a series of four steps 3-phosphoglyceraldehyde is converted to pyruvic acid. Chemical

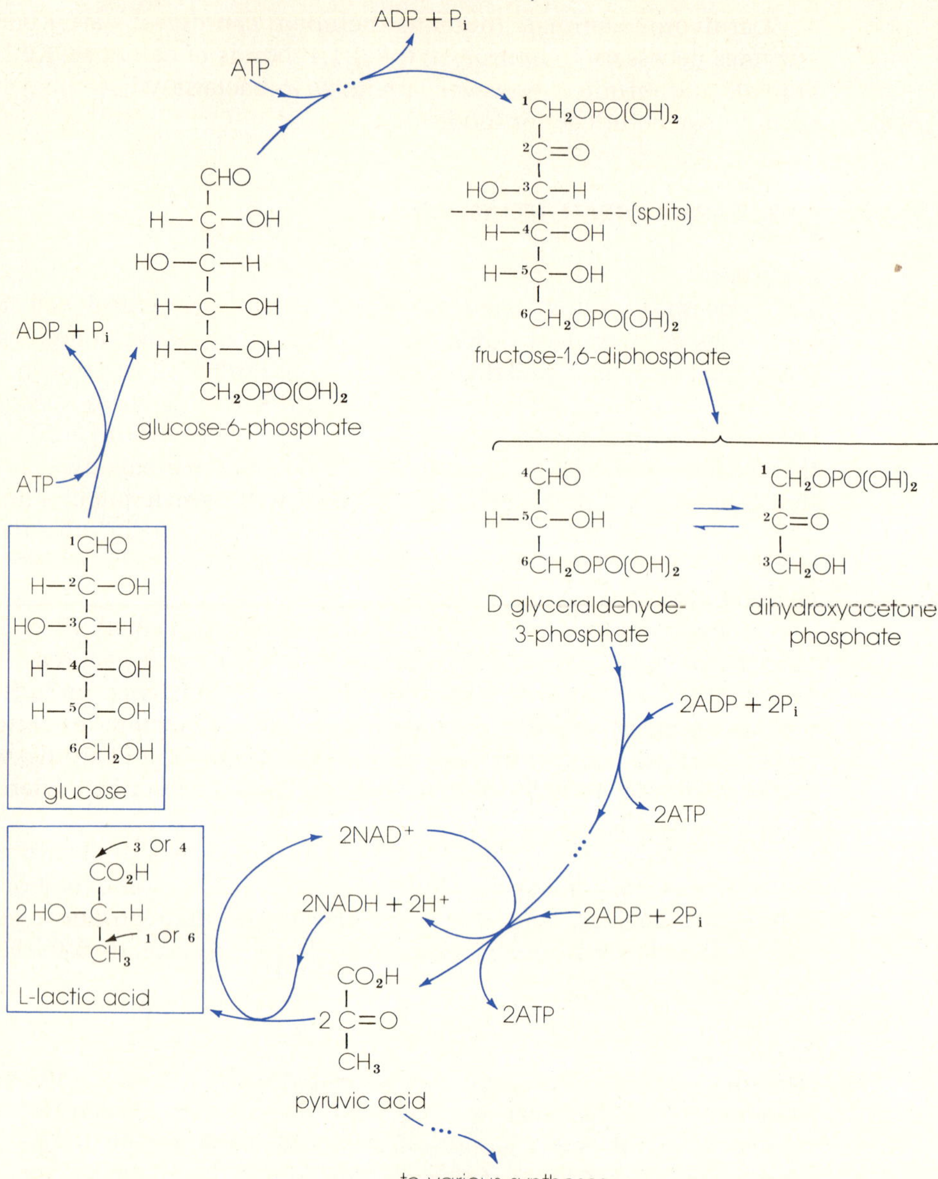

Figure 23-4 Glycolysis, the anaerobic process whereby glucose (or glycogen) is broken down to lactic acid. At the same time energy for cell functions is stored as ATP. Broken arrows indicate places where one or more reactions have been omitted for simplification. The net reaction of glycolysis is

$$C_6H_{12}O_6 + 2ADP + 2P_i = 2H_3CCH(OH)CO_2H + 2ATP$$

Carbon atoms are numbered so they may be followed through the steps.

energy released in these reactions is translated into synthesis of four molecules of the triphosphate ATP from the diphosphate ADP, and reduction of the coenzyme nicotinamide adenine dinucleotide (NAD⁺) to NADH. It is NADH that reduces pyruvic acid to lactic acid in the muscle cells of vertebrates.

Pyruvic acid (or pyruvate ion) crosses cell membranes easily and is distributed throughout the body for various syntheses. If oxygen is available in cells, pyruvic acid is oxidized further. This acid decarboxylates readily. As carbon dioxide is released, acetaldehyde is formed:

$$H_3C-\overset{\overset{O}{\|}}{C}-CO_2H = CO_2 + H_3C-\overset{\overset{O}{\|}}{C}-H$$

Acetaldehyde reacts with *coenzyme A* (abbreviated CoASH, structure with Figure 23-5) which, together with an enzyme, carries the two-carbon unit to its next reaction. Bonding is through coenzyme A's sulfur atom:

$$H_3C-\overset{\diagup\!\!\diagup O}{\underset{\diagdown H}{C}} + CoASH = CoA-S-\overset{\overset{O}{\|}}{C}-CH_3 + 2[H] \qquad (23\text{-}13)$$

The product, acetyl coenzyme A, is the fuel used in the *citric acid cycle* of reactions.

The citric acid cycle

Carbon and hydrogen from the breakdown of carbohydrates, fatty acids (from lipids), and some amino acids (from proteins) enter the final stages of oxidation as the acetyl end of acetyl coenzyme A. The sequence of reactions that follows is variously called the *citric acid cycle*, the *tricarboxylic acid (TCA) cycle,* or the *Krebs cycle* (Figure 23-5). The cycle operates only when oxygen is available (as it is in most tissues), although its compounds are not oxidized by oxygen directly. The acetyl group first becomes attached to a four-carbon compound. As the six-carbon product goes through the cycle, two carbon atoms are oxidized to CO_2. Simultaneously, hydrogen is abstracted and then converted to water in another sequence of reactions called the *respiratory chain.* Finally, the four-carbon acceptor of acetyl is regenerated for another turn of the cycle. Each step is catalyzed by a specific enzyme.

The cycle begins (upper left of Figure 23-5) with oxalacetic acid (C_4) accepting an acetyl group from acetyl coenzyme A, creating citric acid (C_6). Citric acid is isomerized to isocitric acid (C_6, not shown) from which reducing power in the form of two hydrogen atoms and two electrons is removed by the coenzyme NAD^+. Products are NADH, oxalosuccinic acid (C_6, not shown), and H^+, which goes into solution (see (23-14) on page 605). NAD and its reduced form NADH are common to many redox reactions as are the related coenzymes NADP and NADPH. For each mole of NADH formed, 3 moles of ATP are also formed. As oxalosuccinic acid loses CO_2, α-ketoglutaric acid (C_5) is formed.

A second molecule of CO_2 comes from decarboxylation of α-ketoglutaric acid, and two more hydrogens (as $NADH + H^+$) are simultaneously removed. The resulting succinyl group (C_4) attaches coenzyme A, giving succinyl coenzyme A. Removal of coenzyme A from succinyl CoA gives succinic acid (C_4). At the same time another molecule of ATP is produced from ADP.

In a series of three more steps succinic acid is oxidized to the starting material, oxalacetic acid, and four more hydrogen atoms are abstracted via

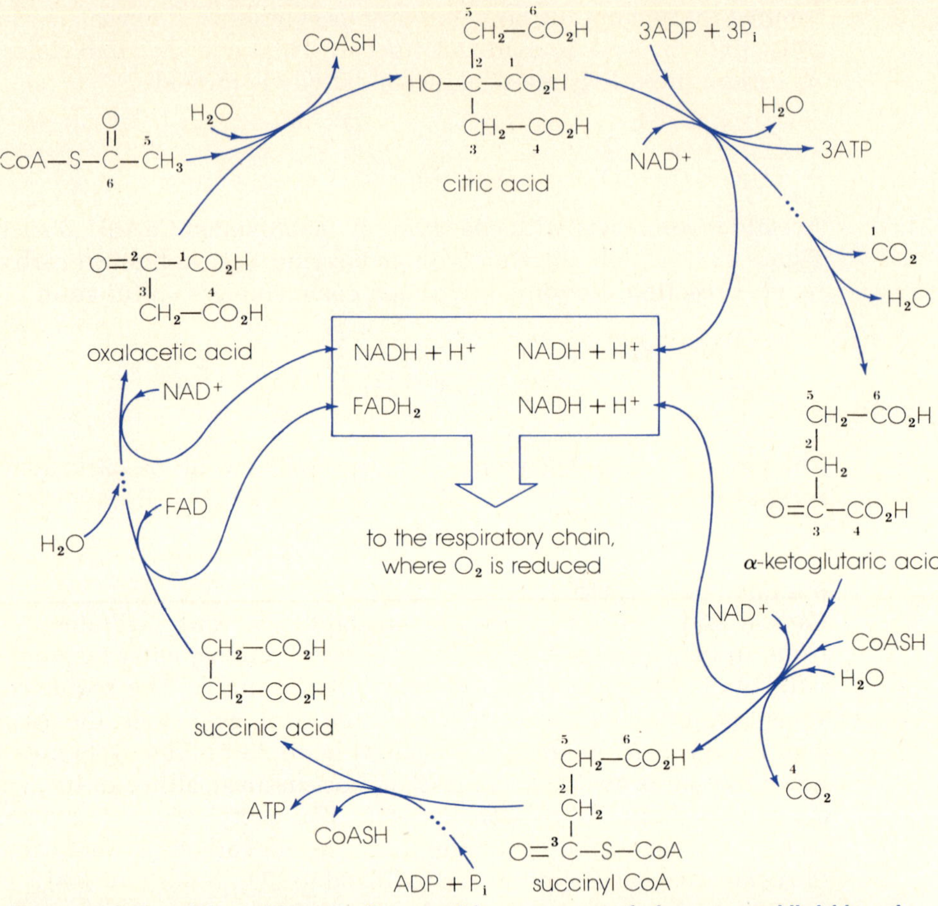

Figure 23-5 The citric acid cycle, the aerobic sequence of glucose, and lipid breakdown. The numbering shows that the acetyl carbons are not the two carbon atoms oxidized to CO_2 in the first turn of the cycle. (For simplification all enzymes are omitted.) Coenzyme A is a combination of a nucleotide and a derivative of pantothenic acid, a B vitamin,

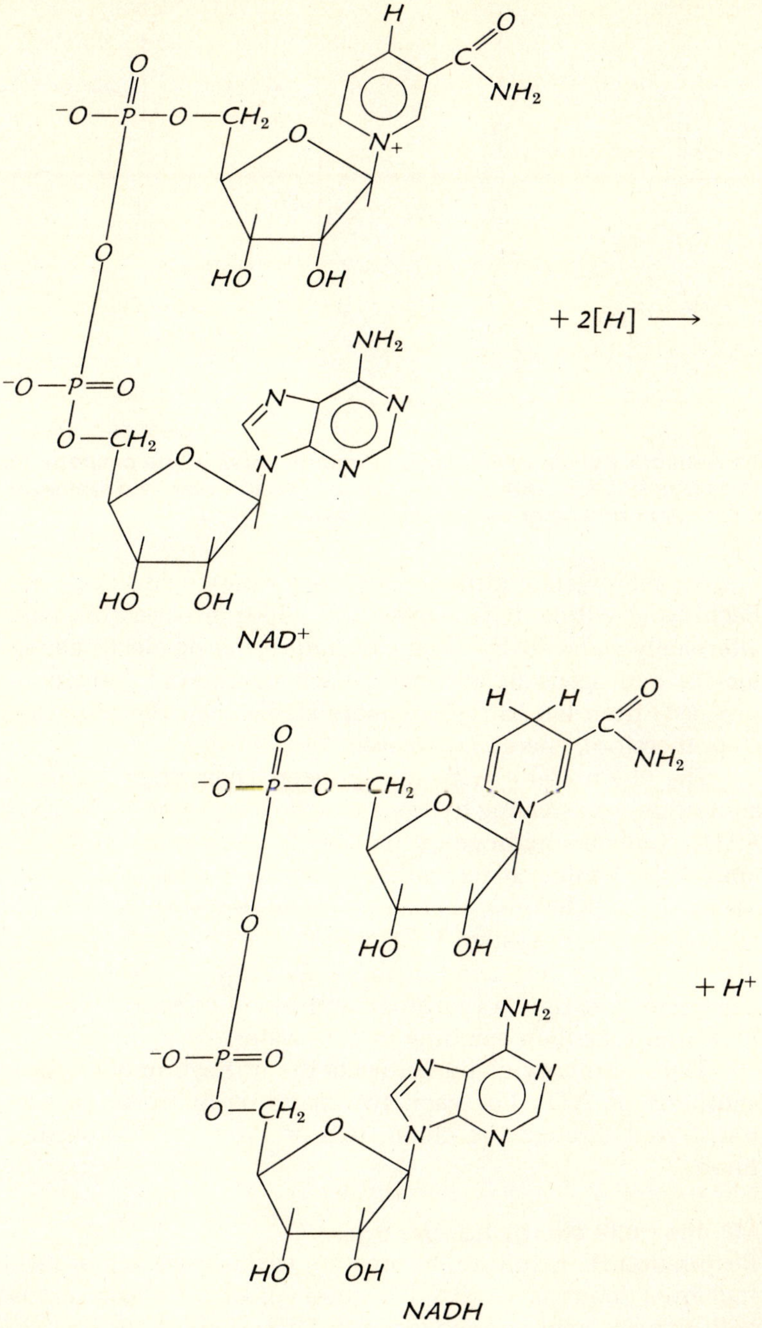

$$+ 2[H] \longrightarrow$$

NAD$^+$

(23-14)

$$+ H^+$$

NADH

coenzymes NAD$^+$ and FAD (flavin adenine dinucleotide), forming more NADH, H$^+$, and FADH$_2$.

The respiratory chain

As noted above, the citric acid cycle oxidations are accomplished by removal of hydrogen atoms in the form of NADH (or NADPH) and FADH$_2$. In a controlled series of steps called the *respiratory chain* or *electron-transfer system* this reducing potential is then used up. The final product is water, produced by reduction of oxygen gas.

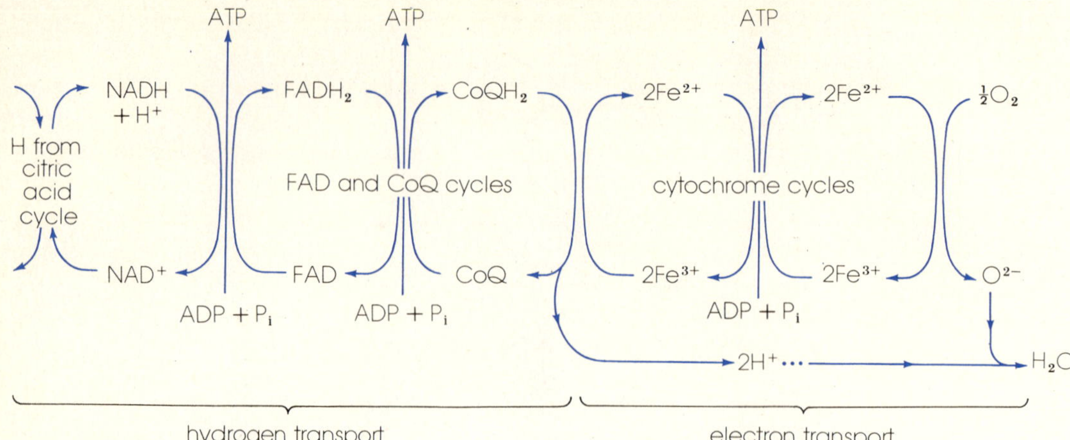

Figure 23-6 The respiratory chain (several cycles and all enzymes are omitted). The net effect is to oxidize NADH to NAD⁺, reduce oxygen gas to water, and synthesize ATP for use in glycolysis and other cellular processes.

An abbreviated outline of the respiratory chain is given in Figure 23-6. Each cycle within it is a reversible redox process, the components being alternately reduced by a gain of hydrogens or electrons, and oxidized by loss of hydrogens or electrons. Each compound formed as the sequence proceeds from left to right is more stable than the preceding one. At each step, therefore, energy is released.

The chain starts with the reduced coenzyme NADH (from the citric acid cycle) transferring hydrogen to the coenzyme FAD, giving FADH₂. The FADH₂ transfers hydrogen to another factor, coenzyme Q. As hydrogen ions are released from coenzyme Q, electrons are transferred to and through a series of cytochromes that are iron porphyrins (see Figure 22-7b). Containing the redox system $Fe^{2+} \rightleftharpoons Fe^{3+} + e^-$, cytochromes are electron carriers. In the final step a cytochrome gives up electrons to oxygen gas which has come to cells via the lungs and blood stream. Oxygen, electrons, and hydrogen ions then combine to give water.

The respiratory chain connects the utilization of oxygen by cells to the synthesis of ATP. For each two pairs of hydrogen atoms and electrons transferred through the chain, two or three ATP molecules are manufactured.

ATP, The cell's energy transfer agent

Respiration (L. *respirare*, to breathe) is the sequence of reactions in which fuel from foods is oxidized by transfer of hydrogen atoms from them to both pyruvic acid and oxygen gas. These oxidations liberate little heat. Instead, an appreciable amount of the energy yield is stored in ATP which carries the chemical energy that makes nonspontaneous reactions go. The importance of ATP should be emphasized by describing the biological use of glucose in the body as

$$C_6H_{12}O_6 + nADP + nP_i \longrightarrow 6CO_2 + 6H_2O + nATP + E \tag{23-15}$$

where E stands for energy not available to the body. If a mole of glucose is burned in a calorimeter, the yield is 684 kcal. To form a mole of ATP from

ADP and inorganic phosphate ion, P_i, requires 7 kcal. The theoretical value for n is therefore 684/7 or 98. The actual number of ATP molecules synthesized biologically per mole of glucose oxidized is about 38, however, indicating a 40-percent efficiency. The way ATP production is linked to the metabolism of both carbohydrates and lipids is diagramed in Figure 23-7.

In cells ATP is made by addition of two inorganic phosphate ions to adenosine monophosphate (AMP, Eq. 22-6). Energy is needed to accomplish the additions because ATP is less stable than AMP and free phosphate ions. Two reasons proposed for this are (1) the hydrolysis products of ATP are freer to resonate and therefore better stablized than ATP, and (2) electronegative phosphate groups that in ATP are bonded close together despite mutual electrostatic repulsion are separated in AMP and $2P_i$.

The energy-rich nature of ATP is located in two oxygen–phosphorus bonds which are called *high-energy bonds*. They are indicated in an abbre-

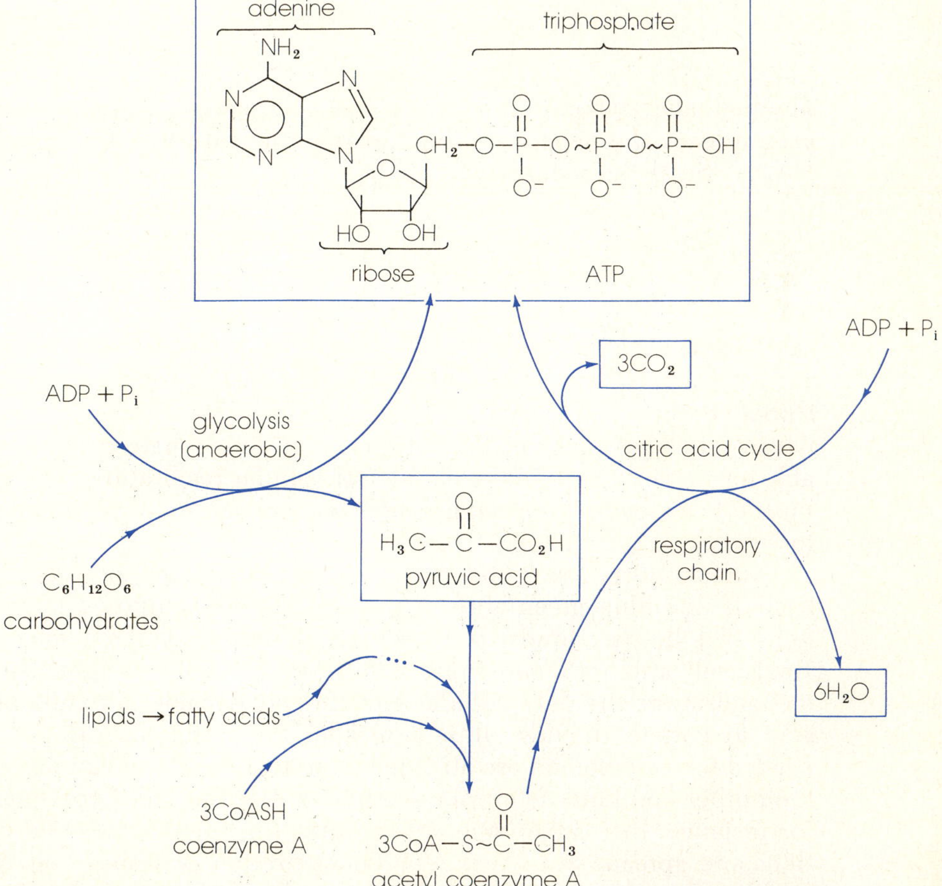

Figure 23-7 Cellular reactions that produce the energy needed to yield the energy-rich nucleotide (base-sugar-phosphate) adenosine triphosphate (ATP). Reactions involving the breakdown of larger molecules and release of energy are called catabolic reactions. (The inverse are called anabolic reactions.)

viated formula for ATP as

$$adenosine\,{-}O{-}P{-}O{\sim}P{-}O{\sim}P$$

Hydrolysis at the O~P positions, giving either an orthophosphate ion $H_2PO_4^-$ and adenosine diphosphate, or a pyrophosphate ion $H_2P_2O_7^{2-}$ and adenosine monophosphate, is accompanied by release of chemical energy. Synthesis of ATP in a cell can be likened to winding a spring in a watch. Once coiled, the spring will gradually unwind, transmitting stored energy to the system attached to it.

Biochemical reactions that require energy input to make them go are called *endergonic*; those that go spontaneously and release energy are called *exergonic*. An endergonic reaction like reduction of a carboxylic acid to an aldehyde can be driven by an exergonic reaction like ATP hydrolysis when the two are related through a common intermediate. In a first step ATP furnishes a phosphate group and the chemical energy needed to attach it:

$$\underset{acid}{R{-}\overset{\overset{O}{\|}}{C}{-}OH} + ATP \xrightarrow{enzyme} \underset{acid\ anhydride}{R{-}\overset{\overset{O}{\|}}{C}{-}O{\sim}\overset{\overset{O}{\|}}{\underset{\underset{OH}{|}}{P}}{-}OH} + ADP \qquad (23\text{-}16)$$

The high-energy bond renders the product unstable. Prone toward hydrolysis, it loses the phosphate ion and the reduced form of a coenzyme like NADH effects the reduction:

$$R{-}\overset{\overset{O}{\|}}{C}{-}O{\sim}\overset{\overset{O}{\|}}{\underset{\underset{OH}{|}}{P}}{-}OH + NADH \xrightarrow[H_2O]{enzyme} \underset{aldehyde}{R{-}\overset{\overset{O}{\|}}{C}{-}H} + NAD^+ + H_2PO_4^-$$

Mitochondria

Glycolysis takes place in the cytoplasm of cells where enzymes are available for its reactions. The remainder of cellular respiration, which includes most of its energy-releasing reactions, occurs in organelles called *mitochondria* (Figure 23-8).

Mitochondria are found in all cells of organisms more complex than bacteria and blue–green algae. Mitochondria are composed mainly of proteins and lipids. Hundreds to several thousand of them may occupy a single cell and may constitute a fifth of the cell's weight. Isolated mitochondria are the only cellular constituents capable of oxidizing pyruvic acid to carbon dioxide. It is probable, therefore, that all the enzymes needed for metabolism are attached to mitochondrial membranes and that their order and kind determine the function of the cell. Experiments show, for instance, that membrane area contracts in high local ATP concentrations and spreads out when ATP concentration is diminished. Such chemical–architectural interdependence may be evidence that by an enfolding process enzyme sites are covered up to regulate the cell's metabolism, or that the changing of cell-wall thickness accomplishes regulation by altering the flow through of water.

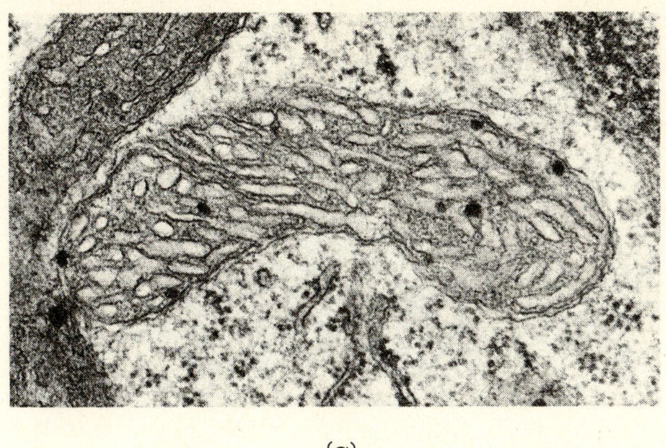

(a)

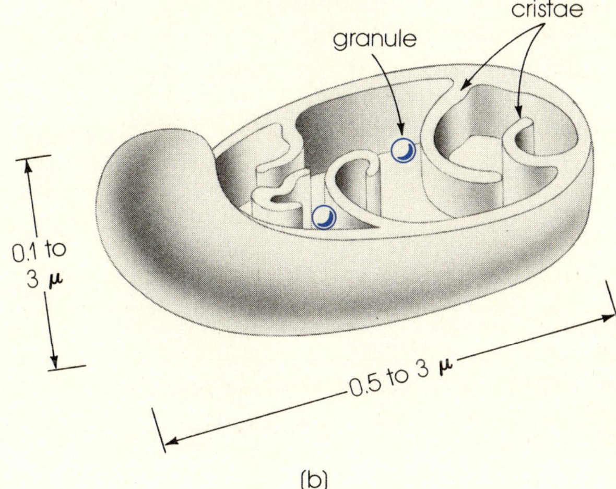

(b)

Figure 23-8 Mitochondria. (a) Electron photomicrograph (×248,500) of a mitochondrion from a salt gland of Tamarix aphylla. The dark spots are granules that may be specific locations where respiratory reactions occur. (b) Drawing of a mitochondrion. The top is removed to show inner membrane folds called cristae. Mitochondria are separated for study by grinding up cells and subjecting the mixture to ultracentrifugation (see Figure 23-15). (Photograph Courtesy W. W. Thomson, University of California, Riverside.)

23-3. LIPIDS

Lipids are a class of natural products extractable from tissues with organic solvents like ether and chloroform. Lipids that are solids at room temperature are called *fats*; lipids that are liquids are called *oils*.

Lipid structure and characterization

Saponification, or the splitting of lipids by hot aqueous alkali, was used for centuries as the method for making soap before it was studied as a chemical process by French chemist M. Chevreul. In the early 1800s he showed that (1) readily available lipids—lard, tallow, and whale oil—were largely glycerol esters (glycerides) of certain weak organic acids, and (2) alkali metal salts of the acids were soaps. A contemporary, Berthelot, proved fatty acids contained even numbers of carbon atoms from C_4 to about C_{30}. Starting with glycerol and the acids, he showed that he could resynthesize glycerides.

Today we know that most natural fatty acids are straight-chain saturated and unsaturated molecules. Those larger than C_4 are, like lipids themselves, water insoluble. Oleic acid is the most widely distributed fatty acid in na-

Table 23-1 Representative fatty acids

Name	C atoms	One source	Formula	M.P. °C
Butanoic	4	Butter	$CH_3(CH_2)_2CO_2H$	−4.7
Lauric	12	Coconuts	$CH_3(CH_2)_{10}CO_2H$	44
Palmitic	16	Palm oil	$CH_3(CH_2)_{14}CO_2H$	63
Stearic	18	Tallow	$CH_3(CH_2)_{16}CO_2H$	70
Oleic	18	Lard	$CH_3(CH_2)_7CH{=}CH(CH_2)_7CO_2H$	14
Linoleic	18	Peanuts	$CH_3(CH_2)_4CH{=}CHCH_2CH{=}CH(CH_2)_7CO_2H$	−5

ture, followed by palmitic acid (Table 23-1). All natural glycerides are mixtures that contain some unsaturation and include mono-, di-, and triesters (1, 2, or 3 —OH groups of glycerol are esterified). Milk fat, one of the more complex mixtures, yields the salts of 14 different fatty acids when saponified.

Table 23-1 suggests that a glyceride can be characterized by the average size of its fatty acid groups and their unsaturation. Saponification and halogenation reactions are used to do that quantitatively (Eq. 23-17).

$$(23\text{-}17)$$

The quantity of potassium hydroxide needed to saponify a given weight of glyceride (break is at dashed line in Eq. 23-17) is related to the glyceride's molecular weight. The smaller the KOH requirement, the larger the average formula weight of the fatty acid portion of the glyceride. The quantity of iodine needed to halogenate a given weight of glyceride is related to the glyceride's unsaturation. The smaller the I_2 requirement, the fewer the carbon–carbon double bonds (see Problems 32, 33).

Gas chromatography; separating fatty acids

In 1952 biochemists A. James and Archer Martin (England 1910–   , Nobel prize in chemistry 1952), seeking a better way of resolving fatty acid mixtures than distillation, developed the technique of gas chromatography. Gas chromatography has become so important for separating difficult mixtures of many kinds, that in recent times each year has seen, worldwide, the introduction of about 10,000 new gas chromatographs and publication of 2000 related scientific papers. (Thus to follow all "g.c." applications one would have to read five articles a day.)

The most versatile and powerful form of chromatography is illustrated in Figure 23-9. It is properly called *gas-liquid or vapor phase chromatography* because a gas is the *moving phase* and a liquid (on a solid support in the column) is the *stationary phase*. The sample, frequently a volatile liquid, is injected with a microliter syringe into the apparatus at a heated port where it vaporizes. The vapor is moved by a continuously flowing stream of carrier gas (helium, for example) into a small-bore chromatographic column. The column is typically packed with brick dust, whose particles have been coated with the stationary phase, an inert nonvolatile oil. Cohesive forces (dipole–dipole attraction, for example—see Section 10-2) acting between the phases cause sample components to be selectively detained. As carrier gas flow continues, components are eventually eluted from the column and emerge as separate volumes of vapor. At the end of the column a detector that senses changes in gas composition feeds a signal to a recorder whose pen traces out the *chromatogram*, a series of peaks above a base line. *Retention time* in the column serves to identify a component, as calibrated previously with known substances. The *area under a chromatographic peak* is proportional to the quantity of that component in the sample.

The first biochemical application of gas chromatography in America dealt with the analysis of lard. The technique used has become routine in lipid analysis: saponifation of the lipid with alkali, addition of methyl alcohol and sulfuric acid to the alkaline mixture to convert the salts of fatty acids to their more volatile methyl esters, followed by chromatography of the esters:

$$R\overset{\overset{\displaystyle O}{\|}}{-C}-O^- Na^+ + H^+ = R\overset{\overset{\displaystyle O}{\|}}{-C}-OH + Na^+ \xrightarrow[H^+]{H_3COH} R\overset{\overset{\displaystyle O}{\|}}{-C}-O-CH_3 + H_2O$$

Lipids other than simple glycerides

Triglycerides (neutral fats) are the most abundant lipids but not the only kind known. Lipids have been found with linkages to phosphorus, sulfur, and nitrogen compounds, as well as to sugars and various alcohols.

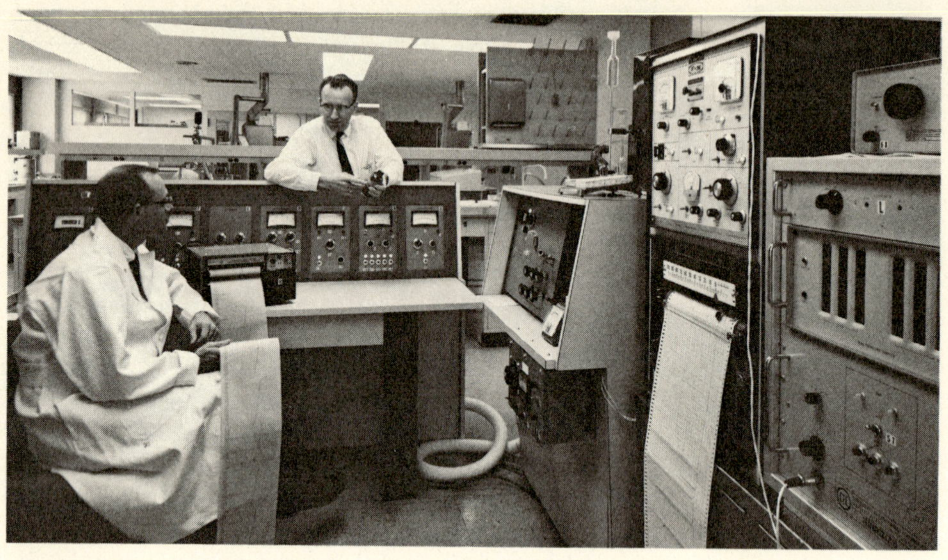

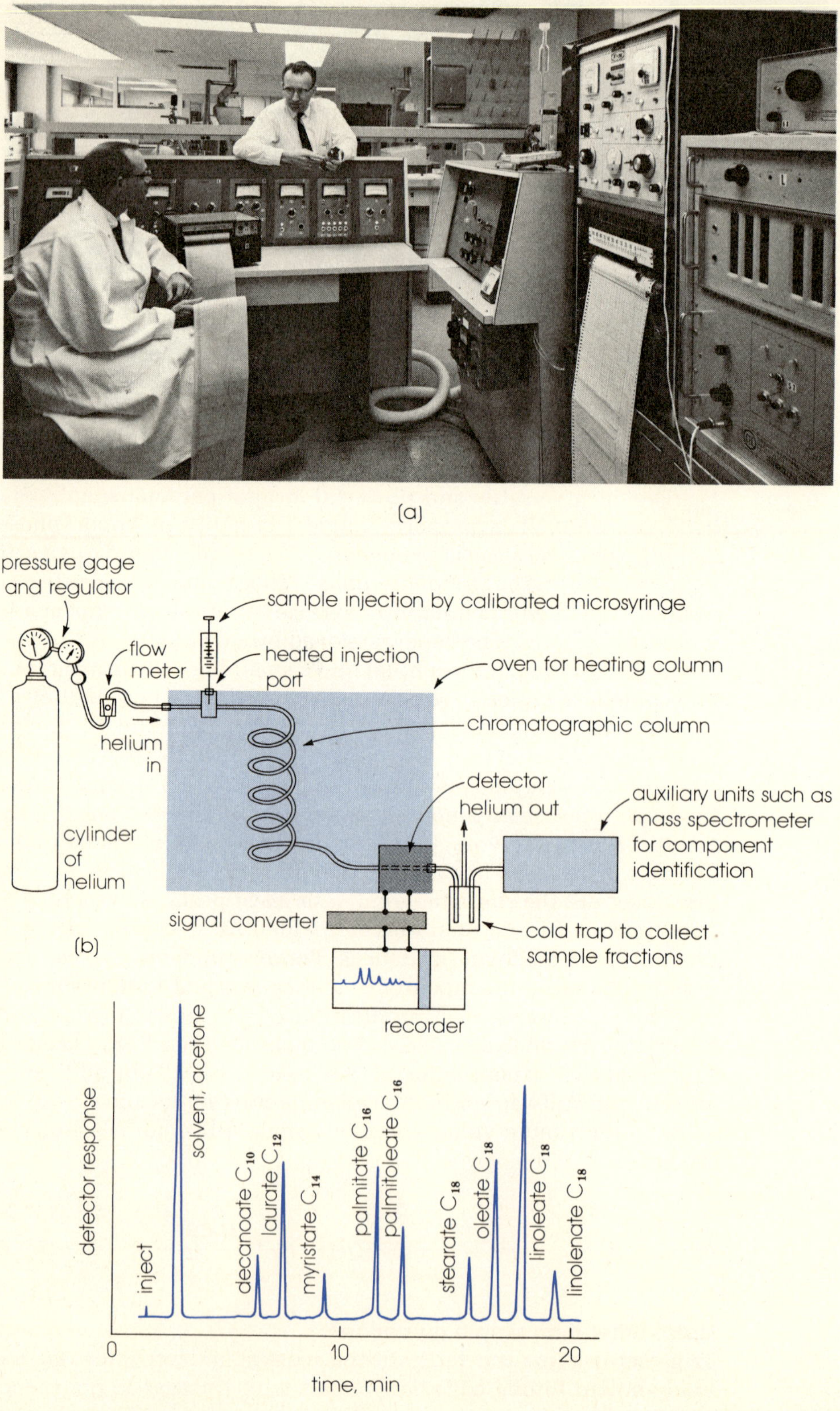
(a)
pressure gage
and regulator
sample injection by calibrated microsyringe
flow
meter
heated injection
port
oven for heating column
chromatographic column
helium
in
detector
helium out
auxiliary units such as
mass spectrometer
for component
identification
cylinder
of
helium
signal converter
cold trap to collect
sample fractions
(b)
recorder
detector response
inject
solvent, acetone
decanoate C_{10}
laurate C_{12}
myristate C_{14}
palmitate C_{16}
palmitoleate C_{16}
stearate C_{18}
oleate C_{18}
linoleate C_{18}
linolenate C_{18}
0
10
20
time, min

(c)

Figure 23-9 Gas chromatography. (a) A gas chromatograph (right, with strip chart) connected to a mass spectrometer. The former separates compounds in a mixture; the latter helps identify them. (A miniaturized gas chromatograph-mass spectrometer combination will be aboard the Viking Lander, an instrument package due for rocket launching by the United States in 1975. Designed to soft land on Mars, it will separate and analyze volatiles in the Martian soil and telemeter the results back to earth. The readout will be compared by computer to similar analyses of earth samples in an attempt to find evidence of life on Mars.) (b) Schematic view of a gas chromatograph. (c) Reproduction of a typical chromatogram. The sample was a mixture of methyl esters of lard fatty acids, separated at 225°C in the chromatographic column. (Photograph courtesy Monsanto Company.)

Phospholipids or *phosphatides* are glyceryl phosphates (see Eq. 22-3) that commonly occupy water–lipid interfaces in cells. Containing both lipidlike and waterlike groups, they bring water and lipids into contact with one another and form stabilized mixtures with the two. Such go-between substances (which include soaps, detergents, and certain natural products) are called *emulsifiers.* Their water–lipid mixtures are *emulsions.* Emulsification facilitates fluid transport through cell membranes. It also makes contact possible between enzymes and the substrates on which enzymes work. A typical phospholipid is *lecithin,* found in nerve sheaths and the walls of mitochondria:

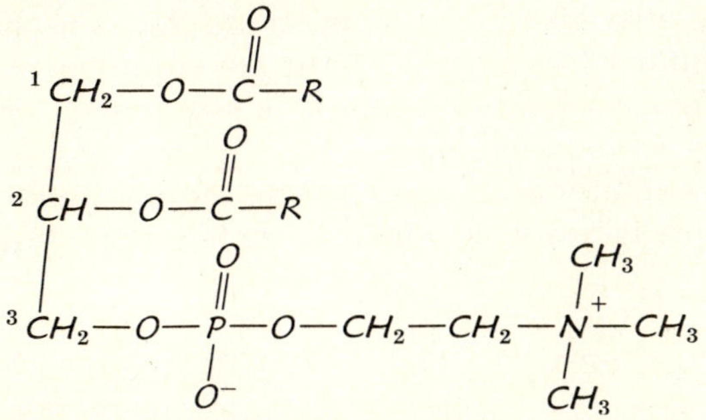

$$^{1}CH_{2}-O-\overset{\overset{\displaystyle O}{\|}}{C}-R$$

$$^{2}CH-O-\overset{\overset{\displaystyle O}{\|}}{C}-R$$

$$^{3}CH_{2}-O-\overset{\overset{\displaystyle O}{\|}}{\underset{\underset{\displaystyle O^{-}}{|}}{P}}-O-CH_{2}-CH_{2}-\overset{\overset{\displaystyle CH_{3}}{|}}{\underset{\underset{\displaystyle CH_{3}}{|}}{N}}\overset{+}{-}CH_{3}$$

The group attached to C-3 in lecithin is *hydrophilic* (water soluble) whereas the chains attached to C-1 and C-2, where R— denotes fatty acid groups, are *lipophilic* (lipid soluble). Lecithin has emulsifying properties.

Steroids are a class of lipids derived from the following C_{17} ring system:

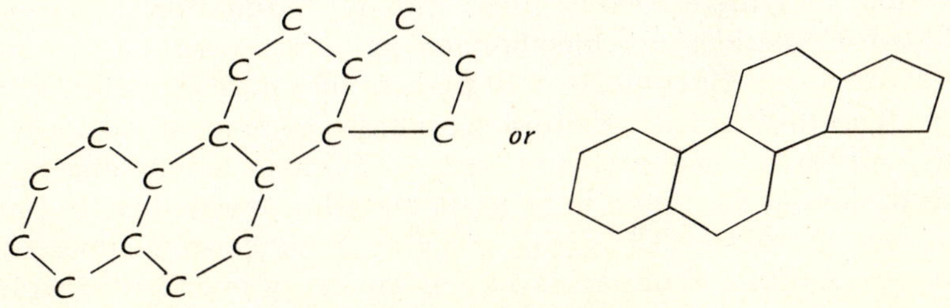

1,2-cyclopentanoperhydrophenanthrene

A steroid whose only functionality is a hydroxyl group is called a *sterol.* As indicated in Figure 23-11 natural steroids perform various and vital physiological functions.

Vitamins belonging to the A, D, E, and K families are lipid related by virtue of a predominantly hydrocarbon structure that makes them lipid soluble. Like all vitamins they are required in small dietary amounts by animals.

Vitamin A, found in cod-liver oil, is necessary to animal growth. First evidence of vitamin A deficiency in a human adult is night blindness. Vitamin A is closely related in structure to the carotenes (see Figure 22-10). The D vitamins, also found in cod-liver oil, aid in the utilization of calcium in the body. Vitamin D_2 specifically prevents rickets, a disease of growing bone:

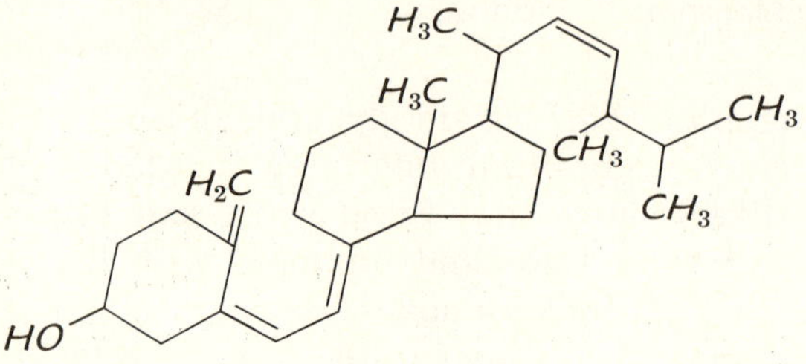

vitamin D₂ (calciferol)

The E vitamins are found mainly in plant oils like wheat-germ oil. In laboratory animals vitamin E deficiency results in several disorders including sterility and muscular dystrophy. In humans vitamin E seems to serve as an antioxidant. It may also be related to other mechanisms but details are lacking. The K vitamins (see Figure 22-10) are supplied continuously to mammals by intestinal bacteria. Vitamin K is essential to synthesis of the compounds that cause blood to clot.

Waxes are esters of fatty acids and alcohols, both generally in the C_{14} to C_{40} range. An example is beeswax, a mixture rich in the following C_{32} ester:

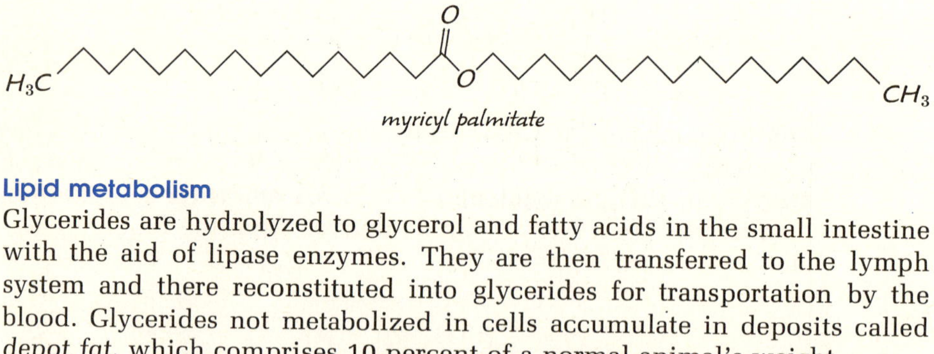

myricyl palmitate

Lipid metabolism

Glycerides are hydrolyzed to glycerol and fatty acids in the small intestine with the aid of lipase enzymes. They are then transferred to the lymph system and there reconstituted into glycerides for transportation by the blood. Glycerides not metabolized in cells accumulate in deposits called *depot fat*, which comprises 10 percent of a normal animal's weight.

The mechanism of lipid oxidation in the animal body began to be explained in 1904 with the work of German biochemist F. Knoop. After feeding dogs synthetic fatty acids tagged at the end with a nondegradable phenyl group (ϕ—$(CH_2)_x$—CO_2H), Knoop analyzed the dogs' urine for breakdown residues. From acids having chains of even-carbon numbers he isolated phenylaceturic acid (ϕ—CH_2—$\overset{\overset{\displaystyle O}{\|}}{C}$—$NHCH_2CO_2H$) which contained intact two original chain carbons. From acids with odd-carbon number chains he found hippuric acid (ϕ—$\overset{\overset{\displaystyle O}{\|}}{C}$—$NHCO_2H$), with but one original

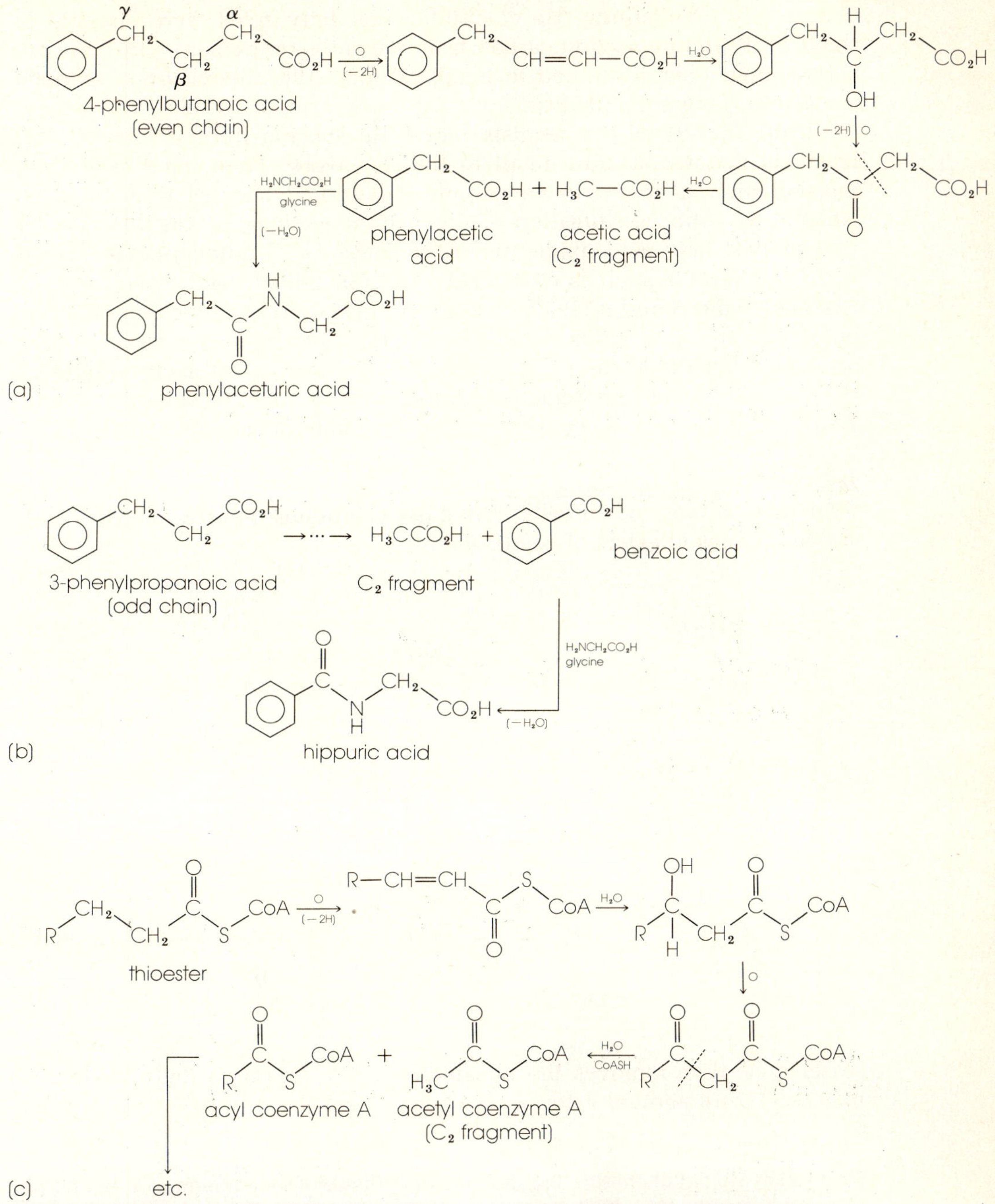

Figure 23-10 Fatty acid metabolism. (a) Knoop's proposal for the digestion of fatty acids. After oxidation at the β position, two-carbon fragments (acetic acid molecules) are broken off until phenylacetic acid, which cannot be oxidized further, is reached. It is eliminated via the urine as phenylaceturic acid, a soluble compound formed by combination with glycine, an amino acid from protein breakdown. (b) The same mechanism, starting with an odd carbon-number chain. End product contains one less carbon. (c) A simplification of the present understanding of fatty acid oxidation. After forming a thioester with coenzyme A, the fatty acid is committed to undergo successive two-carbon degradations in the Krebs cycle. (Necessary enzymes are not shown.)

chain carbon remaining. He concluded that fatty acids were attacked at the β position (the second carbon from the carboxyl carbon) and that two-carbon fragments continued to break off until the phenyl group stopped the action (Figure 23-10a,b).

Until the 1940s biochemists had little success finding intermediate products that would add detail to Knoop's work. Then the site of lipid metabolism was found to be granules (identified later as mitochondria) isolated by ultracentrifugation of finely divided liver. It was then shown that to start fatty acid oxidation in the presence of mitochondria, an oxidizable molecule such as citric acid had to be added. Later it was proved that coenzyme A and ATP also had to be present. Today biochemists know enzymes hydrolyze lipids at ester bonds, and coenzyme A carries the fatty acids through the Krebs cycle (Figure 23-5). Two-carbon acetyl fragments (Knoop's C_2 groups) are detached from fatty acid chains as shown in Figure 23-10c, and the process continues until the chain is used up.

Biosynthesis of glycerides and sterols

Glycerides are formed in cells from L-α-glycerophosphate by reaction with various fatty acid esters of coenzyme A:

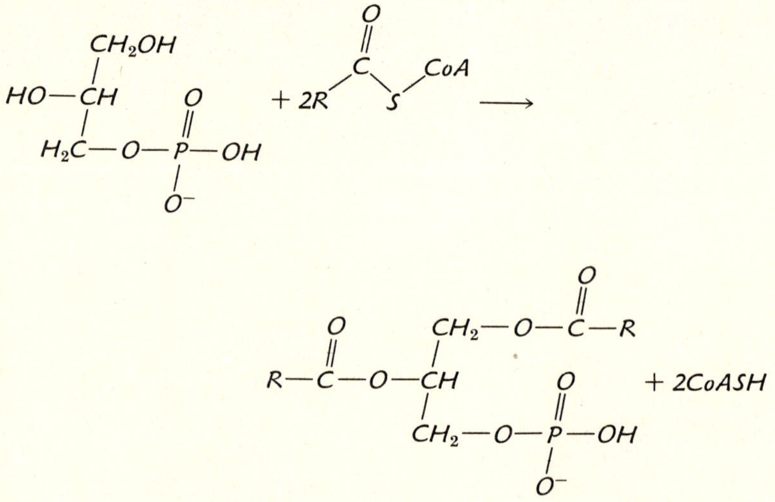

$$
\begin{array}{c}
CH_2OH \\
| \\
HO-CH \\
| \\
H_2C-O-\overset{\overset{O}{\|}}{P}-OH \\
| \\
O^-
\end{array}
\;+\; 2R-\overset{\overset{O}{\|}}{C}-S-CoA \;\longrightarrow
$$

$$
\begin{array}{c}
O \\
\| \\
R-C-O-CH \\
\end{array}
\qquad
\begin{array}{c}
CH_2-O-\overset{\overset{O}{\|}}{C}-R \\
| \\
CH_2-O-\overset{\overset{O}{\|}}{P}-OH \\
| \\
O^-
\end{array}
\;+\; 2CoASH
\qquad (23\text{-}18)
$$

An enzyme then removes the phosphate group, leaving a diglyceride that may react with another ester of coenzyme A to form a triglyceride.

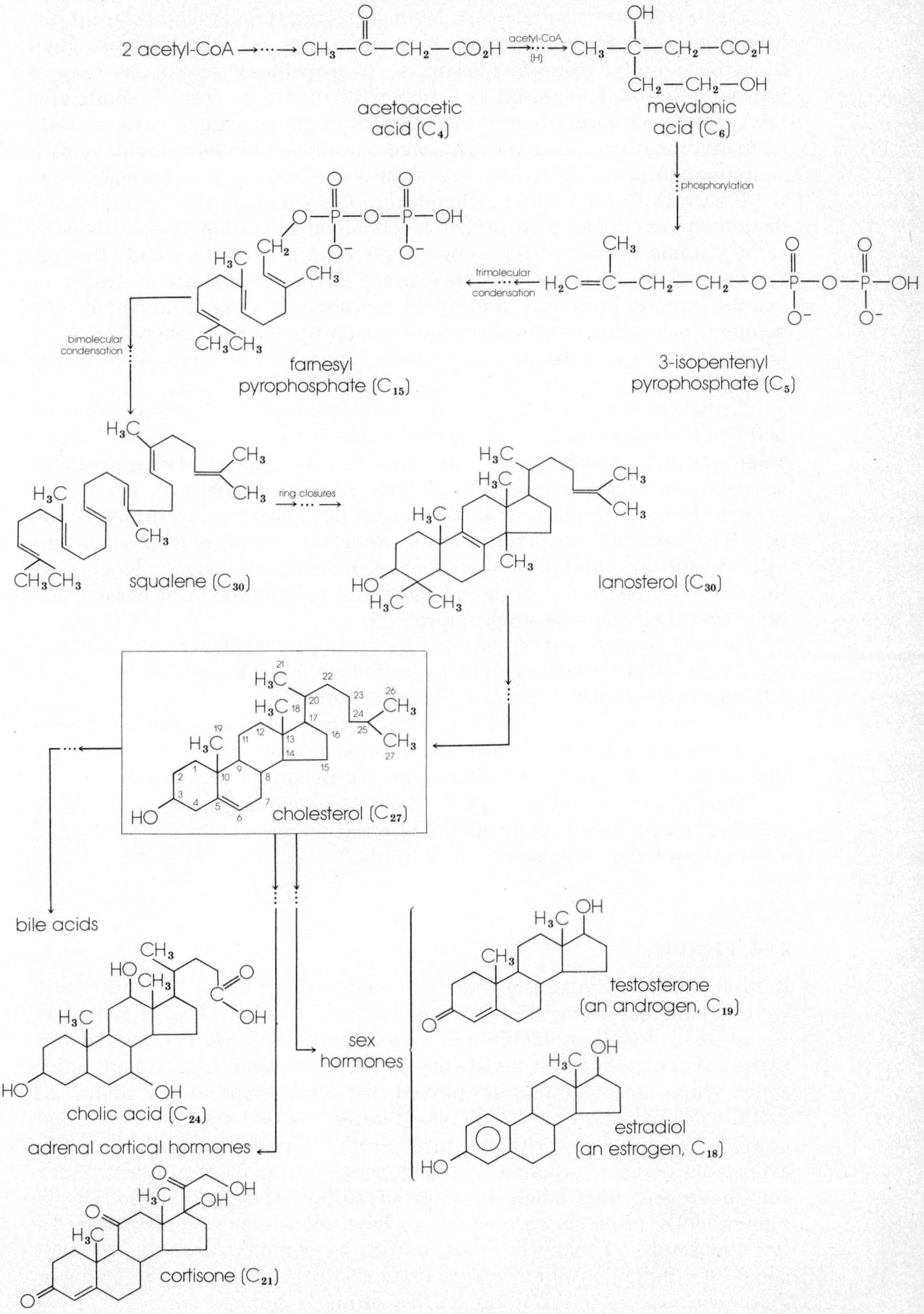

2 acetyl-CoA $\longrightarrow \cdots \longrightarrow$ acetoacetic acid (C_4) $\xrightarrow[\text{(H)}]{\text{acetyl-CoA}}$ mevalonic acid (C_6)

phosphorylation

farnesyl pyrophosphate (C_{15})

$\xleftarrow[\text{condensation}]{\text{trimolecular}} \cdots$ 3-isopentenyl pyrophosphate (C_5)

bimolecular condensation

squalene (C_{30}) $\xrightarrow{\text{ring closures}} \cdots \longrightarrow$ lanosterol (C_{30})

cholesterol (C_{27})

bile acids

cholic acid (C_{24})

sex hormones

testosterone (an androgen, C_{19})

estradiol (an estrogen, C_{18})

adrenal cortical hormones

cortisone (C_{21})

Coenzyme A is also responsible for initiating steroid synthesis and furnishing all the necessary carbon. By a series of condensations and reductions the C_2 acetyl group first becomes a C_5 isopentenyl group, a relative of isoprene, the monomeric unit of rubber (Chapter 21). Three C_5 units give the C_{15} farnesyl group, two of which react to give squalene, a C_{30} unsaturated hydrocarbon. Ring closures and modifications then yield various steroids including *cholesterol,* the most abundant sterol in animal tissue.

Figure 23-11 traces the biochemical routes along which cholesterol and its derivatives appear. Steroids can be classified according to the number of carbon atoms in them. Those containing 18, 19, and 21 carbons that are synthesized in the outer portion (cortex) of the adrenal gland, testes, or ovaries possess hormonal activity. A *hormone* is a compound that after being released directly into the blood stream travels to another part of the body and there activates or deactivates enzymes controlling metabolic processes.

Called *estrogens* the C_{18} steroids are related to secondary sex characteristics in the female and to the ovarian cycle. Birth control pills, at present taken orally by 7 million women in the United States, contain synthetic steroids that create a biochemical false pregnancy. Because a pregnant woman does not produce eggs, she cannot become pregnant in this condition. The ovulation mechanism is incompletely known. One effect of "the pill" is probably inhibition of cholesterol metabolism, hence a blockage of the synthesis of steroid hormones essential to the growth of ovarian follicles and their egg-releasing rupture.

The C_{19} *steroids* are represented by the *androgens* which are responsible for secondary sex characteristics in the male. Androgens and estrogens are found in both sexes but in different amounts.

There are several sorts of C_{21} steroids. One group regulates the body's water and salt balance. Another group helps convert amino acids to carbohydrates and also prevents inflammatory responses in the body.

The C_{24} *steroids* are *bile acids*. About 80 percent of the cholesterol metabolized in the liver is converted to bile acids. Salts of bile acids are emulsifying agents that aid absorption of lipids.

23-4. PROTEINS

Natural nitrogen-containing materials such as egg white and milk curd were at first not recognized by chemists as a general class of nutrients. About 1840, however, Liebig and Dutch chemist G. Mulder proved that hydrolysis of these materials always yielded low-molecular-weight amino acids. When chemical analysis proved that nitrogenous matter comprised half the dry weight of most biological tissues, the two men gave it a name, *protein* (Gk. *proteios,* primary stuff). Protein composition was found to average 54 percent carbon, 21 percent oxygen, 16 percent nitrogen, 7 percent hydrogen, plus small amounts of sulfur, phosphorus and metals. Subsequently proteins were shown to have molecular weights between a few thousand and many millions, and to be polymers derived from only about 20 different α-amino acids (Figure 23-12). A general amino acid synthesis was worked out from the corresponding α–bromoacids:

$$Br-\underset{\underset{R}{|}}{\overset{\overset{CO_2H}{|}}{C}}H \xrightarrow{NH_3} H_2N-\underset{\underset{R}{|}}{\overset{\overset{CO_2NH_4}{|}}{C}}H \xrightarrow{H^+} H_2N-\underset{\underset{R}{|}}{\overset{\overset{CO_2H}{|}}{C}}H \;\; + NH_4{}^+$$

Types of proteins

By 1907 biochemists had enough experimental data to begin describing proteins according to gross composition and behavior in various solutions. Studies on structure still have not provided a more comprehensive scheme, and so we continue to use their original classifications: (1) *simple proteins*, which yield only α-amino acids and their derivatives upon hydrolysis; and (2) *conjugated proteins*, in which a protein molecule is bonded to an organic, nonprotein *prosthetic* (Gk. *prosthetos*, put on) group.

Subclasses of simple proteins include albumins, globulins, prolamines, glutelins, protamines, histones, and scleroproteins. Subclassification is based on solubility and other differences. Albumins, for example, are soluble in water and dilute salt solutions but insoluble in saturated ammonium sulfate. Albumins are found in blood serum (see Figure 23-14b).

Subclasses of conjugated proteins are named to describe the prosthetic groups. Examples are chromoproteins (protein + pigment), mucoproteins (protein + carbohydrate), phosphoproteins (protein + phosphoric acid), and lipoproteins (protein + lipid).

Analysis and character of amino acids

Fractional distillation and crystallization, two classical separation procedures, are not effective in separating mixtures of amino acids because the acids decompose without distilling and form new mixtures when crystallized from solution. Fischer, who began examining protein problems about 1900, avoided these difficulties by esterifying the acids with ethanol and distilling the esters according to their boiling points:

$$H_2NCH_2C\overset{\displaystyle O}{\underset{\displaystyle OH}{\big\backslash\!\!\big/}} + C_2H_5OH \xrightarrow[(-H_2O)]{H^+} H_2NCH_2C\overset{\displaystyle O}{\underset{\displaystyle O-C_2H_5}{\big\backslash\!\!\big/}} \tag{23-19}$$

ethyl glycinate, b.p., 150°C

Today amino acids are separated by chromatographic techniques.

Reluctance of amino acids to volatilize is due to intramolecular reaction between the carboxylate proton and the free electron pair on the amino nitrogen, so that even as crystalline solids amino acids exist in a salt form called a *zwitterion* (Ger. *Zwitter*, hybrid). Zwitterions have high melting points (glycine, 232 to 236°C, with decomposition) and low solubility in organic solvents. The pH at which an amino acid is essentially in the zwitterion form is called its *isoelectric point*. At that pH the substance has its lowest solubility in water. Isoelectric points range from 2.87 to 10.76 (Figure 23-12).

Amino acids may be analyzed quantitatively by colorimetry and titrimetry. Reaction with *ninhydrin*, a mild oxidizer, gives a product whose color intensity (usually dark blue) is proportional to the quantity of amino acid present:

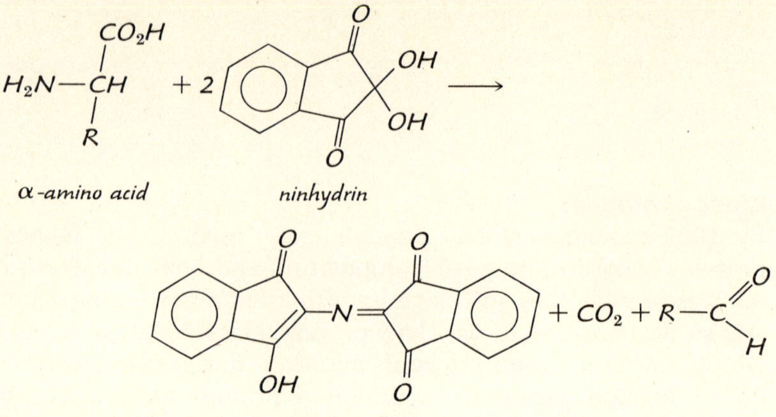

α-amino acid — ninhydrin — colored product

(23-20)

Zwitterions have both basic and acidic character and therefore can be titrated with either strong acid or strong base:

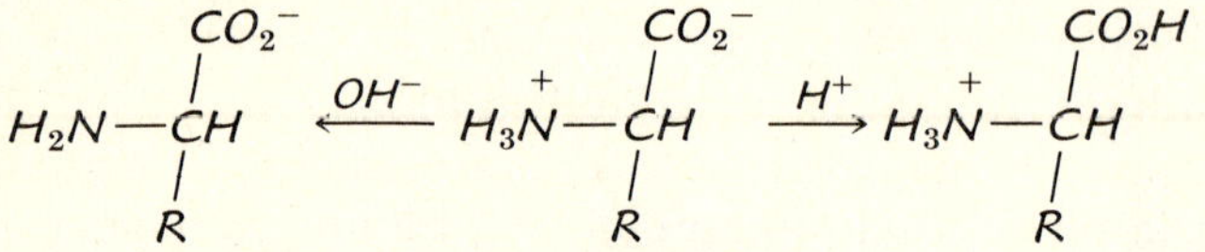

zwitterion

(23-21)

(For another kind of titration see Problem 37.) Amino acids are analyzed for nitrogen content by the procedures described in Problems 38 and 39.

Except for glycine, all the common naturally occurring α-amino acids have an asymmetric α-carbon atom and are optically active. A few D-configuration amino acids are found, especially in certain natural antibiotics and the cell walls of bacteria, but all those obtained by the complete hydrolysis of protein from every known source are configurationally related to L(−)glyceraldehyde:

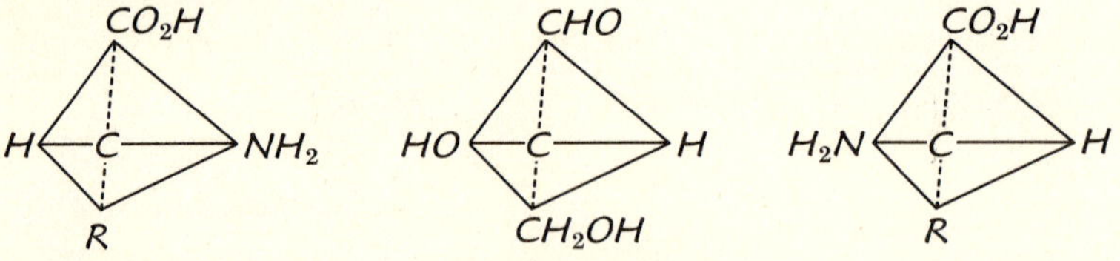

D-amino acid — L(−)glyceraldehyde — L-amino acid

As pointed out in Section 22-2 the origin of asymmetric syntheses in nature is unknown. It is certain that the rest of nature is compatible with the prevalence of L-amino acids, however, since, for example, a diet of D-amino

Figure 23-12 Formulas, common names (L designation is omitted), discovery dates, abbreviations, and isoelectric points of 20 α-amino acids obtained from the complete hydrolysis of proteins. Acids 1 to 5 (the numbering is arbitrary) are aliphatic with no functional group in the side chain; 6 and 7 are hydroxyamino compounds; 8 to 10 contain sulfur; 11 and 12 are called acidic or dicarboxylic; 13 to 18 are called basic for their predominance of amine functions whereas 15 to 18 are also called heterocyclic for their nitrogen-containing rings; 19 and 20 are called aromatic to indicate a benzene ring. Those marked * cannot be synthesized adequately in the animal body. They are called essential amino acids.

1. glycine (1820)
Gly (5.97)

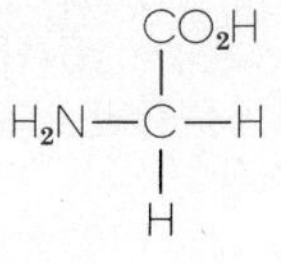

2. alanine (1888)
Ala (6.00)

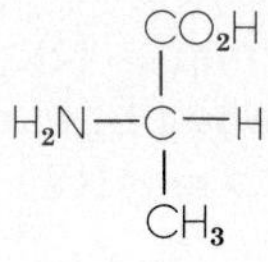

3.* valine (1901)
Val (5.97)

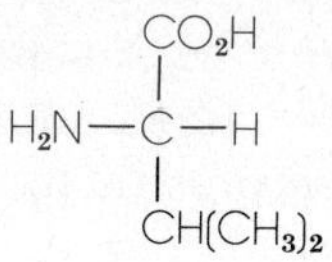

4.* leucine (1820)
Leu (5.98)

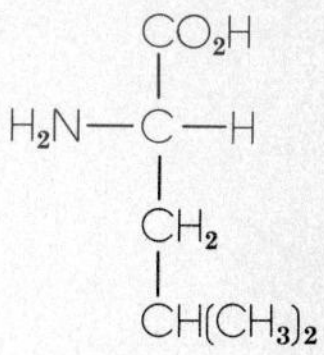

5.* isoleucine (1904)
Ile (6.02)

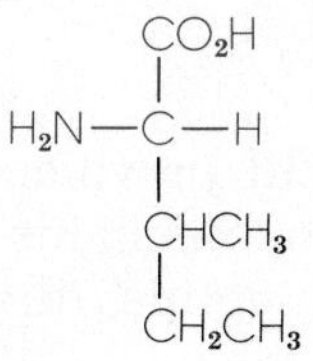

6. serine (1865)
Ser (5.68)

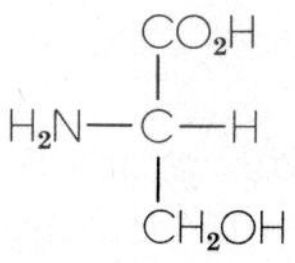

7.* threonine (1935)
Thr (6.53)

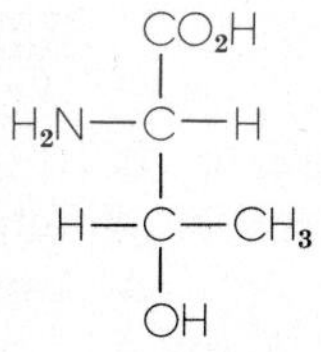

8. cysteine (1899)
CySH (5.02)

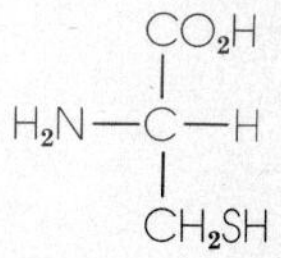

9.* methionine (1922)
Met (5.75)

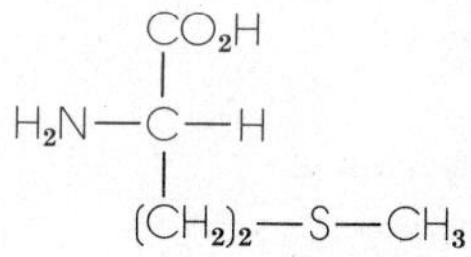

10. cystine (1899)
CyS-SCy (5.06)

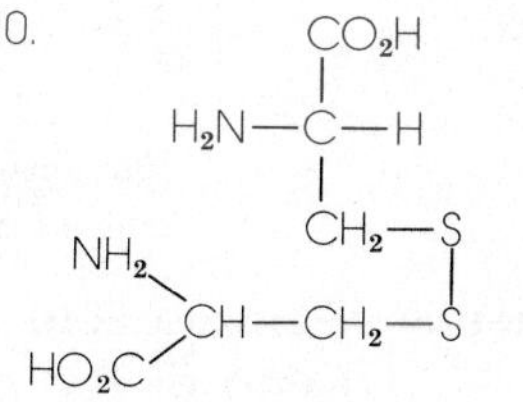

11. aspartic acid (1868)
Asp (2.87)

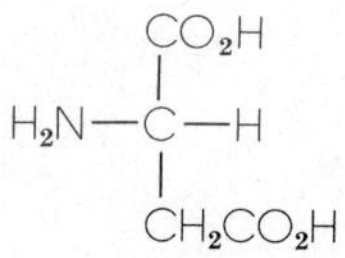

12. glutamic acid (1866)
Glu (3.22)

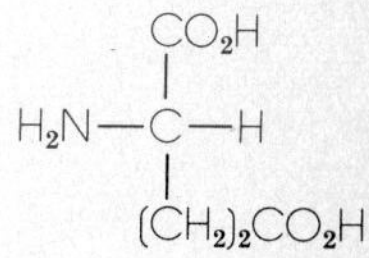

13.* lysine (1889)
Lys (9.74)

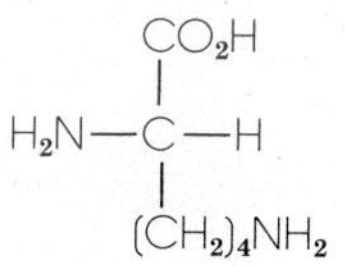

14.* arginine (1895)
Arg (10.76)

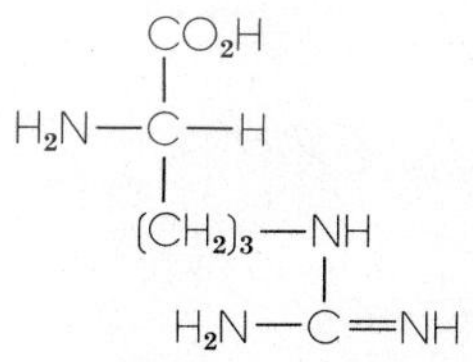

15. proline (1901)
Pro (6.10)

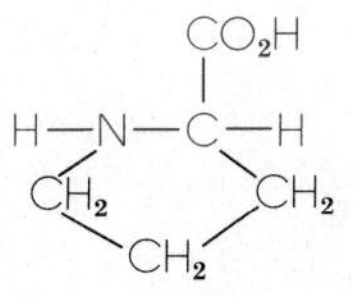

16. hydroxyproline (1902)
Hypro (5.83)

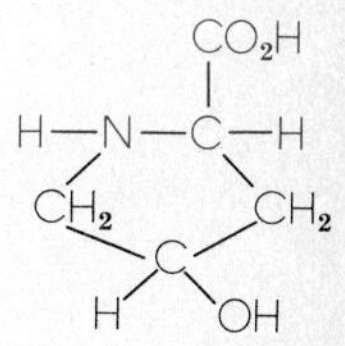

17.* histidine (1896)
His (7.58)

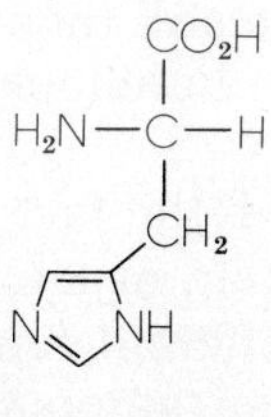

18.* tryptophan (1901)
Try (5.88)

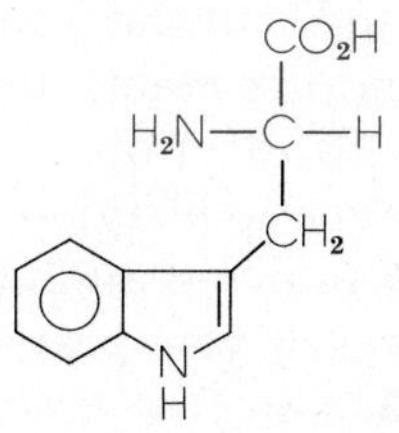

19.* phenylalanine (1881)
Phe (5.98)

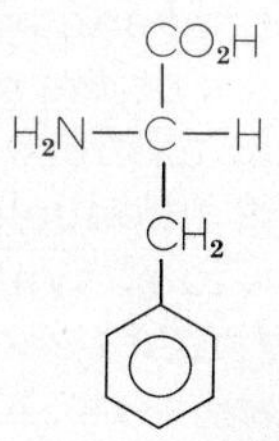

20. tyrosine (1849)
Tyr (5.65)

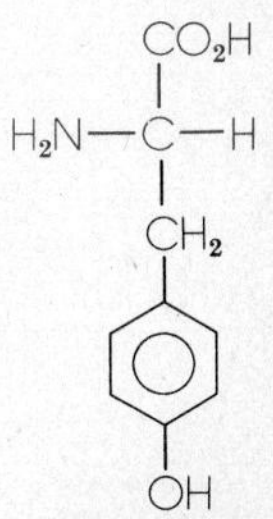

acids cannot be metabolized by animals. The enzymes necessary to the chemical reactions are themselves made of L-amino acid proteins and their surfaces are presumably not shaped to catalyze D-acid reactions.

Peptides

In 1902 Fischer observed that whereas proteins contained few titratable $-NH_2$ and $-CO_2H$ groups, protein hydrolysis produced many such groups in equal numbers. He correctly concluded that proteins were composed of amino acids joined by amide linkages (later called *peptide bonds*),

$$-\overset{\overset{\displaystyle O}{\|}}{C}-\overset{\overset{\displaystyle H}{|}}{N}-.$$

Incomplete hydrolysis of proteins produces amino acid polymers of lower molecular weight. Small proteins or fragments of protein chains are called *peptides*. A combination of two amino acids is a dipeptide, three a tripeptide, many a polypeptide:

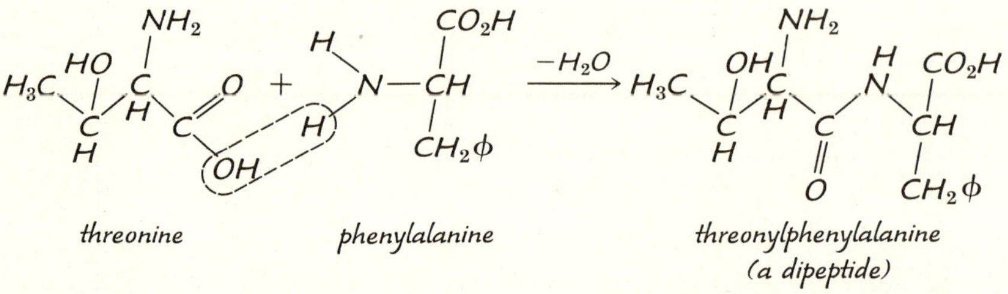

$$\text{(23-22)}$$

threonine   phenylalanine   threonylphenylalanine
(a dipeptide)

Peptides are named as a sequence of amino acid units or *residues*, beginning at the *N-terminal* (free $-NH_2$) end of the chain and proceeding to the *C-terminal* (free $-CO_2H$) end; *-yl* is added to the root name of the amino acids; the C-terminal residue retains its normal name. Thus a tripeptide made from alanine, valine, and serine in that order is

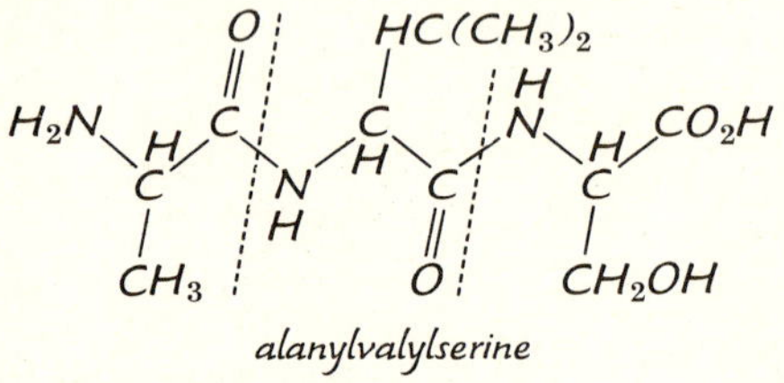

alanylvalylserine

Peptide synthesis

Fischer devised schemes for preparing peptides, linking up in his most successful synthesis 18 amino acid residues. His laborious methods have since been modified and even automated, but the basic problems are still those of blocking one active site while another reacts, then isolating and unblocking the product to prepare for the next reaction.

In 1969 ribonuclease, a pancreatic enzyme was duplicated, the first example of a total synthesis of an enzyme and the largest protein of known structure synthesized to that date. It was prepared by two different routes as described with Figure 23-13. One research group used an automated solid state synthesis technique in which identical peptide chains were built up from molecules of a single amino acid whose carboxyl group was covalently attached to small polystyrene beads. The first acid (lysine) was caused to react with the next L acid (blocked, as needed) in the planned

(a)

(b)

Figure 23-13 Ribonuclease. (a) First total synthesis of an enzyme. Front, Ralph Hirsh-mann, and Robert Denkewalter, leaders of the research team at Merck and Company, with a model of the enzyme. Their approach was preparation of protected peptides containing 6 to 17 residues, followed by joining of the peptides in proper sequence. Back, Rockefeller University biochemists Bernd Gutte and Robert Merrifield, the latter the coinventor of the peptide synthesis machine (rear) described in the text. (b) Primary structure of ribonuclease, a small globular protein that catalyzes the splitting of ribonucleic acid RNA. Nineteen different residues are present in a total of 124. Molecular weight is 13,683. (Photograph, New York Times.)

sequence, giving an attached dipeptide purifiable by filtering out the beads. By machine cycling the blocking group was then removed, making the dipeptide ready for the next amino acid addition. Some 370 consecutive chemical reactions were conducted during a month of continual machine time to add 123 amino acids to the first one according to the structure determined previously by another group of biochemists. When the finished linear molecules were freed from their bead and from the last blocking group and allowed to stand in water, the chains folded themselves and assumed the specific conformation and physiological activity characteristic of the natural enzyme. The same synthesis by older techniques would have taken several years.

Enzyme action

In 1835 when Berzelius discovered that the hydrolysis of potato starch was accelerated by β-amylase, a substance found in barley malt, he proposed that all processes in living cells were aided by catalysts. In a heated controversy that lasted for years, chemists led by Liebig tried to find evidence to support a chemical theory of natural catalysts, whereas microbiologists led by Pasteur tried to prove responsibility of microorganisms like yeasts. The argument was decided in favor of the chemical theory in 1897 when Eduard Buchner (Germany 1860–1917, Nobel prize in chemistry 1907) showed that cell-free yeast juice would ferment sugar. Until the 1920s enzymes were thought to have constitutions different from all other compounds. Then James Sumner (United States 1887–1935, Nobel prize in chemistry 1946) succeeded in preparing the first crystalline enzyme, urease, a catalyst for the hydrolysis of urea. He obtained it from jack-bean meal and proved it was a protein. Since then several hundred enzymes, all proteins, have been isolated in pure form.

The much greater efficiency of enzymes over the best man-made catalysts is partly explained by the special structural relationship that exists between an enzyme and the substrate molecule upon which it works. Experiments indicate that enzyme and substrate form a loose association in lock-and-key fashion. Various kinds of bonding (covalent, hydrogen) are postulated to operate around the enzyme's active site or sites. A site may be a structural hole of correct shape (the lock) containing atoms with ideal placement and electronic characteristics to accommodate the substrate (the key). When the reaction to be catalyzed involves two reacting molecules (a bimolecular reaction), the enzyme probably binds them at active sites in the orientation requiring minimum energy for their interaction, then releases the products. Evidence is also growing that an enzyme can bind the activated complex form (Figure 17-3) particularly effectively, and then "steer" the complex's bonding orbitals toward those of the reaction product. The preciseness of alignment apparently possible here cannot be obtained in other ways. Research in the field promises greater understanding of all catalytic processes.

In many enzyme-catalyzed reactions *coenzymes* like coenzyme A are also required. All living cells utilize the same coenzymes. Biochemists estimate that the human body contains 25,000 different enzymes. Operating at concentrations rarely above 0.001 M, they catalyze 25,000 specific reactions.

A foreign substance that can occupy an enzyme's active site or change its geometry can deactivate the site and inhibit its performance. The sulfa drugs, discovered in 1935 by Gerhard Domagk (Germany 1895–   , Nobel prize in medicine or physiology 1939), are classic examples of enzyme *inhibitors*. Similar in structure to p-aminobenzoic acid (required in the metabolism of certain bacteria but not by man) sulfa compounds apparently compete with that acid by fitting into the active sites of a bacterial enzyme and blocking their action.

p-aminobenzoic acid   *sulfanilamide*   *sulfathiazole*

(sulfa drugs)

Nonspecific enzyme inhibitors include heavy metal poisons such as arsenic and lead. By forming bonds with protein —SH and —CO_2H groups, they alter enzyme shape.

Purification of proteins

Proteins may be roughly separated in water solution by selective precipitation through the addition of other solvents (ethyl alcohol, etc.) or salts (ammonium sulfate, etc.) Proteins may also be separated by *dialysis*. For dialysis the protein solution is sealed in a semipermeable (dialyzing) membrane such as a cellophane bag and the bag placed in a water bath. As small ions and molecules migrate out of the bag, purified proteins are left behind. More exacting protein separations are based on chromatographic principles.

In *gel permeation chromatography*, also called *gel filtration*, the solution of proteins is introduced into a column packed with a polymeric material such as a chemically modified starch that has the ability to absorb solvents and swell up. Depending on gel composition, swelling leaves holes of sizes that permit entry and retention of smaller molecules while denying entry to larger molecules. Proteins thus pass through with excess solvent while smaller species are detained. With gels available for the selective sieving of molecules ranging in molecular weight from thousands to millions, a protein mixture can be separated into molecular-weight fractions by employing a sequence of gel-packed columns.

In the technique called *zone electrophoresis* (Figure 23-14) proteins are separated on a strip of cellulose acetate or paper moistened with inert electrolyte and held in an electric field. Like amino acids, proteins carry charges depending on the numbers of free amino and carboxyl groups per molecule and the pH of surrounding fluid. Representing portions of protein

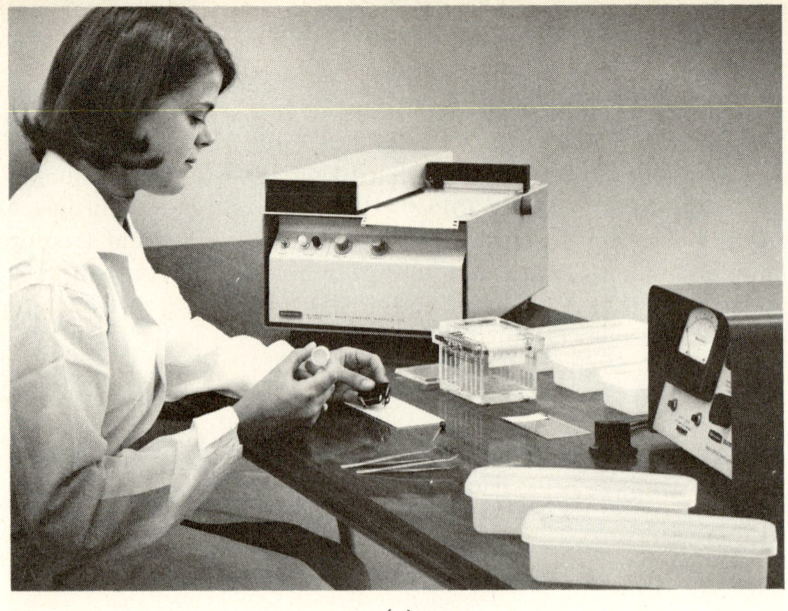

(a)

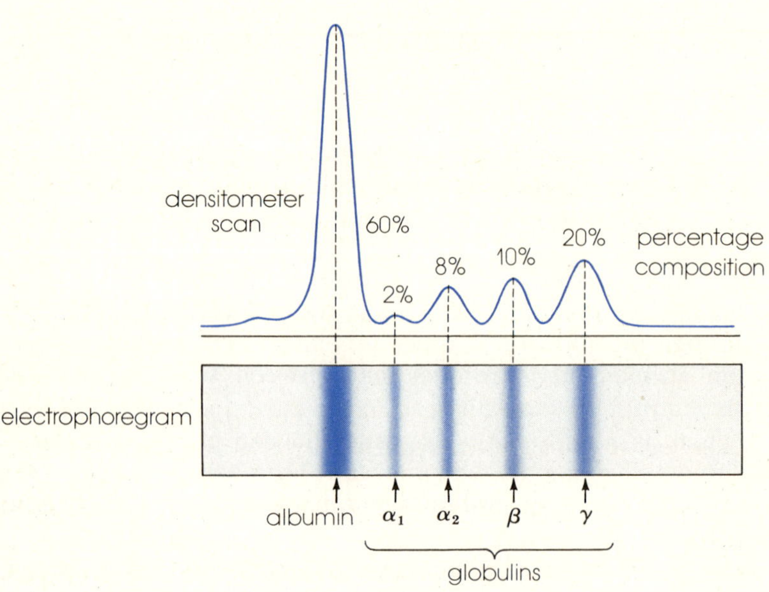

(b)

Figure 23-14 Electrophoresis. The principle, used first in 1892, was made practical in 1937 by Arne Tiselius (Sweden 1902– , Nobel prize in chemistry 1948). (a) Zone electrophoresis equipment. The chemist is spotting sample solution on a strip. The strip will be put in the clear plastic chamber (center) in contact with buffer solution, and the power unit (right) will be connected to chamber electrodes. At a potential of several hundred volts the sample's ions will migrate toward oppositely charged electrodes. After some minutes or hours, the strips are developed with ninhydrin (Eq. 23-20), and the density of the colored zones read and recorded by the densitometer (rear). (b) Electrophoresis of human blood serum protein. Below, the pattern on the strip; above, the color density of each component peak as determined by the densitometer. The area under each peak is proportional to component concentration. Electrophoretic patterns aid physicians in making diagnoses. (Photograph Courtesy Beckman Instruments, Inc.)

chains as zigzag lines the three possible charge conditions are

$$
\begin{array}{ccc}
\text{(a) in acid} & \text{(b) isoelectric point} & \text{(c) in base}
\end{array}
$$

(a) in acid: $^+NH_3$, CO_2H / CO_2H, $^+NH_3$
(b) isoelectric point: $^+NH_3$, CO_2^- / CO_2^-, $^+NH_3$
(c) in base: $:NH_2$, CO_2^- / CO_2^-, $:NH_2$

If the protein mixture is spotted in the middle of the moist strip and oppositely charged electrodes are connected to the ends of the strip, species (a) will migrate toward the negative electrode, species (c) toward the positive electrode, and species (b), which is neutral, will not move from the original position. Migration rate depends on charge, molecular size, and electric field strength.

The purity of isolated proteins can be ascertained by carefully controlled electrophoresis or ultracentrifugation; in either operation a pure protein moves as a well-defined zone. For protein enzymes like ribonuclease, relative purity is measured by the catalytic activity of a given weight of it.

Protein molecular weights

Four methods for finding protein molecular weights are described below.

1. Chemical analysis The method of chemical analysis depends on finding the protein's metal content. For instance, analysis shows vitamin B_{12} (a complex porphyrin) contains 4.35 percent cobalt (at.wt. 58.9) by weight. Because 4.35 percent of the compound's weight is 58.9 (assuming one cobalt atom per molecule), the minimum molecular weight is $58.9 \times (100/4.5) = 1354$. Other methods verify that is the molecular weight.

2. Gel filtration Assuming a gel contains uniformly shaped conical holes, molecules being passed through it will settle into the cones to depths determined by their sizes. The amount of solvent needed to carry the molecules out of the holes and through the column bears an inverse relationship to molecular size. A calibration graph of elution volume versus proteins of known molecular weights allows the molecular weight of an unknown protein to be estimated.

3. Osmotic pressure Molecular weights of proteins up to a million can be found from osmotic pressure measurements of their dilute solutions (see Figure 16-6). With a protein mixture the method yields the average weight of all molecules present (see Problem 36).

4. Ultracentrifugation (sedimentation) A particle settles under centrifugation (Figure 23-15) at a rate dependent on molecular weight, shape, and other factors (see Problem 42.).

Determining residue sequence

The *primary structure* of peptides and proteins is defined by the identity of the amino acid residues and their order of occurrence. The acids can be

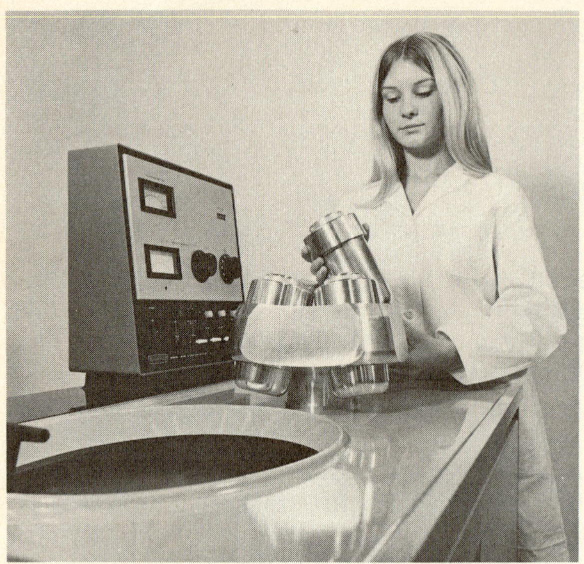

Figure 23-15 The ultracentrifuge, a machine for subjecting samples to large gravitational forces. The preparative instrument illustrated operates up to 40,000 revolutions per minute for the isolation and purification of relatively large quantities of high-molecular-weight substances such as viruses. Rotor, rotor well, controls, and a set of 2-liter sample holders are shown. The ultracentrifuge was developed by Theodore Svedberg (Sweden 1884–   , Nobel prize in chemistry 1926). Very high-speed models exert up to 500,000 times gravity for the sedimentation of smaller molecules. Means are available for viewing and photographing the sedimentation process. (Courtesy Beckman Instruments, Inc.)

readily freed by vigorous hydrolysis and separated and identified by chromatography, but only since 1950 have routines been worked out to discover the acid-by-acid sequence in protein chains. For small proteins (mol. wt. ~ 30,000), which are usually a single unit like ribonuclease, the first step is to determine the residue at each end of the chain. Sanger's reagent, 1-fluoro-2,4-dinitrobenzene, is used to react with the free —NH$_2$ group at the N-terminal end. Then, after all peptide bonds are hydrolyzed with HCl and the mixture separated by chromatography, the yellow-colored dinitro derivative of the residue is identified:

$$(23\text{-}23)$$

The residue at the C-terminal end with its free —CO$_2$H group is found through the use of hot anhydrous hydrazine. As the reagent cleaves all peptide bonds, it converts all residues to hydrazides except the C-terminal residue that emerges as an amino acid, separable and identifiable by chromatography:

$$\text{protein chain, C-terminal end} + H_2N-NH_2 \xrightarrow{100°C} \text{hydrazides of all other acids} + \text{C-terminal acid only} \quad (23\text{-}24)$$

Disulfide bridges, such as the four that crosslink the ribonuclease chain (Figure 23-13b), are oxidized with performic acid so they will not confuse analysis:

$$\text{disulfide link} + \text{performic acid} \longrightarrow \text{bond broken by oxidation} + \text{formic acid}$$

Following these operations, enzymes known to cleave specific kinds of peptide bonds are used to cut up chains. An example is trypsin, which catalyzes the hydrolysis of peptide bonds whose carbonyl group is part of a basic amino acid (examples: lysine and arginine, Figure 23-12). The resulting small polypeptides are investigated by end-group analysis methods and by further cleavages until the whole primary structure can be reconstructed from the jigsaw puzzle of pieces. The molecule may then be synthesized to confirm the analysis.

Chromatographic separations followed by sequencing techniques are being used in brain-chemistry studies. Indications are that learning and memory are related to brain synthesis of peptides believed capable of coding acquired information in the central nervous system. After a peptide is isolated and identified, the next step is to synthesize it and inject it into the brains of untrained laboratory animals to test its behavioral effects. Active peptides identified thus far contain about 10 amino acid residues.

Like peptide synthesis, peptide sequencing has been automated (Figure 23-16).

Protein secondary structure

Protein primary structure is described by amino acid sequence (Figure 23-13b, for example). Protein *secondary structure*, like that modeled in Figure 23-17, is the configuration of folding of the polypeptide chain. A very difficult thing to determine, complete secondary structure is presently known for fewer than a dozen proteins.

Primary links in proteins are the covalent bonds holding amino acid residues together, including sulfur–sulfur bonds that crosslink chains. *Secondary links*, which maintain a protein's three-dimensional shape, consist mainly of *apolar* (also called *hydrophobic*) bonds and hydrogen bonds.

Apolar bonds come from cohesive forces between hydrocarbon (R—) side chains, particularly as they relate to solvent water. Being water repellant the hydrocarbon parts tend to become folded into the interior of the

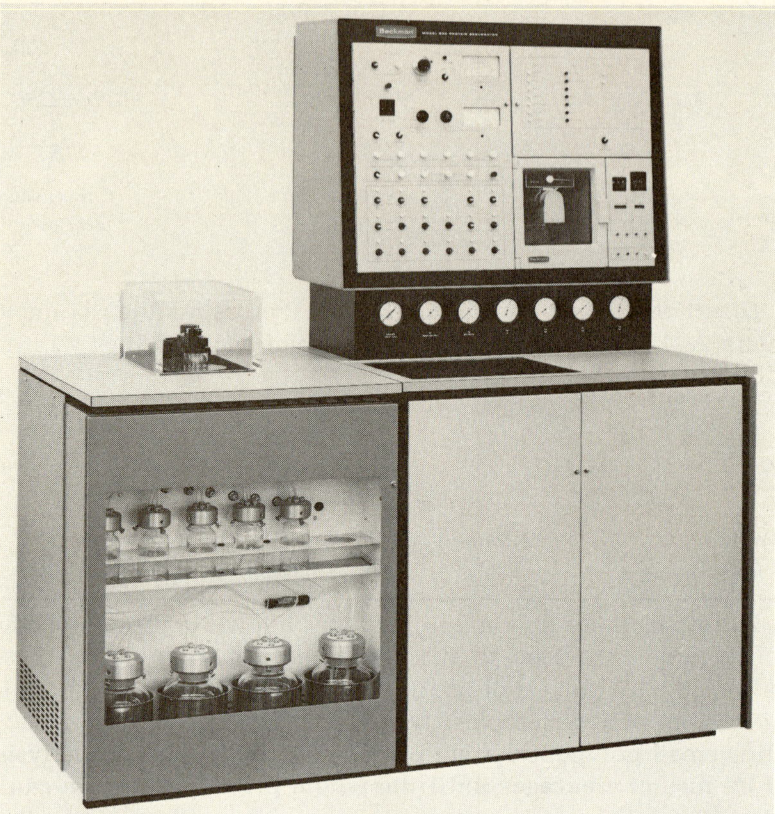

Figure 23-16 A protein and peptide sequencer. One residue at a time is automatically removed from the polypeptide chain and identified by a series of programed operations. The N-terminal end residue is cleaved with specific reagents, converted to a volatile derivative, and identified by gas chromatography. The remainder of the polypeptide chain is recycled for the next set of operations which will remove and identify the new residue at the N-terminal end.

Reagents are delivered from bottles shown in holders in the lower left cabinet. Above them at bench level is the reaction chamber. On the right is the control panel. Inside the control panel window is the punched tape that programs the operations. About a dozen residues can be determined in a 24-hour day. (Courtesy Beckman Instruments, Inc.)

protein structure where water contact is minimized. In opposite fashion polar groups, which are water attracting, are turned outward.

Hydrogen bonds establish themselves spontaneously between the oxygen and nitrogen atoms of favorably situated peptide links. As deduced by Pauling and associates in 1950, maximum stability in polypeptides is attained when all residues are aligned to permit formation of a maximum number of hydrogen bonds. When *intra*molecular hydrogen bonds form parallel to the long axis of a peptide chain, they cause the molecule to assume a helical shape (Figure 23-18a). When *inter*molecular hydrogen bonds form perpendicular to the long axis of peptide chains, they produce sheet structures (Figure 23-18b). Polypeptides of the former type are apt to be

Figure 23-17 British X ray crystallographers with their model of lysozyme, a protein with a molecular weight of 14,400. The group had to take and interpret thousands of X ray diffraction photos before they could deduce the molecule's complete secondary structure. Left to right, J. D. Bernal, John Kendrew (England 1917– , Nobel prize in chemistry 1962), Dorothy Hodgkin (England 1910– , Nobel prize in chemistry 1964), and D. Phillips. (Courtesy Thomson Newspapers, Limited, London.)

globular like ribonuclease and chemically active. Those of the latter type are fibrous, like the keratin of hair, and comparatively inert.

The extent of hydrogen bonding can be determined by *exchange studies.* Hydrogen atoms in free $-NH_2$, $-CO_2H$, and $-SH$ exchange relatively readily with hydrogens from water, whereas hydrogen atoms forming hydrogen bonds exchange slowly. The rate at which a peptide becomes deuterated in deuterium oxide (heavy water) as measured by the decreasing density of the solvent is an inverse measure of the number of hydrogen bonds in the peptide's secondary structure.

Permutations in polypeptide primary structure

The number of proteins possible of synthesis is almost limitless because of variations in the selection of amino acids used and their sequence.

Consider a small protein composed of only 100 residues representing 19 different kinds of L-amino acids apportioned as follows: 14 residues of one kind of acid (glycine for example), 10 each of three other kinds, four each of four other kinds, two each of five other kinds, and five each of six other kinds. In a general case the number of different possible arrangements or *permutations p* is given by the formula

$$p = \frac{n!}{(n_1!)(n_2!)(n_3!) \ldots (n_k!)} \tag{23-25}$$

where n is the total number of amino acid residues, n_1 the number of residues of the first kind of acid, n_2 the number of residues of the second kind of acid, etc., up to the number of residues of the last (kth) kind of acid, and

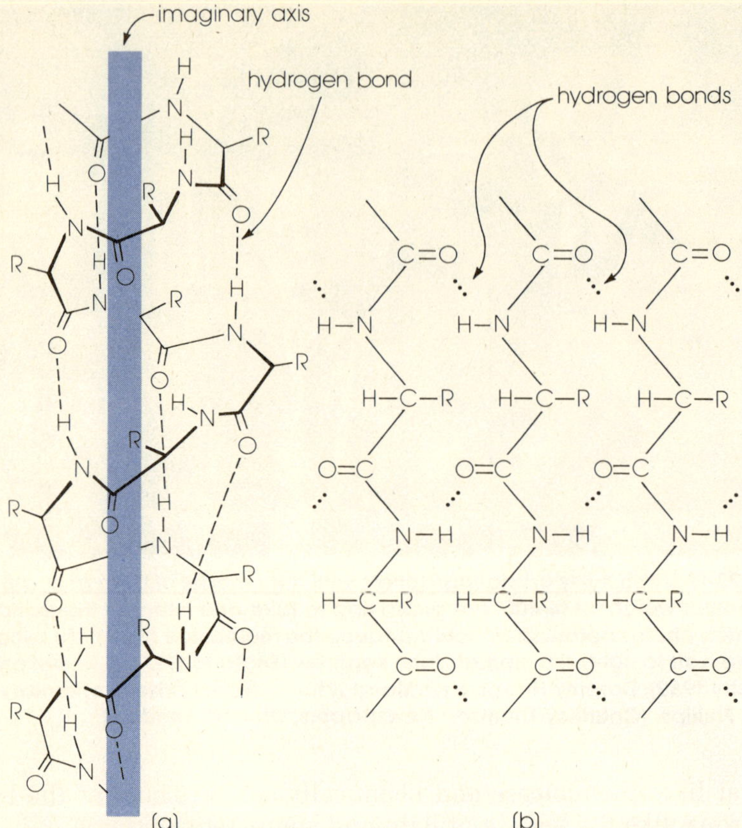

Figure 23-18 Polypeptide secondary structure, as first deduced from X ray diffraction by Pauling, Corey, and co-workers. (a) The right-handed α-helix structure. Hydrogen bonding is parallel to the long axis and within the same molecule. Spacing of the atoms leads to 3.6 residues per turn. (b) A parallel pleated sheet structure. Hydrogen bonding is perpendicular to the long axis and between different molecules.

! is the factorial symbol. Substitution gives

$$p = \frac{100!}{(14!)^1 (10!)^3 (4!)^4 (2!)^5 (5!)^6}$$

The fraction can be evaluated by computer. The answer is 10^{110}. For more typical proteins, containing hundreds or thousands instead of only 100 residues, the permutations are much higher, yet in each case only one of that vast number is the compound synthesized exactly by the organism. In Chapter 24 we will see how this remarkable process takes place.

Protein biodegradation and biosynthesis

Hydrolysis of ingested proteins to polypeptides is begun in the stomach by secretions of hydrochloric acid and the enzyme *pepsin*. Digestion continues in the upper intestine aided by other peptidases which break peptide bonds between specific residues. Liberated amino acids are absorbed into the blood and transported to cells. In the cells they are used in protein synthesis and also oxidized for energy.

The degradation of amino acids, especially in the liver, involves oxidation to β-keto acids such as pyruvic acid which are oxidized to CO_2 via the citric acid cycle. In the process amino nitrogen is converted to ammonia which in mammals is eliminated in urine as urea.

The synthesis of proteins in cells begins by transfer of energy from ATP to amino acid carboxyl groups and the release of pyrophosphate. The process is catalyzed by enzymes and magnesium ions. A different enzyme is required for each acid:

$$ATP + R\!-\!\underset{\underset{NH_2}{|}}{CH}\!-\!CO_2H \xrightarrow[Mg^{2+}]{enzyme} R\!-\!\underset{\underset{NH_2}{|}}{CH}\!-\!\underset{\underset{||}{O}}{C}\!-\!O\!-\!\underset{\underset{O^-}{|}}{\overset{\overset{O}{||}}{P}}\!-\!O\!-\!adenosine + H_2P_2O_7{}^{2-}$$

amino acid   "activated"
amino acid   pyrophosphate

In activated form each amino acid is transferred via molecules of a special kind of ribonucleic acid (RNA) to organelles called *ribosomes* where polymerization into proteins takes place; RNA is genetically related and protein synthesis is under genetic control. To understand the details we must first examine the chemistry of nucleic acids. They are discussed in the next chapter.

QUESTIONS

1. Sugars You have separate solutions of D(−)erythrose and D(−)threose. Tell how to distinguish between them by converting them to the corresponding dicarboxylic acids (see Eq. 23-9 and Figure 20-6.)

2. Sucrose **(a)** What is mutarotation? **(b)** Why does sucrose not exhibit mutarotation? **(c)** Following are formulas as Fischer wrote them:

α-D-glucopyranose β-D-fructofuranose

What do the suffixes pyranose and furanose signify? **(d)** What is a glycosidic bond? **(e)** What is the location of the glycosidic bond that joins the above molecules to give the sucrose molecule?

3. Disaccharides Maltose gives a Fehling's test, sucrose does not. Explain, with specific references to structures of the two sugars.

4. Starch The maltose unit of starch is illustrated with functional groups in equatorial positions. Is that what you would expect from the discussion accompanying Eq. 20-6? Explain.

5. Lactic acid A muscle that is worked finally gets sore due to an accumulation of lactic acid. Briefly without equations give the chemistry of lactic acid's production in cells.

6. Fatty acids Explain: **(a)** The iodine number of a fatty acid may be zero (see Problem 32). **(b)** Lard taken from two hogs is compared. Lard from hog A, fed on corn, has an iodine number of 90. Lard from hog B, fed on another food, has an iodine number of 70. **(c)** As hogs over a period of months are fed on safflower seed cake, gas chromatographic analysis of their lard fatty acids (Figure 23-9c) gives oleate and linoleate peaks of increasing height.

7. Emulsifiers **(a)** Why will soap (Eq. 23-17) lather in dilute NaOH but feel gummy in dilute HCl? **(b)** The following emulsifier is made commercially from oleic acid and sorbitol (a glucose derivative). How is it similar to lecithin? Explain how it can emulsify tiny droplets of oil suspended in water.

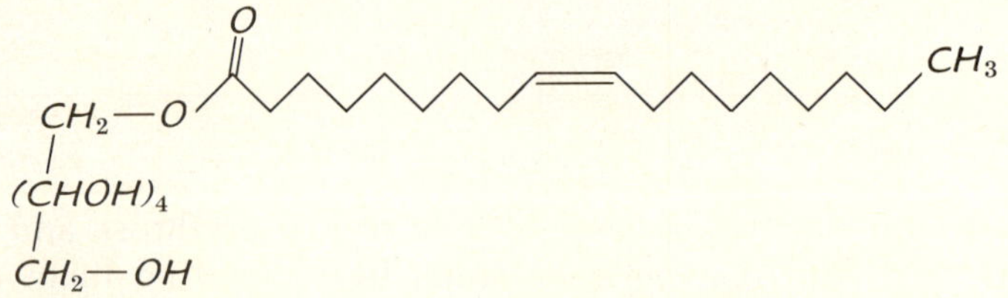

$$CH_2-O-CO-(CH_2)_7-CH=CH-(CH_2)_7-CH_3$$
$$|$$
$$(CHOH)_4$$
$$|$$
$$CH_2-OH$$

sorbitan monooleate

8. Lipids **(a)** What is the difference between a fat and an oil? **(b)** Based on Table 23-1 data, what is the effect of unsaturation on the melting points of fatty acids? **(c)** What is a probable structural difference between a fat and an oil?

9. Amino acids (Use Figure 23-12.) **(a)** A pure amino acid is tested by decomposing it with molten sodium metal, adding water, then adding lead acetate solution. No black precipitate of PbS forms. What conclusions can you make? **(b)** Name two amino acids that contain a benzene ring. **(c)** Name an amino acid that cannot be optically active. Explain. **(d)** Name two amino acids that could exist in four optically active forms (two *dl* pairs). Explain. **(e)** Explain why the isoelectric point of glutamic acid is lower than that of lysine.

10. Hair Disulfide bonds can be reduced and broken by ammonium thioglycollate, the ammonium salt of thioacetic acid:

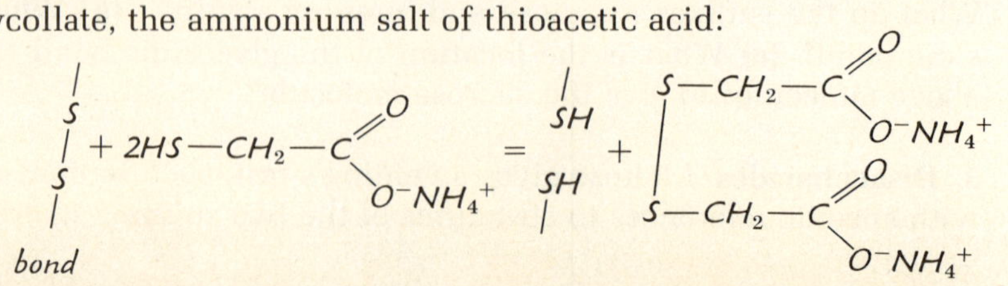

Figure 23-19 Famous biochemists. (See Question 11.) Left, Hans Krebs (Germany and England 1900–   ) and family. Right, Fritz Lipmann (Germany and the United States 1899–   ). The two men shared the 1953 Nobel prize in medicine or physiology. (Courtesy Reportagebild (Sweden), left, and Rockefeller University, right.)

Ammonium thioglycollate is the active ingredient in some mixtures used for the cold "permanent" waving of hair. **(a)** What does this tell you about the composition of hair? **(b)** How do you think the cold waving process works? (A second solution, an oxidizer, is later added to the hair that has been rolled on curlers.)

11. Biochemists (library) Aided by reference to an encyclopedia or other book, briefly describe one important contribution each man in Figure 23-19 made to the chemistry presented in Chapter 23.

12. Molecular weights Why must the molecular weight of a protein be determined by the osmotic pressure method at the isoelectric point rather than at some other pH?

13. D and L Forms **(a)** In 1970 a meteorite that contained small amounts of α-amino acids fell in Australia. The acids were found to be approximately evenly distributed between D and L forms. Did the discovery mean life is possible elsewhere in the universe? Explain. **(b)** L-Amino acids, if protected from heat and light, are converted very slowly to D-amino acids until a state of equilibrium is reached with a 1:1 ratio of L and D acids. In 1971 this was first used to date ancient fossils as old as 1 million years. Explain.

14. Definitions Define or otherwise explain **(a)** peptide link; **(b)** protein; **(c)** carbohydrate; **(d)** unsaturated fat; **(e)** steroid; **(f)** enzyme; **(g)** mitochondria; **(h)** saponification; **(i)** zwitterion; **(j)** essential amino acid.

15. Natural products (library) The men in Figure 23-20 are **(a)** Paul Karrer, **(b)** Adolf Butenandt, **(c)** Richard Kuhn. All won Nobel prizes in chemistry for work with natural products. From library reading prepare a short biography of one of them with emphasis on his contribution to the chemistry of Chapters 22 and 23.

(a)   (b)   (c)

Figure 23-20 Prominent natural products chemists. (See Question 15.) (Courtesy Nobel Foundation.)

PROBLEMS

16. Sugars Use formulas and words to explain the difference between **(a)** an aldotetrose and an aldopentose; **(b)** an aldose and a ketose; **(c)** a D sugar and an L sugar; **(d)** an α-amino acid and a β-amino acid.

17. Sugar identity An aldopentose A, whose number 2, 3, and 4 carbons have the same stereochemical orientation as the 3, 4, and 5 carbons of glucose (see numbering of chain form in Eq. 23-11), is heated with a mild oxidizer, giving compound B, a polyhydroxymonocarboxylic acid; B is reduced with HI, giving C. Write equations, using abbreviations like those shown in Problem 18.

18. Stereochemistry Fischer realized that certain sugars, found optically active in nature, would give optically inactive *meso* diacids on nitric acid oxidation (Eq. 23-9). Will any of the following give *meso* acids? Draw acid formulas and explain.

$$
\begin{array}{ccc}
\text{CHO} & \text{CHO} & \text{CHO} \\
\text{(a) } \text{CH}_2\text{OH} & \text{(b) } \text{CH}_2\text{OH} & \text{(c) } \text{CH}_2\text{OH}
\end{array}
$$

19. Optical activity Consider saccharides with two different asymmetric carbon atoms, a and b, each of whose signs of optical rotation can be $+$ or $-$. The possible combinations are four, giving two pairs of *dl* (mirror-image) isomers:

$$
\begin{array}{cccc}
+a & -a & +a & -a \\
+b & -b & -b & +b \\
\end{array}
$$

dl pair   dl pair

In similar fashion show the combinations and *dl* pairs possible for saccharides containing three different asymmetric centers, a, b, c.

20. Rayon Rayon is made by esterifying —OH groups of cellulose with (in effect) acetic acid. Draw the formula for the repeating unit of cellulose. Show acetic acid reacting with the —OH at C_6 to give cellulose monoacetate.

21. Caloric value Combustion of a mole of sucrose ($C_{12}H_{22}O_{11}$, mol. wt. 342) produces 1350 kcal of heat. If the average adult in the United States consumes 100 lb of sucrose per year, what percentage of his caloric intake of 2800 kcal/day is satisfied by sucrose?

22. Sugar series In the manner of Figure 23-3 show derivation of four L-aldopentoses from L-glyceraldehyde. Give a name for each compound (include sign of rotation) based on those in the figure.

23. Glycoside During energy storage and membrane building, all cells synthesize glycosides, molecules that yield on hydrolysis one sugar molecule and one nonsugar molecule. The glycoside pictured here has been isolated from wheat germ. Write an equation showing the two molecules obtained on hydrolysis, and name them.

24. Chitin The hard shell of crustaceans (lobsters, etc.) and insects (roaches, etc.) is a polysaccharide called *chitin* (ki′ tin). On enzymatic hydrolysis N-acetylglucosamine is obtained. This molecule resembles glucose except that at C-2 an $-N-\overset{\displaystyle O}{\overset{\|}{C}}-CH_3$ is attached instead of —OH. **(a)** Write an open-chain formula for N-acetylglucosamine. **(b)** The structure of chitin is analogous to that of cellulose. Draw a formula containing two joined N-acetylglucosamine units. **(c)** If the molecular weight of chitin is 150,000, how many units are in the polymer?

25. Glycolysis One step in glycolysis (Figure 23-4) is the reaction,

D-glyceraldehyde 3-phosphate $\rightleftharpoons$ dihydroxyacetone phosphate

(a) Draw structural formulas for the two compounds. **(b)** The enzyme catalyzing this reaction is called *isomerase*. Why is the name appropriate?

26. Water Per day an average adult may lose liquid water in the following liter amounts: as urine, 1.0; as sweat, 0.6; as fecal matter, 0.1; in expired air, 0.4. How many gallons of water pass through a human being, assuming these data hold for 60 years?

27. Lipid analysis Linolenic acid is a fatty acid derived from lipids found in seeds. The molecule absorbs iodine equivalent to three carbon–carbon double bonds. To locate the double bonds an ozonolysis (Eq. 21-5d) is run. These products are obtained in 1:2:1 mole ratio on reduction of the ozonide:

$$H_3C-CH_2-\overset{\displaystyle O}{\underset{\displaystyle H}{C}} \qquad \overset{\displaystyle O}{\underset{\displaystyle H}{C}}-CH_2-\overset{\displaystyle O}{\underset{\displaystyle H}{C}} \qquad \overset{\displaystyle O}{\underset{\displaystyle H}{C}}-(CH_2)_7-\overset{\displaystyle O}{\underset{\displaystyle OH}{C}}$$

(a) How many carbon atoms are in linolenic acid? (b) Write a structural formula for the acid.

28. Fatty acid Consider these data on melting points of saturated acids isolated from lipids: C_4, $-5°C$; C_8, $16°C$; C_{16}, $63°C$; C_{26}, $88°C$. Arachidic acid, isolated from a wax, gives a melting point of $75°C$ and does not absorb iodine. What assumptions can you make about its structure? Explain. (Consider graphing the data.)

29. Lipid formulas (a) Draw a structural formula for a mixed glyceride which, on saponification, gives 1 mole of potassium oleate and 2 moles of potassium palmitate. (b) Hiragonic acid is a fatty acid obtainable by the hydrolysis of sardine oil. Its IUPAC name is 6,10,14-hexadecatrienoic acid. Draw the structural formula.

30. Acid isomers (a) Oleic acid (Table 23-1) is a *cis* compound. Draw its structural formula as a carbon–carbon double bond with four groups attached. (b) Elaidic acid (m.p., $51°C$) is the *trans* isomer of oleic acid. Draw its structural formula.

31. Unsaturated acids Cottonseed oil is a glyceride mixture containing oleate and other unsaturated groups. Hydrogenation of the C=C bonds converts the oil to a white fat sold for frying purposes. (a) Write an equation showing reaction of H_2 (nickel catalyst) with glyceryl trioleate. (b) If that product is hydrolyzed, name the fatty acid that is obtained. (c) In excess of a billion pounds of unsaturated lipids like linseed oil is used yearly in the United States to make paints. Propose an equation illustrating interaction between (atmospheric) oxygen atoms and several oil molecules that will lead to the polymer "skin" that a paint develops on drying. Represent an oil molecule simply as a zig-zag line interrupted by one double bond in the middle.

32. Iodination The number of grams of iodine that 100 g of lipid will absorb at its double bonds is called the *iodine number*. Calculate the theoretical iodine number of the lipid in Eq. 23-17.

33. Saponification The milligrams of KOH needed to saponify 1.00 g of neutral lipid is called the *saponification number*. Calculate the theoretical saponification number of the lipid in Eq. 23-17.

34. Molecular weight The boiling point of pure chloroform is $60.19°C$ and K_b for chloroform is $3.63°C–kg$ of solvent/mole of solute (see Eq. 16-7).

When 1.42 g of a certain fatty acid is dissolved in 50.0 g of chloroform, the solution boiling point is 60.55°C. Calculate the acid's molecular weight. Tentatively identify the acid (Table 23-1).

35. Sterol structure Biochemists try to correlate structure with physiological activity. Diethylstilbesterol (DES), shown here, is a powerful synthetic estrogen. **(a)** Copy its formula carefully. Directly on top of it, in another color, draw the formula for estradiol, using the same atom-to-atom spacing. (Begin by matching up the lower ring and —OH group.) Comment on the similarity. **(b)** Beside those formulas draw the formula for the *cis* isomer of DES. It does not have estrogenic activity. Might that be expected if the relationship in (a) is necessary? Explain.

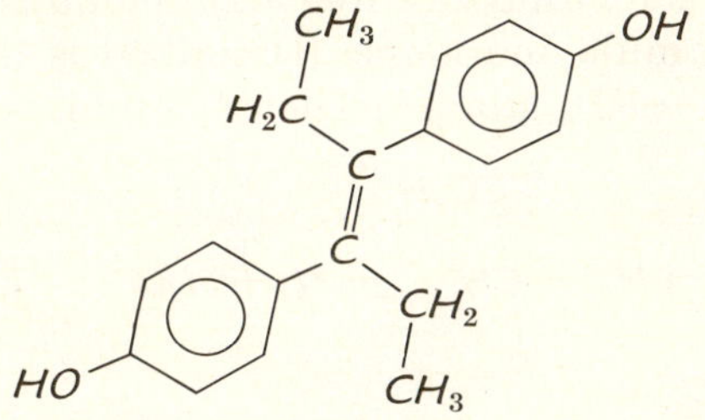

36. Molecular weight Insulin controls the glucose level in the blood. In 1969 Hodgkin (Figure 23-17) announced the secondary structure of insulin, the first protein hormone to be resolved by X ray diffraction. **(a)** Chemical analysis shows insulin is 2.15 percent zinc by weight. Calculate the minimum molecular weight. **(b)** Osmotic pressure studies show the molecular weight of insulin is about 6000. How many zinc atoms does the molecule contain?

37. Acid analysis An amino acid (m.p., 224–225°C with decomposition) is isolated. By the freezing point method its molecular weight is 135 to 155. After formaldehyde is added to a 438-mg sample, 30.0 ml of 0.100 M NaOH is needed to titrate the carboxyl acidity. Find the precise molecular weight of the acid. (Described is the *Sørensen formol titration*.) Reactions are

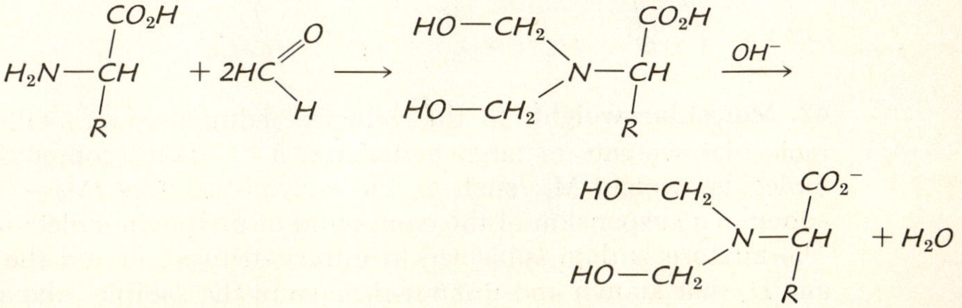

38. Acid analysis Another 438-mg sample of the amino acid described in Problem 37 is decomposed in hot, concentrated H_2SO_4. After cooling, excess NaOH is added. The mixture is boiled and the ammonia liberated is

led into 50.0 ml of 0.200 M HCl. The (excess) HCl not neutralized by NH_3 then requires 40.0 ml of 0.100 M KOH for titration. Calculate the number of nitrogen atoms in each molecule of the amino acid. (Described is the *Kjeldahl* method for total nitrogen.) Reactions are

$$H_2N-\underset{\underset{R}{|}}{\overset{\overset{CO_2H}{|}}{CH}} \xrightarrow[\Delta]{H_2SO_4} NH_4^+ + CO_2 + H_2O \xrightarrow[\Delta]{NaOH} NH_3 \xrightarrow{HCl} NH_4^+$$

39. Acid analysis A 146-mg sample of the amino acid described in Problems 37 and 38 is mixed with nitrous acid. The nitrogen gas liberated is collected in a gas buret. Corrected to standard conditions its volume measures 44.8 ml. Calculate the number of free amino groups in the molecule. Identify the acid by formula and name. (Described is the *Van Slyke* method for determining free $-NH_2$ groups.) The reaction is

$$H_2N-\underset{\underset{R}{|}}{\overset{\overset{CO_2H}{|}}{CH}} + HNO_2 \longrightarrow HO-\underset{\underset{R}{|}}{\overset{\overset{CO_2H}{|}}{CH}} + N_2 + H_2O$$

40. Proteins (a) Use a few examples from Figure 23-12 to estimate the molecular weight of an "average" amino acid. **(b)** A polymer of amino acid residues may be called a polypeptide if mol. wt. < 5000, and a protein if mol. wt. > 5000. About how many amino acid molecules are obtained upon acid hydrolysis of the smallest protein?

41. Peptides (a) Write the formula for the tetrapeptide having the amino acid sequence, tyrosine–glycine–valine–glycine. **(b)** Copy the formula below. Insert dotted lines to show where acid hydrolysis would break the molecule. **(c)** Write formulas and names of the amino acids formed on hydrolysis. **(d)** Name the peptide.

42. Molecular weight In the velocity sedimentation method of getting molecular weights of large molecules, a "marker" compound of known molecular weight M_k, such as the enzyme catalase ($M_k = 2.44 \times 10^5$), is added to a suspension of the compound of unknown molecular weight M_u. The mixture is then subjected to ultracentrifugation and the distances D_k and D_u that known and unknown move in the sample tube are measured. Assuming spherical molecules, the relative molecular weight of the unknown is then calculated from

$$M_u = M_k \left(\frac{D_u}{D_k}\right)^{\frac{3}{2}}$$

(23-26)

Calculate the molecular weight of a protein that moves 0.30 cm while catalase moves 0.40 cm.

43. Acid analysis α-Amino-β-hydroxy acids are determined quantitatively by reaction with periodic acid, a reagent which cleaves them as follows:

$$\underset{\underset{\underset{R}{|}}{\underset{HO-CH}{|}}{\overset{\overset{CO_2H}{|}}{H_2N-CH}} \xrightarrow{HIO_4} NH_4^+ + \underset{\text{glyoxalic acid}}{\underset{H}{\overset{O}{\diagdown}}C-CO_2H} + \underset{\text{aldehyde}}{R-C\overset{O}{\underset{H}{\diagdown}}}$$

(a) Which two amino acids in Figure 23-12 will give this reaction? **(b)** Write their cleavage equations and name the aldehyde products.

44. Analytical reagents Give equations for the reaction of **(a)** Sanger's reagent with glycylalanine; **(b)** performic acid with cystine.

45. Sequencing How many isomers based on sequence alone are possible for **(a)** a dipeptide containing two different amino acid residues? **(b)** a tripeptide containing three different residues? **(c)** Write all the latter, representing the first possibility as ABC.

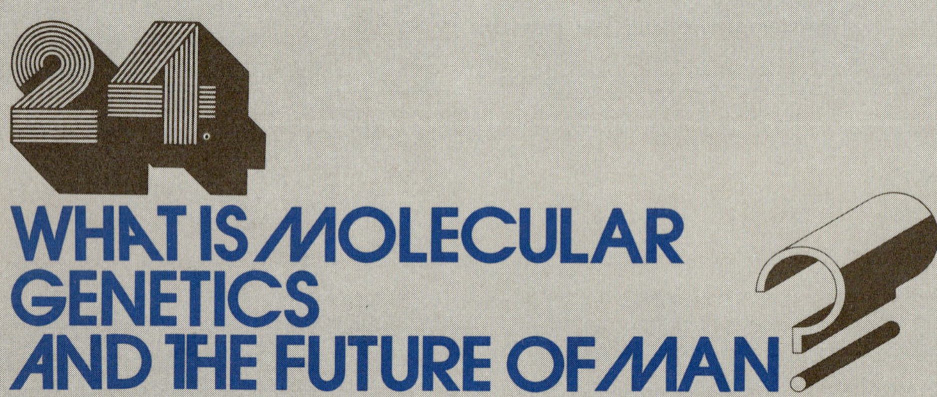

THE IMPORTANT CONCEPTS

24-1. Brief history of genetics

1. Mendel's first law explained the statistics of trait transfer in breeding experiments.
2. Chromosome splitting with daughter cell production may be seen under ordinary magnification.
3. A gene is a chromosome segment that carries a single characteristic.
4. Chromosomes contain two kinds of high polymers: proteins and DNA.
5. Genes are lengths of DNA.
6. Molecular genetics began with the discovery that protein synthesis is genetically controlled.

24-2. The DNA theory

1. Nucleic acids from all organisms are the same.
2. Nucleic acid hydrolysis essentially yields only six different compounds.
    a. Five important bases are known; four occur in a given nucleic acid.
3. The DNA structure was inferred from A:T and C:G ratios, X ray diffraction pictures, and scale models.
4. The Watson-Crick model is a double-stranded polynucleotide helix with base pairing and hydrogen bonding.
    a. The exterior backbone is hydrophilic, the interior stacking is hydrophobic.
5. Replication of DNA occurs as strands unwind and nucleotides from solution build up base-paired complementary strands.

24-3. Biosynthesis of proteins

1. Base sequence of DNA is genetic language.
    a. An alphabet of four letters spells out 64 three-letter codons.
    b. Each amino acid is directed toward peptide assemblage by its specific codon.
    c. Instructions necessary to assemble a human being are possibly contained in 6 million genes.
2. Instructions of DNA are utilized by means of three kinds of RNA.
3. Unlike DNA, RNA contains ribose and uracil.
4. Protein synthesis takes place on ribosomes.
    a. Amino acids in cytoplasm, activated by ATP, react with specific RNA's coded to recognize them.
    b. About 60 different tRNA's are formed in the cell nucleus as complements to DNA segments.
    c. Messenger RNA's, similarly formed in the nucleus on DNA lengths, move to ribosomes for attachment.
    d. Codon-anticodon bonding between mRNA and amino acid tRNA units precedes peptide chain buildup.
    e. Ribosomes carrying the growing peptide as they move along the mRNA chain, reading the codon triplets.
    f. Final protein configuration is established by sulfur–sulfur links and hydrogen bonding.

24-4. New directions in molecular biology

1. Many biochemical problems are interpretable with DNA theory.
    a. A virus can change cell performance by injecting its DNA into the cell.
    b. Only a few genes are active along a DNA chain; inactive genes may be blocked by histones.
    c. In man most genetic mutations are in recessive genes.
    d. Antibiotics may stop bacterial growth by hindering ribosomal action.
2. Gene therapy may some day allow cures for certain inherited diseases.
3. Gene research may allow man to program his own evolution.

24-5. Nutrition and population

1. Proteins, carbohydrates, lipids, vitamins, and minerals are dietary essentials.

2. Carbohydrates and proteins have equal caloric value; fats have more than twice as much.
3. Population increase can be viewed optimistically or pessimistically.
4. The Green Revolution can buy time.
5. Birth control proposals are open to questions.

24-6. Into the next century
1. The world can be modeled for study of its future using five variables.
2. The future of man's use of energy can be divided into three phases.
3. Humanity is very young; a long future in which to resolve problems may lie ahead.

The means by which living organisms preserve their characteristics in reproducing their kind is called *heredity*. The study of heredity in all its aspects is called *genetics*. Genetics includes the physical transfer of the chemical material of heredity from one generation to the next, the reproduction and variation of that material, and the biochemical use of the information it carries.

Hereditary factors passed to progeny are called *genes*. A gene is a unit of a high-polymer nucleic acid. The first nucleic acid was discovered in 1869. The chemical composition of nucleic acids was determined in the 1920s. In the 1940s nucleic acids were confirmed as the carriers of genetic messages.

Mechanisms involving genes are responsible for the synthesis of proteins (including enzymes) in cells. Because enzymes catalyze the reactions that are responsible for cell building and function, genetic material exerts its effect by guiding the development and behavior of all living systems. Studied today from a chemical standpoint, genetics is the molecular basis for understanding biology.

24-1. BRIEF HISTORY OF GENETICS

About a century ago it was generally believed that heredity traits became mixed and lost when they passed from parent to offspring. That misconception was corrected by the work of Gregor Mendel (1822–1884), an Austrian monk. Noting that no one had made an orderly statistical study of heredity, Mendel initiated a series of experiments with garden peas in which he crossbred pure variety parents and tabulated the retention of traits through generations of offspring. In each case he found one of two traits chosen for reference to be *dominant* and therefore capable of masking the other trait which he termed *recessive*. In crossbreeding pure tall and pure short parents, for example, he obtained an all-tall first generation. (Thus tall was dominant, short was recessive.) Fertilizing first-generation plants with one another, he grew a second generation. In it he found 75 percent possessed the dominant feature, 25 percent the recessive feature. The data suggested the following hypothesis, now called *Mendel's first law*.

1. Each trait is determined by a pair of heredity units (later named *genes*).
2. When the genes for a reference trait are alike, the organism is pure bred (or *homozygous*) in that trait; when the genes are different the organism is hybrid (or *heterozygous*).
3. When a reproductive or sex cell (called a *gamete*) forms, the two genes related to any particular trait separate — and the gamete gets only one gene of each pair.

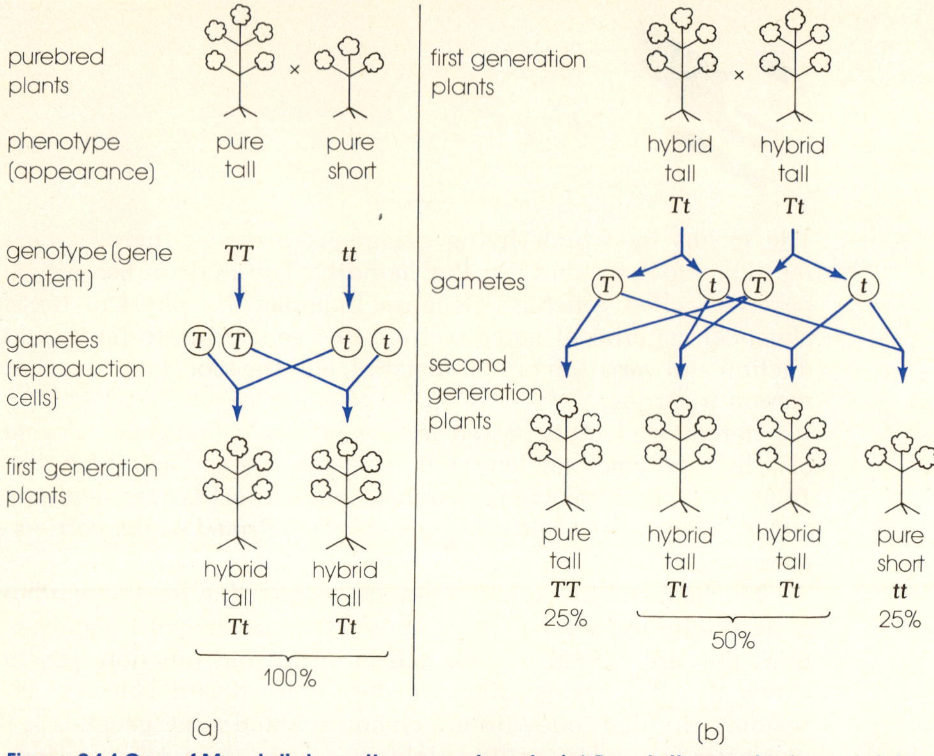

Figure 24-1 One of Mendel's breeding experiments. (a) Pure tall pea plants containing genes for tallness, symbolized TT, mated with pure short pea plants containing genes for shortness, symbolized tt. Before the parent plants matured, Mendel correctly assumed a separation of genes occured, giving gametes with one gene each related to plant height. Mating gave an all-hybrid crop of plants containing one of each gene type, symbolized Tt. (b) Mating of first generation hybrid plants. Gametes formed consisted of two genetically different sperm types and two genetically different egg types. Combining at random they produced a second generation with the frequencies noted. Tallness being dominant and shortness recessive, three-quarters of the second generation was tall. Other experiments gave similar statistics.

4. Fertilization of an egg-type gamete by a sperm-type gamete gives offspring containing two genes for each trait.
5. When the two transmitted genes for a given trait are different, evidence will be seen for the dominant one.

The experimental basis for the law is illustrated in Figure 24-1.

Thirty years after Mendel the theory of the gene in the nucleus of a cell and as part of larger structures called *chromosomes* (so named for an ability to be dyed for microscopic investigation) became one of biology's guiding principles. Alignment and splitting of the threadlike chromosomes with half going to each of two daughter cells was first definitely identified under the microscope in the 1870s. The significance of the observation was not appreciated until 1900, however, when it was correlated with Mendel's laws (the laws had just been rediscovered after having lain in the obscurity of minor publications).

The concept of the gene was introduced in 1911. The existence of genes and their location was demonstrated by Columbia University biologist

Thomas Morgan (United States 1866–1945, Nobel prize in medicine or physiology 1933). Fruit fly chromosomes have banded markings observable under the light microscope which can be related to such features as eye color in the adult fly, and Morgan used them to follow inherited traits. In every case he found chromosome mark and inherited feature transmitted together. Biologists today believe that thousands of genes lie in linear array along the length of the chromosome chain, each one responsible for transmitting a single characteristic when the nucleus of a cell divides.

Chromosomes contain proteins and deoxyribonucleic acid, DNA. Evidence that DNA is the genetic stuff of cells and that genes are segments of DNA molecules came from studies in the last few decades on bacteria. When one kind of bacteria (call it x) was grown in a medium containing DNA from a related bacteria y, bacteria x was found to acquire the ability to synthesize protein characteristic of y. This indicated the invasion of x by y-DNA and incorporation of y-DNA into x's genetic system which was then programed in a way characteristic of y.

The regulation of cellular chemistry through the control of protein synthesis by genes was demonstrated in the late 1940s by Stanford geneticists George Beadle (United States 1903–   ) and Edward Tatum (United States 1909–   )who shared the 1958 Nobel prize in medicine or physiology. They showed mutations could be induced in a certain variety of bread mold that prevented the organism from its normal synthesis of some essential compound such as an amino acid. Inability to perform normal function was traced to lack of an enzyme—and lack of the enzyme was carried from one generation to the next. That meant the mutation was in a particular gene, and thus it was a gene defect that caused the enzyme not to be formed. The science of biochemical genetics began with this work.

24-2. THE DNA THEORY

Structure of DNA

Nucleic acids taken from organisms ranging from one-celled plants to man are structurally the same, meaning that nature in all her experimentation found only one satisfactory genetic mechanism. The polymer DNA is strongly acidic and carries a negative charge at physiological pH (pH $\sim$ 7). It is found in the cell nucleus associated with positive species such as protonated amines and magnesium ions. DNA from various kinds of cells (animal tissue contains about 2 mg of DNA per gram, or 0.2 percent) has molecular weights between 10^6 and 10^9 and therefore includes some of the larger molecules known. Progressive acid hydrolysis followed by chromatographic separations reveals that DNA and other nucleic acids are polymers of subunits which themselves can be broken down as follows:

$$\begin{matrix} nucleic \\ acids \end{matrix} \longrightarrow nucleotides \longrightarrow \begin{cases} phosphoric\ acid \\ nucleosides \longrightarrow \begin{cases} C_5\ sugars \\ purine\ bases \\ pyrimidine\ bases \end{cases} \end{cases} \qquad (24\text{-}1)$$

The structure of nucleic acids, nucleotides, and nucleosides was introduced in Section 22-1 with Eqs. 22-5 and 22-6. Using that information we can define Eq. 24-1 further:

$$(-\text{sugar}\underset{|}{\overset{base}{|}}PO_4-)_n \longrightarrow \text{sugar}\underset{|}{\overset{base}{|}}-PO_4 \longrightarrow H_3PO_4 + \text{sugar}\underset{|}{\overset{base}{|}} \longrightarrow C_5 \text{ sugar} + \text{bases}$$

$$\text{nucleic acid} \qquad\qquad \text{nucleotides} \qquad\qquad \text{nucleosides}$$

The five important bases and related substances are given in Table 24-1.

Despite a molecular weight in the millions, a given nucleic acid contains only four different nucleotides. Complete hydrolysis of DNA, for example, yields essentially only six different compounds:

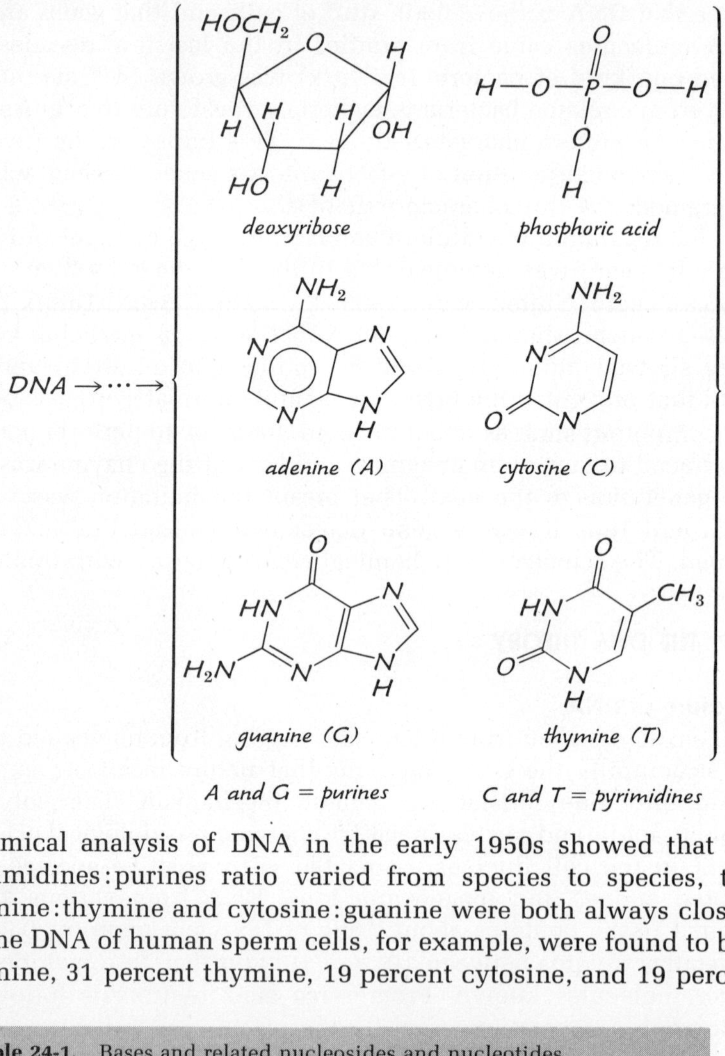

deoxyribose   phosphoric acid

adenine (A)   cytosine (C)

$$DNA \rightarrow \cdots \rightarrow \qquad\qquad (24\text{-}2)$$

guanine (G)   thymine (T)

A and G = purines   C and T = pyrimidines

Chemical analysis of DNA in the early 1950s showed that whereas the pyrimidines:purines ratio varied from species to species, the ratios of adenine:thymine and cytosine:guanine were both always close to 1. Bases in the DNA of human sperm cells, for example, were found to be 31 percent adenine, 31 percent thymine, 19 percent cytosine, and 19 percent guanine.

Table 24-1. Bases and related nucleosides and nucleotides

Base and abbreviation	Nucleoside (base-sugar)	Nucleotide (base-sugar-phosphate)	Nucleic acid source
Adenine, A	Adenosine	Adenylic acid	DNA, RNA
Guanine, G	Guanosine	Guanylic acid	DNA, RNA
Cytosine C	Cytidine	Cytidylic acid	DNA, RNA
Thymine, T	Thymidine	Thymidylic acid	DNA
Uracil, U	Uridine	Uridylic acid	RNA

Examination of DNA under the electron microscope showed some DNA molecules to be long fibers 20 Å thick and 30,000 Å long.

The first X ray diffraction pictures of a nucleic acid were taken in 1938 on a sodium salt of DNA drawn into a semicrystalline thread. The patterns suggested the presence of repeating units of a linear polymer, but it was not until 1953 that enough evidence was accumulated through improved techniques for assignment of the polymer's secondary structure. Leaders in this work shared the Nobel prize in medicine or physiology in 1962: Francis Crick (England 1916–   ), Maurice Wilkins (England 1918–   ), and James Watson (Figure 24-2). They found that

1. DNA from various sources yielded the same diffraction pictures, indicating a structural regularity that superceded differences in base percentages.
2. DNA could take two forms: crystalline in low humidity, and less ordered or "paracrystalline" in high humidity. Pictures of the latter hinted at a slippage of chains past one another along their fiber axis.
3. Although calculations with a preliminary model of a sugar-phosphate chain showed the chemical repeat unit (phosphate-to-phosphate distance) should be 7 A long, X ray pictures gave a repeat distance of 34 Å for the paracrystalline form. That was interpreted to mean that the chemical unit recurred several times before the structural unit repeated. A helical arrangement seemed a logical explanation.
4. Titration data and proton exchange studies inferred that hydrogen bonds played a part in holding polymer chains together. With base ratio figures suggesting bases occurred as adenine–thymine and cytosine–guanine pairs, it seemed likely that hydrogen bonds existed between the pairs (Figure 24-2).

Reasoning from scale models built to conform to accurately determined bond lengths, Watson and Crick in England (and in the United States, Pauling) concluded that DNA was not a single chain but a pair of them coiled in helices about a common axis. Twisting their models in equally spaced turns, Watson and Crick discovered base pairing could only take place when a pyrimidine base was positioned opposite a purine base, as shown on the next two pages. In this arrangement the bases, having the flatness of aromatic rings, required minimum volume for stacking, and the distances between them were optimum for the establishment of hydrogen bonds.

No work since has refuted the Watson-Crick model. DNA is believed to be a double-stranded helix made from the winding of two antiparallel polynucleotide chains as pictured in Figure 24-3. Phosphate and sugar residues form an exterior hydrophilic backbone whereas the bases are turned inward, giving a hydrophobic interior. Base pairs lie approximately parallel to one another and at right angles to the long helical axis. Van der Waals forces and dipole interactions are mainly responsible for holding the helix together; interbase hydrogen bonds add a final stiffening to the alignment. The hydrogen bonds probably form last as the structure tightens and water is squeezed from its interior.

sugar—thymine···H···adenine—sugar (with phosphate groups above and below)

sugar—cytosine···H···guanine—sugar

(a)

(b)

(c)

Figure 24-2 Base pairing through hydrogen bonds in DNA. (a) Simplest representation. (b) Primary structure detail. Sugar–phosphate chains are outside, base pairs inside. (c) James Watson (United States 1928–   ). Watson was working as a postdoctoral fellow with Crick at Cambridge, England, when they deduced the DNA structure. Watson described the work in a frankly written book, The Double Helix. (Photograph Courtesy Harvard University.)

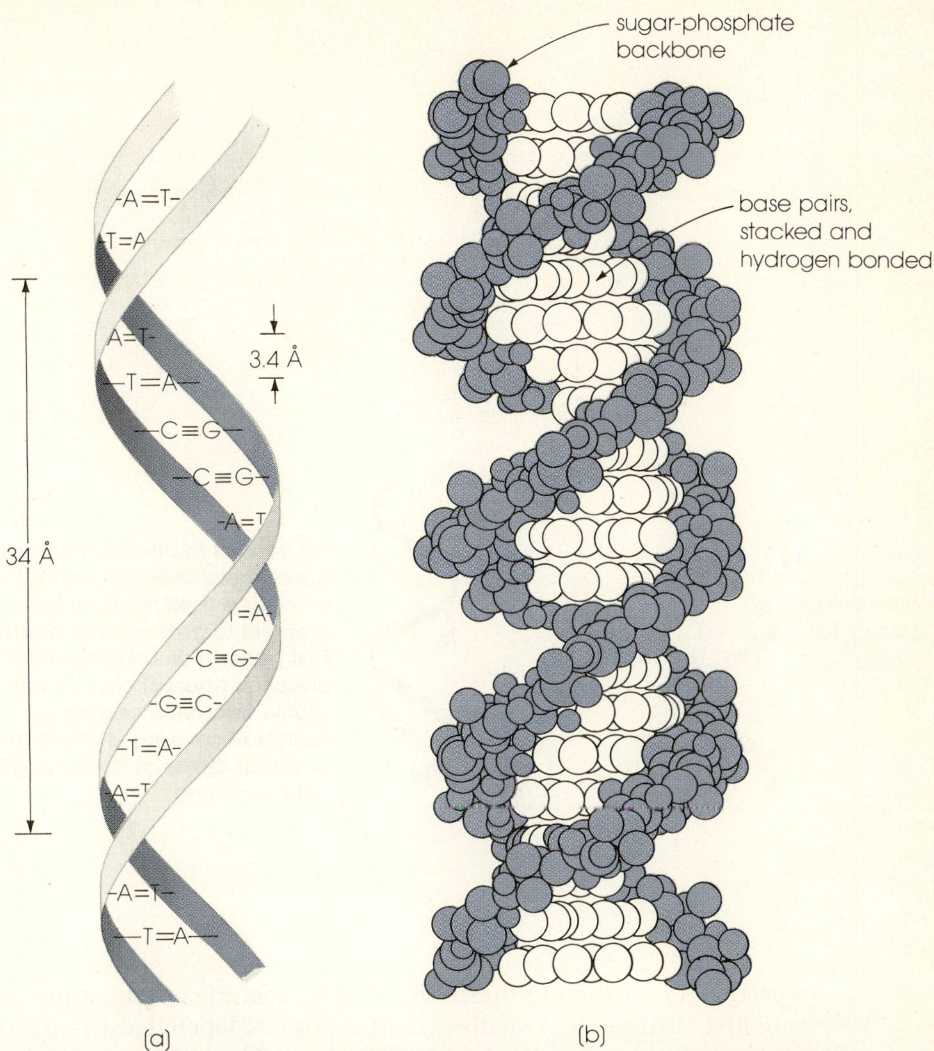

Figure 24-3 Models of DNA. (a) Schematic view of the double helix. Adenine (A) and thymine (T) share two hydrogen bonds; guanine (G) and cytosine (C) share three. Ten base pairs are found in each complete turn (360 degrees, 34 Å). (b) A space filling model.

Replication of DNA

The Watson-Crick model offers an explanation of how DNA can replicate itself. During replication, the two strands of DNA are believed to unwind and add nucleotides present in the surrounding medium. As nucleotides are guided into position by base pairing, each original strand acts as a pattern or template for the construction beside it of a complementary strand. A sequence of —ACGT—, for example, would give a new strand with the sequence complement, —TGCA—. Energy for unwinding the original chains is furnished by the process of new bond formation. As shown in Figure 24-4 the first-generation result is two double helices, each composed of one new and one original chain.

Some of the best support for the replication mechanism has come from

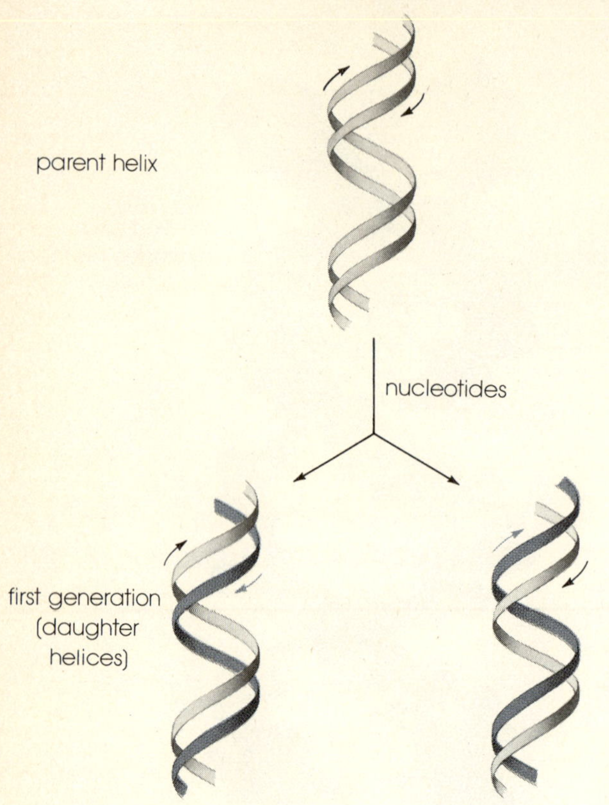

Figure 24-4 Replication of DNA. After the parent helix unwinds, each stand guides alongside itself the synthesis of its complement to give another double helix. First generation DNA therefore contains equal numbers of old (parental) and new atoms. Each succeeding generation contains half as many of the original parental atoms as does the generation just preceding.

experiments on the synthesis of DNA in bacteria. Generations of bacteria were first cultured in a medium containing ^{15}N-labeled nutrients until all nitrogen in the bacterial DNA was ^{15}N isotope. The culture was then placed in a medium containing nutrients whose nitrogen was all ^{14}N isotope. Samples were taken after the population had doubled (first generation) and again after a second doubling (second generation). The DNA from each was isolated and, along with DNA taken from the parental bacteria, was subjected to an ultracentrifugation (see Figure 23-15) procedure in which species were separated according to density. The DNA from the parental bacteria gave a single band in the sample tube, corresponding to DNA completely ^{15}N labeled. The first generation gave a single band corresponding to DNA containing equal amounts of ^{15}N and ^{14}N labels. The DNA from the second generation gave two bands of equal size, one corresponding to 100 percent ^{14}N-DNA, the other to a 1:1 mixture of ^{15}N- and ^{14}N-DNA.

The observations were those expected from the process pictured in Figure 24-4 because in any replication the two daughter helices would contain 50 percent new material. By contrast, if DNA replication involved chain fragmentation followed by recombination, a variety of different densities would be found rather than the clean production of one or two species. A more detailed visualization of DNA replication is given in Figure 24-5.

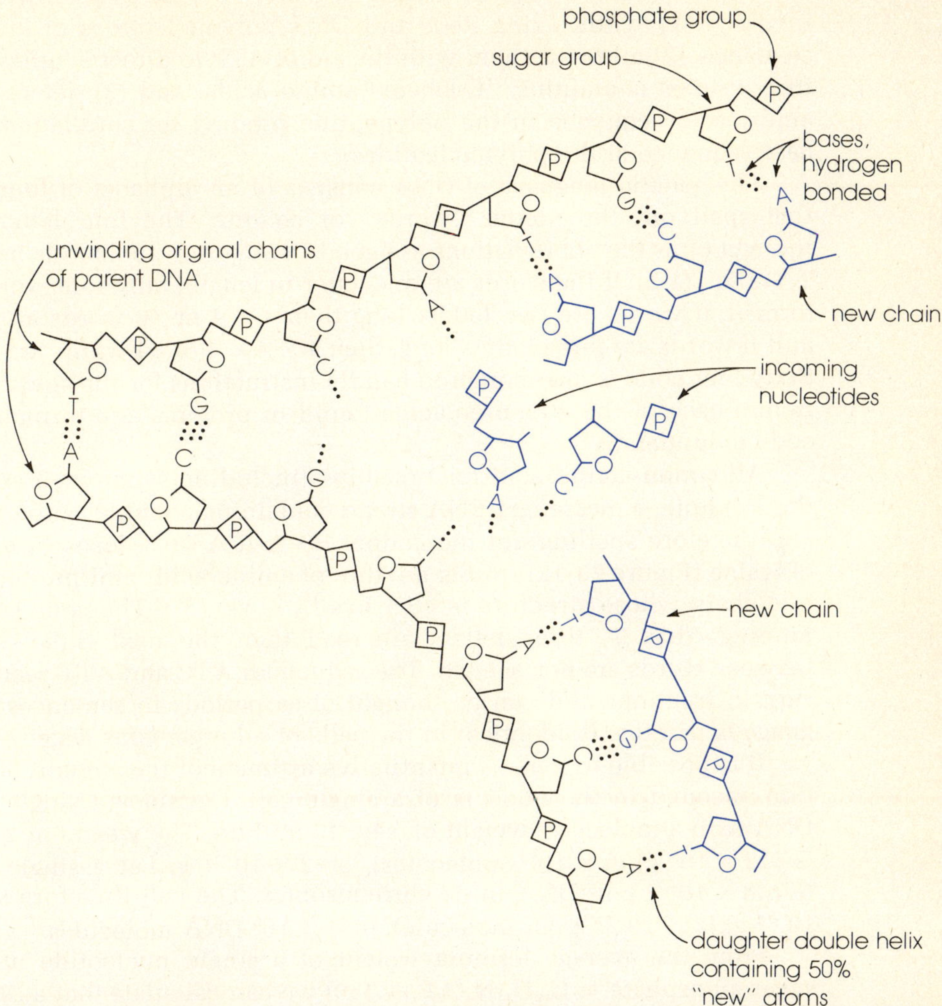

Figure 24-5 Synthesis of DNA (helical twist omitted). As the original helix unwinds, the separate chains begin attaching nucleotides (synthesized in surrounding fluid) according to base-pairing rules. An enzyme, DNA polymerase, catalyzes polymerization of mononucleotides.

24-3. BIOSYNTHESIS OF PROTEINS

Encoding of DNA

The ability of DNA to carry information lies in the sequencing of its four bases. *Base sequence* contains instructions in somewhat the same way as an arrangement of holes in a punched tape comprises instructions that program a machine to perform a repetitive task. The code was explained by the scientists shown in Figure 24-7 and others during the 1960s.

Very briefly, DNA with its coded instructions directs the synthesis of a smaller nucleic acid called *messenger RNA* (mRNA) which carries the genetic message to the cytoplasm. There, with the aid of certain organelles and other types of RNA, messenger RNA directs assemblage of amino acid molecules into peptide chains. Experiments that clarified this mechanism involved (1) synthesizing RNA and DNA polynucleotides of known base sequence, (2) allowing them with the aid of ATP to direct synthesis in cell-free systems containing ^{14}C-labeled amino acids, and (3) determining the amino acid sequence in the polypeptide product for correlation with the base sequence of the polynucleotides.

The genetic language of DNA consists of an alphabet of four "letters" that spell out three-letter "words" or *codons*. The four letters can be thought of as the abbreviations assigned DNA's four nucleotide bases: A, C, T, G (Eq. 24-2). If the words are one letter in length, only four words can be formed. If words are two letters long then 4×4 or 16 words are possible; and if words are three letters long, then $4 \times 4 \times 4$ or 64 words are available. Sixty-four codons are enough to handle instructions for the specific utilization of each of the 20 amino acids found in proteins and to include some code redundancy.

All codon-amino acid relationships (including synonyms) are known. For example, a messenger RNA strand containing adenine as the only base, and therefore spelling out the codon $\cdots$ A-A-A $\cdots$, leads to a selection of lysine (Figure 23-12) from a mixture of amino acids and produces a peptide chain whose structure is $\cdots$ Lys-Lys-Lys $\cdots$ The code is nonoverlapping—that is, three letters are read then the next three—so spaces between words are not needed. The sequences ATC and ATT seem to interrupt instructions and can be thought of as periods in sentences. The language is practically identical in the cells of all organisms tested.

It is possible to make a quantitative estimate of the amount of information encoded in the nucleus of a single cell. Consider a single strand of DNA with a molecular weight of, say, 12 million. The weight of 1 molecule is $(12 \times 10^6$ g)/$(6 \times 10^{23}$ molecules), or 2×10^{-17} g. Let a single cell contain 8×10^{-12} g of DNA in its chromosomes. The cell therefore holds $(8 \times 10^{-12}$ g)/$(2 \times 10^{-17}$ g per molecule), or 4×10^5 DNA molecules.

From the average formula weight of a single nucleotide unit (base–sugar–phosphate $\cong C_{10}H_{11}N_4O_7P \cong 330$) we can calculate that a DNA molecule having a molecular weight of 12 million contains 3.6×10^4 nucleotide units. Letting a sequence of three consecutive nucleotide units on one strand comprise one word of code, each DNA molecule carries $(3.6 \times 10^4$ nuclestides)/(3 nucleotides per word), or 1.2×10^4 words. That gives a total of $(4 \times 10^5$ molecules) $(1.2 \times 10^4$ words per molecule), or 4.8×10^9 words. For comparison, a 1000-page volume of a closely printed encyclopedia contains 106 words. The single-strand DNA code in one cell nucleus is equivalent to almost 5000 volumes of the encyclopedia. Each nucleated cell in the human body contains this much information.

All the instructions necessary to produce a human being are carried in perhaps 6 million genes that originate with union of egg and sperm. The genes are distributed along the DNA molecules that occupy the 46 chromosomes comprising human genetic matter. As already mentioned, the work of Beadle and Tatum indicated that instructions for synthesizing a single protein are carried by a single gene. Consider a protein containing 200 amino acid residues. The gene responsible for its synthesis must contain a sequence of 200 codons or 600 bases. If each base–sugar–phosphate unit takes up a length along the polynucleotide chain of 7.7 Å (the covalently bonded atom sequence, —C—C—O—P—O—), then a 600-base gene would reach about 4.6×10^3 Å or 4.6×10^{-5} cm. The length of 6 million such genes connected end to end would stretch 276 cm, or roughly 11 feet. Each of us is a product of this 11-foot set of instructions.

RNA and the assembly of proteins

The base sequence in DNA determines the amino acid sequence in proteins. As stated previously DNA is not directly involved in protein synthesis but its message is "transcribed," carried, and utilized by three kinds of *ribonucleic acid*. Ribonucleic acid (RNA) is similar to DNA with two exceptions: In RNA the sugar ribose replaces deoxyribose, and the base uracil replaces thymine. Like thymine, uracil is complementary with adenine:

ribose *instead of* *deoxyribose*

uracil *instead of* *thymine*

The site of protein synthesis is the surface of *ribosomes*. Ribosomes are electron-microscopic particles having molecular weights of a few million and consisting of 60 percent ribosomal RNA and 40 percent protein. They are attached to parallel double-membrane protein sheets, collectively called the *endoplasmic reticulum*, which are found in the cytoplasm of almost all animal cells. Ribosomes consist of two subunits. The smaller apparently has protein sites that hold the message to be transcribed; the larger has sites that hold the polypeptide chain being synthesized.

In simplified outline protein synthesis (Figure 24-6) proceeds as follows.

1. In the cytoplasm amino acids react with adenosine triphosphate (ATP), eliminating pyrophosphate ion and forming "activated" amino acids attached by a high-energy bond. A specific enzyme catalyzes the reaction of each different amino acid:

$$\textit{amino acid} + \textit{ATP} \longrightarrow \textit{AMP} \sim \textit{amino acid} + \textit{PP}_\textit{i} \tag{24-3}$$

2. Activated amino acids react in the cytoplasm with a ribonucleic acid called *transfer RNA* (tRNA) and an enzyme molecule that is specific for each amino acid:

$$\textit{AMP} \sim \textit{amino acid} + \textit{tRNA} \longrightarrow \textit{amino acid} \sim \textit{tRNA} + \textit{AMP} \tag{24-4}$$

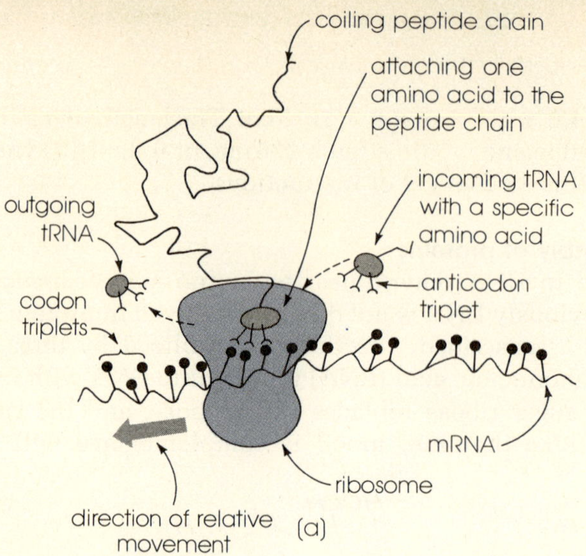

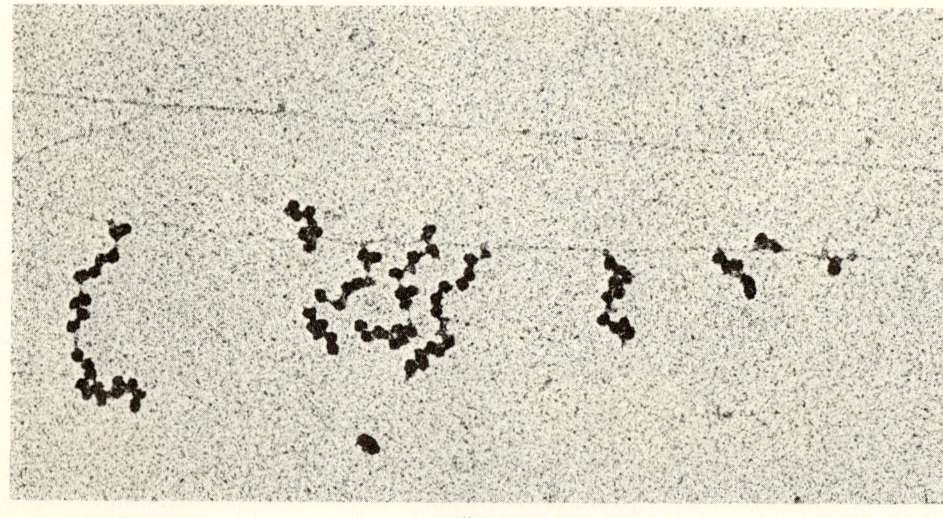

Figure 24-6 Ribosomal synthesis of protein. (a) Schematic view of transfer RNA molecules bringing amino acids to the ribosome where they are joined into polypeptide chains. Assembly follows a coded order as base codons of messenger RNA are interpreted by base anticodons of the transfer RNA. (b) Electron micrograph (×166,000) of chromosome segments (genes) from cells of the bacterium Escherichia coli. The upper fiber is an inactive length of chromosome, the lower fiber an active length. On the latter are strings of messenger RNA in the process of being transcribed. The dark granules attached to the strings are identifiable as ribosomes (they are destroyed if treated with ribonuclease (see Figure 23-13)). The ribosomes have been assembled from molecules made at other segments of the chromosome. At each point of attachment of the mRNA to the gene is a molecule of RNA polymerase, the enzyme that initiates transcription. Presumably attached to each ribosome is a growing thread of protein (a feature unresolvable at the magnification used.) RNA synthesis is proceding from right to left along the chromosome; protein synthesis is proceding from bottom to top along the mRNA strings. In prokaryotic cells like this one transcription of mRNA and translation into protein occur simultaneously at genes. It is interesting that the theory of protein synthesis was worked out biochemically and described in drawings several years before photographs were available for directly examining the mechanism. [(a) Adapted from R. W. Holley in M. Calvin and M. J. Jorgenson (Eds.), Bio-Organic Chemistry, Freeman, San Francisco, 1968. (b) "Visualization of Bacterial Genes in Action," O. C. Miller, Jr., et. al., Science, 169, 392–395 (July 1970). Copyright 1970 by the American Association for the Advancement of Science.]

Transfer RNA is a family of single-stranded polymers that contain about 75 base–sugar–phosphate units (mol. wt. ~ 25,000) called *ribotides*. Transfer RNA molecules are formed in the nucleus against specific lengths of one strand of DNA and carry away their message complements (uracil being substituted for thymine). Some 60 slightly different *t*RNAs have been identified, more than enough so that one or more is available to recognize each of the 20 different L-α-amino acids needed to build proteins.

3. Another kind of ribonucleic acid called *messenger RNA* or mRNA (mol. wt. 3×10^5 to 4×10^6) also forms in the nucleus on portions of the master DNA template. Its molecules then move to the cytoplasm where they attach themselves to ribosomes. Each amino acid~*t*RNA complex is guided by an *anticodon* of three bases on its *t*RNA part which recognizes three bases of a complementary codon on the mRNA chain. Loose anticodon–codon bonding is assumed established until the amino acid becomes tied into the polypeptide chain:

$$amino\ acid \sim tRNA + mRNA \longrightarrow mRNA\text{–}amino\ acid \sim tRNA \qquad (24\text{-}5)$$

4. The amino acids are linked together in the peptide chain using the energy of the ~ bond and enzymatic processes not yet fully understood. For each peptide link that is established, one water molecule is eliminated and one *t*RNA molecule returns to the cytoplasm where it can pick up another amino acid molecule. Linking begins at the N-terminal end of the polypeptide chain and proceeds one amino acid at a time to the C-terminal end:

$$mRNA\text{–}amino\ acid \sim tRNA \longrightarrow protein\ chain + tRNA + mRNA \qquad (24\text{-}6)$$

5. The mRNA codons are brought up to bonding position by movement between ribosome and mRNA. A number of ribosomes may move simultaneously down a single mRNA strand, each "reading" the triplet base code and synthesizing a protein chain (Figure 24-6). As the last ribosome leaves the mRNA, the finished protein is released.

6. Normal bond angles, hydrogen bonding, and other factors cause the protein to coil as it is being formed. On release an enzyme catalyzes sulfur–sulfur crosslinking (as in Figure 23-13b) that gives the protein chain its final characteristic shape. When synthesis is completed, mRNA depolymerizes, and the whole process is ready to start again.

24-4. NEW DIRECTIONS IN MOLECULAR BIOLOGY

Using the DNA theory

Discovery of the structure of DNA is one of the most significant events in modern science. Indispensable to genetics, it has also established a molecular basis for correlating a wide range of observations on living systems. We will briefly consider DNA theory as it relates to some current research aimed at understanding cancer, cell differentiation, cell mutation, antibiotics, genetically linked diseases, and aging.

Cells begin dividing when they start to manufacture DNA. A regulatory mechanism that controls the process in a normal adult keeps cell count fairly constant. In *cancer cells* control over DNA production is somehow lost and the cells multiply wildly. Cancers can be induced by radiation, carcinogenic chemicals, some particulate materials, and certain small *viruses*. Viruses represent DNA and RNA that is found outside cells. A virus is a noncellular parasitic particle composed of a nucleic acid core and a protein shell. Given the ability of injecting its DNA through a cell wall, it can cause the cell machinery to cease normal function and begin manufacturing virus particles. Sol Spiegelman and co-workers at Columbia University have demonstrated that genetic information in certain viruses that cause tumors in mice is similar to the genetic information in the cells of corresponding human tumors. Their experiments indicate certain kinds of cancer are viral diseases, and also support the more general theory which holds that viruses, by continually promoting the exchange of genetic material, may be responsible for the remarkable similarity of the genetic code used by all living species. (See the figure on page 557.)

Other experiments with laboratory animals seem to imply that instructions for the production of cancer-causing viruses may be unavoidably present at birth in genes intermingled with normal genes. Individuals who do not later develop cancer may possess repressor genes that prevent cancer from starting. Further research may identify the specific genes responsible.

The problem of *cell differentiation*—why one cell containing the same genetic information as another becomes, say, a brain cell instead of a muscle cell—will be closer to solution when the complete functions of nucleic acids are known. The specific character of a cell depends largely on the type and amount of protein it contains. Because each differentiated cell in a complex organism contains a complete gene inventory, this means that in each different cell a different combination of genes has been activated and through the RNA system has synthesized a different collection of proteins. The question then is, how are specific genes "turned on" while others are kept inactive? Some answers may have been found in hormones, many of which are able to act on genes. Other answers could lie in *histones*, small basic proteins that fit into grooves around the DNA helix (Figure 24-3b). In some coordinated manner the histones may "switch" genes on or off by allowing or preventing sections of the helix from uncoiling. Their action could regulate the synthesis of messenger RNA and therefore dictate the proteins to be synthesized. We have estimated each human cell contains enough DNA to code for 6 million proteins. Only about 10 percent of the DNA is thought to be functionally active, however.

When base sequence in the DNA of a gene becomes mutated (altered) and the mutation persists in later generations, changes may become noticeable in cell functions and even in whole organisms. Mutation can be caused by radiation (see Figure 4-16). Radiation breaks bonds and also produces reactive hydrogen and OH radicals from cell water. In either case DNA structure can be damaged. Chemicals that cause changes in nucleic acid structure are called *mutagens*. Mutagens may substitute for bases, crosslink base pairs, react with bases to prevent pairing, or in other ways harm DNA function. Pesticides, defoliants, food additives, industrial chemicals, tobacco "tar," air pollutants, and such drugs as tranquilizers, oral contraceptives, antibiotics, hallucinogens, narcotics, and antinausea preparations are among the things recently investigated for potential mutagenicity. Compounds deliberately synthesized to disrupt cellular processes are certain anticancer drugs, which probably destroy cancer cells by producing chromosome breaks in them. Actually no chemical has yet been proved to induce mutations in human sex cells. However, 25,000 chemicals are in common use—the average American ingests 3 pounds of food additives a year, for example—and a few thousand new compounds are being synthesized yearly, so potential for genetic disturbance certainly exists. If damage does occur, the results will not be immediately obvious because in man most genetic mutations involve recessive genes. Recessive genes, as visualized in principle by Mendel (Figure 24-1), are those that produce an observable effect only if both parents transmit the same abnormal recessive gene to the child, and that is an unusual occurrence in humans. Four percent of all babies born annually in the United States (140,000 out of 3.7 million) show significant defects, of which about four-fifths are related to genetic abnormalities.

At present there is no absolute procedure available for evaluating mutagenic hazards in humans. For example, studies indicate caffeine (see caption, Figure 23-1) and sodium nitrite (a meat preservative) cause chromosome breaks in laboratory organisms as well as in human somatic (general-body) cells. (On the other hand, nicotine, ethyl alcohol, and the antibiotic penicillin apparently cause no chromosome breaks.) Caffeine may interfere with the mechanism that patches up DNA ruptures by bonding to sites ordinarily held by adenine. Nitrite may act by removing the —NH_2 groups of adenine, guanine, and cytosine. As removal would prevent normal base pairing, it would also blur the genetic message. Fortunately, nucleic acids are highly resistant to natural disruptive effects such as thermal agitation, and enzymes exist that make quick repairs of breaks. It is estimated that if an undisturbed piece of DNA could carry out its normal copying duties indefinitely, on average it would make only one mistake every million years.

Another kind of interruption in the genetic message system is found in the action of *antibiotics*. An antibiotic is a compound taken from a microorganism that inhibits the metabolism of other microorganisms. Experiments show that streptomycin, for instance, somehow distorts or alters the subunit of ribosomes that holds *m*RNA, and it garbles the reading of *m*RNA codons. In one experiment synthetic *m*RNA was made solely from uridylic acid (Table 24-1). Used to prepare a polypeptide from a mixture of amino acids, the message of its only codon · · · U-U-U · · · yielded a polypeptide chain containing only phenylalanine. With streptomycin present, however, the polypeptide produced contained 40 percent isoleucine. The result suggests that antibiotics stop bacterial growth by causing incorrect protein to be produced.

No one yet knows why living organisms including humans age, possibly because the aging process is too complex to be explained by a single mechanism. Neither has it been possible to extend appreciably the normal lifetime of a human being, although the life span of some cold-blooded organisms has been prolonged by keeping them a few degrees below normal temperature, and the lives of mice have been extended by underfeeding or including in their diets antioxidants such as vitamin E (Figure 22-10c), butylated hydroxyanisole (BHA), and butylated hydroxytoluene (BHT):

A point of obvious agreement is that each organism has its own built-in timetable of aging (insects may last a maximum of a few weeks, man a hundred years or so, the giant sequoia thousands of years). Thus far no "aging genes" that regulate life span have been found but scientists are on the outlook for them. Another theory of aging assumes that DNA simply wears out, develops unrepaired breaks (possibly through natural ionizing radiation) and programs assembly of increasingly defective RNA and proteins, including enzymes. Presumably those building blocks would then be unable to construct normal cells. Also they would not be recognized as familiar molecules by the body's *immune* (self-defense) system and the system would produce *antibodies* to destroy the foreign invaders. By combating its own molecules the body would be accelerating its own aging. Orderly DNA function could likewise be impaired by failure of the histones that work with DNA to do their job. Some experiments show that histones become progressively more tightly bound with time. This may account for the observation that older cell populations contain fewer cells capable of incorporating radio-labeled DNA precursor molecules and dividing as they did in early age. Defects in DNA and other polymers would also be created if the polymers became crosslinked by free radicals produced with radiation, or by reaction of their active groups with small polyfunctional molecules such as malonaldehyde; molecules of this kind can come from incompletely oxidized food:

$$(24\text{-}7)$$

biopolymers   malonaldehyde   superpolymer

Although deterioration of DNA with time sounds logical, protein taken from persons of various age groups seems about the same. Old people apparently do not contain any significant amount of abnormal proteins.

If all causes of death except the general infirmities of old age were eliminated, life expectancy would be extended a decade or more. The added years would be precarious, however, marked by increasing feebleness. But if muscle tone, vigor, and flexibility could be retained into advanced age, people would then have to continue finding useful and interesting things to do. More research is needed in these areas.

Genetic engineering

Within a decade or two molecular geneticists will learn the location and code of some specific genes in human chromosomes. It is also expected they will be able in some cases to eliminate faulty genes by microsurgery, perhaps using as a scalpel the heat of a laser beam. Experiments already have demonstrated the feasibility of introducing new genes (including man-made varieties) into mammalian cells, carrying them via bacterial or animal viruses or simply by fusing two somatic cells together. Perfection of the techniques in the next century could lead to cures for some of the flaws created and inbred during ages of evolution.

Gene therapy will probably first be directed toward alleviating genetically recessive disorders in the newborn. More than 2 percent of humans suffer from genetic disorders. At least 1500 different diseases are known to be inherited according to Mendel's laws, and new examples are being found every year. The list includes diabetes, cystic fibrosis, sickle-cell anemia, and phenylketonuria. In sickle-cell anemia, a serious blood disease, a single defective gene directs synthesis of an abnormally insoluble hemoglobin that differs from normal hemoglobin by only one amino acid residue in a chain of 300. In phenylketonuria, a condition characterized by the non-synthesis of the enzyme responsible for the following oxidation,

$$\text{phenylalanine} \xrightarrow{(O)} \text{tyrosine} \tag{24-8}$$

the physiological result is usually a child with severe mental retardation.

The challenge in such cases is to rewrite defective codes on bits of DNA and position the bits properly in the organism. It might be done by removing groups of somatic cells from the diseased person, correcting their genetic message, then replacing them. Dangers attendant to the procedures are that the new genes might prove to be mutagenic and the viruses used as DNA vectors (carriers) might be cancer inducing. Years of laboratory work lie ahead before gene engineering can hope to become practical.

A synthetic virus indistinguishable from its natural model has been made by Spiegelman and used to infect bacteria and to produce new viruses. Khorana has synthesized a gene, which is a double-stranded DNA molecule containing 77 nucleotides. Research like this hints that perhaps within the next hundred years man will stand at the place where he understands genetic manipulation sufficiently to control his own evolution. That will pose momentous questions. What kinds of desirable traits should man program into himself—reasonableness, physical strength, mental agility, delayed senility? Or will he go further and create superman with the genes to produce, say, a larger brain, a stomach able to digest cellulose should other food become scarce, or ability to regenerate worn-out organs so he can repair himself? What standards should he set for change and who will do the deciding? How will the changes affect society, international rivalries, and such institutions as marriage and the family?

(a)

(b)

(c)

(d)

24-5. NUTRITION AND POPULATION

Dietary essentials

In order for the human body to extract oxidative energy and to build, maintain, and regulate itself, it must be furnished food that contains a variety of elements and molecules. The essentials are as follows.

1. *Proteins* come from meat, soybeans, milk, etc. About half the amino acids known (Figure 23-12) are not synthesized in the body and therefore must be supplied in the diet. Milk protein contains all the essential acids.
2. *Carbohydrates* are the starches and sugars from grains, vegetables, etc. Carbohydrates are the least expensive foods and generally the main source of body energy.
3. *Fats* are found in dairy products, nuts, etc. Linoleic (Table 23-1) and arachidonic acids are regarded as essential because unsaturated acids are not synthesized in the body.
4. *Vitamins* come from many food sources. Vitamins are essential in small amounts to maintain body physiology. Their need is associated with the building of coenzymes such as coenzyme A (see caption with Figure 23-5).
5. *Minerals* required number at least 19 elements other than carbon, hydrogen, oxygen, and nitrogen. Included are iron, sodium, chlorine, potassium, copper, iodine, zinc, magnesium, manganese, vanadium, silicon, sulfur, and cobalt. Approximately 85 percent of the body's minerals are in the calcium phosphate of bones and teeth.

Figure 24-7 Molecular geneticists. (a) Arthur Kornberg (United States 1918–   , Nobel prize in medicine or physiology 1959). Kornberg has worked on the biosynthesis of nucleotides and DNA. In 1967 he succeeded in causing a single-strand DNA of a virus to produce a complementary DNA strand from a nutrient solution. Isolating the synthetic strand he used it in a nutrient solution to prepare a second generation strand. The totally synthetic DNA was found to infect bacteria just as the original virus did. One application of the technique will be duplication of DNA's of cancer-producing viruses so they can be studied. (b) H. Gobind Khorana (India and United States 1922–   , Nobel prize in medicine or physiology 1968). Now at MIT, Khorana is perhaps the foremost expert in nucleotide synthesis, having synthesized all 64 possible trinucleotide codons. He has also prepared high-molecular-weight DNA-like polymers with known sequences and, from analysis of proteins they synthesized, proved that the genetic code is read, consecutively, three bases at a time. (c) Marshall Nirenberg (United States 1927–   , Nobel prize in medicine or physiology 1968). Nirenberg demonstrated the relationship between synthetic RNA and the synthesis of specific proteins. He also worked out an important research technique for synthesizing proteins without cells. (d) Robert Holley (United States 1922–   , Nobel prize in medicine or physiology 1968). Now at Salk Institute for Biological Studies, Holley and co-workers were the first to determine the complete structure of a specific transfer RNA. They did it by enzymatically hydrolyzing the tRNA with ribonuclease (Figure 23-13), separating the fragments by column chromatography, hydrolyzing the fragments with alkali, then separating and identifying the resulting nucleotides by paper chromatography and electrophoresis (Figure 23-14). He is shown holding a twisted wire model representing a possible RNA configuration. Behind him is printed a hypothetical model of a transfer RNA. (Photographs; a, courtesy Stanford University; b, courtesy University of Wisconsin; c, courtesy the National Institutes of Health; d, courtesy Cornell University.)

Food energy needed is measured in kilocalories (also called *large calories*, abbreviated Cal). For normal activity a young American male adult, 6 ft, 1 inch tall (185 cm), and weighing 176 pounds (80 kg), requires about 3000 kcal per day. His female counterpart, 5 ft, 6 inches (168 cm), and 132 pounds (60 kg), requires about 2200 kcal per day. Caloric values of foods are determined with a bomb calorimeter (see Figure 12-3). A correction in heat evolved is made for the oxidation of nitrogen because in the human body nitrogen is voided mainly in reduced form as urea, $(NH_2)_2CO$. Caloric values per gram are lipids 9.3, carbohydrates 4.1, proteins 4.1. If during the day a person consumes 50 g of fat, 500 g of carbohydrate, and 100 g of protein, his caloric intake is $465 + 2050 + 410 = 2925$ kcal.

Undernutrition and mental retardation

Half the world's population is underfed. In Africa, Central and South America, India, and southeast Asia the toll of hunger has historically outdistanced the toll of war. In some countries death by starvation is commonplace. Survivors of famine are stunted physically; many are also stunted mentally.

Prenatal brain cells increase in number as measured by chemical analysis of DNA phosphorus. Postnatal brain cells increase in size as measured by analysis of cell cholesterol. Both kinds of growth depend on adequate diet. Experiments with underfed humans are difficult to make because emotional factors cannot be excluded, but experiments with rats have shown that (1) low *caloric* intake does not affect problem-solving ability; (2) low *protein* intake does affect problem-solving ability; (3) offspring from protein-deprived mothers have lower than normal learning ability; and (4) adequate diet later in life after early deprivation restores neither normal body size nor normal mental ability.

Tragic implications in human biology are observable in undernutritioned youngsters in primitive areas in the tropics. They are slow responding, incurious and anemic, have vitamin-related deficiencies, and often have diarrhea which increases the problem of absorbing nutrients. Their stature is short and their brains give irregular curves with the electroencephalograph (EEG) machine, a device for measuring brain electric potentials. They are people destined for subnormal lives.

Correlation of brain function, structure, and composition has only begun to be made by scientists. Brain nutritional requirements are incompletely known and probably vary among individuals. Pauling has pointed out directions research might take when he said,

> I believe that mental disease is for the most part caused by abnormal reaction rates, as determined by genetic constitution and diet, and by abnormal molecular concentrations of essential substances. Significant improvement in the mental health of many persons might by achieved by the provision of the optimum molecular concentrations of substances normally present in the human body.

An example is furnished by recent experiments with rats which established that mental disorders are likely unless the amino acid tryptophan (from protein) is supplied to the brain via circulating blood. In brain cells tryptophan is used to synthesize serotonin:

$$ \text{(24-9)} $$

tryptophan   serotonin

Serotonin is a powerful hormone involved in the control of sleep and other essential rhythmic functions.

Food and demography

> I think I may fairly make two postulata. First that food is necessary to the existence of man. Secondly that the passion between the sexes is necessary, and will remain nearly in its present state. Therefore . . . the power of population is indefinitely greater than the power in the earth to produce subsistence for man.

The grim prophesy that the human race would helplessly increase to the limit of its agriculture and be kept at that level by famine was published in 1798 by English economist Thomas Malthus. At first widely discussed and feared, it tended to be forgotten as the Industrial Revolution of the nineteenth century promised an escape route through increasing national incomes and decreasing birth rates due to urbanization. More recently, however, the Malthusian specter has been revived by scientists called demographers, who make quantitative deductions from the vital statistics

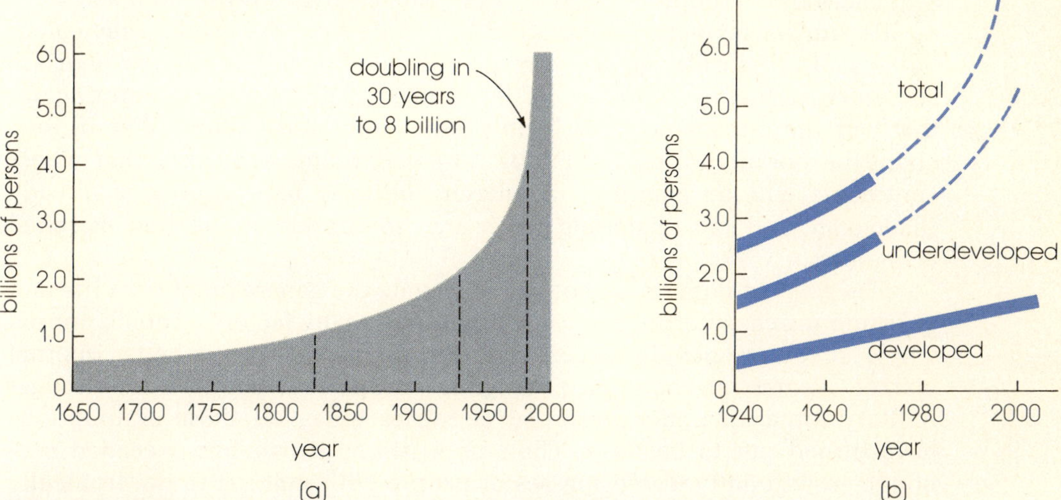

Figure 24-8 Statistics for demographers. (a) Growth of human population. For most of man's existence birth rate has been barely more than death rate. Before 1800 population never grew faster than 0.5 percent per year. In some countries today the growth rate is seven times that figure. From his very beginning to about 1825 modern man increased in number from a few individuals to a total of 1 billion. From 1950 to 1969 he increased from 2.6 to 3.6 billion. Thus what once took 10,000 years to do, recently took 19 years. The next net gain of 1 billion will surely be added by 1980. (b) Comparison of population growth in developed and underdeveloped nations. The latter are growing twice as fast as the former and by the year 2000 will probably contain 78 percent of the world's population.

of populations. Citing trends like those of Figure 24-8, some of them are predicting that hundreds of millions will die of starvation within the next few decades unless extensive efforts are instituted immediately to improve both the output and distribution of food world-wide, and to decrease the rate of population gain. With the majority of mankind existing in the lesser developed countries and multiplying there the fastest, undernutrition in all its ramifications is already a major cause of death. In an average day on earth the balance sheet presently reads: 327,000 human births, 12,000 deaths due to starvation and malnutrition, 124,000 deaths due to other causes; net gain 191,000 people.

Optimists tend to view population growth as a natural prod to progress which should not be interfered with. They feel everyone can be fed if more land is cleared, harmful insects controlled, better grains planted. Using Japan's present nutrition as a standard, they say the earth could support 40 times as many people as it does now. As evidence that technology is keeping ahead of the prospects of famine, they cite harvests of fish meal protein from the sea, experiments on increasing the photosynthetic yield of green plants, research in the microbial production of high-protein edibles from mixtures of petroleum and ammonia, and the *Green Revolution* — the introduction of high-yield wheat and rice developed by American research.

Somewhat less optimistic observers see the population problem as serious but think it can be solved if time (perhaps a decade) bought by the Green Revolution is used intelligently. They call for radical changes in agricultural technology based on new hybrid grains, fertilizers, and intensive use of natural resources. They want the world educated to the consequences of overpopulation and to the use of birth control methods.

Pessimists say world population is already over the brink. They calculate that if all worldly goods were now divided equally, the very poor are so numerous that everyone would be reduced to poverty level. Pointing out that half the human race has been present only since World War II, they claim the population curve (Figure 24-8a) is rising so steeply that whole continents will be engulfed in anarchy because food supply is already inadequate and raw materials too scarce to support more than marginal economies.

The goals of population control to elevate the general quality of life and to secure universal affluence are hindered by many factors — beliefs, for example, that more people assure a country of more wealth, a better internal market, greater international prestige, a cheap labor pool, and a stronger military force. In underdeveloped countries today all these assumptions have turned out to be false. Their population growth has exceeded economic growth and created masses of people with nothing to do. Ironically the Green Revolution itself is creating some of the trouble. Facilities are not available to handle large harvests. And irrigation, mechanization, fertilization, and other costly requirements of the new grains can only be met by big landowners through consolidation of small farms. As farm employment has dropped, mass movement has taken place to the cities where slums lacking services of all kinds have burgeoned into festering sores. Two-fifths of the world's population, many in the circumstances just described, are illiterate, vocationally untrained, and without hope of getting any education or finding employment. When life becomes a daily struggle for existence, political stability vanishes and civil strife follows.

Globally cultivation takes place on 3.3 billion acres of land. Another 4.5 billion acres classed as potentially arable is not being cultivated because the cost of making it usable is too high. Deserts require irrigation, which can only come from expensive desalination and pumping plants (see Figures 4-22 and 15-5). Jungles need clearing and topsoil needs replacement because that which is lateritic (L. *later*, brick) is a mixture of leached oxides of silicon, aluminum, iron, and titanium that becomes rock hard when plowed and exposed to air. New varieties of grains will have to be tailored to virgin soils; they should have high protein yields and be resistant to infestations that invariably appear. Roads and distribution centers must be built, machinery imported, and people taught to manage land continuously without destroying its potential.

Despite such massive efforts, the earth will not accommodate increasing population indefinitely. Neither can technology stay ahead forever in the production of food and energy and disposal of wastes. Inevitably man's dignity must diminish as growth brings increases in diseases, crowding, and crime. (Tokyo, the world's largest city, records 300,000 crimes yearly.) At the indicated rate of multiplication the population of the world will double by 2006, and be seven times today's total in 100 years. Many suggestions have been made for halting the trend. Some of them are (1) development of mass fertility control chemicals for use in public water supply systems; (2) temporary sterilization of all women by implantation of capsules containing slow dissolving steroids (present available effectiveness is two to three years); (3) compulsory sterilization of all men with three or more living children; (4) incentive programs such as paying couples not to have children and taxing those who do; (5) a requirement that women work outside the home with jobs of suitable status to provide interests supplemental to marriage and children; (6) a requirement that birth control programs accompany financial aid to foreign countries; and (7) greater efforts to lower infant mortality rates so fewer births would be considered necessary in lands where children are their parents' only old age security.

These and other proposals are open to questions of social–moral acceptance, cost, availability of technology and trained personnel, and administrative feasibility and effectiveness. When a government vigorously pursues an acceptable birth control policy, population growth has been slowed. Japan's program, based on legalized abortion, has been the world's most effective. With 0.9 million abortions balancing 1.8 million births each year, population growth rate since World War II has been less than 1 percent. Most of the rest of Asia has been unwilling or unable to follow Japan's example. India has had government support for family planning programs for 20 years, but the impact has been small. Already burdened with 15 percent of the world's population on 2.4 percent of the world's land, India is growing by 200,000 people a week. China, with 20 percent of the world's population, is growing even faster. In 25 years there will be 1.2 billion Indians and 2 billion Chinese. Two-thirds of all peoples will then live in Asia.

24-6. INTO THE NEXT CENTURY

Models for world dynamics

Two approaches for discussing the future of mankind have already been introduced: (1) The pessimistic or Malthusian view that claims the earth has limited resources and thus a high birth rate will eventually be balanced by a high death rate because population must eventually exceed productivity, and (2) the optimistic or technological view that says as invention and mass production increase living standards, birth rates will drop and population will level off, being balanced by low death rate.

Until recently the future has been looked at intuitively. Today it is being contemplated with the aid of computer modeling. The essence of the best known study of 1972 is summarized in Figure 24-9. Its authors began with five physical quantities that they considered most important in a world system: population, nonrenewable resources, food, industrial output, and pollution. Next, they wrote mathematical equations interrelating the quantities. Then they set a computer to solve the equations. Given the exponential nature of most changes (world population, for instance, is doubling every 33 years), the model predicts a very difficult future: As nonrenewable resources are exhausted, food supply will decrease, pollution will increase, and population will be be cut drastically by a climbing death rate. Even when the equations are manipulated to include universal birth control, 75 percent recycling of all resources, abundant nuclear power, reduction of pollution to 25 percent that of the 1970 level, and doubled present food production, the model forecasts that a decrease in resources and increase in pollution will compel a downturn in population sometime in the twenty-first century. The only hope seems to lie in halting growth of both population and industrial output now.

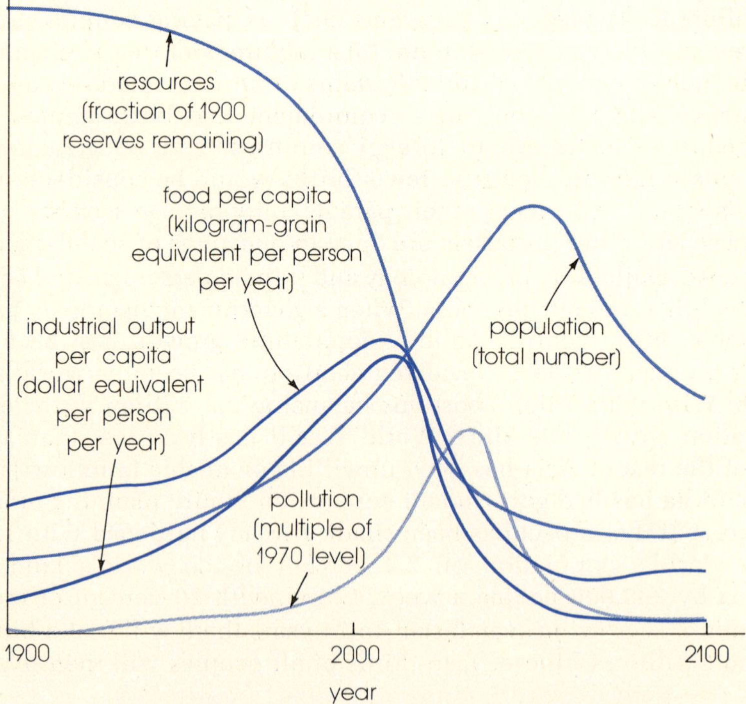

Figure 24-9 A computer-generated dynamic model of global interactions by MIT scientists that included J. Forrester and D. Meadows. The variables, from different charts but in the units indicated, are plotted on the same graph for easy comparison of trends. (From *Limits to Growth,* Universe Books, New York, 1972, as adapted in *Chemical & Engineering News,* March 6, 1972, p. 2.)

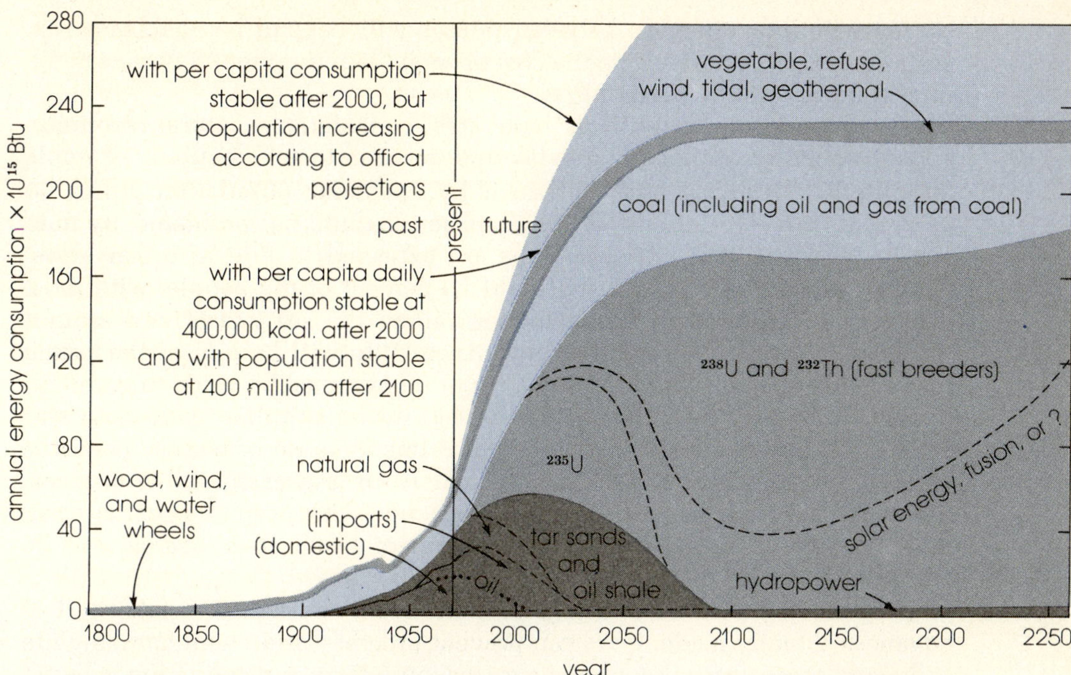

Figure 24-10 Supply and demand of energy in the United States. (After E. Cook, Texas A & M University, and reproduced from Chemical & Engineering News, Jan. 10, 1972, p. 28.)

Critics point out the model does not take into account that (1) resources are not really consumed, only diluted; (2) given unlimited clean energy (solar, hydrogen), "exhausted" resources will be recoverable from dilute sources; (3) when resources become scarce, their rise in price will lead to substitutes; (4) computers cannot predict how human attitudes will change, and people can make value judgments capable of preserving their future; (5) the equations are sensitive to changes in basic assumptions, so that non-catastrophic computer programs can also be written; and (6) the world is, after all, too big and complex to be modeled with only five variables. Further studies will certainly be forthcoming.

We have looked at energy in several contexts and the discussion above reminds us how critically man's future is linked to his use of energy. A final comment is presented in Figure 24-10 which is a model of future energy sources and their usage in the United States. It assumes a rise in our total energy consumption to four times present demand, followed by a leveling off by the year 2100 when presumably population will be stabilized at 400 million. The model's author envisions events proceeding in three stages.

1. A mining stage in which oil and natural gas will disappear by 2025, tar sands and shale by 2090, and uranium-235 by 2080. Coal will last for a few hundred years but finally mining will come to an end with nothing left to mine.
2. A uranium–thorium stage that will also last for a few centuries through use of breeder atomic reactors. Energy output will be limited by the problem of radioactive waste disposal.
3. A hydrogen–solar stage in which hydrogen fusion and solar energy will be used. The latter would last as long as man or the sun is here.

Growth and transition between phases will depend on cost. Most experts agree cost must include the cost of social–environmental impact or future living cannot be harmonious.

The United States, with its large area, considerable natural resources, and advanced technology is in a favored position to set a national example for orderly advance. Nevertheless, if by 2000 our population grows 50 percent (to 310 million), it can generate serious internal problems. By then almost 90 percent of all Americans are expected to live in urban areas. Only 10 percent of the land will hold 70 percent of the people, with most persons concentrated in megalopolises along the sea coasts and around the Great Lakes. How to lead a meaningful life will be one of the prime social concerns. For national health major efforts must be made to maintain livable cities, provide better social services, insure supplies of clean air and water, build mass transit systems, fight crime, dispose of wastes, conserve and reuse natural resources, design products for greater durability and easier repair, save the beauty of nature, and cope with psychological stresses that increase as we change the environment faster than change can be assimilated.

Popular predictions of things to come seem invariably to fall short of advances actually made. If we can prevent progress from dehumanizing life and avoid mass death due to nuclear war, pollution, and starvation, a better world is possible through wise use of the knowledge of science and power of technology. Although science is a newcomer among man's intellectual pursuits, it is a uniquely fundamental and truthful means for understanding both nature and ourselves, and thus a necessary part of the education we need to make judicious decisions. As science goes forward, perhaps humanity will also progress by becoming more intelligent. We are, after all, only newcomers, too, and may have a long future in which to realize a fuller potential.

In the words of Peter Medawar (England 1915–   , Nobel prize in medicine or physiology 1960),

> If we imagine the evolution of living organisms compressed into one year of cosmic time, then the evolution of man occupied one day. Only during the past 10 to 15 minutes of the human day has our life on earth been anything but precarious. . . . Only during that time has there been progress, though, of course, it doesn't amount to very much. We cannot point to a single definitive solution to any of the problems that confront us — political, economic, social, or moral, that is, having to do with the conduct of life. We are still beginners, and for that reason may hope to improve. To deride the hope of progress is the ultimate fatuity, the last word in poverty of spirit and meanness of mind.

QUESTIONS

1. Nucleic acids What is a nucleic acid? What general kinds of chemical units are in it?

2. Nucleic acids What do the abbreviations DNA and RNA stand for? How do the two substances differ from one another?

3. Memory Rats normally prefer a dark place rather than a lighted place. Rats have been trained (by electric shock) to fear the dark. Extracts of their brains injected into untrained rats cause the latter to fear the dark. Pretreatment of the extract with ribonuclease does not change its ability to transfer information. Pretreatment of the extract with trypsin does destroy its ability to transfer information. **(a)** What do the experiments seem to indicate? **(b)** Suppose chemicals responsible for memory become known and are synthesized. Would you favor implanting them in humans to speed up learning? If they tended to normalize the mentally retarded would you favor their use?

4. Evolution Jean Lamark, a French naturalist, developed a theory of evolution about 1800 in which he proposed that acquired characteristics were inheritable. Karl Marx and Friedrich Engels, founders of modern socialism and communism, adopted Lamarkism into their philosophy because it promised that the influences of teaching and environment given to one generation would be preserved in the next. In 1948 Lamarkism was declared the official biology of the U.S.S.R. Does molecular genetics, as you understand the subject, support this theory? Explain.

5. Gene alteration Salvador Luria (Italy and United States 1912–   , Nobel prize in medicine or physiology 1969) has envisioned the possibility of *gene warfare*. Nation *A* in theory could prepare a gene capable of sensitizing humans to a common substance such as carbon dioxide. The gene would be connected to a virus vector and released in the air over nation *B*, scheduling for extinction everyone whom the virus infected. You are a member of a congressional committee charged with funding scientific projects. Would you appropriate money to study gene warfare? Explain.

6. Gene changes If gene manipulation in man becomes possible, do you believe there would be a tendency to lose genetic variability, that is, to homogenize society? What would be the advantages and disadvantages of a society more homogeneous than we now have in the United States?

7. Gene alteration Nirenberg has written, "When man becomes capable of instructing his own cells, he must refrain from doing so until he has sufficient wisdom to use this knowledge for the benefit of mankind." When do you think that will be, and how would you know that time had arrived? What cell instructions do you think would be beneficial?

8. Cloning Cloning is the process in which virtually exact genetic copies of living things are produced asexually from somatic cells (which contain complete sets of chromosomes). Complete trees have been grown from a single tissue slice, for example, and complete frogs from transplanting the nucleus of a frog intestinal cell into an unfertilized frog egg cell. Some time it may be possible to clone humans either in donor mothers or in artificial wombs. A nation then might duplicate its best workers, warriors, scientists, and athletes. Discuss some possible international implications of cloning.

9. Cloning Try to imagine being one of several children in a family cloned from your parents. What problems do you see living in a house of duplicates?

10. Sex A method seems sure to be available soon whereby parents will be able to preselect the sex of their offspring. **(a)** Do you think this will change today's male–female ratio in the United States? **(b)** If so, what might be some of the consequences?

11. Enzymes The cell wall of certain bacteria is a single glycopeptide molecule having a molecular weight upward of 100 billion. Its structure consists of a network of sugar units oriented in one direction with cross-linking peptide units oriented at 90 degrees. The peptides sometimes contain only residues of D-glutamic acid, D-alanine, L-alanine, and L-lysine. Peptide bonds in proteins are digested in animals by enzymes like trypsin and pepsin. The same enzymes do not attack the peptides of bacterial walls. Suggest and explain a reason for this.

12. Demography Explain the shape of the curve in Figure 24-8a, why it has a long low plateau and why it suddenly becomes almost vertical.

13. Birth rates In 1970 the lowest birth rates in the world (about 14 births per 1000 population) were found in Hungary, Rumania, Bulgaria, East Germany, and Czechoslovakia. The highest rates (about 52 births per 1000 population) were in Togo, Sudan, Niger, Ivory Coast, Rwanda, Mali, Zambia, and Kenya. Suggest factors in the two areas that might account for the extreme contrast.

14. Food The food crisis today is most acute in the tropics and subtropics, where plant life may be lush. **(a)** Explain the anomaly. **(b)** Cite reasons why the food problem there is not easily solved.

15. Demography Graph (a) (Figure 24-11) refers to developed countries, graph (b) to underdeveloped countries. **(a)** Give reasons for the trends in each. **(b)** What assumptions can you make from these data?

16. Demography The charts in Figure 24-12 are called age–population graphs. Which represents a rapidly expanding population, which a decreasing population? Explain.

17. Heredity Do you think people with serious hereditary diseases should be sterilized? Do you think that couples whose genetics make deformed offspring likely should refrain from having children? Discuss.

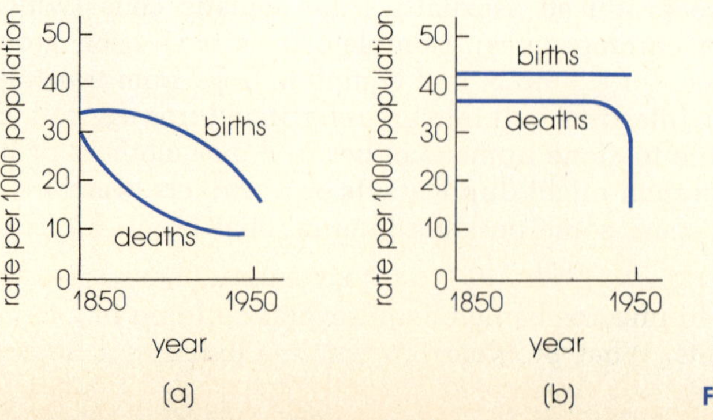

Figure 24-11

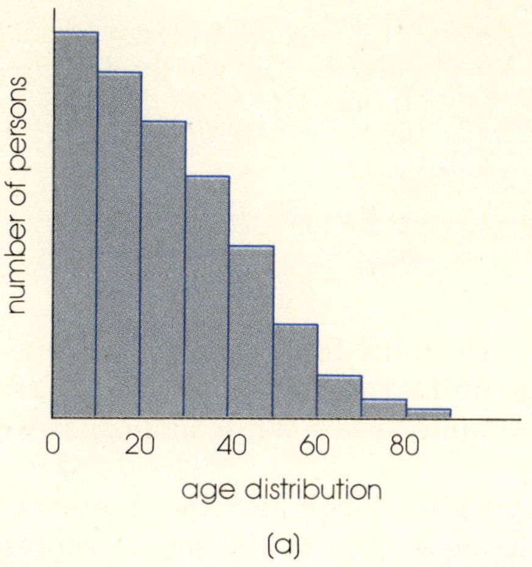

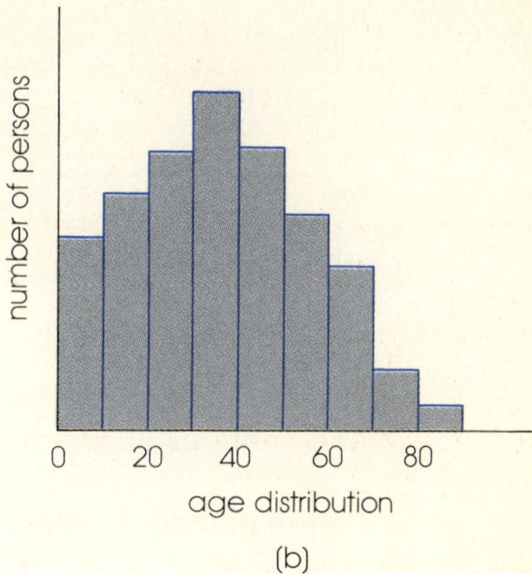

Figure 24-12

18. Futuristics It is sometime in the future. Nation A is industrially poor. It has, however, the largest natural deposits of a mineral that advanced nations critically need and have been buying. Due to unmanageable population growth in nation A, its social and economic systems are disintegrating and anarchy seems near. What do you think will happen? Is your prediction pessimistic or optimistic?

19. Families In 1967, 30 nations within the United Nations agreed to The Universal Declaration of Human Rights, which in effect stated that family size is solely a family matter and not to be dictated by government. What is your opinion of the declaration?

20. World model Is the future, as predicted by Figure 24-9, pessimist-Malthusian or optimist-technological? Explain. Do you concur with its view? Explain.

21. Population It can be argued that the planet should be a *commons* and that every living person should share equally in the available territory and available food. It can also be argued that national boundaries must be maintained to prevent a disastrous breeding race in which the fastest breeders would take over the earth. Develop one side of this argument.

22. Research **(a)** If you could have been laboratory assistant to any of the famous scientists mentioned in this book, whom would you choose? Why? **(b)** If you could do research work on any scientific problem facing man today, which one would you choose? Explain.

PROBLEMS

23. Mendelian genetics In the manner of Figure 24-1b show the results of crossing **(a)** hybrid tall plants with pure short plants; **(b)** hybrid tall plants with pure tall plants.

24. DNA **(a)** Write the formula for a nucleotide unit of DNA in which the

Table 24-2. Birth rate data

Year	Birth rate per 1000 population, United States	Year	Birth rate per 1000 population, United States
1900	32	1960	23
1920	27	1970	18
1940	18	2000	28 (est.)

base is a pyrimidine. Calculate the unit's formula weight. Repeat with a purine base. **(b)** Average the two unit formula weights. **(c)** If a single DNA strand contains 30,000 nucleotide units, estimate its molecular weight.

25. Base pairing In representing the DNA polymer, the abbreviations A, T, C, and G are used. **(a)** What compounds do the letters represent? **(b)** Given a single DNA strand that begins with the sequence ATGGAC-GTATTC, write the sequence of the complementary DNA strand. **(c)** If mRNA is made from the DNA strand described in (b), what will its sequence be?

26. Permutations A certain small polypeptide contains one amino acid residue we will call M, four residues called I, four called S, and two called P. The sequence, as determined with a sequencing machine, is MISSIS-SIPPI. How many permutations of the eleven residues are possible (see Eq. 23-25)?

27. Permutations Silk is a natural protein. Hydrolysis yields only four different amino acids: glycine and alanine in a ratio of 2:1 and comprising 75 percent of the number of residues, and serine and tyrosine in a ratio of 1:1. **(a)** What is the simplest ratio by which the four can combine to form a peptide having the same analysis as silk? Explain your method. **(b)** How many permutations are possible for this peptide? Show calculation.

28. Codon (a) Suppose the genetic code consisted of four nitrogen bases read two at a time. How many different codons would be possible? Why is this not enough to program the synthesis of proteins? **(b)** How many codons would be possible if the four bases were read four at a time? Show calculations.

29. Population Fifty-two births per 1000 persons per year in a general population is close to the maximum that can be expected. List the factors that make this true.

30. Birth rate (a) Graph the data ("year" as abscissa) in Table 24-2. **(b)** Suggest a reason for each peak and valley in the graph, based both on your knowledge of events in American history and on demographic principles.

APPENDIX

MEASUREMENTS

A-1. Origin of the metric system

Length, mass, time, and temperature are the fundamental measures with which the system of scientific measurement is constructed. Systemization began in 1670 with Abbé Gabriel Mouton, a French astronomer and mathematician, who conceived of a decimal system based on a single, unchanging distance measurement of the earth. After more than a century in limbo his plan was finally implemented during the French Revolution. The primary unit was called the *meter* (Gk. *metron*, a measure). It was defined as a certain small segment of the quarter *meridian** passing through Paris. During a 9-year period, a portion of the line extending from Dunkirk, France, south to Barcelona, Spain, was surveyed accurately and the quarter-meridian distance was calculated. The meter, selected to be one

---

* A meridian is a line describing a great circle on the earth which passes through both poles and any other selected point.

ten-millionth of that distance (a length equivalent to 39.37 inches), was then ruled for easy reference on a metal bar. A cubical box 0.1 meter on a side was designated as the volume standard. Its capacity was called 1 *liter*. The mass of 1 liter of distilled water at 4°C was called a *kilogram* and designated the standard of mass. Abbé Mouton's concept was thus fulfilled; the units of length, volume, and mass were interrelated in terms of a fraction of the earth's dimension.

Weight is a measure of the force with which an object is attracted by the earth. Weight therefore varies with distance from the center of the earth. The weight of an object is measured by suspending the object on the pan of a spring balance. *Mass* is the quantity that expresses an object's resistance to any change in its condition of motion or rest. The mass of an object is measured by placing the object on one pan of an equal arm balance and counterbalancing it with a body of equal, known mass on the other pan. Mass is absolute; it does not vary with gravitational attraction. The mass of an object is often also called its *weight*. If doubt in meaning exists, the word *mass* is used.

The unit of *time*, the second, had been in use before proposal of the metric system, and was adopted without change. It was defined in terms of earth–sun position as a portion of a mean solar day.

The Fahrenheit and Celsius *temperature* scales were also already in use in the eighteenth century. In the nineteenth century the Kelvin scale was invented. (The three scales are described in Sections 14-1 and 14-2, and compared in Figure 14-7.)

Table A-1. Four basic units of measure; old and new metric systems compared

Measure	Unit	Definition
Length	Meter, m	*Original:* 1/10,000,000th of the quarter meridian passing through Dunkirk and Barcelona, the length marked off on a platinum-iridium alloy bar kept at Sevres, France. *IS:* 1,650,763.73 wavelengths, in vacuum, of the reddish-orange emission line of krypton-86 excited at −210°C.
Mass	Kilogram, kg	*Original:* Mass of one cubic decimeter (1/10th of a meter, cubed) of water at its maximum density (4°C). *IS:* Mass of the standard kilogram, a platinum-iridium alloy object kept at Sevres, France.
Time	Second, s	*Original:* 1/86,400th of a mean solar day. *IS:* 9,192,631,770 cycles of frequency related to a certain electronic transition in cesium-133.
Temperature	Degree Celsius, °C or kelvin, K or °K	*Original:* (on the Celsius or centigrade scale) 1/100th the interval between the freezing point (0°C) and the boiling point (100°C) of water which is saturated with air. *IS:* (on the Kelvin or thermodynamic scale) 1/273.16th the interval between absolute zero (0°K, −273.15°C) and the triple point of water (273.16°K, 0.01°C). (At the triple point pure liquid water, ice, and water vapor are in equilibrium.) A degree Celsius is called a *kelvin*.

Table A-2.   Fractions and multiples

Fraction	Prefix	Symbol	Multiple	Prefix	Symbol
10^{-1}	Deci	d	10	Deka	da
10^{-2}	Centi	c	10^2	Hecto	h
10^{-3}	Milli	m	10^3	Kilo	k
10^{-6}	Micro	μ	10^6	Mega	M
10^{-9}	Nano	n	10^9	Giga	G
10^{-12}	Pico	p	10^{12}	Tera	T

A-2. The international system

As technology progressed, demands for precise physical measurement exceeded the old standards' capability. The result was development of measurement units definable in terms of invariant atomic properties that could be independently reproduced with high precision in any laboratory having the proper equipment. From these refinements came the modern metric system, called the *International System (IS)*, adopted in 1960.

Six basic IS units are defined. They are the *meter, kilogram, second, ampere, degree Kelvin,* and *candela* (for luminous intensity). Two supplementary units are appended: the plane angle and the solid angle. All other units are derived. Thus the IS unit for area is the square meter, the unit for volume is the cubic meter, the unit for density is the kilogram per cubic meter, etc. Some derived units are given special names. One example, the unit of electric charge, defined as the product, ampere × second, is called the *coulomb.* Details on the four most used units are given in Table A-1.

Table A-3.   Units of length, mass, and volume

Measure	Unit	Equivalent
Length	Millimeter, mm	10^{-3} meter, m
	Centimeter, cm	10^{-2} m
	Kilometer, km	10^3 m
	Ångstrom, Å	10^{-10} m
	Micron, μ	10^{-6} m
	Inch, in.	0.0254 m = 2.54 cm
	Mile, mi	1609 m = 1.609 km
Mass	Milligram, mg	10^{-6} kg = 10^{-3} g
	Gram, g	10^{-3} kg
	Metric ton, t	10^3 kg = 10^6 g = 2205 lb
	Grain, gr	64.80 mg
	Ounce, oz	28.35 g
	Pound, lb	0.4536 kg = 453.6 g
Volume	Milliliter, ml	10^{-3} liter = 10^{-6} m^3 = 10^{-3} dm^3
	Liter	10^3 ml
	Liter	1000.028 cm^3
	Quart, qt (U.S. liquid)	0.9463 liter
	Gallon, gal (U.S. liquid)	3.785 liters
	Cubic foot, ft^3	28.32 liters

A-3. Fractions and multiples

When Mouton proposed the decimal system of measure he also proposed Latin prefixes for fractions and multiples. They are given in Table A-2. A deciliter is $\frac{1}{10}$ liter, a kilogram 1000 g, etc.

A-4. Measurement equivalents

Frequently used units of length, mass, and volume are listed in Table A-3. In each grouping the first three units are IS derived or IS approved; the remainder are examples of equivalents not approved in the IS. The volume unit in the metric system is the liter. It was defined in 1901 as the volume occupied by 1 kilogram of water at 4°C. The liter was intended to be equal to the volume of a cube $\frac{1}{10}$ meter on a side. A small difference, as shown in the table, exists between the two, however. Except in precise work, it is ignored.

Three equivalents worth remembering for problem solving in the text are

$$1 \text{ in.} = 2.54 \text{ cm}$$
$$1 \text{ lb} = 454 \text{ g}$$
$$1 \text{ qt} = 946 \text{ ml} = 946 \text{ cm}^3$$

B-1. Significant figures

Every measurement is limited in its precision—that is, in its reproducibility when the measuring operation is carefully repeated. The digits in a number that express the precision of a measurement are called *significant figures*. If, for example, using a triple-beam balance a flask is found to weigh 60.05 g, the precision of the measurement is 0.01 g and the number 60.05 contains four significant figures. A zero is significant when used to help express a quantity, as do the zeroes in 60.05. A zero is called insignificant when used merely to fix a decimal point, as do the zeroes in 0.01.

The number of significant figures derived from multiplication or division operations can be no more than the least number of significant figures in any of the measurements used. Thus if a piece of copper having a volume of 11.1 cm^3 is found to weigh 99.0502 g, the density may be expressed as a number containing no more than three significant figures, 99.0502 g/ 11.1 cm^3 = 8.92 g/cm^3.

Table B-1. Rules of exponents

Operation	Symbolism
Multiplication	$10^a \times 10^b = 10^{(a+b)}$
Division	$10^a/10^b = 10^{(a-b)}$
Raising to a power	$(10^a)^b = 10^{ab}$
Extracting a root	$\sqrt[b]{10^a} = 10^{a/b}$

In adding significant figures the sum is expressed with the same precision as the least refined measurement in the series being considered. The sum of these milliliter volumes, 21.1, 105, and 7.25 is 133 ml, because the least refined measurement, 105 ml, has been made only to 1 ml.

B-2. Exponential notation

Large and small numbers encountered in scientific measurement are conveniently handled when written in *exponential* form,

$$A \times 10^n$$

where A, a number between 1 and 10 is called the *coefficient*, and n, a positive or negative integer, is called the *exponent*. The exponent is the number of places the decimal point has been moved to write the original number in exponential form. A positive exponent means the decimal point has been moved to the left. A negative exponent means the point has been moved to the right. Thus the charge of one electron in coulombs, 0.000 000 000 000 000 000 160 206, is better written 1.60206×10^{-19}.

Exponential notation facilitates correct interpretation of significant figures. One may find, for instance, that the velocity of light in meters per second, although known to just seven significant figures, is written 299 792 500. The number is better written 2.997925×10^8.

To *add* or *subtract* exponential numbers, rewrite them as needed to make all exponents the same. Then add or subtract coefficients in normal fashion and retain the exponent:

$$4.1 \times 10^3 + 2.0 \times 10^2 = 4.1 \times 10^3 + 0.20 \times 10^3 = 4.3 \times 10^3$$

Rules for handling exponents in other basic operations are given in Table B-1 and illustrated below.

To *multiply* exponential numbers, multiply the coefficients normally and add the exponents:

$$(1.5 \times 10^3)(4.0 \times 10^{-9}) = 6.0 \times 10^{-6}$$

To *divide* exponential numbers, divide the coefficients normally then subtract the exponent of the divisor from the exponent of the dividend. (Subtraction can be accomplished by changing the sign of the divisor's exponent and adding it to the dividend's exponent.)

$$\frac{8 \times 10^{-7}}{2 \times 10^3} = 4 \times 10^{-10}$$

To *raise an exponential number to a power*, treat the coefficient normally then multiply the exponent by the power:

$$(0.030)^3 = (3.0 \times 10^{-2})^3 = 27 \times 10^{-6}$$

APPENDIX

LOGARITHMS

C-1. Logarithms

The power to which the number 10 must be raised to equal another number is its *logarithm* or *log* (Gk. *logos*, proportion + *arithmos*, number). For instance

$$\log 1000 = \log 10^3 = 3$$
$$\log 1 = \log 10^0 = 0$$
$$\log 0.01 = \log 10^{-2} = -2$$

Logarithms are exponents. Therefore, the rules of exponents (Table B-1) also govern the use of logarithms, as shown in Table C-1.

To *extract a root* of an exponential number, treat the coefficient normally, then divide the exponent by the root. As needed rewrite the exponential number to make the exponent divisible by the root:

$$\sqrt[3]{0.0000820} = \sqrt[3]{82.0 \times 10^{-6}} = 4.35 \times 10^{-2}$$

Table C-1. Rules of logarithms	
Operation	Symbolism
Multiplication	$\log (XY) = \log X + \log Y$
Division	$\log (X/Y) = \log X - \log Y$
Raising to a power	$\log X^a = a \log X$
Extracting a root	$\log \sqrt[a]{X} = \log X/a$

On the slide rule use the A and D scales for *square root* (and *squaring*), and the K and D scales for *cube root* (and *cubing*). See a slide rule manual for manipulations.

To find the log of a number, write the number in exponential form with one digit to the left of the decimal point. Rewrite as the sum of the log of the coefficient and the log of the exponent. (Addition of logs is equivalent to multiplying the corresponding numbers (Table C-1).) Find the log of the coefficient with the slide rule (D and L scales; see a manual on the slide rule) or in a log table. The log of the exponential part is simply the exponent itself, as shown above. Add the two numbers (logs) to obtain the log of the original number. For example,

$$\log 2240 = \log (2.240 \times 10^3) = \log 2.240 + \log 10^3 = 0.350 + 3 = 3.350$$

and

$$\log 0.0375 = \log 3.75 \times 10^{-2} = \log 3.75 + \log 10^{-2} = 0.574 + -2 = -1.426$$

C-2. Antilogarithms

An *antilogarithm* (antilog) is the number corresponding to a given log. A log consists of two parts. The integer preceding the decimal point is called the *characteristic*. It determines the placement of the decimal point in the number sought. The integers following the decimal point comprise the *mantissa*. The mantissa corresponds to the numerical part of the number sought. When the antilog is positive, find the mantissa in a log table and read the value of the number having this log. Alternatively locate the mantissa with the L and D scales on the slide rule. Thus

$$10^{2.575} = \text{antilog } 2.575 = 10^2 \times \text{antilog } 0.575 = 10^2 \times 3.76 = 3.76 \times 10^2$$

When a number such as $10^{-3.464}$ is to be expressed in nonexponential form, first rewrite it exponentially with a *positive* mantissa (log tables do not provide for negative mantissas). Do this by adding -1 to the characteristic, subtracting the mantissa from 1.000, and rewriting as a product of 10's to the resulting exponents:

$$10^{-3.464} = \text{antilog } -3.464 = 10^{-4} \times 10^{0.536}$$

As before, addition of exponents is equivalent to multiplication of the numbers, thus $-4 + 0.536 = -3.464$. The number 10^{-4} establishes the decimal point; $10^{0.536}$ is evaluated as in the preceding example:

$$10^{-4} \times 10^{0.536} = 10^{-4} \times \text{antilog } 0.536 = 3.48 \times 10^{-4}$$

C-3. Multiplication by logs

To multiply numbers using logs, first find the logs of the numbers. Add them. Then find the antilog of their sum; it is the desired product. For example, to multiply 0.275 by 358,

$$(0.275)(358) = \log\,(2.75 \times 10^{-1}) + \log\,(3.58 \times 10^{2})$$
$$= (0.440 - 1) + (0.554 + 2) = -0.560 + 2.554 = 1.994$$
$$antilog\ 1.994 = 10^{1} \times antilog\ 0.994 = 9.86 \times 10 = 98.6$$

C-4. Division by logs

To divide numbers using logs, first find the logs of the numbers. Then subtract the log of the divisor from the log of the dividend. The antilog of the remainder is the desired quotient. To divide 15,200 by 783, for example,

$$\frac{15200}{783} = \log\,(1.52 \times 10^{4}) - \log\,(7.83 \times 10^{2})$$
$$= (4 + 0.182) - (2 + 0.894) = 4.182 - 2.894 = 1.288$$
$$antilog\ 1.288 = 10^{1} \times antilog\ 0.288 = 1.94 \times 10 = 19.4$$

C-5. Roots by logs

To find the *root* of a number using logs, first find the log of the number. Divide it by the root. Then find the antilog of the quotient. For example,

$$\sqrt[5]{6000} = \frac{\log\,(6.00 \times 10^{3})}{5} = \frac{3.779}{5} = 0.756$$
$$antilog\ 0.756 = 5.71$$

C-6. p values

The p value of a number is the log of its reciprocal. This is equivalent to the number's negative log. The most common application is in problems dealing with pH, defined in the text by Equation 17-11,

$$pH = \log \frac{1}{[H^{+}]} = -\log\,[H^{+}]$$

Given, for example, the molar hydrogen ion concentration of human blood, 3.89×10^{-8}, find the corresponding pH as follows:

$$pH = \log \left(\frac{1}{3.89 \times 10^{-8}} \right) = \log \left(\frac{10^{8}}{3.89} \right)$$
$$= \log 10^{8} - \log 3.89 = 8 - 0.59 = 7.41$$

Convert a pH value to the corresponding H^{+} concentration as follows:

$$7.41\ pH = -\log\,[H^{+}]$$
$$[H^{+}] = 10^{-7.41} = 10^{-8} \times 10^{0.59}$$
$$= 10^{-8} \times antilog\ 0.59 = 3.89 \times 10^{-8}\ M\ H^{+}$$

APPENDIX

FACTOR-UNIT METHOD

D-1. Units
As already noted, measurements are expressed in numbers and *units*. Examples of units (also called *dimensions*) are given in Table A-3. Measurements to be added or subtracted must have the same units. When measurements are multiplied or divided, both numbers and units are multiplied or divided. Thus

$$5 \text{ cm} \times 6 \text{ cm} = 30 \text{ cm}^2$$
$$10 \text{ kW} \times 2 \text{ hr} = 20 \text{ kWh}$$
$$50 \text{ g}/10 \text{ ml} = 5 \text{ g}/ml$$

D-2. Factor-unit method
A problem that can be solved by means of *proportions* can better be solved by the *factor-unit method*. First the problem is read to find the quantity

sought and the quantity given. The solution to the problem is then set up in this form:

$$quantity\ sought = quantity\ given \times factor(s)$$

Each factor is a fraction. In numerator and denominator of each fraction appear equivalent measurements such as taken from a compilation like Table A-3; thus (2.54 cm/1 in.) or (1 in./2.54 cm). Just as identical numbers cancel in the multiplication of fractions,

$$\frac{\cancel{2}}{3} \times \frac{5}{\cancel{2}} = \frac{5}{3}$$

so do units,

$$\frac{1\ \cancel{in.}}{2.54\ cm} \times \frac{1\ ft}{12\ \cancel{in.}} = \frac{1\ ft}{30.5\ cm}$$

Terms in factors are arranged so that unwanted units cancel and wanted units are retained. The indicated multiplication and division of the numerical parts are then performed. To facilitate approximating the answer before using a slide rule, all numbers should be rewritten in exponential form. Finally, the correct answer is written with the correct units.

Example D-1 Rhodopsin, a visual pigment in the eye, has maximum light absorbance at 500 nm. Express the wavelength in inches.

Solution Two operations are indicated: conversion from nanometers to meters, then from meters to inches. Each conversion requires a factor. Equivalents for the factors come from Appendix Tables A-2 and A-3: 1 nm = 10^{-9} m, 1 in. = 0.0254 m. Factors made of these equivalents are arranged to cause cancellation of all units except inches, the unit wanted. Numbers are written exponentially:

$$?\ in. = 5.00 \times 10^2\ \cancel{nm} \left(\frac{10^{-9}\ \cancel{m}}{1\ \cancel{nm}}\right) \left(\frac{1\ in.}{2.54 \times 10^{-2}\ \cancel{m}}\right) = 1.97 \times 10^{-5}\ in.$$

D-3. Chemical factors

Stoichiometry problems (Chapter 12) are simplified by use of *chemical factors*. Chemical factors, like the fractions above, are derived as needed from table values of atomic and molecular weights and from identities related to the mole concept such as 6.02×10^{23} particles = 1 mole, and 1 mole of gas = 22.4 liters at standard temperature and pressure (STP).

Example D-2 If in the synthesis of silver nitrate,

$$Ag + 2HNO_3 = AgNO_3 + NO_2 + H_2O$$

1000 g of silver is used, theoretically how many grams of pure nitric acid are needed, and how many liters of by-product nitrogen dioxide (STP) are produced?

Solution The atomic weight of silver is 108, the molecular weight of nitric acid is 63.0. The balanced equation above shows that 1 mole of silver reacts with 2 moles of nitric acid to yield 1 mole of each of the three products. These data lead to the following factors and factor-unit solutions:

$$? \text{ g } HNO_3 = 1.00 \times 10^3 \text{ g Ag} \left(\frac{2 \times (6.30 \times 10 \text{ g } HNO_3)}{1.08 \times 10^2 \text{ g Ag}} \right) = 1.17 \times 10^3 \text{ g}$$

$$? \text{ liter } NO_2 = 1.00 \times 10^3 \text{ g Ag} \left(\frac{2.24 \times 10 \text{ liters } NO_2}{1.08 \times 10^2 \text{ g Ag}} \right) = 2.07 \times 10^2 \text{ liters}$$

APPENDIX

SELECTED LISTING OF TEXT DATA

Chapter 1

13. (a) 21.0   (b) 1.97 g
15. 32.0
17. (a) 1.60 g O, 0.10 g H, 3.55 g Cl
18. 80.0

Chapter 2

10. 6560 Å
11. 52 ft
12. (a) 60 g
14. (a) 8.2×10^{-8} newton

Chapter 3

8. (a) 5.48×10^7 cm/sec, 5.32×10^{-7} cm
9. (a) 7.25×10^9 cm/sec
14. 1.18×10^{-32} cm

Chapter 4

17. Q is $^{222}_{86}$Rn
20. ^{256}Md is 8.25 f
23. 1.813 amu, 0.0076 amu/nucleon
27. (a) 1 km;   (b) 10 km

Chapter 5

14. 10.6×10^{22} mi, 1.80×10^{10} lt-yr
16. (a) estimated distance 3300 mi or 5.3×10^8 cm

18. 312 cpm after 4 half-lifes
21. (a) 1.21×10^{-4} yr;   (b) 4.82×10^3 yr

Chapter 6

13. (a) π units^2
14. (a) 0.54 units^2
15. (a) 1 sphere;   (b) 2 spheres
18. 2.81 A

Chapter 7

14. (a) Li 13.1 ml, Na 23.7 ml, K 45.5 ml, Rb 55.9 ml, Cs 70.0 ml
16. (b) Ar $1s^2 2s^2 2p^6 3s^2 3p^6$
17. $n = 3, l = 0, m = 0, s = \pm\frac{1}{2}$
18. (c) 32 electrons

Chapter 8

11. (b) V^{2+} 3 unpaired, Fe^{3+} 5 unpaired, Zn^{2+} 0 unpaired
13. (b) 2+ and 4+
15. (b) $^{249}_{98}Cf + ^{12}_{6}C = ^{257}_{104}Rf + 4^{1}_{0}n$

Chapter 9

12. $2K\cdot + \,:\ddot{O}: \,= K^+ \,:\ddot{O}:^{2-} K^+$
13 (c) $:N: \,::N:$ or $:N\equiv N:$
14. $\ddot{S}: \,:C: \,:\ddot{S}$
18. KBr, 63%: CO_2, 22%

Chapter 10

16. (b) B—F, 1.52; C—Cl, 1.76; Si—Br 2.31
17. Pd^{2+} 10 electrons, 5 ligands
20. (b) 0.414;   (d) 0.82; Cl^- will not touch

Chapter 11

12. (a) $2 + 1 = 2 + 3$;   (b) $2 + 1 = 1 + 1$
13. (a) As_2O_3, Br_2O, I_2O_5
18. (c) $Pb^{2+} + OH^- + H^+ + SO_4^{2-} + PbSO_4 \downarrow \, + H_2O$
20. (a) $S_2O_3^{2-} + 2H^+ = S \downarrow \, + SO_2 \uparrow \, + H_2O$

Chapter 12

16. CuO, As_2O_5, PbO, Sb_2O_3
19. (b) 79.91
20. (b) 35.46
21. (c) La_2O_3
24. (b) MCl_3
28. (b) 2.75 mmole $BaCl_2$, 5.60 mmole H_2O
31. (a) 0.0022 mole $KClO_4$;   (b) 0.0022 mole K;   (c) 0.086 g;
    (d) 6.88% K
34. 6.023×10^{23} particles
36. (a) 29.7 g
40. (b) -74.6 kcal/mole

Chapter 13

13. 1212 g
16. 918 gal, 12.480 mi
17. (a) 5800 lb; (b) 4.41 lb
18. 60.0 g
21. (a) 2.79 mg; (b) 2.00%

Chapter 14

16. (a) Ag, 1234°K; 1761°F
18. (c) about 14,000 ft
20. (b) 341°K, 68.3°C
23. (b) 1.34 g/liter
26. (b) 718 ft^3
29. (a) 0.450 ml; (b) 50.6%
32. 0.387 mg
36. (a) 200 torr or 0.236 atm; (d) 3.86×10^{21} molecules;
 (g) 4.74×10^{27} collisions/cm^3 sec
39. 2.96×10^9 mi
40. 5.09×10^5 cm/sec. Escape velocity of earth is 2.2 times as much; thus gases escape from Mars more readily.

Chapter 15

11. (a) 4.85×10^5 cal/ft^2; (b) 5.27 gal/day
13. (a) 2.09×10^6 gal/min
15. (a) 37,200 cal
16. Na$^+$ 10.6 g/1000 g or 10,600 ppm
 Cl$^-$ 19.5 g/1000 g or 19,500 ppm
22. 2.22×10^4 lb

Chapter 16

17. 79.86 g/100 g
19. (a) 0.301 m
20. 4.92 atm
22. 81.75%
25. (a) 1.46 g HCl; (b) 52.5 mmole; (c) 35.4 ml; (d) 2.00 mmole

Chapter 17

17. (a) 55.7; (b) 0.334 mole/liter
21. (c) 2.4; (d) 10.84; (e) 5.0
24. (b) 9.4
26. (a) 9.11×10^{-6} M in both ions; (b) 8.3×10^{-11} mole^2/liter^2
28. (b) 4.43×10^3 mg Pb^{2+}/liter

Chapter 18

11. (b) 4.40 ohm
12. (a) 1.60 A; (b) 5.4×10^{10} erg
14. 4.06 g Sb, 5.93 g Sn, 19.7 g Au
16. (b) 4.1 ppm
17. (a) 4680 W; (b) 112 kWh; (c) $2.44

Chapter 19

11. Lead cell develops 2.05 V via the reaction,
$$Pb + 2SO_4^{2-} + PbO_2 + 4H^+ = 2PbSO_4 + 2H_2O$$

14. (a) $Br_2 + 6H_2O = 2BrO_3^- + 12H^+ + 10e^-$

18. (c) 1.37 V;   (d) 4.11 V

19. Yes; $E^0_{cell} = 1.54$ V for O_3 to oxidize I^-

22. (a) 250 A;   (b) 100 W/lb;   (c) \$1.35/lb;   (d) 45%

Chapter 20

11. $x = 1, y = 2$

12. C_6H_4S, $C_{12}H_8S_2$

14. (a) 145°

17. $H_3CCHICH_2I$ and $H_3CCIHCH_2I$

18. cyclopentane, cyclohexene

Chapter 21

16. -1664 and -2252 kcal/mole

17. (a) 840

18. (b) $C_nH_{2n}O$

21. (a) A is 1-hexanol

22. B is 3-methylbutanal

27. 2

Chapter 22

15. (a) $CH_4 + H_2O - HCHO + 2H_2$ (d) $H_2NCH_2CN + 2H_2O -$
$H_2NCH_2CO_2H + NH_3$

18. 12,400

Chapter 23

18. Forms (a) and (b) will yield *meso* diacids with a plane of symmetry
through the middle.

19.
+a	−a	+a	−a	+a	−a	+a	−a
+b	−b	−b	+b	−b	+b	+b	−b
+c	−c	−c	+c	+c	−c	−c	+c
dl pair		*dl* pair		*dl* pair		*dl* pair	

23. Products are glucose and *p*-hydroxyphenol.

24. (c) About 680

26. 12,150 gal

27. (a) 18

31. (b) stearic acid

32. 173.5

33. 192

37. 146

38. 2

40. (a) 141 (average of all 20 amino acids);   (b) 35

42. 1.58×10^5

45. (a) 2;   (b) 6

Chapter 24

23. (a) Second generation will be 50% Tt, 50% tt.

24. (b) 324;   (c) 9.72×10^6
25. (c) UACCUGCAUAAG
27. (a) 20 glycine; 10 alanine, 5 serine, 5 tyrosine (total: 40 residues);   (b) 6.42×10^{18}
28. (a) 16;   (b) 256

Boldface numbers refer to figures, n to footnotes, and t to table data.

Androgens, 617–618
-ane (suffix), 528
Anhydrate, 253
Anhydride, acidic, 253, 511, 608
  basic, 253
Aniline, 352, 542, 544
Animals, as heterotrophs, 570
Anions, 249, 250–251t
Anode, 26, 453, 468, **469, 470, 475, 478**
Antiballistic missile (ABM), 95
Antibiotics, 659
Antibodies, 660
Antibonding orbitals, 210–211
Anticancer drugs, 659
Anticodon, 657
Antifreeze, 402
Antiknock, 530
Antilogarithms, 682–683
Antimatter, 57, 76
Antimony, compounds of, 173, 175, 255t, 460, 488
Antioxidants, 614, 660
Apatite, 141t, 174, 365, **366**
Apollo moon missions, 119, 190, 289
Appleton, Edward, 302
Aquinas, St. Thomas, 7
Arabinose, **596**, 597
Arachidonic acid, 663
Archimedes, 192, 368
Arginine, **621, 623**, 629
Argon, 179, **294**, 299, 300t
Aristotle, 5–7
Aromatic compounds, 518, 519, 542–545
  amines, 542–543
  amino acids, 620, **621**
  hydrocarbons, 501
  in oil, **527,** 528
  reactions of, 534–535
Aromatic ring, 502, **519**
Arrhenius, Svante, 200, 406–409, 425
Arsenic, 173, 255t, 429, 481, 625
  compounds of, 175, 253t
Aryl group, 533n
Asbestos, **140,** 141, 309
Aspartic acid, **580,** 621, 623
Astatine, 177–178
Aston, Frances, 67
Asymmetric carbons. See Asymmetric center
Asymmetric center, 507, 598, 636
  in glucose, 594
Atmosphere, 114, 116t, 289–322.
  See also Air
  composition of, 300, **301**
  evolution of, **294, 585**
  mass of, 296
  origin of, 293, 562
  pressure of, 295–296, **301,** 329–330, 335, 353t
  in rock cycle, 126
  structure of, **301**
Atmosphere (unit), 326
Atom, 20, 44–45, 47, 60
Atomic absorption (AA), 127, **130**
Atomic energy, **1,** 78–93, 374, 669
Atomic heat. See Heat capacity
Atomic mass. See Atomic weight
Atomic mass unit, 73n, 82–83
Atomic nucleus, 45, 65
Atomic number (Z), 35–37, 45, 66, 148, 151
Atomic-orbital theory, 211
  and complexes, 226

and covalent bonding, 513–519
Atomic orbitals, 211–213
  of carbon, 515–519
  filling order, **154**
  shapes of, **155,** 156
Atomic reactor, **1,** 78–81, 87–91, **93,** 181, 479, 669
Atomic theory, early, 5, 9–10, 13
Atomic volume, **149**
Atomic weights, 12, 271–272, 275, front and back endpapers
  in chemical factors, 685–686
  Dalton's, 269
  and equivalent weights, 268t
  from heat capacities, 270–271
  and Mendeleev, 148, 149
ATP, 567, 606–608
  and amino acids, 655
  in glycolysis, 601, **607**
  in photosynthesis, 574–575
  from photolysis, 580–581
Aufbau method, **154,** 210
Autocatalytic reactions, 568
Automobile, 171, 306, 315t
  electric, 475–476, **477,** 478–479
Autotrophs, 570

B

B subgroup elements, 179–187
Bacteria, 374, 382–384, 570–571, 601
Baeyer theory, 511–512
Baking soda, 167, 437t. See also Sodium bicarbonate
Balance, analytical, 265, **412**
Balanced equations, 250–251, 486–489
Balancing formulas, 249
Balmer series, 48, 49, 61
Band model, metals, 236–237
Barbiturates, 593
Barium, 78–79, 248t, 255t
  compounds of, 170, 257, 258, 279, 440t
Barkla, Charles, 36
Barometer, **295,** 326, **327,** 328t, **337**
Barton, Derek, 512
Baryons, 76
Base, 13, 257, 435, 437t
  in DNA, RNA, 648t
  and K_b, 433
  theories, 408–411
Base paring, in DNA, 649–651
Base sequence, in DNA, 654
Basic anhydrides, 253
Battery, 25, 199, **451,** 452, 456, 466–479. See also Voltaic cells
  alkaline, 168, 184
  dry cell, **475**
  lead storage, 173, **475, 477**
  solar, 482
  zinc-air, **477**
Bauxite, 171, 460
Beadle, George, 647, 654
Becquerel, Antoine-Henri, 31
Benzaldehyde, resonance in, 545
Benzene, 501–502, 527, 547
  combustion of, 288
  freezing point, 402
  production, 526t
  ring, 534–545
Benzoic acid, 542, 615
Bernoulli's equation, 343
Bertholet, M., 548, 609
Beryllium, compounds of, 70, 108, 168, 231, 255t

Berzelius, Jons, 12, 124, 173, 199, 271, 247, **248,** 427, 497, 520, 624
Beta-decay theory, 65
Beta particle, rays, 32, 65–68
Bethe, Hans, 81
BHA and BHT, 659
Bicarbonate ion, 312, 364, 410
  analysis of, 379, 421
  in blood, 439
  in natural waters, 464t
  in photosynthesis, 578
Bidentate group, 239
Big bang theory, 106
Bile, 601
Bile acids, **617,** 618
Binary compounds, 250
Binding energy, of nucleons, **82**
Biochemical oxygen demand, 382
Biochemistry, 557ff, 589ff, 642ff
Bioelectricity, 482–484, 489
Biological evolution, 570–572
Biopoesis. See Oparin–Haldane
Biopolymers, 568, **569**
  carbohydrates, 600–601
  nucleic acids, 647–653, 661
  proteins, 618–633, 653–657, 661
Biosphere, 116, **126**
Biphenyl, **527**
Birth control pills, 618
Birth rates, 672, 674
Bismuth, compounds of, 173–175, 488
Black hole, 109
Blackbody radiation, **42,** 44, 106
Blackett, Patrick, 69
Blast furnace, 182, 193
Bleach, 13, 178
Bloch, Felix, 515
Bloch, Konrad, 616
Blood, 85, 437t, 439, 626
Boat conformer, 512
Body-centered lattice, **132–135, 230**
Bohr (unit), 216
Bohr, Niels, 46, **47,** 78, 81, 151–152
  atom model, 48–49, 201–202
Bohr radius, 195
Boiling point, 337, 394
  effect of solute on, 398–401
  elevation, 400–401
  of liquids, 257, **337,** 354–355, 535
  of water, **337,** 401
Boltwood, B., 113, 273, 288
Boltzmann's constant, 346
Bond angles, 213, 214t
  in ammonia, **213**
  in hydrocarbons, **516–519**
  in water, **212**
Bond character, 207
Bond energy, 199, 207, 210, 217, 282, 305t
Bond lengths, 204, 230–231, 240
  carbon–carbon, **515–519**
Bond strength, 204. See also Bond energy
Bonding, 49, 199, 214t, 238t. See also Bonds
  from computers, 214–217
Bonding force, electrostatic, 229
Bonding orbital, 210–211
Bonds, 60, 214t, 238t
  apolar (hydrophobic), 629
  in aromatics, 518–519
  axial, 512
  coordinate-covalent, 206

covalent, 203–204, 208, 526. *See
also* Covalent bond
electrovalent. *See* Bonds, ionic
equatorial, 512
high-energy, 607–608
hydrogen, 233. *See also* Hydrogen bond
ionic, 199–203
London, 232
metallic, 235–237, 238t
pi, **517–519**
sigma, 515, **516**
sp, 518
sp², 517–518
sp³, 515–517
van der Waals, 232
Bone, 85, 663
Boranes, 171
Borax, 170, 254t
Born, Max, 56
Boron, compounds of, 125t, 144,
170–171, 293, 481
Boron trichloride, 214t, 411
Boron trifluoride, 205–206, 214t,
530
Bothe, Walter, 70
Boyle, Robert, 328, 342, 408
Boyle's law, 328, 330
Bragg method, 131, 201, 273
Bragg, W. H., and W. L., 132
Brass, bronze, 185, 395
Breeder reactor. *See* Atomic re-
actor
Brine, 167, 178, 369. *See also*
Sodium chloride
Bromide ion, **230–231**
Bromination, 532, 534, 544
Bromine, compounds of, 177–178
isotopic ratio, 286
molecules, **234**
Bromobenzene, 501–502, 544
Brønsted, J., 409
Brownian movement, **342,** 273,
406
Bubble chamber, **75**
Buchner, Eduard, 624
Buffer, 439–440, 626
Buret, 329, **413,** 414
Butane, 372, 523, 528t
reactions of, 530
Butenandt, Adolf, 635, 636
1-Butene, reactions of, 531
2-Butene isomers, 510

C

C-terminal end, 622, 628–629, 657
Cadmium, 186, 227, 249, 255t
in atomic pile, 81
compounds, 186, 224, 227–228
as contaminant, 378
in electrolysis, 457–458
Caffeine, 593, 659
Calcite, 125t, 141t, 504, **505.** *See
also* Calcium carbonate
Calcium, 169, 231t, 255t
in the body, 614
chelation of, 228
in natural waters, 137, 364t
in rocks, 117t, 139t, 364
in shells, 584
in soil, 126
spectral lines, **37,** 102, **105**
Calcium carbonate, 125t, 254,
265, 578
solubility, 441–442
Calcium halides, 169, 454, 497
Calcium hydroxide, 253
Calcium phosphate, 169, 367, 663
Calcium sulfate, 254t, 440t

Calibration, **333,** 353t
Californium, 190, 193
Calomel, 187, **438**
Calorie, 48, 372–373, 664
Calorimeter, **280,** 606, 664
Calvin, Melvin, 562, 578, **579,** 580
Calvin-Bassham cycle, 573, 575,
580, **581**
Cancer, 32, 190, 379, 557, 657
Cannizzaro, S., 13, 147, 270
Capacitor, **208,** 209, 482
Capillarity, **394**
Carbides, 181, 252
Carbohydrates, 565, **569,** 592–601.
See also Saccharides; Sugars
conformations, 513
as food, 601–608, 663–664
Carbon, 134, 171–172, 285
atom, **215**
atomic orbitals, 513–519
bonding ability, 497
fixation of, 580–584
oxidation of, 285
in proteins, 618
reduction with, 255t
tetrahedral structure, 507
valency, 500
Carbon-14, 120, 578, **580**
Carbon cycle, 312, 571
Carbon dioxide, 205, 312–313
in atmosphere, 293, 294, 573,
585
from combustion, 255, 306t,
312–313
ethalpy diagram, **285**
from glucose, 604, 606
krebs cycle, 603–604, **607,** 633
mole of, 272–273
molecular shape, 209
in photosynthesis, 293, 571–
575, 578–584
in water, 364, 571
Carbon-hydrogen analysis, 497
Carbon monoxide, 255, 429
in air, **294,** 313, 315t
enthalpy, 281t, 285
reactions, 85, 317, 432
Carbon oxides, 269t
Carbon tetrachloride, 281t, 288,
499
Carbonic acid, 253t, 257
as blood buffer, 439–440
Carboniferous era, 526, 584, **585**
Carbonium ion, 534–537, 540, 544–
545
Carbonyl group, 233, **308,** 537–
539, 596–597, 599
Carboxyhemoglobin, 313
Carboxyl group, 228, 539–540,
543
Carboxylic acids, 508–511, 533t,
539–540, 542, 569. *See also*
Fatty acids
Carotenes, **576,** 582, 614
Catalyst, 427–430. *See also* Enzymes
for ammonia synthesis, 432–
433
for automobile, 184, 315, **428,**
429
boron trifluoride, 170
and equilibrium, 432
in esterification, 548–549
for hydrogenation, 638
Lewis acid, 530
in oil industry, 168, 181
and reaction rates, 424
in sulfuric acid synthesis, 176,
181

Cathode, 26, 44, 453
in cells, **469, 470, 475, 478**
mercury, 461–462
Cathode rays, 26, 27
Cathodic protection, 486, **490**
Cations, 248–249t
Cavendish, H., 267, 298
Cell, biological, 85–86, 561, 567–
572, 577, 601, 606–608, 645–
647, 654, 657, 658, 661, 664,
671, 672
Cellophane, 177
Cellulose, 176, 564, **569,** 600
as waste, 317
Cellulose acetate, 625, 637
Celsius scale, 325–326, 676
Cement. *See* Portland cement
Centigrade scale, 325, 676
Cerium, 78, 187–188
Cesium, compounds of, 167, 255t
Cesium bromide lattice, **230**
CGS system, 24n
Chadwick, James, 69, 81
Chair conformer, 512, 598–600
Charcoal, 134, 297, 406
in water treating, 384–385
Charge, measurement of, **23**
Charge symmetry, 208
Chelation, 228–229, 378
Chemical change, 9
Chemical equation, 250
Chemical equilibrium. *See* Equi-
librium
Chemical evolution, 562–570
Chemical factor. *See* Factor-unit
Chemical formulas, 11–12, 200,
498–499
writing, 247–250, 500–501
Chemical industry, 13, 165–166,
178, 526–528, 545–551
plants, **447, 493, 551**
Chemical symbols, 11–12, 15,
247–251, end papers
Chemisorption, 428–429
Chemistry, 6
analytical, 9, 247
biological, 592, 609, 618
inorganic, 13, 223
organic, 497, 500
physical, 339
structural, 223, 506, 513
Chitin, 637
Chloric acid, 253t
Chloride ion, 178, 201, 224, 250
analysis of, 379, 415
in biological cells, 483
in natural waters, 364t
radius, 231t
Chlorinated hydrocarbons, 200,
376–377, 501, 503, 530–531, 547
Chlorination, 200, 530–531, 544
Chlorine, 11–14, 166t, 238t, 259
in chlorination, 530–531
compounds of, 177–178
in displacement, 256
production, 259, **447, 458, 462**
in water treatment, 464
Chloroform, 609, 638
Chlorophyll, 571, **574,** 579, 582
Chloroplasts, 573, **576, 577,** 582
Cholesterol, 313, 617–618, 664
Chromatogram, **579, 580,** 611, 612
Chromatography, column, 578,
663. *See also* Electrophoresis
gas, 611–613
gel permeation, 625
liquid–solid, 303
paper, 578, **579, 580**
types of, 611

Chromium, compounds of, **37,** 181–182, 255t
from ocean, 367
supply and demand, 165t
Chromosomes, 572, 646–647, 654, 656
Cis isomer, 225–226, 509, **510,** 532, 638
Citric acid, **580,** 603–604
cycle, 603–605, **606,** 616
Clay, 126, 364, **365**
Cleavage, of crystals, 130
with ozone, 533
with periodic acid, 641
Cloning, 671
Cloud chamber, 33, **34,** 69
Cloud seeding, 202
Coal, 86, **126,** 312
chemicals from, 526–527
future use of, 669
origin of, 526, 585
Cobalt, in complexes, 214t, 226, 239
compounds of, **37,** 179, 182–184, 255t, 367, 627, 663
Cockcroft, John, 71
Codons, 654, 656–657, 663
Coefficients, in equations, 250
in exponential notation, 680
Coenzymes (cofactors), 608, 624
A, 603–605, **607,** 616, 618, 663
FAD, FADH$_2$, 605, **606**
NAD, NADH, 603, **604,** 605, **606,** 608
NADP$^+$, NADPH, 575, 581, 603
Q, 606
Colligative properties, 398–399, 408
Colloids, 405–406
Color, in ceramics, glass, 184
complementary, 582
in complexes, 228
in indicators, **438**
in photosynthesis, 582, 583
in smog, 302, 306
measurement of, 102, 583
in metal ions, 180
wheel, 586, **587**
Colorimeter, 380. See also spectrophotometer
Columbium. See Niobium
Combined gas law, 334
Combining weight. See Equivalent weight
Combustion, analysis by, 497–498
complete, 255, 530
Common ion effect, 441–442
Complexes, 180, 223, 225–226, 228
Complexometric titrations, 414
Compounds, 9, 247–248
Compton, Arthur, **1,** 80, 83
Computer model, bonds, **195,** 214–217
global dynamics, 668
Concentration, as molality, 399–405
as molarity, 411–415
and reaction rates, 424
and temperature, 397–398
as weight percent, 419
Condensation, of Group VIB ions, 182
polymers, 545, 548
of silicates, 137, **138**
Condenser, **549**
Conformation, 512
Conjugate acid–base pair, 409
Conjugated proteins, 619
Conservation, of charge, 67

of mass, 67, 265, 268
Continental shelf, 367
Continents, 112, **363, 365**
Coordinate covalent bond, 206, 411, 515
Coordination number, 223
Coordination theory, 223
Copolymer, 545
Copper, compounds of, 91, 184–185, 214t, 255t, 367, **460,** 663. See also Alloys, copper
as electrode, **460,** 468–469, 471, 473–474
supply-demand, 165t
Coral, dating, 121
reefs, 364
Core, of earth, 114, **115,** 116t, 363
Core electrons, 201, **236–237**
Corrosion, 169, 185, 484–485
Cortisone, 616–617
Cosmic rays, **21,** 57, 72, 301
Cottrell precipitator, 406. See also Electrostatic precipitator
Coulomb (unit), 456, 677
Coulomb attraction, 216, **396.** See also Electrostatic force
Coupling, of diazonium salts, 543
Covalent bond, 203, 232, 238t
angles, 209, 211–214, 507–508, 511–512, **516–519**
and atomic orbital theory, 513–519
and electronegativity, 207
energies, 305t
hybrid orbitals in, 212–213
lengths, 204, **216,** 234, **516–519,** 654
multiple, 204–205, 500
strengths, 204, 305t
Covalent network solds, 235, 238t
Covalent radius, **234**
Crick, Francis, 649
Critical size, 84
Crookes tube, 27
Crude oil, 379, 527–528
future use of, 669
Crust, of earth, 114, **115,** 116t, 365
Cryolite, 178, 460
Crystal, 31, 36, 129–141, 223
bonds, 397
coordination number, 230
energy, 229
lattices, 132–133
radii, 231
Crystal ligancy. See Crystal, coordination number
Cube root and cubing, 682
Cubic lattice, 136, 277
Cupric sulfate, 247, 254t
Curie, Irene, 70
Curie, Marie, and P., 31, **32**
Curie temperature, 183
Current electricity. See Electric current
Cyanide ion, 224, 227
Cyanohydrin, 539
Cyclic transport, 573
Cyclo- (prefix), 529
Cyclohexane conformations, 512–513
NMR spectrum, 513, **514,** 515
Cyclohexanol, 512, 547
Cyclotron, **71,** 81
Cysteine, **621**
Cystic fibrosis, 661

Cystine, 574, **621, 623**
Cytochromes, 574–575, 588, 606
Cytoplasm, 653
Cytosine, 564, 648, **650–651**
Cytosine-guanine pair, 195, 648, **650–651**

D

d- (prefix). See Dextro form
D- (prefix), 594
d-block elements, **157**
d orbitals, 180
Dacron, 546
Dalton, John, 10, 15, **266,** 267–269, 274, 331, 337, 343
Dana mineral classes, 124
Daniell cell, 468, **469,** 473–474
Darwinian evolution, 562
Dating, 113–114, 120, 121
Daughter element, 66–68
Davisson, Clinton, 55
Davisson-Germer experiment, **54**
Davy, Humphry, 168, 199, 408–409, 452, 485
DDT, 276, 376–377
Deamination, 314, 541
de Broglie, Louis, 52, 56–57
de Broglie wavelength, 53
Debye (unit), 209n
Debye, Peter, **209,** 271, 392
Decane, 255, 528t
Decay series, 67, **68**
Decompositions, 254
Degenerate matter, 109
Dehydration, of alcohols, 536
of amides, 542
of hydrates, 254
Delocalization energy, 519
Demography, 665
Denitrifying organisms, **366**
Density, 679
of gases, 327, 336, 339t
of liquids, 392–393
of matter, 393t
Deoxyribonucleic acid. See DNA
Deoxyribose, 648, **650–653,** 655
Derivatives, of hydrocarbons, 530–545
Desalination, 368–372, 667
Detergents, 168, 170, 174
Deuterium, 84, 87
Deuterium oxide. See Heavy water
Dew point, 339
Dextro (d) form, 506–509
Diabetes, 661
Dialysis, 405, 625
Diamagnetic compound, 203
Diamond, 134, **135,** 141t, 235, 237
Diatomic elements, 252
Diazonium salt, 533t, 542–543
Dicarboxylic acid, 546
Dichlorobenzene, 503, 520
Dichloroethane, conformers, 521
Dielectric constant, 229, 396–397
Diesel oil, 528
Dietary essentials, 663
Diethyl ether, **337,** 533t, 536, 609
Diethylstilbesterol (DES), 639
Diffraction, of electrons. See Electron diffraction
of light, 54
of X rays. See X ray diffraction
Diffusion, of gases, 339–340
of liquids, 392, 393t
Digestion, 601
Dimensions, 684
Dimethyl mercury, 378

Dinosaurs, 585
Dipole interactions, 232, 649
Dipole moment, 209, 520
Dirac, Paul, 57
Direct union, 252
Disaccharides, 592, 599–600
Discharge tube, 26, **27**
Dispersed, dispersing phases, 405
Displacement reactions, 536–537. *See also* Replacement reactions
Displacement rule, 66
Dissociation constants. *See* Ionization constants
Distillation, **7**, 368, **369**, 528, 549
Disulfide bonds, 629, 634
DNA, 567, 647–653
  genetic language of, 654
  synthesis, 663
  in a virus, 557, 661
Domagk, Gerhard, 625
Doppler effect, **104**
Double bond, 204–205, 500, 509, 517. *See also* Bonds
Double helix, **652**
Downes cell, **458**, 459
Dry cell, 182, **475**
Dry Ice, 239
Dulong and Petit rule, 270, 275
Dumas, J., 497, 499
Dumas method, **340**, 341
Dustfall, 309
Dyne, 24n
Dysprosium, 187–188

E
E° values, 471ff
$E°_{cell}$, 473–474
Earth, **111**, **363**
  composition, 112–117, 117t
  oarly atmosphere, 562
  formation of, 112–113
  measurements, 116t, 296, 675–676
  minerals in, 138, 139t, 182
  zones, 114–117, 363–365
Effusion, 340
Ehrlich, Paul, 175
Einstein (unit), 305
Einstein, Albert, **45**, 80, 271
  photoelectric effect, 44
Einstein's equation, 82, 283
Elastic collisions, 345
Elastomer, 545
Electric bonding forces, 199
Electric circuits, **484**
Electric conductance, 228, **408**, 452
  of solutions, 200, 225
  of water, 379
Electric current, 451, 479, 484
  in electrolysis, 199
  in Tokamak, **89**, 90
Electric dipole, 208
Electric energy, 457
Electric field, 504
Electric fishes, 482–484
Electric generation, 87t, **88**, 455
Electric potential, 26. *See also* Electromotive force
Electric power, **21**, 456
  for vehicles, 475–479
Electric pressure. *See* Electromotive force
Electric units, 456–457
Electric vehicles, 475–479
Electricity, 23. *See also* Electric current

analogy to water, 456
from nuclear sources, 86
Electrochemical cells, 452–463, 466–486
Electrochemical equivalent, 12, 455–456
Electrochemical series, 451, 472t. *See also* Metal activity series
Electrochemistry, 451–463, 468–486
Electrode potentials (standard), 472
Electrodes, 452, **460**, **475**. See *also* Electrolysis
  in electrophoresis, **626**
  in fuel cells, **480**
  hydrogen (standard), **471**, 472
  molten, **478**
Electrodialysis, **370**
Electroencephelograph (EEG), 664
Electrolysis, 12, 199, 255t, 451, **452**
  for alkaline earths, 168
  in mercury cells, 187, **447**, **462**
  quantitative relationships, 455
  reaction predictions, 453
Electrolytes, 406–408, 452
  and colligative properties, 408
Electrolytic cell, 452–459, 469–470
Electrolytic industries, 459, 461–462
Electrolytic refining, 184, 459, **460**
Electromotive force (emf), 26, 327t, 456, 469, 626
Electron, angular momentum, 47
  beam, 50, 76
  charge, 28–30, 45, 455
  energy, 60
  ground state, 50
  mass, 45
Electron capture, 113t
Electron density maps, 214–217
Electron diffraction, **54–56**, 234
Electron dot formula, 203
Electron-gas model, 235
Electron levels. *See* Electron shells
Electron micrographs, **310**, **428**, **557**, **576**, **609**, **656**
Electron microprobe, 127, **128**
Electron microscope, 56, 134, 309, **310**. *See also* Electron micrographs
  and cellulose, 600
  and colloids, 406
  and DNA, 649
Electron orbits, 47
Electron pairing. *See* covalent bond
Electron-probability contours, 195, 215–216
Electron repulsion, 212
Electron shells, 152–159
Electron spin, 51, 210
Electron transport, in photosynthesis, 573–574, **575**
  in respiration, 605–606
Electron wave, **46**, **58**, 60
Electronegativity, 206–207
Electrons, **75**
  delocalized, 519
  in orbitals, 155
  in outermost level, 153
  in shells, subshells, 156t
  s, *p*, *d*, *f*, 153

Electrophoresis, 625, **626**, 627, 663
Electroplaques, 482–484
Electroplating, 181, 186, 461
Electropositive elements, 167, 472t, 473
Electrostatic force, 229
Electrostatic precipitator, **311**
Electrostatic unit, 24n
Elements, 9. *See also* Elemental abundances; Groups of elements
  A subgroup, 166–179, 248
  abundance, 101, **103**, 117t
  actinoids, 189–190
  B subgroup, 179–187, 248
  discovery dates, **8**
  Group VIII, 182–184
  lanthanoids, 187–189
Empirical formula, 276, 497–498
Emulsifier, 601, 613, 634
End point, 414
End-group analysis, 629
Endothermic reactions, 280, 432, 550
-ene (suffix), 528
Energy, of ATP, 602
  billion electron-volt, 72–73
  in bonding, 216
  consumption of, **669**
  electric, 86n, 457
  electromagnetic, **21**
  microwave, 106
  million electron-volt, 72
  radio-frequency (RF), **21**, 513
  resonance, 519
Energy levels, 48, **77**, **236–237**
Enthalpy, 281–282, **285**, 530, 550
Enzyme inhibition, 175, 625
Enzymes, 599, 608, 624–625
  lack of, 647
  in lipid metabolism, 616
  in mitochondria, 608
  origin of, 566
Epsom salt, 169, 254t
Equilibrium, 398, 424–425, 548–549
Equilibrium constant (K), 424, 430ff, 548
Equilibrium position, 431
Equivalence point, 414
Equivalent, 267, 455
Equivalent weight, 12, 267–269, 275, 455
Equivalents in measurement, 678
Eras, geologic, **585**
Erbium, 187–188
Erg, 71n
Essential amino acids, 620, **621**
Esterification, 540, **549**
Esters, 533t, 539–540, 616
Estrogens, 617–618, 639
Ethane, 281t, 288, 499–500, 528t
  bonding in, 515
  NMR spectrum, 513
Ethanol. *See* Ethyl alcohol
Ether. *See* Diethyl ether
Ethyl acetate, 548–551
Ethyl alcohol, **337**, 498–499, 535, 570, 659
Ethylene, 281t, 500, 526t
Ethylene glycol, 546
Ethyne. *See* Acetylene
Eukaryote cells, 571, **572**
Europium, 187–188
Eutrophication, 375
Evolution, 85
  biological, 567–572, **585**
  chemical, 562–567

Halogenation, 532, 610
Halogens, 177–178
  activity order, 256
Hanford, Washington, 81
Hard water, 137
Hassel, Odd, 512
Haüy, Abbé R., 133–134, 504
Heat, 91, 113, 325, 343
  waste, 374–375
Heat capacity, 270, 271t
Heat content. *See* Enthalpy
Heats of reaction, 280–281, 550
Heavy elements, 109, 170ff
Heavy ion linear accelerator
  (HILAC), **189**
Heavy water, 84, 631
Heisenberg, Werner, **56,** 59, 204
Heisenberg equation, 59
Helium, 32, 88, 179, 300t, 327t
  as alpha particle, 70
  configuration, 204
  discovery of, 299
  fusion, 85, 108
  in gas chromatography, **612**
Helix, 506, 630, **651–653**
Hemoglobin, 313, 420, 661
Hemolysis, 418
Hench, Philip, 616
n-heptane, 528t, 530
Heredity, 645
Hertzsprung-Russell (H-R) dia-
  gram, 101, **104**
Hess's law, 282
Heteronuclear molecule, 206
Heterotrophs, 570
Hexagonal lattice, **132,** 235
n-hexane, 528t, 530
High-energy bonds, 607–608
High-temperature gas-cooled re-
  actor (HTGR), **88**
Hippuric acid, 521, 614–615
Histidine, **621, 623**
Histones, 619, 658, 660
Hodgkin, Dorothy, 631, 639
Hole theory of liquids, 392–393
Holley, Robert, **662,** 663
Holmium, 187–188
Homologous series, 554
Homonuclear molecule, 206
Hormones, **617,** 618, 658
Hubble radius, 119
Hubble's constant (*H*), 105
Huggins, W., 101
Humans, 585, 665-670
Humidity, 338
Hund's rule, 180, 227
Hybrid orbitals, 211–213, 214t
  in carbon, 513–519
Hybridization, 206, 211–213
  geometries, 214t
  sp, 518
  sp^2, 517, **519**
  sp^3, 515, **516**
Hydrated ions, 395–397
Hydrates, 253, 254t
Hydration of alkenes, 531–532
Hydrazine, 629
Hydride, 84–85, 167
Hydriodic acid, 257, 594
Hydrobromic acid, 257
Hydrocarbons, 379, 501, 526–528
  aliphatic, 530–533
  aromatic, 534–535
  from automobile, 315t
  derivatives of, 530–545
  nomenclature of aliphatic,
    528–530
  in smog, 305–309

Hydrochloric acid, 166t, 178,
  256, 257, 281t
  pH, 435–436
  in stomach, 601
Hydrocyanic acid, 257
Hydrofluoric acid, 257, 433
Hydrogen, 11–14, 249, 349
  atomic, 50, 203–204
  computer simulation of, 216
  in early atmosphere, 293, 294,
    562, 563
  economics, 479–480
  in fuel cell, 477
  fusion to helium, 107, 669
  isotopes, 85, 87
  metallic, 110
  molecular orbitals of, 210–211
  in NMR studies, 512–515
  preparation, 296, 297, 462–463
  in proteins, 618
  reduction with, 255t
Hydrogen atom, 30, 48, 58
  spectrum, **22,** 48, 102
Hydrogen bomb, 84
Hydrogen bonding, 232–233, 239
  in cellulose, 600
  in DNA, 649–651
  in polypeptides, 233, 629–631,
    632
  in water, 233
Hydrogen bromide, 207
Hydrogen chloride, 430
Hydrogen cyanide, 204, 520, 562–
  563, 587, 595
Hydrogen economics, 479–480
Hydrogen electrode, 470, **471,**
  472
Hydrogen fluoride, 239, 433
Hydrogen iodide, equilibrium,
  424–427, 431–432
Hydrogen ion, 45, 258, 683
  in acids, 408–411, 433–434
  in buffers, 439–440
  and pH 435–439
  from water, 434–435
Hydrogen peroxide, 209, 300t,
  463
Hydrogen sulfide, 257, 387
  in early atmosphere, 562–563,
    569
Hydrogenation, 317, 531–532
Hydrolysis, 415–416
  and ATP, 608
  and DNA, 648
Hydrometer, 475
Hydronium ion, 409
Hydrophilic character, 578, 613
Hydrophobic character, 578, 629
Hydrosphere, 114, 116t, 564
Hydroxyl ions, 227–228, 485
  from water, 434–437
Hypertension, 378
Hypo, 168, 254t, 287

I

Ice, 238t, 313, 372
Ideal gas, 350
Ideal solutions, 398
Igneous rock, 124–126, **365**
Immune system, 660
Incineration, **316**
Indicators, **413,** 437–439
Indium, 171
Inductive effect, 539
Industrial chemistry. *See* Chem-
  ical industry
Inelastic collision, **75**
Information in a cell, 654

Infrared, **21,** 48, 312, 480
  spectrometry, **307–308,** 319, 579
Inhibitors, 427, 625
Inner transition elements, 157
Inorganic chemicals, production,
  166t
Insecticides, 175, 178, 185, 287
Insoluble salts, **397**
Inspiration, 225
Insulator, 237
Insulin, 639
Interference, **54–55**
International System, 677–678
International Union of Pure and
  Applied Chemistry (IUPAC),
  249
  nomenclature rules of, 528–529
Inversion, atmospheric, **303–304**
Iodine, 177-178, 300t, 318, 610
Iodine number, 638
Ion-exchange resins, 188
Ion product (I.P.), 442
Ion radius, 230–231
Ionic bonding, 202, 207, 238t.
  See also Electrovalence
Ionic compounds, 201, 258–259
Ionic crystals, 229
Ionic equations, 258–259
Ionic equilibria, 433–443
Ionic radii. *See* Crystal radii
Ionization, of air, 31, 311
  constants, 434t
  energy, 44, **158,** 206
  gauge, **327**
  potential, **51**
  of water, 434–437
  of weak acids and bases,
    433–434, 439–440
Ionosphere, 300, **301**
Ions, 200, 248–251t
  in acid–base theory, 409–411
  in discharge tubes, 26
  of electrolytes, 407–409
  in ionic equations, 258–259
  in solution, 249–250, 395–397
Iridium, 182, 184
Iron, compounds of, **37,** 111,
  182–183, 238t, 255t, 376
  corrosion, 485
  Curie point, 119
  in cytochromes, **574, 606**
  in earth, 112, 117t, 126–127,
    139t
  as electrode, 471, 472, 485, **486**
  spectrum, **37,** 39, 105
  supply-demand, 165t, 193
Iron ore, 125, 288
Iso- (prefix), 535
Isobar, 67
Isoelectric points, 619–620, **621**
Isoelectronic species, 217
Isoleucine, **621, 623,** 659
Isomerization, 530
Isomers, benzene, 501–503
  cis-trans, 225–226, 509–511, 532
  dl, 507–509, 537, 568
  DL, 594–598
  of glucose, 598–599
Isoprene, 545, 616–618
Isotopes, 67, 77, 93, 268, 286
  total number, 70
IUPAC. *See* International Union
  of Pure and Applied Chem-
  istry

J

James, A., 611
Jensen, Johannes, 77
Joliot, Frederick, 70

in displacement, 256
in electrolytic cell, 447, 462
Merrifield, Robert, **623**
Meso compounds, 508, **509**, 598
Mesons, 74–76
Messenger ribonucleic acid. *See*
mRNA
Meta directors, 543–545
Meta positions, 501–503
Metabolism, 561, 601, 608
Metallic bonding, 235–237, 238t
Metalloids, 152, 170, 247
Metals, 6–8, 110
 activity, 201, 255t, 451
 alkali, 166–168
 alkaline earth, 168–170
 aluminum group, 170–171
 band model, 236–237
 in complexes, 225–229
 crystal lattices, 235
 as diet essentials, 663
 electron-gas model, 235
 heat capacities, 217t
 inner transition, 187–191
 melting points, back endpaper
 micronutrient, 229, **569**
 oxides of, 253
 properties, 23, 152, 159, 235
 in rocks, 117, 139
 supply-demand, 165t, 192
 toxic, 378
 transition, 179–187
 in waters, 364t
Metamorphic rocks, 124–126
Metathesis, 257, 397
Meteorites, **111**, 191, 562, 635
Methane, 86, 90, 179, 281t, 288,
 367, 528t
 in air, 300t, 313
 in early atmosphere, 562–563,
 569
 from sewage, 383, **384**
 source of, 300, 313
 sp^3 orbitals, 214t, 515, **516**
 supply, 669
Methanol. *See* Methyl alcohol
Methionine, **621, 623**
Methyl alcohol, 218, 416, 526t
 synthesis of, 431
Methyl esters, 598, 611
Methyl halides, 499
Methyl orange, 438
Methylamine, reactions, 541
Metric system, 10, 675–678
MeV, 71n
Meyer, J. L., 147ff
Mica, 139, **140**, 141
Micelles, 405–406
Microwave, **21**, 106
Milk, 437t, 610, 663
Milky way, 102, 106, 112
Miller, Stanley, **563**
Millikan, Robert, **28**, 29, 44, 455
 determination of N, 273, 288
 oil drop apparatus, **30**
Minerals, 124, 125t, 126, 139t, **140**
 as body essentials, 663
 composition of, 136–141
 from ocean, 367
 in waters, 364t
Mirror image, **508, 509**, 598
Mitochondria, 608, **609**, 613
Mohs' scale of hardness, 141t
Moissan, Henri, 177
Molal boiling-point constant,
 400, 403
Molal solutions, 399, 402–405
Molality, 399

Molar heat capacity. *See* Heat
 capacity
Molar solutions, 412–414
Molar volume, 272, 277, 352
Molar weight. *See* Molecular
 weight
Molarity, 411–415
Mole, 272–273, 278, 685
Mole fraction, 399
Molecular compounds. *See* Co-
 ordination compounds
Molecular crystal, 231–232, 238t
Molecular energy, 308, **309**
Molecular formulas, 274–276,
 497–498
Molecular orbitals, 210–211, 236
Molecular sieves, 142
Molecular weight, 270, 272, 278
 from atomic weights, 276
 in chemical factors, 685–686
 from colligative properties,
 402–405
 of gases, 272, 277, 340–342
 of proteins, 627, 635
 by sedimentation, 640
Molecule, 12n, 14, 199, 500, 661
 in chemical evolution, **569**
Molybdenum, compounds of,
 181–182, 380
Monochromatic light, **505**
Monoclinic lattice, **133**
Monomer units, 545
Monosaccharides, 592–599
Moon rocks, 139, **140**
Morgan, Thomas, 647
Moseley, Henry, 36–37, 151, 187
Mouton, Abbé Gabriel, 675–676
Moving phase, 611
mRNA, 653, 657
Muller, Hermann, 85
Muller, Paul, 376
Mulliken, Robert, 210
Multiple bonds, 500. *See also*
 Bonds
Multiple-proportion ratios, 268
Multiples, mathematical, 677–678
Muriatic acid, 178
Muscle, 634
Mutarotation, 598
Mutation, 658
Mylar, 353, 546

N
n (negative) crystal, 481–482
N-terminal end, 622, 628, 630,
 657
NAD+, NADH. *See* Coenzymes
NADP+, NADPH. *See* Coen-
 zymes
Naming, acids, 253t, 257
 alcohols, 535
 aldehydes, 537–538
 aliphatic hydrocarbons, **527,**
 528–529
 amides, 541–542
 amines, 541
 amino acids, **621**
 anions, 249–250t
 aromatic hydrocarbons, **527**
 carbohydrates, 592–600, 655
 carboxylic acids, 539–540, 610
 cations, 248–249t
 elements, front endpaper
 fatty acids, 539–540, 610
 hydrocarbon derivatives, **533**
 hydrates, 254t
 inorganic compounds, 249–250
 ketones, 537–538
 minerals, 125, 138–139

nitrogen bases, 648, 655
nucleic acid components, 648t
peptides, 622
rocks, 124–125, 139–141
silicates, 137–138
Native state, 255t
National Aeronautics and Space
 Administration (NASA), 179,
 289
Natta, Gulio, 546
Natural gas. *See* Methane
Natural selection, 562
Nearest neighbors, crystal, 230
Negative catalyst. *See* Inhibitor
Negative ions. *See* Anions
Neodymium, 187–188
Neon, 179, 299
Neon configuration, 204
Neptunium, **80**, 189
Nerve gas, 175
Nerve net, 484
Network crystals. *See* Covalent
 network solids
Neutral solution, 435
Neutralization, 258, 540
Neutralizing capacity, 267
Neutrinos, 76
Neutron, 70, 75, **76, 77,** 78, 106
 activation, 71
 in fusion power, 90
 stars, 109, 301
Newman notation, 521
Newton, Issac, **9**, 42, 199
 second law, 46
Nickel, compounds of, 111, 117t,
 182–184, 275–276, 367, 454,
 491
 supply-demand, 165t
Nicotinamide adenine dinucleo-
 tide. *See* NAD+
Nicotinamide adenine dinucleo-
 tide phosphate. *See* NADP+
Nicotine, 286, 552, 659
Night blindness, 614
Ninhydrin, 619–620
Niobium (columbium), 181
Nirenberg, Marshall, **662**, 663,
 671
Nitrate, 250t, 364–366
Nitration. *See* Nitric acid
Nitric acid, 166t, 184, 185, 253,
 257, 283, 472t, 489–490
 in nitrations, 503, 534, 544
 in oxidations, 259, 597, 685–686
Nitric oxide, 266, 281t, 283
Nitrile, 533t, 542, **547,** 595–596
Nitrilotriacetic acid (NTA), 375
Nitro group, 542–544
Nitrogen, 92, 173–174, 294, 652
 amino acid, 639–640
 in ammonia synthesis, 432–433
 in atmosphere, **294,** 296, 300t
 compounds, 174
 cycle, 174, 365, **366**
 from deaminations, 541, 543
 from food, 664
 as nonideal gas, **349**
 nuclear disintegration, 69
 orbitals, 213
 preparation of, 174, 298–299
 production, 166t
 protein, 618
 sources of, 294, 375
Nitrogen bases, 566, **569**, 593,
 648, 655
Nitrogen dioxide, 266, 281t, 685.
 See also Nitrogen oxides
 in air, 300t, 306
 and biological systems, 314

Standard conditions, for electrode reactions, 470
for gas calculations, 334
for thermochemical calculations, 282
Standard electrode potentials, 470–474
Standard temperature and pressure. *See* STP
Starch, 564, 592, 593, 600
hydrolysis of, 593, 601, 624
Stark, Johannes, 49
Stars, reactions in, 101–109
Starvation, 571, 666
States of matter, 8–9, 393t
Static electricity, 23, 451
Statistical mechanics, 343, 345
Steam, geothermal, 91, **386,** 669
Stearic acid, 610t
Steel, 165, **181,** 183, 192–193
Stereochemistry, 503–513, 532
Sterilization, 667
Stern-Gerlach experiment, 51
Steroids, 513, 613–618
Sterols, 613, 616–618
Stoichiometry, 265–280, 497–498, 685–686
Storage cell, **475,** 476–478
STP, 326, 334–335
Strain, in rings, 511–512
Stratosphere, 300, **301**
Strecker synthesis, 587–588
Strontium, 78, 169, 231, 255t
Subgroup elements, A, 166–179
B., 179–187
Subscripts, 249
Substitution reactions, 200, 537
of benzene, 534
Succinic acid, 564t, 603–604
Sucrose, **580,** 592, 599, 637
Sugar-phosphate chains, **650–653**
Sugars, 564–565, 569, 592–601. *See also* Carbohydrates
alpha and beta forms, 598
chain expansion of, 595–597
D series, 594–598
nonreducing, 599
reducing, 593
ring structure of, 598–601
synthesis of, 564–565
Sulfa drugs, 625
Sulfate, 125t, 214t, 249, 250, 379
in natural waters, 364t
solubilities, 257
Sulfide ion, **231,** 250, 253
Sulfite ion, 250t
Sulfur, compounds of, 176–178, 214t, 429
cycle, 387
mineral, 117t, 143, 176, 367
in proteins, 623, 657
in rubber, 176, 279
Sulfur dioxide, 176, 281–282
in atmosphere, 293, 313–314
molecular shape, 209
origin of, 313–314
Sulfur trioxide, 281t
Sulfuric acid, 166t, 176, 257–258, 281t, 463
in alkene hydration, 532
in battery, 475
as mine seepage, 376
Sulfurous acid, 257
Sumner, James, 624
Sun, 101, **104,** 108, 110, 117
Sunlight economics, 480–482
Sunlight energy, in photosynthesis, 582
Superheavy elements, **151,** 191

Supernova, 109, **118**
Supersaturated solution, 396
Surface tension, 393, **394**
Svedberg, Theodore, 628
Synchrotron, **72**
Synthetic fibers, 318, 545–548

T
Tantalum, 181, 461
Tar sand, **669**
Tartaric acid, 257, 507
Tatum, Edward, 647, 654
Technetium, 182
Teeth, 173, 228, 663
Teller, Edward, 84
Tellurium, 175–177
Temperature, 84, 325–327, 676
definition of, 331
during nucleogenesis, 106
of the earth, **294**
effect on reaction rate, 424–426
effect on solubility, **397**
Temperature scales, 325, **333**
Terbium, 187–188
Termites, 601
Tetraethyl lead, 167, 173, 287, 427
Tetragonal lattices, **133**
Tetrahedral structure, **132,** 506, **516**
Tetraphosphorus decoxide, 542
Thallium, 171
Theoretical yield, 279
Theory, acid–base, 408–411
aging, 659–660
atmospheric evolution, 293–294
atomic orbital (AO), 226–228
ATP, 606–608
Baeyer's, 511
beta decay, 65
big bang, 106
Bohr's, 44–51
catalyst, 427–429, 624–625
coordination, 223–226
cosmology, 106–113
covalent bond, 203–217, 223–229
Dalton's, 10–12, 266–267
DNA, 647–652
ionic bond, 199–203, 229–231
Kekulé's, 500–502
liquid structure, 392–395
Maxwell's, 61
Mendel's, 645–646
metal structure, 235–238
nuclear structure, 73–78
Oparin-Haldane, 561–567
phlogiston, 296–297
protein synthesis, 655–657
quantum, 51–60
resonance, 204–205, 544–545
tetrahedral C atom, 507–519
Thermal pollution, 374–375
Thermite process, 171
Thermochemistry, 280
Thermocouple, **326,** 353t
Thermometers, 325, 327t
Thomson, George, 55
Thomson, J. J., **28,** 67
apparatus, 29
Thorium, 67, 68, 86, 113t, 669
Thought experiment, 59
Threonine, **621, 623**
Thulium, 187–188
Thymine, 564, 648t, 655
Thyroid gland, 178
Time, measure of, 676

Tin, 86, 173, 256, 367
Tiselius, Arne, 626
Titanium, **37,** 180–181, 488
Titration, 414–416
of amino acids, 622
of DNA, 649
Titration error, 414
Tobacco smoke, 378
Tokamak, **89,** 90
Tollens' test, 539
Toluene, 526t, 534–535
Toothpaste, 173
Torr (unit), 326
Torsion balance, 23, **24**
Total dissolved solids, 379
Trace metals, 378
Trans groups in complexes, 225–226
Trans isomer, 509, **510,** 532
Transfer RNA. *See* tRNA
Transformer, **89,** 90
Transition elements, 157, 179, 205, 223
Transition state. *See* Activated complex
Transmutation, of elements, 69
Trash disposal, **316**
Triads of elements, 147
Tricarboxylic acid cycle. *See* Citric acid cycle
Triclinic lattice, **133**
Triglyceride, 610–611
Triple bond, 204–205, 501, 518
Triple point, water, 332, 676t
Trisaccharides, 592
Tritium, 84, 87, 118
tRNA, 655–657, 663
Troposphere, 300, **301,** 313
Trypsin, 629, 671
Tryptophan, **621, 623,** 664–665
Tumors, 515. *See also* Cancer
Tungsten, compounds of, 101–182, 224, 327t, 367
Turnbull's blue, 183
Twsett, M., 578
Tyndall effect, 406
Tyrosine, **621, 623,** 661

U
Ultracentrifugation, 406, 627, **628,** 640, 652
Ultraviolet, **21,** 48, 293, 308
Uncertainty principle, 59
Undernutrition, 664
Unit cell, 134
Units of measure, 676–677t, 684–686
Universal gas constant (R). *See* Gas constant
Universe, 101, 106
Unsaturated compounds, 500, 531–533, 610
Uracil, 564, 648t, 655
Uranium, 31, 86, 189, 340
decay of, 67, **68,** 113t, **114**
fission, 78ff
in power generation, 669
supply, 86, 669
Urea, 401, 499, 526t, 633, 664
Urey, Harold, **563**
Urey-Miller experiment, **563,** 564t
U.S. Atomic Energy Commission (AEC), 1, 90, 92, **93,** 189

V
Valence, 148–149, 200, 268t
Valence bond formula, 203

Valence number, 247–251, 268.
See also Oxidation state
Valence shell, 201
Valine, **621, 623**
Vanadium, compounds of, **37,** 125t, 181, 367
van der Waals, Johannes, 232, 392
van der Waals forces, 232–234, 238t, 349, 649
van der Waals radius, 234
Van Slyke method, 354, 541, 555, 640
Van't Hoff, Jacobus, 403, **506,** 507–509, 594
Vapor pressure, **337,** 393–394
effect of solute on, 398
lowering, 400–401
Vector, **24,** 37
VIBGYOR, **20–21**
Vibrational energy. *See* Molecular energy
Vibrational modes, **309**
Viking Lander, 613
Vinegar, 414, 539
Viruses, 374, **557,** 628, 658, 661
Viscosity, of liquids, 392, 416
Vitalism, 497, 499
Vitamins, 613–614, 663
A, 614
B, 604, 614
B_{12}, 627
D, 614
E, **576,** 614, 660
K, **576,** 614
Volcanism, 97, 112, 293, 363, **365,** 584
Volt, 26, 456
Voltaic cells, 451, 452, 468–486
Volumetric analysis, 411–416

W
Walton, Ernest, 71
Wastes, 315–318, 382–385
Wastewater, 382–385
Water, analysis of, 6, 7, 364t, 379–382
atomic orbitals, 211–212
boiling point constant, 403
bond angle, 218
cooling, 374
demand, 388t
desalination, **1,** 368–372, 404
dielectric constant, 397
dipole, 209, **396**
dissociation of, 434
electrolysis, 298, 453–454
enthalpy, 281t, **282**
freezing point constant, 401–402
and life, 363, 637
modern use of, 373
molecular shape, 209, 238t
natural, 364t
photolysis of, 573
precambrian, 586
from respiration, 605–606
as solvent, 396
specific heat of, 281
study by NMR, 515
vapor pressure, **337,** 338t, 401
viscosity of, 416
Water constant (K_w), 434
Water of hydration, 254
Water molecule, 209, **212,** 214t, 220, **396**
Water pollution, 373–385
Water softening, 137
Water vapor, 312, 562–563, 584
Watson, James, 649, **650**
Watt (unit), 456
Wave functions. *See* Atomic orbitals
Wave mechanics, 57, 210–211
Wave motion, **52, 53,** 55, **58**
Wavelength, **21,** 43
Waxes, 614
Weak electrolytes, 407, 410
Weathering, 126
Weight, 676
Werner, Alfred, 223–226, 239
White dwarf, **104,** 108–109
Wieland, Heinrich, 616
Wien, Wilhelm, 26
Wilkins, Maurice, 649
Wilson, Charles, 33–34
Windaus, Adolph, 616
Wöhler, F., 499
World dynamics, 667–670
World Health Organization, 376
World War II, 80, 83

X
X rays, **21,** 39t, 85n, 133, 170
X ray diffraction, **36, 131,** 201, 211, 234
and Avogadro's number, 277
and cellulose, 600
and complexes, 226
and crystals, 229–230
and DNA, 649
and liquids, 392
and proteins, 631
X ray spectra, 151
in electron microprobe, 127, **128**
of fourth period metals, **37**
of lanthanoids, 187
X ray tube, **31, 36,** 181
Xenon, 179, 191, 299
Xylene, 526t, **527**

Y
Yeast, 497, 570, 599
Yield. *See* Percentage yield
-yl (suffix), 529
-yne (suffix), 528
Ytterbium, 187–188
Yttrium, 180, 187–188
Yukawa, Hideki, 74

Z
Zeeman, Pieter, 50
Zeolites, 137
Ziegler, Karl, 546
Zinc, in batteries, 451, 469, 472–475, **477**
compounds of, 186, 255t, **310,** 367, 378, 460
ion, 249
supply-demand, 165t
Zirconium, compounds of, 142, 180–181
Zwitter ion, 619–620

PERIODIC TABLE

GROUP IA

IA	IIA	IIIB	IVB	VB	VIB	VIIB	VIII	VIII	VIII
1 1.008 −259.2 1 0.071 **H** Hydrogen									
3 6.941 180.5 1 0.53 **Li** Lithium	**4** 9.012 1277 2 1.85 **Be** Beryllium								
11 22.99 97.8 1 0.97 **Na** Sodium	**12** 24.30 650 2 1.74 **Mg** Magnesium								
19 39.10 63.7 1 0.86 **K** Potassium	**20** 40.08 838 2 1.55 **Ca** Calcium	**21** 44.96 1539 3 3.0 **Sc** Scandium	**22** 47.90 1668 4,3 4.51 **Ti** Titanium	**23** 50.94 1900 5,4,3,2 6.1 **V** Vanadium	**24** 52.00 1875 6,3,2 7.19 **Cr** Chromium	**25** 54.94 1245 7,6,4,3,2 7.43 **Mn** Manganese	**26** 55.85 1536 2,3 7.86 **Fe** Iron	**27** 58.93 1495 2,3 8.9 **Co** Cobalt	**28** 58.71 1453 2,3 8.9 **Ni** Nickel
37 85.47 38.9 1 1.53 **Rb** Rubidium	**38** 87.62 768 2 2.6 **Sr** Strontium	**39** 88.91 1509 3 4.47 **Y** Yttrium	**40** 91.22 1852 4 6.49 **Zr** Zirconium	**41** 92.91 2468 5,3 8.4 **Nb** Niobium	**42** 95.94 2610 6,5,4,3,2 10.2 **Mo** Molybdenum	**43** 98.91 2140 7 11.5 **Tc** Technetium	**44** 101.1 2500 2,3,4,6,8 12.2 **Ru** Ruthenium	**45** 102.9 1966 2,3,4 12.4 **Rh** Rhodium	**46** 106.4 1552 2,4 12.0 **Pd** Palladium
55 132.9 28.7 1 1.90 **Cs** Cesium	**56** 137.3 714 2 3.5 **Ba** Barium	**57** 138.9 920 3 6.17 **La** Lanthanum	**72** 178.5 2222 4 13.1 **Hf** Hafnium	**73** 180.9 2996 5 16.6 **Ta** Tantalum	**74** 183.8 3410 6,5,4,3,2 19.3 **W** Tungsten	**75** 186.2 3180 7,6,4,2,−1 21.0 **Re** Rhenium	**76** 190.2 3000 2,3,4,6,8 22.6 **Os** Osmium	**77** 192.2 2454 2,3,4,6 22.8 **Ir** Iridium	**78** 195.1 1769 2,4 21.4 **Pt** Platinum
87 (223) (27) 1 − **Fr** Francium	**88** 226.0 700 2 5.0 **Ra** Radium	**89** (227) 1050 3** − **Ac** Actinium	**104** (257) **Rf** Rutherfordium	**105** (260) **Ha** Hahnium					

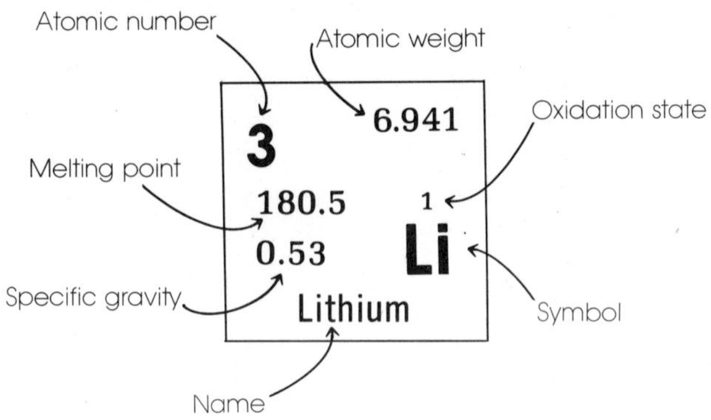

Lanthanoids *

58 140.1 795 3,4 6.67 **Ce** Cerium	**59** 140.1 935 3,4 6.77 **Pr** Praseodymium	**60** 144.2 1024 3 7.00 **Nd** Neodymium	**61** (147) (1027) 3 − **Pm** Promethium	**62** 150.4 1072 3,2 7.54 **Sm** Samarium	**63** 152.0 826 3,2 5.26 **Eu** Europium	**64** 157.2 1312 3 7.89 **Gd** Gadolinium

Actinoids **

90 232.0 1750 4 11.7 **Th** Thorium	**91** 231.0 (1230) 5,4 15.4 **Pa** Profactinium	**92** 238.0 1132 6,5,4,3 19.07 **U** Uranium	**93** 237.0 637 6,5,4,3 19.5 **Np** Neptunium	**94** (242) 640 6,5,4,3 − **Pu** Plutonium	**95** (243) − 6,5,4,3 11.7 **Am** Americium	**96** (247) − 3 − **Cm** Curium

1970 atomic weights, rounded off to 4 significant figures. See also list of elements inside front cover.

For a periodic table extended to element 168, see Figure 7-4, page 151.